Certain Number-Theoretic Episodes in Algebra

PURE AND APPLIED MATHEMATICS

A Program of Monographs, Textbooks, and Lecture Notes

MONOGRAPHS AND TEXTBOOKS IN PURE AND APPLIED MATHEMATICS

Recent Titles

E. Hansen and G. W. Walster, Global Optimization Using Interval Analysis, Second Edition, Revised and Expanded (2004)

M. M. Rao, Measure Theory and Integration, Second Edition, Revised and Expanded (2004)

W. J. Wickless, A First Graduate Course in Abstract Algebra (2004)

R. P. Agarwal, M. Bohner, and W-T Li, Nonoscillation and Oscillation Theory for Functional Differential Equations (2004)

J. Galambos and I. Simonelli, Products of Random Variables: Applications to Problems of Physics and to Arithmetical Functions (2004)

Walter Ferrer and Alvaro Rittatore, Actions and Invariants of Algebraic Groups (2005)

Christof Eck, Jiri Jarusek, and Miroslav Krbec, Unilateral Contact Problems: Variational Methods and Existence Theorems (2005)

M. M. Rao, Conditional Measures and Applications, Second Edition (2005)

A. B. Kharazishvili, Strange Functions in Real Analysis, Second Edition (2006)

Vincenzo Ancona and Bernard Gaveau, Differential Forms on Singular Varieties: De Rham and Hodge Theory Simplified (2005)

Santiago Alves Tavares, Generation of Multivariate Hermite Interpolating Polynomials (2005)

Sergio Macías, Topics on Continua (2005)

Mircea Sofonea, Weimin Han, and Meir Shillor, Analysis and Approximation of Contact Problems with Adhesion or Damage (2006)

Marwan Moubachir and Jean-Paul Zolésio, Moving Shape Analysis and Control: Applications to Fluid Structure Interactions (2006)

Alfred Geroldinger and Franz Halter-Koch, Non-Unique Factorizations: Algebraic, Combinatorial and Analytic Theory (2006)

Kevin J. Hastings, Introduction to the Mathematics of Operations Research with *Mathematica®*, Second Edition (2006)

Robert Carlson, A Concrete Introduction to Real Analysis (2006)

John Dauns and Yiqiang Zhou, Classes of Modules (2006)

N. K. Govil, H. N. Mhaskar, Ram N. Mohapatra, Zuhair Nashed, and J. Szabados, Frontiers in Interpolation and Approximation (2006)

Luca Lorenzi and Marcello Bertoldi, Analytical Methods for Markov Semigroups (2006)

M. A. Al-Gwaiz and S. A. Elsanousi, Elements of Real Analysis (2006)

R. Sivaramakrishnan, Certain Number-Theoretic Episodes in Algebra (2006)

Aderemi Kuku, Representation Theory and Higher Algebraic K-Theory (2006)

Certain Number-Theoretic Episodes in Algebra

R. Sivaramakrishnan

Chapman & Hall/CRC
Taylor & Francis Group

Boca Raton   London   New York

Chapman & Hall/CRC is an imprint of the
Taylor & Francis Group, an informa business

Chapman & Hall/CRC
Taylor & Francis Group
6000 Broken Sound Parkway NW, Suite 300
Boca Raton, FL 33487-2742

© 2007 by Taylor & Francis Group, LLC
Chapman & Hall/CRC is an imprint of Taylor & Francis Group, an Informa business

International Standard Book Number-10: 0-8247-5895-1 (Hardcover)
International Standard Book Number-13: 978-0-8247-5895-0 (Hardcover)

Library of Congress Cataloging-in-Publication Data

Sivaramakrishnan, R., 1936-
  Certain number-theoretic episodes in algebra / R. Sivaramakrishnan.
    p. cm. -- (Pure and applied mathematics ; 286)
  Includes bibliographical references and indexes.
  ISBN 0-8247-5895-1 (alk. paper)
  1. Algebraic number theory. 2. Number theory. I. Title. II. Series.

QA247.S5725 2006
512.7'4--dc22   2006048994

Visit the Taylor & Francis Web site at
http://www.taylorandfrancis.com

and the CRC Press Web site at
http://www.crcpress.com

CATALYZED AND SUPPORTED BY THE

DEPARTMENT OF SCIENCE AND TECHNOLOGY

UNDER ITS

UTILIZATION OF SCIENTIFIC EXPERTISE OF RETIRED SCIENTISTS

SCHEME

(USERS scheme)

PROJECT No: HR/UR/21/98

The author acknowledges with thanks the financial support of the Department of Science and Technology under USERS scheme for undertaking the project for preparation of a monograph/textbook. But for the timely financial help of the DST and the encouraging letters received from Dr. Parveen Farooqui, Head, Human Resources Wing of DST, the task of implementation would not have found fulfilment.

R. Sivaramakrishnan

ACKNOWLEDGEMENT

The author wishes to express his deep sense of gratitude to the authorities of the University of Calicut for having given him the opportunity to utilize the facilities at the Calicut University Campus in general and at the Mathematics Department in particular.

Thanks are due to

(1) Prof. V. Krishnakumar for his help and guidance at a very personal level and in his capacity as Head, Department of Mathematics
(2) Dr. P. T. Ramachandran for valuable discussions and Sri. Kuttappan C, Librarian in the Department of Mathematics for having provided abundant help in the matter of library reference
(3) the Deputy Registrar, Pl.D Branch and his colleagues for matters of official correspondence
(4) the Finance Officer and his staff for processing bills and vouchers and such other transactions.

Many academicians and friends have offered help at various stages in the progress of the project.

The author had the opportunity to visit Mangalore University, Mangalore (Karnataka) during 1996–97. It was during this period that the spade work for the project was done, and the encouragement received from Prof. B. G. Shenoy and Prof. Juliet Britto was of great help. The author thanks the Faculty of the Department of Mathematics, Mangalore University for the invitation to stay and work for a year.

Thanks are due to

(i) Prof. C. S. Seshadri FRS, Director, Chennai Mathematical Institute, Chennai for arranging a visit to the library of the Institute
(ii) Prof. R. Sridharan and Prof. K. R. Nagarajan of Chennai Mathematical Institute for advice about the inclusion of certain topics/papers in the sections on the Pell equation and rings with chain conditions
(iii) Prof. R. Balasubramanian, Director, Institute of Mathematical Sciences (MATSCIENCE) for valuable comments in the choice of topics in Algebraic Number Theory
(iv) Prof. M. Thampan Nair and Prof. C. Ponnuswamy for help in connection with a visit to I.I.T (Madras), Chennai
(v) Prof. T. Thrivikraman and Prof. R. S. Chakravarti of Cochin University of Science and Technology, Kochi for their scholarly suggestions and remarks
(vi) Prof. M. I. Jinnah and Prof. A. R. Rajan for facilitating the author's visit to the Mathematics Department, University of Kerala, Thiruvananthapuram
(vii) Dr. Rajendran Valiaveetil for all the help rendered by way of discussions and for having assisted the author by going through the first draft of the manuscript and for making suggestions about the simplification of proofs of some theorems

(viii) Prof. Pentti Haukkanen of the University of Tampere, Finland and Prof. S. A. Katre of the University of Poona, Pune for sending reprints of their articles to the author

(ix) Prof. Don Redmond of Southern Illinois University, Carbondale for having supplied the references relating to the Goldbach conjecture

(x) Prof. V. K. Balachandran, former Director, Ramanujan Institute for Advanced Study, Chennai for help received by way of discussions and correspondence

(xi) Dr. N. Raju, Head, Department of Statistics, University of Calicut for timely help and advice in the matter of typesetting and format of the manuscript.

PREFACE

This monograph is an attempt to justify the following assertion: "It is desirable to learn algebra via number theory and to learn number theory via algebra". Many concepts in commutative algebra such as Euclidean domains, prime and primary ideals, field of quotients of an integral domain and such others have originated from notions in number theory. For one who goes deeper into the finer aspects of number theory, algebraic techniques would appear to be powerful and elegant. Examples are from the crisp proofs of Gauss's quadratic reciprocity law, Fermat's Two-squares theorem and Lagrange's theorem on the expressibility of a positive integer as a sum of four squares. Though all authors of books on number theory have emphasized this aspect, perhaps, two books that make the algebraic approach explicit are

1. Ethan D. Bolker: *Elementary Number Theory — an Algebraic Approach*
   W. A. Benjamin Inc. NY (1970) and
2. F. Richman: *Number Theory — An Introduction to Algebra*
   Brooks/Cole Monterey/California (1971).

It is true that classical textbooks such as O. Zariski and P. Samuel: *Commutative Algebra* Vols I and II (Springer Verlag GTM Nos. 28, 29 (1982) original version Van Nostrand Edition (1958)) and K. Ireland and M. I. Rosen: *A Classical Introduction to Modern Number Theory*, 2nd Edition, Springer Verlag GTM No. 84 (1985) original version: Bogden and Quigley Inc., Publishers, Tarrytown-on-Hudson, NY (1972) convey the message of doing algebra with full number-theoretic support and vice versa exceedingly well.

The aim of this monograph is to spread this message with greater emphasis. It is for the mathematical community, at large, to pass judgement as to how far the desired goal has been achieved.

This monograph presupposes rudimentary knowledge of elementary number theory as well as algebra on the part of the reader. The main theme is the study of

(i) the ring $\mathbb{Z}$ of integers
(ii) the Chinese Remainder Theorem and reciprocity laws
(iii) finite groups from the point of view of enumeration
(iv) abstract Möbius Inversion
(v) the role of generating functions
(vi) rings of arithmetic functions and
(vii) certain analogues of the Goldbach problem.

Many interesting topics such as p-adic fields, cyclotomy, Emil Artin's conjecture and Fermat's Last Theorem (FLT) have not been discussed in detail. However, the overall picture is what one gets about the nice interconnections between number theory and algebra.

The monograph has been divided into four parts containing 16 chapters in all. Each chapter begins with a 'historical perspective' and closes by giving 'notes with illustrative examples/worked-out example(s)'. Part I dealing with elements of number theory and algebra contains seven chapters. The details are given below.

PART I
ELEMENTS OF NUMBER THEORY AND ALGEBRA

Chapter 1: Theorems of Euler, Fermat and Lagrange

Certain new proofs of classical theorems of number theory are pointed out. Using a counting principle of Melvin Hausner, the theorems of Fermat and Lucas are proved. D. Zagier's proof of Fermat's two-squares theorem is given. Lagrange's four-squares theorem is deduced from the fact that a certain 2×2 matrix with entries from $\mathbb{Z}[i]$ has a factorization of the type BB^* when B^* is the adjoint (conjugate transpose of B). Linear Diophantine equations are also discussed.

Chapter 2: The integral domain of rational integers

$\mathbb{Z}$ is shown as an ordered integral domain. It is proved that an ordered integral domain whose subset of positive elements is well-ordered, is the same as $\mathbb{Z}$, up to isomorphism. Operations on ideals of a commutative ring with unity are described. They give analogues of g.c.d. and l.c.m. of integers. In the case of an integral domain, characterizations of irreducibles and primes are shown. The criterion for an integral domain to satisfy UFD property is given. The notion of a GCD domain is also pointed out.

Chapter 3: Euclidean domains

$\mathbb{Z}$ is a Euclidean domain. The ring of algebraic integers of a quadratic number field $\mathbb{Q}(\sqrt{m})$ is a Euclidean domain when $m = -1, -2, -3, -7$ and -11. 'Almost Euclidean' domains are discussed. It is proved that the ring $R(-19)$ of algebraic integers of $\mathbb{Q}(\sqrt{-19})$ is a PID, but not a Euclidean domain. Further, $\mathbb{Z}$ is shown to be the unique Euclidean domain having 'double-remainder property'.

Chapter 4: Rings of polynomials and formal power series

Polynomial rings are introduced. If F is a field, the uniqueness of the division algorithm in $F[x]$ characterizes $F[x]$ among Euclidean domains. The ring $\mathcal{A}$ of arithmetic functions under the operations of addition and Dirichlet convolution is shown to be a UFD via the ring $\mathbb{C}_\omega$ of formal power series (over the field $\mathbb{C}$ of complex numbers) in countably infinite indeterminates. This significant result is due to E. D. Cashwell and C. J. Everett. See 'The ring of number-theoretic functions', Pacific J. Math 9 (1959) 975–985.

Next, we give a formula for the number of monic irreducible polynomials of degree m (> 0) over the finite field $\mathbb{Z}/p\mathbb{Z}$ (where p is a prime) via Möbius inversion. It is deduced that the number of monic irreducible polynomials over $\mathbb{Z}/p\mathbb{Z}$ is infinite.

Chapter 5: The Chinese Remainder Theorem and the evaluation of number of solutions of a linear congruence with side conditions

The Chinese Remainder Theorem is one of the landmarks of number theory. Its proof along with illustrations is indicated. Direct products and direct sums of rings are discussed. Simultaneous congruences modulo ideals of a commutative ring with unity are considered. This gives a ring-theoretic analogue of the Chinese Remainder Theorem. The theorem holds in a polynomial ring $F[x]$ in which congruences with a set of pairwise relatively prime polynomials provide the desired data for generalization. Next, a class of arithmetical functions called even functions (mod r) is studied with a view to evaluating the number $N(n, r, s)$ of solutions of a linear congruence: $x_1 + x_2 + x_3 + \cdots + x_s \equiv n \pmod{r}$, under the restriction g.c.d $(x_i, r) = 1$ $(i = 1, 2, 3 \ldots s)$. David Rearick's theorem gives $N(n, r, s)$ in terms of Ramanujan Sums, see theorem 39. The Rademacher formula for $N(n, r, s)$ is also derived in corollary 5.6.1.

Chapter 6: Reciprocity laws

Quadratic residues modulo a prime are discussed and Gauss's quadratic reciprocity law is shown by a proof using finite fields. Eisenstein's cubic reciprocity law is proved using primes in the ring $\mathbb{Z}[\omega]$, where ω is an imaginary cube root of unity. As pointed out by W. C. Waterhouse, the genesis of reciprocity laws is in Gauss's lemma.

Chapter 7: Finite groups

This chapter considers various aspects of enumeration vis-a-vis finite groups. Firstly, one notes that the partition function whose value at n is $p(n)$ gives the number of conjugate classes of elements in the symmetric group S_n. Following David Jacobson and Kenneth S. Williams, the number of representations of an element in a finite group G as a product of s 'special elements' possessing a specified property P is considered. A formula for the number $N(D, a, s)$ of representations of $a \in G$ as a product of s elements belonging to D, where $G \setminus D$ is a subgroup of G, is obtained in theorem 51. Some illustrations are shown. Next, as an application of Burnside's lemma, it is shown that the number of cyclic subgroups of a group G of order r is $d(r)$ (the number of divisors of r) if, and only if, G is cyclic. See theorem 54 which is due to I. M. Richards. An identity due to P. Kesava Menon is also deduced. Further, given a positive integer r, a group G of order r is the only cyclic group of order r if, and only if, g.c.d $(r, \phi(r)) = 1$, where ϕ denotes Euler totient. See theorem 55.

Part II comprises four chapters, 8 to 11, and they deal with certain aspects of algebraic structures with reference to (i) partial ordering, (ii) valuation. Abstract Möbius inversion, generating functions and convolutions of functions defined on a finite semigroup are also discussed.

PART II
THE RELEVANCE OF ALGEBRAIC STRUCTURES TO NUMBER THEORY

Chapter 8: Ordered fields, fields with valuation and other algebraic structures

Fields with valuation are discussed. The notion of a normed division domain due to S. W. Golomb is discussed. Properties of modular lattices are pointed out. Jordan-Hölder theorem is described. Unique factorization for elements of a non-commutative ring is made possible via lattices of ideals. An analogue of the fundamental theorem of arithmetic is also noted in the context of a finite Boolean algebra.

Chapter 9: The role of the Möbius function

This chapter is about abstract Möbius inversion. G. C. Rota's idea of Incidence functions defined on a locally finite partially ordered set places Möbius inversion in a general setting. Möbius inversion formula of number theory is obtained as a special case. The incidence algebra of $n \times n$ matrices is described. Considering a vector space $V_n(q)$ of dimension n over a finite field $\mathbb{F}_q$, one obtains a formula for the number of k-dimensional subspaces of $V_n(q)$. The Möbius function of the lattice $L(V_n(q))$ of subspaces of $V_n(q)$ is derived. See theorem 72.

Chapter 10: The role of generating functions

Perhaps the first instance of a generating function was noticed by Euler while studying the partition function $p(n)$, denoting the number of unrestricted partitions of n. Examples of generating functions occur in results relating to Stirling numbers and Bernoulli numbers. While deriving proofs of theorems, the essential analytical background is sketched. Certain generating functions are expressible as a suitable infinite product under given hypotheses. The generating function of Ramanujan's τ-function is an example. Using the notion of binomial posets, we consider an algebra of incidence functions. Its connection with the algebra $\mathbb{C}[[x]]$ of formal power series in x is pointed out. See theorem 77. Dirichlet series of an arithmetic function gives yet another example of a generating function. Properties of Dirichlet series are discussed. Given a field F, the ring $F[[x]]$ of formal power series in x is also considered in order to show that it is an example of a valuation ring. See theorem 81.

Chapter 11: Semigroups and certain convolution algebras

Following E. Hewitt and H. S. Zuckerman (Finite dimensional convolution algebras: Acta Mathematica 93 (1955), 67–119) a convolution algebra of functions defined on a semigroup G is introduced. Denoting the convolution algebra by $\mathcal{L}_1(G)$, it is shown that $\mathcal{L}_1(G)$ is isomorphic to the semigroup algebra $\mathbb{C}G$.

Certain applications to arithmetical convolutions are pointed out. Abstract arithmetical functions defined on a finite semigroup of idempotents give illustrations of generating functions behaving like Dirichlet series. See theorem 86. This generalisation is due to M. Tainiter. See 'Generating functions on idempotent semigroups with applications to combinatorial analysis', J. Comb. Theory 5(1968) 272–288. A subclass of the class of Dirichlet series of arithmetic functions gives rise to a new kind of algebra called a functional-theoretic algebra (F-T.A). We remark that a finite dimensional F-T.A is, indeed, a convolution algebra.

Part III gives a bird's eye view of the fundamentals of algebraic Number Theory. Noetherian and Dedekind domains are discussed in detail. The Pell equation and its solution by the Cakravala method of Brahmagupta are presented in connection with quadratic number fields. Dirichlet's unit theorem is proved. Next, the case of class-number two number fields gives rise to the notion of half-factorial domains. Carlitz's characterization of such number fields is worthy of mention. See theorem 118 (Chapter 13).

PART III
A GLIMPSE OF ALGEBRAIC NUMBER THEORY
Chapter 12: Noetherian and Dedekind domains

This chapter is about the study of Noetherian rings, Artinian rings and Dedekind domains. While discussing Noetherian rings, it is shown that if R is a Noetherian ring in which all maximal ideals are principal, then R is a principal ideal ring (PIR) (see worked-out example b). The Jacobson radical of a ring is introduced. One comes across the class of semisimple rings in which the Jacobson radical is (0). An analogue of Euclid's theorem on infinitude of primes is that a commutative ring R (with unity) is semisimple if, and only if, R is either a field or has an infinite number of maximal ideals (see theorem 92). Properties of Dedekind domains are shown. One meets with an analogue of the Chinese Remainder Theorem in the context of Dedekind domains. Integral domains with finite-norm property are also discussed.

Chapter 13: Algebraic number fields

The ideal class-group is introduced. Number fields having class-number 1 or 2 are discussed. Some properties of cyclotomic fields are pointed out. The Pell equation and its solution are shown. Dirichlet's unit theorem is given with proof. See theorem 124.

Part IV is the concluding part of the monograph. There are three chapters in this section, namely, chapters 14 to 16. These give some more interconnections.

We mention certain classes of periodic functions (mod r) ($r \geq 2$). Various convolutions of arithmetic functions are discussed. Let $B_r(\mathbb{C})$ denote the algebra of even functions (mod r). In the case of the ring $(B_r(\mathbb{C}),+,\cdot)$ under addition and Cauchy multiplication, as $B_r(\mathbb{C})$ is a finite dimensional algebra over $\mathbb{C}$, the ring has no nonzero nilpotent elements. So, divisors of 0 in $(B_r(\mathbb{C},+,\cdot))$ are not nilpotent. However, if one considers the algebra $\mathcal{A}'$ of complex-valued arithmetic functions under the operations of addition and Lucas multiplication, $\mathcal{A}'$ is a ring in which there are zero divisors that are nilpotent. Carlitz Conjecture (1966) says that in $(\mathcal{A}',+,*)$ (the Lucas ring of arithmetic functions $f : \widetilde{\mathbb{Z}} \mapsto F$, where $\widetilde{\mathbb{Z}}$ is the set of non-negative integers and F is a field of characteristic zero) $f \in \mathcal{A}'$ is a zero divisor if, and only if, f is nilpotent. As far as the knowledge of the author goes, this conjecture is yet to be resolved.

A brief account of the well-known Goldbach problem is given. Eckford Cohen obtained a finite analogue of the Goldbach problem in 1954. (See theorems 139 and 140.) An extension to the situation in algebraic number fields is also possible. Two more analogues are known. One is in the context of the ring $M_n(\mathbb{Z})$ of $n \times n$ matrices with entries from $\mathbb{Z}$. This is due to L. N. Vaserstein (1989) with generalisation by Jun Wang (1992). The polynomial 3-primes conjecture due to D. R. Hayes (1966) is narrated along with some of the theorems of G. W. Effinger (1991), which lead to a complete solution of the polynomial analogue of the Goldbach conjecture. These are described in chapter 15. See propositions 15.5.1 and 15.5.2.

Chapter 16 is an epilogue giving some more interconnections. Specifically, we look at a finite group of units of a commutative ring. We also observe that one can make a quadratic reciprocity law in the context of a finite group.

PART IV
SOME MORE INTERCONNECTIONS

Chapter 14: Rings of arithmetic functions

Following Eckford Cohen, if $r \geq 1$, the class $\mathcal{A}_r(F)$ of (r,F)-arithmetic functions $f : \mathbb{Z} \to F$ (a field) is defined. $\mathcal{A}_r(F)$ forms an algebra of dimension r under the operations of addition and Cauchy composition. $\mathcal{A}_r(F)$ is a semisimple algebra that is the direct sum of r fields each isomorphic to F. See proposition 14.2.1. Then, the set $B_r(\mathbb{C})$ of even functions (mod r) ($\mathbb{C}$, the field of complex numbers) is shown to be a semisimple algebra of dimension $d(r)$, the number of divisors of r. The algebra of even functions (mod r) is studied in section 14.3. Next, Carlitz conjecture about $\mathcal{A}'$ (defined earlier) is mentioned. This conjecture is about the structure of the Lucas ring $\mathcal{A}'$ (of arithmetic functions) that is a commutative ring with unity. Defining a 'primary ring' as a ring in which there is a proper minimal prime ideal, we show that a commutative ring S having unity element is a primary ring if, and only if, every zero divisor of S is nilpotent (see Remark 14.8.3).

When the set of $\mathcal{A}$ of arithmetic functions is considered as a vector space over $\mathbb{C}$, certain linear operators on $\mathcal{A}$ (which are norm-preserving) yield interesting number-theoretic identities. Examples are given.

Chapter 15: Analogues of the Goldbach problem

The Goldbach problem of number theory is that every even number greater than 4 is a sum of two odd primes. This problem is about 260 years old. I. M. Vinogradov (1891–1983) proved in 1937 that when n is an odd integer which is 'sufficiently large', n could be expressed as a sum of three primes. Experimental results, using super computers show that the Goldbach conjecture is true for all even numbers up to 4.10^{11}. An interesting connection with algebra is that every element of the residue class ring $\mathbb{Z}/r\mathbb{Z}$ ($r \geq 2$) (considered in terms of a least non-negative residue system (mod r)) is a sum of two primes of $\mathbb{Z}/r\mathbb{Z}$. This is due to Eckford Cohen (1954). He has also extended this result to a residue class ideal of a number ring.

$M_n(\mathbb{Z})$ denotes the ring of $n \times n$ matrices with entries from $\mathbb{Z}$. L. N. Vaserstein has shown that given an integer p and $A \in M_2(\mathbb{Z})$, one can find matrices $X, Y \in M_2(\mathbb{Z})$ such that $A = X + Y$ with det $X =$ det $Y = p$. Next, let n be even and q be an arbitrary positive integer. Then, given $A \in M_n(\mathbb{Z})$, there exist matrices $X, Y \in M_n(\mathbb{Z})$ such that $A = X + Y$ with det $X =$ det $Y = q$. See theorem 142. Theorem 143 covers the case of $M_n(\mathbb{Z})$ with n odd.

Let $\mathbb{F}_q$ be a finite field of characteristic p (a prime). Suppose that $M(x) \in \mathbb{F}_q[x]$. $M(x)$ is called an even polynomial, if $q = 2$ and if x or $x + 1$ divides $M(x)$. $M(x)$ is called odd, if it is not even.

Let $M(x) \in \mathbb{F}_q[x]$, monic with deg $M(x) = r$. $M(x)$ is called a 3-primes polynomial, if there exist irreducible monic polynomials $P_1(x)$, $P_2(x)$ and $P_3(x) \in \mathbb{F}_q[x]$ such that deg $P_1(x) = r$, deg $P_2(x) < r$, deg $P_3(x) < r$ and $M(x) = P_1(x) + P_2(x) + P_3(x)$.

We examine the polynomial 3-primes conjecture given below:

Every odd monic polynomial $M(x) \in \mathbb{F}_q[x]$ is a 3-primes polynomial except for the case q even and $M(x) = x^2 + a \in \mathbb{F}_q[x]$.

Certain particular cases are given with proofs.

Chapter 16: An epilogue: More interconnections

A journey through the adjacent lanes of number theory and algebra is, indeed, an experience beyond theorem-proving. One is tempted to believe that Gauss was more an algebraist than a number-theorist. The various types of integral domains that have appeared are a PID, a Dedekind domain, a Bézout domain, a valuation domain and a Prüfer domain. The final observation is that $\mathbb{Z}$ finds a place in many of them.

Four more interesting situations that arise are

(i) There exist commutative rings without maximal ideals. See theorem 157.

(ii) Fabrizio Zanello (2004) looks at 'infinitude of primes' in a principal ideal domain R in terms of a property of maximal ideals of $R[x]$ thus: If R is a PID, R has an infinite number of pairwise nonassociated irreducible elements if, and only if, every maximal ideal of $R[x]$ has height 2 (see theorem 158).

(iii) We mention about the structure of the group G of units of a commutative ring R when G is finite and of odd order. Further, if the order of G (the number of units in R) is of the form p^m, either p equals 2 or p is a Mersenne prime $M_q = 2^q - 1$, where q is a prime (see theorem 160 and corollary 16.4.2).

(iv) An analogue of quadratic reciprocity law due to William Duke and Kimberly Hopkins in the context of a finite group G is made possible using the notion of a discriminant d of G. See theorem 162.

Some of the theorems presented are adaptations from journal articles and other known sources. They are duly acknowledged with proper references. It was kind of the referee to have suggested that the polynomial analogue of the Goldbach problem be included. This has improved the original version of chapter 15 in the present form for which the author is thankful to the referee.

Before concluding, the author wishes to remark that selected chapters from Parts I to IV may be chosen as the course material for a one-semester programme for senior undergraduate students and for beginning research scholars entering the areas of number theory and algebra. Some suitable combinations of chapters are

(i) chapters 1,2,3,4,5 and 6   (v) chapters 1,3,4,9,10 and 11

(ii) chapters 2,3,4,5,11 and 14   (vi) chapters 4,5,6,7,14 and 16

(iii) chapters 1,3,4,5,6 and 7   (vii) chapters 2,8,12,13,15 and 16.

(iv) chapters 2,3,5,6,12 and 13

It goes without saying that an instructor could select chapters of his/her choice.

This monograph was originally planned to be published by Marcel Dekker, Inc., New York. However, due to certain unforeseen circumstances, there was a delay for completion of the final draft of the manuscript, on the part of the author. He is grateful to Ms. Maria Allegra, Mr. Kevin Sequeira, Mr. Fred Coppersmith, Mr. David Grubbs, Ms. Theresa Delforn and Mrs. Gerry Jaffe of the Taylor & Francis Group for all the help received in connection with the publication of this manuscript.

Thanks are also due to Dr. T. R. Aggarwal, Principal Scientific Officer, Department of Science and Technology, Ministry of Science and Technology, Government of India, New Delhi for the timely help received in the matter of a release of the Grant for writing the book.

The author expresses his sincere thanks to M/s Srividya Computers, Chenakkal, Calicut University P. O. and M/s Beeta Computers, XXII/20, Rajendra Nivas, Fort, Tripunithura P. O. (both located in Kerala) for their valuable help and assistance in typesetting work done extremely well. In particular, the author is indebted to Mr. Sanjai Varma and to Mr. K. Manu for their unfailing courtesy and efficiency in the execution of LATEX.

Mistakes, if any, found in the narration of proofs or statements may kindly be pointed out to the author. None of the persons who helped the author in this venture should be held responsible for mistakes/errors that the reader may find in this monograph.

25th June 2006   R. Sivaramakrishnan

ABOUT THE AUTHOR

R. Sivaramakrishnan (b. 1936) has served the University of Calicut, India as a Faculty member of the Mathematics Department since 1977. He retired from the university service in 1996, while holding the post of Professor and Head. He earned the Ph.D Degree of the University of Kerala in 1972, while working in the Kerala State Collegiate Department. His research interests are in arithmetic function theory, enumerative combinatorics and commutative algebra. He has published many research articles individually and under joint-authorship. His earlier textbook entitled: *Classical theory of arithmetic functions* appeared in the series of Monographs and Textbooks (No: 126), Marcel Dekker, Inc., New York in 1989. He has held visiting positions at the University of Kansas, Lawrence, KS 66045, U.S.A. (1987–88) and at Mangalore University, Mangalore, DK 574199, India (1996–97).

This monograph is the outcome of the author's effort to bring out some of the interesting interconnections between number theory and commutative algebra. His project on the manuscript bearing the title of this volume got the approval of the Department of Science and Technology, Ministry of Science and Technology, Government of India, New Delhi, and the project was partially funded by the Department of Science and Technology under the scheme of 'Utilization of Scientific Expertise of Retired Scientists' (USERS).

Sivaramakrishnan is a life-member of the Allahabad Mathematical Society, Allahabad, India and is also a member of the American Mathematical Society, Providence, RI 02940 (since 1988).

CONTENTS

Part I

ELEMENTS OF NUMBER THEORY AND ALGEBRA

CHAPTER 1

Theorems of Euler, Fermat and Lagrange

Historical perspective

Number theory has a long and interesting history. It deals with the study of properties of integers. The fact that there are infinitely many primes was noted by Euclid (300 B.C.) in Euclid's Elements (Book IX, theorem 20). The result: 'For p a prime, if p divides ab (a, b integers), either p divides a or p divides b' is found in Euclid's Elements (Book VII, theorem 30). Further, Euclid noted that every natural number is divisible by at least one prime p (see Euclid's Elements, Book VII, theorem 31). Every positive integer $n(> 1)$ is a product of primes and apart from rearrangement of factors, n can be expressed as a product of primes uniquely. This is known as the fundamental theorem of arithmetic (F.T.A). F.T.A does not seem to have been stated in this form before Carl Friedrich Gauss (1777– 1855). As pointed out in [4], it was familiar to earlier mathematicians, but Gauss was the first to develop arithmetic as a "systematic science". Problems in number theory led to many important developments in other branches of mathematics: for instance, Gauss's construction of a regular polygon of 17 sides. Over the years, many results of significance sprang up.

A positive integer is said to be 'representable' if it can be expressed as the sum of two squares of integers (including zero). In fact, a perfect square r^2 is representable in the sense that $r^2 = r^2 + 0^2$. It is known that the least integer which is representable in three ways is

$$325 = 18^2 + 1^2 = 17^2 + 6^2 = 15^2 + 10^2.$$

Representable numbers were first studied by Diophantos in 250 A.D. Equations for which solutions are sought in integers are called Diophantine equations. In the case of the equation $2x + 5y = 100$, a solution by inspection gives $\langle x, y \rangle = \langle 10, 16 \rangle$, as are many others. Pierre de Fermat (1601–1665) who was a lawyer by profession, took interest in mathematics while reading a translation by Bachet (1581–1638) of Diophantos' 'Arithmetica'. Fermat gave a formula for the number of solutions of the Diophantine equation

$$x^2 + y^2 = r.$$

This follows from Fermat's Two-squares theorem. It may be remarked that Fermat merely stated theorems and many of his theorems were codified with proofs by Leonhard Euler (1707–1783), the way we learn number theory from textbooks.

The 'congruence' notation was introduced and used extensively by Gauss. The famous 'Four-squares theorem' which states that every positive integer is the sum of four integer squares, was guessed by Fermat and proved by Joseph Louis Lagrange (1736–1813). Diophantos probably knew the Four squares theorem.

Among the mathematicians who contributed to number theory during the 17th and the 18th centuries, the name of John Wilson (1741–1793) is never missed as he proved only one theorem of number theory in his lifetime and he is known by that lone theorem.

1.1. Introduction

The set $\{0, \pm 1, \pm 2, \ldots\}$ of integers is denoted by $\mathbb{Z}$. This notation probably got established due to the fact that the German word zahlen means 'number'.

The purpose of this chapter is to point out certain proofs of classical theorems of elementary number theory using ideas arising from

(1) an elementary enumeration principle of Melvin Hausner [6]

(2) a map on $\mathbb{Z} \times \mathbb{Z} \times \mathbb{Z}$ due to D. Zagier [17] and

(3) Newman's [11] factorisation of 2×2 matrices over the ring of Gaussian integers.

We also give conditions necessary and sufficient for the existence of solutions of Diophantine equations of first degree in s unknowns. A formula for solutions of

$$a_1 x_1 + a_2 x_2 + \ldots + a_s x_s = r, \quad a_i \geq 1, \quad i = 1, 2, \ldots, s; \quad r \geq 1,$$

when solutions exist, is pointed out.

1.2. The quotient ring $\mathbb{Z}/r\mathbb{Z}$

Let r be an arbitrary element of $\mathbb{Z}$. We write

$$r\mathbb{Z} = \{rk : k \in \mathbb{Z}\}.$$

It suffices to choose r to be either 0 or a positive integer. $r\mathbb{Z}$ consists of integers congruent to 0 (mod r). $r\mathbb{Z}$ is an ideal of the ring $(\mathbb{Z}, +, \cdot)$, where $+$ denotes addition and $\cdot$ denotes multiplication. We know that for $a, b \in \mathbb{Z}$,

$$a \equiv b \pmod{r}$$

if, and only if, $a - b \in r\mathbb{Z}$, that is, if, and only if, a and b come from the same coset of $(r\mathbb{Z}, +)$ in $(\mathbb{Z}, +)$. Congruence (modulo r) is an equivalence relation on $\mathbb{Z}$ and a congruence class is a coset of $(r\mathbb{Z}, +)$ in $(\mathbb{Z}, +)$. Each coset of $(r\mathbb{Z}, +)$ is a subset of $\mathbb{Z}$ containing integers which form an arithmetic progression with common difference r. It is this connection between congruence (modulo r) and cosets of $(r\mathbb{Z}, +)$ that motivates quotient groups and quotient rings.

It is easy to check that the quotient group $(\mathbb{Z}/r\mathbb{Z}, \oplus)$ under addition (modulo r) is cyclic and is of order r. The quotient ring $(\mathbb{Z}/r\mathbb{Z}, \oplus, \otimes)$ under addition and multiplication (modulo r) is an example of a finite ring which is a field if, and only if, r is a prime. (The abstract definition of a field was given by

Heinrich Weber (1842–1913) in a paper of 1893). The group of units in $\mathbb{Z}/r\mathbb{Z}$ is the set of cosets

$$\{ [a] : a \in \mathbb{N}, \text{g.c.d}\, (a,r) = 1) \},$$

where $\mathbb{N}$ denotes the set of positive integers. The cardinality of the group of units in $\mathbb{Z}/r\mathbb{Z}$ is $\phi(r)$, the Euler totient. To find x such that

(1.2.1) $ax \equiv b \pmod{r}$, where a, b are fixed integers,

is to obtain the congruence class $[t]$ such that $at \equiv b \pmod{r}$. When t is obtained, $[t]$ is a solution of (1.2.1). It is known that (1.2.1) is solvable if, and only if, g.c.d (a,r) divides b. If this happens, the congruence (1.2.1) is equivalent to the equation

(1.2.2) $AX = B$

in $\mathbb{Z}/r\mathbb{Z}$, where $A = [a]$ and $B = [b]$. If $X \in \mathbb{Z}/r\mathbb{Z}$ is obtained from (1.2.2), any $t \in X$ solves (1.2.1). Conversely, if t is a solution of (1.2.1) then $[t]$ solves (1.2.2). In short, what are congruences in number theory are equations in algebra.

Next, we note that (1.2.1) could be written as

(1.2.3) $ax = b + ry$

which is an example of a Diophantine equation in x, y. That is, (1.2.1), (1.2.2) and (1.2.3) are three formulations of the same problem, namely,

(1) solving a linear congruence
(2) solving an equation in $\mathbb{Z}/r\mathbb{Z}$ and
(3) solving a linear Diophantine equation in two unknowns x, y.

Now, let us denote the group of units in $\mathbb{Z}/r\mathbb{Z}$ by $U(r)$. For $r \geq 1$, $U(r)$ has $\phi(r)$ elements. One notes that

(1.2.4) $a^{\phi(r)} \equiv 1 \pmod{r}$, whenever g.c.d $(a,r) = 1$.

which is Euler's theorem.

In the case $r = p$, a prime, one sees that

(1.2.5) $a^{p-1} \equiv 1 \pmod{p}$

(whenever p does not divide a) which is Fermat's little theorem.

Next we point out that Wilson's theorem follows as a corollary of

Lemma 1.2.1 : *Let $G = \{a_1, a_2 \ldots, a_r\}$ be an abelian group of order r under multiplication, with a_1 serving as the identity element. If a_j ($j \neq 1$) is the only element of order 2 in G, then $a_1 \cdot a_2 \cdot \ldots \cdot a_r = a_j$.*

Proof : We observe that a_j is its own inverse. The remaining elements a_i ($i \neq 1, i \neq j$) have distinct inverses. That is, the product of all these elements is the identity a_1.

$$a_1 a_2 \ldots a_{j-1} a_{j+1} \ldots a_r = a_1.$$

Multiplication by a_j gives the desired result.   □

Corollary 1.2.1 (Wilson's theorem)**:** *If p is a prime, then, $(p-1)! + 1 \equiv 0 \pmod{p}$.*

Proof : For, the case $p = 2$ is obvious. So, let p be an odd prime. If we consider the multiplicative group of $\mathbb{Z}/p\mathbb{Z}$, it is cyclic and so abelian. Its $(p-1)$ elements are such that $[(p-1)]$ is the only element of order 2 in the group. So, by lemma 1.2.1, we see that $(p-1)! \equiv -1 \pmod{p}$. $\square$

Remark 1.2.1 : Wilson's theorem has a valid converse which is stated below:

If r is a positive integer > 1 such that $(r-1)! + 1 \equiv 0 \pmod{r}$, then r is a prime.

This by itself is not an efficient way to determine primes. But, Wilson's theorem is relevant to applications in certain contexts. For instance, if p is a prime, one knows that the quadratic congruence

$$(1.2.6) \qquad\qquad x^2 \equiv -1 \pmod{p}$$

has a solution if, and only if, p is a prime of the form $4k+1$. To prove (1.2.6) one way, Wilson's theorem is used.

Before concluding this section, we point out an interconnection between algebra and number theory.

Definition 1.2.1 : *Let $r > 1$ be a composite number. Suppose that d is an arbitrary but fixed integer such that $1 \leq d < r$. We write*

$$(1.2.7) \qquad\qquad A_r(d) = [\{x \in \mathbb{Z} : xd \equiv 0 \pmod{r}\}]$$

It is easily verified that

$$A_r(d) = \{[x] \in \mathbb{Z}/r\mathbb{Z} : [x]\,[d] = [0]\}.$$
$$A_r(d) = A_r(dt), \text{ if g.c.d } (t,r) = 1$$
$$A_r(d) \text{ is an ideal of the ring } \mathbb{Z}/r\mathbb{Z}.$$
$$A_r(d) \text{ is the annihilator of } [d] \text{ in } \mathbb{Z}/r\mathbb{Z}.$$

As an illustration, when $r = 12, d = 7$, one has

$$A_{12}(7) = [\{x \in \mathbb{Z} : 7x \equiv 0 \pmod{12}\}] = [0].$$

When $r = 12, d = 4$, we get

$$A_{12}(4) = [\{x \in \mathbb{Z} : 4x \equiv 0 \pmod{12}\}] = \{[0], [3], [6], [9]\}.$$

$A_{12}(4)$ is contained in $\mathbb{Z}/12\mathbb{Z}$ and is an ideal of $\mathbb{Z}/12\mathbb{Z}$. Next, suppose that d divides r. When $x = r/d, 2r/d, \ldots, (d-1)r/d, r$; $xd \equiv 0 \pmod{r}$ and so, there are d elements in $A_r(d)$.

Lemma 1.2.2 (Charles Green [5]) **:** *Let d be a proper divisor of r. Suppose that $A_r(d)$ is as defined in (1.2.7). Then, $A_r(d)$ is a field if, and only if, d is a prime such that g.c.d $(d, r/d) = 1$. If $p_1, p_2, \ldots, p_s$ are prime divisors of r such that g.c.d $(p_i, r/p_i) = 1$, $(i = 1, 2, \ldots, s)$, there are s fields contained in $\mathbb{Z}/r\mathbb{Z}$.*

Proof : $:\Rightarrow$ Let $A_r(d)$ be a field. Suppose that $d = d_1d_2$. We write $r = d_1d_2t$. We note that $[d_1t]$ and $[d_2t]$ are both nonzero elements in $A_r(d)$. But, $[d_1t][d_2t] = [d_1d_2t^2] = [0]$. So, if $d^2|r$, then $r = d^2s$ and $[d][s]$ is a nonzero element of $A_r(d)$ such that $[ds][ds] = [d^2s^2] = [0]$. As $A_r(d)$ is a field, d has to be a prime p such that $p^2 \nmid r$; that is, d is a prime and g.c.d $(d, \frac{r}{d}) = 1$.

$\Leftarrow$: Suppose that d is a prime p and $p^2 \nmid r$. Let $q = \frac{r}{p}$ and g.c.d $(q, p) = 1$. If $m \in A_r(p)$, then $m = [nq]$ for some integer n. Since $a \equiv b \pmod{r}$ implies $aq \equiv bq \pmod{r}$, we see that an element $[nq]$ of $A_r(p)$ is $[0]$ if, and only if, $p|n$. Hence, if $[aq]$ and $[bq]$ are nonzero elements of $A_r(p)$, we have

$$[aq][bq] = [abq^2] \text{ is nonzero.}$$

p is a prime such that $p \nmid a$, $p \nmid b$ and $p \nmid q$. So $p \nmid abq$. Therefore, $A_r(p)$ is an integral domain. Since $A_r(p)$ is finite, $A_r(p)$ is a field.

Further, there are as many fields $A_r(p)$ as there are primes p for which $p \mid r$ and g.c.d $(p, \frac{r}{p}) = 1$. This proves the lemma 1.2.2. $\qquad\square$

Corollary 1.2.2 : *Let p be a prime such that $p^2 \nmid r$. Then, the identity element of the field $A_r(p)$ is $[cc']$ where $c \neq 0$, $[c] \in \mathbb{Z}/p\mathbb{Z}$ and $cc' \equiv 1 \pmod{p}$.*

Proof : Given $A_r(p)$ is a field, we can consider a nonzero element $[c] \in \mathbb{Z}/p\mathbb{Z}$, where $\mathbb{Z}/p\mathbb{Z}$ is a field. Since $p \nmid c$, there exists $[c'] \in \mathbb{Z}/p\mathbb{Z}$ such that

$$cc' \equiv 1 \pmod{p}.$$

Then $[cc']$ is a nonzero element of $A_r(p)$. Further,

$$[cc'][cc'] = [c^2c'c'] \text{ and } c^2c' \equiv c \pmod{p}.$$

$$\text{Also, } c' \equiv c' \pmod{\frac{r}{p}}.$$

So,

$$c^2c'c' \equiv cc' \pmod{\text{r}} \text{ or } [cc'][cc'] = [cc'] \text{ in } A_r(p).$$

That is, $[cc']$ is the identity element in $A_r(p)$. $\qquad\square$

Example 1.2.1 : Taking $r = 12$, $p = 3$, we observe that $A_{12}(3)$ is a field, as g.c.d$(3,4) = 1$. Further,

$$A_{12}(3) = \{[0], [4], [8]\}$$

$[2] \in \mathbb{Z}/3\mathbb{Z}$ is such that $2^2 \equiv 1 \pmod{3}$. Clearly, $[4]$ is the identity element of $A_{12}(3)$.

Example 1.2.2 : Taking $r = 14$, $p = 7$, we have

$$A_{14}(7) = \{[0], [2], [4], [6], [8], [10], [12]\}.$$

7^2 does not divide 14. Further, 2 and 4 are such that $8 \equiv 1 \pmod{7}$. Then, $[8]$ is the identity element in $A_{14}(7)$.

Next, we note that every ideal of $\mathbb{Z}$ is principal. That is, an ideal of $\mathbb{Z}$ is generated by a single element. For, let I be an ideal of $\mathbb{Z}$. If $I = (0)$, I is generated by 0. Suppose $I \neq (0)$. I contains a nonzero element say n. If $n \in I$, $-n \in I$ and so I contains positive integers. Let m be the least positive integer contained in I. The division algorithm in $\mathbb{Z}$ says that given $a \in \mathbb{Z}$,

$$a = mq + r, \text{ where } 0 \leq r < m.$$

Since $a \in I$ and $mq \in I$, $r = a - mq \in I$.

As $r < m$, if $r \neq 0$, we arrive at a contradiction to the minimality of $m \in \mathbb{Z}$. This forces r to be zero. So, $a = mq$.

We write

$$(m) = \{km : k \in \mathbb{Z}\}.$$

(m) is the ideal of $\mathbb{Z}$ generated by m. So $I \subseteq (m)$. As $m \in I$, $(m) \subseteq I$. Thus, $I = (m)$, or every ideal of $\mathbb{Z}$ is principal. In fact, $\mathbb{Z}$ is an example of principal ideal domain written PID.

Definition 1.2.2 : *The set of Gaussian integers is defined by*

$$\mathbb{Z}[i] = \left\{a + bi : a, b \in \mathbb{Z}, \ i \text{ denotes } \sqrt{-1}\right\}.$$

The function $g : \mathbb{Z}[i] \to \mathbb{Z}$ given by

(1.2.8) $g(a + bi) = a^2 + b^2 \quad a, b \in \mathbb{Z}$

serves as a Euclidean Norm on $\mathbb{Z}[i]$ and so $\mathbb{Z}[i]$ is a PID.

We look at Euclidean domains more closely in Chapter 3.

1.3. An elementary counting principle

X denotes a finite set. Let p be an arbitrary but fixed prime. $|S|$ denotes the number of elements of S when S is finite. There exist functions $f : X \to X$ such that

$$f^p = f \circ f \circ \cdots \circ f \text{ (p times) } = j, \text{ the identity map on } X.$$

For example, let G be a finite group. We define $f : G \to G$ by $f(x) = x^{-1}$ (the inverse of x) for each $x \in G$. It is clear that $f \circ f = f^2 = j$.

If T is defined by

$$T = \{x \in G : f(x) = x\},$$

then,

$$T = \left\{x \in G : x^{-1} = x\right\}$$
$$= \left\{x \in G : x^2 = e, \text{ the identity element in G}\right\}.$$

We note that $|T|$ gives the number of elements of order 2 in G. If $|G|$ is odd, $|T| \equiv 1 \pmod 2$ and we get

(1.3.1) $|G| \equiv |T| \pmod 2$.

If $|G|$ is even, elements x for which $x \neq x^{-1}$ can be paired off and so $|T| \equiv 0 \pmod 2$. In this case also (1.3.1) holds. As $e \in T$, it follows that if G is a group of even order, then G has an element $a \neq e$ such that $a^2 = e$ (see I. N. Herstein [8]). This is precisely what we did in the proof of Wilson's theorem for odd primes p. The sort of argument used to obtain (1.3.1) could be given in a general set-up.

Theorem 1 (Melvin Hausner (1983)) **:** *Let $f : X \to X$ be such $f^p = j$, the identity map. Suppose T is a subset of X defined by*

$$T = \{x \in X : f(x) = x\}.$$

Then,

(1.3.2) $|X| \equiv |T| \pmod{p}$.

Proof : For $x \in X$, we define $A(x)$ by

$$A(x) = \{x, f(x), f^2(x), \cdots, f^{p-1}(x)\}.$$

$A(x)$ is called the orbit of x under f, as $f^p(x) = x$ for all $x \in X$. The orbits of the elements of X give rise to a partition of set X. Further,

$$|A(x)| = 1 \text{ if, and only if, } f(x) = x.$$

That is, $|A(x)| = 1$ if, and only if, $x \in T$.

Claim : If $|A(x)| > 1$, then $|A(x)| = p$. If $A(x)$ is such that, for some s, t; $0 \le s < t < p$; $f^s(x) = f^t(x)$, then, $f^{t-s}(x) = x$. Since $f^p(x) = x$ and g.c.d $(t-s, p) = 1$, it follows that $t - s = 1$ and $f(x) = x$. So, $|A(x)| = 1$. Thus, the elements of $A(x)$ are all distinct if $|A(x)| > 1$. Then, $|A(x)| = p$. Now, there are $|T|$ orbits of length 1 in X. As X is a disjoint union of orbits, we get

$$|X| = |T| + mp,$$

where m is the number of orbits of length p. (1.3.2) follows.   □

Theorem 2 (Fermat) **:** *For $n \in \mathbb{N}$ and p a prime*

(1.3.3) $n^p \equiv n \pmod{p}$.

Proof : Let $\mathbb{R}$ denote the field of real numbers.

$$\mathbb{R}^n = \{(x_1, x_2, \cdots, x_n) : x_i \in \mathbb{R}, i = 1, 2, \cdots, n\}$$

is referred to as an n–dimensional vector space over $\mathbb{R}$.
Points $(x_1, x_2, \cdots, x_n)$ where $x_i \in \mathbb{Z}$ $(i = 1, 2, \cdots, n)$ are called lattice points in $\mathbb{R}^n$.
    From $x_1, x_2, \cdots, x_n$, $1 \le x_i \le n$ $(i = 1, 2, \cdots, n)$, we choose a lattice point $(x_1, x_2, \cdots, x_p) \in \mathbb{R}^p$. We define

$$X = \{(x_1, x_2, \cdots, x_p) : 1 \le x_i \le n, (i = 1, 2, \cdots, p)\}.$$

A function $f : X \to X$ is defined by

$$f(x_1, x_2, \cdots, x_p) = (x_2, x_3, \cdots, x_p, x_1).$$

Then $f^p = j$, the identity map. We note that $|X| = n^p$.

If $(x_i, x_i, \cdots, x_i) \in X, i = 1, 2, \cdots, n;$

$$f(x_i, x_i, \cdots, x_i) = (x_i, x_i, \cdots, x_i), \; i = 1, 2, \ldots n.$$

So, if

$$T = \{(x_1, x_2, \cdots, x_p) : f(x_1, x_2, \cdots, x_p) = (x_1, x_2, \cdots, x_p)\},$$
$$|T| = n.$$

By theorem 1, we get the desired congruence (1.3.3). $\qquad\qquad\square$

Remark 1.3.1 : We note that this argument works with $\mathbb{R}^p$, whatever be the value of $n \neq 0$.

Theorem 3 (Lucas' Theorem) **:** *Let p be any prime. Suppose that*

$$\left. \begin{array}{ll} n & = n_0 + n_1 p + \cdots + n_k p^k \quad 0 \leq n_i < p \; ; \\ and \; r & = r_0 + r_1 p + \cdots + r_k p^k \quad 0 \leq r_i < p. \end{array} \right\} i = 0, 1, \cdots, k.$$

Then,

(1.3.4)
$$\binom{n}{r} \equiv \binom{n_0}{r_0} \binom{n_1}{r_1} \cdots \binom{n_k}{r_k} \pmod{p},$$

where

$$\binom{a}{b} = \begin{cases} 0, & \text{if } b > a; \\ \frac{a!}{b!(a-b)!}, & \text{if } b < a; \\ 1, & \text{if } b = a. \end{cases}$$

Proof : n and r are such that

$$n = n_0 + Np, \; n_0 \geq 0$$
$$r = r_0 + Rp, \; r_0 \geq 0.$$

Once we prove that

(1.3.5)
$$\binom{n}{r} = \binom{n_0}{r_0} \binom{N}{R} \pmod{p},$$

we will get from

$$n = n_0 + n_1 p + n_2 p^2$$
$$r = r_0 + r_1 p + r_2 p^2$$
$$n = (n_0 + n_1 p) + n_2 p \cdot p$$
$$r = (r_0 + r_1 p) + r_2 p \cdot p$$

So,

$$\binom{n}{r} \equiv \binom{n_0 + n_1 p}{r_0 + r_1 p} \binom{n_2 p}{r_2 p} \pmod{p},$$

Or,

$$\binom{n}{r} \equiv \binom{n_0}{r_0} \binom{n_1}{r_1} \binom{n_2 p}{r_2 p} \pmod{p}.$$

But,

$$\binom{n_2 p}{r_2 p} \equiv \binom{n_2}{r_2} \pmod{p}$$

This would imply that

$$\binom{n}{r} \equiv \binom{n_0}{r_0} \binom{n_1}{r_1} \binom{n_2}{r_2} \pmod{p}.$$

Therefore, it suffices to show that (1.3.5) holds. We write

$$A_i = \{(i,1),(i,2),\cdots,(i,N)\}; \quad i = 1,2,\cdots,p.$$

and

$$B = \{(0,1),(0,2),\cdots,(0,n_0)\}.$$

$A_1, A_2, \cdots, A_p$ and B are sets of ordered pairs. $\mid A_i \mid = N$, $i = 1, 2, \cdots, p$; $\mid B \mid = n_0$. There are $(p+1)$ sets under consideration. We write

(1.3.6) $$A = A_1 \cup A_2 \cup \cdots \cup A_p \cup B.$$

Then,

(1.3.7) $$\mid A \mid = N p + n_0.$$

We define $f : A \to A$ by moving A_i's cyclically and keeping B fixed. That is,

(1.3.8) $$\begin{cases} f((i,x)) = (i+1,x), \ 1 \le i \le p-1, \ x = 1,2,\cdots,N. \\ f((p,x)) = (1,x), \ x = 1,2,\cdots,N. \\ f((0,x)) = (0,x), \ x = 1,2,\cdots,n_0. \end{cases}$$

From (1.3.8), we get

(1.3.9) $$\begin{cases} f(A_i) = A_{i+1} \ (1 \le i \le p-1), \\ f(A_p) = A_1, \\ f(B) = B. \end{cases}$$

It is seen that $f^p(A) = A$ or $f^p = j$, the identity map. We take X as the collection of subsets C of A with $\mid C \mid = r$.

$$f(C) = \{f(x) : x \in C\}$$

As f is one-to-one, $|f(C)| = |C|$. $|X|$ is the number of r-element subsets of A, where $|A| = n$. So, $|X| = \binom{n}{r}$.

Any subset C of A can be uniquely written as

$$C = C_1 \cup C_2 \cup \cdots C_p \cup C_0$$

where $C_i \subseteq A_i$ and $C_0 \subseteq B$.

Since f sends A_i cyclically around and keeps B fixed, we see that

$$f(C) = C$$

if, and only if, $C_i = f^{i-1}(C_1)$ $i = 1, 2, \cdots, p$. For, $f(C_1) = C_2$, $f^2(C_1) = C_3, \cdots$ and

$$f(C_i) = C_{i+1},$$
$$f(C_p) = C_1,$$
$$f(C_0) = C_0.$$

Then $f(C) = C$.

For C contained in X, we have $|C| = r$. If C is in T,

$$r = |C| = p|C_1| + |C_0|.$$

Also, $|C_0| \geq 0$, $r_0 < p$. The cardinality restriction on C is satisfied if, and only if, $|C_1| = R$, $|C_0| = r_0$. So, $|C| = Rp + r_0$. There are $\binom{N}{R}$ such choices for C_1 and $\binom{n_0}{r_0}$ independent choices for C_0. So,

$$|T| = \binom{N}{R}\binom{n_0}{r_0}$$

and $|X| \equiv |T| \pmod{p}$ yields (1.3.5). $\square$

Remark 1.3.2 : Proofs of theorems 1, 2 and 3 have been adapted from [6].

1.4. Fermat's two squares theorem

It is known [4] that a prime p of the form $4k + 1$ can be expressed as a sum of two squares. Many proofs are available. One is based on the fact that a prime of the form $4k + 1$ splits in $\mathbb{Z}[i]$. (It can be proved without appealing to $\mathbb{Z}[i]$). In 1984, D. R. Heath-Brown [7] gave a new proof based on an involutory map on a finite set. D. Zagier [17] gave a proof in 1990 on the lines of proof given by Heath-Brown. D. Zagier's proof is presented below. We need a definition and a special case of theorem 1.

Definition 1.4.1 : *Let X be a finite set. Suppose $f : X \to X$ is a well-defined map. f is called an involution, if $f \circ f = j$, the identity map.*

The set

$$T = \{y \in X : f(y) = y\}$$

is called the set of fixed points of f.

From theorem 1, we see that we could take the case $p = 2$ in the congruence $|X| \equiv |T| \pmod{p}$.

So, when $f : X \to X$ is an involution, we deduce that $|X|$ and $|T|$ have the same parity, or

(1.4.1) $|X| \equiv |T| \pmod{2}$.

D. Zagier remarks that the above congruence is the combinatorial analogue of the

Proposition 1.4.1 : *The Euler characteristic of a topological space and its fixed-point set under any continuous involution have the same parity.*

The above theorem is from algebraic topology. See [2].

Theorem 4 (Fermat's Two-squares theorem) **:** *Any prime $p \equiv 1$ (mod 4) is a sum of two squares.*

Proof : We define a set

$$S = \{(x,y,z) \in \mathbb{N}^3 : x^2 + 4yz = p\}$$

where $\mathbb{N}$ denotes the set of positive integers. p is an arbitrary but fixed odd prime, where $p \equiv 1 \, (\text{mod } 4)$.

If $p = 4k + 1$ taking $x = 1$, $k = yz$, the different values of y, z are obtained from the different factorizations of k into the form yz. So, there are only a finite number of solutions $\langle x, y, z \rangle$ to the equation

$$x^2 + 4yz = p.$$

So, S is a finite set. For $(x, y, z) \in S$, let

$$(1.4.2) \qquad f(x,y,z) = \begin{cases} (x+2z, z, y-x-z), & \text{if } x < y-z \\ (2y-x, y, x-y+z), & \text{if } y-z < x < 2y \\ (x-2y, x-y+z, y), & \text{if } 2y < x. \end{cases}$$

The image of (x, y, z) under f is the triple (x', y', z') where x', y', z' are positive integers. Three cases arise.

(a) Suppose that (x, y, z) is such that $x < y-z$.
Then,

$$f(x,y,z) = (x+2z, z, y-x-z).$$

We have $y-x-z > 0$, $z > 0$.
As $x > 0$, $x+2z > 0$.
We write

$$x' = x+2z, y' = z, z' = y-x-z.$$

$2y' = 2z$ and $2y' < x' = x+2z$.
So,

$$f(x',y',z') = (x'-2y', x'-y'+z', y') = (x,y,z)$$

So,

$$f^2(x,y,z) = (x,y,z).$$

(b) Suppose that x, y, z are such that $y-z < x < 2y$. Then,

$$f(x,y,z) = (2y-x, y, x-y+z).$$

$$2y-x > 0, y > 0, x-y+z > 0 \text{ or } x+z > y \text{ or } y-z < x < 2y.$$

If $2y-x = x'$, $y = y'$ and $x-y+z = z'$.

$$2y'-x' = 2y-(2y-x) = x > 0.$$

Also,

$$2y' > x', y'-z' = y-(x-y+z) = 2y-x-z < x', \text{ as } x' = 2y-x.$$

Therefore,

$$y' - z' < x' < 2y'.$$

That is,

$$f(x', y', z') = (2y' - x', y', x' - y' + z') = (x, y, z).$$

So,

$$f^2(x, y, z) = (x, y, z).$$

(c) Suppose that, x, y and z are such that $2y < x$.

$$f(x, y, z) = (x - 2y, x - y + z, y).$$

Take

$$x' = x - 2y, y' = x - y + z, z' = y,$$

$$x' > 0, \ y' - z' = (x - y + z) - y = x - 2y + z = x' + z > x' \text{ or } x' < y' - z'.$$

So,

$$f(x', y,' z') = (x' + 2z', z', y' - x' - z') = (x, y, z).$$

Or,

$$f^2(x, y, z) = (x, y, z).$$

This shows that f is an involution on S. To obtain T, we note that $f(x, y, z) = (x, y, z)$ if, and only if, we get positive integral values for x, y, z such that $2y - x = x$, as $x + 2z > x$ and $x - 2y < x$ with $y - z = 0 < x < 2y$. Then,

$$x^2 + 4yz = x^2 + 4xz = p \Rightarrow x \mid p. \text{ So, } x = 1 \text{ or } p.$$

$x \neq p$. So, $x = 1$. Then, $(1, 1, k)$ is the only fixed point under f. Next, let $g : S \to S$ be given by $g(x, y, z) = (x, z, y)$. g is an involution and g has a fixed point (x, y, y). Then, $p = x^2 + 4y^2$ gives the required property of p. $\square$

Remark 1.4.1 :

(1) The above type of argument could be applied to prove the expressibility of a prime of the form $8k + 3$ as $x^2 + 2y^2$.
    See Terrence Jackson [9].

(2) For a recent but a different proof of the Two-squares theorem, see John A. Ewell [3].

(3) Counting the number of solutions of $x^2 + y^2 = p$ requires the study of the nature of primes in $\mathbb{Z}[i]$ where $\mathbb{Z}[i]$, the ring of Gaussian integers, is a unique factorization domain.

If $p_1, p_2, \ldots, p_r$ are primes congruent to 1 (mod 4) and $q_1, q_2, \ldots, q_r$ are primes congruent to 3 (mod 4), and $n = 2^a p_1^{a_1} \cdots p_r^{a_r} \cdots q_1^{2b_1} q_2^{2b_2} \cdots q_p^{2b_p}$, $x^2 + y^2 = n$ has $[\frac{(a_1+1)(a_2+1)\cdots(a_r+1)+1}{2}]$ solutions where solution $< x, y >$ and $< x', y' >$ are considered equivalent whenever $x = \pm x', y = \pm y'$. See [1]. ([x] denotes the greatest integer not exceeding x).

We metion that one can count the number $r(n)$ of solutions of $x^2 + y^2 = n$ without appealing to $\mathbb{Z}[i]$. Considering $(\pm x, \pm y)$ as distinct solutions we note that $r(n) = 8$, when n is the prime of the form $4k+1$, where as $r(n) = 0$ when n is a prime of the form $4k+3$. It can be shown that

$$(1.4.3) \qquad r(n) = 4 \sum_{d^2 | n} t\left(\frac{n}{d^2}\right),$$

where d runs through positive integers whose squares divide n and

$$t(m) = \begin{cases} 0, & \text{if 4 divides } m \text{ or if } m \text{ is divisible by a prime} \\ & \text{of the form } 4k+3 \\ 2^q, & \text{if 4 does not divide } m, m \text{ is not divisible by a prime} \\ & \text{of the form } 4k+3 \text{ and } q \text{ is the number of distinct primes} \\ & \text{of the form } 4k+1 \text{ dividing } m. \end{cases}$$

For proof, see E. Landau [10] or Don Redmond [12].

1.5. Lagrange's four squares theorem

It was Lagrange who proved in 1770 that every positive integer is a sum of four squares. The proof given below is originally due to M. Newman [11]. The simplification of the proof is due to Charles Small [16].

We begin with

Lemma 1.5.1 : *Let p be a prime. Every element of the field $\mathbb{Z}/p\mathbb{Z}$ is a sum of two squares.*

Proof : We may assume that $p \neq 2$. We write

$$S_1 = \left\{x^2 : x \in \mathbb{Z}/p\mathbb{Z}\right\}, \quad S_2 = \left\{r - y^2 : y \in \mathbb{Z}/p\mathbb{Z}\right\}$$

where r is an arbitrary representative element of $\mathbb{Z}/p\mathbb{Z}$. As x and $kp-x$ both give the same square x^2, S_1 has $\frac{p-1}{2} + 1 = \frac{p+1}{2}$ elements. S_2 also has $\frac{p+1}{2}$ elements. So, $S_1 \cap S_2$ is non empty and so r is a sum of two squares. $\qquad\square$

Remark 1.5.1 : The equation $ax^2 + cy^2 = r$ $(a, c \neq 0)$ has solutions in any finite field $\mathbb{F}_p$ (a field of characteristic p). For, one has to consider

$$S_1 = \left\{ax^2 : x \in \mathbb{F}_p\right\}.$$

and

$$S_2 = \left\{r - cy^2 : y \in \mathbb{F}_p\right\}.$$

To see that a positive integer r is a sum of four squares, without loss of generality, we may take r to be square-free. For, if $r = a^2 r'$, (r' square-free) and if $r' = w^2 + x^2 + y^2 + z^2$, then

$$r = (aw)^2 + (ax)^2 + (ay)^2 + (az)^2.$$

If $r = p_1 p_2 \cdots p_k$ where $p_1, p_2, \cdots, p_k$ are distinct primes,

$$\mathbb{Z}/r\mathbb{Z} \cong \mathbb{Z}/p_1\mathbb{Z} \times \mathbb{Z}/p_2\mathbb{Z} \times \cdots \times \mathbb{Z}/p_k\mathbb{Z} \text{ (see [8]).(See also section 5.3, chapter 5)}$$

As each element of $\mathbb{Z}/p_i\mathbb{Z}$ ($i = 1, 2, \cdots, k$) is a sum of two squares, each element of $\mathbb{Z}/r\mathbb{Z}$ is a sum of two squares. This is clear from the fact that writing $t \in \mathbb{Z}/p_i\mathbb{Z}$ as a sum of two squares for each p_i, t is a sum of two squares in $\mathbb{Z}/r\mathbb{Z}$.

So, -1 is a sum of two squares in $\mathbb{Z}/r\mathbb{Z}$, where r is square-free. We could write -1 as

(1.5.1) $-1 = c^2 + d^2 - rs$

We define a 2×2 matrix A by

(1.5.2) $A = \begin{bmatrix} r & c+di \\ c-di & s \end{bmatrix}$

where i denotes $\sqrt{-1}$. Then,

$$\det A = rs - c^2 - d^2 = 1 \text{ by (1.5.1)}.$$

Theorem 5 : *Let the matrix A be given by (1.5.2) where $c, d, r, s \in \mathbb{Z}$ and $r > 0$. Assume that det $A = 1$. Then,*

$$A = BB^*$$

where B is a 2×2 matrix over $\mathbb{Z}[i]$ and B^* is the conjugate transpose of B.

Proof : To prove the theorem, we apply induction on $c^2 + d^2$. If $c^2 + d^2 = 0$, we will have

$$A = \begin{bmatrix} 1 & 0 \\ 0 & 1 \end{bmatrix}$$

and then $B = A$ will do.

Therefore, we assume that $c^2 + d^2 > 0$. So, c and d are not both zero.

As $rs = 1 + c^2 + d^2$, s is a positive integer. There are two cases to be considered: (i) $0 < r \le s$ (ii) $0 < s \le r$

Case (i): $0 < r \le s$. Let $A' = MAM^*$, where

$$M = \begin{bmatrix} 1 & 0 \\ x-yi & 1 \end{bmatrix}$$

and x, y are integers to be specified. Then,

$$
\begin{aligned}
A' &= \begin{bmatrix} 1 & 0 \\ x-yi & 1 \end{bmatrix} \begin{bmatrix} r & c+di \\ c-di & s \end{bmatrix} \begin{bmatrix} 1 & x+yi \\ 0 & 1 \end{bmatrix} \\
&= \begin{bmatrix} 1 & 0 \\ x-yi & 1 \end{bmatrix} \begin{bmatrix} r & r(x+yi)+c+di \\ c-di & (c-di)(x+yi)+s \end{bmatrix} \\
&= \begin{bmatrix} r & c'+d'i \\ c'-d'i & * \end{bmatrix}
\end{aligned}
$$

where
$$c' = c + rx, \ d' = d + ry$$
and $\det A' = 1$, as $\det M = \det M^* = 1$.

We choose x, y such that
$c'^2 + d'^2 < c^2 + d^2$. Applying induction, $A' = CC^*$. Further,

$$
\begin{aligned}
A &= M^{-1} A' (M^*)^{-1} \\
&= M^{-1} CC^* (M^*)^{-1} \\
&= (M^{-1} C)(M^{-1} C)^* \\
&= BB^*, \text{where } B = M^{-1} C.
\end{aligned}
$$

Now, when $c > \frac{r}{2}$, we choose $x = -1, y = 0$. Then, $c'^2 = (c - r)^2 < c^2$ and $d'^2 = d^2$. Therefore, $c'^2 + d'^2 < c^2 + d^2$. Similarly, if $c < -\frac{r}{2}$, we take $x = 1, y = 0$.

If $d > \frac{r}{2}$, take $x = 0, y = -1$.

If $d < \frac{r}{2}$, take $x = 0, y = 1$.

If $r = 1, |c| > \frac{1}{2}$, since c and d are not both zero and $s = 1 + c^2 + d^2$, as $s \geq 1, |d| > \frac{1}{2}$.

If $r > 1$, we claim that $|c| > \frac{r}{2}, |d| > \frac{r}{2}$. Suppose on the contrary, $|c| \leq \frac{r}{2}, |d| \leq \frac{r}{2}$. Since $0 < r \leq s$, we have $r^2 \leq rs = c^2 + d^2 + 1 \leq (\frac{r}{2})^2 + (\frac{r}{2})^2 + 1 = \frac{r^2}{2} + 1 < r^2$, a contradiction. Therefore, we will only have $|c| > \frac{r}{2}, |d| > \frac{r}{2}$ for $r \geq 1$. Therefore, in all the above possibilities involving r, s, c, d; x and y are determinable and so the proof is okay in case (i).

Case (ii): $0 < s \leq r$. We write $A' = MAM^*$, where

$$
M = \begin{bmatrix} 1 & x + yi \\ 0 & 1 \end{bmatrix}
$$

Thus

$$
A' = \begin{bmatrix} * & c' + d'i \\ c' - d'i & s \end{bmatrix}
$$

with $c' = c + sx, \ d' = d + sy$. It suffices to find x and y such that $c'^2 + d'^2 < c^2 + d^2$. As in case (i), the values $x = \pm 1, y = 0; x = 0, y = \pm 1$ yield the desired condition $c'^2 + d'^2 < c^2 + d^2$, since here $|c| > \frac{s}{2}, |d| > \frac{s}{2}$. Proof by induction is complete. $\qquad \square$

Corollary 1.5.1 : *If r is square-free, r is a sum of four squares.*
For, writing

$$
A = \begin{bmatrix} r & c + di \\ c - di & s \end{bmatrix}
$$

with $\det A = 1$, we obtain $A = BB^*$ for some B. We take

$$
B = \begin{bmatrix} w + xi & y + zi \\ * & * \end{bmatrix}
$$

then,

$$
B^* = \begin{bmatrix} w - xi & * \\ y - zi & * \end{bmatrix}
$$

and $A = BB^*$ yields $r = w^2 + x^2 + y^2 + z^2$. $\qquad \square$

Remark 1.5.2 : Lagrange's theorem is a consequence of corollary 1.5.1.

1.6. Diophantine equations

A linear Diophantine equation in two unknowns is of the form $ax + by = c$ where $a, b, c \in \mathbb{Z}$.

It is known that

$$(1.6.1) \qquad ax + by = c, \ \langle a, b \rangle \neq \langle 0, 0 \rangle,$$

has solutions if, and only if, g.c.d $(a, b) \mid c$. When solutions exist, they are all given by

$$(1.6.2) \qquad \begin{cases} x = \dfrac{c}{d}\, x_0 + \dfrac{b}{d}\, r \\[2mm] y = \dfrac{c}{d}\, y_0 - \dfrac{a}{d}\, r \end{cases}$$

where g.c.d $(a, b) = d$, $ax_0 + by_0 = d$ and r is any integer. See [1].

Remark 1.6.1 : (a) The pair $\langle x_0, y_0 \rangle$ is obtained in view of the fact that d is expressible as a linear combination of a and b.

(b) In the case of the linear congruence $ax \equiv b \pmod{r}$, a solution exists if, and only if, g.c.d $(a, r) \mid b$. When solutions exist, they are unique modulo $\frac{r}{d}$ where $d = $ g.c.d (a, r).

Definition 1.6.1 : *Let $a_1, a_2, \cdots, a_t$ be integers, at least one of which is not zero. A greatest common divisor d of $a_1, a_2, \cdots, a_r$ is a common divisor which is a multiple of every common divisor.*

The existence of a g.c.d $(a_1, a_2, \cdots, a_t)$ follows from the fact that $\mathbb{Z}$ is a PID.

Lemma 1.6.1 : *The Diophantine equation*

$$(1.6.3) \qquad a_1 x_1 + a_2 x_2 + a_3 x_3 + \cdots + a_s x_s = c. \quad (some\ a_i \neq 0)$$

has a solution if, and only if, g.c.d $(a_1, a_2, \cdots, a_s) \mid c$.

Proof : When $s = 2$ the result is known. We apply induction on s. Suppose that $s > 2$. We assume that the result is true for all Diophantine equations having $s - 1$ unknowns. Now,

$$\text{g.c.d } (a_1, a_2, \cdots, a_s) = \text{ g.c.d } (t, a_3, a_4, \cdots, a_s),$$

where $t = $ g.c.d (a_1, a_2).

By the linear expressibility of g.c.d, there exist integers y_1, y_2 such that $a_1 y_1 + a_2 y_2 = t$ and $tx + a_3 x_3 + \cdots + a_s x_s = c$ is solvable, by induction hypothesis. We write $x_1 = y_1 x$, $x_2 = y_2 x$. Then, $a_1 x_1 + a_2 x_2 = tx$. So, $\langle x_1, x_2, \cdots, x_s \rangle$ solves (1.6.3) if, and only if, g.c.d $(a_1, a_2, \cdots, a_s) \mid c$. $\qquad \Box$

Remark 1.6.2 : Given a, b, c ; the Diophantine equation

$$ax + by = c, \quad \langle a, b \rangle \neq \langle 0, 0 \rangle$$

where g.c.d (a,b) divides c, has solutions stated in (1.6.2). To pick solutions for which $x > 0$, $y > 0$, plot points (x,y) (which satisfy the given Diophantine equation) in the cartesian plane and locate the pairs $\langle x,y \rangle$ corresponding to the points which fall in the first quadrant.

1.7. Notes with illustrative examples

The division algorithm in the ring $\mathbb{Z}$ of integers has many applications. The Euclidean algorithm is a technique of iterating the division algorithm. It provides an efficient way to find the greatest common divisor of two positive integers.

Suppose $a = 288$, $b = 51$. The division algorithm gives

$$288 = 51 \times 5 + 33, 0 < 33 < 51$$

Applying the division algorithm with $a = 51$ and $b = 33$, we have

$$51 = 33 \times 1 + 18, 0 < 18 < 33.$$

One more iteration yields

$$33 = 18 \times 1 + 15, 0 < 15 < 18.$$

Once again iterating, we get

$$18 = 15 \times 1 + 3, 0 < 3 < 15$$

and

$$15 = 5 \times 3 + 0.$$

The algorithm tells us that 3 is the g.c.d. of 51 and 288.

In a general setting, one would have the following iterations:

$$a = bq_1 + r_1, \qquad 0 < r_1 < b$$
$$b = r_1 q_2 + r_2, \qquad\qquad 0 < r_2 < r_1$$
$$r_1 = r_2 q_3 + r_3, \qquad\qquad 0 < r_3 < r_2$$
$$\dots\dots \qquad\qquad\qquad \dots$$
$$r_{k-2} = r_{k-1} q_k + r_k, \qquad\qquad 0 < r_k < r_{k-1}$$
$$r_{k-1} = r_k q_{k+1} + 0 \qquad \text{say at the } k^{th} \text{ iteration.}$$

Then r_k is the g.c.d of a and b. For, clearly, $r_k \mid r_{k-1}$. $r_{k-1} \mid r_{k-2}$. So, $r_k \mid r_{k-2}$. Proceeding thus, $r_k \mid r_1$ and $r_k \mid b$. So, $r_k \mid a$. If $t \mid a$ and $t \mid b, t \mid r_1$, so, $t \mid r_2$, $\cdots, t \mid r_k$. So, any common divisor a and b divides r_k.

Next, if l is the least common multiple of a and b and g is a g.c.d (a,b), one has

$$lg = ab$$

If we consider $\mathbb{N}$ as a partially ordered set under the divisibility relation, the g.c.d and l.c.m of a,b are their greatest lower bound and least upper bound respectively in the lattice $(\mathbb{N}, \leq)$ where $\leq$ is to mean 'divides'. Lattices are discussed in chapter 8.

A number of the form $f_n = 2^{2^n} + 1$ is called a Fermat number when $n \geq 0$. Fermat conjectured that f_n is a prime for all positive integers n. Euler proved that 641 divides f_5. So, f_5 is not a prime. It is likely that all Fermat numbers other than f_0, f_1, f_2, f_3 and f_4 are composite. It is true that Fermat aimed at generating the primes belonging to $\mathbb{N}$. He believed that $\{f_n\}$ would generate primes.

If $p_1, p_2, \ldots, p_t$ are Fermat primes, that is, primes of the form $2^{2^n} + 1$ and $r = 2^a p_1, p_2, \cdots, p_t$ ($a \geq 0$), then the Euler function value $\phi(r)$ is a power of 2.

Now, we write $M_p = 2^p - 1$. M_p is called a Mersenne prime (Marin Mersenne (1588–1648)). If M_p is to be a prime, it is necessary but not sufficient to say that p is a prime.

The 30th Mersenne prime

$$2^{2,16,091} - 1$$

was discovered in 1985. Prof. Curtis Cooper and Prof. Steven Boone of Central Missouri State University, Warrensburg, MO have reported (Dec 2005) that the 43rd Mersenne prime M_p is $2^p - 1$ where $p = 30402457$ and M_p has 9152052 digits.

Next, we compare the nature of solutions of a linear Diophantine equation and a linear congruence.

In the case of Diophantine equation

(1.7.1) $ax + by = c$

if $d = $ g.c.d (a,b) divides c, writing $a = da_1$, $b = db_1$, $c = dc_1$, we get

(1.7.2) $a_1 x + b_1 y = c_1$

It is easy to check that the equations in (1.7.1) and (1.7.2) have exactly the same solutions.

In the case of congruence

(1.7.3) $18x \equiv 30 \pmod{66}$,

since g.c.d $(18,66) = 6$ and 6 divides 30, the congruence (1.7.3) has six incongruent solutions. From (1.7.3) division by 6 yields

(1.7.4) $3x \equiv 5 \pmod{11}$.

Since the g.c.d $(3,11) = 1$ which divides 5, the congruence (1.7.4) has a unique solution (modulo 11). The inverse of 3 (modulo 11) is 4 (modulo 11). Multiplying both sides of (1.7.4) by 4 (mod 11), we get the unique solution as

(1.7.5) $x_0 \equiv 9 \pmod{11}$.

But, (1.7.3) is given modulo 66 so that its incongruent solutions are integers among the numbers $0, 1, 2, \cdots, 65$. The incongruent solutions of (1.7.3) are given by

$$y = 9 + 11k, \quad k = 0, 1, 2, 3, 4, \text{ and } 5.$$

So the solutions of (1.7.3) are [9], [20], [31], [42], [53] and [64] and they are six distinct residue classes satisfying (1.7.3) and incongruent modulo 66. The summary is that a unique solution of (1.7.4) gives rise to six distinct solutions of (1.7.3).

The next observation is about Fermat's little theorem. A direct converse of theorem 2 is false. It is not true that if r does not divide a and

$$(1.7.6) \qquad a^{r-1} \equiv 1 (\text{mod } r)$$

then r is necessarily a prime. As noted in Hardy and Wright [4], it is easy to check that if a is prime to $561 = 3.11.17$, then

$$a^{560} = 1 \ (\text{mod } 561)$$

However, 561 is not a prime!

A valid converse of theorem 2 is the following.

Theorem 6 : *For $a \geq 2$ and prime to r, if $a^{r-1} \equiv 1 (\text{mod } r)$ and $a^d \not\equiv 1 (\text{mod } r)$ for any divisor $d(1 \leq d < r-1)$ of $(r-1)$, then r is a prime.*

Proof : As g.c.d $(a, r) = 1$, if d is the order of a in the group of units of $\mathbb{Z}/r\mathbb{Z}$, then, $d | \phi(r)$. We are given that

$$a^d \not\equiv 1 \ (\text{mod } r)$$

for any divisor d of $r-1$ ($1 \leq d < r-1$). So, as $a^{r-1} \equiv 1 (\text{mod } r)$, $(r-1)$ divides $\phi(r)$. If r is composite, then r has a divisor t such that $1 < t < r$. Further, we note that there are at least two integers t_1 and t_2 among the numbers $1, 2, \cdots, r$ which are not relatively prime to r, namely t and r themselves. So, $\phi(r) \leq (r-2) < (r-1)$. So, if $(r-1)$ divides $\phi(r)$, $(r-1)$ has to be equal to $\phi(r)$ and in that case r is a prime. $\qquad \square$

Remark 1.7.1 : For an account of Euler's ϕ-function and its generalizations, see the expository articles [11], [12] and [13].

Definition 1.7.1 : *Let b be a positive integer. If r is a composite number and*

$$b^r \equiv b (\text{mod } r)$$

then r is called a pseudoprime to the base b.

For example $341 = 11 \times 31$ is a pseudoprime to the base 2, as

$$2^{340} = 1 (\text{mod } 341)$$

Definition 1.7.2 : *Let r be a positive composite integer such that*

$$a^{r-1} \equiv 1 (\text{mod } r) \text{ for all } a \text{ with g.c.d } (a, r) = 1.$$

Then, r is called a Carmichael number.

It is verified that 561 is a Carmichael number and it is the smallest such number. We state, without proof, a

Proposition 1.7.1 : *An odd composite number r is a Carmichael number if, and only if, r is square-free and $p-1$ divides $r-1$ for every prime p dividing r.*

In 1994, Alford, Granville and Pomerance [A1] have shown that there are infinitely many Carmichael numbers. While discussing sums of squares, we gave certain number-theoretic aspects only. For an advanced level reading of the theme on sums of squares, see Olga Taussky [A4].

1.8. Worked-out examples

a) What is meant by a 'primality test' ?
   Answer: The problem of distinguishing primes from composites has attracted the attention of those who are interested in some kind of a numerical computation. By a 'primality test', we mean a test that will check whether a given number is composite or not. To say that a positive integer r passes a primality test is to conclude that when the test is executed, r is shown to be composite. If r fails a primality test, then r is a prime. The basic idea is that on account of Fermat's little theorem, if we can find an integer a such that

(1.8.1) $a^r \not\equiv a \pmod{r}$,

   then, r is composite.

   Remark 1.8.1 : Suppose that r is composite. We write $r = r_1 r_2$; where $1 < r_1, r_2 < r$. So, we will have $r_1 \leq \sqrt{r}$, since, otherwise, $r_2 \geq r_1 > \sqrt{r}$ implies that $r_1 r_2 > \sqrt{r}.\sqrt{r} = r$, which is impossible. As every positive integer has at least one prime factor, r_1 has a prime factor $p \leq \sqrt{r}$. For instance, every composite number < 100 has a prime factor $< \sqrt{100} = 10$. Since the only primes < 10 are $2, 3, 5$ and 7, we have only to check each number < 100 for divisibility by $2, 3, 5$ and 7. Crossing out such multiples of $2, 3, 5$ and 7 we arrive at integers > 1, which are primes < 100. This procedure is known as the sieve of Eratosthenes (276–196 B.C.) who belonged to the school of Alexandria. He devised a systematic method (the sieve method) for attaining all primes up to a given number r.

   Remark 1.8.2 : A recent efficient algorithm known as AKS algorithm (Aggarwal, Kayal and Saxena (2002)) is known to determine whether a given integer is prime. See Andrew Granville [A2].

                                                                        □

b) (Ralph G. Archibald) It is known that the polynomial $x^2 + x + 41$ yields a prime for $x = 0, 1, 2, \ldots, 39$, but is composite for $x = 40$ and 41. Show that there does not exist a polynomial $f(x)$ of degree $m > 0$ and having integer coefficients such that $f(x)$ yields primes for every integer value of x or every integer $x > n_0$ (a specified integer).

Answer: Let $g(x)$ be a polynomial of degree n and having real coefficients. We write

$$g(x) = a_0 x^n + a_1 x^{n-1} + \ldots + a_{n-1} x + a_n$$

which is the same as

(1.8.2) $$g(x) = x^n (a_0 + \frac{a_1}{x} + \ldots + \frac{a_{n-1}}{x^{n-1}} + \frac{a_n}{x^n}).$$

for sufficiently large x, in numerical value, $g(x)$ has the same sign as that of $a_0 x^n$. Further,

$$\text{as } x \to \infty, \quad |g(x)| \to \infty.$$

We write

(1.8.3) $$f(x) = b_0 x^m + b_1 x^{m-1} + \ldots + b_{m-1} x + b_m.$$

Assume that $b_0 > 0$. Let x_0 be an integer such that

(1.8.4) $$f(x_0) = q > 1.$$

For $x > t$, suppose that $f(x) - q > 0$. We use Taylor expansion of f at $x = x_0 + sq$ where s is arbitrary.

(1.8.5) $$f(x_0 + sq) = f(x_0) + sqf'(x_0) + \frac{s^2 q^2}{2!} f''(x_0) + \ldots + \frac{s^m q^m}{m!} f^{(m)}(x_0).$$

Now, $\frac{1}{r!} f^{(r)}(x_0)$ has integer coefficients for $1 \leq r \leq m$. So, for s an integer, as $f(x_0) = q$ (1.8.4)

$$f(x_0 + sq) - q = q\{sf'(x_0) + \frac{s^2 q}{2!} f''(x_0) + \ldots + \frac{s^m q^{m-1}}{m!} f^{(m)}(x_0)\}$$
$$= qM(\text{ say}).$$

So, when $x = x_0 + sq$, $f(x) - q$ is a multiple of q and is positive when $x > t$. So,

(1.8.6) $$f(x) = q(1 + M).$$

Therefore, we have exhibited $q(> 1)$ as a divisor of $f(x)$ for $x_0 + sq > t$. So, $f(x)$ is composite for $sq > t - x_0$. We have only to take $x_0 = t$. $\square$

c) (Ethan D. Bolker) (i) For $r \geq 2$, let $U(r)$ denote the group of units in $\mathbb{Z}/r\mathbb{Z}$. If $U(r)$ contains an element of order $r - 1$, show that r is a prime.
(ii) Given $r \geq 2$, show that r is a prime if, and only if, every linear polynomial with coefficients in $\mathbb{Z}/r\mathbb{Z}$ has at most one zero in $\mathbb{Z}/r\mathbb{Z}$.
Answer: (i) Let $[a]$ be an element of order $(r-1)$ in $\mathbb{Z}/r\mathbb{Z}$. It follows that

$$[a]^{r-1} = [1].$$

That is, $a^{r-1} \equiv 1 \pmod{r}$. As $\phi(r)$ is the order of $U(r)$, $r - 1$ divides $\phi(r)$. But, $\phi(r) \leq (r-1)$. Hence, $\phi(r) = r - 1$ and so, r is a prime.
(ii) :$\Rightarrow$. Let r be a prime. Then, $U(r)$ has order $(r-1)$. A linear equation having coefficients in $\mathbb{Z}/r\mathbb{Z}$ has the form

(1.8.7) $$[a]x + [b] = [0].$$

When $a \neq 0$, $[a]$ has a multiplicative inverse $[a]^{-1}$. So, solving (1.8.7), we get

$$x = -[a]^{-1} \otimes [b]$$

which is a unique solution.

$\Leftarrow$: To prove the converse, we use contrapositive argument. Suppose that r is composite. As $r \geq 2$, the ring $(\mathbb{Z}/r\mathbb{Z}, \oplus, \otimes)$ has divisors of zero. Suppose that a, b are divisors of r such that $1 < a \leq b \leq r$ such that $ab \equiv 0 (\mathrm{mod}\ r)$. Let $[t]$ be a unit in $\mathbb{Z}/r\mathbb{Z}$. We consider the polynomial equation

(1.8.8) $[t]x + [b] = [0]$.

One solution of (1.8.8) is $x = -[t]^{-1} \otimes [b]$. However, multiplying both sides of (1.8.8) by $[a]$, we get

$$[a] \otimes ([t]x + [b]) = [0].$$

As $[a] \otimes [t] \neq [0]$, we see that $x = [0]$, which is another solution of (1.8.8). Or, we have exhibited a linear equation having more than one solution. So, when r is composite, we can find a linear polynomial having more than one zero. In other words, if every linear polynomial having coefficients in $\mathbb{Z}/r\mathbb{Z}$, has at most one zero, r cannot be composite. $\square$

d) (Nicol and Vandiver) Given $r \geq 1$ and d a divisor of r, we consider the set

$$S = \{n_1, n_2, \ldots, n_c\}$$

where S is the set of positive integers less than and relatively prime to r. $c = \phi(r)$, the Euler ϕ-function. Show that the number of elements of S which are congruent to $t(\mathrm{mod}\ \frac{r}{d})$ with g.c.d $(t, \frac{r}{d}) = 1$ is $\frac{d}{b}\phi(b)$ where b denotes the greatest divisor of d such that g.c.d $(b, \frac{r}{d}) = 1$.
Answer: Let

(1.8.9) $T = \{t + j(\dfrac{r}{d}) : j = 0, 1, 2, \ldots, (d-1)\}$.

Elements of T are integers which are relatively prime to $\frac{r}{d}$, but not necessarily prime to r.
Case: (i) Suppose that $b > 1$. We write the d integers $0, 1, 2, \ldots, (d-1)$ as

(1.8.10) $< m + hb >$

where $m = 0, 1, \ldots, (b-1)$ and $h = 0, 1, 2, \ldots, (\frac{d}{b}) - 1$.

The values of j in (1.8.9) are reduced modulo b using (1.8.10) and we obtain

(1.8.11) $T' = \{t + m(\dfrac{r}{d}) : m = 0, 1, 2, \ldots, (b-1)\}$

corresponding to each value of $h = 0, 1, 2, \ldots, (\frac{d}{b}) - 1$.
Since g.c.d $(b, \frac{r}{d}) = 1$, the set T' (reduced modulo b) is the same as $\{0, 1, 2, \ldots, (b-1)\}$.

By definition, d can be written as $d = d_1 b$ where each prime factor of d_1 divides $\frac{r}{d}$. To obtain those elements of T which are prime to d (and so prime to r), it is enough if we select those which are prime to b, since they are prime to $\frac{r}{d}$ and hence to d_1. This subset of T becomes a subset W of T' in which the elements are relatively prime to b. So $|W| = \phi(b)$. Now, there are $\frac{d}{b}$ values of b corresponding to each set T' (1.8.11). Hence, the number of integers relatively prime to r is $(\frac{d}{b})\phi(b)$, as required.   □

Remark 1.8.3 : If $b = 1$, each prime factor of d occurs in $\frac{r}{d}$ so that each element in S is prime to r. There are $\frac{d}{1}\phi(1) = d$ such numbers in this case.

Remark 1.8.4 : The result in worked out example (d) is fruitfully employed to evaluate Ramanujan's sum $C(n,r)$ (specifically, relation (5.4.4) of chapter 5). See [A3].

EXERCISES

1. *Mark the following statements true (T) or false (F) justifying your answer briefly.*
   a) *Let $N(> 1)$ be a given positive integer. All the positive divisors of N can be determined from the prime factorization of N.*
   b) *Let p be an odd prime and $n > 1$. Then, $p^n + 1$ is not a square.*
   c) *Let t be an even integer. Suppose that a,b,c are integers having no common factor > 1, then, it is impossible to choose a,b,c such that*
   $$t^2 = a^2 + b^2 + c^2.$$
   d) *If $n \equiv 3$ or $6 \,(\mathrm{mod}\, 9)$, then n is not representable as a sum of two squares.*
   e) *Let r be a square-free integer. It is certain that an abelian group of order r is cyclic.*
   f) *The matrix ring $M_2(\mathbb{Z}/2\mathbb{Z})$ has proper two-sided ideals. That is, $M_2(\mathbb{Z}/2\mathbb{Z})$ is not a simple ring.*
2. *For any $a \in \mathbb{Z}$, show that $a^5 - a \equiv 0 \,(\mathrm{mod}\, 5)$.*
3. *Find the integral solutions of $6x + 4y = 14$.*
4. *Let $a_1, a_2, \cdots, a_t$ be integers. We define*
   $$I = \{a_1 x_1 + a_2 x_2 + \cdots + a_t x_t : x_i \in \mathbb{Z}, \quad i = 1, 2, \cdots, t\}.$$
   Show that I is an ideal of $\mathbb{Z}$, generated by g.c.d $(a_1, a_2, \cdots, a_t)$.
5. *Find all integers a such that for $0 < a < 13$,*
   $$x^2 = a \,(\mathrm{mod}\, 13)$$
   has a solution.
6. *Let $\{c_1, c_2, \cdots, c_t\}$ where $t = \phi(r)$ be a reduced residue system mod r. Show that*
   $$c_1 + c_2 + \cdots + c_t \equiv 0 \quad (\mathrm{mod}\, r).$$
7. *Solve the congruence: $45x \equiv 36 \,(\mathrm{mod}\, 54)$.*
8. *Solve the Diophantine equation: $6x + 15y = 6$.*

9. Solve the Diophantine equation: $8x + 3y = 28$.

10. Let p be an odd prime. Suppose that g.c.d $(a, p) = 1$. Show that the congruence

$$x^2 \equiv a \pmod{p^t}$$

has either no solutions or exactly two solutions modulo p^t.

11. A Pythagorean triple is a triple (a, b, c) of positive integers such that $a^2 + b^2 = c^2$. Let $c > 0$. Show that there is a Pythagorean triple (a, b, c) if, and only if, c is divisible by some prime p with $p \equiv 1 \pmod 4$.

12. We write $A_{700}(7) = \{[x] \in \mathbb{Z}/700\mathbb{Z} : [7][x] = [0]\}$.
    Show that $A_{700}(7)$ is a field isomorphic to $\mathbb{Z}/7\mathbb{Z}$.

13. Find the least positive residue of (i) 2^{32} (ii) 2^{47} modulo 47.

14. Let $f_n = 2^{2^n} + 1$. f_n is called a Fermat number. If f_n is a prime, it is called a Fermat prime. For $m \neq n$, show that f_m and f_n are relatively prime to one another.
    (Gauss showed that a regular p-gon can be constructed with a ruler and compass for those values of prime p for which $p = f_n$; $n = 2$ gives $p = 17$.)

15. Solve the congruence $71x \equiv 4 \pmod{55}$.

16. Let n be a product of four consecutive positive integers. Prove or disprove: $(n + 1)$ is a perfect square.

17. (Landau) Let $m \equiv 5 \pmod{12}$ and $m > 17$. Show that m is expressible as a sum of three distinct positive squares.

18. Let $r = x^2 + y^2$ where g.c.d $(x, y) = 1$. If $d(\geq 1)$ is a divisor of r, show that d is also a sum of two squares.

19. (Ethan D. Bolker). Let $r = p_1^{a_1} p_2^{a_2} \cdots p_k^{a_k}$ where $p_1, p_2, \ldots, p_k$ are distinct primes. We denote the (symmetric) group of permutations on n symbols by S_n. Show that the least value of n for which S_n contains an element of order r is $n = p_1^{a_1} + p_2^{a_2} + \cdots + p_k^{a_k}$.

20. (Thue) Suppose that r is not a perfect square. Let $t \in \mathbb{Z}$. Show that the congruence $tx \equiv y \pmod r$ has a solution $< x, y >$ in which $|x|$ and $|y|$ are both $< \sqrt{r}$ and $< x, y > \neq < 0, 0 >$.

REFERENCES

[1] Ethan D. Bolker : Elementary Number theory-an algebraic approach, Chapters 1,2 and 4, pp 1–22, 34–59 and 107–119, W. A. Benjamin Inc NY (1970).

[2] Fred H. Croom : Basic concepts of Algebraic Topology, Chapters 1 & 2, pp 1–30, UTM. Springer Verlag NY (1978).

[3] John A. Ewell : A simple proof of Fermat's Two-square theorem, Amer. Math. Monthly 90 (1983) 635–637.

[4] G. H. Hardy and E. M. Wright : An introduction to the theory of numbers, Chapters V & VI pp 48–81, Oxford at the Clarendon Press, Fourth Edn (1965).

[5] Charles Green : Adv. Problem 5469. Amer. Math. Monthly 74 (1967) p 208. Solution by L. D. Crowson: Amer. Math. Monthly 75 (1968) p30.

[6] Melvin Hausner : Applications of a simple counting technique, Amer. Math. Monthly 90 (1983) 127–129.

[7] D. R. Heath-Brown : Fermat's Two-square theorem, Invariant (1984) 3–5.

[8] Herstein I. N. : Topics in Algebra, Wiley Eastern Ltd. Second Edn. (15th reprint 1993), New Delhi.

[9] Terrence Jackson : A short proof that every prime $p \equiv 3 \pmod 8$ is of the form $x^2 + 2y^2$. Amer. Math. Monthly 107 (2000) p 447.

[10] E. Landau : Elementary Number Theory, Chelsea Pub. Co: New York 2nd edition (1966) pp. 135–140.

[11] M. Newman : Integral Matrices, Chapter XI pages 201–205, Academic Press NY, London (1972).

[12] Don Redmond: Number Theory, Monographs and Textbooks in Pure and Applied Mathematics, No. 201 (1996) pp.312–319.

[13] R. Sivaramakrishnan : The many facets of Euler's Totient I, A general perspective, Niew Archief voor Wiskunde (1986) serie IV, 4 No. 3, 175–190.

[14] R. Sivaramakrishnan : The many facets of Euler's Totient II, Generalizations and analogues, Niew Archief voor Wiskunde (1990) serie VIII, No. 2, 169–187.

[15] J. Sándor and R. Sivaramakrishnan : The many facets of Euler's Totient III, An assortment of miscellaneous topics, Niew Archief voor Wiskunde (1993) Vol XI, No. 2, 97–130.

[16] Charles Small : A simple proof of the four-square theorem. Amer. Math. Monthly 89 (1982) 59–61.

[17] D. Zagier : A one sentence proof that every prime $p \equiv 1 \pmod 4$ is a sum of two squares. Amer. Math. Monthly 97 (1990) 144.

ADDITIONAL REFERENCES

[A1] W. R. Alford, A. Granville and C. Pomerance : There are infinitely many Carmichael numbers. Annals of Math 140 (1994) 703–722.

[A2] Andrew Granville : It is easy to determine whether a given integer is prime, Bull. Amer. Math. Soc., 42 (2005) 3–38.

[A3] C. A. Nicol & H. S. Vandiver : A Von Sterneck arithmetical function and restricted partitions with respect to a modulus. Proc. Nat. Acad. Sci. (USA) 40 (1954) 825–835.

[A4] Olga Taussky : Sums of squares, Amer. Math. Monthly 77 (1970) 805–830.

CHAPTER 2

The integral domain of rational integers

Historical perspective

Commutative algebra owes its development during the last two hundred years or so, to the discovery of various algebraic properties of the set $\mathbb{Z}$ of rational integers (as the set of integers forms a ring contained in the field $\mathbb{Q}$ of rational numbers). Attempts to prove Fermat's last theorem (FLT) — namely, if $n > 2$, there exist no positive integers x, y and z such that $x^n + y^n = z^n$— led to the birth of a new branch of number theory called 'algebraic number theory'. It was found out that the set of rational integers could be considered in a larger set-up giving the notion of 'rings of algebraic integers'. For many of these rings it was possible to show that for very large values of n, $x^n + y^n = z^n$ has no integer solutions. For many values of n, nothing could be said about the truth or falsity of FLT. Andrew Wiles (1994) of Princeton University proved FLT in the affirmative establishing a conjecture in algebraic geometry, the so-called Shimura-Tanyama-Weil conjecture. This was a remarkable achievement of Twentieth Century.

In 1843, Ernst Eduard Kummer (1810–1893) extended the definition of an integer and gave a 'proof 'of FLT. Peter Gustav Lejeune Dirichlet (1805–1859) found that Kummer had made an incorrect assumption/statement about factorization of numbers. Then, Kummer worked again and took pains to correct the error in his 'proof'. In order to rectify the incorrect factorizations, Kummer introduced the notion of 'ideal numbers'which enabled him to complete his proof in certain special cases, as in the ring of integers of the number field $\mathbb{Q}(\omega)$ where $\omega = \exp(2\pi i/23)$. His 'ideal numbers' were sets of numbers. Richard Dedekind (1831–1916) used this latter observation to invent the notion of an ideal. It was at this point of time that ring theory was born in a disguised manner as part of algebraic number theory. Leopold Kronecker (1823–1891) gave the name 'order' to the ring of algebraic integers contained in a number field. However, it was David Hilbert (1862–1943) who coined the word 'ring' for an algebraic structure with two binary operations + (addition) and · (multiplication) satisfying the appropriate axioms as we know. We owe this axiomatic approach to David Hilbert.

It was not until 1900 that a subject called ring theory appeared to have been born. Commutative ring theory covers number systems and polynomials. Various types of rings and their properties were established during the beginning of the Twentieth Century. Factorial rings/unique factorization domains come from a generalisation of the fundamental theorem of arithmetic.

2.1. Introduction

The ring $\mathbb{Z}$ of rational integers is known to be an integral domain. We first show that $\mathbb{Z}$ is an ordered integral domain.

The aim of this chapter is to discuss some properties of commutative rings with unity. The ideal-theoretic analogues of the notions of l.c.m and g.c.d are pointed out. The fact that $\mathbb{Z}$ is a Principal Ideal Domain (PID) makes it a Unique Factorization Domain (UFD). The role of irreducibles and primes is important in the context of uniqueness of factorization of a nonzero, non-unit in an integral domain. The notion of a GCD domain due to I. Kaplansky is introduced. Not all GCD domains are unique factorization domains. Further, $\mathbb{Z}[\sqrt{-5}]$ is shown as an example of an integral domain which is not a GCD domain.

2.2. Ordered integral domains

The elements of $\mathbb{Z}$ could be exhibited in an ascending order as follows:

$$\cdots, -3, -2, -1, 0, 1, 2, 3, \cdots.$$

For $a, b \in \mathbb{Z}$, we see that $a \geq b$ if, and only if, $a - b \geq 0$. One distinguishes the positive elements of $\mathbb{Z}$ by writing $\mathbb{N}$ for the subset of positive integers.

Definition 2.2.1 : *An integral domain $(D, +, \cdot)$ is said to be an ordered integral domain, if D contains a subset D_P with the following properties:*

   i. If $a, b \in D_P$, $a + b \in D_P$ (closure under addition)
   ii. If $a, b \in D_P$, $a \cdot b \in D_P$ (closure under multiplication)
   iii. For each element $a \in D$, exactly one of the following is true:

$$a = 0, \quad a \in D_P \text{ or } -a \in D_P \quad \text{(law of trichotomy)}.$$

D_P is called the set of positive elements of D. The nonzero elements of D which are not in D_P are called the negative elements of D.

This is a clear generalization of $\mathbb{Z}$.

Definition 2.2.2 : *Let D be an ordered integral domain with D_P denoting the subset of positive elements. If $a, b \in D$, we say that $a > b$ to mean $a - b \in D_P$ and $a < b$ if $a - b \notin D_P$.*

Clearly $a > 0 \Rightarrow a \in D_P$. $a \geq b$ means either $a = b$ or $a - b \in D_P$.
   The absolute value $|a|$ of $a \in D$ is such that

   (1) If $a \geq 0$, $|a| = a$
   (2) If $a < 0$, $|a| = -a$. So, if $a \neq 0$, $|a| > 0$

Definition 2.2.3 : *Let S be subset of an ordered integral domain D. S is said to be well-ordered, if each non-empty subset T of S contains a least element. That is, for each subset T of S there exists an element a in T such that $a \leq x$ for all $x \in T$.*

We deduce that $\mathbb{N}$ is well-ordered. However, if $\mathbb{Q}^+$ denotes the set of positive rational numbers. $\mathbb{Q}^+$ is not well-ordered, as the set of all positive rational numbers has no least element.

Theorem 7 : *Let D be an integral domain in which the set D_P of positive elements is well-ordered. If 1_D denotes the unity element of D, then,*

$$D_P = \{m1_D : m \in \mathbb{N}\}$$

$$and\ D = \{n1_D : n \in \mathbb{Z}\}.$$

Proof : If $a \in D_P$, $a > 0_D$ (the zero element of D).
For $a, b \in D$, $(-a) \cdot (-b) = ab$, since $a \cdot (-b) = -a \cdot b$ and $(-a) \cdot (-b) = -(a \cdot (-b)) = -(-(a \cdot b)) = a \cdot b$.
If $a > 0_D$, $a^2 > 0_D$. If $-a > 0_D$, $(-a)^2 > 0_D$ or $a^2 > 0_D$. So, as $1_D > 0_D$, $1_D^2 = 1_D$, $1_D^2 > 0_D$. For each positive integer n, let S_n be the assertion that $n1_D > 0_D$.
As $1_D = 1_D$, we note that S_1 is true. Let $k \in \mathbb{N}$ such that $k1_D > 0$. Then $(k+1)1_D = k1_D + 1_D > 0$.
So, S_{k+1} is true if S_k is true. So, by induction on n, S_n is true for all $n \in \mathbb{N}$. So, $n1_D \in D_P$ for all positive integers n.

Claim : All elements of D_P are of the form $m1_D$ for $m \in \mathbb{N}$.

Since D_P is well-ordered, D_P has a least element. The least element is going to be 1_D. For, suppose that c is the least element of D_P and that

$$1_D > c > 0_D$$

So $1_D \cdot c > c^2 > 0_D$ or $c > c^2 > 0_D$, since $c \cdot 1_D = c$. So $c^2 \in D_P$ and $c^2 < c$. This violates the assumption that c is the least element of D_P.
So, c is not smaller than 1_D, that is, $c = 1_D$.

To show that every element of D_P is of the form $m1_D$ where $m \in \mathbb{N}$, we assume the contrary. We pick some elements which are not of the form $m1_D$.
Let T be a non-empty subset of D_P where each element of T is not of the form $m1_D$. T has a least element say d. Since 1_D is the least element of D_P, we should have $d > 1_D$.
So, $d - 1_D > 0_D$. But then, $d - 1_D \in D_P$ and since $1_D > 0_D$, $d > d - 1_D$. So $d - 1_D \notin T$. Therefore, by hypothesis $d - 1_D = m'1_D$ or $d = (m' + 1)1_D$ and $m' + 1 \in \mathbb{N}$. This contradicts the fact that $d \in T$. So T is empty. So every element of D_P is of the form $m1_D$ where $m \in \mathbb{N}$.

Next, if $a \in D$ and $a \notin D_P$, either $a = 0_D$ or $-a \in D_P$.
If $a = 0_D$, $a = 0 \cdot 1_D$. If $-a \in D_P$, $-a = m_2 1_D$, $m_2 \in \mathbb{N}$. So, $a = -m_2 1_D$ where $m_2 \in \mathbb{N}$.
So every element of D is of the form $n1_D$ where $n \in \mathbb{Z}$.
Finally, if $n_1, n_2 \in \mathbb{Z}$ and $n_1 1_D = n_2 1_D$ we will have $n_1 = n_2$. For, suppose not.
That is, we assume that $n_1 \neq n_2$. Without loss of generality, we could take $n_1 > n_2$.
Then, $n_1 - n_2 > 0$. So, $(n_1 - n_2)1_D \in D_P$. Then, $(n_1 - n_2)1_D > 0_D$ or $n_1 1_D \neq n_2 1_D$, a contradiction to the assumption $n_1 1_D = n_2 1_D$. So, $n_1 1_D = n_2 1_D \Rightarrow n_1 = n_2$. This proves theorem 7.   □

Theorem 8 (A Characterization of $\mathbb{Z}$) : *Let D, D' be two ordered integral domains in which the sets of positive elements are well-ordered. Then, D and D' are isomorphic.*

Proof : Let 1_D, $1_{D'}$ be the unity elements of D, D' respectively. By theorem 7,

$$D = \{n1_D : n \in \mathbb{Z}\}, D' = \{n1_{D'} : n \in \mathbb{Z}\}$$

We define a map $\psi : D \to D'$ given by

$$\psi(n1_D) = n1_{D'} \quad n \in \mathbb{Z}.$$

ψ is a homomorphism of D into D' which is one-one. For,

$$\psi(n_1 1_D + n_2 1_D) = \psi((n_1 + n_2)1_D)$$
$$= (n_1 + n_2)1_{D'}$$
$$= n_1 1_{D'} + n_2 1_{D'}$$
$$= \psi(n_1 1_D) + \psi(n_2 1_D)$$

Similarly, $\psi(n_1 n_2 1_D) = \psi(n_1 1_D)\psi(n_2 1_D)$

Further, given $n_1 1_{D'} \in D'$, there exists $n_1 1_D \in D$ such that

$$\psi(n_1 1_D) = n_1 1'_D$$

Also $n_1 1'_D = n_2 1'_D \Rightarrow n_1 = n_2$ and so $n_1 1_D = n_2 1_D$. So D and D' are isomorphic ring-theoretically.   □

Corollary 2.2.1 : $\mathbb{Z}$ *is the only ordered integral domain up to isomorphism, as* $\mathbb{N}$, *the subset of positive integers is well-ordered.*

Remark 2.2.1 : Theorems 7 and 8 have been adapted from N. H. McCoy and Thomas Berger [4].

2.3. Ideals in a commutative ring

Commutative rings with unity are in plenty. $\mathbb{Z}$ is one such. Let $(R, +, \cdot)$ be a commutative ring with unity 1_R. We recall that by an ideal I of $(R, +, \cdot)$ we mean an additive subgroup $(I, +)$ of $(R, +)$ satisfying the property that for $a \in I$, $r \in R$, $r \cdot a = a \cdot r \in I$. Henceforth, we denote $(R, +, \cdot)$ by R.

Definition 2.3.1 : *If I, J are ideals of R, their sum $I + J$ is defined by*

$$I + J = \{a + b : a \in I, b \in J\}.$$

We notice that $I + J$ is the smallest ideal of R containing I and J. For, if T is an ideal containing I and J, then T contains $I + J$.

Definition 2.3.2 : *The product IJ of two ideals $I, J \in R$ is defined by*

$$IJ = \{\sum_{finite} a_i \cdot b_i : a_i \in I, b_i \in J\}.$$

It is easy to check that IJ is an ideal of R. In fact, IJ is the additive subgroup of $(R,+)$ generated by all products $x \cdot y$ where $x \in I, y \in J$.

Next, the intersection $I \cap J$ of two ideals I, J of R is the largest ideal contained in I as well as J.

In the case of $\mathbb{Z}$, for $a, b \in \mathbb{Z}$, if $I = (a)$, $J = (b)$; the principal ideals generated by a and b respectively,

(2.3.1) $I + J = (a) + (b) = (g)$ where $g = \text{g.c.d } (a,b)$

(2.3.2) $IJ = (a)(b) = (ab)$ the ideal generated by ab

(2.3.3) $I \cap J = (a) \cap (b) = (l)$ where $l = \text{l.c.m } (a,b)$.

For I, J, K ideals of R, one has

(2.3.4) $I(J + K) = IJ + IK$

(2.3.5) $IJ \subseteq I \cap J$

(2.3.6) $I \cap (J + K) = I \cap J + I \cap K$,   if $J \subseteq I$ or $K \subseteq I$ (modular law)

(2.3.7) $(I + J)(I \cap J) \subseteq IJ$.

Details of verification of (2.3.1) to (2.3.7) are omitted.

Definition 2.3.3 : *An ideal I of the ring R is said to be a nil ideal if each element a in I is nilpotent. That is, there exists a positive integer n such that $a^n = 0_R$. (n depends on the particular element a).*

Definition 2.3.4 : *An ideal I of the ring R is called a nilpotent ideal if $I^n = (0_R)$, for same positive integer n.*

We remark that I^n denotes the set of all finite sums of products of n elements taken from I. It means that for every choice of n elements $a_1, a_2, \ldots, a_n \in I$, one has $a_1 \cdot a_2 \cdot \ldots \cdot a_n = 0_R$. So, $a_n = 0$ for all $a \in I$. Thus every nilpotent ideal of R is a nilideal.

Definition 2.3.5 : *Let $\{I_\lambda\}$ ($\lambda \in \Lambda$, an index set) be a family of ideals of R. Their sum $\sum_\lambda I_\lambda$ is the ideal consisting of elements which are all possible finite sums of elements drawn from the family $\{I_\lambda\}$.*
That is,

$$\sum_\lambda I_\lambda = \{ \sum_{finite} a_\lambda : a_\lambda \in I_\lambda, \text{all but a finite number of } a_\lambda \text{ are zero} \}.$$

Further, $\sum_\lambda I_\lambda$ is the smallest ideal containing every $I_\lambda, \lambda \in \Lambda$.

In the case of intersection and product of members of a family of ideals, one obtains

(2.3.8) $\bigcap_\lambda I_\lambda = \{ a_\lambda : a_\lambda \in I_\lambda \text{ for each } \lambda \in \Lambda \}$, the largest ideal of R containing each $I_\lambda (\lambda \in \Lambda)$, and

$$(2.3.9) \qquad \prod_{i=1}^{n} I_i = \{\sum_{\text{finite}} a_1 \cdot a_2 \cdots a_n : a_i \in I_i, i = 1, 2, \cdots, n\}$$

In particular,

$$(2.3.10) \qquad I^n = \{\sum_{\text{finite}} a_{i_1} \cdot a_{i_2} \cdots a_{i_n} : a_{i_k} \in I \quad (k = 1, 2, \cdots n)\}$$

One easily checks that

$$I \supseteq I^2 \supseteq I^3 \supseteq \cdots \supseteq I^n \supseteq \cdots$$

In $\mathbb{Z}$, one has

$$(2) \supseteq (4) \supseteq (8) \supseteq \cdots\cdots, \text{ as } (2)^n = (2^n), n \geq 1.$$

Definition 2.3.6 : *For ideals I, J of R, the quotient of I by J denoted by $I : J$ is defined by*

$$I : J = \{a \in R : aJ \subseteq I\}$$

Here, aJ means the set $\{as : s \in J\}$. As R is commutative, $aJ = Ja$. If $a, b \in I : J$, we get

$$ax - bx = (a - b)x \in I, \text{ whenever } x \in J.$$

So, $a - b \in I : J$.

For $r \in R$, $raJ \subseteq rI \subseteq I$ and so $ra \in I : J$, whenever $a \in I : J$.

That is, $I : J$ is an ideal of R.

In the case of $\mathbb{Z}$, for principal ideals (a), (b) in $\mathbb{Z}$, one verifies that

$$(2.3.11) \qquad (a) : (b) = (c), \text{ where } c = \frac{a}{\text{g.c.d } (a, b)}.$$

2.4. Irreducibles and primes

R denotes a commutative ring with unity. The notion of 'divisibility' is viewed from two aspects: (i) through irreducible elements and (ii) through primes.

Definition 2.4.1 : *A nonzero element $q \in R$ is called an irreducible if, and only if, q is not a unit (divisor of the multiplicative identity 1_R) and in every factorization $q = b \cdot c$ with $b, c \in R$, either b or c is a unit.*

Definition 2.4.2 : *A nonzero element $p \in R$ is called a prime if, and only if, p is not a unit and for $0 \neq a$, $0 \neq b$ elements in R whenever $p \mid a \cdot b$, either $p \mid a$ or $p \mid b$.*

For $x, y \in R$, we write $x \mid y$ to denote that x divides y. When x does not divide y, we express it as $x \nmid y$. We recall that two elements $a, b \in R$ are said to be associates if a is of the form $b \cdot u$ where u is a unit.

We write $a \sim b$ to say that a and b are associates. $\sim$ is an equivalence relation on R. The equivalence classes are sets of associated elements. The associates of 1_R are units. In $\mathbb{Z}$, the associates of $n \in \mathbb{Z}$ are $\pm n$. If a and b are associates in

R, the statements: $a \mid b$ and $b \mid a$ are valid and the principal ideals (a) and (b) are equal.

We note that any element which is an associate of an irreducible (prime) is also an irreducible (prime). It is easy to check that if D is an integral domain, a prime p in D is always an irreducible. The converse of this statement is not true, in general. In $\mathbb{Z}[\sqrt{-5}] = \{a+b\sqrt{-5} : a,b \in \mathbb{Z}\}$ one has $2 \cdot 3 = (1+\sqrt{-5})(1-\sqrt{-5})$ and 2 is an irreducible. However, 2 divides $(1+\sqrt{-5})(1-\sqrt{-5})$ and $2 \nmid (1+\sqrt{-5})$, $2 \nmid (1-\sqrt{-5})$. So, 2 is not a prime in $\mathbb{Z}[\sqrt{-5}]$.

In the context of a PID, the notions of an irreducible and a prime coincide.

Theorem 9 : *Let D be a PID. A nonzero element p in D is an irreducible if, and only if, p is a prime.*

Proof : $\Leftarrow$: If p is a prime, suppose that $p = a \cdot b$, where $a,b \in D$, then $p \mid a \cdot b$. Therefore, $p \mid a$ or $p \mid b$. For definiteness, we shall take it as $p \mid a$. Then, by definition of a divisor, we have $a = p \cdot c$ where $c \in D$. Then, $p = (p \cdot c) \cdot b = p \cdot (c \cdot b)$. By cancellation law, $c \cdot b = 1_D$. So, c and b are units in D. Thus, $p = a \cdot b$ where b is a unit. That is, p is an irreducible.

$\Rightarrow$ Suppose that p is an irreducible such that $p \mid a \cdot b$ where a and b are nonzero elements of D. Then, $p \cdot c = a \cdot b$ for some element $c \in D$. As D is a PID the ideal generated by p and a written (p,a) is another principal ideal say (d), where $d \in D$. But, then, $p = r \cdot d$ where $r \in D$, for some choice of $r \in D$. Now, p is an irreducible. So, either r or d is a unit. If d is a unit, $(p,d) = D$. There exist elements $s,t \in D$ such that

$$1_D = s \cdot p + t \cdot d.$$

Then, $b = b \cdot 1_D = b \cdot (s \cdot p + t \cdot d) = b \cdot s \cdot p + b \cdot t \cdot d.$

Or, $b = b \cdot s \cdot p + b \cdot t \cdot d.$

Further, there exist $x,y \in D$ such that $d = x \cdot p + y \cdot a$. So,

(2.4.1) $b = b \cdot s \cdot p + b \cdot t \cdot (x \cdot p + y \cdot a)$

(2.4.2) $b = b \cdot s \cdot p + b \cdot t \cdot x \cdot p + t \cdot y \cdot (a \cdot b).$

Or, $p \mid a \cdot b$. So, $p \mid b$. So, $b = p$ so $p \mid b$. If, on the other hand, r is a unit, one could similarly arrive at $p \mid a$. So, whenever $p \mid a \cdot b$ either $p \mid a$ or $p \mid b$. Thus, p is a prime. $\square$

Next, we mention about prime and maximal ideals in a commutative ring with identity 1_R.

Definition 2.4.3 : *An ideal I of R is called a prime ideal if for all $a,b \in R$ with $a \cdot b \in I$, one has either $a \in I$ or $b \in I$.*

It is known [1] that an ideal I of R is a prime ideal $\Leftrightarrow$ R/I is an integral domain.

Definition 2.4.4 : *An ideal M of R is called an maximal ideal if $M \neq R$ and whenever J is an ideal of R such that $M \subseteq J \subseteq R$, then, either $J = M$ or $J = R$.*

In other words, it is not possible to squeeze in a proper ideal of R between a maximal ideal M and the whole ring R. If M is a maximal ideal and if $a \in R$ and $a \notin M$ is chosen, the ideal generated by the set $M \cup \{a\}$ written $(M,a) = R$.

It is known [1] that M is a maximal ideal of $R \Leftrightarrow R/M$ is a field.

In $\mathbb{Z}$, an ideal (r) is a maximal ideal $\Leftrightarrow r$ is a prime. In fact, ideals generated by primes in $\mathbb{Z}$ are maximal ideals. Clearly, in R, a maximal ideal M of R is a prime ideal, as a field is an integral domain.

Proposition 2.4.1 : *[Krull-Zorn theorem] In a commutative ring R with unity, every proper ideal is contained in a maximal ideal.*

For proof see D. M. Burton [1].

We deduce that $u \in R$ is a unit if, and only if, u belongs to no maximal ideal of R. The ideal-theoretic version of theorem 9 is given in

Theorem 10 : *Let R be a PID. A non-trivial ideal (a) of R is a prime ideal if, and only if, it is a maximal ideal.*

Proof : $\Leftarrow$: As a maximal ideal is also a prime ideal, (a) is a maximal ideal implies that (a) is a prime ideal.
:$\Rightarrow$ Let (a) be a prime ideal. Suppose that J is an ideal of R such that $(a) \subseteq J \subseteq R$. Since R is a PID, there exists an element $0 \neq b \in R$ such that $J = (b)$. Now, $a \in (a) \subseteq (b)$ implies that $a = r \cdot b$ for some choice of $r \in R$. Since (a) is a prime ideal, $r \cdot b \in (a)$ implies that either $r \in (a)$ or $b \in (a)$. If $b \in (a)$ one has $(b) \subseteq (a)$ and so $(b) = (a)$. If $r \in (a)$, $r = s \cdot a$ for some $s \in R$. So, $a = r \cdot b = s \cdot a \cdot b = a \cdot (s \cdot b)$. Therefore, $s \cdot b = 1_R$ or b is a unit. Then, $(b) = R$. So, $(a) \subseteq J \subseteq R \Rightarrow$ either $J = (a)$ or $J = R$.
Thus, (a) is a maximal ideal of R. $\qquad\qquad\square$

Theorem 11 : *Let p be a nonzero non-unit element of an integral domain D.*
(i) p is an irreducible element of D if, and only if, the principal ideal (p) is a maximal principal ideal, that is, (p) is a maximal in the set of proper principal ideals of D.
(ii) p is a prime element of D if, and only if, the principal ideal $(p) \neq D$ is a prime ideal.

Proof : (i) :$\Rightarrow$ Suppose that p is an irreducible element in D. Let (a) be a principal ideal such that

$$(p) \subseteq (a) \subseteq D.$$

As $p \in (a)$, we can write $p = t \cdot a$ for some $t \in D$. As p is an irreducible, t or a is a unit in D. If t is a unit,

$$a = t^{-1} \cdot p \in (p). \text{ So, } (a) \subseteq (p), \text{ or, } (p) = (a).$$

If a is a unit, then $(a) = D$. So, (p) is a maximal principal ideal in D.

$\Leftarrow$: Conversely, let (p) be a maximal ideal in the set of proper principal ideals of D. Suppose that p is not an irreducible element. Then, $p = a \cdot b$, where $a, b \in D$ and neither a nor b is a unit.

If $a \in (p)$, $a = s \cdot p$ for some choice of $s \in D$. Then $p = a \cdot b = s \cdot p \cdot b = p \cdot (s \cdot b)$. Using cancellation law (of D), $s \cdot b = 1_D$. That is, b is a unit—a contradiction to the assumption that a, b are non-units. So, $a \notin (p)$. Hence, $(p) \subset (a)$, as $a \mid p$.

If $(a) = D$, a is a unit contrary to the assumption that a is a non-unit. So, $(p) \subset (a) \subset D$. But, then, (p) fails to be maximal in the set of principal ideals of D—a contradiction. Thus, $p = a \cdot b \Rightarrow$ either a or b is a unit. That is, p is an irreducible element in D.

(ii) $: \Rightarrow$ Suppose that p is a prime in D. To prove that (p) is a prime ideal in D, assume that $a \cdot b \in (p)$ for $a, b \in D$. Then, $a \cdot b = x \cdot p$ for some choice of $x \in D$. So, $p \mid a \cdot b$. As p is a prime, either $p \mid a$ or $p \mid b$. That is, either $a \in (p)$ or $b \in (p)$. Consequently, (p) is a prime ideal.

$\Leftarrow$: Conversely, let (p) be a prime ideal of D. Let $p \mid a \cdot b$ for $a, b \in D$. Then, $a \cdot b \in (p)$. Since (p) is a prime ideal, either $a \in (p)$ or $b \in (p)$. That is, either $p \mid a$ or $p \mid b$. Hence p is a prime in D. $\qquad \square$

Fact 2.4.1 : Given a PI Domain D, a non-trivial ideal (p) of D is a maximal ideal $\Leftrightarrow p$ is an irreducible element. (p) is a prime ideal of $D \Leftrightarrow p$ is a prime in D.

Theorems 9, 10 and 11 give us the following:

Fact 2.4.2 : (a) In a PID, an irreducible element is a prime and vice versa. Consequently a prime ideal is a maximal ideal. Therefore, if D is a PID, every nonzero non-unit in D is divisible by some prime $p \in D$.

(b) In an integral domain, once we locate irreducible elements in the sense that there exist nonzero non-units that are irreducible, one can have an ascending chain of principal ideals that terminates. One takes this as the ascending chain condition on principal ideals, briefly written as ACCP. This idea will be explained when we deal with GCD domains in section 2.5 of this chapter.

Definition 2.4.5 : *An integral domain D is called a unique factorization domain (briefly written as UFD) if*
(i) every nonzero non-unit element $a \in D$ can be factorized into a (product of) a finite number of irreducibles and
(ii) $0 \neq a \in D$ has two factorizations into irreducibles, namely,

$$a = p_1 p_2 \cdots p_n = q_1 q_2 \cdots q_m$$

$n = m$ and there is a permutation π of the suffix set $\{1, 2, \cdots, n\}$ such that p_i and $q_{\pi(i)}$ are associates $(i = 1, 2 \cdots, n)$.

Indeed, $\mathbb{Z}$ is a UFD. In fact, we know that every PID is a UFD. An example of an integral domain which is not a UFD, is given by

$$\mathbb{Z}[\sqrt{-5}] = \{a + b\sqrt{-5} : a, b \in \mathbb{Z}\}$$

For, 3 is an irreducible in $\mathbb{Z}[\sqrt{-5}]$. To see this, suppose that

$$3 = (a + b\sqrt{-5}) \cdot (c + d\sqrt{-5})$$

gives (taking conjugates)

$$3 = (a - b\sqrt{-5}) \cdot (c - d\sqrt{-5})$$

and on multiplication,

$$3^2 = (a^2 + 5b^2) \cdot (c^2 + 5d^2)$$

So, $a^2 + 5b^2 \mid 9$. But $a^2 + 5b^2 \neq 1$ and $c^2 + 5d^2 \neq 1$, if $a + b\sqrt{5}$ and $c + d\sqrt{5}$ are both non-units. So, $a^2 + 5b^2 \neq 9$, as in that case $c^2 + 5d^2$ would be 1. We are left with $a^2 + 5b^2 = 3$. This equation has no solutions $\langle a, b \rangle$ as $a^2 + 5b^2 \geq 5$ for nonzero integers a and b. Similarly $c^2 + 5d^2 = 3$ is not permissible. So, a non-trivial factorization of 3 in $\mathbb{Z}[\sqrt{-5}]$ is not possible. Therefore, 3 is an irreducible. Similarly it can be shown that $2 + \sqrt{-5}$ and $2 - \sqrt{-5}$ are irreducibles. But then, one has

$$9 = 3 \cdot 3 = (2 + \sqrt{-5}) \cdot (2 - \sqrt{-5}).$$

This gives two different factorizations of 9 into irreducibles and so $\mathbb{Z}[\sqrt{-5}]$ is not a UFD.

We state below, without proof, a consequence of theorem 9.

Proposition 2.4.2 : *Let D be an integral domain in which every nonzero non-unit is expressible as a finite product of irreducibles. Then D is a UFD if, and only if, every irreducible is a prime.*

For proof, see I. Stewart & D. O. Tall [5].

2.5. GCD domains

Following I. Kaplansky [2], an integral domain D is called a GCD domain, if every pair of nonzero elements in D has a greatest common divisor. It follows that in a *GCD* domain, any finite number of elements have a g.c.d.

Let D be a *GCD* domain.

Suppose a, b, c are elements of D.

The g.c.d of $c \cdot a$ and $c \cdot b$ exists and g.c.d $(c \cdot a, c \cdot b) = c \cdot$ g.c.d(a, b).

If g.c.d $(a, b) = 1_D =$ g.c.d (a, c) then g.c.d $(a, b, c) = 1_D$.

For $a_1, a_2, \cdots a_n \in D$,

g.c.d $(a_1, a_2, \cdots, a_n) =$ g.c.d(g.c.d$(a_1, a_2, \cdots, a_{n-1}), a_n)$, $n \geq 2$.

Fact 2.5.1 : In a *GCD* domain D, every irreducible is a prime.

For, if $p \in D$ is an irreducible and $p \mid a \cdot b$ $(a, b, \in D)$ as g.c.d (p, a) divides p, the g.c.d (p, a) is either p or 1_D. Similarly, g.c.d (p, b) is either p or 1_D.

Now, g.c.d $(p, a) =$ g.c.d$(p, b) = 1_D$ contradicts g.c.d $(p, ab) = p$. So when $p \mid ab$, either $p \mid a$ or $p \mid b$, thereby asserting that p is a prime.

From the definition of a GCD domain, we deduce that in a GCD domain D, the ideal generated by a finite number of elements is a principal ideal. Further, in a GCD domain every irreducible element is a prime. Since in $\mathbb{Z}[\sqrt{-5}]$, one has irreducible elements which are not primes, we see that $\mathbb{Z}[\sqrt{-5}]$ is not a GCD domain.

Claim : In $\mathbb{Z}[\sqrt{-5}]$, the elements 9 and $3 \cdot (2 + \sqrt{-5})$ do not permit a g.c.d. We proceed as follows. In $\mathbb{Z}[\sqrt{-5}]$, we have

$$9 = 3 \cdot 3 = (2 + \sqrt{-5}) \cdot (2 - \sqrt{-5})$$

(see the example following definition 2.4.5).

The common divisors of 9 and $3 \cdot (2 + \sqrt{-5})$ and $1, 3$ and $(2 + \sqrt{-5})$. However, 3 is not divisible by $2 + \sqrt{-5}$ and vice versa. Therefore, a unique g.c.d of 9 and $3(2 + \sqrt{-5})$ fails to exist. Further, though 3 and $2 + \sqrt{-5}$ are relatively prime to one another, the equation

(2.5.1) $\text{g.c.d}(9, 3(2 + \sqrt{-5})) = 3 \cdot \text{g.c.d}(3, 2 + \sqrt{-5})$

is not valid in $\mathbb{Z}[\sqrt{-5}]$. See D. M. Burton [1]. We remark that if $\mathbb{Z}[\sqrt{-5}]$ is a *GCD* domain, we should have g.c.d $(c \cdot a, c \cdot b) = c \cdot \text{g.c.d}\,(a, b)$ for any three elements a, b, c in $\mathbb{Z}[\sqrt{-5}]$. This, again, shows that $\mathbb{Z}[\sqrt{-5}]$ is not a *GCD* domain.

Definition 2.5.1 : *Let R be a commutative ring with unity* 1_R. *R is said to satisfy the ascending chain condition on principal ideals written ACCP, if every ascending chain of principal ideals terminates. That is, given a chain of principal ideals*

$$(a_1) \subseteq (a_2) \subseteq \cdots \subseteq (a_n) \subseteq \cdots$$

there exists $m \in \mathbb{N}$ *such that* $(a_n) = (a_m)$ *for all* $n \geq m$.

It is clear that $\mathbb{Z}$ satisfies ACCP. We ask the question : When is a GCD domain a UFD?
The answer is in

Theorem 12 : *An integral domain D is a UFD if, and only if, it is a GCD domain in which ACCP is satisfied.*

Proof : $\Leftarrow$: Let D be a GCD domain. We have seen that D has the property given in Fact 2.5.1. That is, every irreducible is a prime.

Also, here, D satisfies ACCP. Let a be a nonzero non-unit in D. Then, there is a chain

$$(a) \subseteq (a_1) \subseteq (a_2) \subseteq \cdots \subseteq (a_n) \subseteq \cdots$$

which terminates.

So, there exists $m \in \mathbb{N}$ such that

$$(a_m) = (a_{m+1}) = \cdots .$$

So then, a_m is irreducible say p_1.

Writing $a = p_1 \cdot b_1$ we check whether b_1 is an irreducible or not. If b_1 is an irreducible, a is a product of irreducibles. Otherwise, we write $b_1 = p_2 \cdot c_2$ where p_2 is an irreducible. Continuing in this manner, we obtain,

$$(a) \subseteq (b_1) \subseteq (b_2) \cdots .$$

This breaks off with an irreducible element $b_n = p_{n+1}$.
Thus, we get

$$a = p_1 \cdot b_1 = p_1 \cdot p_2 \cdot b_2 = \cdots = p_1 \cdot p_2 \cdots p_{n+1}.$$

That is, a is finite product of irreducibles. By Proposition 2.4.2, we conclude that D is a UFD as every irreducible in a GCD domain (see Fact 2.5.1) is a prime.
$:\Rightarrow$ Conversely, if D is a UFD, we write $a, b \in D$ as

$$a = \mu \cdot p_1^{a_1} \cdot p_2^{a_2} \cdots p_r^{a_r}, \quad a_i \geq 0, i = 1, 2, \cdots, r\,;$$

$$b = \nu \cdot p_1^{b_1} \cdot p_2^{b_2} \cdots p_r^{b_r}, \quad b_i \geq 0, i = 1, 2, \cdots, r$$

where μ, ν are units in D. p_i $(i = 1, 2, \cdots, r)$ are pairwise non-associated primes. The g.c.d of a and b is given by

$$\text{g.c.d } (a, b) = \omega p_1^{c_1} \cdot p_2^{c_2} \cdots p_r^{c_r}, \text{ where } c_i = \min\{a_i, b_i\}\,(i = 1, 2, \cdots, r)$$

and ω is a unit.

So, D is a GCD domain. Further, every nonzero non-unit $t \in D$ is a finite product of irreducibles. We can make an ascending chain of principal ideals of the form

$$(t) \subseteq (t_1) \subseteq (t_2) \cdots \subseteq (t_n) \subseteq \cdots$$

where t_n is an irreducible for an appropriate choice of $n \geq 1$. Then, the ascending chain terminates, as whenever $b \mid a$, $(a) \subseteq (b)$.
So, D is a GCD domain in which ACCP holds. $\square$

Remark 2.5.1 : The criterion for a GCD domain to be a UFD has been adapted from G. Karpilovsky [3].

2.6. Notes with illustrative examples

We have observed that the set of integers has the properties of an ordered integral domain in which the set of positive elements is well-ordered. It is possible to assume simple properties of $\mathbb{N}$, the set of positive integers and to derive all other properties in a logical way. The idea is due to the Italian mathematician G. Peano (1858–1932) who stated a few axioms to define $\mathbb{N}$. They are called

PEANO'S AXIOMS:
Axiom1: $1 \in \mathbb{N}$
Axiom 2: To each element n of $\mathbb{N}$ there corresponds a unique element n', called the successor of n in $\mathbb{N}$.
Axiom 3: For each $n \in \mathbb{N}$, $n' \neq 1$. Or 1 is not the successor of any element of $\mathbb{N}$.
Axiom 4: If $m, n \in \mathbb{N}$, $m' = n' \Rightarrow m = n$.
Axiom 5: Let K be a set of elements belonging to $\mathbb{N}$. Then $K = \mathbb{N}$, if
(a) $1 \in K$
(b) Given $k \in K$, then $k' \in K$.
Axiom 5 is the basis of proof by mathematical induction. Using the five axioms one could define addition and multiplication in $\mathbb{N}$ and to prove that $\mathbb{N}$ has all the properties of an integral domain except that $0 \notin \mathbb{N}$ and $n \in \mathbb{N}$ has no additive inverse. For addition, define $n + 1 = n'$ and proceed to show that $m + n' = (m + n)'$. For $n \in \mathbb{N}$, define $n \cdot 1 = n$ and for $n \in \mathbb{N}$, we get $n \cdot m' = n \cdot m + n$. We have to introduce negative integers and the zero element in $\mathbb{Z}$.

We write

$$S = \{(a,b) : a,b \in \mathbb{N}\}$$

We are to frame (a,b) as $a-b$. For $(a,b),(c,d) \in S$, we say that $(a,b) \sim (c,d)$ if $a+d = c+b$. $\sim$ is an equivalence relation on S. The equivalence class $[a,b]$ is given by

$$[a,b] = \{(x,y) : x,y \in \mathbb{N} \text{ and } x+b = y+a\}.$$

Let us define

(2.6.1) $$[a,b] + [c,d] = [a+c, b+d]$$

(2.6.2) $$[a,b] \cdot [c,d] = [ac+bd, ad+bc]$$

$\mathbb{Z}$ is realised as an integral domain with the new notation for addition and multiplication given in (2.6.1) and (2.6.2). The zero element of $\mathbb{Z}$ is $[c,c]$ for every $c \in \mathbb{N}$. The additive inverse of $[a,b]$ is $[b,a]$, that is

$$-[a,b] = [b,a].$$

Let $\mathbb{N}'$ be the set of elements of $\mathbb{Z}$ of the form $[x+1,1]$, $x \in \mathbb{N}$
The map $\phi : \mathbb{N} \to \mathbb{N}'$ defined by

$$\phi(n) = [n+1,1], n \in \mathbb{N}$$

is one-one and onto $\mathbb{N}'$. Addition and multiplication are preserved under ϕ. We can identify $\mathbb{N}'$ with $\mathbb{N}$ and take n to represent $[n+1,1] \in \mathbb{N}'$. So, $\mathbb{Z}$ contains $\mathbb{N}$.
If $[a,b] \in \mathbb{Z}$,

$$[a,b] = [a+1,1] + [1,b+1]$$
$$= [a+1,1] - [b+1,1]$$

$[a+1,1]$ and $[b+1,1]$ are respectively elements $a,b \in \mathbb{N}$.
So then,

$$[a,b] = a-b$$

as desired. Thus, $\mathbb{Z}$ contains zero and $\pm n$ for each $n \in \mathbb{N}$.

This argument parallels the method of construction of the set of rational numbers starting from the integral domain $\mathbb{Z}$ of rational integers.

Let D be a PID. Suppose $a(\neq 0) \in D$. $\lambda(a)$ is defined as the number of irreducible factors of a. $\lambda(a)$ is called the length of a. $\lambda(a) = 0$ if, and only if, a is a unit in D. If $a \mid b$, $\lambda(a) \le \lambda(b)$. If a and b do not divide one another, there exist $p,q \in D$ such that

$$\lambda(pa+qb) \le \min\{\lambda(a), \lambda(b)\}$$

If a has length n, it can be shown that there are at the most 2^n ideals containing a, in D. For, suppose

$$a = p_1 p_2 \cdots p_n \text{ where } p_i \text{ is irreducible}$$

$p_i = p_j$ for $i = j$ is allowed.
The ideals generated by p_i are maximal ideals of D. When $p_i \neq p_j$, $i \neq j$, the set $S = \{p_1, p_2, \cdots, p_n\}$ has 2^n subsets. The subset containing k irreducibles

generates a principal ideal containing a. Therefore, there are at the most 2^n ideals (of D) containing a.

Recalling the example of $\mathbb{Z}[\sqrt{-5}]$, we note that g.c.d (a,b) may exist but g.c.d (ta,tb) may not exist in an integral domain. Therefore, it is wrong to assume always

$$\text{g.c.d } (ta,tb) = t \text{ g.c.d } (a,b)$$

in an arbitrary integral domain, unless it is a GCD domain.

Next, the dual concept of g.c.d is that of l.c.m.

Definition 2.6.1 : *Let $a_1, a_2, \cdots, a_n$ be nonzero elements of an integral domain D. $l \in D$ is called a least common multiple (l.c.m) of $a_1, a_2, \cdots, a_n$, if*
(i) $a_i | l$, for $i = 1, 2, \cdots n$.
(ii) $a_i | c$, for $i = 1, 2, \cdots n$ implies that $l | c$.

One could verify that $a_1, a_2, \cdots, a_n$ have an l.c.m if, and only if, $\bigcap_i (a_i)$ is principal. If D is a *GCD* domain, it follows that any finite number of nonzero elements of D admit a l.c.m. Every principal ideal domain is a *GCD* domain possessing the l.c.m property also.

This chapter has touched upon the structure of $\mathbb{Z}$. To supplement the content with related material bearing on $\mathbb{Z}$, see A. G. Hamilton [A2]. See also Aaboe [A1].

2.7. Worked-out examples

a) A ring R is said to be ordered when there is a non-empty subset P of R, called the set of positive elements of R satisfying
(i) $a \in P$ and $b \in P \Rightarrow a+b \in P$ and $a \cdot b \in P$.
(ii) for each $a \in R$, either $a \in P$, $a = 0_R$ or $(-a) \in P$. (law of trichotomy (see definition 2.2.1 relating to an integral domain D.))
Prove that (i) In any ordered ring R, all squares of nonzero elements are positive (ii) Any ordered commutative ring R is an integral domain of characteristic zero.

Answer: (i) Let P be the set of positive elements of R. Suppose that $0_R \neq a \in R$. By the law of trichotomy, either $a \in P$ or $-a \in P$. Since P is closed under multiplication, $a^2 = (-a)^2 \in P$ in either case, as asserted.

(ii) R is an ordered commutative ring. Suppose that $0_R \neq a$, $0_R \neq b$ where a, b make $a \cdot b = 0_R$. Then, $(\pm a) \cdot (\pm b) = 1_R$. By trichotomy law, one of $\pm a$ and one of $\pm b \in P$. So, some one of the four products $(\pm a) \cdot (\pm b)$ belongs to P—a contradiction to $0_R \notin P$. So, R is an ordered integral domain. Since $1_R \in P$,

$$1_R + 1_R + \cdots + 1_R (n \text{ times }) \in P.$$

Therefore $1_R + 1_R + \cdots + 1_R (n \text{ times }) \neq 0_R$. That is, R has characteristic zero. □

Remark 2.7.1 : If $1_R \in R$, $1_R^2 = 1_R$ is always positive. -1_R is never positive.

b) Discuss the nature of $\mathbb{Z}, \mathbb{R}, \mathbb{Z}[\sqrt{2}]$ and $\mathbb{Q}$ as ordered integral domains.

Answer: (i) There is only one way to make $\mathbb{Z}$ an ordered integral domain. (See theorem 8.)

(ii) In the case of $\mathbb{R}$, the field of real numbers, we make the observation that by example (a), any nonzero square is positive, (in any ordering of the ring). The positive real numbers have real square-roots, so real numbers remain positive in any ordering of $\mathbb{R}$. So, there is only one way to make $\mathbb{R}$, an ordered integral domain. Further, the monomorphism $\phi : \mathbb{Z} \to \mathbb{R}$ given by $\phi(n) = n.1$ is order-preserving.

(iii) In the case of $\mathbb{Z}[\sqrt{2}]$, we have

$$\mathbb{Z}[\sqrt{2}] = \{a + b\sqrt{2} : a, b \in \mathbb{Z}\}$$

$\mathbb{Z}[\sqrt{2}]$ can be made an ordered integral domain by at least two choices of P (the set of positive elements in $\mathbb{Z}[\sqrt{2}]$).

One ordering is given by the condition:

$$a + b\sqrt{2} \in P, \text{ if } a > 0 \text{ and } a^2 > 2b^2.$$

The other ordering is given by the condition:

$$a + b\sqrt{2} \in P, \text{ if } b > 0 \text{ and } 2b^2 > a^2.$$

Further details are to be verified, without any difficulty.

(iv) $\mathbb{Q}$ is the field of quotients of $\mathbb{Z}$. Let P denote the set of positive elements in $\mathbb{Z}$.

For $a, b \in \mathbb{Z}$ with $b \neq 0$, we say that $\frac{a}{b} \in P_1$, the set of positive elements of $\mathbb{Q}$, if, and only if, $a \cdot b \in P$ in $\mathbb{Z}$. Now,

$$\frac{a}{b} = (a \cdot b) \cdot \frac{1}{b^2}. \quad \frac{1}{b^2} \text{ is } (\frac{1}{b})^2 \text{ is necessarily in } P_1.$$

So, $\frac{a}{b} \in P_1$, if, and only if, $a \cdot b \in P_1$.

If $\phi : \mathbb{Z} \to \mathbb{Q}$ is given by $\phi(n) = n \cdot 1 \in \mathbb{Q}$, ϕ is order-preserving and so $a \cdot b \in P_1$ if, and only if, $a \cdot b \in P$. So, P_1 is the set of positive elements of $\mathbb{Q}$, provided P_1 is closed under addition and multiplication. This can be checked easily. $\square$

Remark 2.7.2 : There is one and only one way of making $\mathbb{Q}$ an ordered field.

c) Let $\mathbb{Z}[i]$ denote the integral domain of Gaussian integer. For $\alpha, \beta \in \mathbb{Z}[i]$, we define $\alpha \equiv \beta \pmod{\eta}$ if $\frac{\alpha - \beta}{\eta} \in \mathbb{Z}[i]$, (where $\eta \in \mathbb{Z}[i]$). Congruences are clearly additive and multiplicative as for rational integers. Write down the residue classes of integers of $\mathbb{Z}[i]$ modulo $(2 + i)$. Show that they form a field having 5 elements.

Answer: The zero residue class $[0]$ contains multiples of $(2 + i)$. The other residue classes are determined by the rational integers $1, 2, 3$ and 4.

The residue classes are $\{[1], [2], [3], [4], [0]\}$. They form a field having 5 elements. $\square$

Remark 2.7.3 : $\mathbb{Z}[i]$ is a U.F.D. Further, $(2+i)$ is a prime in $\mathbb{Z}[i]$, as norm of $(2+i)$ equal to $2^2+1^2=5$ is a prime. So, $\mathbb{Z}[i]/(2+i)$ is an integral domain which is finite.

EXERCISES

1. **Mark the following statements true (T) or false (F) justifying your answer briefly.**
   a) The ring $\mathbb{Z}/r\mathbb{Z}$ is a principal ideal ring ($r \geq 1$)
   b) Let I be an nonzero ideal of the ring $\mathbb{Z}$ of integer. The following statement are equivalent:
   (i) I is a prime ideal,
   (ii) I is a maximal ideal,
   (iii) $I = (p)$, where p is a prime.
   c) $\mathbb{Z}[\sqrt{3}]$ is not a GCD domain.
   d) A commutative ring with unity can be made an ordered ring.
   e) $D = \{x+y\sqrt[3]{2}: x,y \in \mathbb{Q}\}$ is an integral domain under ordinary addition and multiplication.
   f) A prime subdomain is of an integral domain D is one containing no proper subdomains. Let 1_D denote the unity element in D. Then,

   $$D' = \{n1_D : n = 0, \pm1, \pm2, \ldots\}$$

   is a prime subdomain of D with unity 1_D.
   (Note: Every integral domain D contains a unique prime subdomain which is either isomorphic to $\mathbb{Z}/p\mathbb{Z}$ (p a prime) or isomorphic to $\mathbb{Z}$).

2. $\mathbb{Z}$ denotes the integral domain of rational integers. I, J denote the principal ideals of $\mathbb{Z}$ generated by 36 and 49 respectively. Find $I+J$, IJ, $I \cap J$.

3. $\mathbb{Z}_e$ denotes the ring of even rational integers. ($\mathbb{Z}_e$ does not have a unity element). Show that the principal ideal generated by 4 is a maximal ideal in $\mathbb{Z}_e$ though $\mathbb{Z}_e/4\mathbb{Z}_e$ is not a field. What is your inference?

4. (a) What are the prime and maximal ideals in $\mathbb{Z}/36\mathbb{Z}$?
   (b) Discuss the general case when 36 is replaced by an integer $r > 1$.

5. Show that $\mathbb{Z}/r\mathbb{Z}$ is a principal ideal ring for all $r \geq 1$ (it need not be a PID).

6. Give an example of a ring possessing a non-trivial prime ideal which is not a maximal ideal.

7. Let P be proper prime ideal of a non-trivial commutative ring R with unity 1_R. If R/P is finite, show that P is a maximal ideal of R.

8. Show that $\mathbb{Z}[\sqrt{-5}]$ is not a PID.

9. Show that $\mathbb{Z}[\sqrt{-6}] = \{a+b\sqrt{-6} : a,b \in \mathbb{Z}\}$ is not a UFD.

10. Give an example of a GCD domain that is not a UFD.

11. Let R be the ring of $n \times n$ matrices with entries from a field F. Show that R has no right or left ideals other than $[0]$ and R.

12. *Prove or disprove : In an ordered field, the set of positive elements is not well-ordered.*

13. *Let F be a field. The intersection of all subfields of F is called the prime subfield of F. It is the subfield generated by 1_F. Show that in an ordered field F, the prime subfield is isomorphic to $\mathbb{Q}$.*

14. *Make the integral domain $\mathbb{Z}[\sqrt{3}]$ into an ordered integral domain, by defining P, the set of its positive elements.*

15. *By considering $\mathbb{Z}[\omega]$ where $\omega = \exp(\frac{2\pi i}{3})$. Show that a rational prime of the form $3k+1$ is expressible in the form $a^2 - ab + b^2$.*

16. *Let $\theta = \exp(\frac{2\pi i}{5})$. We write*

$$\mathbb{Z}[\theta] = \{a_0 + a_1\theta + a_2\theta^2 + a_3\theta^3 : a_0, a_1, a_2, a_3 \in \mathbb{Z}\}.$$

Show that $\mathbb{Z}[\theta]$ is a PID.

17. *(Charles Vanden Eynden) Given an integer $r > 1$, determine the set of integers which can be written as a sum of two integers relatively prime to r. (Ref: Problem 10338, Amer.Math.Monthly, 104 (1997) p 75).*

REFERENCES

[1] D. M. Burton : A first course in rings and fields, Chapters 2, 4, 5, and 6, pp 16–111. Addison-Wesley Pub. Co., Reading, Mass USA (1970).

[2] I. Kaplansky: Commutative rings, Allyn and Bacon Inc, Boston, 1970.

[3] G. Karpilovsky : Commutative group algebras, Chapter 4, pp 83–107, Monographs & Textbooks in Pure & App. Math. No. 78, Marcel Dekker Inc. NY (1983).

[4] Neal H. McCoy and Thomas R. Berger : Algebra : Groups, Rings & Other topics, Chapter 4 pp 160–170, Allyn Bacon Inc Boston USA (1977).

[5] Ian Stewart and D. O. Tall: Algebraic Number Theory, Chapter 4, pp 71–104, Chapman & Hall Inc., New York (1987), Second Edn.

ADDITIONAL REFERENCES

[A1] A. Aaboe : Episodes from the early history of mathematics: The New Mathematical Library-13. The Mathematical Association of America, Washington DC (1981).

[A2] A. G. Hamilton : Numbers, sets and axioms - The apparatus of Mathematics, Cambridge University Press, New York (1980).

CHAPTER 3

Euclidean domains

Historical perspective

The making of algebra (where symbols replaced numbers) took roots in India and Arabia around 400 A.D. Arithmetic became a discipline as a branch of mathematics at the hands of mathematicians in India and the Middle East. The origin of the word 'algorithm' is traced back to the Arab mathematician Al-Khowarizmi (825 A.D.) who computed via algebra areas of rectangles and used them to represent algebraic quantities and vice versa. But, it was not until 1637 when Rene Descartes (1596–1650) invented Analytical Geometry, a mixing of algebraic and geometric concepts took place. (Descartes made use of 'x' and other letters near the end of the alphabet to represent an 'indeterminate'.) In fact, the ideas of algebra and geometry were brought together in a recognizable way. That was a great event in the history of mathematics. For, it helped the invention of Calculus by Isaac Newton (1642–1727) and Gottfried Wilhelm von Leibnitz (1646–1716) during the early years of eighteenth century.

As years passed by, one could interpret the set $\mathbb{Z}$ of integers in many ways. The efforts to prove Fermat's last theorem paved the way for the development of new concepts in Ring theory. A close examination of certain types of integral domains which are number rings corresponding to algebraic number fields (which are finite extensions of the field $\mathbb{Q}$ of rational numbers) was done. Thanks to the efforts of Kummer, Kronecker, Dedekind and others, factorization of elements in commutative rings was analysed in depth. It was found that uniqueness of factorization of elements was possible in certain integral domains such as a Euclidean domain. The structure of a Euclidean domain (whose properties are discussed in this chapter) arises from the notion of the 'division algorithm' property occurring in the integral domain $\mathbb{Z}$. The idea of a 'Euclidean norm' works out beautifully well in a general set-up.

3.1. Introduction

The purpose of this chapter is to highlight the properties of

(1) Euclidean domains
(2) the ring of integers of $\mathbb{Q}(\sqrt{m})$, m an integer,

and to point out the well-known example of $\mathbb{Z}[\sqrt{\theta}]$ where $\theta = \frac{1+\sqrt{-19}}{2}$, which is a PID, but not a Euclidean domain. We also observe that $\mathbb{Z}$ is characterised by its 'double-remainder property'.

3.2. $\mathbb{Z}$ as a Euclidean domain

We denote the set of non-negative integers by $\widetilde{\mathbb{Z}}$. We observe that $\mathbb{Z}$ is an example of a Euclidean domain whose definition is shown below:

Definition 3.2.1 : *A Euclidean domain is an integral domain D together with a function* $g : D^* \to \widetilde{\mathbb{Z}}$ *(D* is the set of nonzero elements of D) such that*

(1) $g(ab) \geq g(b)$ for all $a,b \in D^*$,
(2) if $a \in D, b \in D^*$, there exist elements $q, r \in D$ such that

$$a = bq + r$$

where either $r = 0$ or $g(r) < g(b)$.

We write (D,g) to say that D is a Euclidean domain with the associated function g. We assume that D is not a field. As D is commutative, g satisfies $g(ba) \geq g(a)$ also.

It is easily verified that when (D,g) is a Euclidean domain, the following statements hold:

(3.2.1)   For each $a \in D^*$, $g(a) \geq g(1_D)$.

(3.2.2)   If a,b are associates in D, $g(a) = g(b)$.

(3.2.3) $u \in D$ is a unit if, and only if, $g(u) = g(1_D)$.

Theorem 13 : *Let (D,g) be a Euclidean domain. For $a \in D, b \in D^*$, one has $a = bq + r$ where either $r = 0$ or $g(r) < g(b)$. Then, q and r are unique if, and only if,*

(3.2.4) $g(a+b) \leq \max\{g(a), g(b)\}$ *for* $a, b \in D^*$.

Proof : $\Leftarrow$: Suppose that (3.2.4) holds. We claim that q, r are unique. On the contrary, assume that q, r are not unique. That is, let

$$a = bq' + r' \text{ where } r' = 0 \text{ or } g(r') < g(b)$$

for another representation of a in terms of q', r'. Also, let $r \neq r'$, $q \neq q'$. Then,

$$g(b(q - q')) \geq g(b)$$

But, $g(b(q - q')) = g(r' - r) < \max\{g(r'), g(-r)\}$.

That is,

(3.2.5) $g(b) < \max\{g(r'), g(-r)\}$, by definition (3.2.1)

(3.2.5) is possible only when one of $r' - r$ or $q - q'$ is zero. For,

$$g(r') < g(b), \quad g(-r) < g(b)$$

If $r' - r \neq 0$, $g(r)$ and $g(r')$ are nonzero and so one of $g(r)$, $g(r')$ has to be less than $g(b)$. (3.2.5) contradicts this situation. So, when once $r = r'$, $q = q'$ follows and vice versa. Therefore q, r are unique.

$:\Rightarrow$ We are given that in $a = bq + r$ with $r = 0$ or $g(r) < g(b)$, q and r are unique. We have to establish (3.2.4).

If possible, assume that

$$g(a+b) > \max\{g(a), g(b)\}$$

Then,

(3.2.6) $\qquad \begin{cases} b & = 0_D(a+b) + b \\ b & = 1_D(a+b) - a \end{cases}$

Also, from (3.2.6) $g(-a) = g(a) < g(a+b)$ and

$$g(b) < g(a+b).$$

We observe that we get a contradiction to the uniqueness of 'q' and 'r'. So, we must have

$$g(a+b) \leq \max\{g(a), g(b)\}, \quad \text{for } a, b \in D^*.$$

$\qquad\qquad\qquad\qquad\qquad\qquad\qquad\qquad\qquad\qquad\qquad\qquad\qquad\qquad \Box$

Remark 3.2.1 : (i) The above criterion for uniqueness of 'quotient' and 'remainder' has been adapted from D. M. Burton [1].
(ii) A commutative ring R satisfying the conditions in definition 3.2.1 is called a Euclidean ring.

Remark 3.2.2 : $(\mathbb{Z}, g)$ is a Euclidean domain with $g(a) = |a|$ for all $a \in \mathbb{Z}$. But (3.2.4) does not hold for the absolute value function. In the division algorithm $a = bq + r$ with $r = 0$ or $|r| < b$, q and r are not unique. However, in $a = bq + r$ with $0 \leq r < |b|$, q and r are unique. This is what we use in number-theoretic situations.

We state, without, proof

Fact 3.2.1 : Every Euclidean domain is a PID.

For proof, see I. N. Herstein [6]. The converse is not true in general. See theorem 19 below. See also [A2].

3.3. Quadratic number fields

$\mathbb{Q}$ denotes the field of rational numbers. Let $\mathbb{C}$ denote the field of complex numbers. A subfield K of $\mathbb{C}$ which is finite extension of $\mathbb{Q}$ is called an algebraic number field. It is known that K is of the form $\mathbb{Q}[\alpha]$ for some $\alpha \in \mathbb{C}$. If the degree of K over Q (written $[K : \mathbb{Q}]$) is equal to n, then,

(3.3.1) $\qquad Q[\alpha] = \{a_0 + a_1\alpha + \cdots + a_{n-1}\alpha^{n-1} : a_i \in \mathbb{Q}, 0 \leq i \leq n-1\}.$

In fact, $\{1, \alpha, \alpha^2, \cdots, \alpha^{n-1}\}$ forms a basis for $\mathbb{Q}[\alpha]$ considered as a vector space over $\mathbb{Q}$. Also, α is a root of an irreducible polynomial of degree n having coefficients from $\mathbb{Q}$.

Quadratic number fields are of the form $\mathbb{Q}[\sqrt{m}]$ where m is not a perfect square. Further, $[\mathbb{Q}[\sqrt{m}] : \mathbb{Q}] = 2$. $\{1, \sqrt{m}\}$ is a basis for $\mathbb{Q}[\sqrt{m}]$ as a vector space over $\mathbb{Q}$. We have

(3.3.2) $\mathbb{Q}[\sqrt{m}] = \{a + b\sqrt{m} : a, b \in \mathbb{Q}\}$

$\mathbb{Q}[\sqrt{m}]$ is called a real quadratic field when $m > 0$. It is called an imaginary quadratic field when $m < 0$.

We recall that if $\omega = \exp(\frac{2\pi i}{m})$ (m any positive integer), $\mathbb{Q}[\omega]$ is called the mth cyclotomic field. The irreducible polynomial of ω is the cyclotomic polynomial of degree $\phi(m)$, where ϕ is the Euler ϕ-function. If $\rho = \exp(\frac{2\pi i}{3})$

$$\mathbb{Q}[\rho] = \{a + b\rho : a, b \in \mathbb{Q}\}$$

and the irreducible polynomial of ρ is $x^2 + x + 1$, as $\rho^2 + \rho + 1 = 0$.

Definition 3.3.1 : *A complex number α is called an algebraic integer if, and only if, α is a zero of a monic polynomial (leading coefficient 1) with coefficients from $\mathbb{Z}$.*

As examples, we have algebraic integers such as $1 + \sqrt{2}$, $\frac{-1+\sqrt{3}\,i}{2}$, $\exp(\frac{2\pi i}{3})$ etc. However, if t is a rational number of the form $\frac{a}{b}$, $b \neq 0$, g.c.d $(a, b) = 1$, $\frac{a}{b}$ is not an algebraic integer.

Fact 3.3.1 : Let f be a monic polynomial with coefficients from $\mathbb{Z}$. Assume that $f = gh$, where g and h are monic polynomials with coefficients from $\mathbb{Q}$. Then, g and h have coefficients from $\mathbb{Z}$.

For proof, see D. A. Marcus [8, lemma p 14].

Lemma 3.3.1 : *Let α be an algebraic integer. Suppose that α satisfies a monic polynomial f (of lowest degree) with coefficients from $\mathbb{Z}$. Then, f is irreducible over $\mathbb{Q}$.*

Proof : On the contrary, suppose that f is reducible. Take $f = gh$ where g and h are non-constant polynomials with coefficients from $\mathbb{Q}$. g and h can be considered as monic polynomials (division by a suitable rational number will help). From fact 3.3.1, we see that g and h are monic polynomials with coefficients from $\mathbb{Z}$. But then, α is a zero of either g or h which are of lower degree than that of f. This contradicts the status of f as a monic polynomial of lowest degree and having α as a zero. So, f is irreducible. $\square$

Remark 3.3.1 : The notation : $\mathbb{Q}(\alpha)$.
In general, parentheses are used when one wishes to indicate that the set is a field, although no harm would be done by using $\mathbb{Q}[\alpha]$ to denote

$$\{a_0 + a_1\alpha + \cdots + a_{n-1}\alpha^{n-1} : a_i \in \mathbb{Q}, 0 \leq i \leq n-1\}.$$

$\mathbb{Z}[\sqrt{2}]$ is the ring of integers of $\mathbb{Q}(\sqrt{2})$. $\mathbb{Z}[\sqrt{2}]$ is merely a ring where as $\mathbb{Q}(\sqrt{2})$ is a field. Usage of a single bracket rather than a square bracket, conveys a bit more information about the set.

Let F be a field. By convention, we write $F[x]$ to denote the ring of polynomials in x. $F[x]$ is, indeed an integral domain. The field of quotients of $F[x]$ is denoted by $F(x)$. If $f(x)$ is the irreducible polynomial of α (an algebraic number) of degree n, we get an extension of $\mathbb{Q}$ by considering $\mathbb{Q}[x]/(f(x))$. We write $\mathbb{Q}(\alpha)$ to denote the finite extension of $\mathbb{Q}$ by adjoining α to $\mathbb{Q}$. It can be shown [6] that

$$\mathbb{Q}(\alpha) \cong \mathbb{Q}[x]/(f(x)).$$

$$\mathbb{Q}(\alpha) = \{a_0 + a_1\alpha + \cdots + a_{n-1}\alpha^{n-1} : a_i \in \mathbb{Q}, i = 0, 1, 2, \cdots, n-1\}.$$

Evidently, $\mathbb{Q}(\alpha)$ and $\mathbb{Q}[\alpha]$ are one and the same, in the present context.

Remark 3.3.2 : The monic irreducible polynomial of α has coefficients from $\mathbb{Z}$ when α is an algebraic integer.

Remark 3.3.3 : If α is an algebraic integer and $\alpha \in \mathbb{Q}$, then $\alpha \in \mathbb{Z}$. That is, if a rational number is an algebraic integer, it is a rational integer.

When the context is clear, the ring of algebraic integers contained in a number field K is referred to as the ring of integers of K. Properties of the ring of integers of a number field are treated in chapter 13.

Theorem 14 : *Let m be a square-free integer. Then the set $R(m)$ of algebraic integers in $\mathbb{Q}[\sqrt{m}]$ is given by*

$$R(m) = \begin{cases} \{a + b\sqrt{m} : a, b \in \mathbb{Z}\}, & \text{if } m \equiv 2 \text{ or } 3 \pmod 4 \\ \{\frac{a+b\sqrt{m}}{2} : a, b \in \mathbb{Z}, a \equiv b \pmod 2\}, & \text{if } m \equiv 1 \pmod 4 \end{cases}$$

Proof : $\alpha \in \mathbb{Q}[\sqrt{m}]$ is of the form $\alpha = \frac{a}{t} + \frac{b}{t}\sqrt{m}$ where $a, b, t \in \mathbb{Z}$ and $t > 0$. We may assume that $\frac{a}{t}$ and $\frac{b}{t}$ are in their lowest terms. That is, a, b, t have no common divisor > 1. When $\alpha \in R(m)$, α satisfies a monic polynomial f of degree 2 with coefficients from $\mathbb{Z}$. The zeros of f are $\alpha, \bar{\alpha}$ where $\bar{\alpha}$ is the conjugate of α. Then,

$$f(x) = (x - \alpha)(x - \bar{\alpha}) = x^2 - \frac{2a}{t}x + \frac{a^2 - b^2 m}{t^2}.$$

The coefficients of x and x^2 in f are rational integers. So, $\frac{2a}{t} \in \mathbb{Z}$ and $\frac{a^2-b^2m}{t^2} \in \mathbb{Z}$. As g.c.d $(t, a) = 1$, $t | 2$. So, $t = 2$ or $t = 1$.

If $t = 2$, $a^2 - b^2 m \equiv 0 \pmod 4$. As $t = 2$ and g.c.d $(t, a) = 1$, a is odd. This forces b to be odd. So, a and b are odd. Then,

$$a^2 \equiv 1 \pmod 4, \quad b^2 \equiv 1 \pmod 4.$$

Therefore, $a^2 - b^2 m \equiv 0 \pmod 4$ happens when $m \equiv 1 \pmod 4$ only.

Then, $a^2 - b^2 \equiv 0 \pmod 4$. This shows that a and b have the same parity. That is, $a \equiv b \pmod 2$. Thus, $\alpha = \frac{a+b\sqrt{m}}{2}$ as required.

Conversely, if $m \equiv 1 \pmod 4$, when a and b are odd, $a^2 - b^2 m \equiv 0 \pmod 4$ and

$a \in \mathbb{Z}$. So, $\alpha = \frac{a+b\sqrt{m}}{2}$ satisfies a monic polynomial with coefficients from $\mathbb{Z}$. So, $R(m) = \{ \frac{a+b\sqrt{m}}{2} : a,b \in \mathbb{Z}, a \equiv b \pmod 2 \}$ when $m \equiv 1 \pmod 4$.

If $m \equiv 2$ or $3 \pmod 4$, from $\frac{2a}{t} \in \mathbb{Z}$, when $t \neq 2$, we get $t = 1$ and $a^2 - b^2 m \in \mathbb{Z}$. a and b could take odd or even values in any manner. In all such cases, $\alpha = a + b\sqrt{m}$ is an algebraic integer, whenever $a,b \in \mathbb{Z}$ and $m \not\equiv 1 \pmod 4$. Hence, $R(m)$ has the structure as prescribed in the theorem. $\qquad\square$

Remark 3.3.4 : The spirit of theorem 14 is that $R(m)$ is expressible in the form

$$R(m) = \begin{cases} \mathbb{Z} + \sqrt{m}\mathbb{Z}, & \text{if } m \not\equiv 1 \pmod 4 \\ \mathbb{Z} + \theta\mathbb{Z}, & \text{if } m \equiv 1 \pmod 4 \text{ where } \theta = \frac{1+\sqrt{m}}{2}. \end{cases}$$

In other words,

(3.3.3) $$R(m) = \begin{cases} \mathbb{Z}[\sqrt{m}] & \text{if } m \not\equiv 1 \pmod 4 \\ \mathbb{Z}[\theta] & \text{if } m \equiv 1 \pmod 4. \end{cases}$$

Definition 3.3.2 : *For $\alpha \in \mathbb{Q}[\sqrt{m}]$, if $\alpha = a + b\sqrt{m}$, $a,b \in \mathbb{Q}$, we define the norm of α, written $N(\alpha)$, as $N(\alpha) = \alpha\bar{\alpha} = a^2 - b^2 m$.*

It is clear that for $\alpha, \beta \in \mathbb{Q}[\sqrt{m}]$,

(3.3.4) $$N(\alpha\beta) = N(\alpha)N(\beta).$$

We observe that $R(m)$ can be made a Euclidean domain by defining a function $g : R^*(m) \to \widetilde{\mathbb{Z}}$ by $g(\alpha) = | N(\alpha) |$ where $R^*(m) = R(m) \setminus \{0\}$; $\alpha \epsilon R^*(m)$.

Let $\alpha, \beta \in R(m)$. If $\alpha \mid \beta$, one has $g(\alpha) \leq g(\beta)$. Further, given $\alpha, \beta \in R(m)$ with $\beta \neq 0$, one can find γ and δ in $R(m)$ such that

(3.3.5) $$\alpha = \beta\gamma + \delta, \text{ where either } \delta = 0 \text{ or } g(\delta) < g(\beta),$$

provided one shows that for $\theta \in \mathbb{Q}[\sqrt{m}]$ there exists $\eta \in R(m)$ such that

(3.3.6) $$g(\theta - \eta) < 1.$$

If (3.3.6) is satisfied, $R(m)$ becomes a Euclidean domain.

Theorem 15 : $(R(m), g)$ *is a Euclidean domain if given $\theta \in \mathbb{Q}[\sqrt{m}]$, we can find $\eta \in R(m)$ such that (3.3.6) holds.*

Proof : In order to show that $(R(m), g)$ is a Euclidean domain, we need to show that the division algorithm shown in (3.3.5) holds for any pair α, β of elements of $R(m)$ with $\beta \neq 0$.

We are given that $\theta \in \mathbb{Q}[\sqrt{m}]$. We write $\theta = \frac{\alpha}{\beta}$ where $\alpha, \beta \in R(m)$, $\beta \neq 0$. In fact, one has only to multiply θ by an algebraic integer $\beta(\neq 0)$ and take $\beta\theta = \alpha$.

(i) If $\theta \in R(m)$, from $\beta\theta = \alpha$ we see that β divides α and so the 'remainder' on division of α by β is zero.

(ii) If $\theta \notin R(m)$, we are assuming that we can choose an $\eta \in R(m)$ such that

$$g(\theta - \eta) = | N(\theta - \eta) | < 1.$$

Now, $g(\beta(\theta-\eta)) = g(\beta)g(\theta-\eta) = g(\alpha-\beta\eta)$ as $\theta = \frac{\alpha}{\beta}$. So, writing $\delta = \alpha - \beta\eta$, we see that

$$g(\delta) = g(\alpha-\beta\eta) = g(\beta)g(\theta-\eta) < g(\beta) \text{, as } g(\theta-\eta) < 1.$$

So, $\alpha = \beta\eta + \delta$ holds and either $\delta = 0$ or $g(\delta) < g(\beta)$. This makes $(R(m), g)$ a Euclidean domain. $\qquad\square$

Remark 3.3.5 : The multiplicativity of g is essential in the context of the number ring $R(m)$.

Remark 3.3.6 : The essence of theorem 15 is that in order to show that $R(m)$ is a Euclidean domain, one need only establish the sufficiency condition stated in (3.3.6).

Theorem 16 : $R(m)$ *is a Euclidean domain for $m = -1, -2, -3, -7$ and -11 with associated function $g : R^*(m) \to \tilde{\mathbb{Z}}$ given by*

$$g(\alpha) = |N(\alpha)|, \text{ for } \alpha \in R^*(m).$$

Proof : We have only to show that given $\theta \in \mathbb{Q}[\sqrt{m}]$, one can find $\eta \in R(m)$ such that

$$(3.3.7) \qquad\qquad |N(\theta-\eta)| < 1.$$

Let $\theta = r + s\sqrt{m}$, $r, s \in \mathbb{Q}$. If $m \not\equiv 1 \pmod 4$, take $\eta = x + y\sqrt{m}$, $x, y \in \mathbb{Z}$ satisfying (3.3.7).

That is, $|(r-x)^2 - m(s-y)^2| < 1$.

Case 1: $m = -1$ or $m = -2$.

Take x, y to be integers nearest to r and s respectively.

Then, $|r-x| \le \frac{1}{2}|$ and $|s-y| \le \frac{1}{2}$.

Therefore, $|(r-x)^2 - m(s-y)^2| \le |(\frac{1}{2})^2 + 2(\frac{1}{2})^2| = \frac{3}{4} < 1$. $\eta \in R(m)$ exists and is such that $|r-x| \le \frac{1}{2}$ and $|s-y| \le \frac{1}{2}$. Further, (3.3.7) is satisfied.

So, $R(m)$ is a Euclidean domain.

Case 2: $m = -3, -7$ or -11.

We have $\theta = r + s\sqrt{m}$, $m \equiv 1 \pmod 4$. Let us take $\eta = x + y(\frac{1+\sqrt{m}}{2})$ with $x, y \in \mathbb{Z}$. Then,

$$\left|\left(r-x-\frac{y}{2}\right)^2 - m\left(s-\frac{y}{2}\right)^2\right| \text{ should be } < 1.$$

Take y as the integer nearest to $2s$ so that $|2s-y| \le \frac{1}{2}$. We can find $x \in \mathbb{Z}$ such that $|r-x-\frac{y}{2}| \le \frac{1}{2}$ for given y. Then,

$$\left|\left(r-x-\frac{y}{2}\right)^2 - m\left(s-\frac{y}{2}\right)^2\right| \le \left|\frac{1}{4} + \frac{11}{6}\right| = \frac{15}{16} < 1.$$

So, for $m = -3, -7, -11$; given $\theta \in \mathbb{Q}[\sqrt{m}]$, we can find $\eta \in R(m)$ such that (3.3.7) holds. This proves theorem 16. $\qquad\square$

Remark 3.3.7 : For $m = -1$,

$$R(-1) = \mathbb{Z}[i] = \{a + bi : a, b \in \mathbb{Z}, i = \sqrt{-1}\}$$

is the familiar ring of Gaussian integers and it is a Euclidean domain. It is a PID and so a UFD.

Remark 3.3.8 : When m is negative, it can be shown that $R(m)$ is not Euclidean for $m < -11$. We conclude that for m negative, $R(m)$ is a Euclidean domain if, and only if, $m = -1, -2, -3, -7$ and -11.

Remark 3.3.9 : For $m > 0$, it is known that $R(m)$ is a Euclidean domain if, and only if, $m = 2, 3, 5, 6, 7, 11, 13, 17, 19, 21, 29, 33, 37, 41, 55$ and 73. See Chatland and Davenport [3] and K. Inkeri [7].

Remark 3.3.10 : C. F. Gauss had stated that for $m < 0$, $R(m)$ would be a UFD for $m = -1, -2, -3, -7, -11, -19, -43, -67$ and -163. In 1968, H. M. Stark [10] proved that when $m < 0$, $R(m)$ is a UFD precisely for these values of m. For a study of other significant contributions of Stark, H. M. see W. Narkeiwicz [9]. As we have seen, for $m = -1, -2, -3, -7$ and -11, $R(m)$ is Euclidean domain and so a UFD. The case $m = -19$ is discussed separately. See Section 3.4.

3.4. Almost Euclidean domains

There exist integral domains which resemble a Euclidean domain in certain respects. The following theorem is due to Helmut Hasse (1898–1979). It is also contained in a paper of R. Dedekind under a slightly different version. We will call it Dedekind-Hasse theorem.

Theorem 17 (Dedekind-Hasse) **:** *Given an integral domain D, suppose that a function $g : D \to \tilde{\mathbb{Z}}$ satisfies the following conditions:*
(i) $g(a) = 0 \Leftrightarrow a = 0_D$, $a \in D$, (0_D being the additive identity),
(ii) $g(ab) = g(a)g(b)$ for all $a, b \in D$,
(iii) Whenever b does not divide a and $0 < g(b) \leq g(a)$, there exists a pair $\langle x, y \rangle$ of elements in D with

$$0 < g(ax - by) < g(b),$$

then D is a PID.

Proof : Let I be a nonzero ideal of D. We consider the set

$$S = \{x : x = g(a), a \in I, a \neq 0_D\}.$$

As S is a subset of $\tilde{\mathbb{Z}}$, S has a minimal element say x_0. If $g(a_0) = x_0$ is the least of the g-values of nonzero elements of I, then $0 < g(a_0) \leq g(a)$, if a_0 does not divide a.
By (iii) for $0_D \neq a \in I$, there exists a pair $\langle x, y \rangle$ of elements in D such that

$$0 < g(ax - a_0 y) < g(a_0).$$

This contradicts the choice of a_0, since $ax - a_0y \in I$. Therefore, $a_0 \mid a$ or $a = ta_0$ for some $t \in D$. Thus, $I = (a_0)$, the principal ideal generated by a_0. That is, D is a PID. $\qquad\qquad\qquad\qquad\qquad\qquad\qquad\qquad\qquad\qquad\qquad\qquad\qquad\qquad$ □

Remark 3.4.1 : K. R. Nagarajan has rightly asked the following question: should the function g be multiplicative? Clearly, we have not used the condition of multiplicativity in the proof of theorem 17. In the case of rings of integers of algebraic number fields, multiplicativity of norm of an element holds. So, the same is retained. Moreover, the existence of a function $g : D^* \rightarrow \tilde{\mathbb{Z}}$ satisfying the conditions of the theorem is not always guaranteed. Theorem 17 has been adapted from W. Narkeiwicz [9].

Our aim is show that $R(-19)$ is a PID.

Theorem 18 : $R(-19)$ *is a PID.*

Proof : We write $\theta = \frac{1+\sqrt{-19}}{2}$. As $-19 \equiv 1 \pmod 4$, $R(-19) = \mathbb{Z}[\theta]$. For $\alpha = a + b\theta \in R(-19)$, the norm of α written $N(\alpha)$ is given by

$$(3.4.1) \qquad N(\alpha) = (a + b\theta)(a + b\bar{\theta}) = a^2 + ab + 5b^2.$$

$N : R(-19) \rightarrow \tilde{\mathbb{Z}}$ satisfies
(i) $N(\alpha\beta) = N(\alpha)N(\beta)$ for all $\alpha, \beta \in R(-19)$.
(ii) $N(\alpha) = 0 \Leftrightarrow \alpha = 0$.
(iii) $N(\alpha) > 0$ for $\alpha \neq 0$ in $R(-19)$.

In view of theorem 17, we get through if the following condition is established: For $\alpha, \beta \in R(-19)$, where β does not divide α and $0 < N(\beta) \leq N(\alpha)$ there exist, $\gamma, \delta \in R(-19)$ such that

$$(3.4.2) \qquad 0 < N(\alpha\gamma - \beta\delta) < N(\beta)$$

To arrive at (3.4.2), we proceed as follows: We assume that $\beta \neq 0$. Then,

$$(3.4.3) \qquad \frac{\alpha}{\beta} = a + b\theta; a, b \in \mathbb{Q}$$

and at least one of a, b is not an element of $\mathbb{Z}$. This is okay as the inverse of θ (as a complex number) is in $\mathbb{Q}[\theta]$.

The following situations are to be handled:
Case 1: $b \in \mathbb{Z}$, $a \notin \mathbb{Z}$.
Let $\{x\}$ denote the integer nearest to x.
Take $\gamma = 1$ and $\delta = \{a\} + b\theta$.

$$\frac{\alpha}{\beta}\gamma - \delta = \frac{\alpha}{\beta} - \delta = a - \{a\}$$

As $\mid a - \{a\} \mid \leq \frac{1}{2}$, $0 < N(\frac{\alpha}{\beta}\gamma - \delta) \leq \frac{1}{4} < 1$. This makes $N(\alpha\gamma - \beta\delta) < N(\beta)$.

Case 2(a): $a \in \mathbb{Z}$, $5b \notin \mathbb{Z}$.
As $\bar{\theta} = 1 - \theta$,

$$\frac{\alpha}{\beta}\bar{\theta} = \frac{\alpha}{\beta}(1-\theta) = (a+b\theta)(1-\theta) \quad \text{from (3.4.3)}$$

$$= a - a\theta + b\theta - b\theta^2 = a(1-\theta) + b\theta - b(\theta - 5)$$

or,

$$\frac{\alpha}{\beta}\bar{\theta} = a + 5b - a\theta$$

Take $\gamma = \bar{\theta}$, $\delta = (a+5b) - a\theta$
Then,

$$\frac{\alpha}{\beta}\gamma - \delta = \frac{\alpha}{\beta}\bar{\theta} - (a+5b) + a\theta = 0.$$

So,

$$N(\frac{\alpha}{\beta}\gamma - \delta) < 1.$$

That is, $N(\alpha\gamma - \beta\delta) < N(\beta)$ as in case 1.
Case 2(b): $a \in \mathbb{Z}$, $5b \in \mathbb{Z}$. Take $\gamma = 1$, $\delta = a + \{b\}\theta$. Then,

$$\frac{\alpha}{\beta}\gamma - \delta = \frac{\alpha}{\beta} - \delta = a + b\theta - (a + \{b\}\theta)$$

$$= (b - \{b\})\theta$$

$$\text{So,} \quad N(\frac{\alpha}{\beta}\gamma - \delta) = N((b - \{b\})\theta) = N(b - \{b\})N(\theta)$$

$$= (b - \{b\})\theta\bar{\theta}$$

$$= 5(b - \{b\}), \text{ as } \theta\bar{\theta} = 5.$$

Or, $N(\frac{\alpha}{\beta}\gamma - \delta) = 5b - \{5b\} = 0 < 1$. So, $N(\alpha\gamma - \beta\delta) < N(\beta)$ as in case 1.
Case 3(a) a and b are not elements of $\mathbb{Z}$, but $2a, 2b \in \mathbb{Z}$.
It can be shown that

$$\frac{\alpha}{\beta}\theta = -5b + (a+b)\theta \text{ and } a+b \in \mathbb{Z}.$$

Taking $\gamma = \theta$ and $\delta = \{-5b\} + (a+b)\theta$, we get

$$N(\frac{\alpha}{\beta}\gamma - \delta) = N(\frac{\alpha}{\beta}\theta - \delta)$$

$$= N(-5b - \{-5b\})$$

$$\leq \frac{1}{4} < 1.$$

So, $N(\alpha\gamma - \beta\delta) < N(\beta)$ as in case 1.
Case 3(b): a, b are not elements of $\mathbb{Z}$, $2a$ and $2b$ are also not elements of $\mathbb{Z}$. Then, either $\mid b - \{b\} \mid \leq \frac{1}{3}$ or $\mid 2b - \{2b\} \mid \leq \frac{1}{3}$. In the former case, we take $\gamma = 1$,

$\delta = \{a\} + \{b\}\theta$. We arrive at

$$0 < N(\frac{\alpha}{\beta}\gamma - \delta) = N(a - \{a\} + (b - \{b\})\theta)$$
$$= (a - \{a\})^2 + (a - \{a\})(b - \{b\}) + 5(b - \{b\})^2$$
$$\leq \frac{1}{4} + \frac{1}{6} + \frac{5}{9} \ (= \frac{35}{36})$$
$$< 1.$$

In the latter case, take $\gamma = 2$ and $\delta = \{2a\} + \{2b\}\theta$
we get $N(\frac{\alpha}{\beta}\gamma - \delta) \leq \frac{35}{36} < 1$.
Case 3(c): a, b are not elements of $\mathbb{Z}$, but $2a \in \mathbb{Z}$ and $2b \notin \mathbb{Z}$.
When $5b \in \mathbb{Z}$, take $\gamma = 5$ and $\delta = \{5a\} + 5b\theta$.
When $5b \notin \mathbb{Z}$, take $\gamma = 2\bar{\theta}$ and $\delta = \{2a + 10b\} - 2a\theta$. This leads to

$$N(\frac{\alpha}{\beta}\gamma - \delta) < 1 \text{ as in the earlier cases.}$$

Case 3(d): a and b are not elements of $\mathbb{Z}$, $2a \notin \mathbb{Z}$, but $2b \in \mathbb{Z}$.
Take $\gamma = 2$, $\delta = \{2a\} + 2b\theta$. Then,

$$\frac{\alpha}{\beta}\gamma - \delta = 2(a + b\theta) - \{2a\} - 2b\theta = 2a - \{2a\}.$$

So, $N(\frac{\alpha}{\beta}\gamma - \delta) \leq \frac{1}{4} < 1$ and hence $N(\alpha\gamma - \beta\delta) < N(\beta)$ as in the earlier cases. This exhausts all possibilities for the choices of a, b in $\frac{\alpha}{\beta} = a + b\theta$. Thus, (3.4.2) holds and the proof is complete. □

Theorem 19 : $R(-19)$ *is not a Euclidean domain.*

Proof : Let $R(-19)^* = R(-19) \setminus \{0\}$. To show that $R(-19)$ is not a Euclidean domain, one has to establish that $R(-19)$ does not admit a function

$$g : R(-19)^* \to \widetilde{\mathbb{Z}}$$

where (i) $g(\alpha) \leq g(\alpha\beta)$ for $\alpha \neq 0, \beta \neq 0$ in $R(-19)$
(ii) given α, β nonzero elements in $R(-19)$, there exist $\gamma, \delta \in R(-19)$ such that $\alpha = \beta\gamma + \delta$ with either $\delta = 0$ or $g(\delta) < g(\beta)$.

Assume the contrary. That is, assume that there exists a function $g : R(-19)^* \to \widetilde{\mathbb{Z}}$ with the properties given above.

Take $g(\alpha) = N(\alpha) = \alpha\bar{\alpha}$ for $\alpha \in R(-19)$. The group of units in $R(-19)$ is $\{+1, -1\}$. We have seen in (3.4.1) that if $\alpha = a + b\theta$ where $\theta = \frac{1 + \sqrt{-19}}{2}$.

$$N(\alpha) = \alpha\bar{\alpha} = (a + b\theta)(a + b\bar{\theta}) = a^2 + ab + 5b^2.$$

Also, the norm is multiplicative. Further,

(3.4.4) $$N(\alpha) = N(\bar{\alpha}) = (a + b)^2 - ab + 4b^2$$

If α is a unit, then $N(\alpha) = 1$.

We introduce two sets S and T defined by

(3.4.5) $\qquad\qquad S = \{\alpha \in R(-19) : \alpha \neq 0, 1 \text{ or } -1\}$

(3.4.6) $\qquad\qquad T = \{N(\alpha) : \alpha \in S\}$

T has a minimal element say $n > 0$.

We claim that 2 and 3 stay as primes in $R(-19)$. Applying the division algorithm to 2 with n as divisor (in $\mathbb{Z}$) we get

$$2 = qn + r \text{ with } |r| < |n|.$$

r is one of $0, 1, -1$. So, either $n|2$ or $n|3$. We show that $n = \pm 2$ or $n = \pm 3$. If 2 is not a prime in $R(-19)$ we could write

$$2 = (a + b\theta)(c + d\theta)$$

where $a + b\theta$, $c + d\theta$ are in $R(-19)$ and are nonzero non-units.

$$N(2) = 4 = N(a + b\theta)N(c + d\theta) = ((a+b)^2 - ab + 4b^2)((c+d)^2 - cd + 4d^2)$$

From $((a+b)^2 - ab + 4b^2) = 2$, considering the cases $ab \geq 0$ and $ab < 0$, we note that $b = 0$. In the same manner, $d = 0$. So then, $2 = ac$ is a factorization in $\mathbb{Z}$ —a contradiction. So, 2 stays prime in $R(-19)$. Similarly, it could be shown that 3 stays prime in $R(-19)$.

Applying the division algorithm in $R(-19)$ to θ with ± 2 or ± 3 as a divisor, we see that if $\theta = 2\gamma + \delta$ either $\delta = 0$ or $N(\delta) < N(2)$.

$\delta = 0 \Rightarrow \theta$ is divisible by 2. $\delta = \pm 1$ would give $\theta \pm 1$ is divisible by 2. Similarly, either θ is divisible by 3 or $\theta \pm 1$ is divisible by 3. But this is impossible since

$$N(\theta) = 5 = N(\theta - 1) \text{ and } N(\theta + 1) = 7,$$

$$N(2) = 4 \text{ and } N(3) = 9.$$

The minimal element of T (3.4.6) being 2 or 3 is not allowed. This contradiction asserts that our assumption about the existence of a function $g : R(-19)^* \to \tilde{\mathbb{Z}}$ with the stated properties is wrong. So, $R(-19)$ is not a Euclidean domain. $\qquad\square$

Remark 3.4.2 : Theorems 18 and 19 have been drawn from Oscar A. Campoli [2].

The Dedekind-Hasse theorem (theorem 17) says that a PID is near to a Euclidean domain. John Greene [5] remarks that the structure of $R(-19)$ along with the norm $N(\alpha)$, $\alpha \in R(-19)$ makes us suggest that $R(-19)$ is 'almost Euclidean'.

Definition 3.4.1 : *An integral domain D is called 'almost Euclidean' if there is a function $g : D \to \tilde{\mathbb{Z}}$ satisfying the following conditions:*
(i) $g(0_D) = 0$ and $g(a) > 0$ for $0 \neq a \in D$. (0_D being the additive identity)
(ii) for $0 \neq a, b$ elements in D, one has

$$g(ab) \geq g(a) \text{ for all } a \in D$$

and

(iii) either a = bq, for some q ∈ D or

(iv) $0 < g(ax+by) < g(b)$, for some $x, y \in D$.

Theorem 20 : *An integral domain D is a PID if, and only if, it is 'almost Euclidean'.*

Proof : ⟸: Dedekind-Hasse theorem (theorem 17) shows that if D is 'almost Euclidean', then it is a PID.

For, $g(ab) = g(a)g(b)$ satisfies $g(ab) \geq g(a)g(b)$ for all $a, b \in D$. Let I be a nonzero ideal in D. Among the elements $x \in I$, let $x = b$ be an element such that $g(b)$ is minimal. For $a \in I$, since $ax+by \in I$ for $x, y \in D$ one has

$$0 < g(ax+by) < g(b).$$

This is not acceptable as $g(b)$ is minimal. So, $a = bq$ must hold for some $q \in D$. Thus, $I = (b)$ and so D is PID.

:⟹ Suppose that D is a PID. We define a function $g : D \to \widetilde{\mathbb{Z}}$ as follows: $g(0_D) = 0$. For $0_D \neq a \in D$, let $a = \epsilon \, p_1 p_2 \cdots p_n$ where ϵ is a unit and $p_1, p_2, \cdots, p_n$ are irreducibles in D.

Let $g(a) = 2^n$, n denoting the total number of irreducible factors of a. (repetitions being allowed)

Since $g(ab) = g(a)g(b)$, g satisfies conditions (i) and (ii) of definition 3.4.1.

Let $a, b \in D$ with $b \neq 0_D$. We write

$$I = \{ax+by : x, y \in D\}$$

Since I is an ideal in D, $I = (d)$, the principal ideal generated by d (say an element of I).

If $a = bq$ for some $q \in D$, $I = (b)$. Otherwise, $I \neq (b)$. Since $b \in I$, $b = td$ for some $t \in D$. So, $g(b) \geq g(d)$. Since $I \neq (b)$, t is not a unit in D. So $g(t) > 1$. So, $g(b) > g(d)$. If $d = ax_0 + by_0$ for some $x_0, y_0 \in D$, $0 < g(d) < g(b)$ gives $0 < g(ax_0 + by_0) < g(b)$.

This shows that either $a = bq$ or there exist $x_0, y_0 \in D$ such that

$$0 < g(ax_0 + by_0) < g(b).$$

That is, D is 'almost Euclidean'. □

Remark 3.4.3 : Theorem 20 has been adapted from John Greene [5].

Next, the division algorithm in $\mathbb{Z}$ says that for $a, b \in \mathbb{Z}$ ($b \neq 0$), we can find q, r such that

(3.4.7) $a = bq + r$ with $0 \leq |r| < |b|$

In (3.4.7), we have considered the absolute-value norm. If $r > 0$, we see that $0 < r < |b|$. When $q \neq -1$ we also have

(3.4.8) $a = b(q+1) - t$

with $r = b - t$ or $t = b - r$ where $|t| < |b|$. If $q = -1$, $a = -b + r$ is the same as $a = 0 \cdot b - t$ where $t = -a = (b - r)$ and $|t| < |b|$.

(3.4.7) and (3.4.8) together suggest that the remainder on division of a by b is not unique and it is said to have a 'double-remainder property' abbreviated d.r.p. Following Steven Galovich [4], we make the

Definition 3.4.2 (Steven Galovich) : *A Euclidean domain* (D, g) *is said to have the double-remainder property (d.r.p) if, for each pair of elements* a, b $(a \in D, b \in D^*)$ *such that* b *does not divide* a, *there exist exactly two pairs* (q_i, r_i) $(i = 1, 2)$ *such that*

$$a = q_i b + r_i \ (i = 1, 2)$$

where $g(r_i) < g(b)$ $(i = 1, 2)$.
Further, $a \equiv r_1 \pmod{b}$ *and* $a \equiv r_2 \pmod{b}$.

Next, we show that d.r.p characterizes $\mathbb{Z}$ among Euclidean domains. See Steven Galovich [4].

Let (R, g) be a Euclidean domain possessing d.r.p. R is assumed to be an infinite set. $U(R)$ denotes the group of units of R. As usual, we write $R^* = R \setminus \{0_R\}$. Let 1_R denote the multiplicative identity in R.

Definition 3.4.3 : *Given a Euclidean domain* (R, g), *a subset* R_1 *of* R^* *is defined by*

$$R_1 = \left\{ x \in R^* : g(x) \le g(y) \, \text{for all} \, y \in R^* \right\}$$

If $u \in U(R)$, $g(u) = g(1_R)$. Further, $g(1_R) \le g(a)$ for all $a \in R^*$. Also, if $t \in R_1$, $g(t) \le g(1_R)$. But, $g(1_R) \le g(t)$. So, $g(t) = g(1_R)$ or $t \in U(R)$. That is, R_1 is precisely the set $U(R)$ of units of R.

Definition 3.4.4 : *For* $n \ge 2$, *we define*

$$R_n = \left\{ x \in R^* : g(x) \le g(y) \, \text{for all} \, y \in R^* \setminus R_{n-1} \right\}$$

We need a series of lemmas (nine of them) to reach the desired goal.

Lemma 3.4.1 : $R^* = \bigcup_{n=1}^{\infty} R_n$.

Proof : We have observed that $R_1 = U(R)$. Now,

$$R_2 = \left\{ x \in R^* : g(x) \le g(y) \quad \text{for all} \, y \in R^* \setminus R_1 \right\}.$$

If $t \in R_1$, $g(t) \le g(y)$ for all $y \in R^*$. So $g(t) \le g(y)$ for all $y \in R^* \setminus R_1$ also. So, $t \in R_2$. That is, $R_1 \subseteq R_2$.

We prove by induction on n that $R_n \subseteq R_{n+1}$ for all $n \ge 1$. The result is true for $n = 1$. Assume that there exists $m > 2$ such that

$$R_{m-1} \subseteq R_m.$$

$R_{m-1} \subseteq R_m \Rightarrow R^* \setminus R_m \subseteq R^* \setminus R_{m-1}$.

Let $t \in R^*$ and $t \in R_m$ where $g(t) \leq g(y)$ for all $y \in R^* \backslash R_m$. Then, $t \in R_{m+1}$ or $R_m \subseteq R_{m+1}$. It follows that $R_n \subseteq R_{n+1}$ for all $n \geq 1$. Thus, $\{R_n\}$ is an enlarging sequence of subsets of R^*. If $M = \bigcup_{n=1}^{\infty} R_n$, $M \subseteq R^*$.

Now, for $s \in R^*$, there exists R_q contained in M such that $s \in R_q$ for some $s \in \mathbb{N}$. That is, $R^* \subseteq M$ and this shows that $R^* = M$ which proves lemma 3.4.1. $\square$

Lemma 3.4.2 : *If $u \in U(R)$ and $u \neq \pm 1_R$, then $1_R + u \in U(R)$.*

Proof : We note that if $v \in U(R)$, then

$$g(v) < g(a) \text{ for all } a \in R^* \setminus U(R).$$

No two of $1_R, -u, u^2$ are equal. However,

$$1_R \equiv -u \equiv u^2 \pmod{(1_R + u)}.$$

So, if $1_R + u \notin U(R)$ we get the congruences

$$(3.4.9) \qquad \begin{cases} 1_R & \equiv -u \pmod{(1_R + u)} \\ 1_R & \equiv u^2 \pmod{(1_R + u)} \end{cases}$$

(3.4.9) is due to the fact that (R, g) possesses d.r.p. Further,

$$(3.4.10) \qquad \begin{cases} g(-u) & < g(1_R + u) \\ g(u^2) & < g(1_R + u) \end{cases}$$

Next, when $u \in U(R)$, there exists $v \in U(R)$ such that $uv = 1_R$. Let I denote the ideal generated by $1_R + u$ in R. Then, as I is a principal ideal, $g(a) \geq g(1_R + u)$ for all $a \in I$. $1_R + u$ is a nonzero non-unit and so I is a proper ideal of R, provided we assume that $1_R + u \notin U(R)$. The nonzero elements of I are contained in R_m for some $m \geq 1$.

Now, $1_R + u = uv + u = u(1_R + v)$. $1_R + u \neq 0_R$, as $u \neq \pm 1_R$. So, $g(1_R + u) \leq g(y)$ for every $y \in R^* \backslash R_{m-1}$, as $1_R + u \in R_m$. As $U(R) \subset R_m$ and $u \in U(R)$, $g(1_R + u) \leq g(u)$. Further, $g(-u) \leq g(u)$ and so $g(1_R + u) \leq g(-u)$. As $g(u) \leq g(u^2)$, we have $g(1_R + u) \leq g(u^2)$. Therefore, d.r.p does not work if $1_R + u \notin U(R)$. Thus, $1_R + u \in U(R)$. $\square$

Corollary 3.4.1 : *If $u \in U(R)$, $u \neq \pm 1_R$, $1_R - u \in U(R)$.*

Proof follows on lines of proof of lemma 3.4.2.

Lemma 3.4.3 : *Let $S(R) = U(R) \cup \{0_R\}$. Then, $S(R)$ does not form a field under the ring operations in R.*

Proof : Assume the contrary. That is, we suppose that $S(R)$ is a field. Let $r \in R_2 \setminus R_1$. Since (R, g) possesses d.r.p, there exists $u \in U(R)$, $u \neq 1_R$ and $q \in R$ such that

$$1_R = qr + u$$

By the assumption that $S(R)$ is a field, $qr = 1_R - u \in U(R)$, by Corollary 3.4.1. Then, q and r would be units. But, we have taken r to be a non-unit. This contradicts the field structure (assumed) of $S(R)$. So, $S(R)$ is not a field. $\quad\square$

Lemma 3.4.4 : *Taking $1_R + 1_R$ as 2_R, 2_R considered as an element of R is a nonzero non-unit.*

Proof : If $2_R \in S(R) = U(R) \cup \{0\}$, then for all $u, v \in U(R)$ either $u + v = 0$ or $u + v = u(1_R + u^{-1}v) \in U(R)$ by lemma 3.4.2. So, then $S(R)$ becomes a field which is not true by lemma 3.4.3. That is, 2_R is a nonzero non-unit in R. $\quad\square$

Corollary 3.4.2 : *The characteristic of R is not equal to 2 and $2_R \notin U(R)$. For, 2_R is a nonzero non-unit of R and so $2_R \neq 0_R$.*

We deduce that as R is assumed to be an infinite set, characteristic of R is zero.

Lemma 3.4.5 : *If R_n is as given in definition 3.4.4, R_n is a finite set.*

Proof : We first show that $R_1 = U(R) = \{1_R, -1_R\}$. Suppose that there exists $u \in R$ such that $u \in U(R)$ and $u \neq \pm 1_R$. Then,

$$u \equiv u - 2_R \pmod{2_R}$$

$$u \equiv u + 2_R \pmod{2_R}$$

Now, $2_R + u = 1_R + (1_R + u)$. Let $v = 1_R + u \in U(R)$. Then, $1_R + v = 1_R + (1_R + u)$ is again an element of $U(R)$, by lemma 3.4.2. So, $2_R + u \in U(R)$. Also,

$$u - 2_R = -(1_R + (1_R - u)) \in U(R).$$

Therefore, by d.r.p,

$$g(u - 2_R) < g(2_R)$$

$$g(2_R + u) < g(2_R)$$

As $2_R \neq 0_R$, 2_R belongs to R_m for some m. So,

$$g(2_R) \leq g(y) \quad \text{for all } y \in R^* \backslash R_{m-1}$$

Next, $R_1 \subseteq R_m$. So, if a property is true for all $y \in R_m$, it is true for all $y \in R_1$.

That is, $g(2_R) \leq g(y)$ for all $y \in R_1 = u(R)$

In particular, $g(2_R) \leq g(u + 2_R)$ and $g(2_R) \leq g(u - 2_R)$ as $u + 2_R, u - 2_R$ are in $U(R)$. This violates *d.r.p*. So, $R_1 = \{1_R, -1_R\}$.

Next, we assume that R_{n-1} is finite. Let $x \in R_n \backslash R_{n-1}$. By d.r.p, each nonzero coset of the ideal (x) (generated by x) contains exactly two elements of R_{n-1}. By hypothesis, R_{n-1} is finite. We write

$$k = 1 + \frac{1}{2}(\# R_{n-1}).$$

Then, $R/(x)$ is a finite ring having k elements. So, if $\oplus$ denotes addition modulo (x), $(R/(x), \oplus)$ is a finite group with k elements. In $R/(x)$,

$$k(1_R \oplus (x)) = k1_R \oplus (x) = (x).$$

So, $k1_R \in (x)$. That is $k1_R$ is a multiple of x. Since R is a UFD with a finite group of units,

$$k1_R = (1 + \frac{1}{2}(\#R_{n-1})1_R$$

has only a finite number of divisors in R. So, R_n is finite whenever R_{n-1} is finite. Since R_1 is finite, the induction is complete. $\square$

Corollary 3.4.3 : *If $x \in R^*$, $R/(x)$ is a finite ring.*

Remark 3.4.4 : Since R is infinite and for each $n \geq 1$, R_n is a finite set, we note that R_n is strictly contained in R_{n+1} for $n \geq 1$.

Definition 3.4.5 : *For $x \in R^*$, $\#(R/(x))$ denoted by $N(x)$ is a positive integer and $N(x)$ is called the norm of $x \in R^*$.*

Lemma 3.4.6 : *The norm $N(x)$ of $x \in R^*$ is multiplicative. That is, for $x, y \in R^*$,*

$$N(xy) = N(x)N(y).$$

Proof : For $x, y \in R^*$, we consider the quotient rings $R/(x)$ and $(y)/(xy)$. This is meaningful as R is a PID. We define

$$\psi : R/(x) \longrightarrow (y)/(xy)$$

by $\psi(a + (x)) = ay + (xy)$ where $a \in R$. If $b \in R$ and $ay + (xy) = by + (xy)$, we get $(a - b)y \in (xy)$. This yields $a - b = 0_R$ or $a = b$. That is, ψ is one-one. Therefore,

$$N(x) = \#((y)/(xy)).$$

Further, $R/(y) \cong R/(xy)\big/(y)/(xy)$. Thus, $N(y) = \frac{N(xy)}{N(x)}$ which proves the desired property of N. $\square$

Lemma 3.4.7 : *For $n \geq 2$,*

$$R_n \setminus R_{n-1} = \{x \in R^* : N(x) = 1 + \frac{1}{2}(\#R_{n-1})\}.$$

Proof : While proving lemma 3.4.5, we have observed that if $x \in R_n \setminus R_{n-1}$, $R/(x)$ has $k = 1 + \frac{1}{2}(\#R_{n-1})$ elements.
    We set

$$A_n = R_n \setminus R_{n-1}$$

and

$$B_n = \{x \in R^* : N(x) = 1 + \frac{1}{2}(\#R_{n-1})\}.$$

$A_m \cap A_n = \phi$ for $m \neq n$. Similarly $B_m \cap B_n = \emptyset$ for $m \neq n$. However, we have

$$\bigcup_{n=2}^{\infty} A_n = \bigcup_{n=2}^{\infty} B_n = R^* \setminus U(R).$$

Also, $A_n \subseteq B_n$. For, if $x \in R_n \setminus R_{n-1}$, $N(x) = 1 + \frac{1}{2}(\#R_{n-1})$ and so $x \in B_n$.

We claim that A_n as a proper subset of B_n is impossible. For if $y \in B_n$, $y \notin A_n$. So, $y \in A_j \subset B_j$ is a contradiction as $B_j \cap B_n = \emptyset$ for $j \neq n$. So, $A_n = B_n$ for $n \geq 2$. This completes the proof of lemma 3.4.7. $\square$

Lemma 3.4.8 : *For $x, y, \in R^*$, $g(x) < g(y) \Leftrightarrow N(x) < N(y)$. In other words, the Euclidean domain (R, N) also possesses d.r.p.*

Proof : $\Leftarrow$: Let $x, y, \in R^*$ such that

$$N(x) = 1 + \frac{1}{2}(\#R_{n-1})$$

$$N(y) = 1 + \frac{1}{2}(\#R_{m-1})$$

$x \neq y \Rightarrow n \neq m$. If $N(x) < N(y)$, $\#R_{n-1} < \#R_{m-1}$. As $R_n \subset R_{n+1}$ for $n \geq 1$, $R_n \subset R_m$ when $x \in R_n \setminus R_{n-1}$; $y \in R_m \setminus R_{m-1}$. So, when $n < m$, $g(x) < g(y)$.
:$\Rightarrow$ Suppose that $g(x) < g(y)$. Choose n, m the least positive integers such that $x \in R_n$, $y \in R_m$. Then, $n < m$.

$$R_m = \{ y \in R^* : g(y) \leq g(t) \text{ for all } t \in R^* \setminus R_{m-1} \}$$

$$R_m - R_{m-1} = \{ y \in R^* : N(y) = 1 + \frac{1}{2}(\#R_{m-1}) \}$$

Now,

$$N(x) = 1 + \frac{1}{2}(\#R_{n-1}) \leq N(y) = 1 + \frac{1}{2}(\#R_{m-1}).$$

Inequality is strict. For, otherwise, $x \in R_m \setminus R_{m-1}$.
So, $g(x) < g(y) \Rightarrow N(x) < N(y)$. As N is multiplicative, (R, N) has also d.r.p. $\square$

Remark 3.4.5 : Lemma 3.4.8 connects the function g with the norm N.

Remark 3.4.6 : $\#R_{n-1} = 2(N(y) - 1)$, whenever $y \in R_n \setminus R_{n-1}$.

Lemma 3.4.9 :
$$R_2 \setminus R_1 = \{ \pm 2_R \}.$$

Proof : By lemma 3.4.7,

$$R_2 \setminus R_1 = \{ x \in R^* : N(x) = 1 + \frac{1}{2}(\#R_1) \}.$$

As $\#R_1 = 2$, we have
$$R_2 \setminus R_1 = \{ x \in R^* : N(x) = 2 \}.$$

$N(x) = 2 \Rightarrow \#R/(x) = 2$. Therefore, $2(1_R + (x)) = (x)$ or $2_R \in (x)$. There exists $y \in R^*$ such that $xy = 2_R$ (as 2_R is a multiple of x.) We claim that $N(y) = 1$. This implies that y is a unit and $x = \pm 2_R$. We proceed as follows:
We observe that

$$N(x^2) = N(x)N(x) = 4.$$

$$1_R \equiv x \equiv x^2 \pmod{(x - 1_R)}.$$

For,

$$1_R = -(x - 1_R) + x$$
$$1_R = -(x + 1_R)(x - 1_R) + x^2$$
$$1_R = 0_R(x - 1_R) + 1_R.$$

If $N(x - 1_R) > 4$, then $N(x) < N(x - 1_R)$, $N(x^2) = 4 < N(x - 1_R)$. Also, $N(1_R) < N(x - 1_R)$. As R satisfies d.r.p, there exist exactly two elements x_1, x_2 with $(x - 1_R)$ not dividing 1_R such that

$$N(x_1) < N(x - 1_R)$$
$$N(x_2) < N(x - 1_R)$$

and

$$1_R \equiv x_1 \pmod{(x - 1_R)}$$
$$1_R \equiv x_2 \pmod{(x - 1_R)}$$

This shows that $\# (R/(x - 1_R)) = N(x - 1_R)$ cannot exceed 4. That is,

(3.4.11) $$N(x - 1_R) \leq 4.$$

Arguing similarly, we show that $N(\pm 1_R \pm x) \leq 4$.

(3.4.12) Now, no two of $\pm 1_R \pm x$ are equal.

For,

If $1_R + x = -1_R + x$, we have $2_R = 0$.

If $1_R + x = 1_R - x$, we have $2_R x = 0$.

If $-1_R - x = 1_R + x$, we have $2_R x = -2R$ or $x = -1_R$.

But, x is a non-unit. So (3.4.11) holds good.
We observe that

$$1_R + x = 2_R x + (1_R - x)$$
$$1_R + x = 2 \cdot 1_R + (-1_R + x)$$
$$1_R + x = 2 \cdot 0_R + (1_R + x)$$

So, they are congruent modulo 2_R. So, by d.r.p, $N(2_R) \leq N(\pm 1_R \pm x) \leq 4$ or, $4 \geq N(2_R) = N(x)N(y) = 2N(y)$ as $N(x) = 2$. So, $N(y) \leq 2$. If all of $(\pm 1_R \pm x)$ have norm < 4, then,

$$N(2_R) < 4 \quad \text{or else, d.r.p is violated.}$$

If $N(2_R) < 4$, $2N(y) < 4$ and so $N(y) = 1$ as $N(y) < 2$.
From the argument shown above, we arrive at the fact that $N(y) = 1$.
So, we can assume without loss of generality that

$$N(1_R - x) = 4 \text{ and } N(y) = 1.$$

$N(a) = 4 \Rightarrow \#R/(a) = 4$. Since any element of norm 4 is a product of irreducibles dividing 2_R, we suppose that $ab = 4_R$ for some $b \in R$. As R is a UFD, $ab = 2_R \cdot 2_R$.

Suppose that $xy = 2_R$. Then, $N(1_R - x) = 4 \Rightarrow x$ divides $1_R - x$ or y divides $1_R - x$. If x divides $1_R - x$, $1_R - x \in (x)$. So $1_R - x + x = 1_R \in (x)$. This shows that x has to be a unit which is not the case. So, y divides $1_R - x$ and $N(y) = 2$. So, $1 - x = \pm y^2$.

Now,

$$R_2 \setminus R_1 = \{x \in R^* : N(x) = 2\} = \{\pm x, \pm y\}$$

Furthermore, if $N(1_R - x) = 4$, $1_R - x$ must be of the form $a_1 a_2$ where a_1, a_2 are xy or yx.

$$R_3 \setminus R_2 = \{x \in R^* : N(x) = 4\} = \{\pm x^2, \pm xy, \pm y^2\}$$

Since $N(1_R + x) \leq 4$, x does not divide $1_R + x$ and $1_R + x \neq 1_R - x$. So, $1_R + x \neq \pm y^2$ or $1_R + x = \pm y$. Therefore,

$$y^2 = (1_R + x)^2 = \pm(1_R - x).$$

If $(1_R + x)^2 = (1_R - x)$, then, $1_R + 2_R x + x^2 = 1_R - x$ or $x(x + 3_R) = 0$. So, $x = -3_R$ and $y = \pm 2_R$ which implies that $x = \pm 1_R$, a contradiction. So,

$$(1_R + x)^2 = -1_R + x$$

$$\text{or } x^2 + x + 2_R = 0.$$

So, $x = \frac{-1_R \pm \sqrt{1_R - 8_R}}{2}$. That is,

(3.4.13) $2x + 1_R = \sqrt{-7_R} \in R.$

Also, $y = -1_R - x = -(\frac{1_R \pm \sqrt{-7_R}}{2})$, $x - y = 1_R + 2x$.
Now, $\#R_3 = \#(R_3 - R_2) + \#(R_2 - R_1) + \#R_1 = 12$. So, by lemma 3.4.7

$$R_4 \setminus R_3 = \{t \in R^* : N(t) = 7\}$$

If $z \in R_4 \setminus R_3$, z divides $\sqrt{-7_R}$. Thus z divides $\sqrt{-7_R} = 1_R + 2x = x - y$.

Therefore, $x^2 \equiv xy \equiv y^2 \pmod{z}$ which contradicts d.r.p. So, we conclude that $N(y) = \pm 1$ and so $R_2 \setminus R_1$ has elements 2_R and -2_R as desired. This proves Lemma 3.4.9. $\square$

Theorem 21 (Steven Galovich (1978).) **:** *If (R, g) is a Euclidean domain with d.r.p, then $R \cong \mathbb{Z}$.*

Proof : Proof is by induction on the norm function N given in definition 3.4.5. We have seen that $R_1 = U(R) = \{\pm 1_R\}$. By lemma 3.4.9, $R_2 \setminus R_1 = \{\pm 2_R\}$. Further,

$$U(R) = \{x \in R^* : N(x) = 1\}.$$

Let

$$n_R = 1_R + 1_R + \cdots + 1_R (\text{n times})$$

$N(n_R) = n$ is true for $n = 1$ and 2. Suppose that $N(k_R) = k$ for all $k \leq (n-1)$. If n is composite, $N(n_R) = n$ by induction on n, as $N : R^* \to \mathbb{N}$ is multiplicative.

Let n be an odd prime. Then,

$$N((n+1)_R) = N\left(\frac{2(n+1)_R}{2}\right)$$

$$= N(2_R)N\left(\frac{(n+1)_R}{2}\right)$$

$$= 2 \cdot \left(\frac{(n+1)}{2}\right)$$

$$= (n+1).$$

Since $1_R \equiv 1_R - n_R \equiv (n+1)_R \pmod{n_R}$,

$$N(n_R) \leq (n+1) \text{ as } N((n+1)_R) = (n+1).$$

But, 1_R has additive order n in $R/(n_R)$ and so n divides $N(n_R)$. Therefore, $N(n_R) = n$. Next, let $y \in R^*$

(3.4.14) $\qquad \#\{x \in R^* : N(x) < N(y)\} = 2(N(y) - 1).$

For, if $y \in R_n \setminus R_{n-1} = \{y \in R^* : N(y) = 1 + \frac{1}{2}(\#R_{n-1})\}$

$$\{x \in R^* : N(x) < N(y)\} = R_{n-1}$$

and $\#R_{n-1} = 2(N(y)) - 1$. Writing $y = n_R$ in (3.4.14) and noting that $N(n_R) = n$, we obtain

$$\{x \in R^* : N(x) < n\} = \{\pm 1_R, \pm 2_R, \cdots, \pm(n-1)_R\}.$$

As $\bigcup_{n=1}^{\infty}\{x \in R^* : N(x) < n\} = R^*$, $R \cong \mathbb{Z}$ or $R = \mathbb{Z}$ up to isomorphism. This shows that the double-remainder property characterizes $\mathbb{Z}$ among Euclidean domains. $\qquad \Box$

Remark 3.4.7 : Proof of theorem 21 has been adapted from [4].

3.5. Notes with illustrative examples

One need not confine oneself to integral domains while discussing the 'division algorithm'. Euclidean rings could be studied. So, then, there are principal ideal rings.

The study of the ring of algebraic integers of $\mathbb{Q}[\sqrt{m}]$ has arisen from a class of Diophantine equations

$$x^2 - my^2 = n$$

where $|m|$ is given and n is arbitrary. In the case of

$$x^2 - y^2 = p \text{ (a prime)}$$

we have $(x-y)(x+y) = p$. One has $x - y = 1$ and $x + y = p$ and so $x = \frac{p+1}{2}$, $y = \frac{p-1}{2}$. So, a prime p is a difference of two squares if, and only if, it is odd. For an odd integer n, one can write

$$n = \left(\frac{n+1}{2}\right)^2 - \left(\frac{n-1}{2}\right)^2$$

and hence every odd integer is the difference of two consecutive squares.

So, we are interested in studying rings contained in $\mathbb{Q}[\sqrt{m}]$ where $m \in \mathbb{Z} \setminus \{0,1\}$, m square free. In the case of the ring $\mathbb{Z}[i]$ of Gaussian integers, it can be shown that for any ideal I of $\mathbb{Z}[i]$, the quotient ring $\mathbb{Z}[i]/I$ is finite. As I is a principal ideal, $I = (\alpha)$, $\alpha \in \mathbb{Z}[i]$. The cosets $I + \beta$ are those determined by $\theta = \alpha\gamma + \beta$ where $\theta \in \mathbb{Z}[i]$. There are only a finite number of β for which $g(\beta) < g(\alpha)$.

Now, given m_1 and m_2 square-free integers such that one does not divide the other, it is easy to check that $\mathbb{Q}[\sqrt{m_1}]$ and $\mathbb{Q}[\sqrt{m_2}]$ are not isomorphic as fields. Further, for a given m, one can determine all the subfields of the quadratic field $\mathbb{Q}[\sqrt{m}]$ as $[\mathbb{Q}[\sqrt{m}] : \mathbb{Q}] = 2$. Regarding units in $\mathbb{Z}[\sqrt{m}]$, $m > 1$, $\mathbb{Z}[\sqrt{m}]$ has infinitely many units.

If $\langle a_1, b_1 \rangle$ is a solution of $a^2 - mb^2 = \pm 1$, $\langle a_k, b_k \rangle$ is also a solution where $\langle a_k, b_k \rangle = (a_1 + b_1\sqrt{m})^k$, $k \in \mathbb{Z}$. If $\langle a_1, b_1 \rangle$ is a solution of $a^2 - mb^2 = \pm 1$, then $a_1 + b_1\sqrt{m}$ is a unit. It follows that $\pm(a_1 + b_1\sqrt{m})^k$, $k \in \mathbb{Z}$ are all units. In particular, the elements of the form $\pm(1 + \sqrt{2})^k$, $k \in \mathbb{Z}$ are units of $\mathbb{Q}[\sqrt{2}]$. See section 13.7, chapter 13.

If $m < -1$ $\mathbb{Z}[\sqrt{m}]$ has units $+1, -1$. Further, $\mathbb{Z}[\sqrt{-5}]$ is not a PID, as the ideal generated by 3 and $2 + \sqrt{-5}$ is not a principal ideal. $\mathbb{Z}[\sqrt{-5}]$ deserves greater attention. If $p_1, p_2, \cdots, p_k$ and $q_1, q_2, \cdots, q_m$ are irreducible elements of $\mathbb{Z}[\sqrt{-5}]$, such that $p_1, p_2, \cdots, p_k = q_1, q_2, \cdots, q_m$, then $k = m$. An integral domain which satisfies this kind of property is referred to as a half-factorial domain (H.F.D). In general, if H is a H.F.D, it happens that a finite set of elements of H need not have a greatest common divisor. See Section 2.5, chapter 2. Half-factorial domains are considered in Section 13.4, chapter 13.

3.6. Worked-out examples

a) Let $K = \mathbb{Q}[\sqrt{m}]$ where $m > 0$ and $m \equiv 2$ or $3 \pmod 4$. Show that the number of fields K for which the ring $R(m)$ is a Euclidean domain is finite.

Answer: Let us suppose that $R(m)$ is Euclidean for $m > 0$ and $m \not\equiv 1 \pmod 4$. Given $\theta = \mathbb{Q}[\sqrt{m}]$, there exists $\eta \in R(m)$ such that

$$(3.6.1) \qquad\qquad g(\theta - \eta) < 1$$

where $g : R^*(m) \to \widetilde{\mathbb{Z}}$ given by $g(\alpha) = |N(\alpha)|, \alpha \in R^*(m)$.

Let $\theta = r + s\sqrt{m}$, $r, s \in \mathbb{Q}$. $\eta = x + y\sqrt{m}$, $x, y \in \mathbb{Z}$ and θ, η satisfy (3.6.1). Then, by (3.6.1),

$$|(r-x)^2 - m(s-y)^2| < 1.$$

Take $r = 0$, $s : t/m$ where t is an integer to be chosen. Then,

$$|x^2 - m(y - t/m)^2| < 1.$$

That is, $|(my-t)^2 - mx^2| < m$. But, $(my-t)^2 - mx^2 \equiv t^2 \pmod m$. So, there exist rational integers x, z such that

$$(3.6.2) \qquad z^2 - mx^2 \equiv t^2 \pmod m, \quad |z^2 - mx^2| < m.$$

If $m \equiv 3 \pmod 4$, we choose t such that $5m < t^2 < 6m$. This is possible when m is large enough. By (3.6.2) $z^2 - mx^2$ is equal to $t^2 - 5m$ or $t^2 - 6m$. Therefore, one of

(3.6.3) $$t^2 + z^2 = m(5 \cdot x^2), \quad t^2 - z^2 = m(6 \cdot x^2)$$

holds. Considering integers t^2, z^2, x^2, m modulo 8, we have

$$t^2 \equiv 1 \pmod 8$$

$$z^2, x^2 \equiv 0, 1 \text{ or } 4 \pmod 8, \quad m \equiv 3 \text{ or } 7 \pmod 8$$

$$t^2 - z^2 \equiv 0, 1 \text{ or } 5 \pmod 8$$

$$5 - x^2 \equiv 1, 4 \text{ or } 5 \pmod 8$$

$$6 - x^2 \equiv 2, 5 \text{ or } 6 \pmod 8$$

$$m(5 - x^2) \equiv 3, 4 \text{ or } 7 \pmod 8$$

$$m(6 - x^2) \equiv 2, 3, 6 \text{ or } 7 \pmod 8.$$

whatever be the choice of m, each of (3.6.3) is impossible.

If $m \equiv 2 \pmod 4$, we choose t odd with the property

(3.6.4) $$2m < t^2 < 3m,$$

when m is large enough. (3.6.4) holds for large values of m. In this case, one of

(3.6.5) $$t^2 - z^2 = m(2 - x^2), \quad t^2 - z^2 = m(3 - x^2)$$

holds. But, to the modulus 8, $m \equiv 2$ or $6 \pmod 8$. Then,

$$2 - x^2 \equiv 1, 2 \text{ or } 6 \pmod 8$$

$$3 - x^2 \equiv 2, 3 \text{ or } 7 \pmod 8$$

$$m(2 - x^2) \equiv 2, 4 \text{ or } 6 \pmod 8$$

$$m(3 - x^2) \equiv 2, 4 \text{ or } 6 \pmod 8.$$

Then, each of (3.6.5) is impossible.

Hence, if $m \equiv 2$ or $3 \pmod 4$ and m is large enough, the ring $R(m)$ of integers of $\mathbb{Q}(\sqrt{m})$ is not Euclidean.

This completes the answer. □

Remark 3.6.1 : Worked out example (a) is applicable to the case $m \equiv 1 \pmod 4$. As noted in [A1], the proof is more difficult and so not attempted.

b) (Godement) Let R be a PID having F for its field of quotients. Suppose that $t \in F$ is given by

(3.6.6) $$t = \frac{a}{p_1^{a_1}, p_2^{a_2}, \dots, p_k^{a_k}}$$

where $p_1, p_2, \ldots, p_k$ are distinct primes of R; $a \in R$, $a_i \geq 1 (i = 1, 2, \ldots, k)$. Show that there exist elements $t_1, t_2, \ldots t_k$ in R such that

(3.6.7)
$$t = \frac{t_1}{p_1^{a_1}} + \cdots + \frac{t_k}{p_k^{a_k}}.$$

(that is, every element of F can be written as a sum of fractions of the form $\frac{b}{p^r}$, where $b \in R$, p a prime in R, $r \geq 0$).

Answer: We make the following

Observation: (i) If $p_1, p_2, \ldots, p_k$ are nonassociated primes in R, then, $p_1^{a_1}, p_2^{a_2}, \ldots, p_k^{a_k}$ ($a_i \geq 1; i = 1, 2, \ldots, k$) are relatively prime to one another.

Observation: (ii) Suppose that $b_1, b_2, \ldots, b_k$ are elements of R which are relatively prime to one another. If

$$s = \frac{c}{b_1, b_2, \ldots, b_k}$$

where $c \in R$, $s \in F$ and s can be expressed as

(3.6.8)
$$s = \frac{c_1}{b_1} + \frac{c_2}{b_2} + \cdots + \frac{c_k}{b_k}$$

where $c_i \in R (i = 1, 2, \ldots, k)$.

Observation (i) follows from the fact that if d is a g.c.d of $p_i^{a_i}$ and $p_j^{a_j}$, $i \neq j$ and if d is a non-unit, d has a prime divisor p which divides $p_i^{a_i}$ and $p_j^{a_j}$. So, p is an associate of p_i as well as p_j — a contradiction to the fact that p_i and p_j are not associated. So, d is a unit and hence $p_i^{a_i}$ and $p_j^{a_j}$ are relatively prime to one another.

Observation (ii) can be proved by induction on k, starting with $k = 2$. (3.6.7) follows from (3.6.8).   □

Remark 3.6.2 : Example (b) is useful when we consider $\mathbb{R}[x]$, the polynomial ring with coefficients from $\mathbb{R}$, the field of real numbers. $\mathbb{R}[x]$ is a PID. Its field of quotients $\mathbb{R}(x)$ contains rational functions of the form $\frac{f(x)}{g(x)}$, $g(x) \neq 0$. By the UFD property of $\mathbb{R}[x]$, we express $g(x)$ as a product of irreducible elements of $\mathbb{R}[x]$. Then, $\frac{f(x)}{g(x)}$ is expressible as a sum of partial fractions.

Illustration 3.6.1 :

(3.6.9)
$$\frac{x^3 + 2x + 1}{x^4 - 1} = \frac{1}{x - 1} + \frac{1}{2(x + 1)} - \frac{(x + 1)}{2(x^2 + 1)}.$$

EXERCISES

1. **Mark the following statements true (T) or false (F) justifying your answer briefly.**

   a) d is a g.c.d of nonzero elements $a_1, a_2, \ldots, a_n$ of a PID R. Then, d is expressible as

   $$d = x_1 a_1 + x_2 a_2 + \cdots + x_n a_n$$

   where $x_1, x_2, \ldots, x_n$ are elements of R. It is correct to say that $x_1, x_2, \ldots, x_n$ are pairwise relatively prime to one another.

   b) Let R be a PID. Suppose that p denotes an irreducible element of R. Then the ideal $I(= (p))$ generated by p need not be a prime ideal.

   c) (Godement) Let d be an element of a commutative ring R with unity. In $R \times R$, we define addition and multiplication by

   $$(r,s) + (r',s') = (r+r', s+s')$$

   (3.6.10) $$(r,s) \cdot (r',s') = (rr' + dss', rs' + r's)$$

   (where r, r'; s, s' are elements of R) involving the given element d and the laws of composition in R. We denote the new ring obtained from $R \times R$ using (3.6.10) by $R[\sqrt{d}]$.

   If $R = \mathbb{F}_p = \mathbb{Z}/p\mathbb{Z}$, where p is a prime, $\mathbb{F}_p[\sqrt{d}]$ is a field having 121 elements, for $d = [7]$ (element of $\mathbb{F}_p$) and $p = 11$.

   d) The g.c.d of $11 + 7i$ and $18 - i$ in $\mathbb{Z}[i]$ is $2 + i$.

   e) Let $R(3)$ denote the ring of integer of $Q(\sqrt{3})$. The equation $2.11 = (5 + \sqrt{3})(5 - \sqrt{3})$ does not contradict the UFD property of $R(3)$.

   f) Let $\theta = \exp(\frac{2\pi i}{5})$. Then, $\mathbb{Z}[\theta]$ is a Euclidean domain.

2. Let (D, g) be a Euclidean domain. For nonzero $a, b \in D$, show that $g(a) < g(b)$ if, and only if, b is not a unit in D.

3. (Picavet) [A3] Let (D, g) be a Euclidean domain whose algorithm g satisfies the following properties :

   (a) $g(ab) = g(a)g(b)$ for all $a, b \in D^*$ with $a + b \neq 0$

   (b) $g(a) = g(b)$ if, and only if, a and b are associates.

   Show that $D = \mathbb{Z}$, the ring of integers.

4. Prove that 1 and -1 are the only units in $R(-19)$.

5. If $\omega = \exp(\frac{2\pi i}{3})$, show that $\mathbb{Z}[\omega]$ is a Euclidean domain.

6. Show that $\mathbb{Z}[\sqrt{-10}]$ is not a unique factorization domain.

7. Prove that for each of the number fields $\mathbb{Q}(\sqrt{m})$ where $m = -13, -14, -15, -17, -21, -22, -23$ and -26, the associated ring $R(m)$ of integers is not a U.F.D and so not a PID.

8. Show that $R(10)$ and $R(15)$ are not unique factorization domains. (Recall that $R(m)$ denotes the ring of integers of $\mathbb{Q}[\sqrt{m}]$).

9. What is the irreducible polynomial of $\alpha = 1 + \sqrt[3]{5}$ determining a number field K such that $[K : \mathbb{Q}] = 3$. Determine the ring of integers of K.

10. Let $a \in \mathbb{Q}$. Suppose that α is a zero of $x^2 - a$. Find the number field K of degree 2 over $\mathbb{Q}$ such that $\alpha \in K$. Describe the ring of integers of K.

11. Let K be equal to $\mathbb{Q}(\sqrt{3}, i)$ where i denotes $\sqrt{-1}$. Determine the ring of integers of K. Is it a Euclidean domain?

REFERENCES

[1] D. M. Burton : A first Course in rings and ideals, Chapter 6, pp 90–111, Addison-Wesley Pub. Co., Reading Mass (1970).

[2] Oscar A. Campoli : A Principal Ideal Domain that is not a Euclidean domain, Amer. Math. Monthly 95 (1988) 868–871.

[3] H. Chatland and H. Davenport : Euclid's algorithm in real quadratic fields, Canadian J. Math 2 (1950) 289–296.

[4] Steven Galovich : A Characterization of integers among Euclidean domains, Amer. Math. Monthly 85 (1978) 572–575.

[5] John Greene : Principal Ideal Domains are almost Euclidean, Amer. Math. Monthly 104 (1997), 154–156.

[6] Herstein I.N. : Topics in Algebra, Wiley Eastern Ltd. Second Edn. (15th reprint 1993), New Delhi.

[7] K. Inkeri: Über den Euklidischen algorithms in quadratischen Zahlkorpern, Annals Academic Sinentiarum Fenrical 41 (1947), 35–40.

[8] D. A. Marcus : Number fields, Universitext : Chapter 2, pp 12–54, Springer Verlag (1977).

[9] W. Narkeiwicz: Elementary and Analytical theory of algebraic numbers, Chapters 3 & 4, PWN Polish Scientific Publishers Warsawa, 1990 (Original edition). Third revised and extended edition (English) 2004. Springer Monographs in Mathematics, Chapter 1 pp. 1–41.

[10] H. M. Stark : A Complete determination of Complex quadratic fields of class number one, Michigan Math J 14 (1967) 1–27.

ADDITIONAL REFERENCES

[A1] G. H. Hardy and E. M. Wright : An introduction to the theory of numbers, Chapter XIV, Theorem 249, Oxford at the Clarendon Press, Fourth Edn (1965) pp 204–217.

[A2] T. Motzkin : 'The Euclidean algorithm', Bulletin Amer Math. Soc., 55 (1949) 1142–1146.

[A3] G. Picavet : caracterisation de certains types d' anneaux euclidiens: L' Ensignment Math. 18 (1972) 245–254.

CHAPTER 4

Rings of polynomials and formal power series

Historical perspective

Polynomials have played a major role in the study of algebra and geometry. Even before the time of Euclid, Babylonians (586 B.C.) knew that a right-angled triangle with legs each of unit length must have a hypotenuse $\sqrt{2}$ units in length. (The ancient Babylonian Empire was situated in Euphrates valley about 100 kms south of Baghdad (Capital of Iraq which was formerly known as Mesopotamia). Pythagoras (580–500 B.C.) is known by the theorem that bears his name, although the theorem was known to Babylonians. It is reported that Egyptians were familiar with methods of solving polynomial equations in special cases. From the time of Hippocrates (c. 460 B.C.) till the time of Diophantos (c 250 A.D.), Greeks attempted numerical problems which could be stated in terms of polynomials. In Greek algebra, magnitudes were represented by line segments and the problem of finding the roots of a quadratic equation meant a solution in the form of a straight-edge and compass construction for line segments representing the roots. By 1100 A.D., Arabs developed algebra to the extent where they were conscious of tackling cubic equations. In the sixteenth and seventeenth centuries, methods of finding the roots of a quadratic, cubic and biquadratic equations were found out in the form of 'formulae for roots'. The contributions of Euler, Lagrange, Niels Henrik Abel (1802–1829) and Everiste Galois (1811–1832) to the 'insolvability of the quintic' are well-known. As mentioned in chapter 3, when Rene Descartes (1596–1650) invented analytic geometry, the door was opened for a fusion of ideas of Calculus which unfolded itself at the hands of Newton and Leibnitz.

A major breakthrough was in sight when Poncelot (1788–1867) published his work on projective geometry. Poncelot was in prison during 1813–1814 and it was during this period he invented many outstanding results in geometry—mainly his concept of central projection. Polynomials are studied in the context of central projections. If we consider curves whose equations are given by polynomials, central projections change these curves into other curves given by ratios of polynomials. We examine the properties of a curve which do not change by such projections. Those properties are called invariants. Invariants provide a 'bridge' between algebra and geometry. The idea of central projections gave place to one-one onto mappings θ of an n-dimensional space so that θ and θ^{-1} were polynomial maps. They are birational transformations. The question that was asked is the following:

*What properties of curves remain invariant under certain kinds of transfor-
mations? In the general situation, polynomials with coefficients from a field were
considered. The ring $F[x]$ is a PID. Polynomials in two indeterminates x, y belong to the integral domain $F[x, y]$. But $F[x, y]$ is not a PID. We look at an ideal
I of the ring $F[x_1, x_2, \cdots, x_n]$ in n indeterminates. The question is : Is I finitely
generated? Hilbert's Basis theorem says that I is finitely generated. It was P.
Gordon (1837–1912) who first proved this theorem for ideals of $F[x, y]$ in 1868.
For the next 20 years or so, Gordon and others were able to extend this result to
certain special kinds of ideals in $F[x_1, x_2, x_3]$ and $F[x_1, x_2, x_3, x_4]$. Even though the
work was about the structure of ideals, they studied certain kinds of polynomials.
Indeed, D. Hilbert's contributions are remarkable and his influence on modern
mathematics is profound.*

4.1. Introduction

Polynomial rings are introduced. If F is a field, $F[x]$ is a Euclidean domain
and the uniqueness of the division algorithm characterizes $F[x]$ among Euclidean
domains.

The ring $\mathcal{A}$ of arithmetic functions under the operations of addition and
Dirichlet convolution is shown to be a UFD via the ring $\mathbb{C}_\omega$ of formal power
series in countably infinite indeterminates. This was proved by E. D. Cashwell
and C. J. Everet [2] in 1959. This is given in theorem 27.

When we consider polynomials over a finite field, say $\mathbb{Z}/p\mathbb{Z}$, p, a prime it is
possible to find a formula for the number of monic irreducible polynomials of a
given degree say m, via Möbius inversion. It is deduced that the number of monic
irreducible polynomials over $\mathbb{Z}/p\mathbb{Z}$ is infinite. This is an analogue of Euclid's
theorem on the infinitude of primes (done in arithmetic) in algebra. Noting that $\mathbb{Z}$
is a PID, another analogue involving a PID (due to Fabrizio Zanello) is considered
in Section 16.3, Chapter 16. See Theorem 158.

4.2. Polynomial rings

Let R be a commutative ring with unity 1_R. A polynomial in x (an indeterminate) with coefficients from R is of the form $\sum_{i=0}^{n} a_i x^i$. It is a sequence
$(a_0, a_1, a_2, \cdots)$ where $a_i \in R$ and all the a_i except a finite number of them are zero.
$(a_0, a_1, a_2, \cdots)$ and $(b_0, b_1, b_2, \cdots)$ are equal if, and only if, $a_i = b_i$ for $i = 0, 1, 2, \cdots$.
Addition and multiplication of polynomials are defined by

(4.2.1) $(a_0, a_1, a_2, \cdots) + (b_0, b_1, b_2, \cdots) = (a_0 + b_0, a_1 + b_1, a_2 + b_2, \cdots)$

and

(4.2.2) $(a_0, a_1, a_2, \cdots)(b_0, b_1, b_2, \cdots) = (c_0, c_1, c_2, \cdots)$

where

$$c_n = \sum_{j=0}^{n} a_j b_{n-j}$$

The set of polynomials with coefficients from R is denoted by $R[x]$. $R[x]$ forms a commutative ring under operations of addition and multiplication given above. The zero polynomial is

$$(0_R, 0_R, 0_R, \cdots)$$

and x is denoted by $(0_R, 1_R, \cdots .)$. It is a polynomial. The symbol x multiplied by x, t times is the polynomial

$$(0_R, 0_R, \cdots, \underset{\substack{\downarrow \\ (t+1)^{th}\,place}}{1_R}, 0_R, \cdots).$$

0_R is the additive identity and 1_R, the unity element in R. With this notation, we get

(4.2.3) $$(a_0, a_1, a_2, \cdots, a_n, 0_R, 0_R, \cdots) = \sum_{i=0}^{n} a_i x^i.$$

It is easy to check that when R is an integral domain so is $R[x]$.
If $f(x) = \sum_{i=0}^{n} a_i x^i \in R[x]$, $\deg f(x) = n$ when $a_n \neq 0_R$. By convention, the degree of zero polynomial is defined to be $-\infty$ (negative infinity). Let R be an integral domain. We observe that

(4.2.4) $$\deg(f+g)(x) \leq \max\{\deg f(x), \deg g(x)\}$$

(4.2.5) $$\deg(fg)(x) = \deg f(x) + \deg g(x)$$

When R is a commutative ring with unity, division algorithm holds in $R[x]$. We state this property without proof.

Proposition 4.2.1 : *Suppose that $f(x)$, $g(x)$ ($\neq 0$) are elements of $R[x]$ such that the leading coefficient of $g(x)$ is a unit in R. Then, there exist unique elements $q(x), r(x) \in R[x]$ such that*

$$f(x) = q(x)g(x) + r(x)$$

where either $r(x) = 0$ or $\deg r(x) < \deg g(x)$.

For proof, see D. M. Burton [1].

Remark 4.2.1 : When $R = F$, a field, the leading coefficient in $g(x)$ is invertible and so the condition that the leading coefficient should be a unit is not needed.

Corollary 4.2.1 : $F[x]$ *is a Euclidean domain.*

For, one could define $\delta : F[x] \to \tilde{\mathbb{Z}}$ (the set of non-negative integers), by

$$\delta(f) = \deg f, \quad f \in F[x].$$

Let $f, g \in F[x]$.

(i)

$$\begin{aligned}
\delta(f(x)g(x)) &= \deg(f(x)g(x)) \\
&= \deg f(x) + \deg g(x) \\
&\geq \deg f(x), \quad \text{as } \deg g(x) \geq 0 \\
&= \delta(f(x))
\end{aligned}$$

(ii) Given $f, g \in F[x]$, $g(x) \neq 0$, one has, by proposition 4.2.1,

$$f(x) = q(x)g(x) + r(x)$$

where either $r(x) = 0$ or $\delta(r(x)) < \delta(g(x))$. This proves corollary 4.2.1.

We prove in theorem 23 below that the division algorithm for polynomials in $F[x]$ gives a unique quotient and remainder. Such uniqueness characterises polynomial rings among Euclidean domains.

Let D be a Euclidean domain with $1_D \neq 0_D$. Let $D^* = D \setminus \{0_D\}$. The function $g : D^* \to \tilde{\mathbb{Z}}$ has the property:

(4.2.6)   If $a, b \in D$ and a is a proper divisor of b then $g(a) < g(b)$

and $g(a) = g(b)$ if, and only if, a and b are associates.

Further by theorem 13, (chapter 3), the quotient and remainder on application of division algorithm are unique if, and only if,

(4.2.7) $g(a+b) \leq \max\{g(a), g(b)\}$ for every pair $\langle a, b \rangle$ in D^*.

Now, we may assume that $g(1) = 0$. If not, we could write

(4.2.8) $g' : D^* \to \tilde{\mathbb{Z}}$ such that $g'(a) = g(a) - g(1)$.

g' also gives a Euclidean norm (function) in D^* just as g does give.

Theorem 22 : *If $g' : D^* \to \tilde{\mathbb{Z}}$ is as given in (4.2.8) and for $a \in D^*$, $g'(a)$ is a positive minimum. Then for each $b \in D^*$, there exist unique elements*
$q_0, q_1, \cdots, q_k \in U(D) \cup \{0_D\}$ *such that*

$$b = q_k a^k + q_{k-1}a^{k-1} + \cdots + q_1 + q_0, \ q_k \neq 0_D \ \text{(the zero element in } D\text{)},$$

where $U(D)$ denotes the set of units in D. Further, the map

$$\phi : D \to F[x]$$

given by $b = \displaystyle\sum_{j=1}^{k} q_j x^j$ is an isomorphism, where $F = U(D) \cup \{0_D\}$.

Proof : The sequence $\{g'(a^k)\}$, $k = 0, 1, 2, \cdots$ is a strictly increasing sequence as a^k is a proper divisor of a^{k+1}. $(k \geq 1)$.

(4.2.9)   If $b \neq 0_D, g'(a^k) \leq g'(b) \leq g'(a^{k+1})$ for some $k \geq 1$.

Then,

$$b = q_k a^k + r \text{ where } r = 0 \text{ or } g'(r) < g'(a^k).$$

$q_k \neq 0_D$. We claim that q_k is a unit. If q_k were not a unit, we will have

$$q_k = la + m, m \in F \text{ and } l \neq 0_D \text{ as } g'(a) \text{ is a positive minimum.}$$

So, $q_k a^k = la^{k+1} + ma^k$. From

$$b = q_k a^k + r,$$

we have

$$b = la^{k+1} + ma^k + r.$$

Now, l and m are unique, as m is a unit.

$$\text{So, by (4.2.7), } g'(b - ma^k - r) = g'(b + (-ma^k - r))$$

$$\leq \max\{g'(b), g'(ma^k + r)\}$$

Further,

$$g'(ma^k + r) \leq \max\{g'(ma^k), g'(r)\}$$

$$\leq \max\{g'(a^k), g'(r)\}, \text{ as } m \text{ is a unit.}$$

$$= g'(a^k).$$

By (4.2.9), $g'(b) \geq g'(a^k)$. So, $\max\{g'(b), g'(ma^k + r)\} = g'(b)$.
Also, $g'(b - ma^k - r) \leq g'(b)$. But, $b - ma^k - r = la^{k+1}$.
Therefore, $g'(b) \geq g'(la^{k+1}) \geq g'(a^{k+1})$ which contradicts the inequality
$g'(b) < g'(a^{k+1})$ in (4.2.9). So, q_k is a unit and $b = q_k a^k + r$.

If $r \neq 0$, we proceed as before and arrive at

$$(4.2.10) \qquad b = \sum_{j=0}^{k} q_j a^j, \ q_j \in F, \ j = 0, 1, 2, \cdots k.$$

The representation of b in (4.2.10) is unique.
For, suppose that

$$\sum_{j=m}^{n} s_j a^j = 0_D, s_j \in F, \ s_m \neq 0_D.$$

Then,

$$-s_m = a \sum_{j=m+1}^{n} s_j a^{j-m-1}$$

Therefore, $g'(s_m) = g'(1) \geq g'(a)$ which is impossible as, $g'(1) = 0$. That is,

$$\sum_{j=m}^{n} s_j a^j = 0_D, \Rightarrow s_j = 0_D, \ m \leq j \leq n.$$

It follows that the representation (4.2.10) for b is unique. $\phi : D^* \to F[x]$ is such
that

$$\phi(b) = \sum_{j=0}^{k} q_j x^j$$

provides a one-one map from D^* into $F[x]$.

The question is : Is F a well-defined ring?
We answer this in the affirmative, once we claim that the sum of two units in D is again a unit.
If u_1, u_2 are in $u(D)$

$$0 \le g'(u_1 + u_2) \le \max\{g'(u_1), g'(u_2)\} = g'(1) = 0.$$

So $u_1 + u_2$ is a unit whenever u_1, u_2 are units.
If $D = u(D) \cup \{0_D\}(= F), D$ is a field. This completes the proof. $\qquad \square$

Theorem 23 (M. A. Jodiet (1967)) : *Let D be a Euclidean domain in which $1_D \ne 0_D$. If the quotient and remainder on applying division algorithm to every pair $\langle a, b \rangle$ in D^* are unique, the set F of units together with 0_D forms a field. If $F \ne D$, $D \cong F[x]$.*

Proof : If the quotient and remainder are unique, the function $g' : D^* \to \tilde{\mathbb{Z}}$ could be so defined as to make $g'(1_D) = 0$. In this case $D = F$ by theorem 22.

If $D \ne F$, there exist nonzero non-units in D and then, by theorem 22, we have a one-one map $\phi : D^* \to F[x]$ given by $\phi(b) = \sum_{j=0}^{k} q_j x^j$, $q_j \in F$, $q_k \ne 0$, $b \in D^*$. Let us map the zero element 0_D into the zero polynomial in $F[x]$. We get a map $\phi' : D \to F[x]$ which is a homomorphism and which is one-one onto $F[x]$. For,

$$\phi'(b_1 + b_2) = \phi'(\sum_{j=0}^{k_1} q_j a^j + \sum_{j=0}^{k_2} q'_j a^j)$$

$$= \phi'(\sum_{j=0}^{\max\{k_1, k_2\}} t_j a^j) \text{, where } t_j = q_j + q'_j \in F;$$

$$= \sum_{j=0}^{\max\{k_1, k_2\}} t_j x^j.$$

Or, $\phi'(b_1 + b_2) = \phi'(b_1) + \phi'(b_2)$.
Also,

$$\phi'(b_1 b_2) = \phi'(\sum_{j=0}^{k_1} q_j a^j \sum_{j=0}^{k_2} q'_j a^j)$$

$$= \phi'(\sum_{j=0}^{k_1+k_2} c_j a^j \text{, where } c_j = \sum_{s=0}^{j} q_s q'_{j-s})$$

$$= \sum_{j=0}^{k_1+k_2} c_j x^j.$$

Or, $\phi'(b_1 b_2) = \phi'(b_1)\phi'(b_2)$.

So, $D \cong F[x]$. □

Remark 4.2.2 : Theorems 22 and 23 have been adapted from M. A. Jodiet [6].

Theorem 24 : *Let D be an integral domain. Suppose that a function $\phi : D \to \tilde{\mathbb{Z}}$ satisfies the conditions:*
(a) for $a, b \in D$ whenever $a|b$. $\phi(a) \leq \phi(b)$ with equality if, and only if, a and b are associates.
(b) If $a, b \in D^* (= D\backslash\{0\})$ such that neither of them divides the other, then there exist elements k, l, m with the property $m = ka + lb$ and $\phi(m) < \min\{\phi(a), \phi(b)\}$. Then D is a PID and conversely.

Proof : $:\Rightarrow D$ is an integral domain and $\phi : D \to \tilde{\mathbb{Z}}$ is a function satisfying conditions (a) and (b) of the theorem.

We claim that D is a *PID*.

To achieve this, we consider a proper ideal I of D. $I \neq (0)$. We take an element x among the nonzero elements of I such that $\phi(x)$ is a minimum. We show that I is the principal ideal generated by x.

Suppose not. Let $y \in I$. If y is not a multiple of x, x cannot be a multiple of y as $g(x) \leq g(y)$ by condition (a). Applying (b) we note that there exist elements k, l, m such that

$$m = kx + ly \text{ and } \phi(m) < \min\{\phi(x), \phi(y)\} = \phi(x)$$

This contradicts the minimality of $\phi(x)$. So I is a principal ideal and $I = (x)$. Thus, D is a PID.

$\Leftarrow$: Conversely, if D is a PID we introduce a function $\phi : D^* \to \tilde{\mathbb{Z}}$ as follows: As D is a UFD, every nonzero element of D is a product of a finite number of irreducibles.

For $a \in D^*$, take $\phi(a)$ = the number of irreducible factors of a and write $\phi(0_D) = 0$. Then (a) holds. For (b), take m to be a g.c.d of a and b. As D is a PID, D satisfies the criteria for the function ϕ defined as above. □

Remark 4.2.3 : An integral domain D for which a function $\phi : D^* \to \tilde{\mathbb{Z}}$ satisfying conditions (a) and (b) of theorem 24 does not make it a Euclidean domain. For, condition (b) is not the equivalent of the 'division algorithm' property of a Euclidean domain. So, theorem 24 does not hold good for Euclidean domains though they are principal ideal domains. For instance, as in theorem 18, chapter 3, $R(-19)$ is a PID and is not a Euclidean domain. Theorem 24 indicates that an integral domain D could be 'almost Euclidean' as indicated in theorem 17, chapter 3 (see definition 3.4.1).

4.3. Elementary arithmetic functions

By an arithmetic function f we mean a mapping $f : \mathbb{N} \to \mathbb{C}$ where $\mathbb{C}$ denotes the field of complex numbers. We denote the set of arithmetic functions by $\mathcal{A}$. An arithmetic function is merely a complex-valued sequence: $f(1)$, $f(2)$, $f(3)$, $\cdots$. For $f, g \in \mathcal{A}$, we define $f + g$ by

(4.3.1) $(f + g)(r) = f(r) + g(r)$, $r \geq 1$

Definition 4.3.1 : *The Dirichlet convolution $f \cdot g$ of $f, g \in \mathcal{A}$ is defined by*

$$(f \cdot g)(r) = \sum_{t \mid r} f(t) g(\frac{r}{t})$$

where the summation on the right is over all positive divisors t of r.

Addition and Dirichlet convolution are associative. They are also commutative. Further, for $f, g, h \in \mathcal{A}$

(4.3.2) $f \cdot (g + h) = f \cdot g + f \cdot h$

The function z given by

(4.3.3) $z(r) = 0$, $\quad r \geq 1$

serves as the identity element for addition. The function e_0 given by

(4.3.4) $e_0(r) = [\frac{1}{r}]$

where $[x]$ denotes the greatest integer not greater than x serves as the identity for Dirichlet convolution (multiplication). It follows that $(\mathcal{A}, +, \cdot)$ is a commutative ring with unity e_0. It is known [7] that $f \in \mathcal{A}$ is a unit if, and only if, $f(1) \neq 0$.

Definition 4.3.2 : $f \in \mathcal{A}$ *is called a multiplicative function if, whenever* g.c.d $(r, s) = 1$, $f(r)f(s) = f(rs)$.

The function d where $d(r)$ denotes the number of divisors of r and the Euler totient ϕ are examples of multiplicative functions.

Definition 4.3.3 : *For $f \in \mathcal{A}$, $f \neq z$, $N(f)$, the norm of f is defined as the least positive integer a such that $f(a) \neq 0$. $N(z) = 0$, where z is the zero function.*

Fact 4.3.1 : Norm of an arithmetic function satisfies

$$N(f \cdot g) = N(f)N(g) \quad \text{for all } f, g \in \mathcal{A}.$$

For proof, see Sivaramakrishnan [8].

It is clear that $N(f) = 1$ if, and only if, f is a unit. If $f \in \mathcal{A}$ is a unit, there exists $g \in \mathcal{A}$ such that $f \cdot g = e_0$. We denote g by f^{-1} and call it the Dirichlet inverse of f.

If $e(r) = 1$ for all $r \geq 1$, e, is a unit in $\mathcal{A}$. The Dirichlet inverse of e is called the Möbius function which is denoted by μ. That is, $e^{-1} = \mu$. Accordingly, we give

Definition 4.3.4 : *The Möbius function μ is defined by*

$$\mu(r) = \begin{cases} 1, & r = 1; \\ (1)^k, & \text{if } r = p_1 p_2 \cdots p_k, \text{ where } p_1, p_2, \cdots p_k \text{ are distinct primes}; \\ 0, & \text{if } a^2 | r, \quad a > 1. \end{cases}$$

μ is multiplicative.

The relation $e \cdot \mu = e_0$ translates into

$$\sum_{t|r} \mu(t) = \begin{cases} 1, & r = 1 \\ 0, & \text{otherwise.} \end{cases}$$

Further, we note that for $f, g \in \mathcal{A}$

$$f \cdot e = g \Leftrightarrow f = g \cdot e^{-1}$$

This is the well-known Möbius Inversion formula namely: for $f, g \in \mathcal{A}$

(4.3.5) $$\sum_{t|r} f(t) = g(r) \Leftrightarrow f(r) = \sum_{t|r} g(t)\mu(\tfrac{r}{t})$$

Let $U(\mathcal{A})$ denote the group of units in $(\mathcal{A}, +, \cdot)$. As remarked earlier, $f \in U(\mathcal{A})$ if, and only if, $f(1) \neq 0$.

Lemma 4.3.1 : $(\mathcal{A}, +, \cdot)$ *is an integral domain.*

Proof : We have already seen that $(\mathcal{A}, +, \cdot)$ is a commutative ring with unity e_0. We claim that $\mathcal{A}$ has no divisors of zero. As the norm $N(f)$ of an arithmetic function f satisfies

$$N(f) = N(g \cdot h) = N(g)N(h), \text{ whenever } f = g \cdot h;$$

$N(f) = 0$ if, and only if, $N(g)$ or $N(h)$ is zero. So, if $g \neq z$, $h \neq z$, $f = g \cdot h \neq z$. So, $\mathcal{A}$ is an integral domain.   □

We, next, mention some identities occurring in elementary number theory. Let $I \in \mathcal{A}$ be defined by

(4.3.6) $$I(r) = r, \quad r \geq 1.$$

Then, if ϕ denotes Euler's totient, $\sum_{t|r} \phi(t) = r$ is expressed by

$$\phi \cdot e = I,$$

$$\text{or } \phi = I \cdot e^{-1} = I \cdot \mu$$

That is, $\phi(r) = \sum_{t|r} t\, \mu(\tfrac{r}{t})$.

The number of divisors function $d = e \cdot e$, or $d \cdot e^{-1} = e$. That is,

$$\sum_{t|r} d(t)\mu(\tfrac{r}{t}) = 1.$$

If $\omega(r)$ denotes the number of distinct prime divisors of r

$$\sum_{t|r} |\mu(t)| = 2^{\omega(r)} \Leftrightarrow |\mu(r)| = \sum_{t|r} 2^{\omega(t)} \mu(\tfrac{r}{t}).$$

It is, in effect, $|\mu| \cdot e = \lambda$ where $\lambda(r) = 2^{\omega(r)}$. Or

$$|\mu| = \lambda \cdot e^{-1}.$$

Many of the known number-theoretic identities can be brought under Dirichlet convolution, using the calculus of multiplicative functions introduced by R. Vaidyanathaswamy (1894–1960) [9].

4.4. Polynomials in several indeterminates

We begin with a commutative ring R having unity element 1_R. Let S be a subring of R and T a subset of R. The smallest subring of R containing S and T is the intersection of all subrings of R which contain both S and T. We denote it by $S[T]$ and $S[T]$ is the subring of R generated by S and T.

If $T = \{x_1, x_2, \cdots, x_n\}$ we get the ring $S[x_1, x_2, \cdots, x_n]$. It is the subring of R generated by S and the x_i ($i = 1, 2, \cdots n$). $S[x_1, x_2, \cdots x_n]$ contains finite sums of monomials

$$sx_1^{a_1} x_2^{a_2} \cdots x_n^{a_n}, \text{ where } a_i \geq 0 \ (i = 1, 2, \cdots n), s \in S.$$

So, $S[x_1, x_2, \cdots, x_n]$ consists of all polynomials in $x_1, x_2, \cdots, x_n$ with coefficients from S. A typical polynomial in n indeterminates is of the form

$$(4.4.1) \qquad f(x_1, x_2, \cdots, x_n) = \sum_{a_1, a_2, \cdots, a_n \geq 0} s(a_1, a_2, \cdots, a_n) \ x_1^{a_1} x_2^{a_2} \cdots x_n^{a_n};$$

where $s(a_1, a_2, \cdots, a_n)$ is written to represent an element of S which occurs as a coefficient of $x_1^{a_1} x_2^{a_2} \cdots x_n^{a_n}$. $f(x_1, x_2, \cdots, x_n)$ can also be written in the form

$$(4.4.2) \qquad f(x_1, \cdots, x_n) = a + \sum_{i=1}^{n} a_i x_i + \sum_{i,j=1}^{n} a_{i,j} x_i x_j + \sum_{i,j,k=1}^{n} a_{i,j,k} x_i x_j x_k + \cdots$$

with coefficients $a, a_i, a_{i,j}, a_{i,j,k}, \cdots$ all but a finite number of them being zero.

To obtain (4.4.2), we have considered the terms in (4.4.1) for which

$$a_1 + a_2 + \cdots + a_n = 0,$$
$$a_1 + a_2 + \cdots + a_n = 1,$$
$$a_1 + a_2 + \cdots + a_n = 2,$$
$$\cdots \cdots \cdots \cdots \cdots$$

Now, $S[x_1, x_2, \cdots, x_n]$ contains all polynomials shown in (4.4.1) and conversely, every element of this ring is such a polynomial.

That is, $S[x_1, x_2, \cdots, x_n]$ is precisely the set of elements $y \in R$ which are of the form $f(x_1, x_2, \cdots, x_n)$ given in (4.4.1). For, since the product of two monomials in the x_i ($i = 1, 2, \cdots, n$) is another monomial in x_i ($i = 1, 2, \cdots, n$), the set S'

of polynomials (4.4.1) is a subring of R which contains S and $x_1, x_2, \cdots x_n$. So, $S' \supseteq S[x_1, x_2, \cdots, x_n]$. But $S[x_1, x_2, \cdots, x_n] \supseteq S'$ and so,

$$S' = S[x_1, x_2, \cdots, x_n].$$

When $T = \{x\}$ we get $S[x]$ to be the ring of polynomials in x with coefficients from S. For example, $\mathbb{C} = \mathbb{R}[i]$ where $i^2 = -1$ and $\mathbb{R}$ is the set of reals. Suppose $x = \sqrt[3]{2}$. The successive powers of x are

$$1, x, x^2, 2, 2x, 2x^2, 4, \cdots$$

So, the subring $\mathbb{Q}[\sqrt[3]{2}]$ of $\mathbb{R}$ is the set of all real numbers of the form $a + b\alpha + c\alpha^2$ where $\alpha = $ real cube root of 2; $a, b, c \in \mathbb{Q}$.

Definition 4.4.1 : *Let R be a commutative ring and S a subring of R. Suppose $x_1, x_2, \cdots, x_n$ are elements of R. An algebraic relation between $x_1, x_2, \cdots, x_n$ with coefficients in S is of the form*

$$(4.4.3) \qquad \sum_{a_1, a_2, \cdots a_n \geq 0} s(a_1, a_2, \cdots, a_n) x_1^{a_1} x_2^{a_2} \cdots x_n^{a_n} = 0$$

where $s(a_1, a_2, \cdots, a_n)$ are elements of S and all but a finite number of coefficients $s(a_1, a_2, \cdots, a_n)$ are zero.

If (4.4.3) implies that each $s(a_1, a_2 \cdots, a_n) = 0$, then $x_1, x_2, \cdots, x_n$ are said to be *algebraically independent* over S. On the other hand, if there exists at least one non-trivial relation (4.4.3), $x_1, x_2, \cdots, x_n$ are said to be *algebraically dependent*.

In the particular case $S[x]$, if x is algebraically dependent over S, that is, if there exists a relation of the form

$$(4.4.4) \qquad a_0 + a_1 x + \cdots + a_t x^t = 0$$

for some $t \geq 1$, and the coefficients $a_i \in S$ not all of which are zero, then, x is said to be algebraic over S. $i = \sqrt{-1}$ is algebraic over $\mathbb{Q}$.

Next, let R be a commutative ring containing S as a subring and generated by S and $(n-1)$ elements $x_1, x_2, \cdots x_{n-1}$ which are algebraically independent over S. Let T be the polynomial ring in one indeterminate x_n with coefficients from R, then,

$$(4.4.5) \qquad T = R[x_n] = S[x_1, x_2, \cdots, x_{n-1}][x_n]$$

(1) $S[x_1, x_2, \cdots, x_{n-1}]$ is a subring of T.
(2) $x_1, x_2, \cdots, x_n$ are algebraically independent over S.
(3) T is generated by S and $x_1, x_2, \cdots, x_n$.

Definition 4.4.2 : T given in (4.4.5) *is called the ring of polynomials in n indeterminates with coefficients from S.*

$f \in T$ can be written as

$$(4.4.6) \qquad f = \sum s(a_1, a_2, \cdots, a_n) x_1^{a_1} x_2^{a_2} \cdots x_n^{a_n}$$

with the coefficients $s(a_1, a_2, \cdots, a_n) \in S$, all but a finite number of them being zero.

For $n = 2$, one has

(4.4.7)
$$f = \sum_{a_1, a_2 \geq 0} s(a_1, a_2) x^{a_1} y^{a_2}$$

$s(a_1, a_2) \in S$, all but a finite number being zero. f in (4.4.7) can be rewritten as

$$f = s(0,0) + (s(1,0)x + s(0,1)y)$$
$$+ (s(2,0)x^2 + s(1,1)xy + s(0,2)y^2)$$
$$+ (s(3,0)x^3 + s(2,1)x^2y + s(1,2)xy^2 + s(0,3)y^3) + \cdots$$

with only a finite number of nonzero terms.

Fact 4.4.1 : In the case of $S[x]$, if S is a UFD, so is $S[x]$.
In the case of $S[x_1, x_2, \cdots, x_n]$, we have

Fact 4.4.2 : If S is an integral domain, so is $S[x_1, x_2, \cdots, x_n]$.

Fact 4.4.3 : If S is a UFD, so is $S[x_1, x_2, \cdots, x_n]$.

For proofs of facts 4.4.1, 4.4.2, see R. Godement [3] and T. W. Hungerford [4].

4.5. Ring of formal power series

Let R be a commutative ring with unity 1_R. We consider an infinite sequence

(4.5.1) $f = (a_0, a_1, a_2, \cdots, a_n, \cdots)$ where $a_i \in R, i \in \widetilde{\mathbb{Z}}$

f is called a formal power series over R. The set of all such infinite sequences is denoted by seq R. We introduce suitable operations of addition and multiplication in seq R in order to make it a ring containing R.

For $f, g \in$ seq R, when f is as given in (4.5.1) and

(4.5.2) $g = (b_0, b_1, b_2, \cdots, b_n, \cdots); b_i \in R, i \in \widetilde{\mathbb{Z}}$

$f = g$ if, and only if, $a_n = b_n$ for all $n \geq 0$.

Definition 4.5.1 : *When f and g are as given in (4.5.1) and (4.5.2) respectively, $f + g$ and fg are given by*

$$f + g = (a_0 + b_0, a_1 + b_1, \cdots)$$
$$fg = (c_0, c_1, c_2, \cdots)$$

where, for each $n \geq 0$, $c_n = \sum_{j=0}^{n} a_j b_{n-j}$.

It can be checked that for $f, g, h \in$ seq R

$$f(g + h) = fg + fh$$

The zero sequence is given by

(4.5.3) $z = (0_R, 0_R, 0_R, \cdots)$

where 0_R is the additive identity in R and additive inverse of f is given by

(4.5.4) $-f = (-a_0, -a_1, -a_2, \cdots, -a_n, \cdots)$

Thus, the seq R forms a commutative ring with unity

(4.5.5) $j = (1_R, 0_R, 0_R, \cdots)$

The set $S = \{(a, 0_R, 0_R, \cdots) : a \in R\}$ is a subring of seq R isomorphic to R. Let x be an indeterminate. We define

(4.5.6) $ax = (0_R, a, 0_R, \cdots), \quad a \in R$

and then,

(4.5.7) $ax^t = (0_R, 0_R, \cdots, 0_R, \underset{\underset{(t+1)^{th}place}{\downarrow}}{a}, 0_R, \cdots), t \geq 2.$

If

$$f = (a_0, 0_R, \cdots) + (0_R, a_1, 0_R, \cdots) + (0_R, 0_R, a_2, 0_R, \cdots)$$
$$+ (0_R, 0_R, \cdots, 0_R, \underset{\underset{(n+1)^{th}place}{\downarrow}}{a_n}, 0_R, \cdots) + \cdots$$

(4.5.8) $$f = \sum_{n=0}^{\infty} a_n x^n$$

f is now in the form of a formal power series, the form we want. When we take

$$x = (0_R, 1_R, 0_R, 0_R, \cdots)$$
$$ax = (a, 0_R, \cdots)(0_R, 1_R, 0_R, \cdots)$$
$$\text{or, } ax = (0_R, a, 0_R, 0_R, \cdots) \text{ as in (4.5.6)}$$

Notation 4.5.1 : seq R is denoted by $R[[x]]$, even though x need not be considered as an element of $R[[x]]$ when R does not have 1_R. As an example, we take $R = \mathbb{Z}$ to obtain $\mathbb{Z}[[x]]$.

It is easily verified that

$$(1, 0, 1, 0, 1, 0, \cdots) = 1 + x^2 + x^4 + \cdots + x^{2n} + \cdots$$

Definition 4.5.2 : *If $f(x) = \sum a_n x^n$ is a nonzero formal power series, then the smallest integer n such that $a_n \neq 0$ is called the order of $f(x)$ and is denoted by* ord $f(x)$.

Suppose $f(x), g(x) \in R[[x]]$ and ord $f(x) = s$, ord $g(x) = t$ then,

$$f(x) = a_s x^s + a_{s+1} x^{s+1} + \cdots \quad a_s \neq 0$$
$$g(x) = b_t x^t + b_{t+1} x^{t+1} + \cdots \quad b_t \neq 0$$

Then,

$$f(x)g(x) = a_s b_t x^{s+t} + (a_{s+1} b_t + a_s b_{t+1}) x^{s+t+1} + \cdots$$

If a_s or b_t is not a zero divisor in R, then, $a_s b_t \neq 0_R$ and so,

(4.5.9)   ord $(f(x)g(x)) = s+t =$ ord $f(x) +$ ord $g(x)$

So, if, by chance, $a_s b_t = 0$, we can very well write

(4.5.10)   ord $(f(x)g(x)) \geq$ ord $f(x) +$ ord $g(x)$

with equality if R is an integral domain.
It can happen that $f(x)g(x) = z$ (4.5.3).

In the same manner, one notes that if $f(x)$ and $g(x)$ are nonzero formal power series,

(4.5.11)   ord $(f(x) + g(x)) \geq \min\{$ord $f(x),$ ord $g(x)\}$

Or, it can happen that $f(x) + g(x) = z$ (4.5.3). We also deduce that

(4.5.12)   if R is an integral domain, so is $R[[x]]$.

Lemma 4.5.1 : *Let R be a commutative ring with 1_R. A formal power series $f(x) = \sum_{n=0}^{\infty} a_n x^n$ is a unit in $R[[x]]$ if, and only if, the constant term a_0 is invertible in R.*

Proof : $:\Rightarrow$ Let j denote the formal power series $(1_R, 0_R, 0_R, \cdots)$ (4.5.5).
If $f(x)g(x) = j$ where $g(x) = \sum b_n x^n$, then $a_0 b_0 = 1_R$ and so a_0 is invertible in R.
$\Leftarrow$: Conversely, suppose that a_0 is invertible in R. We proceed inductively to define the coefficients of a power series $\sum b_n x^n$ in $R[[x]]$, given $f(x) = \sum a_n x^n$. Take $b_0 = a_0^{-1}$. Assuming that $b_1, b_2, \cdots, b_{n-1}$ are already obtained, let

$$b_n = -a_0^{-1}(a_1 b_{n-1} + a_2 b_{n-2} + \cdots + a_n b_0)$$

Then, $a_0 b_0 = 1_R$ and for $n \geq 1$

$$c_n = \sum_{i=0}^{n} a_i b_{n-i} = a_0 b_n + a_1 b_{n-1} + \cdots + a_n b_0 = 0. \text{ It follows that}$$

$$\sum a_n x^n \sum b_n x^n = j = (1_R, 0_R, 0_R, \cdots)$$

and so $\sum a_n x^n$ has an inverse in $R[[x]]$. $\square$

Remark 4.5.1 : If F is a field, $F[[x]]$ is such that $f(x) = \sum a_n x^n \in F[[x]]$ has an inverse in $F[[x]]$ if, and only if, $a_0 \neq 0_F$.

Theorem 25 : *Given a field F, $F[[x]]$ is a PID. In fact, the nontrivial ideals of $F[[x]]$ are of the form (x^t), $t \in \mathbb{N}$.*

Proof : Let I be a proper ideal of $F[[x]]$. Either $I = (z)$ in which case it is the zero ideal or else, I has nonzero elements. Let $f(x) \in I$ with minimal order, say k. Then

(4.5.13) $f(x) = a_k x^k + a_{k+1} x^{k+1} + \cdots = x^k (a_k + a_{k+1} x + \cdots), \quad a_k \neq 0.$

So, the series

$$a_k + a_{k+1}x + \cdots \text{ is a unit in } F[[x]], \text{ by lemma 4.5.1}$$

So, $f(x) = x^k g(x)$ where $g(x)$ is a unit in $F[[x]]$.
Then, $x^k = f(x)g^{-1}(x) \in I$.
So the ideal (x^k) generated by x^k is contained in I.

On the other hand, let $h(x)$ be a nonzero power series such that $h(x) \in I$ and $h(x)$ is of order m. Since $f(x)$ is assumed to have the least order k, $k \leq m$. Thus, $h(x)$ can be written as

$$h(x) = x^k(b_m x^{m-k} + b_{m+1}x^{m-k+1} + \cdots) \in (x^k)$$

So, $I \subseteq (x^k)$ and thus $I = (x^k)$. □

Definition 4.5.3 : *A commutative ring R with unity 1_R is said to be a quasi-local ring, if it has a unique maximal ideal.*

On the basis of definition 4.5.3, we note that $F[[x]]$ is a quasi-local ring as (x) is its unique maximal ideal. We also note that any element $f(x) \in F[[x]]$ can be written as $f(x) = x^k g(x)$ where $g(x)$ is a unit in $F[[x]]$ and $k \geq 0$.

If we consider $\mathbb{C}[[x]]$, there is a one-one correspondence between $f(x) = \sum a_n x^n$ and the arithmetic function

$$(4.5.14) \qquad f = (a_0, a_1, a_2, \cdots)$$

when the domain of definition of f is taken as $\tilde{\mathbb{Z}}$. The set $\mathcal{A}'$ of arithmetic functions f (4.5.14) forms a commutative ring under the operations of ordinary addition and Cauchy multiplication given by

$$(f * g)(r) = \sum_{i=0}^{r} f(i)g(r-i).$$

As $\mathbb{C}[[x]]$ is a UFD, we note that $\mathcal{A}'$ is also a UFD. But Dirichlet multiplication of arithmetic functions (see definition 4.3.1) is not that easy to handle. We need a formal power series in countably many variables.

Let $\omega = \{x_n : n \in \mathbb{N}\}$ be a countably infinite set of indeterminates. We write

$$(4.5.15) \qquad \mathbb{C}_\omega = \mathbb{C}[[x_1, x_2, x_3, \cdots]]$$

to denote the ring of formal power series in the indeterminates x_n, $n \in \mathbb{N}$.

We connect the set $\mathcal{A}$ of arithmetic functions having domain $\mathbb{N}$ with $\mathbb{C}_\omega$. Our aim is to show that $(\mathcal{A}, +, \cdot) \cong \mathbb{C}_\omega$ and $\mathbb{C}_\omega$ is going to be a UFD.

Definition 4.5.4 : *Let $r = p_1^{a_1} p_2^{a_2} \cdots p_k^{a_k}$, $a_i \geq 1$, $i = 1, 2, \cdots, k$ be the prime factorization of r with $p_1 < p_2 < \cdots < p_k$. Given $f \in \mathcal{A}$, we define a formal power series relating to f by*

$$(4.5.16) \qquad P(f) = \sum_{a_1, a_2, \cdots} f(r)x_1^{a_1} x_2^{a_2} \cdots$$

where the summation extends over all r of the form $p_1^{a_1} p_2^{a_2} \cdots p_k^{a_k}$.

$P(f)$ is a formal power series in a countably infinite number of indetermi-nates $x_1, x_2, \cdots$ and having coefficients in $\mathbb{C}$, the field of complex numbers. This gives a one-one correspondence between the set $\mathcal{A}$ of arithmetic functions and $\mathbb{C}_\omega$ (4.5.15). Actually, only a finite number of x_i will occur in any term of $P(f)$. But, infinitely many x_i may also occur in the terms with non zero coefficients of terms in $P(f)$.

We examine $\mathcal{A}$ more closely. For $f, g \in \mathcal{A}$, suppose that there exists $h \in \mathcal{A}$ such that $f = g \cdot h$ (Dirichlet convolution of g and h). We say that g divides f and we write $g \mid f$ to express this fact. Two elements $f, g \in \mathcal{A}$ are called associates written $f \sim g$, if, and only if, $f \mid g$ and $g \mid f$. Further, $\sim$ is an equivalence relation on $\mathcal{A}$ partitioning the set of arithmetic functions into mutually disjoint classes of associates. The class $[0]$ contains only z (4.5.3). The class $[u]$ of units is the group of units in $\mathcal{A}$. $[u]$ consists of all the arithmetic functions f for which $N(f) = 1$. Primes in $\mathbb{N}$ are called rational primes.

Lemma 4.5.2 : *An arithmetic function f for which $N(f) = p$, a rational prime, is an irreducible in $\mathcal{A}$.*

Proof : Since the norm is multiplicative, if $\pi \in \mathcal{A}$ is such that $N(\pi) = p$, $\pi = \epsilon_1 \cdot \epsilon_2$ will imply that either ϵ_1 or ϵ_2 is a unit. So π is an irreducible in $\mathcal{A}$. $\qquad\square$

Definition 4.5.5 : *$f \in \mathcal{A}$ is called a composite function, if f is not an irreducible in $\mathcal{A}$.*

Lemma 4.5.3 : *Every composite function $f \in \mathcal{A}$ can be written as a finite product of irreducibles.*

Proof : Suppose that $f \neq z$ is given. If $f_1 \mid f$, we can write $f = f_1 \cdot g_1$ where g_1 is not a unit. Then, we call f_1 a proper divisor of f. So every composite function $f \in \mathcal{A}$ can be written as $f = g \cdot h$ where g and h properly divide f. As g, h are not units, $N(g) > 1$, $N(h) > 1$. Also,

$$N(f) = N(g)N(h)$$

where $N(g)$ and $N(h)$ divide $N(f)$ properly. If $N(f) = p$ a prime in $\mathbb{N}$, f would be an irreducible by lemma 4.5.2. So, $N(g) < N(f)$ and $N(h) < N(f)$ with $N(g) \neq 1$, $N(h) \neq 1$. Therefore, if g and h are not irreducibles they could be split further. Also, $N(f)$ has only a finite number of prime factors which are rational primes. So, every chain of proper divisors of f terminates at an irreducible element of $\mathcal{A}$. Therefore, if $f \neq z$ and $f \notin U(\mathcal{A})$, where $U(\mathcal{A})$ is the group of units of $\mathcal{A}$, f has an irreducible factor. Thus, a composite f can be expressed as a finite product of irreducibles. $\qquad\square$

Next, suppose that the uniqueness of factorization of an arithmetic function into irreducibles is false. We will divide the set of nonzero non-units of $\mathcal{A}$ into two mutually disjoint subsets:

(i) the subset E of normal elements whose factorization into irreducibles is unique,

(ii) the subset E' of abnormal elements whose factorization is such that there are essentially two different ways of factorization of an abnormal element into irreducibles.

$E \neq \emptyset$, as irreducible elements of $\mathcal{A}$ belong to E. What about E'?
We have

Theorem 26 : *If f is an abnormal element of minimum norm $N(f)$, then f can be expressed as*

$$f = g_1 \cdot g_2 = h_1 \cdot h_2$$

where g_1, g_2; $h_1 h_2$ are distinct irreducibles having the same norm say N.

Proof : If f is an abnormal element, suppose that

$$f = g_1 \cdot g_2 \cdots g_m = h_1 \cdot h_2 \cdots h_n$$

where $g_1 \cdot g_2 \cdots g_m$ and $h_1 \cdot h_2 \cdots h_n$ are essentially two decompositions of f into irreducibles. Now, $m \neq 1$, $n \neq 1$, as an irreducible is a normal element. No g_j is an associate of an h_l. For, if so, cancellation will produce an element of norm $< N(f)$. Without loss of generality, let us assume that

$$N(g_1) \leq N(g_2) \leq \cdots \leq N(g_m)$$
$$N(h_1) \leq N(h_2) \leq \cdots \leq N(h_m)$$

and $N(g_1) < N(h_1)$.
Then, $N(g_1 \cdot h_1) = N(g_1)N(h_1) \leq N(h_1)N(h_2) = N(h_1 \cdot h_2) \leq N(f)$.
If any of $\leq$ is a strict inequality, we will have

(4.5.17) $N(g_1 \cdot h_1) < N(f)$.

We claim that (4.5.17) leads to a contradiction.
For, let

$$y = f - g_1 \cdot h_1,$$

$y \neq z$ as $g_2 \cdot g_3 \cdots g_m = h_1$ is false. Also, $y \notin U(\mathcal{A})$, since $g_1 | y$. From the definition of norm and the assumption (4.5.17) it follows that $N(y) < N(f)$. This implies that y is normal, since $N(f)$ is minimal. g_1 and h_1 are not associates. They divide y. So,

$$g_1 \cdot h_1 | f = g_1 \cdot g_2 \cdots g_m$$
$$f = g_1 \cdot h_1.s_1 \text{ (say)}$$

or $g_2 \cdot g_3 \cdot g_m = h_1 \cdot s_1$. Now, $N(g_2 \cdot g_3 \cdots g_m) < N(f)$.
So, $g_2 \cdot g_3 \cdots g_m = h_1 \cdot s_1$ is normal by the minimality of $N(f)$. So, h_1 is an associate of some g_j $(j = 2, 3, \cdots m)$—a contradiction. So, we are forced to arrive at $N(g_1 \cdot h_1) \geq N(f)$. So, together with $N(g_1 \cdot h_1) \leq N(f)$, we get

(4.5.18) $N(f) = N(g_1)N(h_1)$.

From $N(g_1 \cdot h_1) \leq N(h_1)N(h_2)$, as we have $N(g_1)N(h_1) \leq N(h_1)N(h_2)$ or, $N(g_1) \leq N(h_2)$. From $N(g_1 \cdot h_1) \geq N(f) \geq N(h_1 \cdot h_2)$ we get $N(g_1)N(h_1) \geq N(h_1)N(h_2)$ or $N(g_1) \geq N(h_2)$. So, $N(g_1) = N(h_2)$.
From (4.5.18) we get $N(f) = N(h_1)N(h_2)$. As $N(g_1) < N(h_1) \leq N(h_2)$ and as

$N(g_1) = N(h_2)$ we should have $N(g_1) = N(h_1) = N(h_2)$.
So, if $N(g_1) = N(h_1) = N(h_2) = N$, we get

$$N^2 = N(f) = N(h_1)N(h_2) = N(g_1)N(g_2)\cdots N(g_m) \geq N^m.$$

As $m > 1$, $m = 2$. So, $N(g_2) = N$. Hence, if the unique factorization property fails in $\mathcal{A}$, we should have an element of the form $g_1 \cdot g_2 = h_1 \cdot h_2$ where g_1, g_2, h_1, h_2 are irreducible elements of identical norm N. $\qquad\square$

Lemma 4.5.4 : *The integral domain $\mathcal{A}$ of arithmetic functions is isomorphic to* $\mathbb{C}_\omega$.

Proof : $\mathbb{C}_\omega$ is the ring of formal power series in countably infinite indeterminates $x_1, x_2, \cdots$, (4.5.15). For $f \in \mathcal{A}$, the associated formal series $P(f)$ is as given in (4.5.16). The correspondence $f \mapsto P(f)$ preserves addition. For, if $f, g \in \mathcal{A}$

$$P(f+g) = \sum_{a_1, a_2, \cdots} (f(r) + g(r))x_1^{a_1} x_2^{a_2} \cdots$$
$$= \sum_{a_1, a_2, \cdots} f(r)x_1^{a_1} x_2^{a_2} \cdots + \sum_{a_1, a_2, \cdots} g(r)x_1^{a_1} x_2^{a_2} \cdots$$

or,

(4.5.19) $P(f+g) = P(f) + P(g)$

Dirichlet convolution (4.3.1) of f and g corresponds to multiplication of $P(f)$ and $P(g)$ where 'like terms' are collected and arranged as a formal power series. For,

$$P(f)P(g) = \left(\sum_{a_1, a_2, \cdots} f(r)x_1^{a_1} x_2^{a_2} \cdots \right)\left(\sum_{a_1, a_2, \cdots} g(r)x_1^{a_1} x_2^{a_2} \cdots \right).$$

If $t = p_1^{\delta_1} p_2^{\delta_2} \cdots p_k^{\delta_k}$, $\dfrac{r}{t} = p_1^{a_1-\delta_1} p_2^{a_2-\delta_2} \cdots p_k^{a_k-\delta_k}$.

$$f(t)g(\tfrac{r}{t})(x_1^{\delta_1} x_2^{\delta_2} \cdots x_k^{\delta_k})(x_1^{a_1-\delta_1} x_2^{a_2-\delta_2} \cdots x_k^{a_k-\delta_k}) = f(t)g(\tfrac{r}{t})x_1^{a_1} x_2^{a_2} \cdots x_k^{a_k}$$

So, like terms add up to $\left(\displaystyle\sum_{t|r} f(t)g(\tfrac{r}{t})\right) x_1^{a_1} x_2^{a_2} \cdots x_k^{a_k}$.

We have

(4.5.20) $P(f)P(g) = \displaystyle\sum_{a_1, a_2, \cdots} h(r)x_1^{a_1} x_2^{a_2} \cdots$, where $h(r) = \displaystyle\sum_{t|r} f(t)g(\tfrac{r}{t})$.

Thus, $\mathcal{A} \cong \mathbb{C}_\omega$. $\qquad\square$

Definition 4.5.6 : $\mathbb{C}_\ell = \mathbb{C}[[x_1, x_2, \cdots, x_\ell]]$ *is the ring of formal power series in ℓ indeterminates $x_1, x_2, \cdots, x_\ell$.*

By Krull's theorem [7], $\mathbb{C}_\ell$ is a UFD. The units of $\mathbb{C}_\ell$ are formal power series with nonzero constant terms.

We use the notation

(4.5.21) $P(f)_\ell = \mathbb{C}[[x_1, x_2, x_3, \cdots, x_\ell, 0, 0, 0, \cdots]]$

$P(f) \mapsto P(f)_\ell$ gives a ring homomorphism of $\mathbb{C}_\omega$ or $\mathbb{C}_m$ onto $\mathbb{C}_\ell$ when $m \geq \ell$. We also note that

$$(P(f)P(g))_\ell = P(f)_\ell P(g)_\ell.$$

Let $\mathbb{C}[[0, 0, 0, \cdots]] = 0$. If a series $P(f)$ in $\mathbb{C}_\omega$ is neither zero nor a unit, there exists a minimal $L = L(P(f))$ for which $P(f)_\ell$ is neither zero nor a unit of $\mathbb{C}_\ell$ for $\ell \geq L$. Since $P(f) \neq 0$, $P(f)$ must contain a nonzero term containing $x_1^{a_1} x_2^{a_2} \cdots$ with $(a_1, a_2, \cdots) \neq (0, 0, 0, \cdots)$. There is a minimal $L = L(P(f))$ with $P(f)_L \neq 0, L \geq 1$. But then, $P(f)_\ell$ is a nonzero non-unit for any $\ell \geq L$.

We come to the crucial point. If $P(f)$ is neither zero nor a unit in $\mathbb{C}_\omega$, then, for any $P(f)_\ell$ irreducible in $\mathbb{C}_\ell$ where $\ell \geq L$, $P(f)_m$ is an irreducible in $\mathbb{C}_m$ for all $m \geq \ell$ implies that $P(f)$ is an irreducible in $\mathbb{C}_\omega$. For such a $P(f)$, there is a minimal $q = q(P(f)) \geq L(P(f))$ such that $P(f)_\ell$ is an irreducible in $\mathbb{C}_\ell$ for all $\ell \geq q(P(f))$. We say that such irreducibles in $\mathbb{C}_\omega$ are 'finitely irreducible'. The other possibility is that for some $P(f)$ nonzero, non-unit, one has $P(f)_\ell$ as a composite power series in $\mathbb{C}_\ell$ for all $\ell \geq L(P(f))$. Next, we need a lemma called the principal lemma. See [2].

Lemma 4.5.5 (PRINCIPAL LEMMA) : *All irreducibles in $\mathbb{C}_\omega$ are finitely irreducible.*

Proof : Let $P(f)$ be a fixed nonzero non-unit in $\mathbb{C}_\omega$ with $L = L(P(f))$. $P(f)_\ell$ is a nonzero non-unit for any $\ell \geq L$. Let $P(f)_\ell = P(h_1)_\ell P(h_2)_\ell$ where $P(h_1)_\ell$ and $P(h_2)_\ell$ are non-units in $\mathbb{C}_\ell$. We say that $P(h_1)_\ell$ and $P(h_2)_\ell$ are true factors of $P(f)_\ell$ and $P(h_1)_\ell P(h_2)_\ell$ is a true factorization of $P(f)_\ell$. A true factor of $P(f)_\ell$ is thus a non-unit proper divisor of $P(f)_\ell$ in $\mathbb{C}_\ell$ and so has a companion of the same kind.

We call any chain $[P(h_1)_L, P(h_1)_{L+1}, \cdots, P(h_1)_M]$ of true factors of the corresponding $P(f)_\ell, \ell = L, L+1, \cdots, M$ telescopic if each

$$P(h_1)_{\ell-1} \in \mathbb{C}[[x_1, x_2, \cdots, x_{\ell-1}, 0, \cdots]]$$

induces a true factorization of

$$P(f)_{m-1} = (P(f)_m)_{m-1} = (P(h_1)_m)_{m-1}(P(h_2)_m)_{m-1}$$
$$\equiv P(h_1)_{m-1} P(h_2)_{m-1}$$

and so down to $(P(f))_L = P(h_1)_L P(h_2)_L$, where the chain of true factors is telescopic (gradual reduction of number of indeterminates). In the above notation $(P(f)_m)_{m-1}$ means $\mathbb{C}[[x_1, x_2, \cdots x_{m-1}, 0, 0, \cdots]]$ considered as isomorphic to a subring (containing $m-1$ indeterminates) of $\mathbb{C}[[x_1, x_2, \cdots, x_{m-1}, x_m, 0, 0, \cdots]]$.

The assumption on $P(f)$ is as follows:
We assume the existence of sequence $K_0, K_1, K_2, \cdots$ defined via the true factors

$P(h_1)_{i,j}; j = 0, 1, \cdots, i$ of $(P(f))_{L+j}$.

$$K_0 = [P(h_1)_{0,0}]$$
$$K_1 = [P(h_1)_{1,0}, P(h_1)_{1,1}]$$
$$K_2 = [P(h_1)_{2,0}, P(h_1)_{2,1}, P(h_1)_{2,2}]$$
$$\cdots\cdots\cdots\cdots\cdots$$

We want to prove the existence of an infinite chain of true factors

(4.5.22) $K^* = [P(h_1)_0^*, P(h_1)_1^*, P(h_1)_2^*, \cdots]$

where $P(h_1)_i^* = [P(h_1)_{i,0}, P(h_1)_{i,1}, \cdots, P(h_1)_{i,i}]$ which is telescopic throughout. If this is achieved, we would have

$$(P(f))_{L+j} = P(h_1)_j^* P(h_2)_j^*.$$

Clearly, the chain $[P(h_1)_0^*, P(h_1)_1^*, P(h_1)_2^*, \cdots]$ is also telescopic since

$$(P(h_1)_{j-1}^* P(h_2)_j^*)_{L+j-1} = (P(h_1)_j^*)_{L+j-1} (P(h_2)_j^*)_{L+j-1}$$
$$= P(h_1)_{j-1}^* (P(h_2)_j^*)_{L+j-1}$$

But, any infinite telescopic chain defines unambiguously a series belonging to $\mathbb{C}_\omega$. If $P(h_1)^*$ and $P(h_2)^*$ are the non-unit series defined by $P(h_1)_j^*$ and $P(h_2)_j^*$ chains, we will have

$$P(f) = P(h_1)^* P(h_2)^*,$$

since we can prove the left and right coefficients of any term by considering

$$(P(f))_{L+j} = P(h_1)_j^* P(h_2)_j^*   \text{for suitable } j.$$

and then we are done with the principal lemma.

Since unique factorization holds in $\mathbb{C}_\ell$ there are only a finite number of classes of associates into which the true factors of any $P(f)_\ell$ can fall. Hence, by the pigeon-hole principle, an infinite set of the chains K_i are such that they have their first entry equivalent to some one true factor $P(h_1)_0$ of $P(f)_L$. Choose one of these and call it K_0'. Belonging to this infinite set, there is an infinite subset of K_i whose second entry is equal to some one true factor $P(h)_1$ of $((P(f))_{L+1}$. Choose one and call it K_1'.

Continuing in this way, we are led to a subsequence of telescopic chains

$$K_0' = [P(h_1)_{0,0}', \cdots,]$$
$$K_1' = [P(h_1)_{1,0}', P(h_1)_{1,1}', \cdots,]$$
$$K_2' = [P(h_1)_{2,0}', P(h_1)_{2,1}', P(h_1)_{2,2}', \cdots,]$$

each of which extends at least to the main diagonal such that the entries of this diagonal and below have the property that for each $j = 0, 1, 2, \cdots$, $P(h_1)_{i,j}' \sim P(h_1)_j$ for all $i \geq j$.

We can now construct the telescopic infinite chain K^* working only with the main diagonal next below it as follows:

Define $P(h_1)_0^* = P(h_1)_{0,0}'$.

Since $P(h_1)_{1,0}' \sim P(h_1)_0 \sim P(h_1)_0^*$ in $\mathbb{C}_\ell$, there is a unit U_L of $\mathbb{C}_\ell$ such that

$$P(h_1)_0^* = P(h_1)_{1,0}' U_L = (P(h_1)_{1,1}' U_L)_L.$$

Define $P(h_1)_1^* = P(h_1)_{1,1}' U_L$ in $\mathbb{C}_{L+1}$. We note that $P(h_1)_1^*$ is a true factor of $P(f)_{L+1}$,

$$P(h_1)_1^* = P(h_1)_0^* \quad \text{and}$$
$$P(h_1)_1^* \sim P(h_1)_1 \text{ in } \mathbb{C}_{L+i}.$$

To make the process clear and to avoid a formal induction we carry out the construction through one more step.

Since $P(h_1)_{2,1}' \sim P(h_1)_1 \sim P(h_1)_1^*$ in $\mathbb{C}_{L+1}$, there is a unit U_{L+1} of $\mathbb{C}_{L+1}$ such that

$$P(h_1)^* = P(h_1)_{2,1}' U_{L+1} = (P(h_1)_{2,2}' U_{L+1})_{L+1}$$

We define $P(h_1)_2^* = P(h_1)_{2,2}' U_{L+1}$ in $\mathbb{C}_{L+2}$ and we observe that $P(h_1)_2^*$ is a true factor of $P(f)_{L+2}$. Also $P(h_1)_2^* \sim P(h_1)_2$ in $\mathbb{C}_{L+2}$. $P(f)$ reducible in $\mathbb{C}_\ell$ for $\ell \geq L$ implies that $P(f)$ is reducible in $\mathbb{C}_\omega$. This completes the proof of the principal lemma. $\square$

Theorem 27 (Cashwell and Everett (1959)) : $\mathbb{C}_\omega$ *is a UFD.*

Proof : We have seen that by lemma 4.5.4, $\mathcal{A} \cong \mathbb{C}_\omega$. Suppose that unique factorization into irreducibles fails in $\mathcal{A}$. Then, it fails in $\mathbb{C}_\omega$. By theorem 26 an abnormal element q has two factorizations $f \cdot g$ and $h \cdot k$ where f, g, h and k have the same norm. Accordingly, we have a formal power series $P(q)$ in $\mathbb{C}_\omega$ of the form

$$P(q) = P(f)P(g) = P(h)P(k)$$

where $P(f)$, $P(g)$, $P(h)$, $P(k)$ are irreducibles in $\mathbb{C}_\omega$ and $P(f)$ is not an associate of $P(h)$ or $P(k)$. Since all irreducibles in $\mathbb{C}_\omega$ are finitely irreducible, there exists an integer t such that in the equations

$$(P(f)P(g))_\ell = P(f)_\ell P(g)_\ell = (P(h)P(k))_\ell = P(h)_\ell P(k)_\ell;$$

$P(f)_\ell$, $P(g)_\ell$, $P(h)_\ell$, $P(k)_\ell$ are distinct irreducibles in $\mathbb{C}_\ell$ for each $\ell \geq t$. Since factorization in each $\mathbb{C}_\ell$ is unique, $P(f)_\ell$ must be an associate of $P(h)_\ell$ or $P(k)_\ell$ in $\mathbb{C}_\ell$ for each $\ell \geq t$. Hence, there must exist an infinite increasing subsequence $\sigma = \{m\}$ of $\mathbb{N}$ $(m \geq t)$ such that

$$P(f)_m \text{ is an associate of } P(h)_m \text{ in } \mathbb{C}_m \text{ or}$$

$$P(f)_m \text{ is an associate of } P(k)_m \text{ in } \mathbb{C}_m,$$

for all $m \in \sigma$.

We shall use the notation $\sim$ to mean 'an associate of', in the context of $\mathbb{C}_m$ or $\mathbb{C}_\omega$. To fix ideas, we take

$$P(f)_m \sim P(h)_m.$$

Then,

$$P(f)_m = U_m P(h)_m$$

where U_m is a unit of $\mathbb{C}_m$ for each $m \in \sigma$. If m, n are chosen from σ such that $m < n$,

$$U_m P(h)_m = P(f)_m = (P(f)_n)_m,$$

where $(P(f)_n)_m$ is obtained from $P(f)_n$ by substituting

$$x_{m+1} = x_{m+2} = \cdots = x_n = 0.$$

However,

$$(P(f)_n)_m = (U_n)_m (P(h)_n)_m = (U_n)_m P(h)_m,$$

where U_n is an extension of U_m by terms each of which involves the indeterminate x_i with $i > m$ and so does not occur in U_m. Thus, $\{U_m\}_{m \in \sigma}$ defines a unit U of $\mathbb{C}_\omega$ and by an argument used to show that irreducibles in $\mathbb{C}_\omega$ are 'finitely irreducible', we arrive at $P(f) = U P(h)$. Thus, $P(f) \sim P(h)$ in $\mathbb{C}_\omega$. This leads to a contradiction. Hence, the factorization of an element into irreducibles in $\mathbb{C}_\omega$ exists and is unique.

$\square$

Corollary 4.5.1 : $(\mathcal{A}, +, \cdot)$ *is a UFD.*

This follows from the fact that $\mathcal{A} \cong \mathbb{C}_\omega$ which is a UFD by theorem 27.

Remark 4.5.2 : The ring $\mathcal{A}$ also forms an algebra over $\mathbb{C}$. It is called the 'Dirichlet algebra' of arithmetic functions, as multiplication is Dirichlet convolution.

4.6. Finite fields and irreducible polynomials

It is known that the characteristic of a field is either 0 or a prime p. Given a field K, if char $K = 0$, K has a subfield isomorphic to $\mathbb{Q}$, the field of rational numbers. If char $K = p$ (a prime), K has a subfield isomorphic to $\mathbb{Z}/p\mathbb{Z}$ denoted by $\mathbb{F}_p$. If K is a finite field, K has p^n elements for some prime p and $n \in \mathbb{N}$. Further, *char* $K = p$. K can be considered as an extension of $\mathbb{F}_p$ of degree n. We write $n = [K : \mathbb{F}_p]$. Further, every element of K with p^n elements is the zero of a polynomial $f(x) = x^{p^n} - x \in \mathbb{F}_p[x]$. We express this by saying that K is the splitting field of $f(x) = x^{p^n} - x \in \mathbb{F}_p[x]$. Since any two splitting fields of a given nonconstant polynomial are isomorphic, any two finite fields having the same number of elements are isomorphic.

Now, $x^p - x$ has all its zeros in $\mathbb{F}_p[x]$. So, every nonzero element of $\mathbb{F}_p$ is a zero of $x^{p-1} - 1 \in \mathbb{F}_p[x]$. Therefore,

$$x^{p-1} - 1 \equiv (x-1)(x-2) \cdots (x-(p-1)) \pmod{p}$$

Putting $x = \alpha \equiv 0 \pmod{p}$, we obtain

$$-1 \equiv (-1)^{p-1} (p-1)! \pmod{p}.$$

So, for $p \geq 2, (p-1)! + 1 \equiv 0 \pmod{p}$ which is Wilson's theorem. (See corollary 1.2.1, chapter 1).

Lemma 4.6.1 : *Given a prime p, let $\mathbb{F}_d(x)$ denote the product of monic irreducible polynomials of degree d over $\mathbb{F}_p$. Then,*

$$(4.6.1) \qquad\qquad x^{p^n} - x = \prod_{d \mid n} F_d(x).$$

Proof : $t(x)$ denotes a factor of $x^{p^n} - x$. We claim that $t(x)$ can occur only to the first power. That is, if $t(x) \mid (x^{p^n} - x)$, $t^2 \nmid (x^{p^n} - x)$. For, if $x^{p^n} - x = t^2(x)s(x)$, by formal differentiation, we obtain

$$p^n(x^{p^n-1}) - 1 = 2t(x)t'(x)s(x) + t^2(x)s'(x).$$

As char $\mathbb{F}_p = p$, we get

$$-1 = 2t(x)t'(x)s(x) + t^2(x)s'(x).$$

This implies that $t(x)$ divides -1 which is not correct.

Next, we show that if $t(x)$ is a monic irreducible polynomial of degree d, then

$$t(x) \mid x^{p^n} - x \Leftrightarrow d \mid n.$$

Let α be a zero of $t(x)$. Adjoining α to $\mathbb{F}_p$ we obtain an extension $\mathbb{F}_p(\alpha)$ of degree d over $\mathbb{F}_p$. That is, the extension $\mathbb{F}_p(\alpha)$ is such that $[\mathbb{F}_p(\alpha) : \mathbb{F}_p] = d$. Each element of $\mathbb{F}_p(\alpha)$ is a zero of

$$(4.6.2) \qquad\qquad x^{p^n} - x = t(x)s(x)$$

As α is a zero of $t(x)$, from (4.6.2), $\alpha^{p^n} - \alpha = 0$. Moreover, $\mathbb{F}_p(\alpha)$ is a vector space (over $\mathbb{F}_p$) of dimension d and has a basis $\{1, \alpha, \alpha^2, \cdots, \alpha^{d-1}\}$. $v \in \mathbb{F}_p(\alpha)$ can be written as

$$(4.6.3) \qquad v = b_1\alpha^{d-1} + b_2\alpha^{d-2} + \cdots + b_{d-1}\alpha + b_d, \ b_i \in \mathbb{F}_p \quad (i = 1, 2, \cdots d).$$

Then, as $(a+b)^{p^n} = a^{p^n} + b^{p^n}$ for $a, b \in \mathbb{F}_p(\alpha)$ $(n \geq 1)$

$$(b_1\alpha^{d-1} + b_2\alpha^{d-2} + \cdots + b_d)^{p^n} = b_1(\alpha^{p^n})^{d-1} + \cdots + b_d$$
$$= b_1\alpha^{d-1} + \cdots + b_d$$

or $v^{p^n} = v$ by (4.6.3), for every $v \in \mathbb{F}_p(\alpha)$. Therefore, the elements of $\mathbb{F}_p(\alpha)$ satisfy $x^{p^n} - x = 0$. Since the elements of $\mathbb{F}_p(\alpha)$ also satisfy $x^{p^d} - x = 0$, we see that

$$x^{p^d} - x \mid x^{p^n} - x$$

or $x(x^{p^d-1} - 1)$ divides $x(x^{p^n-1} - 1)$.
Further, for $a \in \mathbb{N}$, $a^\alpha - 1$ divides $a^\beta - 1$ if, and only if, $\alpha \mid \beta$.
So, $x^{p^d-1} - 1$ divides $x^{p^n-1} - 1 \Leftrightarrow p^d - 1 \mid p^n - 1$. This, in turn, implies $d \mid n$.
Conversely, suppose that $d \mid n$.
Then $x^{p^d} - x \mid x^{p^n} - x$ and so $t(x) \mid x^{p^n} - x \Leftrightarrow d \mid n$. For each d, a divisor of n, there is at least one monic irreducible polynomial dividing $x^{p^n} - x$. Let $d_1 = 1, d_2, \cdots, d_t = n$ be the divisors of n. By the definition of $F_d(x)$, $d \mid n$,

$$x^{p^n} - x = F_{d_1}(x)F_{d_2}(x) \cdots F_{d_t}(x) \text{ which is } (4.6.1).$$

□

Theorem 28 : *If $N(m)$ denotes the number of monic irreducible polynomials of degree m in $\mathbb{F}_p[x]$, then*

(4.6.4) $$N(m) = \frac{1}{m} \sum_{d|m} p^d \mu(\frac{m}{d})$$

where μ is the Möbius function. (See definition 4.3.1).

Proof : By lemma 4.6.1, we have

$$x^{p^n} - x = \prod_{d|n} F_d(x)$$

where $F_d(x)$ denotes the product of monic irreducible polynomials of degree d in $\mathbb{F}_p[x]$. $N(m)$ denotes the number of monic irreducible polynomials of degree m in $\mathbb{F}_p[x]$. Replacing n by m in (4.6.1), we have

(4.6.5) $$x^{p^m} - x = \prod_{d|m} F_d(x).$$

p^m occurs as the highest degree of x on the right side of (4.6.5). So, we get

$$p^m = \sum_{d|m} d\, N(d).$$

Using Möbius inversion (4.3.5), we obtain

$$mN(m) = \sum_{d|m} p^d \mu(\frac{m}{d})$$

from which (4.6.4) follows.   □

Corollary 4.6.1 : *For every integer $m \geq 1$, there exists a monic irreducible polynomial of degree m in $\mathbb{F}_p[x]$.*

Proof : From (4.6.3), we have

(4.6.6) $$N(m) = \frac{1}{m}\{p^m - \cdots + p\mu(m)\}.$$

We note that $\mu(r) = 0$ if r contains a squared factor > 1. So, the terms inside the bracket on the right side of (4.6.6) will be a sum of distinct powers of p with coefficients $+1$ and -1. So, $N(m) \neq 0$.   □

Remark 4.6.1 : The above theorem and corollary are adapted from K. Ireland and M. I. Rosen [5].

Remark 4.6.2 : For each $m \geq 1$, there is a monic irreducible polynomial of degree m with coefficients from $\mathbb{F}_p$. This enables us to state the following analogue of Euclid's theorem on infinitude of primes.

Fact 4.6.1 : The number of monic irreducible polynomials over $\mathbb{F}_p$ is infinite.

For a detailed study of finite fields and their applications, see Lidl and Niederreiter [A2].

4.7. More about irreducible polynomials

Let $p(x)$ be an irreducible polynomial in $F[x]$ where F is a field. If $p(x)$ divides a product $f(x)\,g(x)$ where $f(x)$, $g(x)$ are elements of $F[x]$, $p(x)$ divides either $f(x)$ or $g(x)$. Further, $F[x]$ is a PID. If $f(x) \in \mathbb{Z}[x]$ and

(4.7.1) $\qquad f(x) = a_0 + a_1 x + a_2 x^2 \cdots + a_n x^n, \quad a_n > 0$

$f(x)$ is called a primitive polynomial, if $a_0, a_1, \ldots a_n$ have no common factor other than ± 1.

Fact 4.7.1 : $f(x)$ given by (4.7.1) is irreducible in $\mathbb{Z}[x]$ if, and only if, either
(i) $f(x)$ is a prime number or
(ii) $f(x)$ is a primitive polynomial which is irreducible in $\mathbb{Q}[x]$ ($\mathbb{Q}$, being the field of rational numbers).

Fact 4.7.2 : Let $f(x) = a_0 + a_1 x + \cdots + a_n x^n \in \mathbb{Z}[x]$. Let p be a prime not dividing a_n. If $f(x)$ is reduced modulo p, say $\overline{f}(x)$, and $\overline{f}(x)$ is irreducible over $\mathbb{Z}/p\mathbb{Z}$, then $f(x)$ is irreducible in $\mathbb{Q}[x]$.

Fact 4.7.3 : (Eisenstein criterion) Let $f(x) = a_0 + a_1 x + \cdots + a_n x^n \in \mathbb{Z}[x]$. For a prime p, suppose that
(i) $p^2 \nmid a_0$, (ii) $p | a_i$ ($i = 0, 1, 2, \ldots, (n-1)$) and (iii) $p \nmid a_n$, then $f(x)$ is irreducible in $\mathbb{Q}[x]$. If $f(x)$ is also primitive, $f(x)$ is irreducible in $\mathbb{Z}[x]$.

For proofs of Facts 4.7.1 to 4.7.3, see M. Artin [A1, chapter 11, pp 390–404].

In the case finite fields $\mathbb{F}_q$, where $q = p^m$, p a prime; $m \geq 1$, the elements of $\mathbb{F}_q$ are the zeros of the polynomial $x^q - x$. $x^q - x$ factors into linear factor in $\mathbb{F}_q$, as remarked earlier.

Suppose that E denotes a field of characteristic p, a prime. We consider

$$F = \{\alpha \in E : \alpha \text{ is a zero of } x^q - x, \text{ where } q = p^m; m \geq 1\}.$$

It is verified that $x^q - x$ has no multiple roots in E and F is a subfield of E.

Let $f(x)$ be an irreducible polynomial of degree t in $\mathbb{F}_p[x]$. Suppose that $f(x)$ has a zero α in an extension E of F. We write $E' = \mathbb{F}_p(\alpha)$ where $[E' : \mathbb{F}_p] = t$. Then, $|E'| = p^t$. Elements of E' are the zeros of $x^{q'} - x$ where $q' = p^t$. If $|E| = q = p^m$, ($m \geq 1$), α is a zero of $x^q - x$ also. So, if $f(x)$ is an irreducible polynomial of degree t in $\mathbb{F}_p[x]$, $f(x)$ divides $x^q - x$.

Next, suppose that $f(x)$ is an irreducible polynomial of degree t and $t | m$. Then $f(x)$ is a factor of $x^q - x$. For, as $f(x)$ divides $x^{q'} - x$ where $q' = p^t$, if $m = ts$, then, as $x^{q'} - x$ divides $x^q - x$, we note that an irreducible polynomial $f(x)$ whose degree

t divides m is such that $f(x)$ divides $x^q - x$. However, if $f(x)$ is irreducible and its degree t does not divide m, since $[E : \mathbb{F}_p] = m$, $f(x)$ has no zero in E and so, $f(x)$ is not a factor of $x^q - x$. We arrive at

Fact 4.7.4 : (a) Every irreducible polynomial of degree t in $\mathbb{F}_p[x]$ is a factor of $x^q - x$, for some $q = p^m$ $(m \geq 1)$.

(b) A field $\mathbb{F}_q (q = p^m; m \geq 1)$ contains a subfield $\mathbb{F}_{q'}$ $(q' = p^t; t \geq 1)$ if, and only if, t divides m.

Proofs are omitted.

Remark 4.7.1 : In the case of $\mathbb{F}_8$, having 2^3 elements, $\mathbb{F}_4$ is not a subfield of $\mathbb{F}_8$, as $[\mathbb{F}_8 : \mathbb{F}_2] = 3$, $[\mathbb{F}_4 : \mathbb{F}_2] = 2$ and 2 does not divide 3. But, in the case of $\mathbb{F}_{16}$, $\mathbb{F}_{16}$ contains $\mathbb{F}_4$.

Remark 4.7.2 : If $\mathbb{F}_q$ is a field having $q = p^m$ elements, to obtain an irreducible polynomial of degree t in $\mathbb{F}_q[x]$, we have only to consider an irreducible polynomial of degree t in $\mathbb{F}_p[x]$.

4.8. Notes with illustrative examples

We begin with a UFD say R. K denotes the field of quotients of R. Suppose that $f(x) = a_0 + a_1 x + \cdots + a_n x^n$ is a nonconstant polynomial in $R[x]$. Eisenstein criterion says: For some prime $p \in R$, suppose that $p \nmid a_n, p | a_k (k = 0, 1, \cdots, n - 1)$ and $p^2 \nmid a_0$. Then, $f(x)$ is irreducible in $K[x]$. As an example, one could prove that for $a \neq \pm 1$, a nonzero square-free integer, $x^n + a \in \mathbb{Z}[x]$ is irreducible over $\mathbb{Q}$ for $n \geq 2$. Using the example of $\mathbb{Z}[x]$, one can also show that $\mathbb{Z}[x]$ has a prime ideal which is not a maximal ideal. For, if

$$(x) = \{a_1 x + a_2 x^2 + \cdots + a_n x^n : a_i \in \mathbb{Z}, 1 \leq i \leq n\}$$

(x) is a prime ideal of $\mathbb{Z}[x]$. For $\mathbb{Z}/(x) \cong \mathbb{Z}$ is an integral domain. One also notes that $\mathbb{Z}[x]$ is not PID, though $\mathbb{Z}[x]$ is a UFD. If we consider the ideal generated by 2 and x, $(2, x)$ is the maximal ideal consisting of polynomials with constant terms equal to an even integer. Further, $(x) \subset (2, x)$.

Let F be a field. If $n > 1$, $F[x_1, x_2, \cdots, x_n]$ is neither a PID nor a Euclidean domain. In fact, the ideal (x_1, x_2) is not a principal ideal.

In the case of the ring $\mathcal{A}$ of arithmetic functions, we have thrown the problem of uniqueness of factorization into the ring $\mathbb{C}_\omega$ of formal power series in countably infinite indeterminates $x_1, x_2, \cdots$. $\mathbb{C}_\omega$ is shown to be a UFD by an ingenious method and so $\mathcal{A}$ is shown to be a UFD. As $\mathcal{A}$ satisfies ACCP (ascending chain condition on principal ideals), it suffices to show that $\mathcal{A}$ is a GCD domain in order to show that it is a UFD. The difficulty arises because of the fact that the linear expressibility of the g.c.d is not to be presumed though g.c.d property holds.

Finite fields are easy to handle. Counting monic irreducible polynomials of degree m in $\mathbb{F}_p[x]$ involves the use of Möbius inversion. This is not the only place in algebra where we use Möbius inversion. This inversion technique is powerful and can be used in many other contexts. Abstract Möbius inversion is an idea due to G. C. Rota and it will be taken up in detail in chapter 9.

As an illustration of theorem 28, we observe that $N(2)$, the number of monic irreducible polynomials of degree 2 over $\mathbb{F}_p$ is given by

$$(4.8.1) \qquad N(2) = \tfrac{1}{2} \sum_{d|2} p^d \mu(\tfrac{2}{d}) = \tfrac{1}{2}(p^2 - p) = \tfrac{p(p-1)}{2}$$

In the same manner,

$$(4.8.2) \qquad N(3) = \tfrac{1}{3} \sum_{d|3} p^d \mu(\tfrac{3}{d}) = \tfrac{1}{3}(p^3 - p) = \tfrac{(p-1)p(p+1)}{3}$$

$$(4.8.3) \qquad N(4) = \tfrac{1}{4} \sum_{d|4} p^d \mu(\tfrac{4}{d}) = \tfrac{1}{4}(p^4 - p^2) = \tfrac{p^2(p-1)(p+1)}{4}$$

In the formula for $N(4)$, we note that $(p-1)p$ and $p(p+1)$ are even numbers and so $p^2(p-1)(p+1)$ is exactly divisible by 4.

4.9. Worked-out examples

a) R denote a commutative ring with unity. J_{n+1} denotes the ideal of $R[x]$, generated by x^{n+1} $(n \geq 0)$. Show that $S_{n+1} = R[x]/J_{n+1}$ is generated by R and λ where $\lambda^{n+1} = 0$. Describe the units in S_{n+1}.

Answer: S_{n+1} is the ring of polynomials in λ of degree $\leq n$, when they are reduced modulo λ^{n+1}. The set $\{1_R, \lambda, \lambda^2, \ldots, \lambda^n\}$ generates S_{n+1}. The units in S_{n+1} are polynomials $u(\lambda), v(\lambda)$ of degree $\leq n$ such that

$$(4.9.1) \qquad u(\lambda)v(\lambda) \equiv 1_R \,(\text{mod } \lambda^{n+1}).$$

$\square$

b) Let $\mathbb{F}_q$ denote a field having $q = p^m$ elements. $(p$, a prime; $m \geq 1)$. Let $F^r = \{x^r : x \in \mathbb{F}_q : r$, a positive integer$\}$. If $s = \text{g.c.d } (r, q-1)$, show that $F^s = \{x^s : x \in \mathbb{F}_q\}$ and F^r are identical sets.

Answer: It is known that $\mathbb{F}_q^* = \backslash \{0\}$ is a cyclic group of order $(q-1)$. Let y be a generator of $\mathbb{F}_q^*$, so that $y^{q-1} = 1$. If $x \in \mathbb{F}_q$, $x = y^t$ for some integer $t \geq 1$. So,

$$x^r = y^{rt}. \text{ Also, } x^s = y^{st} \text{ and } r = (\tfrac{r}{s}s).$$

So, $x^r = y^{(\frac{r}{s})st} = (y^{st})^{\frac{r}{s}} = (y^{\frac{rt}{s}})^s = (x')^s$ for some $x' \in \mathbb{F}_q$. So, $F^r = F^s$ where $s = \text{g.c.d}(r, q-1)$. $\square$

c) (Mowaffaq Hajja) Find all infinite sequences $\vec{c} = (c_0, c_1, c_2, \ldots)$ of integers for which the set

$$I_{\vec{c}} = \{\sum_{i=0}^{n} a_i x^i \in \mathbb{Z}[x] : \quad \sum_{i=0}^{n} a_i c_i = 0\}$$

is an ideal of $\mathbb{Z}[x]$.
Answer: We prove

Proposition 4.9.1 : $I_{\bar{c}}$ is an ideal of $\mathbb{Z}[x]$ if, and only if, there exist integers r, s such that $c_i = r^i s$, for all i.

Proof : $:\Rightarrow$ Given: $I_{\bar{c}}$ is an ideal. If $c_0 = 0$ and $1 \in I_{\bar{c}}$, we must have $1 \cdot x^i \in I_{\bar{c}}$. Then, $c_i = 0$ for all i. (We choose $s = 0$ and r is arbitrary). On the other hand, if $c_0 \neq 0$, then $c_1 - c_0 x \in I_{\bar{c}}$, as $a_0 c_0 + a_1 c_1 = 0$ for $a_1 = -c_0$ and $a_0 = c_1$. This requires that $(c_1 - c_0 x) x^i \in I_{\bar{c}}$. We conclude that

(4.9.2) $c_i c_1 - c_{i+1} c_0 = 0$

Therefore, $c_{i+1} = r c_i$ where $r = \frac{c_1}{c_0}$. This gives

(4.9.3) $c_i = r^i c_0 \quad (i \geq 1)$

r has to be an integer, since otherwise c_i will not be an integer. When i is sufficiently large, we choose $s = c_0$. This shows that the condition $c_i = r^i s$ for all i, is necessary.

$\Leftarrow:$ Suppose we are given that $c_i = r^i s$ for fixed integers r and s. If $c_0 = 0$, then $I_{\bar{c}} = \mathbb{Z}[x]$.
If $c_0 \neq 0$, then $f \in I_{\bar{c}}$ if, and only if, $f(r) = 0$, where $f(x) = \sum_{i=0}^n a_i x^i \in \mathbb{Z}[x]$. Thus, $I_{\bar{c}}$ is the set of polynomials in $\mathbb{Z}[x]$ that vanish at r. Clearly, $I_{\bar{c}}$ is an ideal of $\mathbb{Z}[x]$. To make $I_{\bar{c}}$ an ideal of $\mathbb{Z}[x]$, it is sufficient that $c_i = r^i s$ for some $r, s \in \mathbb{Z}$. $\square$

Remark 4.9.1 : The above example has been adapted from problem 10399 in Amer. Math. Monthly 104 (1997) pp 279–280 for which a composite solution was provided by John H. Lindsey II and Nasha Komanda.

EXERCISES

1. **Mark the following statements true (T) or false (F) justifying your answer briefly.**
   a) Let $F[x]$ be a ring of polynomials with coefficients from a field F. Consider $f(x) \in F[x]$. If $f(x) = 0$ has a nonzero root in F that is twice another root, then all the roots of $f(x) = 0$ are in F.
   b) Let $p, q \in \mathbb{Z}$. Consider $x^2 + 3x - pq \in \mathbb{Z}[x]$. Then, $x^2 + 3x - pq$ is a prime element of $\mathbb{Z}[x]$ for all primes p and q.
   c) Let $f(x) = x^3 - 3x^2 - 2x + 6$ be an element of $\mathbb{R}[x]$ ($\mathbb{R}$, the field of real numbers). Then, $f(x)$ is a product of 3 distinct primes in $\mathbb{R}[x]$.
   d) Let
   $$f(x) = \sum_{j=0}^n a_j x^j, \quad g(x) = \sum_{j=0}^n a_j (x+1)^j \ (a_j \in \mathbb{Q}; \ j=0,1,2,\ldots,n).$$
   $f(x)$ is composite in $\mathbb{Q}[x]$ if, and only if, $g(x)$ is composite in $\mathbb{Q}[x]$.
   e) $f(x) = x^2 + 3x + 2 \in \mathbb{Z}[x]$ as well as $\mathbb{Z}[[x]]$. The assertion is:
   $f(x)$ is reducible in $\mathbb{Z}[x]$, but not in $\mathbb{Z}[[x]]$.

    f) *The formal power series ring* $F[[x]]$ *(F, a field) is such that* $F[[x]]$ *is a PID having the only ideals* (0_F) *and* (x^k), $k \geq 0$.

2. $p(x) = x^2 - 2x - 1$ *is irreducible over* $\mathbb{Q}$, *the field of rational. However, it is reducible for the finite fields* $\mathbb{Z}/3\mathbb{Z}$, $\mathbb{Z}/5\mathbb{Z}$ *and* $\mathbb{Z}/7\mathbb{Z}$. *Is it reducible over* $\mathbb{Z}/p\mathbb{Z}$, *where p is a prime other than* 3, 5 *or* 7 *?*

3. *Let* $\sigma(r)$, $d(r)$, *denote respectively the sum and number of divisors of* $r(> 1)$. *Prove that*

$$\sigma(r) + \phi(r) = r d(r);$$

if, and only if, r is a prime. (ϕ denotes the Euler ϕ-function).

4. *Let* $\mathbb{F}_p$ *denote a finite field of p elements. (p a prime). Let* $X = \mathbb{F}_p \times \mathbb{F}_p$. *Make X a field of* p^2 *elements by defining suitable laws of composition of addition and multiplication in the cartesian product for X.*

5. *Let R be a commutative ring. If I denotes an ideal of R, show that* $I[x]$ *is an ideal of* $R[x]$.

6. *Decompose* $10x^2 + 5x - 5$ *into prime factors in* $\mathbb{Z}[x]$.

7. $1 - x$ *is a unit in* $\mathbb{C}[[x]]$. *Find its inverse.*

8. *Let R be a commutative ring.* $t \in R$ *is called a nilpotent element if there exists an integer* $n \geq 1$ *such that* $t^n = 0_R$.

    (a) If t is a nilpotent element of R, show that $1 - t$ *is a unit in R.*

    (b) Show that the polynomial $1 - tx$ *($t \in R$) is a unit in* $R[x]$ *if, and only if, t is a nilpotent in R.*

    (c) Show that the intersection of all prime ideals of R is the set of nilpotent elements of R.

9. *Let p be a prime. Show that the cyclotomic polynomial*

$$f(x) = x^{p-1} + x^{p-2} + \cdots + x + 1$$

is irreducible in $\mathbb{Z}[x]$.

10. *Let D be an integral domain. Show that a prime ideal P of D can be described as an ideal whose complement is a multiplicatively closed set. (That is, whenever* $x, y \in D \setminus P, xy \in D \setminus P$).

11. *Let R be a commutative ring with unity* 1_R. *An ideal J of R is called a primary ideal if whenever* $ab \in J$ *($a, b \in R$) and* $a \notin J$, *there exists an integer* $n \in \mathbb{N}$ *such that* $b^n \in J$.

    In $\mathbb{Z}$, *the primary ideals are the principal ideals* (p^k) *($k \geq 1$, p a prime). Considering* $F[x, y]$ *(F a field), give an example to show that a primary ideal need not be a power of a prime ideal. [Hint : Let* $I = (x, y)$. *I is a prime ideal of* $F[x, y]$. *Also*

$$I^2 = (x^2, xy, y^2) \subseteq (x^2, y) \subseteq I$$

$J = (x^2, y)$ *is a primary ideal but* $J \neq I^n$ *for* $n \geq 1$.]

12. *Let R be a commutative ring with unity* 1_R. *If R is a quasi-local ring, (i.e., a ring in which there is a unique maximal ideal) show that* $R[[x]]$ *is also a quasi-local ring. Check that the converse need not be true, in general.*

13. *Show that the ideal* $M = (2, x, y)$ *is a maximal ideal in* $\mathbb{Z}[x, y]$.

14. *[R. Sridharan]* $\mathbb{R}$ *denotes the field of real numbers. Analogous to theorems 18 and 19 of chapter 3, if* $I(x,y)$ *denotes the ideal generated by* $x^2 + y^2 + 1$ *in* $\mathbb{R}[x,y]$, *show that* $\mathbb{R}[x,y]/I(x,y)$ *is a PID, but not a Euclidean domain.*

15. *[R. Sridharan] Let* $\mathbb{C}$ *denote the field of complex numbers. If* $J(x,y)$ *denotes the ideal generated by* $x^2 + y^2 + 1$ *in* $\mathbb{C}[x,y]$, *show that*

$$\mathbb{C}[x,y]/J(x,y) \cong \mathbb{C}[t,t^{-1}]$$

and that $\mathbb{C}[t,t^{-1}]$ *is a Euclidean domain. (Note the difference when* $\mathbb{R}$ *is replaced by* $\mathbb{C}$, *in exercise 14).*

16. *Let A be a set and G be a group. If* $\phi : G \to S$ *is a homomorphism of G into the symmetric group S upon A,* ϕ *is called an action of G on A.*

   Let K be a finite field. The set $\mathcal{G}$ *of all automorphisms of K is called the Galois group of K. When* $K = GF(p^n)$, *the Galois group* $\mathcal{G}$ *is cyclic and is of order n.*

   Let $p = 2^m + 1$ $(m \geq 1)$

   (a) Show that the group U of units of $K = GF(2^{2m})$ *contains an element u of order p.*

   (b) Show that $K = \mathbb{F}_2(u)$ *where* $\mathbb{F}_2 = \mathbb{Z}/2\mathbb{Z}$.

   (c) Let $\mathcal{G}$ *be the Galois group of K. Show that* $\mathcal{G}$ *has an action upon the subgroup* U' *of U generated by u and that this action is an isomorphism of* $\mathcal{G}$.

   (d) Show that $2m|(p-1)$ *and that m has to be a power of 2.*
   (Exercise 16 says that if $2^m + 1$ *is a prime, m has to be a power of 2. Then,* $2^m + 1$ *is referred to as a Fermat prime. See section 1.7 Chapter 1. Ref: N H McCoy & T R Berger : Algebra : Groups, rings and other topics: Allyn & Bacon Inc. Boston (1977), Chapter 12 (problem 12) page 468.)*

REFERENCES

[1] D. M. Burton : A first course in rings and fields, Chapter 7 pp 121–122. Addison-Wesley Pub. Co., Reading Mass USA (1970).

[2] E. D. Cashwell and C. J. Everett : The ring of number-theoretic functions, Pacific J. Math 9 (1959) 975–985.

[3] R. Godement : Algebra, Hermann Paris, Houghton Mifflin Co, Boston (1968), Chapter 27 pp 423–431.

[4] Thomas W. Hungerford: Algebra GTM No. 73, Chapter 3 Section 5 pp 149–165, Springer Verlag (1986).

[5] K. Ireland and M. I. Rosen : A classical introduction to Modern Number Theory. GTM No. 84, Springer Verlag NY, 2nd Edn. (1990). Chapter 7, pp 76–83.

[6] M. A. Jodiet : Uniqueness in division algorithm, Amer. Math. Monthly 74 (1967) 835–836.

[7] W. Krull : Beitrage zur Arithmetik kommutative integritātis beireiche III Zum Dimensions begniff der ideal theorie, Math. Zeit 42 (1937) 745–766.

[8] R. Sivaramakrishnan: Classical theory of arithmetic functions. Monographs and Textbooks in Pure and Applied Mathematics No.126, Marcel Dekker Inc. NY (1989), Chapter I, 3–9.

[9] R. Vaidyanathaswamy: The theory of multiplicative arithmetic functions: Trans. Amer. Math. Soc. 33 (1931), 579–662.

ADDITIONAL REFERENCES

[A1] M. Artin: Algebra, Prentice Hall of India (P) Ltd, New Delhi (1994).

[A2] R. Lidl and H. Niederreiter: Introduction to finite fields and their applications; Cambridge University Press (1986).

The Chinese Remainder Theorem and the evaluation of number of solutions of a linear congruence with side conditions

Historical perspective

The premise is a pair of linear congruences having the same modulus r. Let $a_1, a_2; b_1, b_2; c_1, c_2$ be integers. One considers

(A)
$$a_1 x_1 + a_2 x_2 \equiv c_1 \ (\text{mod } r)$$
$$b_1 x_1 + b_2 x_2 \equiv c_2 \ (\text{mod } r).$$

If

$$D = \begin{vmatrix} a_1 & a_2 \\ b_1 & b_2 \end{vmatrix}, D_1 = \begin{vmatrix} c_1 & b_1 \\ c_2 & b_2 \end{vmatrix}, D_2 = \begin{vmatrix} a_1 & c_1 \\ a_2 & c_2 \end{vmatrix}, one \ gets$$

(B)
$$Dx_1 \equiv D_1 \ (\text{mod } r)$$
$$Dx_2 \equiv D_2 \ (\text{mod } r)$$

If g.c.d $(D, r) = 1$, the congruences (B) have solution $\langle t_1, t_2 \rangle$ where t_1, t_2 are unique modulo r. These give a solution of the system (A) above. In the same manner, if there are m congruences in m unknowns all taken to the same modulus r, they can be reduced to m independent congruences involving only one unknown as in (B) above. These give a unique solution (mod r). A different context is when we have a system of m congruences in a single unknown but taken to different moduli. This problem was solved by Chinese mathematicians as early as the first century A.D. The earliest reference is that of Sun Tsu [18]. But, at about the same period Nichomachus (born in Gerasa, Palestine c 100 A.D.) is known to have solved the problem in his "Introductio to Arithemeticae". An example is:

Find the least positive integer which upon division by 3 leaves a remainder 2, upon division by 5 leaves a remainder 3 and upon division by 7 leaves a remainder 2. In symbols, one has

$$x \equiv 2 \, (\text{mod } 3), \quad x \equiv 3 \, (\text{mod } 5) \quad and \ x \equiv 2 (\text{mod } 7).$$

A common solution is $x \equiv 23$ (mod $3 \cdot 5 \cdot 7$). The gist of the idea is that the solutions to

$$x \equiv b_i (\text{mod } r_i) \quad (i = 1, 2, \cdots, k)$$

form a progression with period r_i. *The Chinese Remainder Theorem says that k arithmetic progressions with pairwise relatively prime moduli have a non-empty intersection. It is just an assertion of the fact that the cosets of the ideals* $r_i\mathbb{Z}$ ($i = 1, 2, \cdots, k$) *fit nicely into a particular coset of the ideal* $N\mathbb{Z}$ *where* $N = r_1 r_2 \cdots r_k$. *See T. W. Hungerford [11]. The 13th century Chinese algebraist Ch'in Chiu-Shao used the Euclidean algorithm in his solution of the Chinese Remainder Theorem (1247) (published in Shu-shu Chiu Chang, a mathematical treatise in 9 sections).*

The second problem we consider is that of determining the number of solutions of the congruence

$$n \equiv x_1 + x_2 + \cdots + x_s \pmod{r},$$

when x's are such that g.c.d $(x_i, r) = 1$ ($i = 1, 2, \cdots, s$). *If* $N(n, r, s)$ *denotes the number of solutions as specified above, H. Rademacher [14] gave the evaluation of* $N(n, r, s)$ *in 1925. Alfred Theodor Brauer (1884–1985) [2] verified it in 1926. An application of the Chinese Remainder Theorem shows that* $N(n, r, s)$ *is a multiplicative function of r.*

5.1. Introduction

The Chinese Remainder Theorem is one of the landmarks of Number theory. It is shown as theorem 29 given below. There are analogues of the theorem in algebra. We give two of them: one in terms of direct sums of rings and the other replacing $\mathbb{Z}$ by the ring $F[x]$ of polynomials with coefficients from a field F. See [3] and [9].

$N(n, r, s)$ denotes the number of solutions of a linear congruence

$$(5.1.1) \qquad\qquad x_1 + x_2 + \cdots + x_s \equiv n \pmod{r}$$

under the restriction g.c.d $(x_i, r) = 1$, ($i = 1, 2, \cdots, s$).

The formula for $N(n, r, s)$ is derived using elementary methods. We remark that Ramanujan sums defined by

$$(5.1.2) \qquad\qquad C(n, r) = \sum_{h(\mathrm{mod}\ r),\ \mathrm{g.c.d}\ (h,r)=1} \exp(\frac{2\pi i h n}{r})$$

where the summation is over a reduced residue system (mod r) plays an important role in the derivation of formulae involving $N(n, r, s)$. The notion of even functions (mod r) due to Eckford Cohen [5] is discussed. The arithmetical representation of an even function (mod r) is obtained by using an orthogonal property of Ramanujan sums. See theorem 34. The Rademacher formula for $N(n, r, s)$ is deduced from David Rearick's theorem which is shown as theorem 39.

5.2. The Chinese Remainder Theorem

We observe that the number of solutions of the linear congruence

$$(5.2.1) \qquad\qquad ax \equiv b \pmod{r}, \quad a, b \in \mathbb{Z}, r \geq 1$$

is the number of incongruent solutions t (mod r) such that $at - b \equiv 0 \pmod{r}$.

Fact 5.2.1 : (5.2.1) has a solution $\Leftrightarrow$ g.c.d $(a,r) \mid b$. When a solution exists, it has $d = $ g.c.d (a,r) solutions. If $x \equiv x_0 \pmod{r}$ is a solution, so is $x = x_0 + (\frac{r}{d})t$, $(t = 0, 1, 2, \cdots (d-1))$.

(This has been noted in an example in Section 1.7, chapter 1).

So, we replace (5.2.1) by $x \equiv d' \pmod{r'}$ where $r' = \frac{r}{\text{g.c.d}(a,r)}$, $d' = \frac{bu}{\text{g.c.d}(a,r)}$ for a suitable u which satisfies

$$(5.2.2) \qquad \frac{au}{\text{g.c.d}(a,r)} \equiv 1 \, (\text{mod} \, \frac{r}{\text{g.c.d}(a,r)}).$$

Fact 5.2.1 is needed for later reference. See illustration 7.4.1, chapter 7.

Example 5.2.1 : The congruence $24x \equiv 6 \pmod{15}$ is equivalent to $x \equiv 2 \pmod{5}$. For, g.c.d $(24, 15) = 3$. Further application of (5.2.2) to $24x \equiv 6 \pmod{15}$ with $a = 24$, $b = 6$, $r = 15$ yields

$$(5.2.3) \qquad 8u \equiv 1 \, (\text{mod} \, 5).$$

The unique solution of (5.2.3) is $u \equiv 2 \pmod{5}$. Also, $\frac{6u}{3} = 2u$ where $u \equiv 2 \pmod{5}$. So, the three solutions of $24x \equiv 6 \pmod{15}$ are given by $x \equiv d' \pmod{5}$ where $d' = 2u \equiv 4 \pmod{5}$. That is, $x \equiv 4 \pmod{15}$, $x \equiv 9 \pmod{15}$, $x \equiv 14 \pmod{15}$.

Next, we consider simultaneous linear congruences taken to different moduli.

Theorem 29 (The Chinese Remainder Theorem) : *The system of simultaneous congruences*

$$(5.2.4) \qquad \begin{cases} x \equiv c_1 \, (\text{mod } r_1) \\ x \equiv c_2 \, (\text{mod } r_2) \\ \qquad \cdots\cdots\cdots \\ x \equiv c_k \, (\text{mod } r_k) \end{cases}$$

is solvable if, and only if, g.c.d $(r_i, r_j) \mid c_i - c_j$ for every pair of subscripts i, j satisfying $1 \leq i < j \leq k$, any two solutions of the system are incongruent modulo l.c.m of the moduli, written $[r_1, r_2, \cdots, r_k]$.

Proof : $:\Rightarrow$ In order that the system (5.2.4) is solvable, it is certainly necessary that every pair

$$x \equiv c_i \, (\text{mod } r_i)$$
$$x \equiv c_j \, (\text{mod } r_j) \, (1 \leq i < j \leq k)$$

is solvable. So, we must have,

$$x = c_i + tr_i, x = c_j + sr_j, \, t, s \in \mathbb{Z}$$

or $tr_i \equiv (c_j - c_i) \pmod{r_j}$. Such a linear congruence in t is solvable if, and only if, g.c.d (r_i, r_j) divides $(c_j - c_i)$. So, the conditions g.c.d $(r_i, r_j) \mid (c_i - c_j)$ $(1 \leq i < j \leq k)$ are necessary.

$\Leftarrow$: Conversely, suppose that g.c.d $(r_i, r_j)|(c_i - c_j)$ $(1 \le i < j \le k)$. We choose a pair

$$x \equiv c_i \pmod{r_i}$$
$$x \equiv c_j \pmod{r_j}$$

and claim that the system (5.2.4) is solvable. Starting from

$$x \equiv c_1 \pmod{r_1}$$
$$x \equiv c_2 \pmod{r_2}$$

we get a congruence $tr_1 \equiv (c_2 - c_1) \pmod{r_2}$. This congruence in t has g.c.d (r_1, r_2) solutions modulo r_2. But, it is uniquely determined modulo $\frac{r_2}{\text{g.c.d}\,(r_1,r_2)}$. Therefore,

$$x \equiv c_1 + tr_1$$

is uniquely determined modulo $\frac{r_1 r_2}{\text{g.c.d}\,(r_1,r_2)} = [r_1, r_2]$, l.c.m of r_1 and r_2.
We write $x \equiv c_{12} \pmod{[r_1, r_2]}$ where c_{12} is uniquely determined modulo $[r_1, r_2]$.
We, next, show that every pair of the congruences

(5.2.5)
$$x \equiv c_{12} \pmod{[r_1, r_2]}$$
$$x \equiv c_j \pmod{r_j}, \ (3 \le j \le k)$$

is solvable. This needs the requirement

$$\text{g.c.d}\,(r_j, [r_1, r_2]) \mid (c_j - c_{12}) \ (3 \le j \le k)$$

This implies that g.c.d $(r_i, r_j) \mid (c_i, c_j)$, $1 \le i < j \le k$. This proves the sufficiency condition.

Now, c_{12} is uniquely determined modulo $[r_1, r_2]$.
Solving the congruences

(5.2.6)
$$x \equiv c_{12} \pmod{[r_1, r_2]}$$
$$x \equiv c_3 \pmod{r_3}$$

simultaneously, we arrive at

$$x \equiv c_{123} \pmod{[r_1, r_2, r_3]}$$

where c_{123} is uniquely determined modulo $[r_1, r_2, r_3]$. Repeating the procedure a finite number of times, we prove that the system (5.2.4) is solved simultaneously and the solution is unique modulo $[r_1, r_2, \cdots, r_k]$. $\quad \square$

Corollary 5.2.1 : *The system of congruences*

(5.2.7)
$$\begin{cases} x \equiv c_1 \pmod{r_1} \\ x \equiv c_2 \pmod{r_2} \\ \cdots\cdots\cdots \\ x \equiv c_k \pmod{r_k} \end{cases}$$

is solvable, if g.c.d $(r_i, r_j) = 1$ $(i \ne j; i, j = 1, 2, \cdots k)$ and any two solutions of the system (5.2.7) are congruent modulo the product $r_1 r_2 \cdots r_k$.

Proof follows from the fact that r_i and r_j are relatively prime to one another and conditions of theorem 29 are satisfied and $[r_1, r_2, \cdots, r_k] = r_1 r_2 \cdots r_k$, when g.c.d $(r_i, r_j) = 1$ $i \neq j$, $i, j = 1, 2, \cdots, k$. See Hugh M. Edgar [10] also.

Remark 5.2.1 : The Chinese Remainder Theorem is about the existence of solution when a certain condition is satisfied and when the solution exists, the uniqueness of solution is from a particular residue class modulo $[r_1, r_2, \cdots, r_k]$.

Remark 5.2.2 : The constructive proof of the corollary emerges from the following observation:

As g.c.d $(r_i, r_j) = 1$ for $i \neq j$ we write

$$M = r_1 r_2 \cdots r_k \text{ and so, g.c.d } (\frac{M}{r_i}, r_i) = 1. \quad i = 1, 2, \cdots, k.$$

Let t_i denote the solution of

(5.2.8) $$\frac{M}{r_i} x \equiv 1 (\text{mod } r_i); \quad i = 1, 2, \cdots, k$$

Then, as

(5.2.9) $$\frac{M}{r_i} t_i \equiv 1 \ (\text{mod } r_j) \quad \text{for } j \neq i$$

we write $x_0 = \sum_{i=1}^{k} \frac{M}{r_i} t_i c_i$. Then, for $1 \leq j \leq k$, $x_0 \equiv \frac{M}{r_j} t_j c_j \equiv c_j \ (\text{mod } r_j)$, by (5.2.8) and (5.2.9). This shows that x_0 is a solution of the system (5.2.7) and is unique modulo $r_1 r_2 \cdots r_k$.

Illustration 5.2.1 : *Solve the system of simultaneous congruences*

$$x \equiv 2 \ (mod \ 3)$$
$$x \equiv 4 \ (mod \ 5)$$
$$x \equiv 6 \ (mod \ 7)$$

Solution : Here, $M = 3 \cdot 5 \cdot 7 = 105$.

$$35x \equiv 1(\text{mod } 3) \Rightarrow x \equiv 2(\text{mod } 3),$$
$$21x \equiv 1(\text{mod } 5) \Rightarrow x \equiv 1(\text{mod } 5),$$
$$15x \equiv 1(\text{mod } 7) \Rightarrow x \equiv 1(\text{mod } 7).$$

So, $x_0 = 35 \times 2 \times 2 + 21 \times 4 \times 1 + 15 \times 6 \times 1 = 104 \equiv -1 \ (\text{mod } 105)$ leads to the unique solution $x_0 \equiv -1 \ (\text{mod } 105)$ which satisfies each of the given congruences.

5.3. Direct products and direct sums

G and G' are two groups with identity elements e, e' respectively.

Definition 5.3.1 : *The direct product of G and G' is a group whose underlying set is $G \times G'$ and whose binary operation is given by*

$$(a,a') \cdot (b,b') = (a \cdot b, a' \cdot b')$$

where $a, b \in G$ and a', $b' \in G'$.

Observation 5.3.1 : *The identity element of the direct product, written $G \times G'$, is (e,e'). The inverse of (a,a') in $G \times G'$ is (a^{-1}, a'^{-1}) where a^{-1} and a'^{-1} are respective inverses of a and a' in G and G'.*

Observation 5.3.2 : *If the group operations are 'addition' in each of the groups G, G', we express $G \times G'$ as $G \oplus G'$.*

Let $\{G_\lambda : \lambda \in \Lambda\}$ be a family of groups indexed by the set Λ. The direct product of the groups G_λ, $(\lambda \in \Lambda)$ written $\prod_\lambda G_\lambda$ or $\sum_\lambda \oplus G_\lambda$ is defined as follows:

$$\text{Let } a : \Lambda \to \bigcup_\lambda G_\lambda, \ b : \Lambda \to \bigcup_\lambda G_\lambda$$

be two functions. Then, $ab : \Lambda \to \bigcup_\lambda G_\lambda$ is the function given by

(5.3.1) $\qquad\qquad ab(\lambda) = a(\lambda)b(\lambda)$ for all $\lambda \in \Lambda$

where $a(\lambda) \in G_\lambda$, $b(\lambda) \in G_\lambda$.

Definition 5.3.2 : *The direct product (or the complete direct sum) of the groups $G_\lambda, \lambda \in \Lambda$ is the set*

$$\prod_\lambda G_\lambda (= \sum_\lambda \oplus G_\lambda) = \{a : a \text{ is a function from } \Lambda \text{ to } \cup G_\lambda \text{ such that } a(\lambda) \in G_\lambda\}.$$

Remark 5.3.1 : $\prod_\lambda G_\lambda$ is a group under the operation of multiplication given in (5.3.1) and for each $\theta \in \Lambda$, the map $\pi_\theta : \Pi_\lambda G_\lambda \to G_\theta$ given by $a \mapsto a(\theta)$ is a surjective homomorphism of groups (homomorphism onto).

Example 5.3.1 : If $\Lambda = \{1, 2, \cdots, n\}$ and G_i $(i \in \Lambda) = (\mathbb{Z}, +)$

$$\mathbb{Z}^n = \{(a_1, a_2, \cdots, a_n) : a_i \in \mathbb{Z}, \quad (i = 1, 2, \cdots, n)\}$$

and for $(a_1, a_2, \cdots, a_n)$, $(b_1, b_2, \cdots, b_n) \in \mathbb{Z}^n$,

$$(a_1, a_2, \cdots, a_n) \oplus (b_1, b_2, \cdots, b_n) = (a_1 + b_1, a_2 + b_2, \cdots, a_n + b_n)$$

$\mathbb{Z}^n$ is the complete direct sum of n groups each equal to $\mathbb{Z}$. $(\mathbb{Z}, +)$ is abelian. So is $(\mathbb{Z}^n, \oplus)$.

Definition 5.3.3 : *A group* $(G, \cdot)$ *is said to be decomposable into the direct product of groups A, B if, and only if,* $G \cong A \times B$ *and* $A \cap B = (e)$, *the subgroup of G containing the identity.*

It follows that if a group G is decomposable into the direct product of two normal subgroups A and B of G, then every element of A commutes with every element of B and an element $x \in G$ is uniquely expressed as $x = a \cdot b$ where $a \in A, b \in B$. G is said to be indecomposable if G admits no non-trivial direct decomposition.

Definition 5.3.4 : *Let S be an arbitrary, but fixed, set. By a free abelian group G on the set S, we mean an abelian group G together with a function* $f : S \rightarrow G$ *such that for every function* $g : S \rightarrow H$ *where H is some abelian group, there is a unique homomorphism* $h : G \rightarrow H$ *such that the relation* $h \circ f = g$ *holds.*

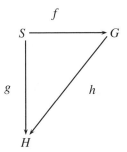

Figure 1

We express this in symbols as in figure 1 above. If for each $s \in S$, $h(f(s)) = g(s)$, the diagram above is said to be commutative. There are two sets of arrows from S to H. The composition of mappings h and f is the same as g. The composition of maps depends only on the initial point and the final point and not on the path chosen.

Fact 5.3.1 : If G is an abelian group and $f : S \rightarrow G$ makes it a free abelian group on S, then $f : S \rightarrow G$ is injective (one-one) and its image $f(S)$ generates G. For proof, see Chih-Han Sah [17].

Observation 5.3.3 : *Let* $\{A_\lambda : \lambda \in \Lambda\}$ *be a collection of sets indexed by* Λ. *Then, there exists a set B together with a collection of maps* $\theta_\lambda : B \rightarrow A_\lambda$ *which satisfies the following universal mapping property namely:*

Let C be any set and let $\phi_\lambda : C \rightarrow A_\lambda$ *be any collection of maps. Then, there exists a unique map* $\phi : C \rightarrow B$ *such that*

$$\theta_\lambda \circ \phi = \phi_\lambda.$$

In figure 2, ϕ is denoted by a dotted arrow and ϕ is not given in advance.

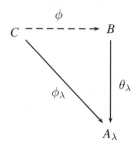

Figure 2

Universal mapping property characterises a cartesian product via maps.
Now, the definition of direct product of a family of groups can be recast as follows:

Definition 5.3.5 : *Let* $\{G_\lambda : \lambda \in \Lambda\}$ *be an indexed family of groups. A group G is called a direct product of the groups* G_λ, $\lambda \in \Lambda$ *if, and only if, there exist homomorphisms* $\theta_\lambda : G \to G_\lambda$ $(\lambda \in \Lambda)$ *with the following universal mapping property:*
    For any group H and group homomorphisms $\phi_\lambda : H \to G_\lambda$ $(\lambda \in \Lambda)$, *there exists a unique homomorphism* $\phi : H \to G$ *such that*

$$\theta_\lambda \circ \phi = \phi_\lambda, \ \lambda \in \Lambda$$

That is, the diagram given in figure 3 (shown below) is commutative.

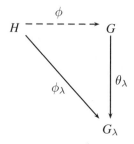

Figure 3

Fact 5.3.2 : Let I be the finite set $\{1, 2, \cdots, n\}$. A group G is a direct product of the groups G_i $(i \in I)$ if, and only if, there exist groups $H_i \subseteq G_I$, $i \in I$ such that the following conditions hold :

  (i) H_i is normal in G for all $i \in I$
  (ii) $H_i \cong G_i$, for all $i \in I$.

(iii) each element $g \in G$ has a unique representation as

$$g = g_1 g_2 \cdots g_n \text{ where } g_i \in H_i \quad (i = 1, 2, \cdots n).$$

If G contains a finite number of subgroups H_i that satisfy the conditions stated above, G is called the internal direct product of the subgroups H_i (i varying over a finite set).

Remark 5.3.2 : A finite abelian group G is the internal direct product of uniquely determined Sylow p-subgroups in which p runs through the distinct prime factors of $n = | G |$, (the order of G).

That is, if $n = \prod_{i=1}^{k} p_i^{a_i}$, p_i primes, $a_i \geq 1$ ($i = 1, 2, \cdots k$), then $G = \prod_{i=1}^{k} H_i$ where H_i is a Sylow p_i-subgroup (of G) of order $p_i^{a_i}$. It is worthwhile noting that

$$(5.3.2) \qquad \mathbb{Z}/(n) \cong \prod_{i=1}^{R} \mathbb{Z}/(p_i^{a_i}).$$

Next, for technical reasons, the 'dual' of direct product of groups is obtained by reversing the arrows in the commutative diagram mentioned in definition 5.3.3 and the resulting group is called the direct sum of groups $G_\lambda (\lambda \in \Lambda)$. Definition 5.3.5 is restated for abelian groups in the following manner:

Definition 5.3.6 : *Let $\{G_\lambda : \lambda \in \Lambda\}$ be an indexed family of abelian groups. An abelian group G is called a direct sum of the abelian groups G_λ, $\lambda \in \Lambda$ if, and only if, there are homomorphisms $\theta_\lambda : G_\lambda \to G$ ($\lambda \in \Lambda$) with the following universal mapping property: For any abelian group H and homomorphisms $\phi_\lambda : G_\lambda \to H$, $\lambda \in \Lambda$, there exists a unique homomorphism $\phi : G \to H$ such that $\phi \circ \theta_\lambda = \phi_\lambda$ for all $\lambda \in \Lambda$.*

That is, the diagram given in figure 4 (shown below) is commutative.

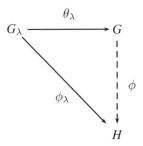

Figure 4

Remark 5.3.3 : Existence of direct sum :

Let G be the subgroup of the direct product $\prod_{\lambda \in \Lambda} G_\lambda$ which consists of the elements g_λ, $\lambda \in \Lambda$ such that $g_\lambda \neq e_\lambda$ (the identity in G_λ) for at most a finite number of indices $\lambda \in \Lambda$. (the finite number may vary from element to element). The map $\theta_\lambda : G_\lambda \to G$ is such that

$$(5.3.3) \qquad g_\lambda \mapsto (1,1,\cdots,1,g_\lambda,1,1,\cdots) \text{ of } G,$$

where 1 represents the identity element of the component of G. So, every element of G is expressed uniquely as a finite product of elements of the form $\theta_\lambda(g_\lambda)$, $\lambda \in \Lambda$ and only finite products occur in G. This proves the existence of direct sum of groups G_λ, $\lambda \in \Lambda$.

Further, $\phi((1,1,\cdots,g_\lambda,1,1,\cdots)) = \phi_\lambda(g_\lambda) \in H$.

Theorem 30 (Fundamental theorem of arithmetic) : *Let $\mathbb{Q}^* = \mathbb{Q} \setminus \{0\}$ be the group of non-zero rational numbers under multiplication. Then, $\mathbb{Q}^*$ is the direct sum of a cyclic group of order 2 and a countable number of infinite cyclic groups.*

Proof : Let G_0 be a cyclic group of order 2. That is, $G_0 = \{1,-1\}$. Let p_k be the kth prime in $\mathbb{N}$. $p_k^\lambda \mathbb{Q}^*$ denotes an infinite cyclic subgroup of $\mathbb{Q}^*$ generated by $p_k^\lambda (\lambda \in \mathbb{Z})$. We consider a countable family of cyclic subgroups of $\mathbb{Q}^*$ given in

$$\{G_0, p_1^\lambda \mathbb{Q}^*, p_2^\lambda \mathbb{Q}^*, \cdots\}, \lambda \in \mathbb{Z}.$$

Every element of $\mathbb{Q}^*$ is of the form $\pm \prod_{j=1}^{s} p_j^{a_j}$, p_j are primes, a_j is a positive or negative integer. Then $\mathbb{Q}^*$ is the direct sum of a cyclic group of order 2 and a countable number of infinite cyclic groups generated by p^λ, p a prime and λ a nonzero element in $\mathbb{Z}$. □

Remark 5.3.4 : Theorem 30 has been adapted from Chih-Han Sah [17].

Next, we shall extend these ideas to the case of a family of rings.

Let $\{R_\lambda : \lambda \in \Lambda\}$ be an indexed family of rings. $\prod_\lambda R_\lambda$ is the direct product of the additive abelian groups $(R_\lambda,+)$, $\lambda \in \Lambda$.

(i) $\prod_\lambda R_\lambda$ is a ring with multiplication given by

$$(5.3.4) \qquad \{a_\lambda\} \cdot \{b_\lambda\} = \{a_\lambda b_\lambda\}$$

(ii) If R_λ ($\lambda \in \Lambda$) has multiplicative identity, then, $\prod_\lambda R_\lambda$ also has multiplicative identity.

(iii) If R_λ ($\lambda \in \Lambda$) is commutative, so is $\prod_\lambda R_\lambda$.

(iv) For each $\theta \in \Lambda$, the canonical projection $\pi_\theta : \prod_\lambda R_\lambda \to R_\theta$ given by $\{a_\lambda\} \mapsto a_\theta$ is an epimorphism of rings
(onto homomorphisms)

(v) For each $\theta \in \Lambda$, the canonical injection $i_\theta : R_\theta \to \prod_\lambda R_\lambda$ given by $a_\theta \mapsto \{a_\lambda\}$ where $a_\lambda = 0$ for $\lambda \neq \theta$ is a monomorphism of rings
(one-one homomorphisms).

Now, we translate definition 5.3.5 of direct product of groups to direct product of rings.

Definition 5.3.7 : *Let $\{R_\lambda : \lambda \in \Lambda\}$ be a non-empty family of rings. A ring R is called a direct product of rings R_λ, $\lambda \in \Lambda$ written $\prod_\lambda R_\lambda$ if, and only if, there exist homomorphisms $\{\phi_\lambda : R \to R_\lambda\}$ ($\lambda \in \Lambda$) with the following universal mapping property:*

For any ring S and any ring homomorphisms $\psi_\lambda : S \to R_\lambda$ ($\lambda \in \Lambda$), there exists a unique ring homomorphism $\phi : S \to R$ such that

$$\phi_\lambda \circ \phi = \psi_\lambda \quad (\lambda \in \Lambda).$$

Theorem 31 : *$\{A_i : 1 \leq i \leq n\}$ is a collection of ideals of a commutative ring R with unity 1_R such that*

$$R = A_1 + A_2 + \cdots + A_n.$$

Suppose that for k ($k = 1, 2, \cdots, n$),

$$A_k \cap (A_1 + A_2 + \cdots + A_{k-1} + A_{k+1} + \cdots + A_n) = (0_R).$$

Then, there is a ring isomorphism $R \cong \prod_{i=1}^n A_i$.

Proof : From the given data, we are led to consider

$$\theta : \prod_{i=1}^n A_i \to R$$

given by $\theta((a_1, a_2, \cdots, a_n)) = a_1 + a_2 + \cdots + a_n$, $(a_i \in A_i, i = 1, 2, \cdots, n)$. θ is an isomorphism of additive abelian groups.

Claim : θ is a ring homomorphism.

If $i \neq j$ and $a_i \in A_i$, $a_j \in A_j$, $a_i a_j \in A_i \cap A_j = (0_R)$. So, for all $a_i, b_i \in A_i$,

$$(a_1 + a_2 + \cdots + a_n)(b_1 + b_2 + \cdots + b_n) = (a_1 b_1 + a_2 b_2 + \cdots + a_n b_n).$$

So, $\theta((a_1 b_1, a_2 b_2, \cdots, a_n b_n)) = \theta((a_1, a_2, \cdots, a_n)) \theta((b_1, b_2, \cdots, b_n))$. So, θ is a ring homomorphism, which is both an epimorphism and a monomorphism. Thus, $R \cong \prod_{i=1}^n A_i$. $\qquad\square$

Remark 5.3.5 : Theorem 31 suggests that the ring R (with unity 1_R) is the internal direct product of the ideals A_i ($i = 1, 2, \cdots, n$).

Remark 5.3.6 : The idea of a direct product of k finite fields has already been mentioned while expressing -1 as a sum of squares in $\mathbb{Z}/r\mathbb{Z}$ in Section 1.7 of chapter 1. See (5.3.2) also.

Definition 5.3.8 : *Given a commutative ring R with unity 1_R, and ideal $I \subset R$, we say that for $a, b, \in R$,*

$$a \equiv b \pmod{I} \Leftrightarrow a - b \in I.$$

That is, $a \equiv b \, (\text{mod } I) \Leftrightarrow$ the cosets $a+I$ and $b+I$ are identical. Since R/I can be made a ring, we deduce that :

Given $a_1 \equiv a_2(\text{mod } I)$, $b_1 \equiv b_2(\text{mod } I)$

$$(5.3.5) \qquad \begin{cases} a_1+b_1 & \equiv a_2+b_2 \ (\text{mod } I) \\ a_1 b_1 & \equiv a_2 b_2 \ (\text{mod } I). \end{cases}$$

Definition 5.3.9 : *Two ideals I, J of a ring R are said to be comaximal, if $I+J = R$.*

As an example, if we consider two ideals (a), (b) of $\mathbb{Z}$ where g.c.d $(a,b) = 1$, one gets $(a)+(b) = \mathbb{Z}$ and so (a) and (b) are comaximal. For, when g.c.d $(a,b) = 1$ there exist integers x, y such that

$$ax + by = 1.$$

'Comaximality' among ideals corresponds to the notion of 'relatively prime' among integers.

Theorem 32 (analogue of the Chinese Remainder Theorem.) **:** *R is a commutative ring with unity 1_R. Let $\{I_1, I_2, \cdots, I_n\}$ be a set of n ideals (of R) which are pairwise comaximal. That is, $I_i + I_j = R$ $(i \neq j)$. Let $c_1, c_2, \cdots, c_n$ be elements of R. Then, the system of simultaneous congruences*

$$x \equiv c_i(\text{mod } I_i) \quad (i = 1, 2, \cdots, n)$$

has a unique solution b satisfying

$$x \equiv b \, (\text{mod } I)$$

where

$$I = \cap_{i=1}^n I_i.$$

This implies that

$$(5.3.6) \qquad R/\cap_{i=1}^n I_i \cong \prod_{i=1}^n R/I_i.$$

Proof : We define a mapping $f : R \to \prod_{i=1}^n R/I_i$ by

$$(5.3.7) \qquad f(x) = (x+I_1, x+I_2, \cdots, x+I_n), \quad x \in R$$

Then,

$$\begin{aligned} f(x+y) &= f(x)+f(y) \\ f(xy) &= f(x)f(y) \end{aligned} \Bigg\} \, x,y \in R.$$

If $f(t) = (I_1, I_2, \cdots, I_n)$, $t \in I_i$ for each i and so

$$t \in \cap_{i=1}^n I_i$$

Therefore, $\ker f = \cap_{i=1}^n I_i$.

Our aim is to show that f is an epimorphism. Since $I_i + I_j = R$ for $i \neq j$, there exist elements $a_i \in I_i$, $b_i \in I_j$ such that $a_i + b_i = 1_R$.

If $r_j = a_1 a_2 \cdots a_{j-1} a_{j+1} \cdots a_n$, $r_j \in \cap_{i \neq j} I_i$.

Since $1_R - a_i \in I_j$ the coset $a_i + I_j = 1_R + I_j$ for all $i \neq j$. So,

$$a_i \equiv 1_R(\text{mod } I_j), \quad i \neq j.$$

Let $x_1, x_2, \cdots, x_n \in R$. If $x = \sum_{i=n}^{n} r_i x_i$, we will have

(5.3.8) $$f(x) = (x_1 + I_1, x_2 + I_2, \cdots, x_n + I_n)$$

For,

$$x + I_j = \sum_{i \neq j} (r_i + I_j)(x_i + I_j) + (r_j + I_j)(x_j + I_j)$$

Since $r_i \in I_j$ for $i \neq j$ and $r_j + I_j = 1_R + I_j$ (by comaximality among I_i, I_j), we get

$$x + I_j = x_j + I_j \quad (j = 1, 2, \cdots n)$$

or,

$$x \equiv x_j (\operatorname{mod} I_j) \quad j = 1, 2, \cdots, n$$

Therefore, from (5.3.8) $x = \sum_{i=1}^{n} r_i x_i$ gives (5.3.7) thereby showing that f is an epimorphism. Now, $r_1, r_2, \cdots, r_n$ are such that

$$r_i \equiv 0 (\operatorname{mod} I_j) \text{ for } i \neq j$$
$$\text{and } r_j \equiv 1_R (\operatorname{mod} I_j)$$
$$\text{So, } c_i r_i \equiv 0 (\operatorname{mod} I_j) \text{ for } i \neq j$$
$$\text{and } c_j r_j \equiv c_j (\operatorname{mod} I_j).$$

If $b = \sum_{i=1}^{n} c_i r_i$, $b \equiv c_j (\operatorname{mod} I_j)$ for $j = 1, 2, \cdots n$. So, the system

$$x \equiv c_i (\operatorname{mod} I_i) \quad i = 1, 2, \cdots, n$$

has a solution b which is the preimage of $(c_1 + I_1, c_2 + I_2, \ldots, c_n + I_n)$ under f. If b' is another solution of the given system of congruences,

$$b - b' \equiv 0 (\operatorname{mod} I)$$

where $I = \cap_{i=1}^{n} I_i$. So, the solution is unique modulo I. Since ker $f = I$, this is also expressed by the isomorphism shown in (5.3.6) which is a consequence of the fundamental homomorphism theorem. $\qquad \square$

Observation 5.3.4 : *The corollary 5.2.1 of theorem 29 (the Chinese Remainder Theorem) is a special case of theorem 32.*

    For, if $x \equiv c_i (\operatorname{mod} r_i)$ $(i = 1, 2, \cdots, k)$ with g.c.d $(r_i, r_j) = 1$, these congruences admit a simultaneous solution modulo $r_1 r_2 \cdots r_k$.

    We consider principal ideals $(r_1), (r_2), \cdots, (r_k)$ which are pairwise comaximal and note that

$$\cap_{i=1}^{k} (r_i) = \text{ the ideal generated by } r = r_1 r_2 \cdots r_k.$$

Remark 5.3.7 : The form in which theorem 32 is stated gives the natural extension of the Chinese Remainder Theorem to commutative rings having a given set of pair-wise comaximal ideals. See Thomas W. Hungerford [11] and David M. Burton [3].

Now, we observe that $\mathbb{Z}$ is a U.F.D. So, we replace $\mathbb{Z}$ by a U.F.D, say, $F[x]$, the integral domain of polynomials with coefficients from a field F. Let $a(x)$ and $r(x)$ be polynomials in $F[x]$. We look for a polynomial $u(x)$ modulo $r(x)$ such that

(5.3.9) $a(x)u(x) \equiv b(x) \pmod{r(x)}$,

where $b(x) \in F[x]$ is already given. (5.3.9) is a congruence in which we solve for $u(x)$. (We can compare (5.3.9) with $ax \equiv b \pmod{r}$ of number theory). We state below an analogue of the Chinese Remainder Theorem in the context of $F[x]$.

Theorem 33 : $r_i(x)$ $(i = 1, 2, \cdots k)$ are given polynomials in $F[x]$ and $r_i(x)$ and $r_j(x)$ are relatively prime whenever $i \neq j$. Suppose $a_i(x)$ $(i = 1, 2, \cdots k)$ are polynomials for which $a_i(x)$ and $r_i(x)$ are relatively prime to one another $(i = 1, 2, \cdots k)$. If $b_i(x)$ $(i = 1, 2, \cdots k)$ are arbitrary polynomials in $F[x]$, then, the system of congruences

(5.3.10) $\begin{cases} a_1(x)u(x) & \equiv b_1(x) \pmod{r_1(x)} \\ a_2(x)u(x) & \equiv b_2(x) \pmod{r_2(x)} \\ \cdots\cdots\cdots & \cdots\cdots\cdots\cdots \\ a_k(x)u(x) & \equiv b_k(x) \pmod{r_k(x)} \end{cases}$

has a unique solution modulo $r(x)$ where $r(x) = r_1(x)r_2(x)\cdots r_k(x)$.

Outline of proof : On account of the linear expressibility of g.c.d, since $(a_i(x), r_i(x)) = 1_F$, we can find a polynomial $c_i(x) \in F[x]$ such that

(5.3.11) $c_i(x)a_i(x) \equiv 1 \pmod{r_i(x)}$ $(i = 1, 2, \cdots, k)$

So, (5.3.10) can be restated as

$u(x) \equiv c_i(x)b_i(x) \pmod{r_i(x)}$, $(i = 1, 2, \cdots, k)$

It is of the form

$x \equiv c_i \pmod{I_i}$ $(i = 1, 2, \cdots, k)$

where I_i $(i = 1, 2, \cdots, k)$ are ideals of the ring $F[x]$. Further, I_i, I_j are comaximal for $i \neq j$. Applying theorem 32 we obtain the unique solution $u(x) \equiv h(x) \pmod{r(x)}$ where $r(x) = r_1(x)r_2(x)\cdots r_k(x)$. $\square$

Remark 5.3.8 : Theorem 33 has been adapted from C. Ding, D. Pei and A. Salomaa [9].

Illustration 5.3.1 : *We take $F = GF(2)$, the Galois field having two elements.*

$$r_1(x) = x^3 + x + 1$$
$$r_2(x) = x^3 + x^2 + 1$$

Let $a_1(x) = x^2 + x + 1$ and $a_2(x) = x + 1$.

$$a_1(x)u(x) \equiv 1 \pmod{x^3 + x + 1}$$

has the solution

$$t(x) = x^2$$

as $x^2(x^2+x+1)+(x+1)(x^3+x+1) = 1$. That is, we have

(5.3.12) $$(x^2+x+1)x^2 \equiv 1 \,(\mathrm{mod}\,(x^3+x+1))$$

So, here, $c_i(x)$ of (5.3.11) is $c_i(x) = x^2$. Now,

$$a_2(x)u(x) \equiv 1 \,(\mathrm{mod}\,x^3+x+1)$$

has the solution

$$s(x) = x^2,$$

as $x^2(x+1)+x^3+x^2+1 = 1$. That is ,

(5.3.13) $$(x+1)x^2 \equiv 1 \,(\mathrm{mod}\,(x^3+x+1)).$$

We seek a solution of simultaneous congruences

(5.3.14) $$\begin{cases} (x^2+x+1)u(x) & \equiv b_1(x)\,(\mathrm{mod}\,(x^3+x+1)) \\ (x+1)u(x) & \equiv b_2(x)\,(\mathrm{mod}\,(x^3+x^2+1)) \end{cases}$$

By theorem 33, (5.3.14) has a unique solution modulo $r(x)$ where

$$r(x) = (x^3+x+1)(x^3+x^2+1).$$

This is due to the fact that x^3+x+1 and x^3+x^2+1 are relatively prime to one another, as

(5.3.15) $$(x+1)(x^3+x^2+1)+x(x^3+x+1) = 1$$

From (5.3.12), (5.3.13) and (5.3.14), we obtain

(5.3.16) $$\begin{cases} u(x) \equiv x^2 b_1(x)(\mathrm{mod}\,(x^3+x+1)), \\ u(x) \equiv x^2 b_2(x)(\mathrm{mod}\,(x^3+x^2+1)). \end{cases}$$

So, if

(5.3.17) $$d(x) = (x+1)(x^3+x^2+1)\,x^2 b_1(x)+x(x^3+x+1)x^2 b_2(x),$$

(5.3.18) $$u(x) \equiv d(x)\,(\mathrm{mod}\,(x^3+x+1)(x^3+x^2+1)).$$

From (5.3.16), we notice that $u(x)$ satisfies the congruence in (5.3.18) and as

$$(x^2+x+1)x^2 \equiv 1(\mathrm{mod}\,(x^3+x+1)),$$
$$(x^2+x+1)u(x) \equiv (x+1)(x^3+x^2+1)b_1(x)\,(\mathrm{mod}\,(x^3+x+1)).$$

But,

$$(x+1)(x^3+x^2+1) \equiv 1(\mathrm{mod}\,(x^3+x+1)).$$
$$\text{So, } (x^2+x+1)u(x) \equiv b_1(x)\,(\mathrm{mod}\,(x^3+x+1)).$$

Similarly,

$$(x+1)u(x) \equiv b_2(x)\,(\mathrm{mod}\,(x^3+x+1)).$$

Thus, (5.3.18) gives a common solution to either of the congruences in (5.3.14).

5.4. Even functions (mod r)

Let r denote an arbitrary but fixed positive integer ≥ 1. If $n \in \mathbb{Z}$, Ramanujan's sum $C(n,r)$ (5.1.2) has the arithmetical representation [4] given by

$$(5.4.1) \qquad C(n,r) = \sum_{d|(n,r)} \mu(\tfrac{r}{d})d, \ n \geq 1, r \geq 1$$

where d runs through the common divisors of n and r and μ is the Möbius function and summation is over $a(\bmod r)$ such that g.c.d $(a,r) = 1$.

$C(n,r)$ is known to be multiplicative in r. That is,

$$(5.4.2) \qquad C(n,r)C(n,r') = C(n,rr') \text{ whenever g.c.d. } (r,r') = 1.$$

If ϕ denotes Euler ϕ-function, the Hölder relation [4] for $C(n,r)$ is

$$(5.4.3) \qquad C(n,r) = \frac{\mu(\tfrac{r}{g})\phi(r)}{\phi(\tfrac{r}{g})}$$

μ and ϕ being the Möbius and Euler ϕ-function respectively and $g = $ g.c.d (n,r).

Definition 5.4.1 : *By an arithmetic function f of two variables n, r, we mean a map $f : \mathbb{Z} \times \mathbb{N} \to \mathbb{C}$, the field of complex numbers.*

Definition 5.4.2 : *An arithmetic function f (of two variables n, r) is said to be periodic (mod r) if*

$$f(n,r) = f(n',r) \text{ whenever } n' \equiv n(\bmod r).$$

If f is periodic (mod r) and if $d \mid r$, then f is periodic (mod d), since $n \equiv n'(\bmod r) \Rightarrow n \equiv n'(\bmod d)$

We note that $C(n,r)$ is periodic (mod r).

Definition 5.4.3 : *An arithmetic function f is called an even function (mod r) or briefly 'even (mod r)' if $f((n,r),r) = f(n,r)$ for all n; (n,r) being g.c.d (n,r).*

Evidently an even function (mod r) is also periodic (mod r). An orthogonal property of $C(n,r)$ due to Eckford Cohen [4] states that if d, e are divisors of r,

$$(5.4.4) \qquad \sum_{t|r} C(\tfrac{r}{t},d)C(\tfrac{r}{e},t) = \begin{cases} r, & \text{if } d = e, \\ 0, & \text{otherwise.} \end{cases}$$

For proof, see [5].

(5.3.5) is exploited to obtain an arithmetical representation of an even function (mod r) via Ramanujan Sums. See [6]. This also enables one to consider the set $\mathbb{B}_r(\mathbb{C})$ of even functions (mod r) as a finite dimensional algebra over $\mathbb{C}$, the field of complex numbers. This aspect of the algebra of even functions (mod r) will be considered in chapter 14.

Definition 5.4.4 : *Let f, g be periodic functions (mod r). The Cauchy product [7] of f and g is the arithmetic function defined by*

$$h(n,r) = \sum_{n \equiv a+b(\bmod r)} f(a)g(b)$$

where the sum on the right side is over all solutions $\langle a,b \rangle \,(\mathrm{mod}\ r)$ of the congruence

$$x+y \equiv n\,(\mathrm{mod}\ r).$$

It follows from definition 5.4.4 that h is also periodic $(\mathrm{mod}\ r)$.
Analogous to (5.4.4), one gets the orthogonal relation: If d,e are divisors of r,

$$(5.4.5) \qquad \sum_{n \equiv a+b(\mathrm{mod}\ r)} C(a,d)C(b,e) = \begin{cases} rC(n,d), & \text{if } e=d, \\ 0, & \text{otherwise.} \end{cases}$$

For proof, see Eckford Cohen [7].

Next, we go to the arithmetical representation of an even function $(\mathrm{mod}\ r)$. The two theorems that follow are due to Eckford Cohen [5], [6], [7] and they are fundamental in nature.

Theorem 34 (Eckford Cohen (1955)) : *Let f be an even function $(\mathrm{mod}\ r)$. Then, f can be uniquely expressed as*

$$(5.4.6) \qquad f(n,r) = \sum_{d|r} \alpha(d,r)C(n,d), \text{ for all } n;$$

where $\alpha(d,r)$ is given by

$$(5.4.7) \qquad \alpha(d,r) = \frac{1}{r}\sum_{t|r} f(\frac{r}{t},r)C(\frac{r}{d},t), \ d|r.$$

Proof : Suppose that $\alpha(d,r)$ are given by the formula (5.4.7) for all divisors d of r. Then,

$$\sum_{d|r} \alpha(d,r)C(n,d) = \frac{1}{r}\sum_{d|r}\sum_{t|r} f(\frac{r}{t},r)C(\frac{r}{d},t)C(n,d)$$

$$= \frac{1}{r}\sum_{t|r} f(\frac{r}{t},r)\sum_{d|r}C(\frac{r}{d},t)C(n,d).$$

Now, $C(n,d) = C((n,d),d)$ for all $d|r$, by virtue of (5.4.3). g.c.d $(n,d)|n$, for $d|r$. So, g.c.d $(n,d)|$g.c.d (n,r). So, d runs through those divisors common to n and r. Therefore, $C(n,d) = C(d,d) = C((n,r),d)$ as $d|$ g.c.d (n,r).

Consequently,

$$(5.4.8) \qquad \sum_{d|r} \alpha(d,r)C(n,d) = \frac{1}{r}\sum_{t|r} f(\frac{r}{t},r)\sum_{d|r}C(\frac{r}{d},t)C((n,r),d)$$

Let g.c.d $(n,r) = g$. Then the right side of (5.4.8) becomes

$$\frac{1}{r}\sum_{t|r} f(\frac{r}{t},r)\sum_{d|r}C(\frac{r}{d},t)C(\frac{r}{g},d).$$

The inner sum shown is zero when $\frac{r}{g} \neq t$ and is r if $\frac{r}{g} = t$ by (5.4.4). So,

$$\sum_{d|r} \alpha(d,r)C(n,d) = f(g,r) = f(n,r).$$

This proves (5.4.6).

To show that the representation is unique, we prove that $\sum_{d|r} \alpha(d,r)C(n,d) = 0$ for all $n \Rightarrow \alpha(d,r) = 0$ for each divisor of r. For, choosing $n = 0$ one gets

$$\sum_{d|r} \alpha(d,r)C(0,d) = 0.$$

For divisors d, e of r, one has, using (5.4.5)

$$rC(0,d) = \sum_{a+b \equiv 0 (\text{mod } r)} C(a,d)C(b,e) \text{ when } d = e.$$

So,

$$\frac{1}{r} \sum_{d|r} \alpha(d,r)rC(0,d) = \frac{1}{r} \sum_{d|r} \alpha(d,r) \sum_{a+b \equiv 0 (\text{mod } r)} C(a,d)C(b,e) = 0$$

That is,

$$\frac{1}{r} \sum_{d|r} \alpha(d,r) \sum_{a+b \equiv 0 (\text{mod } r)} C(a,d)C(b,e) = 0$$

As $C(0,d) = \phi(d)$, the Euler totient, $C(0,d) \neq 0$ and so $\alpha(d,r) = 0$ for each $d|r$. $\square$

Remark 5.4.1 : The coefficients $\alpha(d,r)$ given in (5.4.7) are called the Fourier coefficients of f. (5.4.6) gives a finite Fourier expansion of the function which is even (mod r).

Theorem 35 : *If f and g are even functions* (mod r) *with Fourier coefficients $\alpha(d,r)$ and $\beta(d,r)$ respectively, their Cauchy product h is even* (mod r) *with Fourier coefficients $r\alpha(d,r)\beta(d,r)$ for all d dividing r.*

Proof : Let n be arbitrary but fixed. By definition,

$$h(n,r) = \sum_{n \equiv a+b (\text{mod } r)} f(a,r)g(b,r).$$

But, $f(a,r) = \sum_{s|r} \alpha(s,r)C(a,s)$ and $g(b,r) = \sum_{t|r} \beta(t,r)C(b,t)$. So,

$$h(n,r) = \sum_{n \equiv a+b (\text{mod } r)} \sum_{s|r} \alpha(s,r)C(a,s) \sum_{t|r} \beta(t,r)C(b,t)$$

$$= \sum_{s|r} \sum_{t|r} \alpha(s,r)\beta(t,r) \sum_{n \equiv a+b (\text{mod } r)} C(a,s)C(b,t)$$

Or,

(5.4.9) $h(n,r) = \sum_{s|r} r\alpha(s,r)\beta(s,r)C(n,s)$ for $s = t$(by (5.4.5)).

As h is even (mod r), h has a finite Fourier expansion with Fourier coefficients $r\alpha(d,r)\beta(d,r)$ for d dividing r, as stated in the theorem. $\square$

Illustration 5.4.1 : *If $C(n, r)$ denotes Ramanujan's sum*

(5.4.10)
$$\sum_{g.c.d\,(a,r)=1} C(n-a,r) = \mu(r)C(n,r),$$

where μ is the Möbius function and summation is over a reduced-residue system (mod r).

Proof of (5.4.10) is by finding the Cauchy product h of ρ and C where ρ is Kronecker function:

(5.4.11)
$$\rho(n,r) = \begin{cases} 1, & g.c.d\,(n,r) = 1 \\ 0, & g.c.d\,(n,r) \neq 1. \end{cases}$$

and C is Ramanujan's sum (5.1.2). ρ and C are even functions (mod r).

If $\rho(n,r) = \sum_{d|r} \alpha(d,r)C(n,d)$

$$\alpha(d,r) = \frac{1}{r}\sum_{t|r} \rho(\frac{r}{t},r)C(\frac{r}{d},t)$$

$$= \frac{1}{r}\sum_{\substack{t|r \\ g.c.d\,(\frac{r}{t},r)=1}} C(\frac{r}{d},t)$$

For t dividing r, $(\frac{r}{t},r) = 1$ if, and only if, $r = t$. So,

(5.4.12)
$$\alpha(d,r) = \frac{1}{r}C(\frac{r}{d},r).$$

Also, when

$$C(n,r) = \sum_{d|r} \beta(d,r)C(n,d)$$

$$\beta(d,r) = \begin{cases} 1, & \text{if } d = r, \\ 0, & \text{if } d \neq r. \end{cases}$$

Or,

(5.4.13)
$$\beta(d,r) = P(d,r),$$

where P is the 'principal function' whose value is 1 when $d = r$ and zero, otherwise. By theorem 35,

$$\sum_{\substack{a\,(\text{mod }r) \\ (a,r)=1}} C(n-a,r) = r\sum_{\substack{d|r \\ d=r}} \frac{1}{r}C(\frac{r}{d},r)C(n,d) = C(1,r)C(n,r)$$

As $C(1,r) = \mu(r)$, (see (5.4.3)), (5.4.10) follows.

5.5. Linear congruences with side conditions

A linear congruence in s unknowns $x_1, x_2, \cdots, x_s$ is of the form

$$(5.5.1) \qquad a_1x_1 + a_2x_2 + \cdots + a_sx_s \equiv n \pmod{r} \; ; \; a_i \in \mathbb{Z}, i = 1, 2, \cdots, s.$$

By a solution of (5.5.1), we mean a solution (mod r) namely, an ordered s-tuple of integers $\langle x_1, x_2, \cdots, x_s \rangle$ that satisfies (5.5.1). Two s-tuples $\langle x_1, x_2, \cdots, x_s \rangle$ and $\langle x_1', x_2', \cdots, x_s' \rangle$ that satisfy (5.5.1) are counted as the same solution if, and only if,

$$x_i \equiv x_i' \pmod{r}, \; i = 1, 2, \cdots, s.$$

We will count either all solutions of (5.5.1) or all the solutions that are restricted in some way. A problem that is relevant to our context is counting those solutions $\langle x_1, x_2, \cdots, x_s \rangle$ in which g.c.d $(x_i, r) = 1$, $i = 1, 2, \cdots s$. We may call this as a 'side-condition' in respect of solutions of (5.5.1).

Theorem 36 (unrestricted case) : *The congruence* (5.5.1) *has a solution if, and only if, $d | n$ where $d = $ g.c.d $(a_1, a_2, \cdots, a_s, r)$ and when this condition is satisfied, it has dr^{s-1} solutions.*

Proof : $:\Rightarrow$ (5.5.1) has a solution implies that there is an s-tuple $\langle t_1, t_2, \cdots t_s \rangle$ such that

$$a_1t_1 + a_2t_2 + \cdots + a_st_s \equiv n \pmod{r}$$

If $d = $ g.c.d $(a_1, a_2, \cdots, a_s, r)$, d has to divide n. So, this condition is necessary for the congruence to have a solution.

$\Leftarrow$: Suppose that $d | n$. We prove that the congruence (5.5.1) has a solution. We show that it has dr^{s-1} solutions by induction on s. For the case $s = 1$

$$(5.5.2) \qquad \frac{a_1x_1}{d} \equiv \frac{n}{d} \pmod{\frac{r}{d}}$$

yields that (5.5.2) has a unique solution (mod $\frac{r}{d}$) as g.c.d $(\frac{a_1}{d}, \frac{r}{d}) = 1$. So, $a_1x_1 \equiv n \pmod{r}$ has exactly d solutions namely

$$x_1, x_1 + \frac{r}{d}, x_1 + \frac{2r}{d}, \cdots, x_1 + \frac{(d-1)r}{d}.$$

Next, suppose that $s > 1$ and that the result is true for linear congruences with $(s-1)$ unknowns.

Let $b = $ g.c.d $(a_2, a_3, \cdots, a_s, r)$. Since $d = $ g.c.d (a_1, b) divides n, the congruence

$$a_1x_1 \equiv n \pmod{b}$$

has d solutions. Also, in every complete residue system (mod r) there are $(\frac{r}{b})d$ solutions of (5.5.1).

Let t_1 be a solution of $a_1x_1 \equiv n \pmod{b}$. We take

$$a_2x_2 + \cdots + a_sx_s \equiv (n - a_1t_1) \pmod{r}.$$

Since $b | (n - a_1t_1)$, it has br^{s-2} solutions by induction hypothesis. Therefore, (5.5.1) with s unknowns has $(\frac{r}{b})dbr^{s-2} = dr^{s-1}$ solutions. This completes the proof. $\qquad \square$

Next, we proceed to obtain a formula for the number $N(n, r, s)$ of solutions of the congruence.

(5.5.3) $$x_1 + x_2 + \cdots + x_s \equiv n \pmod{r}$$

under the restriction g.c.d $(x_i, r) = 1$ for $i = 1, 2, \cdots s$.

When $s = 2$, we have the special case of counting the number a of integers $1 \le a \le r$ such that

$$\text{g.c.d } (a, r) = \text{g.c.d } (n - a, r) = 1.$$

We denote this number by $\theta(n, r)$. One can show that

(5.5.4) $$\theta(n, r) = r \prod_{p | \text{g.c.d } (n, r)} (1 - \frac{1}{p}) \prod_{\substack{p | r \\ p \nmid n}} (1 - \frac{2}{p})$$

See [6]. $\theta(n, r)$ is referred to as Nagell's totient in the literature. The first step in the derivation of a formula for $N(n, r, s)$ is to show that it is an even function (mod r).

Lemma 5.5.1 : $N(n, r, s)$ *is an even function* (mod r).
That is, $N(n, r, s) = N((n, r), r, s)$*, where* $(n, r) = \text{g.c.d } (n, r)$.

Proof : Let $n = mn'$ with g.c.d $(n', r) = 1$. We consider

(5.5.5) $$x_1 + x_2 + \cdots + x_s \equiv m \pmod{r}$$

under the restriction g.c.d $(x_i, r) = 1$.

Let $\langle y_1, y_2, \cdots, y_s \rangle$ be a solution of (5.5.5).
Then $\langle n'y_1, n'y_2, \cdots, n'y_s \rangle$ is a solution of (5.5.3). So, to each solution of (5.5.5) there corresponds a solution of (5.5.3) and vice-versa. That is , there is a one-one correspondence between the solutions of (5.5.3) and (5.5.5). Also, g.c.d $(y_i, r) = 1$ $(i = 1, 2, \cdots, s)$ holds if, and only if, g.c.d $(n'y_i, r) = 1$ $(i = 1, 2, \cdots, s)$. Therefore, we have

$$N(n, r, s) = N(m, r, s).$$

Using the Chinese Remainder Theorem, we also note that $N(n, r, s)$ is multiplicative in r. That is,

(5.5.6) $$N(n, r, s)N(n, r', s) = N(n, rr', s)$$

whenever g.c.d $(r, r') = 1$.

So, it will suffice to show that $N(n, r, s) = N((n, r), r, s)$ when n and r are powers of same prime p.

Let $n = p^b$, $r = p^a$, $a \ge 1$, $b \ge 1$. If $b \le a$, g.c.d $(n, r) = p^b = n$. So

$$N(p^b, p^a, s) = N(\text{g.c.d}(p^b, p^a), p^a, s), \quad \text{when } b \le a.$$

When $b > a$, $p^a + p^b = p^a t$ where $p \nmid t$. Further,

$$N(p^b, p^a, s) = N(p^b + p^a, p^a, s) = N(p^a t, p^a, s) = N(\text{ g.c.d } (p^a t, p^a), p^a, s)$$
$$= N(p^a, p^a, s) \text{ , as g.c.d } (p^a, t) = 1.$$

This establishes the claim of the lemma (5.5.1). $\qquad\qquad \square$

Theorem 37 : *For all $n \in \mathbb{Z}$ and $r \geq 1$, the arithmetical evaluation of $N(n,r,s)$ is given by*

(5.5.7)
$$N(n,r,s) = \frac{1}{r}\sum_{d|r} C(\frac{r}{d},r)^s C(n,d)$$

where $C(n,r)$ denotes Ramanujan's sum (5.1.2).

Proof : Proof is by induction on s. The congruence $x_1 \equiv n(\text{mod } r)$ with g.c.d $(x_1,r) = 1$ is such that

$$N(n,r,1) = \begin{cases} 1, & \text{if g.c.d } (n,r) = 1, \\ 0, & \text{otherwise.} \end{cases}$$

That is, $N(n,r,1) = \rho(n,r)$ (5.4.11).

We have seen that

(5.5.8)
$$\begin{cases} \rho(n,r) & = \sum_{d|r} \alpha(d,r)C(n,d) \\ \text{where } \alpha(d,r) & = \frac{1}{r}C(\frac{r}{d},r), \text{ whenever } d|r \text{ (see (5.4.12)).} \end{cases}$$

So, (5.5.7) holds for $s = 1$. Suppose that $s > 1$. Assume that the result holds for $s = s' - 1$. Then,

$$N(n,r,s'-1) = \frac{1}{r}\sum_{d|r} C(\frac{r}{d},r)^{s'-1}C(n,d), n \in \mathbb{Z}.$$

By virtue of Cauchy multiplication (see definition (5.4.4)) we see that

$$N(n,r,s') = \sum_{n \equiv a+b \,(\text{mod } r)} N(a,r,1)N(b,r,s'-1).$$

By induction hypothesis, (5.5.7) holds for $s = s' - 1$. Then, by theorem 34 and the expression for $\rho(n,r)$ as in (5.5.8), we deduce that (5.5.7) holds for $s = s'$ as the Fourier coefficients for $\rho(n,r)$ are as given in (5.5.8). This completes the proof. □

Remark 5.5.1 : The evaluation (5.5.7) was discovered by K. G. Ramanathan [15] in 1944. It was rediscovered by Nicol and Vandiver [13] in 1954.

5.6. The Rademacher formula

Though theorem 37 gives an evaluation of $N(n,r,s)$ in terms of $C(n,r)$, it is desirable to give it an explicit form. For this, we use the fact that $N(n,r,s)$ is multiplicative in r and that when $r = p^a$ (p a prime and $a \geq 1$)

(5.6.1)
$$C(n,p^a) = \begin{cases} p^{a-1}(p-1), & \text{if } p^a|n \\ -p^{a-1}, & \text{if } p^{a-1}|n \text{ and } p^a \nmid n \\ 0, & \text{otherwise.} \end{cases}$$

If ϕ denotes Euler's totient, $C(n,p^a) = \phi(p^a)$ whenever $p^a|n$. This implies that

(5.6.2)
$$C(n,r) = \phi(r) \text{ whenever } r|n.$$

Notation 5.6.1 : We write $e(x)$ to represent $\exp(2\pi i x)$, i denotes $\sqrt{-1}$.

Theorem 38 (David Rearick (1963)) : *Let $C(n,r)$ denote Ramanujan's sum. If $N(n,r,s)$ represents the number of solutions of*

$$x_1 + x_2 + \cdots + x_s \equiv n(\bmod\ r)$$

under the restriction g.c.d $(x_i, r) = 1$ $(i = 1, 2, \cdots, s)$, then

$$(5.6.3) \qquad\qquad C^s(n,r) = \sum_{h(\bmod\ r)} N(h,r,s) e(\tfrac{hn}{r})$$

and

$$(5.6.4) \qquad\qquad N(n,r,s) = \frac{1}{r} \sum_{h(\bmod\ r)} C^s(h,r) e(\tfrac{-hn}{r}).$$

Proof : From the definition of $C(n,r)$, we note that

$$C^s(n,r) = \sum_{x_1(\bmod\ r)}^{*} e(\tfrac{x_1 n}{r}) \sum_{x_2(\bmod\ r)}^{*} e(\tfrac{x_2 n}{r}) \cdots \sum_{x_s(\bmod\ r)}^{*} e(\tfrac{x_s n}{r})$$

where $\displaystyle\sum_{x}^{*}$ means that the summation is over $x(\bmod\ r)$ with g.c.d $(x,r) = 1$. (that is, $\displaystyle\sum_{x}^{*}$ is over a reduced-residue system $(\bmod\ r)$). Therefore,

$$C^s(n,r) = \sum_{x_1(\bmod\ r)}^{*} \sum_{x_2(\bmod\ r)}^{*} \cdots \sum_{x_s(\bmod\ r)}^{*} e\left(\tfrac{n(x_1 + x_2 + \cdots + x_s)}{r}\right)$$

$$= \sum_{h(\bmod\ r)} e(\tfrac{nh}{r}) \sum_{\lambda}^{**} 1,$$

where $\displaystyle\sum_{x_i}^{**}$ stands for summation over s reduced-residues $x_i(\bmod\ r)$ $(i = 1, 2, \cdots, s)$ such that

$$x_1 + x_2 + \cdots + x_s \equiv h\,(\bmod\ r).$$

We infer that $\displaystyle\sum_{x_i}^{**} 1 = N(h,r,s)$. This proves (5.6.3). (5.6.4) is the same as the result given in theorem 37. It is proved by the familiar orthogonal methods of Fourier coefficients. More specifically,

$$\frac{1}{r}\sum_{h(\text{mod } r)} C^s(h,r)e(\tfrac{-hn}{r}) = \frac{1}{r}\sum_{h(\text{mod } r)} e(\tfrac{-hn}{r})\sum_{k(\text{mod } r)} N(k,r,s)e(\tfrac{kh}{r})$$

$$= \frac{1}{r}\sum_{h(\text{mod } r)}\sum_{k(\text{mod } r)} N(k,r,s)e(\tfrac{h(k-n)}{r})$$

$$= \frac{1}{r}\sum_{k(\text{mod } r)} N(k,r,s)\sum_{h(\text{mod } r)} e(\tfrac{h(k-n)}{r}).$$

The inner sum is zero unless $r|(k-n)$. As k runs through a complete residue system (mod r), $(k-n)$ also runs through a complete residue system (mod r) and $|k-n| < r$ is what we need. It is so. Therefore, $r|(k-n)$ if, and only if, $k-n=0$ and in such a situation the inner sum is r. This proves (5.6.4). $\qquad\square$

Theorem 39 (David Rearick (1963)) **:** *An evaluation of $N(n,r,s)$ is given by*

(5.6.5) $$N(n,r,s) = \frac{\phi^s(r)}{r}\prod_{p|r}(1+\frac{(-1)^s C(n,p)}{(p-1)^s})$$

where ϕ denotes Euler's totient and $C(n,r)$ is Ramanujan's sum.

Proof : From theorem 37, using Hölder relation for $C(n,r)$ given in (5.4.3), we see that

(5.6.6) $$N(n,r,s) = \frac{\phi^s(r)}{r}\sum_{d|r}\frac{\mu^s(d)}{\phi^s(d)}C(n,d)$$

$C(n,r)$ is multiplicative in r. So are ϕ and μ.

It is easy to check that if f is multiplicative

(5.6.7) $$\sum_{d|r}\mu^s(d)f(d) = \prod_{p|r}(1+(-1)^s f(p)),$$

where p runs through the primes dividing r on the right side of (5.6.7). Taking $f(r) = \frac{C(n,r)}{\phi^s(r)}$, we get from (5.6.6)

$$N(n,r,s) = \frac{\phi^s(r)}{r}\prod_{p|r}(1+\frac{(-1)^s C(n,p)}{\phi^s(p)}).$$

As $\phi(p) = p-1$, we arrive at (5.6.5). $\qquad\square$

Corollary 5.6.1 : *Rademacher's formula for $N(n,r,s)$ is*

$$N(n,r,s) = r^{s-1}\prod_{p|g.c.d\,(n,r)}\frac{(p-1)\{(p-1)^{s-1}-(-1)^{s-1}\}}{p^s}\prod_{p|r,p\nmid n}\frac{(p-1)^s-(-1)^s}{p^s}.$$

For, it is easy to note that $C(n, p) = \phi(p) = (p-1)$ if $p|n$ and $C(n, p) = (-1)$ if $p \nmid n$. Separating the prime factors p of r as those p for which $p|n$ and those p for which $p \nmid n$ and using the fact that

$$\phi^s(r) = r^s \prod_{p|r}(1 - \tfrac{1}{p})^s = r^s \prod \frac{(p-1)^s}{p^s},$$

we get

$$N(n, r, s) = r^{s-1} \prod_{p|r} \frac{(p-1)^s}{p^s} \prod_{p|\text{g.c.d}\,(n,r)} (1 + \frac{(-1)^s(p-1)}{(p-1)^s}) \prod_{p|r, p\nmid n} (1 + \frac{(-1)^s(-1)}{(p-1)^s})$$

$$= r^{s-1} \prod_{p|\text{g.c.d}\,(n,r)} \{\frac{(p-1)\{(p-1)^{s-1}-(-1)^{s-1}\}}{p^s}\} \prod_{p|r, p\nmid n} \frac{(p-1)^s - (-1)^s}{p^s}$$

which is as stated. □

Remark 5.6.1 : Theorems 38 and 39 have been adapted from David Rearick [16].

Remark 5.6.2 : In [1], Henry L. Alder defines the function $\phi(n, r)$ as the number of solutions $\langle x, y \rangle$ of the equation

$$x + y = n + r$$

satisfying $1 \le x \le r$ and g.c.d $(x, r) = $ g.c.d $(y, r) = 1$. $\phi(n, r)$ is the same as $N(n, r, 2)$ which is $\theta(n, r)$ (5.5.4).

For more results of this kind, see Eckford Cohen [6], [8]. See also Paul J. McCarthy [12] for a beautiful exposition of counting linear congruences with specified restrictions.

The use of computers for doing problems in elementary number theory especially in topics dealing with primality test, factorization and solving congruences is recommended for practical training. Choosing PASCAL for programs, Peter Giblin [A1] gives an interesting study of primes and programming touching upon number-theoretic aspects of cryptography.

5.7. Notes with illustrative examples

During the 7th century A.D. the Indian mathematician Brahmagupta (598–665 A.D.) posed the following problem:

Find a positive integer such that when divided by $3, 4, 5$ and 6 it leaves the remainders $2, 3, 4$ and 5 respectively. The system of congruences would mean

$$x \equiv 2 \,(\text{mod } 3)$$
$$x \equiv 3 \,(\text{mod } 4)$$
$$x \equiv 4 \,(\text{mod } 5)$$
$$x \equiv 5 \,(\text{mod } 6)$$

Here, the moduli are not relatively prime in pairs. However, one could apply the Chinese Remainder Theorem to solve the first three congruences simultaneously. Taking $M = 3 \cdot 4 \cdot 5 = 60$, $r_1 = 3$, $r_2 = 4$, $r_3 = 5$ we have $M_i = \frac{M}{r_i}$ and so

$$M_1 = 20, \ M_2 = 15, \ M_3 = 12$$

$M_1 t_1 \equiv 1 \, (\text{mod } r_1)$ gives

$$t_1 \equiv 1 \, (\text{mod } 3), \text{ or } t_1 = 2,$$
$$15 t_2 \equiv 1 \, (\text{mod } 4), \text{ or } t_2 = 3,$$
$$12 t_3 \equiv 1 \, (\text{mod } 5), \text{ or } t_3 = 3.$$

So, $x_0 = 20 \cdot 2 \cdot 2 + 15 \cdot 3 \cdot 3 + 12 \cdot 3 \cdot 4 = 359 \equiv -1 \, (\text{mod } 60)$ So, the general solution of the first three congruences is $x = 60k - 1$. But, then, $60k - 1 \equiv 5 \, (\text{mod } 6)$. So, the general solution is $60k - 1$ and the least positive integer satisfying all the four congruences is 59. When the moduli are not relatively prime in pairs, one has to do a 'splitting' of the moduli into relatively prime numbers and consider the minimum number of congruences for forming a system. Let us look at the simultaneous congruences

$$5x \equiv 2 \ (\text{mod } 24),$$
(5.7.1) $\qquad\qquad\qquad 3x \equiv 62 \ (\text{mod } 88),$
$$x \equiv 28 \ (\text{mod } 99).$$

In (5.7.1), the first is equivalent to $x \equiv 10 \, (\text{mod } 24)$. The second is equivalent to $x \equiv 50 \, (\text{mod } 88)$. So, we get $x \equiv 10 \, (\text{mod } 24)$, $x \equiv 50 \, (\text{mod } 88)$ and $x \equiv 28 \, (\text{mod } 99)$. We factorize $24, 88$ and 99 and obtain a system of non-repeated simultaneous congruences

$$x \equiv 2 \, (\text{mod } 8),$$
$$x \equiv 6 \, (\text{mod } 11)$$
$$\text{and } x \equiv 1 \, (\text{mod } 9)$$

in which the moduli are relatively prime in pairs. The solution is $x \equiv 226 \ (\text{mod } 792)$.

When properly set, the Chinese Remainder Theorem holds in any commutative ring with unity. Theorem 32 is one such.

A generalization of $N(n, r, s)$ was given by Paul J. McCarthy [12] in 1977. It covered several other congruences under suitable side-conditions. For instance, if $P_k(n, r, s)$ (5.6.4) denotes the number of solutions of

$$x_1 + x_2 + \cdots + x_s \equiv n \, (\text{mod } r)$$

under restriction g.c.d (x_i, r) is a $k^{\text{th}}-$ power $(i = 1, 2, \cdots, s)$, then,

(5.7.2) $\qquad\qquad P_k(n, r, s) = \frac{1}{r} \sum_{d \mid r} (H_k(\frac{r}{d}, r))^s \, C(n, d)$

where

(5.7.3) $$H_k(n,r) = \sum_{d^k|r} C(n, \tfrac{r}{d^k})$$

See Eckford Cohen [8] also.

In connection with the evaluation of $N(n,r,s)$ (5.6.4), David Rearick [16] obtained a 'cross-correlation function' of $N(n,r,s)$ by showing that

(5.7.4) $$\sum_{m(\bmod\ r)} N(m,r,s)N(m+k,r,t) = N(k,r,s+t).$$

The summation on the left of (5.7.4) runs over a complete residue system (mod r). For each value of $m(\bmod\ r)$ inside the sum, we note that $N(m,r,s)$ gives the number of solutions of

$$x_1 + x_2 + \cdots + x_s \equiv m\,(\bmod\ r)$$

under the restriction g.c.d $(x_i, r) = 1$, $i = 1,2,\cdots,s$.

$N(m+k,r,t)$ gives the number of solutions of

$$y_1 + y_2 + \cdots + y_t \equiv m+k\ (\bmod\ r)$$

under the restriction g.c.d $(y_j, r) = 1$, $j = 1,2,\cdots,t$.

Accordingly, there are $N(m,r,s)N(m+k,r,t)$ solutions of the simultaneous congruences

$$x_1 + x_2 + \cdots + x_s \equiv m\,(\bmod\ r)$$
$$y_1 + y_2 + \cdots + y_t - x_1 - x_2 - \cdots - x_s \equiv k\,(\bmod\ r)$$

with g.c.d $(x_i, r) = 1$ $(i = 1,2,\cdots,s)$, g.c.d $(y_j, r) = 1$, $(j = 1,2,\cdots,t)$. So, $\sum_{m(\bmod\ r)} N(m,r,s)N(m+k,r,t)$ yields the number of solutions of

(5.7.5) $$y_1 + y_2 + \cdots + y_t - x_1 - x_2 - \cdots - x_s \equiv k\,(\bmod\ r)$$

under the restriction g.c.d $(x_i, r) = 1$ $(i = 1,2,\cdots,s)$, g.c.d $(y_j, r) = 1$, $(j = 1,2,\cdots,t)$. The number of solutions of the restricted congruence (5.7.5) is evidently $N(k,r,s+t)$ as g.c.d $(r-u,r) = 1$ for $1 \le u < r \Rightarrow$ g.c.d $(u,r) = 1$. This proves (5.7.4).

As $N(n,r,1) = \rho(n,r)$ (Kronecker function (5.4.11)), we get, from (5.7.4), the relation

(5.7.6) $$\sum_{\substack{m(\bmod\ r) \\ \text{g.c.d}\ (m,r)=1}} N(m+k,r,t) = N(k,r,t+1)$$

(5.7.6) was also obtained by A. Brauer in [2].

5.8. Worked-out examples

a) (Underwood Dudley) Construct linear congruences modulo 20 that have
   (i) no solutions
   (ii) exactly one solution
   (iii) more than one solution
   (iv) exactly 20 solutions

Answer:

(i) $15x \equiv 14 \pmod{20}$ has no solution as g.c.d $(15, 20) = 5$ does not divide 14.

(ii) $13x \equiv 14 \pmod{20}$ has exactly one solution, as g.c.d $(13, 20) = 1$.

(iii) $12x \equiv 4 \pmod{20}$ has exactly 4 solutions, as g.c.d $(12, 20) = 4$ and 4 divides 4.

(iv) $20x \equiv 0 \pmod{20}$ has 20 solutions. $\square$

b) (Ralph G. Archibald) Let p be an odd prime. Suppose that $f(x)$, $g(x) \in \mathbb{Z}[x]$ are of degrees m and n respectively and $x^{p-1} - 1 \equiv f(x)g(x) \pmod{p}$ identically. Then, show that $f(x) \equiv 0 \pmod{p}$ and $g(x) \equiv 0 \pmod{p}$ have m and n incongruent solutions, respectively, modulo p.

Answer: Two polynomials in $\mathbb{Z}[x]$ are said to be identically congruent to one another modulo k, if the coefficients of like terms in the two polynomials are congruent to one another modulo k. For instance, $f(x) = 5x^3 - 2x^2 + x + 5$ and $g(x) = 6x^4 - x^3 + 10x^2 - 5x - 1$ are such that $f(x) \equiv g(x) \pmod{6}$ identically.

By Fermat's little theorem, $x^{p-1} - 1 \equiv 0 \pmod{p}$ has precisely $(p-1)$ incongruent solutions modulo p, namely,

(5.8.1) $x \equiv 1, 2, 3, \ldots, (p-1) \pmod{p}$

if $x^{p-1} - 1$ is factorized modulo p into polynomials $f(x)$, $g(x) \in \mathbb{Z}[x]$, $f(x) \equiv 0 \pmod{p}$ cannot have more than m incongruent solutions (as deg $f(x) = m$). In the same manner, $g(x) \equiv 0 \pmod{p}$ cannot have more than n incongruent solutions. However, since

$$f(x)g(x) \equiv 0 \pmod{p}$$

has exactly $p - 1 = m + n$ solutions, $f(x) \equiv 0 \pmod{p}$ cannot have fewer than m incongruent solutions and $g(x) \equiv 0 \pmod{p}$ cannot have fewer than n incongruent solutions, modulo p. Moreover, $f(x) \equiv 0 \pmod{p}$ and $g(x) \equiv 0 \pmod{p}$ cannot have a solution in common.

This completes the answer. $\square$

c) (Nicol and Vandiver). Given $r > 1$ and

$$\Phi(n, r) = \frac{\mu(\frac{r}{g})\phi(r)}{\phi(\frac{r}{g})}; \quad g = g.c.d(n, r)(\text{see } (5.4.3)),$$

show that

(5.8.2) $$\sum_{s=1}^{r} s\Phi(s, r) = \frac{r\phi(r)}{2}.$$

Answer: Let $\alpha = \exp(\frac{2\pi i}{r})$. From (5.1.2) and (5.4.3), we note that

(5.8.3) $$\Phi(n, r) = \sum_{\text{g.c.d }(k, r) = 1} \alpha^{nk}.$$

So,

$$\sum_{s=1}^{r} s\Phi(s,r) = \sum_{s=1}^{r} s \sum_{\text{g.c.d } (k,r)=1} \alpha^{ks}$$

or,

(5.8.4) $$\sum_{s=1}^{r} s\Phi(s,r) = \sum_{\text{g.c.d } (k,r)=1} (\alpha^{k} + 2\alpha^{2k} + \cdots + r\alpha^{rk})$$

It is verified that

(5.8.5) $$\alpha^{k} + 2\alpha^{2k} + \cdots + r\alpha^{kr} = r\left(\frac{1}{\alpha^{k}-1} + 1\right)$$

The cyclotomic polynomial $F_r(x)$ is given by

(5.8.6) $$F_r(x) = \prod_{\text{g.c.d } (k,r)=1} (x - \alpha^{k}).$$

Differentiating and letting $x = 1$, we have

(5.8.7) $$\frac{F_r'(1)}{F_r(1)} = -\sum_{\text{g.c.d } (k,r)=1} \frac{1}{(\alpha^{k}-1)}$$

A formula due to Hölder says

(5.8.8) $$F_r'(1) = \begin{cases} \frac{1}{2}\phi(r), & r \text{ not a power of a prime,} \\ p^{m}(p-1), & \text{if } r = p^{m}. \quad m \geq 1. \end{cases}$$

Further, it is known that

(5.8.9) $$F_r(1) = \begin{cases} 1, & \text{if } r \text{ contains two prime factors;} \\ p, & \text{otherwise.} \end{cases}$$

From (5.8.5), (5.8.7), (5.8.8) and (5.8.9), we deduce that

$$\sum_{s=1}^{r} s\,\Phi(s,r) = -\frac{r\phi(r)}{2} + r\phi(r)$$

which yields (5.8.2). $\qquad\qquad\square$

Remark 5.8.1 : Worked-out example (c) has been drawn from [13, Theorem III].

EXERCISES

1. *Mark the following statements true (T) or false (F) justifying your answer briefly.*
   a) *The number of solutions of* $24x \equiv 18 \pmod{21}$ *is four.*
   b) *In order that* $3^{k} \equiv 1 \pmod{10}$*, one should have* $k \geq 4$.
   c) *The congruence* $4x^{2} + x \equiv 14 \pmod{13}$ *is solvable.*

d) *The only solution t of the simultaneous congruences*

$$5x \equiv 6 \,(\text{mod}\,7)$$
$$4x \equiv 5 \,(\text{mod}\,9)$$

for which $0 < t \le 100$ *is* $t = 53$.

e) *Let* $\Phi(n,r) = \dfrac{\mu(\frac{r}{g})\phi(r)}{\phi(\frac{r}{g})}$; $g = g.c.d\,(n,r)$. *For* $r > 1$, *one gets*

$$\sum_{d|r} \Phi(d,r) = r \prod_{p|r} (1 - 2/p).$$

f) *Let* $f(x) = a_n x^n + a_1 x^{n-1} + \cdots + a_1 x + a_0 \in \mathbb{Z}[x]$, $a_n \neq 0$. *Suppose that* $m_1, m_2, \ldots, m_k$ *are pairwise relatively prime positive integers. The number of solutions of*

$$f(x) \equiv 0 \,(\text{mod}\, m_1 m_2 \ldots m_k)$$

equals the product of the numbers of solutions of

$$f(x) \equiv 0 \,(\text{mod}\, m_1),$$
$$f(x) \equiv 0 \,(\text{mod}\, m_2),$$
$$\cdots\cdots$$
$$f(x) \equiv 0 \,(\text{mod}\, m_k).$$

2. *Find the least positive integer which simultaneously satisfies*

$$5x \equiv 2 \,(\text{mod}\, 13)$$
$$x \equiv 1 \,(\text{mod}\, 25)$$
$$3x \equiv 4 \,(\text{mod}\, 11)$$
$$x \equiv 7 \,(\text{mod}\, 20).$$

3. *Solve the congruence :*

$$71x \equiv 4 \,(\text{mod}\, 55).$$

4. *(Landau) Let* $m \equiv 5 \,(\text{mod}\, 12)$ *and* $m > 17$. *Show that* m *is expressible as a sum of three distinct square numbers.*

5. *Find the least positive integer* N *that satisfies*

$$N \equiv 9 \,(\text{mod}\, 11),$$
$$N \equiv 13 \,(\text{mod}\, 28),$$
$$N \equiv 7 \,(\text{mod}\, 45).$$

6. *Solve the congruences*

$$5x \equiv 2 \,(\text{mod}\, 13),$$
$$x \equiv 2 \,(\text{mod}\, 35),$$
$$3x \equiv 13 \,(\text{mod}\, 77),$$
$$x \equiv 7 \,(\text{mod}\, 20).$$

7. *Solve the congruences simultaneously*

$$x \equiv 2 \,(\text{mod } 6),$$
$$x \equiv 3 \,(\text{mod } 5),$$
$$x \equiv 5 \,(\text{mod } 11).$$

8. *Examine whether the following pair of congruences can be solved simultaneously*

$$5x \equiv -2 \,(\text{mod } 10),$$
$$x \equiv 1 \,(\text{mod } 4).$$

9. *Suppose that a, b are positive integers which are relatively prime to one another. Given an integer n, show that there exists an integer m for which g.c.d $(ma+b,n) = 1$.*

10. *Suppose that $\{a_1, a_2, \cdots, a_n\}$ is a set of nonzero elements of a P.I.D say D. Assume that a_i and a_j are relatively prime to one another when $i \neq j$. If $a = [a_1, a_2, \cdots, a_n]$ (l.c.m), show that*

$$D/(a) \cong \oplus \sum_{i=1}^{n} \left(D/(a_i) \right)$$

11. *Let F be a field of 3 elements. Solve the system of congruences*

$$(x^2+1)u(x) \equiv b_1(x) \,(\text{mod } x^2+2),$$
$$(x+1)u(x) \equiv b_2(x) \,(\text{mod } x^2+1),$$

where the polynomials are from $F[x]$.
(Hint : $GF(3) = \{0,1,\alpha\}$ where $\alpha^2 = 1$, $\alpha+1 = 0$.
$x^2+2 = (x+1)(x+\alpha)$ and g.c.d $(x^2+1, x^2+2) = 1$.)

12. *(Eckford Cohen) Let $M(n,r,s)$ denote the number of solutions of*

$$x_1 + x_2 + \cdots + x_s \equiv n \,(\text{mod } r)$$

under the restriction g.c.d $((x_1, x_2, \cdots x_s), r) = 1$. Show that

$$M(n,r,s) = (\tfrac{r}{g})^{s-1} \phi_s(g), \quad g = g.c.d\,(n,r)$$

where $\phi_s(r) = \sum_{d|r} \mu(\tfrac{r}{d}) d^s$.

13. *(Eckford Cohen) Let $M'(n,r,s)$ denote the number of solutions $x_i \,(\text{mod } r)$, $y_i \,(\text{mod } r)$ $(i = 1,2,3,\cdots,s)$ of the congruence*

$$x_1 y_1 + x_2 y_2 + \cdots + x_s y_s \equiv n \,(\text{mod } r).$$

Show that

$$M'(n,r,s) = r^{s-1} \sum_{d|g} d\phi_s(\tfrac{r}{d}); \quad g = g.c.d\,(n,r)$$

and $\phi_s(r)$ is as given in exercise 12.

14. *(Eckford Cohen) Let $D(n,r,s)$ denote the number of solutions (mod r) of the congruence*

$$x + y_1 + y_2 + \cdots + y_s \equiv n \ (\text{mod } r)$$

under the restriction g.c.d $(x,r) = g.c.d(g.c.d(y_1, y_2, \cdots y_s), r) = 1$. Show that

$$D(n,r,s) = r^{s-1} \phi(r) \sum_{\substack{d \mid r \\ g.c.d \ (d,n)=1}} \frac{\mu(d)}{\phi(d) d^{s-1}},$$

where the summation on the right is over those divisors d (of r) such that g.c.d $(d,n) = 1$. For $s = 1$, deduce that Nagell's totient $\theta(n,r)$ is given by

$$\theta(n,r) = \phi(r) \sum_{\substack{d \mid r \\ g.c.d \ (d,n)=1}} \frac{\mu(d)}{\phi(d)}.$$

REFERENCES

[1] Henry L. Alder : A generalization of the Euler $\phi-$ function, Amer. Math. Monthly 65 (1958) 690–692.

[2] A. Brauer : Lösungen der Aufgaben 30 Deutsh math-Verein 35 (1926) 2 Abteilung 92–94.

[3] David M. Burton : A first course in rings and ideals, Chapter 10 pp 204–216. Addison Wesley Pub Co., Reading, Mass. USA (1970).

[4] Eckford Cohen : A class of arithmetical functions, Proc. Nat. Acad. Sci. (USA) 41 (1955) 939–944.

[5] Eckford Cohen : An extension of Ramanujan's Sum II Duke Math. J. 23 (1956) 121–126.

[6] Eckford Cohen : Representations of even functions (mod r) I, Arithmetical Identities, Duke Math. J. 25 (1958) 401–422.

[7] Eckford Cohen : Representations of even functions (mod r) II, Cauchy products, Duke Math. J. 26 (1959) 165–182.

[8] Eckford Cohen : A class of residue systems (mod r) and related arithmetical functions I. A generalization of Möbius inversion, Pacific J. Math. 9 (1959) 13–23.

[9] C. Ding, D. Pei and A. Salomaa : Chinese Remainder Theorem, Applications in computing, coding and cryptography, chapters 1 & 2 pp 1–32, World Scientific Pub Co., (P) Ltd. Singapore (1996).

[10] Hugh M. Edgar: A first course in number theory, Sections 3.1 to 3.4 and notes to chapter 3, Wadsworth Pub Co., (P) Ltd., Belmont, California USA (1988).

[11] Thomas W. Hungerford : Algebra (original edition), chapter III pp 109–133, Holt Rinehart & Winston Inc (1974). Reprinted Springer Verlag GTM 73 (1986).

[12] Paul J. McCarthy : Counting restricted solutions of a linear congruence, Nieuw Archief voor Wiskunde (3) XXV (1977) 133–147.

[13] C. A. Nicol & H. S. Vandiver : A Von Sterneck arithmetical function and restricted partitions with respect to a modulus. Proc. Nat. Acad. Sci. (USA) 40 (1954) 825–835.

[14] H. Rademacher : Aufgabe 30, Jahr. Deutch Math verein 34 (1925) 2, Abteihung 158.

[15] K. G. Ramanathan: Some applications of Ramanujan's trigonometrical sum $C_m(n)$: Proc. Ind. Acad. Sci. (A) 20 (1944) 62–69.

[16] David Rearick: A linear congruence with side conditions. Amer. Math. Monthly 70 (1963) 837–840.

[17] Sah, Chih Han: Abstract Algebra, Chapter III section III-5 pp 89–94, Academic Press Inc NY (1967).

[18] Sun Tsu: "Suan-ching" (arithmetic) edited by Y. Mikaini Abhandlungen-Geichichte de Mathemetischen Wissenchaften 30 (1912) p 32.

ADDITIONAL REFERENCE

[A1] Peter Giblin : Primes and Programming– An Introduction to Number Theory and Programming, Cambridge University Press (1993).

CHAPTER 6

Reciprocity laws

Historical perspective

By a quadratic congruence, we mean a congruence of the form
$x^2 \equiv a \pmod{m}$ where a, m are integers and $m \geq 2$. If it has a solution we say that
a is a quadratic residue of m, written aRm. Otherwise, a is said to be a quadratic
non-residue of m written aNm. More precisely, we give

Definition 6.0.1 : Let $m > 1$ and g.c.d $(a, m) = 1$. a is a quadratic residue of m if
$x^2 \equiv a \pmod{m}$ has a solution. If $x^2 \equiv a \pmod{m}$ has no solution, a is a quadratic
non-residue of m.

It is convenient to suppose that the modulus m is a prime, say p. If $x^2 \equiv$
$a \pmod{p}$ has a solution, there are two distinct solutions $x \equiv \alpha \pmod{p}$ and $x \equiv$
$-\alpha \pmod{p}$. It is possible to show that there are two solutions of $x^2 \equiv a \pmod{p^k}$,
$k \geq 1$. The Legendre symbol $(a|p)$ is given by

$$(6.0.1) \qquad (a|p) = \begin{cases} 1, & \text{if } aRp, \\ -1, & \text{if } aNp, \\ 0, & \text{if } p|a. \end{cases}$$

If p and q are odd primes, the quadratic reciprocity law stated by Adrien-Marie
Legendre (1752–1833) in 1785 says that

$$(6.0.2) \qquad (p|q)(q|p) = (-1)^{\frac{(p-1)(q-1)}{4}}$$

See Gauss [5]. This was proved by Gauss. In fact, Gauss succeeded in discover-
ing eight different demonstrations of (6.0.2). Gauss made use of a lemma which
goes by his name. Many other mathematicians have given proofs. According
to Emil Grosswald [7], Paul Bachmann (1837–1920) counted 45 proofs. More
number of proofs were also coming up abundantly. One of Gauss's own proofs
is due to his student F. G. Eisenstein (1822–1852). The shortest known proof is
due to Georg Frobenius (1849–1917). For a detailed account of different proofs
of quadratic reciprocity law, see F. Lemmermeyer [A2]. A new elementary proof
is found in Sey Y. Kim [A1].
It is known that $x^2 \equiv -1 \pmod{p}$ has a solution if, and only if, p is a prime of the
form $(4k + 1)$. That is,

$$(-1|p) = 1 \iff p \equiv 1 \pmod 4$$

For, if $p \equiv 1 \pmod 4$, $-1 \equiv \{(\frac{p-1}{2})!\}^2 \pmod p$. That is, two of the square roots of -1 in Z/pZ are $\pm(\frac{p-1}{2})!$ Conversely, if $(-1|p) = 1$, there exists $t \in Z$ such that

$$t^2 \equiv -1 \pmod p$$

But, $t^{p-1} \equiv 1 \pmod p$. So, $(t^2)^{\frac{p-1}{2}} \equiv 1 \pmod p$ or $(-1)^{\frac{p-1}{2}} \equiv 1 \pmod p$. That is, $\frac{p-1}{2}$ is even and so $p \equiv 1 \pmod 4$. Thus, there is an odd prime p for which $(-1|p) = 1$ and that $p \equiv 1 \pmod 4$. Each of the infinitely many odd primes p for which $(-1|p) = 1$ is of the form $(4k+1)$. So, there are an infinite number of primes of the form $(4k+1)$. Legendre's quadratic reciprocity law (also referred to as Gauss reciprocity law) enables one to show that certain arithmetic progressions contain infinitely many primes. Dirichlet's theorem [1] states that there are infinitely many primes of the form $ax+b$, where a and b are relatively prime to one another. Undoubtedly, Legendre's (Gauss) reciprocity law occupies a pivotal place in the elementary theory of numbers.

6.1. Introduction

The aim of this chapter is to prove the quadratic reciprocity law using finite fields. Eisenstein's cubic reciprocity law is discussed by considering the ring $Z[\omega]$ where ω is an imaginary cube root of unity. Reciprocity laws are also viewed in a general setting. As pointed out by W. C. Waterhouse [13], the mode of formation of reciprocity laws is suggested by Gauss lemma (stated in theorem 41).

6.2. Preliminaries

We begin with a polynomial $f(x) = c_0 x^n + c_1 x^{n-1} + \ldots + c_n \in Z[x]$. An integer t which satisfies

(6.2.1) $f(x) \equiv 0 \pmod r$ ($r \in \mathbb{N}$, and r arbitrary, but fixed)

is said to be a root of the congruence (6.2.1). If t is a root, so is any integer congruent to $t \pmod r$. Congruent roots are considered to be equivalent. When the congruence has m incongruent roots, we say that the congruence has m roots. There is no analogue of the fundamental theorem of algebra for polynomial congruences such as (6.2.1). However, given a prime p, if $f(x) \equiv 0 \pmod p$ with $c_0 \not\equiv 0 \pmod p$, the congruence has at most n roots or n solutions modulo p. For proof, see Tom Apostol [1]. Further, if $f(x) \equiv 0 \pmod p$ has more than n roots, then every coefficient c_i ($i = 0, 1, 2, \ldots, n$) is divisible by p.

Fermat's little theorem says that

$$x^{p-1} - 1 \equiv 0 \pmod p$$

has $(p-1)$ roots namely $\{t_1, t_2, \ldots, t_{p-1}\}$ where t_i ($i = 1, 2 \ldots (p-1)$) are nonzero residues $\pmod p$ from a complete set of residues modulo p.

If $d \mid (p-1)$, the congruence

$$x^d - 1 \equiv 0 \pmod p$$

has exactly d roots. See [1] or [8].

Definition 6.2.1 : *An integer a is called a primitive root* (mod *p*), *if* [a] *generates the group U of nonzero elements of* $\mathbb{Z}/p\mathbb{Z}$ *(The group U has order* $\phi(p) = (p-1)$*)*.

Thus, if *a* is a primitive root (mod *p*), $(p-1)$ is the smallest positive integer such that

(6.2.2) $$a^{p-1} - 1 \equiv 0 (\text{mod } p)$$

and vice versa.

More generally, *a* is called a primitive root (mod *r*) $(r \geq 2)$ if $\phi(r)$ is the least positive integer such that

$$a^{\phi(r)} - 1 \equiv 0 (\text{mod } r);$$

that is, if [a] generates the group *U* of units in $\mathbb{Z}/r\mathbb{Z}$.

6.2.1. EMIL ARTIN'S CONJECTURE: If *a* is not a square and $a \neq -1$, there are infinitely many primes *p* for which *a* is a primitive root (mod *p*).
For a general exposition of this conjecture see L. J. Goldstein [6].

The conjecture remains unproven. Moving on to certain classes of number fields, E. Artin conjectured that *a* 'form of degree *d*' in $n > d^2$ variables has a non-trivial zero. This is proved by L. Carlitz for the special case where the form is defined in relation to the field *K* of rational functions over a finite field $\mathbb{F}_q$.

It is known [1] that an integer *r* possesses primitive roots if, and only if, *r* is of the form 2, 4, p^t or $2p^t$ where *p* is an odd prime and $t \geq 1$. It amounts to saying that the group *U* of units in $\mathbb{Z}/r\mathbb{Z}$ is cyclic if, and only if, $r = 2, 4, p^t$ or $2p^t$.

Definition 6.2.2 : *Let r, n* $\in \mathbb{N}$, *a* $\in \mathbb{Z}$ *and g.c.d* $(a,r) = 1$, *a is called an nth-power residue* mod *r, if*

$$x^n \equiv a (\text{mod } r)$$

has a solution.

Theorem 40 : *Suppose that r* $\in \mathbb{N}$ *possesses primitive roots. Let g.c.d* $(a,r) = 1$. *Then a is an* n^{th}*-power residue* (mod *r*) *if, and only if,*

(6.2.3) $$a^{\frac{\phi(r)}{d}} - 1 \equiv 0 (\text{mod } r)$$

where d = g.c.d $(n, \phi(r))$.

Proof : $:\Rightarrow$ Let *g* be a primitive root (mod *r*).
Then, the numbers $g, g^2, \ldots, g^{\phi(r)}$ form a reduced residue system (mod *r*). So, as g.c.d $(a,r) = 1$, we could write

$$a = g^b \text{ (say) and}$$

when g.c.d $(x,r) = 1, x = g^y$ (say).
The congruence $x^n - a \equiv 0$ (mod *r*) is equivalent to $g^{ny} - g^b \equiv 0$ (mod *r*). As *g* is a primitive root (mod *r*), $\phi(r)$ is the least positive integer for which

$$g^{\phi(r)} \equiv 1 (\text{mod } r)$$

and so, $ny \equiv 0$ (mod $\phi(r)$) and $b \equiv 0$ (mod $\phi(r)$).

Thus,

(6.2.4) $$ny \equiv b \,(\mathrm{mod}\ \phi(r))$$

(6.2.4) is solvable if, and only if, $d = \mathrm{g.c.d}\,(n, \phi(r))$ divides b. Also, if there is one solution for (6.2.4), there are d solutions for (6.2.4). If $d|b$,

$$a^{\frac{\phi(r)}{d}} \equiv g^{\frac{b\phi(r)}{d}} \equiv 1 \,(\mathrm{mod}\ r).$$

That is, for a to be an nth power residue $(\mathrm{mod}\ r)$, it is necessary that (6.2.3) holds. $\Leftarrow$: If $a^{\frac{\phi(r)}{d}} \equiv 1 \,(\mathrm{mod}\ r)$, then,

$$g^{\frac{b\phi(r)}{d}} \equiv 1 \,(\mathrm{mod}\ r).$$

As g is a primitive root $(\mathrm{mod}\ r)$, $\phi(r) | \phi(r)\frac{b}{d}$ or $d|b$. So, $ny \equiv b \,(\mathrm{mod}\ \phi(r))$ has a solution. That is, $x^n - a \equiv 0 \,(\mathrm{mod}\ r)$ has a solution or a is an nth-power residue $(\mathrm{mod}\ r)$. $\qquad\square$

Corollary 6.2.1 (Euler's criterion) : *Let p be an odd prime.*

$$aRp \Leftrightarrow a^{\frac{p-1}{2}} \equiv 1 \,(\mathrm{mod}\ p).$$

This follows from the fact that $\mathrm{g.c.d}\,(2, \phi(p)) = 2$.

6.3. Gauss lemma

Given $r = 2^b p_1^{b_1} \ldots p_k^{bk}$ where $b \geq 0$, $b_i \geq 1$ $(i = 1, 2, \ldots, k)$ and $p_1, p_2 \ldots p_k$ are distinct odd primes dividing r, we attempt to solve

(6.3.1) $$x^2 \equiv a(\mathrm{mod}\ r), \quad \mathrm{g.c.d}\,(a, r) = 1.$$

By the Chinese Remainder Theorem, we know that (6.3.1) is equivalent to the system of congruences:

$$x^2 \equiv a \,(\mathrm{mod}\ 2^b)$$
$$x^2 \equiv a \,(\mathrm{mod}\ p_1^{b_1})$$
$$\cdots\cdots\cdots\cdots\cdots$$
$$\cdots\cdots\cdots\cdots\cdots$$
$$x^2 \equiv a \,(\mathrm{mod}\ p_k^{b_k}).$$

For $x^2 \equiv a \,(\mathrm{mod}\ 2^b)$, we dispose of the cases $b = 1, 2$ and 3. $x^2 \equiv a \,(\mathrm{mod}\ 2)$ gives $x \equiv 1 \,(\mathrm{mod}\ 2)$. $x^2 \equiv a(\mathrm{mod}\ 4)$ gives $a = 1$ and $x^2 \equiv a \,(\mathrm{mod}\ 8)$ gives $a = 1$.

The values $b = 2$, $b = 3$ are such that 1 is the only quadratic residue modulo 4 as well as modulo 8. For $b \geq 3$, the structure of the group U of units of $\mathbb{Z}/2^b\mathbb{Z}$ is such that U is not a cyclic group.

Lemma 6.3.1 : *If a is odd and $b \geq 3$,*

$$a^{2^{b-2}} \equiv 1 \,(\mathrm{mod}\ 2^b)$$

Proof : When $b = 3$, each element of the group U of units in $\mathbb{Z}/8\mathbb{Z}$ is of order 2 and so U is not cyclic. Moreover, one has

$$1^2 \equiv 3^2 \equiv 5^2 \equiv 7^2 \equiv 1 \,(\text{mod } 8)$$

proving the lemma for $b = 3$.

To prove the lemma for $b > 3$, we shall apply induction on b. Suppose that the lemma is true for $b = q$ (say) $q \geq 3$.

Then $a^{2^{q-2}} \equiv 1 \,(\text{mod } 2^q)$ or $a^{2^{q-2}} = 1 + t\, 2^q, t \in \mathbb{Z}$. Squaring

$$\left(a^{2^{q-2}}\right)^2 = 1 + 2(t\, 2^q) + t^2 2^{2q}.$$

Therefore, $a^{2^{q-1}} \equiv 1 \,(\text{mod } 2^{q+1})$. As the result is true for $b = 3$, induction on b is complete. $\square$

Remark 6.3.1 : Let G_2 be the cyclic subgroup of the group U of units in $\mathbb{Z}/2^b\mathbb{Z}$, generated by the element [5]. By lemma 6.3.1, one has $5^{2^{b-2}} \equiv 1 \,(\text{mod } 2^b)$. Also it is verified that $5^{2^{b-3}}$ is not congruent to 1 (mod 2^b). The order of G_2 is 2^{b-2}. But, the order of U is 2^{b-1}.

As 5^f is not congruent to $-1 \,(\text{mod } 2^b)$ for any $f \geq 0$, -1 is not congruent to any power of 5 modulo 2^b. That is, $-1 \notin G_2$. As $|G_2|$ is one half of $|U|$, the index of G_2 in U is 2. Therefore, U is a disjoint union of the cosets G_2 and $(-1)\, G_2$ (of G_2 in U). So the set $S = \{[-1], [5]\}$ generates U. So, we could write U as isomorphic to $C(2) \times G_2$ where $C(2)$ is cyclic of order 2 and G_2 is cyclic of order $2^{b-2} (b \geq 3)$.

The structure of the group U of units of $\mathbb{Z}/2^b\mathbb{Z}$ ($b \geq 3$) suggests that any element y of U is of the form:

(6.3.2) $y = \pm 5^\alpha$ chosen from the appropriate residue class (mod 2^b),

$$1 \leq \alpha \leq 2^{b-2}.$$

Now, if 5^α is a solution of (6.3.1) with $r = 2^b$, $2^b - 5^\alpha$ is also a solution. And, $5^{2\alpha} \equiv a \,(\text{mod } 2^b)$ is such that $5^{2\alpha} - a = m\, 2^b$, $m \in \mathbb{Z}$ or $(1+4)^{2\alpha} - a = m\, 2^b$ or $1 + 4^{2\alpha} + 2 \cdot 4^\alpha - a \equiv m\, 2^b$ with $b \geq 3$. Then, $a \equiv 1 \,(\text{mod } 8)$.

The conclusion is that $x^2 \equiv a \,(\text{mod } 2^b)$ has a solution

$$t \equiv 5^\alpha \,(\text{mod } 2^b) \Rightarrow a \equiv 1 \,(\text{mod } 8), (1 \leq \alpha \leq 2^{b-2})$$

Fact 6.3.1 : $x^2 \equiv a \,(\text{mod } 2^b)$, $b \geq 3$ is solvable if, and only if, $a \equiv 1 \,(\text{mod } 8)$. When solutions exist, there are four of them (modulo 2^b). Further, $x^2 \equiv a \,(\text{mod } 8)$ is solvable if, and only if, $x^2 \equiv a \,(\text{mod } 2^b)$ is solvable for all $b \geq 3$.

Next, in the case of $x^2 \equiv a \,(\text{mod } p_i^{b_i})$ $i = 1, 2, \ldots, k$, we proceed as follows: As $p_i^{b_i}$ possesses primitive roots, by theorem 40, a is a quadratic residue (mod p_i)

if, and only if,

$$(6.3.3) \qquad a^{\frac{\phi(p_i^{b_i})}{2}} - 1 \equiv 0 \pmod{p_i^{b_i}}$$

This is okay as g.c.d $(2, \phi(p_i - 1)) = 2$. But $\phi(p_i^{b_i}) = p_i^{b_i-1}(p_i - 1)$. So, (6.3.3) reads

$$(6.3.4) \qquad (a^{\frac{p_i-1}{2}})^{p_i^{b_i-1}} \equiv 1 \pmod{p_i^{b_i}}.$$

If aRp_i, by Euler's criterion (corollary 6.2.1) $a^{\frac{p_i-1}{2}} \equiv 1 \pmod{p_i}$.
So (6.3.4) is satisfied and thus $aRp_i^{b_i}$.

Conversely, if $aRp_i^{b_i}$, from (6.3.4) letting $m = a^{\frac{p_i-1}{2}}$.

$$(6.3.5) \qquad m^{p_i^{b_i-1}} \equiv 1 \pmod{p_i^{b_i}}$$

As the group of units of $\mathbb{Z}/p_i^{b_i}\mathbb{Z}$ is cyclic, $m^{\phi(p_i^{b_i})} \equiv 1 \pmod{p_i^{b_i}}$ or

$$(6.3.6) \qquad m^{(p_i^{b_i} - p_i^{b_i-1})} \equiv 1 \pmod{p_i^{b_i}}.$$

From (6.3.5) and (6.3.6), we see that $m^{p_i^{b_i}} \equiv 1 \pmod{p_i^{b_i}}$. This implies that $m \equiv 1 \pmod{p_i}$ or Euler's criterion holds.

Fact 6.3.2 : $x^2 \equiv a \pmod{p^b}$ ($b \geq 1$), p, an odd prime, is solvable if, and only if, $x^2 \equiv a \pmod{p}$ is solvable. Thus, it suffices to consider quadratic congruences modulo a prime p. For proof, see [12].

Theorem 41 (Gauss lemma) **:** *Let p be an odd prime. Suppose that*

$$S = \{-\frac{(p-1)}{2}, \frac{-(p-3)}{2}, \ldots, -1, 1, 2, \ldots, \frac{(p-1)}{2}\}$$

represents the set of nonzero least residues (mod p). *Further, assume that $p \nmid a$. Let μ be the number of least negative residues of the integers $1a, 2a, \cdots, (\frac{p-1}{2})a$. Then $(a|p) = (-1)^{\mu}$.*

Proof : We consider the products $ta(\text{mod } p)$, $1 \leq t \leq \frac{p-1}{2}$. Let $\pm b_t$ be the least residue of $ta(\text{mod } p)$ where b_t is positive. As $1 \leq t \leq \frac{p-1}{2}$, μ is the number of negative signs arising in this manner.

Claim : $b_t \neq b_s$ if $t \neq s$ and t, s lie between 1 and $\frac{p-1}{2}$.

Assume the contrary. If $b_t = b_s$, then $ta \equiv \pm sa \pmod{p}$. Since p does not divide a, $p|(t \pm s)$. This is impossible since $t \neq s$ and $|t \pm s| \leq |t| + |s| \leq (p-1)$. It follows that the sets $\{1, 2, \ldots, \frac{p-1}{2}\}$ and $\{b_1, b_2, \ldots, b_{\frac{p-1}{2}}\}$ coincide. Now,

$$1a \equiv \pm b_1 \pmod{p}$$
$$2a \equiv \pm b_2 \pmod{p}$$
$$\ldots\ldots\ldots\ldots$$
$$\ldots\ldots\ldots\ldots$$
$$\left(\frac{p-1}{2}\right) a \equiv \pm b_{\frac{p-1}{2}} \pmod{p}.$$

Multiplying the left sides and right sides, we get

$$\left(\frac{p-1}{2}\right)! \, a^{\left(\frac{p-1}{2}\right)} \equiv (-1)^{\mu} \left(\frac{p-1}{2}\right)! \, (\bmod \, p).$$

This yields $a^{\frac{p-1}{2}} \equiv (-1)^{\mu}$ (mod p) and Euler's criterion (corollary 6.2.1) gives the desired result.   □

From the definition of Legendre symbol $(a|p)$ (6.0.1), we note that

(6.3.7) $\qquad\qquad\qquad a^{\frac{p-1}{2}} \equiv (a|p) \, (\bmod \, p)$

(6.3.8) $\qquad\qquad\qquad (ab|p) = (a|p)(b|p)$, for all $a, b \in \mathbb{Z}$

(6.3.9) $\qquad\qquad\qquad (a|p) = (b|p)$, if $b \equiv a \, (\bmod \, p)$

Now,

(6.3.10) $\qquad\qquad x^{p-1} - 1 \equiv (x^{\frac{p-1}{2}} - 1)(x^{\frac{p-1}{2}} + 1) \, (\bmod \, p)$

But $x^{p-1} - 1 \equiv 0$ (mod p) has $(p-1)$ solutions $[1], [2], \ldots, [p-1]$. As $x^{\frac{p-1}{2}} \equiv 1$ (mod p) has $\left(\frac{p-1}{2}\right)$ solutions, there are $\left(\frac{p-1}{2}\right)$ quadratic residues (mod p). So, half of the nonzero residues $[1], [2], \ldots, [p-1]$ are quadratic residues and the remaining half quadratic non-residues (mod p).

Using (6.3.7), one deduces that
$$(2|p) = 1 \text{ if, and only if, } p \equiv 1 \text{ or } 3 \, (\bmod \, 8).$$

6.4. Finite fields and quadratic reciprocity law

This section is meant to obtain Gauss quadratic reciprocity law, using Gauss's quadratic sum defined in terms of primitive p^{th} roots of unity, where p is a prime. We have seen properties of finite fields in Section 4.6 of chapter 4.

Let p be a prime and $s \geq 1$. There exists a finite field F having $q = p^{s}$ elements. F is a vector space of dimension s over $\mathbb{Z}/p\mathbb{Z}$. $F^{*} = F \setminus \{0\}$ is a cyclic group of order $q-1$ and having $\phi(q-1)$ generators. Now,

$$x^{q} - x = \prod_{a \in F}(x-a),$$

where $x^{q} - x \in F[x]$. Let $b \in K$, where K is an extension of F. Then, $b \in F$ if, and only if, $b^{q} = b$, as the zeros of the polynomial $x^{q} - x$ are precisely the q elements of F.

Next, we take g to be a generator of the cyclic group F^{*}. Then, g^{j} ($1 \leq j \leq q-1$) is an nth-root of unity, if $g^{jn} = 1_{F}$. This happens in a cyclic group of order $(q-1)$. So, $g^{jn} = 1_{F} \Leftrightarrow jn \equiv 0$ (mod $q-1$). That is, the number of nth-roots of unity contained in F^{*} is equal to the g.c.d $(n, q-1)$, as the number of solutions of

$$nx \equiv 0 \, (\bmod \, q-1)$$

is equal to g.c.d $(n, q-1)$. If $n|(q-1)$, the powers of g namely

$$g^{j}, g^{2j}, \ldots, g^{(n-1)j}, g^{nj} = 1_{F}, \quad (\text{where } nj = (q-1))$$

run through the nth-roots of unity. If g.c.d $(n, q-1) = 1$, 1_F is the only n^{th}-root of unity contained in F^*. See Neal Koblitz [9].

We recall that a complex number ζ is a primitive n^{th}-root of unity, if n is the least positive integer such that $\zeta^n = 1$. ζ is, in fact, a generator of a cyclic group of order n.

Given an odd prime p, suppose that ζ denotes a complex p^{th}-root of unity. We could take $\zeta = \exp(\frac{2\pi i}{p})$. For $p = 5$, we have $\zeta = \exp(\frac{2\pi i}{5})$. ζ is not an element of $\mathbb{Z}/5\mathbb{Z}$ as the nonzero elements of $\mathbb{Z}/5\mathbb{Z}$ are $\{1, g, g^2, g^3\}$ with $g^4 = 1$. However, we could adjoin ζ to $\mathbb{F}_5 = \mathbb{Z}/5\mathbb{Z}$ and get an extension E of $\mathbb{F}_5$ where $|E| = 5^4$. A basis for E is $\{1, \zeta, \zeta^2, \zeta^3\}$ where $\zeta^5 = 1$. When $p = 7$ and $\omega = \exp(\frac{2\pi i}{3})$, as $\omega^3 = 1$, $2 \cdot 3 \equiv 0 \pmod 6$ yields $\omega = g^2$ where g generates the cyclic group of order 6, namely $\mathbb{F}_7^* = \mathbb{Z}/7\mathbb{Z} \setminus \{0\}$. The set $\{1, \omega, \omega^2\}$ forms a subgroup of $\mathbb{F}_7^*$ (of order 3). Indeed, ω and ω^2 are in $\mathbb{F}_7^*$.

We fix an odd prime p. $\zeta = \exp(\frac{2\pi i}{p})$ is a primitive p^{th}-root of unity.

Definition 6.4.1 : *A quadratic Gauss sum is defined by*

$$G(a, \zeta) = \sum_{j=0}^{p-1} (j|p)\zeta^{aj}.$$

In particular, we write

(6.4.1) $$G(1, \zeta) = G(\zeta) = \sum_{j=0}^{p-1} (j|p)\zeta^j$$

Lemma 6.4.1 : $G(a, \zeta) = (a|p)G(\zeta)$

Proof : If $p|a$, $\zeta^{aj} = 1$ for $0 \leq j \leq (p-1)$.

Then, $G(a, \xi) = \sum_{j=0}^{b-1}(j|p) = 0$, since the numbers 1 to $(p-1)$ are such that half of them are quadratic residues (mod p) and the remaining quadratic non-residues (mod p).

And $(a|p) = 0$ when $p|a$. So, the lemma is true for the case where $p|a$.

Next, assume that p does not divide a.

Then,

$$(a|p)\, G(a, \zeta) = \sum_{j=0}^{p-1} (a|p)(j|p)\zeta^{aj} = \sum_{j=0}^{p-1} (aj|p)\zeta^{aj},$$

as the Legendre symbol is multiplicative. But, then, since $p \nmid a$, aj ($j = 0, 1, \ldots, (p-1)$) constitute a complete residue system (mod p). So,

$$(a|p)\, G(a, \zeta) = \sum_{k=0}^{p-1} (k|p)\, \zeta^k = G(\zeta).$$

Since $(a|p)^2 = 1$, the lemma, for the case p not dividing a, holds.   □

Lemma 6.4.2 :

$$G^2(\zeta) = (-1)^{\frac{p-1}{2}} p.$$

Proof :

We evaluate $G^2(\zeta)$ in two different ways. Firstly, summing for $j = 0, 1, \ldots, (p-1)$ and secondly for $-j = p, (p-1), \ldots, 1$ in the reverse order. So,

$$G^2(\zeta) = \sum_{j=0}^{p-1} (j|p)\zeta^j \sum_{k=p}^{1} (-k|p)\zeta^{-k}$$

$$= \sum_{j,k=0}^{p-1} (-1|p)(kj|p)\zeta^{j-k}$$

$$= (-1|p) \sum_{j=0}^{p-1} \sum_{k=0}^{p-1} (kj|p)\zeta^{j-k}.$$

It is known that $(-1|p) = (-1)^{\frac{p-1}{2}}$. For each value of j we make a change of variable in the inner sum namely $k \mapsto kj$. That is, for each j, kj gives a set of residues mod p, as k does. The summands depend only on the residues (mod p). So,

$$G^2(\zeta) = (-1)^{\frac{p-1}{2}} \sum_{j=0}^{p-1} \sum_{k=0}^{p-1} (j^2 k|p)\zeta^{j(1-k)}$$

$$= (-1)^{\frac{p-1}{2}} \sum_{k=0}^{p-1} (k|p) \sum_{j=0}^{p-1} \zeta^{j(1-k)}$$

For $k \neq 1$, the inner sum gives the sum of the jth powers of $\theta = \zeta^{1-k}$, a complex p^{th}-root of unity. Therefore,

$$\sum_{j=0}^{p-1} \zeta^{j(1-k)} = 0 \quad \text{for } k \neq 1.$$

When $k = 1$ $\sum\limits_{j=0}^{p-1} \zeta^0 = p$ and so

$$G^2(\zeta) = (-1)^{\frac{p-1}{2}} (1|p)p = (-1)^{\frac{p-1}{2}} p,$$

as desired. $\qquad\qquad\qquad\qquad\qquad\qquad\qquad\qquad\qquad\qquad\qquad\qquad\qquad\square$

Theorem 42 (Quadratic reciprocity law) **:** *If p and q are distinct odd primes, then*

(6.4.2) $$(p|q)(q|p) = (-1)^{\frac{(p-1)(q-1)}{4}}$$

Proof : Since p and q are distinct odd primes, they are relatively prime to one another. So, there exists a positive integer n such that

$$q^n \equiv 1 \, (\text{mod } p)$$

We remark that n could be $(p-1)$. Let $\mathbb{F}_q$ be a field having q^n elements. $\mathbb{F}_q^* = \mathbb{F}_q \setminus \{0\}$ is a cyclic group of order $q^n - 1$. Let g be a generator of $\mathbb{F}_q^*$.

We set $h = g^\lambda$ where $\lambda = \frac{q^n-1}{p}$, an integer. $h^p = g^{q^n-1} = 1$. In fact, p is the least positive integer such that $h^p = 1$. So, order of h in $\mathbb{F}_q^*$ is p.

Analogous to the definition of a quadratic Gauss sum (definition 6.4.1), we write

$$(6.4.3) \qquad\qquad \chi(a,h) = \sum_{j=0}^{p-1} (j|p)h^{aj}, \quad a \in \mathbb{Z}.$$

As $h \in \mathbb{F}_q^*$, $\chi(a,h) \in \mathbb{F}_q^*$.
If $p|a$, $h^{aj} = 1$. Then,

$$\chi(a,h) = \sum_{j=0}^{p-1} (j|p) = 0 = (a|p),$$

(as there are $\frac{b-1}{2}$ quadratic residues and $\frac{b-1}{2}$ quadratic non-residues mod p). If $p \nmid a$,

$$(a|p)\chi(a,h) = \sum_{j=0}^{p-1} (aj|p)h^{aj}$$

As j runs through a complete residue system (mod p), so is the case with aj, as $p \nmid a$. Therefore,

$$(a|p)\chi(a,h) = \sum_{k=0}^{p-1} (k|p)h^k = \chi(1,h) = \chi(h) \text{ (say)}.$$

Thus,

$$(6.4.4) \qquad\qquad \chi(a,h) = (a|p)\chi(h)$$

As in the case of lemma 6.4.2, one has

$$(6.4.5) \qquad\qquad \chi^2(1,h) = \chi^2(h) = (-1)^{\frac{p-1}{2}}[p],$$

where $[p]$ is the coset of p in $\mathbb{Z}/q\mathbb{Z}$. From (6.4.5), we note that

$$(6.4.6) \qquad\qquad ((-1)^{\frac{p-1}{2}}p|q) = 1 \text{ if, and only if, } \chi(h) \in \mathbb{Z}/q\mathbb{Z}.$$

Next, let K be a finite extension of $\mathbb{Z}/q\mathbb{Z}$. Then, $\alpha \in K$ is in $\mathbb{Z}/q\mathbb{Z}$ if, and only if, $\alpha^q = \alpha$, since all the zeros of $t^q - t$ are in the field $\mathbb{Z}/q\mathbb{Z}$. Therefore,

$$(6.4.7) \qquad\qquad ((-1)^{\frac{p-1}{2}}p|q) = 1 \text{ if, and only if, } \chi^q(h) = \chi(h).$$

Now,

$$\chi^q(h) = \Big(\sum_{j=0}^{p-1} (j|p)h^j\Big)^q, \quad \text{where } h \in \mathbb{F}_q^* \text{ and } h^p = 1.$$

In a field F of characteristic q, when $a, b \in F$, the relation $(a+b)^q = a^q + b^q$ holds. So,

$$\chi^q(h) = \sum_{j=0}^{p-1} (j|p) h^{qj}, \text{ as char } \mathbb{F}_q = q$$

Thus,

(6.4.8) $$\chi^q(h) = \chi(q, h), \quad q \in \mathbb{Z} \text{ by (6.4.3)}.$$

Using (6.4.4), we get

(6.4.9) $$\chi(q, h) = (q|p)\chi(h).$$

Therefore, from (6.4.7) to (6.4.9), we note that

$$((-1)^{\frac{p-1}{2}} p|q) = 1 \text{ if, and only if, } (q|p) = 1.$$

That is,

(6.4.10) $$((-1)^{\frac{p-1}{2}} |q)(p|q) = 1 \text{ if, and only if, } (q|p) = 1$$

But, $(-1|q) = (-1)^{\frac{q-1}{2}}$, as q is an odd prime. So,

$$((-1)^{\frac{p-1}{2}} |q) = ((-1)^{\frac{q-1}{2}})^{\frac{p-1}{2}} = (-1)^{\frac{(p-1)(q-1)}{4}}$$

Thus, from (6.4.10), we deduce that

(6.4.11) $$(-1)^{\frac{(p-1)(q-1)}{4}} (p|q) = 1 \text{ if, and only if, } (q|p) = 1$$

Also,

(6.4.12) $$(-1)^{\frac{(p-1)(q-1)}{4}} (p|q) = -1 \Rightarrow (q|p) = -1$$

(by the contrapositive argument applied to (6.4.11)). So,

$$(-1)^{\frac{(p-1)(q-1)}{4}} (p|q)(q|p) = 1,$$

which completes the proof of theorem 42. $\qquad\qquad \square$

Remark 6.4.1 : In the proof of theorem 42, (6.4.5) mentions about $[p]$, the coset of p in $\mathbb{Z}/q\mathbb{Z}^* = \mathbb{F}_q^*$. It refers to a residue class modulo q. As h is a pth-root of unity in $\mathbb{F}_q^*$, it is advisable to consider congruence modulo q in the ring $\mathcal{A}$ of algebraic integers. (We recall that an algebraic integer is a zero of a monic polynomial with coefficients from $\mathbb{Z}$. It is true that $\omega \in \mathcal{A}$ can be a rational number and an algebraic integer. If $\omega \in \mathbb{Q}$, ω is necessarily an element of $\mathbb{Z}$. That is, ω is a rational integer.)

We introduce a congruence relation modulo ω in $\mathcal{A}$ by writing

(6.4.13) $$a \equiv b \,(\mathrm{mod}\, \omega), \text{ where } a, b \in \mathbb{Z}.$$

The above congruence means that $a - b = c\omega$ when $c \in \mathbb{Z}$. So, while considering residue classes modulo q, we are actually working with congruences modulo q in the ring $\mathcal{A}$ of algebraic integers. This is a point to be noted, while giving the proof of the quadratic reciprocity law using finite fields.

Remark 6.4.2 : Quadratic reciprocity law is an essential tool to find out primes p for which $x^2 \equiv a$ (mod p) has a solution (of course, for a given a). We have pointed out that -1 is a quadratic residue of primes p of the form $4k+1$. Similarly, by using the formula

$$(2|p) = (-1)^{\frac{p^2-1}{8}}$$

one obtains that 2 is a quadratic residue of a prime p of the form $8k+1$ or $8k+7$.

Fact 6.4.1 : -3 is a quadratic residue of primes p of the form $12k+1$ or $12k-5$.

For proof, see Emil Grosswald [7].

For more examples and illustrations, see Tom M. Apostol [1], K. Ireland and M. I. Rosen [8] or Don Redmond [12].

6.5. Cubic residues (mod p)

a denotes a fixed integer and p a prime. When $x^3 \equiv a$ (mod p) has a solution, a is said to be a cubic residue modulo p.

A reciprocity law similar to that of theorem 42 is obtainable in the context of cubic residues. The result is due to Eisenstein [3]. The candidate for cubic reciprocity is the ring $\mathbb{Z}[\omega]$ of Eisenstein integers where $\omega = \exp(\frac{2\pi i}{3})$. This ring is described as

$$(6.5.1) \qquad \mathbb{Z}[\omega] = \{a+b\omega : a,b \in \mathbb{Z}\}$$

Lemma 6.5.1 : $\mathbb{Z}[\omega]$ *is a Euclidean domain.*

Proof : It is easy to check that $\mathbb{Z}[\omega]$ is an integral domain. For $\alpha \in \mathbb{Z}[\omega]$, we define the norm $N(\alpha)$ of α by

$$N(\alpha) = \alpha\bar{\alpha} = a^2 - ab + b^2, \quad \text{where } \alpha = a+b\omega.$$

Let $0 \neq \beta \in \mathbb{Z}[\omega]$.

$$\frac{\alpha}{\beta} = \frac{\alpha\bar{\beta}}{\beta\bar{\beta}} = s+t\omega \text{ where } s,t \in \mathbb{R}.$$

$\alpha\bar{\beta} \in \mathbb{Z}[\omega]$ and $\beta\bar{\beta}$ is a positive integer.
We find integers m,n such that $|s-m| \leq 1/2$ and $|t-n| \leq 1/2$.
As $m,n \in \mathbb{Z}$, $\gamma = m+n\omega \in \mathbb{Z}[\omega]$.

Also, $N(\frac{\alpha}{\beta} - \gamma) = (s-m)^2 - (s-m)(t-n) + (t-n)^2 \leq 1/4 + 1/4 + 1/4 < 1$.
If $\delta = \alpha - \beta\gamma$, then either $\delta = 0$, or

$$N(\delta) = N(\beta(\frac{\alpha}{\beta} - \gamma))$$

$$= N(\beta)N(\frac{\alpha}{\beta} - \gamma)$$

$$< N(\beta).$$

Thus, $N : \mathbb{Z}[\omega] \to \mathbb{N}$ determines a Euclidean norm in $\mathbb{Z}[\omega]$ satisfying the conditions for making $\mathbb{Z}[\omega]$ a Euclidean domain. ((3.3.6) of chapter 3 gives a criterion for Euclidean domain property). □

Remark 6.5.1 : $\alpha \in \mathbb{Z}[\omega]$ is a unit if, and only if, $N(\alpha) = 1$.
For, if $\alpha = a + b\omega$, $N(\alpha) = 1 \Rightarrow a^2 - ab + b^2 = 1$
    or, $4(a^2 - ab + b^2) = 4$
    or, $(2a - b)^2 + 3b^2 = 4$.
The following possibilities arise:
    (i) $2a - b = \pm 1$, $b = \pm 1$
    (ii) $2a - b = \pm 2$, $b = 0$

Solving the above six pairs of equations, we get $a = 1$, $b = 0$; $a = -1$, $b = 0$; $a = 0$, $b = 1$; $a = 0$, $b = -1$; $a = 1$, $b = 1$ and $a = -1$, $b = -1$.
As $1 + \omega + \omega^2 = 0$, $\omega^2 = -1 - \omega$, the units in $\mathbb{Z}[\omega]$ form the set $\{1, -1, \omega, -\omega, \omega^2, -\omega^2\}$. For each of these, the norm N takes the value 1.

Facts 6.5.1 :

a) If π is a prime in $\mathbb{Z}[\omega]$, $N(\pi) = p$ or p^2 where p is a rational prime.
b) If $N(\pi) = p^2$, π is an associate of p.
c) If $\pi \in \mathbb{Z}[\omega]$ and $N(\pi) = p$, then π is a prime in $\mathbb{Z}[\omega]$.
d) $1 - \omega$ is a prime in $\mathbb{Z}[\omega]$ as $N(1 - \omega) = 3$, a prime in $\mathbb{Z}$.

For, if $N(\pi) = n$ $(n > 1)$. $\pi\bar{\pi} = n \in \mathbb{Z}$ as n is a product of primes, π divides a rational prime p. There exists $\eta \in \mathbb{Z}[\omega]$ such that $\pi\eta = p$. Then, $N(\pi)N(\eta) = N(p) = p^2$. So, $N(\pi)$ is either p or p^2. If $N(\eta) = 1$, π and p are associates.
If $N(\pi) = N(\eta) = p$, $\pi = up'$ where u is a unit and p' is a rational prime. Then, $p = p'^2$ leads to a contradiction. So, π is not an associate of a rational prime. Then, π is a prime in $\mathbb{Z}[\omega]$, since $p = \pi\eta$ with $N(\pi)$, $N(\eta) > 1$ is impossible.
Next, $1 - \omega$ is a prime in $\mathbb{Z}[\omega]$ as $N(1 - \omega) = (1 - \omega)(1 - \omega^2) = 3$ a rational prime.
In what follows, we denote $\mathbb{Z}[\omega]$ by D. We can introduce the notation of congruence among the elements of D.

Definition 6.5.1 : *For α, β, $\gamma \in D$ with $\gamma \neq 0$, we say that $\alpha \equiv \beta \pmod{\gamma}$, if γ divides $(\alpha - \beta)$.*

As in $\mathbb{Z}$, the congruence classes $\pmod{\gamma}$, form a ring called the residue class ring modulo γ, denoted by $D/\gamma D$.

Lemma 6.5.2 : *(a) If p is a prime $\equiv 2 \pmod 3$, then p is a prime in D.*
(b) if p is a prime $\equiv 1 \pmod 3$, then $p = \pi\bar{\pi}$ where π is a prime in D.

Proof : (a) We are given that $p \equiv 2 \pmod 3$.
If p were not a prime in D, $p = \pi\eta$, where $N(\pi), N(\eta)$ are greater than 1. Then, $p^2 = N(\pi)N(\eta)$. So, $N(\pi) = N(\eta) = p$. Writing $\pi = a + b\omega$, $a, b \in \mathbb{Z}$, we

get $p = a^2 - ab + b^2$ or $4p = (2a-b)^2 + 3b^2$. That is $p \equiv (2a-b)^2 \pmod 3$. If 3 does not divide p, as 1 is the only square such that $1 \equiv (2a-b)^2 \pmod 3$, p has to be congruent to 1 (mod 3). So, $p = \pi\eta$ with $N(\pi), N(\eta) > 1$ is impossible when $p \equiv 2 \pmod 3$. This proves (a).

(b) If $p \equiv 1 \pmod 3$, using quadratic reciprocity law,

$$(-3|p) = (-1|p)(3|p) = (-1)^{\frac{p-1}{2}}(-1)^{(\frac{p-1}{2})(\frac{3-1}{2})}(p|3)$$
$$= (p|3)$$
$$= 1, \text{ as } p \equiv 1 \pmod 3$$

Therefore, we can find t such that $t^2 \equiv -3 \pmod p$.

That is, p divides $(t + \sqrt{-3})(t - \sqrt{-3}) = (t + 1 + 2\omega)(t - 1 - 2\omega) = \alpha\beta$ (say) where $\omega = \frac{1+\sqrt{-3}}{2}$. If p were a prime in D, p will divide either α or β. As $p \neq 2$ and p is odd, if $p \geq 7$, p divides neither α nor β. So, $p = \alpha\beta$ where $\alpha\beta$ are non-units in D. Therefore,

$$p^2 = N(\alpha)N(\beta)$$

gives $p = N(\alpha) = \alpha\bar\alpha$ or $p = N(\beta) = \beta\bar\beta$. This proves (b). □

Remark 6.5.2 : From lemma 6.5.2, we get the structure of a prime (in D) other than $1 - \omega$. See [8, chapter 9].

Theorem 43 : *Let* $\pi \in D$ *be a prime. Then* $D/\pi D$ *is a finite field with* $N(\pi)$ *elements.*

Proof : As D is a Euclidean domain, D is a PID. As π is a prime in D, the principal ideal πD is a prime ideal of D. So, $D/\pi D$ is an integral domain. We get through if we show that $D/\pi D$ is finite.

Let $\alpha \in D$ and α is not congruent to 0 (mod π). It means that $\pi \nmid \alpha$. So, there exist $\beta, \gamma \in D$ such that

$$\beta\alpha + \gamma\pi = 1_D \quad (1_D = 1, \text{ here})$$

So, $\beta\alpha \equiv 1 \pmod \pi$. So the residue class of α is a unit in $D/\pi D$.

Claim : $D/\pi D$ has $N(\pi)$ elements.

Case 1. Suppose that $\pi = p_1$, a rational prime congruent to 2 (mod 3).

Let $\mu = m + n\omega \in D$.

We apply the division algorithm in $\mathbb{Z}$ to m and n taking p_1 as the divisor. Then,

$$m = p_1 s + a,$$
$$n = p_1 t + b;$$

$s, t, a, b \in \mathbb{Z}$ and $0 \leq a < p_1, 0 \leq b < p_1$. Then $\mu \equiv a + b\omega \pmod{p_1}$. Suppose that $a + b\omega \equiv a' + b'\omega \pmod{p_1}$.

$$0 \leq a < p_1, 0 \leq b < p_1; 0 \leq a' < p_1, 0 \leq b' < p_1.$$

Then, $\dfrac{a-a'}{p_1} + \dfrac{(b-b')\omega}{p_1}$ is in D.

So, then,

$$\frac{a-a'}{p_1} \in \mathbb{Z}, \frac{b-b'}{p_1} \in \mathbb{Z}.$$

This is possible only if $a = a'$, $b = b'$.

So, the set $S = \{a+b\omega : 0 \le a < p_1, 0 \le b < p_1\}$ is a complete set of coset representatives in D/p_1D. So, D/p_1D has $p_1^2 = N(p_1)$ elements.

Case 2. Let p_2 be a rational prime $\equiv 1 \pmod 3$. Then, $p_2 = \pi\bar{\pi}$ where π is a prime in D. See (b) of lemma 6.5.2.

Now, let $\pi = a + b\omega$. Since $p_2 = a^2 - ab + b^2$, it follows that $p_2 \nmid b$. Let $\mu = m + n\omega$. There exists an integer s such that $bs \equiv n \pmod{p_2}$. Then, $\mu - s\pi \equiv (m-sa) \pmod{p_2}$. Therefore,

$$\mu \equiv (m-sa) \pmod{\pi}$$

So, every element such as μ is congruent to a rational integer $\pmod{\pi}$. If $c \in \mathbb{Z}$, we write

$$c = qp_2 + r \text{ where } q, r \in \mathbb{Z}, 0 \le r < p_2.$$

Then, $c \equiv r \pmod{p_2}$ and as $\pi | p_2$,

$$c \equiv r \pmod{\pi}.$$

So every element of D, by reduction modulo p_2, is congruent to an element of $\{0, 1, 2, \ldots, (p_2-1)\}$ modulo π.

If $r \equiv r' \pmod{\pi}$ with $r, r' \in \mathbb{Z}$, we have

$$0 \le r < p_2, \quad 0 \le r' < p_2$$

and then $(r - r') = \pi\delta$ for some $\delta \in D$. Therefore,

$$(r - r')^2 = N(\pi)N(\delta) = p_2N(\delta) \text{ which implies that } p_2 | (r - r').$$

So, then, $r = r'$. Thus, $D/\pi D$ has $N(\pi) = p_2$ elements.

Case 3. $\pi = 1 - \omega$. If $\pi = 1 - \omega$, $N(\pi) = 3$ and by an argument similar to that of case 2, $D/\pi D$ has 3 elements.

Hence $D/\pi D$ is a finite integral domain and so $D/\pi D$ is a field. □

Remark 6.5.3 : When π is a prime, the multiplicative group of $D/\pi D$ has $N(\pi) - 1$ elements. The multiplicative group is cyclic.

When $[\alpha] \in D/\pi D$, if $\pi \nmid \alpha$, one has

(6.5.2) $$\alpha^{N(\pi)-1} \equiv 1 \pmod{\pi}$$

which is an analogue of Fermat's little theorem relating to $\mathbb{Z}[\omega]$.

Since $\{1, \omega, \omega^2\}$ is a cyclic group of order 3 contained in $D/\pi D$, we also see that 3 divides $N(\pi) - 1$.

Lemma 6.5.3 : *Let π be a prime such that $N(\pi) \ne 3$ and that $\pi \nmid \alpha$ ($\alpha \in D$). Then, there exists a unique integer m belonging to the set $\{0, 1, 2\}$ such that*

$$\alpha^{\frac{N(\pi)-1}{3}} \equiv \omega^m \pmod{\pi}.$$

Proof : By (6.5.2), $\pi|(\alpha^{N(\pi)-1}-1)$.

However,

(6.5.3) $\qquad \alpha^{N(\pi)-1}-1 = (\alpha^{\frac{N(\pi)-1}{3}}-1)(\alpha^{\frac{N\pi-1}{3}}-\omega)(\alpha^{\frac{N(\pi)-1}{3}}-\omega^2)$.

Since π is a prime, π must divide one of the factors on the right side of (6.5.3). It can divide at the most one factor only. For, if π were to divide two of the factors, it would divide their difference which would mean that π will divide either $1-\omega$, $1-\omega^2$ or $\omega(1-\omega)$. As $1-\omega$ is a prime in D and ω or $1+\omega$ is a unit, π divides one of the factors on the right side of (6.5.3). So $\alpha^{\frac{N(\pi)-1}{3}} \equiv 1$, ω or ω^2 (mod π), as claimed. $\qquad\qquad\square$

Definition 6.5.2 : *Let π be a prime in D. If $N(\pi) \neq 3$ and $\alpha \in D$, the cubic character of α modulo π denoted by $(\alpha|\pi)_3$ is defined by*
  (i) $(\alpha|\pi)_3 = 0$ if $\pi|\alpha$,
  (ii) $\alpha^{\frac{N(\pi)-1}{3}} \equiv (\alpha|\pi)_3 \,(\text{mod }\pi)$,
where $(\alpha|\pi)_3 = 1, \omega$ or ω^2.

Remark 6.5.4 : $(\alpha|\pi)_3$ is the analogue of the Legendre symbol $(a\,|\,p)$ where $a \in \mathbb{Z}$ and p is a rational prime.

Lemma 6.5.4 : *The following statements hold in respect of $(\alpha|\pi)_3$.*
  (i) $(\alpha|\pi)_3 = 1$ if, and only if, $x^3 \equiv \alpha\,(\text{mod }\pi)$ has a solution in D.
  (ii) $\alpha^{\frac{N(\pi)-1}{3}} \equiv (\alpha|\pi)_3\,(\text{mod }\pi)$
  (iii) For $\alpha, \beta \in D$, $\quad (\alpha|\pi)_3(\beta|\pi)_3 = (\alpha\beta|\pi)_3$
  (iv) If $\alpha \equiv \beta\,(\text{mod }\pi)$, $\quad (\alpha|\pi)_3 = (\beta|\pi)_3$

Proof : (i) We appeal to the following particular case of theorem 40. Suppose that F is a finite field having q elements (q a prime). Let $\alpha \in F^*$. Then, $x^n = \alpha$ has a solution in F if, and only if,

$$\alpha^{\frac{q-1}{d}} = 1 \text{ where } d = \text{g.c.d }(n, q-1)$$

In the case of $D/\pi D$ with $N(\pi)$ elements, as $3\,|\,(N(\pi)-1)$, $x^3 = \alpha$ has a solution in $D/\pi D$ if, and only if,

$$\alpha^{\frac{N(\pi)-1}{3}} = 1$$

This proves (i).
  (ii) is a consequence of lemma 6.5.3.
  (iii) For $\alpha, \beta \in D$,

$$(\alpha\beta|\pi)_3 = (\alpha\beta)^{\frac{N(\pi)-1}{3}} \equiv \alpha^{\frac{N(\pi)-1}{3}} \cdot \beta^{\frac{N(\pi)-1}{3}} \equiv (\alpha|\pi)_3(\beta|\pi)_3\,(\text{mod }\pi)$$

  (iv) If $\alpha \equiv \beta\,(\text{mod }\pi)$

$$(\alpha|\pi)_3 = \alpha^{\frac{N(\pi)-1}{3}} = \beta^{\frac{N(\pi)-1}{3}} \equiv (\beta|\pi)_3\,(\text{mod }\pi)$$

  So, $(\alpha|\pi)_3 = (\beta|\pi)_3$. $\qquad\qquad\square$

Notation 6.5.1 : If a is a complex number, $\bar{a}$ denotes the conjugate of a. The cubic character of α (mod π), namely, $(\alpha|\pi)_3$ is hereafter denoted by $\chi(\alpha,\pi)$.

We verify that

(6.5.4) $\qquad \overline{\chi(\alpha,\pi)} = \chi(\alpha,\pi)^2 = \chi(\alpha^2,\pi)$ as $(\alpha|\pi)_3 = 1, \omega$ or ω^2.

(6.5.5) $\qquad \overline{\chi(\alpha,\pi)} = \chi(\bar{\alpha},\bar{\pi})$

(6.5.5) is a consequence of the fact that

$$\alpha^{\frac{N(\pi)-1}{3}} \equiv \chi(\alpha,\pi) \,(\text{mod } \pi)$$

implies

$$\bar{\alpha}^{\frac{N(\pi)-1}{3}} \equiv \overline{\chi(\alpha,\pi)} \,(\text{mod } \pi)$$

But, $N(\bar{\pi}) = N(\pi)$ and so, $\chi(\bar{\alpha},\bar{\pi}) = \overline{\chi(\alpha,\pi)}$ which is (6.5.5).

Lemma 6.5.5 : *Let q be a prime congruent to 2 (mod 3) and suppose that $n \in \mathbb{Z}$ is such that $q \nmid n$. Then n is a cubic residue* (mod q).

For,

$$\chi(\bar{n}|q) = \chi(n,q) = \chi(n^2,q) = \chi(n,q)^2$$

Since $\chi(n,q) \neq 0$, we get $\chi(n,q) = 1$.

Corollary 6.5.1 : *If q and q_2 are distinct primes each of which is congruent to* 2 (mod 3), $\chi(q_2,q_1) = \chi(q_1,q_2)$ *as each one of them is 1.*

Remark 6.5.5 : Corollary 6.5.1 is a special case of Eisenstein's cubic reciprocity law which says that if π, and π_2 are primes (in D) of a special kind, then

$$\chi(\pi_1,\pi_2) = \chi(\pi_2,\pi_1)$$

This will be proved in theorem 47 after going through a few lemmas providing the necessary preparation.

Since the group of units in D is of order 6, every element of D has six associates. We need to pick a special kind of an associate of a prime π in D.

Definition 6.5.3 : *Let π be a prime in D. π is called 'primary', if $\pi \equiv 2$ (mod 3).*

By lemma 6.5.2 a rational prime $q \equiv 2$ (mod 3) is a 'primary' prime in D. In the case of a prime $\pi = a + b\omega$, $\pi \equiv 2$ (mod 3) if, and only if, $a \equiv 2$ (mod 3) and $b \equiv 0$ (mod 3).

Lemma 6.5.6 : *Let π be a prime in D such that $N(\pi) = p \equiv 1$ (mod 3) (p a rational prime). Then, π has an associate π' which is primary and π' is unique.*

Proof : We take $\pi = a + b\omega$ where $a,b \in \mathbb{Z}$. The associates of π are $\pi, -\pi, \pi\omega$, $-\pi\omega, \pi\omega^2$ and $-\pi\omega^2$. These are respectively $a+b\omega, -a-b\omega, -b+(a-b)\omega, b+(b-a)\omega, (b-a)-a\omega$. Since $N(\pi) = p = a^2 - ab + b^2$, a and b are not both divisible by 3. From the associates of π, we have to pick the one which satisfies the conditions for being 'primary'.

If $a \equiv 2 \pmod 3$, as $p = a^2 - ab + b^2$, we get

$$p \equiv 1 \equiv 4 - 2ab + b^2 \pmod 3,$$

or,

$$b(b-2) \equiv 0 \pmod 3.$$

If $3|b$, $a+b\omega$ is primary. If $b \equiv 2 \pmod 3$, $b+(b-a)\omega$ is primary. So, $a+b\omega$ or $b+(b-a)\omega$ is primary.

To prove the uniqueness, let us assume that $a+b\omega$ is primary. Then none of the others is primary. If $a+b\omega$ is not primary, $-\omega\pi = b+(b-a)\omega$ is primary. □

Example 6.5.1 : (a) $3+\omega \in D$ is such that $N(3+\omega) = 7$, a rational prime. So $3+\omega$ is a prime in D.

$$-\omega^2(3+\omega) = -3\omega^2 - 1 = 3(1+\omega) - 1 = 2 + 3\omega$$

is such that $2+3\omega \equiv 2 \pmod 3$. So $2+3\omega$ is the primary prime associated to $3+\omega$.

(b) $7+3\omega \in D$ is such that $N(7+3\omega) = 37$, a rational prime of the from $4k+1$, $7+3\omega$ is a prime in D, and $37 = \pi\bar\pi = (7+3\omega)(7+3\omega^2)$.

Now, $-(7+3\omega)$ is the unique primary prime associated to $7+3\omega$.

6.6. Group characters and the cubic reciprocity law

$(\mathbb{R},+)$ denotes the additive group of real numbers. It is an abelian group. $(\mathbb{Z},+)$ is a normal subgroup of $(\mathbb{R},+)$. Let T be the quotient group $\mathbb{R}/\mathbb{Z}$.

$$T = \mathbb{R}/\mathbb{Z} = \{\mathbb{Z}+t : t \in \mathbb{R}, 0 < t \le 1\}$$

$\mathbb{Z}+t \mapsto \exp(2\pi it)$ gives a mapping of T onto the circle group S_1, the multiplicative group of complex numbers of absolute value 1. It is easy to check that $T \cong S_1$.

Definition 6.6.1 : *Let R be a ring with unity 1_R. An R-module A is an additive abelian group together with a function $s : R \times A \to A$ written $(r,a) \mapsto ra$ and subject to the following axioms:*
For all $r, r' \in R$, $a, b \in A$,

$$r(a+b) = ra + rb$$
$$(r+r')a = ra + r'a$$
$$(rr')a = r(r'a)$$
$$1_R a = a$$

As 'scalar' multiplication by elements of R is on the left, A is called a left R-module.

When r is fixed, $a \mapsto ra$ is a homomorphism of $(A,+)$ into $(A,+)$.

Definition 6.6.2 : *Given two R-modules A, A', an R-module homomorphism of R-modules is a mapping $f : A \to A'$ such that*

$$f(a+b) = f(a) + f(b)$$
$$f(ra) = r f(a)$$

for all $a, b \in A$ and all scalars $r \in R$.

The above two conditions are equivalent to the following requirement:

(6.6.1) $$f(ra + r'b) = rf(a) + r'f(b), \quad r, r' \in R$$

Fact 6.6.1 : If A and A' are two R-modules, the set

$$\text{Hom }_R(A, A') = \{f \mid f : A \to A' \text{ is an } R\text{-module homomorphism}\}$$

is an abelian group under point-wise addition of R-module homomorphisms.

Any abelian group A is a module over the ring $\mathbb{Z}$ of integers. For, if $a \in A$, $n \in \mathbb{Z}$, we take na to be a multiple of a by n that is, $a + a + \ldots + a$ (n times) if n is positive and $-a + -a + \ldots + -a$ ($-n$ times), if n is negative. $na = 0_A$, if n is zero. As $1a = a$, $(n + 1)a = na + a$, $(-n)a = -(na)$, the abelian group A is a $\mathbb{Z}$-module.

Definition 6.6.3 : *Let A be an abelian group and $T = \mathbb{R}/\mathbb{Z}$. The character group of A is defined to be the group $Hom_{\mathbb{Z}}(A, T)$. A character χ of A is an element of $Hom_{\mathbb{Z}}(A, T)$. Further, χ is multiplicative. That is, $\chi(a)\chi(b) = \chi(ab)$ for $a, b \in A$. $\chi_0 : A \to T$ given by $\chi_0(a) = 1$ for all $a \in A$ is called the principal character of A.*

The multiplication of characters $\chi_i, \chi_j \in \text{Hom}_{\mathbb{Z}}(A, T)$ is defined by

(6.6.2) $$(\chi_i \chi_j)(a) = \chi_i(a)\chi_j(a), \quad \text{for all } a \in A.$$

$\text{Hom}_{\mathbb{Z}}(A, T)$ is the abelian group of characters of A, denoted by $ch(A)$.

Lemma 6.6.1 : *For each positive integer n, the group $T = \mathbb{R}/\mathbb{Z}$ has exactly one cyclic subgroup of order n.*

Proof : Let $G = \{e, a_2, \ldots, a_n\}$ be a subgroup of T of order n. For $a_i \in G$, let $K_i(d)$ be the cyclic subgroup (of order d) generated by a_i. It implies that

$$a_i^d = 1.$$

So, a_i is a dth-root of unity, $K_i(d)$ is the cyclic subgroup (of order d) generated by a complex d^{th}-root of unity. This is true for every divisor d of n. In particular, there is an nth-root of unity say ζ contained in G such that $\zeta^n = 1$. So, G is cyclic of order n.   □

Theorem 44 : *A finite abelian group G of order n has exactly n distinct characters.*

Proof : Let H be a proper subgroup of G. Suppose that $a \in G$ and $a \notin H$. If a is of order m in G, then $a^m = e \in H$. It happens that among the powers a^j, $(j \geq 0)$, there is a smallest positive integer h such that $a^h \in H$. We call h the indicator of a in H. The subgroup $\langle H, a \rangle$ generated by H and a is of the form

$$\langle H, a \rangle = \{ya^k : y \in H \text{ and } 0 \leq k < h\}$$

As H is finite,

$$|\langle H, a \rangle| = h|H|.$$

Let us denote the trivial subgroup (e) by H_0. We assume that $H_0 \neq G$. Let $a_1 \in G$ and $a_1 \neq e$. Suppose that $H_1 = \langle H_0, a_1 \rangle$. Having got H_1, if $H_1 \neq G$, take $a_2 \notin H_1$ and let $H_2 = \langle H_1, a_2 \rangle$. Continuing this process, we get an ascending sequence of subgroups:

$$H_0 \subset H_1 \subset H_2 \subset \cdots \subset H_n = G, \text{ as } G \text{ is finite.}$$

We prove the theorem by showing that if it is true for i, $0 \leq i < n$, it is true for $i+1$.

For $H_0 = (e)$, there is one character for H_0 namely the function which is identically 1. For H_i, assume that H_i has order d (where d divides n) and H_i has d distinct characters.

In $H_{i+1} = \langle H_i, a_{i+1} \rangle$, let h_{i+1} be the indicator of a_{i+1}. That is, h_{i+1} is the smallest integer such that $a_{i+1}^{h_{i+1}} \in H_i$. We prove that there are h_{i+1} different ways to extend each character of H_i to obtain a character of H_{i+1}. This will show that H_{i+1} has exactly dh_{i+1} characters and dh_{i+1} is the order of H_{i+1}.

A typical element of H_{i+1} is of the form ya_{i+1}^k, where $y \in H_i$ and $0 \leq k < h_{i+1}$.

Let $\bar{\chi}$ be an extension of χ defined on H_i. The multiplicative property of χ requires that

$$\bar{\chi}(ya_{i+1}^k) = \bar{\chi}(y)\bar{\chi}(a_{i+1})^k$$

But, $y \in H_i$. So, $\bar{\chi} = \chi(y)$. So, we have

(6.6.3) $$\bar{\chi}(ya_{i+1}^k) = \bar{\chi}(y)\bar{\chi}(a_{i+1})^k$$

So $\bar{\chi}(ya_{i+1}^k)$ is known when $\bar{\chi}(a_{i+1})$ is known. Let $c = a_{i+1}^{h_{i+1}} \in H_i$. Then, $\bar{\chi}(c) = \chi(c)$ as $c \in H_i$.

So,

$$\bar{\chi}(a_{i+1}^{h_{i+1}}) = \chi(c) \quad \text{or} \quad (\bar{\chi}(a_{i+1}))^{h_{i+1}} = \chi(c)$$

$\bar{\chi}_i(a_{i+1})$ is one of the h_{i+1}th-roots of $\chi(c)$. There are h_{i+1} choices for $\bar{\chi}_i(\alpha_{i+1})$. Each one of these gives rise to a character of H_{i+1}. From (6.6.3), we also note that

$$\bar{\chi}(xa_{i+1}^k \cdot ya_{i+1}^j) = \bar{\chi}(xy)\bar{\chi}(a_{i+1})^{k+j}$$
$$= \chi(xy)(\bar{\chi}(a_{i+1}))^{k+j}$$
$$= \chi(x)\chi(y)\bar{\chi}(a_{i+1})^k\bar{\chi}(a_{i+1})^j$$
$$= \bar{\chi}(xa_{i+1}^k)\bar{\chi}(ya_{i+1}^j)$$

and so $\bar{\chi}$ is multiplicative. No two extensions $\bar{\chi}_1$, $\bar{\chi}_2$ can be identical on H_{i+1}, as the characters χ_1 and χ_2 on H_i which they extend would then be the same. So each of the d characters of H_i can be extended in h_{i+1} different ways to produce a character of H_{i+1}. Also, if χ' is any character on H_{i+1}, its restriction to H_i will be a character on H_i and so this process of extension produces all the characters on H_{i+1}. So the number of characters on H_{i+1} is equal to $|H_{i+1}|$. This way, we produce n distinct characters on G. $\qquad\square$

Remark 6.6.1 : Proof of theorem 44 has been adapted from Tom Apostol [1]. [1] gives more properties of characters of a finite abelian group.

Lemma 6.6.2 : *For any positive integer n, $Ch(\mathbb{Z}/n\mathbb{Z}) \cong \mathbb{Z}/n\mathbb{Z}$.*

Proof : By theorem 44, $\mathbb{Z}/n\mathbb{Z}$ has n distinct characters. $(\mathbb{Z}/n\mathbb{Z})$ is cyclic of order n. So if $\chi \in Ch(\mathbb{Z}/n\mathbb{Z})$,

$$\chi([1])^n = 1$$

So $\chi([1])$ is an nth-root of unity. So $Ch(\mathbb{Z}/n\mathbb{Z})$ is also cyclic generated by a primitive nth-root of unity. This proves the lemma. $\qquad\square$

Fact 6.6.2 : Given a finite abelian group A, $Ch(A) \cong A$.
For, if G, G' are finite abelian groups, $Ch(G \oplus G') \cong Ch\ G \oplus Ch\ G'$. As G is a direct product of cyclic subgroups, $Ch\ G \cong G$.

Let p be a prime. We consider $\mathbb{F}_p = \mathbb{Z}/p\mathbb{Z}$. $\mathbb{F}_p^*$ is cyclic of order $p-1$. $Ch F_p^* \cong F_p^*$.
If g is a generator of $\mathbb{F}_p^*$, we define

$$\lambda(g) = \exp(\frac{2\pi i}{p-1})$$

λ is a character on $\mathbb{F}_p^*$ and the set of characters of F_p^* is given by

$$\{\varepsilon, \lambda, \lambda^2, \ldots, \lambda^{p-2}\}, \quad \lambda^{p-1}(g) = 1.$$

(6.6.4) $\varepsilon(g) = 1, g \in \mathbb{F}_p^*$.

ε is called the principal character on $\mathbb{F}_p^*$.

Next, we need to consider sums over elements of $\mathbb{F}_p$. So, it is convenient to extend the domain of definition of a character χ (on $\mathbb{F}_p^*$) to $\mathbb{F}_p$ by writing

$$\chi(0) = \begin{cases} 1, & \text{if } \chi = \varepsilon \\ 0 & \text{if } \chi \neq \varepsilon \end{cases}$$

(where ε is the principal character on $\mathbb{F}_p^*$).
By doing this, we make χ a character on $\mathbb{F}_p$.

Definition 6.6.4 : *Let χ be a character on $\mathbb{F}_p$ and let $a \in \mathbb{F}_p^*$. We write $\zeta = \exp(\frac{2\pi i}{p})$. The sum*

$$G(a, \chi) = \sum_{t \in \mathbb{F}_p} \chi(t)\zeta^{at}$$

is called a Gauss sum on $\mathbb{F}_p$ belonging to the character χ.

Theorem 45 : *Let $a \in F_p$. Suppose that ε is the principal character on $\mathbb{F}_p$. That is, $\varepsilon(b) = 1$ for all $b \in \mathbb{F}_p$. Then,*

(6.6.5) $G(a, \chi) = \begin{cases} \chi(a^{-1})G(1, \chi), & \text{if } \chi \neq \varepsilon, a \neq 0 \\ 0, & \text{if } \chi = \varepsilon \text{ and } a \neq 0 \\ 0, & \text{if } \chi \neq \varepsilon \text{ and } a = 0 \\ p, & \text{if } \chi = \varepsilon \text{ and } a = 0 \end{cases}$

Proof : Case 1. $a \neq 0$, $\chi \neq \varepsilon$.

$$\chi(a)G(a,\chi) = \chi(a)\sum_{t\in\mathbb{F}_p}\chi(t)\zeta^{at} = \sum_{t\in\mathbb{F}_p}\chi(at)\zeta^{at} = G(1,\chi),$$

as $p \nmid a$ and 'at' runs through a complete residue system (mod p).

$$\text{or}\quad G(a,\chi) = \chi(a)^{-1}G(1,x) = \chi(a^{-1})G(1,\chi).$$

Case 2. $a \neq 0$, $\chi = \varepsilon$

$$G(a,\varepsilon) = \sum_{t\in\mathbb{F}_p}\varepsilon(t)\zeta^{at} = \sum_{t\in\mathbb{F}_p}\zeta^{at} = 0.$$

Case 3. $a = 0$, $\chi \neq \varepsilon$

$$G(0,\chi) = \sum_{t\in\mathbb{F}_p}\chi(t)\zeta^{0t} = \sum_{t\in\mathbb{F}_p}\chi(t) = \begin{cases} p, & \text{if } \chi = \varepsilon \\ 0, & \text{otherwise.} \end{cases}$$

For, if we choose $a \in \mathbb{F}_p^*$ such that $\chi(a) \neq 1$,

$$\chi(a)\sum_{t\in\mathbb{F}_p}\chi(t) = \sum_{t\in\mathbb{F}_p}\chi(at).$$

If $S = \sum_{t\in\mathbb{F}_p}\chi(t)$, we get $\chi(a)S = S$, as 'at' runs through a complete residue system (mod p).

This implies that $S = 0$.

Case 4. $a = 0$, $\chi = \varepsilon$

$$G(0,\varepsilon) = \sum_{t\in\mathbb{F}_p}\varepsilon(t)\zeta^{0t} = p.$$

Cases 1 to 4 above yield (6.6.5).   □

Theorem 46 : *If* $\chi \neq \varepsilon$, $|G(1,\chi)| = \sqrt{p}$.

Proof : We evaluate the sum

$$\sum_{a\in\mathbb{F}_p}G(a,\chi)\overline{G(a,\chi)} \text{ in the different ways,}$$

If $a \neq 0$, by theorem 45, $\overline{G(a,\chi)} = \overline{\chi(a^{-1})G(1,\chi)} = \chi(a)\overline{G(1,\chi)}$.
So,

(6.6.6) $G(a,\chi)\overline{G(a,\chi)} = \chi(a^{-1})\chi(a)G(1,\chi)\overline{G(1,\chi)} = |G(1,\chi)|^2$.

Also,

$$G(a,\chi)\overline{G(a,\chi)} = \sum_t\sum_{t'}\chi(t)\overline{\chi(t')}\zeta^{at-at'}.$$

Summing over $a \in \mathbb{F}_p$ on the left and right sides, we obtain

(6.6.7) $\displaystyle\sum_{a\in\mathbb{F}_p}G(a,\chi)\overline{G(a,\chi)} = \sum_{a\in\mathbb{F}_p}(\sum_t\sum_{t'}\chi(t)\overline{\chi(t')}\zeta^{a(t-t')})$

Now,

$$\sum_{a\in\mathbb{F}_p} \zeta^{a(t-t')} = p\,\delta(t,t'),$$

where

$$\delta(t,t') = \begin{cases} 1, & \text{if } t \equiv t' \pmod{p} \\ 0, & \text{otherwise.} \end{cases}$$

So, from (6.6.7)

$$\sum_{a\in\mathbb{F}_p} G(a,\chi)\overline{G(a,\chi)} = \sum_{t}\sum_{t'} \chi(t)\overline{\chi(t')}p\,\delta(t,t')$$

$$= \sum_{t\in\mathbb{F}_p} |\chi(t)|^2 p.$$

Now, as $G(a,\chi) = 0$ for $\chi \neq \varepsilon$, $a = 0$; noting that $|\chi(t)| = 1$, for $t \in F_p^*$, we get

(6.6.8) $$\sum_{a\in\mathbb{F}_p} G(a,\chi)\overline{G(a,\chi)} = p(p-1), \text{ as } \chi(0) = 0.$$

Further, from (6.6.6),

$$\sum_{a\in\mathbb{F}_p} G(a,\chi)\overline{G(a,\chi)} = \sum_{a\in\mathbb{F}_p^*} |G(1,\chi)|^2.$$

That is,

(6.6.9) $$\sum_{a\in\mathbb{F}_p} G(a,\chi)\overline{G(a,\chi)} = (p-1)|G(1,\chi)|^2.$$

From (6.6.8) and (6.6.9), we obtain $|G(1,\chi)|^2 = p$ which proves theorem 46. $\qquad\square$

Definition 6.6.5 : *Let χ and λ be two characters of $\mathbb{F}_p$. The Jacobi sum in terms of χ and λ is defined by*

$$J(\chi,\lambda) = \sum_{a+b=1} \chi(a)\lambda(b)$$

where the summation is over the elements $a,b \in \mathbb{F}_p$ such that $a+b = 1$.

It is easy to check that

(6.6.10) $$J(\varepsilon,\varepsilon) = p,$$

(6.6.11) $$J(\varepsilon,\chi) = 0.$$

Now,

$$J(\chi,\chi^{-1}) = \sum_{a+b=1} \chi(a)\chi^{-1}(b) = \sum_{\substack{a+b=1\\b\neq 0}} \chi(\frac{a}{b}) = \sum_{a\neq 1} \chi(\frac{a}{1-a}).$$

Let $\frac{a}{1-a} = t$. If $t \neq -1$, $a = \frac{t}{1+t}$.

As a varies over the elements of $\mathbb{F}_p$, except 1, t varies over the elements of $\mathbb{F}_p$ except -1.

But,

$$\sum_{t\in\mathbb{F}_p} \chi(t) = 0, \text{ for } \chi \neq \varepsilon.$$

Therefore, $J(\chi,\chi^{-1}) = \sum_{t\neq-1} \chi(t))$ or

$$(6.6.12) \qquad\qquad J(\chi,\chi^{-1}) = -\chi(-1).$$

Next, suppose χ and λ are such that $\chi\lambda \neq \varepsilon$.

$$G(1,\chi)G(1,\lambda) = \left(\sum_a \chi(a)\zeta^a\right)\left(\sum_b \lambda(b)\zeta^b\right)$$

$$= \sum_{a,b} \chi(a)\lambda(b)\zeta^{a+b}$$

$$= \sum_t \left(\sum_{a+b=t} \chi(a)\lambda(b)\right)\zeta^t.$$

If $t = 0$,

$$\sum_{a+b=0} \chi(a)\lambda(b) = \sum_{a\in\mathbb{F}_p} \chi(a)\lambda(-a) = \lambda(-1)\sum_{a\in\mathbb{F}_p} \chi\lambda(a) = 0, \text{ as } \chi\lambda \neq \varepsilon.$$

If $t \neq 0$, let a' be such that $a = ta'$. Let b' be such that $b = tb'$. If $a+b=t$, $a'+b' = 1$ and so

$$\sum_{a+b=t} \chi(a)\lambda(b) = \sum_{a'+b'=1} \chi(ta')\lambda(tb') = \chi\lambda(t)J(\chi,\lambda).$$

So,

$$G(1,\chi)G(1,\lambda) = \sum_t \chi\lambda(t)J(\chi,\lambda)\zeta^t = J(\chi,\lambda)G(1,\chi\lambda),$$

or

$$(6.6.13) \qquad\qquad J(\chi,\lambda) = \frac{G(1,\chi)G(1,\lambda)}{G(1,\chi\lambda)}, \quad \chi\lambda \neq \varepsilon.$$

We deduce that if χ, λ and $\chi\lambda$ are not equal to ε, as $|G(1,\chi)| = \sqrt{p}$ (theorem 46)

$$(6.6.14) \qquad\qquad |J(\chi,\lambda)| = \sqrt{p}.$$

Fact 6.6.3 : If $p \equiv 1 \pmod 3$ and χ is a cubic character $(\bmod\ p)$, that is, $\chi^3 = \varepsilon$

$$(6.6.15) \qquad\qquad G(1,\chi)^3 = pJ(\chi,\chi).$$

Proof : Using (6.6.13), we have $G(1,\chi)^2 = G(1,\chi^2)J(\chi,\chi)$. Multiplying both sides by $G(1,\chi)$, we obtain

$$(6.6.16) \qquad\qquad G(1,\chi)^3 = G(1,\chi)G(1,\chi^2)J(\chi,\chi).$$

But $\chi^2 = \chi^{-1}$.

Also, $G(1,\chi^{-1}) = G(1,\bar{\chi})$ as χ^{-1} is the character which takes $t \in \mathbb{F}_p$ to $\bar{\chi}(t)$.

Now,

$$\chi(-1)G(1,\bar\chi) = \sum_{t\in\mathbb{F}_p} \overline{\chi(-t)}\zeta^t$$

$$= \sum_{s\in\mathbb{F}_p} \overline{\chi(s)}\zeta^{-s}$$

$$= \overline{G(1,\chi)}$$

So,

$$\chi(-1)G(1,\chi)G(1,\bar\chi) = G(1,\chi)\overline{G(1,\chi)}.$$

$$= |G(1,\chi)|^2.$$

$$= p, \quad \text{by theorem 46.}$$

As χ is a cubic character (mod p), $\chi(-1) = \chi((-1)^3) = \chi^3(-1) = 1$. So,

(6.6.17) $\qquad\qquad G(1,\chi)G(1,\chi^2) = p$

From (6.6.16) and (6.6.17), we arrive at (6.6.15). $\qquad\qquad\square$

Lemma 6.6.3 : *Let $p \equiv 1$ (mod 3). If χ is a cubic character (mod p) and if $J(\chi,\chi) = a + b\omega$ where $\omega = \exp(\frac{2\pi i}{3})$, then $a \equiv 2$ (mod 3) and $b \equiv 0$ (mod 3). That is, $J(\chi,\chi)$ is a primary prime.*

Proof : We work with congruences in the ring $\mathcal{A}$ of algebraic integers.

(6.6.18) $\qquad G(1,\chi)^3 = (\sum_{t\in\mathbb{F}_p}\chi(t)\zeta^t)^3 = \sum_{t\in\mathbb{F}_p}\chi(t)^3\zeta^{3t} \text{ (mod 3)}$

We observe that $\chi(0) = 0$ and $\chi^3(t) = 1$ for $t \neq 0$.
Therefore,

$$\sum_{t}\chi(t)^3\zeta^{3t} = \sum_{t\neq 0}\zeta^{3t} = -1.$$

So, from (6.6.15) and (6.6.18), we have

$$G(1,\chi)^3 = pJ(\chi,\chi) \equiv a + b\omega \equiv -1 \text{ (mod 3)}.$$

Working with $\bar\chi$ instead of χ and remembering that $\overline{G(1,\chi)} = G(1,\bar\chi)$, we get

$$G(1,\bar\chi)^3 = pJ(\bar\chi,\bar\chi) \equiv a + b\bar\omega \equiv -1 \text{ (mod 3))}.$$

Subtraction yields $b(\omega - \bar\omega) \equiv 0$ (mod 3)).
But, $\omega - \bar\omega = (\frac{-1+\sqrt{-3}}{2}) - (\frac{-1-\sqrt{-3}}{2}) = \sqrt{-3}$.
Or,

$$b\sqrt{-3} \equiv 0 \text{(mod 3)}.$$

It follows that $-3b^2 \equiv 0$ (mod 9) or 3 divides b.
Since $b \equiv 0$ (mod 3), $a + b\omega \equiv -1$ (mod 3) gives $a \equiv -1$ (mod 3), as desired. $\qquad\square$

Observation 6.6.1

By (6.6.14), $|J(\chi,\chi)| = \sqrt{p}$. So,

$$N(J(\chi,\chi)) = J(\chi,\chi)\overline{J(\chi,\chi)}$$
$$= |J(\chi,\chi)|^2$$
$$= p.$$

If p is a prime congruent to 1 (mod 3) and χ is a cubic character (mod p), by definition 6.5.3 and lemma 6.6.3, $J(\chi,\chi)$ is a primary prime in D. By lemma 6.5.6, $J(\chi,\chi)$ is the associate of a prime π in D such that

$$N(\pi) = p \equiv 1 \,(\text{mod } 3).$$

Theorem 47 (cubic reciprocity law) : *If π_1 and π_2 are primary primes in D and $N(\pi_1)$, $N(\pi_2) \neq 3$ and $N(\pi_1) \neq N(\pi_2)$, then*

$$\chi(\pi_2,\pi_1) = \chi(\pi_1,\pi_2)$$

where $\chi(\alpha,\pi)$ stands for $(\alpha|\pi)_3$ as per Notation 6.5.1.

Proof : If π_1 and π_2 are distinct rational primes congruent to 2 (mod 3), take $\pi_1 = a$, $\pi_2 = b$ (say).

As $\bar{a} = a$, $\bar{b} = b$ and $\chi(\bar{a},\bar{b}) = \overline{\chi(a,b)} = (\chi(a,b))^2$ (by 6.6.4)
$\chi(a,b) = (\chi(a,b))^2 \Rightarrow \chi(a,b) = 1$ as $\chi(a,b) \neq 0$. Similarly, $\chi(b,a) = 1$ and so the result holds for distinct rational primes $\equiv 2$ (mod 3).

The other cases to be considered are

(i) π_1, a rational prime $q \equiv 2$ (mod 3) and π_2 complex with $N(\pi_2) = p$ a prime $\equiv 1$ (mod 3).

(ii) π_1 and π_2 complex with $N(\pi_1) = p_1 \equiv 1$ (mod 3) and $N(\pi_2) = p_2 \equiv 1$ (mod 3), where p_1 and p_2 are distinct primes.

Case 1. We consider the case

$$\pi_1 = q \equiv 2\,(\text{mod } 3) \text{ and } \pi_2 = \pi \text{ with } N(\pi) = p.$$

$\chi(\alpha,\pi)$ is a cubic character modulo p. By lemma 6.6.3

(6.6.19) $J(\chi(\alpha,\pi),\chi(\alpha,\pi)) = \pi'$ a primary prime.

Since $\pi\bar{\pi} = p = \pi'\bar{\pi}'$, we have $\pi|\pi'$ or $\pi|\bar{\pi}'$.
Since all primes are primary, we claim that $\pi = \pi'$ or $\pi = \bar{\pi}'$.
From the definition of J, one obtains

$$J(\chi(\alpha,\pi),\chi(\alpha,\pi)) = \sum_{a+b=1} \chi(a,\pi)\chi(b,\pi)$$
$$= \Big(\sum_{t\in\mathbb{F}_p} t^{\frac{p-1}{3}}(1-t)^{\frac{p-1}{3}}\Big)\,(\text{mod } p).$$

The polynomial $t^{\frac{p-1}{3}}(1-t)^{\frac{p-1}{3}}$ is of degree $\frac{2}{3}(p-1) < (p-1)$.

So,
$$\sum_{t\in\mathbb{F}_p} t^{\frac{p-1}{3}}(1-t)^{\frac{p-1}{3}} \equiv 0 \,(\mathrm{mod}\ p).$$

For,
$$\sum_{i=1}^{p-1} i^k \equiv \begin{cases} 0\,(\mathrm{mod}\ p), & \text{if } (p-1)\nmid k \\ -1\,(\mathrm{mod}\ p), & \text{if } (p-1)\,|\,k \end{cases}$$

So,
$$J(\chi(\alpha,\pi),\chi(\alpha,\pi)) \equiv 0 \ (\mathrm{mod}\ \pi).$$

From (6.6.19), we have $\pi\,|\,\pi'$ and so $\pi = \pi'$.

Using Fact 6.6.3, we obtain $G(1,\chi(\alpha,\pi))^3 = p\pi$.

In particular,

(6.6.20) $\qquad G(1,\chi(q,\pi))^3 = p\pi$ where $\pi_1 = q \equiv 2\,(\mathrm{mod}\ 3)$

Raising both sides to the power $\frac{q^2-1}{3}$, we get
$$G(1,\chi(q,\pi))^{q^2-1} = (p\pi)^{\frac{q^2-1}{3}}$$

Taking congruences (modulo q), we get
$$G(1,\chi(q,\pi))^{q^2} = \chi(p\pi,q)G(1,\chi(q,\pi))\,(\mathrm{mod}\ q), \text{ as } N(q)=q^2.$$

Now, $\chi(p,q) = 1$, as $p^{\frac{q^2-1}{3}} \equiv 1\,(\mathrm{mod}\ q)$. We have

(6.6.21) $\qquad G(1,\chi(q,\pi))^{q^2} = \chi(\pi,q)G(1,\chi(q,\pi))\,(\mathrm{mod}\ q).$

$$\begin{aligned}
G(1,\chi(q,\pi))^{q^2} &= (\sum_{t\in\mathbb{F}_p} \chi(t,\pi)\zeta^t)^{q^2} \\
&= \sum_{t\in\mathbb{F}_p} \chi(t,\pi)^{q^2}\zeta^{tq^2}\,(\mathrm{mod}\ q).
\end{aligned}$$

As $q^2 \equiv 1\,(\mathrm{mod}\ 3)$ and $\chi(t,\pi)$ is a cube root of unity, we have
$$G(1,\chi(q,\pi))^{q^2} \equiv G(q^2,\chi(q,\pi))\,(\mathrm{mod}\ q).$$

By theorem 45, $G(q^2,\chi(q,\pi)) = \chi(q^{-2},\pi)G(1,\chi(q,\pi))$,

(6.6.22) $\qquad$ or $\quad G(1,\chi(q,\pi))^{q^2} = \chi(q,\pi)G(1,\chi(q,\pi)),$

or, from (6.6.21) and (6.6.22) we get
$$\chi(\pi,q)G(1,\chi(q,\pi)) \equiv \chi(q,\pi)G(1,\chi(q,\pi))\,(\mathrm{mod}\ q).$$

Multiplying both sides by $\overline{G(1,\chi(q,\pi))}$ and noting that
$$G(1,\chi(q,\pi))\overline{G(1,\chi(q,\pi))} = p, \quad \text{(by theorem 46)}$$

we have
$$\chi(q,\pi)p \equiv \chi(\pi,q)p\,(\mathrm{mod}\ q),$$

or,
$$\chi(q,\pi) \equiv \chi(\pi,q)\,(\mathrm{mod}\ q).$$

This implies that $\chi(q,\pi) = \chi(\pi,q)$.

Case 2. π_1 and π_2 are complex primes and

$$N(\pi_1) = p_1 \equiv 1 \,(\mathrm{mod}\ 3),$$
$$N(\pi_2) = p_2 \equiv 1 \,(\mathrm{mod}\ 3).$$

Let $\delta_1 = \bar{\pi}_1$ and $\delta_2 = \bar{\pi}_2$.

Then, δ_1, δ_2 are primary primes and $p_1 = \pi_1\delta_1$, $p_2 = \pi_2\delta_2$. Now, using (6.6.15) and observation (6.6.1),

$$G(1, \chi(1,\delta_1))^3 = p_1\delta_1.$$

Raising to the power $\frac{N(\pi_2)-1}{3} = \frac{p_2-1}{3}$ and taking congruences modulo π_2, we get

(6.6.23) $$\chi(p_2^2, \delta_1) = \chi(p_1\delta_1, \pi_2)$$

Starting from $G(1, \chi(1,\pi_2))^3 = p_2\pi_2$ and raising both sides to the power $\frac{p_1-1}{3}$ and taking congruences modulo π_1, we obtain

(6.6.24) $$\chi(p_1^2, \pi_2) = \chi(p_2\pi_2, \pi_1)$$

(since $\chi(\alpha, \pi_1) = \overline{\chi(\alpha, \bar{\pi}_1)} = \chi(\alpha^2, \bar{\pi}_1)$ with $\alpha = \bar{p}_2 = p_2$ and $\delta_1 = \bar{\pi}_1$). We also have

(6.6.25) $$\chi(p_2^2, \delta_1) = \chi(p_2, \pi_1)$$

(since $\overline{\chi(\alpha, \bar{\pi}_1)} = \chi(\alpha, \bar{\pi}_1)^2 = \chi(\alpha^2, \bar{\pi}_1)$ with $\bar{\pi}_1 = \delta_1$ and $\alpha = \bar{p}_2 = p_2$).

Next by, (6.6.23),

$$\begin{aligned}
\chi(\pi_2, \pi_1)\chi(p_1\delta_1, \pi_2) &= \chi(\pi_2, \pi_1)\chi(p_2^2, \delta_1) \\
&= \chi(\pi_2, \pi_1)\chi(p_2, \pi_1) \text{ by (6.6.25)} \\
&= \chi(p_2\pi_2, \pi_1) \\
&= \chi(p_1^2, \pi_2) \text{ by (6.6.24)} \\
&= \chi(p_1\pi_1\delta_1, \pi_2) \text{ as } p_1 = \pi_1\delta_1.
\end{aligned}$$

Or,

$$\chi(\pi_2, \pi_1)\chi(p_1\delta_1, \pi_2) = \chi(\pi_1, \pi_2)\chi(p_1\delta_1, \pi_2).$$

Cancelling $\chi(p_1\delta_1, \pi_2)$ from both sides, we get $\chi(\pi_2, \pi_1) = \chi(\pi_1, \pi_2)$. $\square$

Remark 6.6.2 : Proof of theorem 47 has been adapted from [8]. For an overview of reciprocity laws, see B. F. Wyman [A4]. K. S. Williams gives an Euler criterion for cubic non-residues in [A3].

6.7. Notes with illustrative examples

The Legendre symbol is in respect of primes.

Let b be an odd positive integer. Suppose that $b = p_1p_2\ldots p_r$ where p_i $(i = 1,2,\ldots r)$ are primes not necessarily distinct. Jacobi (1804–1851) defined the symbol $(a|b)$ by

(6.7.1) $$(a|b) = (a|p_1)(a|p_2)\cdots(a|p_r)$$

where $(a|p_i)$ is the Legendre symbol $(i = 1,2,\ldots,r)$.

$(a|b)$ is referred to as the Jacobi symbol. If g.c.d $(a,b) > 1$, we take $(a|b)$ to be zero. Now,

$$(2|175) = (2|5)^2(2|7) = 1$$

However, 2 is not a quadratic residue mod 175. That is, $(a|b) = 1$ does not mean that a is a quadratic residue (mod b). However, if $(a|b) = -1$ then a is a non-residue (mod b). The Jacobi symbol could be used to prove the following.

Theorem 48 : *Let a be square-free integer. There are infinitely many primes p for which a is a quadratic non-residue.*

Proof : We write

$$a = 2^k p_1 p_2 \ldots p_r \text{ where } k = 0 \text{ or } 1 \text{ and } p_i \ (i = 1 \ldots r)$$

are distinct odd primes, $(r \geq 1)$.

Let $\{q_1, q_2, \ldots, q_s\}$ be a finite set of odd primes not containing any p_i. Let t be a quadratic non-residue of p_r.

We solve the congruences

$$x \equiv 1 \,(\text{mod } q_j), \quad j = 1, 2, \ldots, s;$$
$$x \equiv 1 \,(\text{mod } 8),$$
$$x \equiv 1 \,(\text{mod } p_i), \quad i = 1, 2, \ldots (r-1);$$
$$x \equiv t \,(\text{mod } p_r).$$

Suppose that $x \equiv b \bmod (8p_1 p_2 \ldots p_r q_1 q_2 \ldots q_s)$ be the common solution. Let $b = m_1, m_2, \ldots, m_n$ where m_i are primes.

Since $b \equiv 1 \,(\text{mod } 8)$, $x^2 \equiv 2 \,(\text{mod } b)$ has a solution. For,

$$(2|b) = (2|m_1)(2|m_2) \ldots (2|m_n) = 1.$$

The generalized quadratic reciprocity law says:

(6.7.2) $(a|b)(b|a) = (-1)^{\frac{(a-1)(b-1)}{4}}$ when a, b are positive and odd.

By the generalized quadratic reciprocity law, $(p_i|b) = (b|p_i) \ i = 1, 2, \ldots, r$.

So,

$$(a|b) = (2|b)^k (p_1|b) \ldots (p_r|b)$$
$$= (b|p_1)(b|p_2) \ldots (b|p_r)$$
$$= (1|p_1)(1|p_2) \ldots (1|p_{r-1})(t|p_r).$$

This gives

(6.7.3) $(a|b) = -1$

By definition of $(a|b)$, we have

$$(a|b) = (a|m_1)(a|m_2) \ldots (a|m_n).$$

From (6.7.3), we have $(a|m_i) = -1$ for some i, $1 \le i \le n$. The primes $q_1, q_2, \ldots q_n$ are so chosen that none of them divides b. So $m_i \notin \{q_1, q_2 \ldots q_s\}$. Also a is square-free and divisible by an odd prime. We are able to pick a prime m_i outside the set $\{2, q_1, \ldots, q_s\}$ such that $(a|m_i) = -1$. The theorem is okay for such a's.

If $a = 2$, let $\{q_1, q_2, \ldots, q_s\}$ be a finite set of primes ($\neq 3$) for which $(2|q_j) = -1$ ($j = 1, \ldots s$), we write

$$b = 8q_1 q_2 \ldots q_s + 3.$$

Since $b \equiv 3 \pmod 8$, we have $(2|b) = -1$.
As before, take $b = m_1, m_2, \ldots, m_n$ where m_i are primes ($i = 1, \ldots, n$).
Then $(2|m_i) = -1$ for some i (as before) and $m_i \notin \{3, q_1, q_2, \ldots, q_s\}$. This disposes of the case $a = 2$ also.

$\square$

Remark 6.7.1 : Theorem 48 has been adapted from [8].

Next, let p be an odd prime and $a \in \mathbb{N}$, $p \nmid a$. The Legendre symbol $(a|p)$ takes values $+1$ or -1 according as a is a quadratic residue (mod p) or a quadratic non-residue (mod p). $(a|p)$ is a character on $\mathbb{F}_p^*$ where $\mathbb{F}_p = \mathbb{Z}/p\mathbb{Z}$ and $\mathbb{F}_p^* = \mathbb{F}_p \setminus \{0\}$. The Legendre symbol depends only on the residue class of a (mod p) and is multiplicative.
Now,

(6.7.4) $\mathbb{F}_p^{*2} = \{a^2 : a \in \mathbb{F}^*\}$ = the set of quadratic residues (mod p).

It follows that $\mathbb{F}_p^{*2}$ is a subgroup (of $\mathbb{F}_p^*$) of order $\frac{p-1}{2}$. So $\mathbb{F}_p^{*2}$ is a subgroup of index 2 in $\mathbb{F}_p^*$. It is a normal subgroup of $\mathbb{F}_p^*$. One has $\mathbb{F}_p^*/\mathbb{F}_p^{*2} \cong \{1, -1\}$. Further,

(6.7.5) $x^2 \equiv a \pmod p \Leftrightarrow x^2 - a = [0]$ in $\mathbb{F}_p$.

This is possible if, and only if, $a \in \mathbb{F}_p^{*2}$. That is,

(6.7.6) $x^2 \equiv a \pmod p \Leftrightarrow x^2 - a$ splits into a product of linear factors in $\mathbb{F}_p[x]$.

We go to a general problem. Let K be an arbitrary field. Suppose that $f(x) \in K[x]$. We have to check whether $f(x)$ splits into a product of linear factors in $K[x]$ and in $L[x]$ when L is the smallest extension of K. That is, we obtain an extension L of K such that L is the splitting field of $f(x)$. Using results on field extensions [4], one knows that a splitting field of $f(x)$ exists and is unique up to isomorphism.

Is it possible to characterize a splitting field of $f(x)$ in terms of certain invariants depending only on K and the polynomial $f(x)$? When $k = \mathbb{R}$, the field of real numbers, every polynomial in $\mathbb{R}[x]$ splits into a product of irreducible factors of degree ≤ 2. Also, a polynomial $t(x)$ of degree 2 (over $\mathbb{R}$) is irreducible if, and only if, the discriminant of $t(x)$ is negative. In that case, the splitting field of $t(x)$ is $\mathbb{C}$, the field of complex numbers.

If $K = \mathbb{F}_q$, a finite field having $q = p^m$ ($m \ge 1$) elements (where p is a prime), for every $m \in \mathbb{N}$, there exists up to isomorphism a unique field extension L (of

K) such that the degree of L over K is m. L can be viewed as a vector space of dimension m over K. Further, every polynomial $f(x)$ of degree $s \leq m$ splits into a product of linear factors in $L[x]$. The situation is tougher when $K = \mathbb{Q}$, the field of rational numbers or more generally an algebraic number field (a finite extension of $\mathbb{Q}$).

A situation that could be tackled better is the class of finite fields $\mathbb{Z}/p\mathbb{Z}$ (p a prime) and the set of polynomials $f(x) \in \mathbb{Z}[x]$. The question is whether a given monic irreducible polynomial $t(x) \in \mathbb{Z}[x]$ splits into a product of linear factors in $\mathbb{F}_p[x]$ where $\mathbb{F}_p = \mathbb{Z}/p\mathbb{Z}$.

A more difficult question is:
Given a monic irreducible polynomial in $\mathbb{Z}[x]$, find the set of primes p for which $f(x)$ splits into a product of linear factors in $\mathbb{F}_p[x]$?

Definition 6.7.1 : *The set spl. $f(x)$ is defined by*

$$spl. f(x) = \{ p : p \text{ is a prime and } f(x) \text{ splits into a product of}$$
$$\text{linear factors in } \mathbb{F}_p[x] \}$$

The most general reciprocity law due to Emil Artin establishes a curious connection between the splitting field of $f(x)$ and the set spl. $f(x)$ when the splitting field L of $f(x)$ is such that L is abelian over K in the sense that the Galois group $G(L/K)$ is an abelian group.
Let $f(x) = x^2 - a \in \mathbb{Z}[x]$.
Suppose that $f(x)$ is irreducible over $\mathbb{Q}$.

spl. $f(x) = \{ p : f(x) \text{ splits into a product of linear factors over } \mathbb{Z}/p\mathbb{Z} \}$.
For any $a \in \mathbb{Z}$, spl. $f(x)$ contains the prime 2 as

$$x^2 - a \equiv x^2 - 1 \text{ or } x^2 \in \mathbb{F}_2[x]$$

So, $x^2 - a$ splits into linear factors over $\mathbb{Z}/2\mathbb{Z}$. To determine spl. $(x^2 - a)$ we need to compute $(a|p)$ for fixed a and all odd primes p such that $p \nmid a$. The problem, thus, reduces to that of computing $(a|p)$.

$$\text{As } a = \pm 2^b p_1^{a_1} p_2^{a_2} \cdots p_r^{a_r}; \quad b \geq 0, a_i \geq 1 \, (i = 1, 2, \ldots r)$$

and as $(a|p)$ is multiplicative, we need to consider only $(2|p)$, $(-1|p)$ and $(q|p)$ where $q \neq p$ are odd primes. Quadratic reciprocity law comes in handy. We need only to compute the Legendre symbol for finitely many primes.

We have observed that -1 is a quadratic residue of primes of the form $4k+1$. So,

$$spl. (x^2 + 1) = \{ 2, 5, 13, 17, 29, \ldots \}$$

Also, -3 is a quadratic residue (mod p) if, and only if, $p \equiv 1$ (mod 12) or $p \equiv -5$ (mod 12). Therefore,

$$spl. (x^2 + 3) = \{ 2, 7, 13, 19, 37, \ldots \}$$

In the case of $f(x) = x^3 - a$, $a \in \mathbb{Z}$, suppose that $x^3 - a$ is irreducible over $\mathbb{Q}$. If p is a prime $\neq 3$, $x^3 - 1$ splits into a product of linear factors over $\mathbb{F}_p$, as $\mathbb{F}_p$ contains a primitive cube root of unity when $p \equiv 1$ (mod 3). That is, if

$p \equiv 1 \pmod{3}$ $\mathbb{F}_p^*/\mathbb{F}_p^{*3}$ is of order 3. If $p \equiv -1 \pmod{3}$ $\mathbb{F}_p^*/\mathbb{F}_p^{*3}$ is of order 1. For, the map $\theta : \mathbb{F}_p^* \to \mathbb{F}_p^{*3}$ defined by $\theta(t) = t^3$ is surjective and ker θ is the set of all cube roots of unity in $\mathbb{F}_p^*$.

It follows that the cubic residue symbol should take values in a group of order 3 and it is to be expected that such a symbol should be an isomorphism of $\mathbb{F}_p^*/\mathbb{F}_p^{*3}$ onto a group of order 3. But, the primes $p \equiv -1 \pmod{3}$ do not satisfy this condition, as $\mathbb{F}_p^*/\mathbb{F}_p^{*3}$ is the trivial group of order 1 in such a case. For such primes, $\mathbb{F}_p^*$ does not contain primitive cube roots of unity. So, in order to get an analogy with the Legendre symbol, we should work in a ring larger than $\mathbb{Z}$, since $\mathbb{Z}$ does not contain the imaginary cube roots of unity. This led Eisenstein to consider $D = \mathbb{Z}[\omega]$, (6.5.1).

The group of units in D is $\{\pm 1, \pm \omega, \pm \omega^2\}$.

Let p be a rational prime. If π is a prime in D,

$$N(\pi) = \pi \bar{\pi} = p, p^2 \text{ or } 9$$

according as $p \equiv 1 \pmod{3}$, $p \equiv 2 \pmod{3}$ or $p = 3$.

(6.7.7) $1 - \omega$ is a prime in D with $N(1 - \omega) = 3$

Given a prime $\pi \in D$, $D/\pi D \cong \mathbb{F}_p$ or $\mathbb{F}_{p^2}$ according as $N(\pi) = p$ or p^2.

Since $p^2 \equiv 1 \pmod{3}$ for all primes $p \neq 3$, it follows that for all primes π in D for which $N(\pi) \neq 3$, $D/\pi D \cong \mathbb{F}_q$ $(q = N(\pi))$ contains all cube roots of unity (see remark 6.5.3).

Further, given $\alpha \in \mathbb{Z}[\omega]$ and a prime $\pi \in \mathbb{Z}[\omega]$ where $\pi \nmid \alpha$ there exists a unique cube root of unity namely 1, ω or ω^2 such that

(6.7.8) $\alpha^{\frac{N(\pi)-1}{3}} \equiv \omega^i \pmod{\pi}$ $(i = 0, 1 \text{ or } 2)$ (see Lemma 6.5.3)

Next, the cubic residue symbol $(\alpha|\pi)_3$ induces an isomorphism of $F_{N(\pi)}^*/F_{N(\pi)}^{*3}$ onto the group $H = \{1, \omega, \omega^2\}$. $(\alpha|\pi)_3 = 1$ if, and only if, $x^3 - \alpha$ splits into a product of linear factors in $\mathbb{F}_{N(\pi)}[x]$.

To determine the primes π for which $x^2 - \alpha$ splits into a product of linear factors in $\mathbb{F}_{N(\pi)}[x]$, it is enough to compute $(\alpha|\pi)_3$ for $\alpha = -1, \omega, 1 - \omega$ and all primary primes π' co-prime with π. For, $\alpha \in \mathbb{Z}[\omega]$ can be uniquely expressed as

$$\alpha = (-1)^a \omega^b (1 - \omega)^c \pi_1^{a_1} \pi_2^{a_2} \dots \pi_t^{a_t}$$

where $a, b, c; a_1, a_2, \dots a_t$ are integers ≥ 0 and $\pi_i (i = 1, 2, \dots t)$ are primary primes.

David Hilbert's ninth problem asks for the most general reciprocity law in the context of algebraic number fields. Emil Artin (1898–1962) gave the solution and it is now known as Artin's general reciprocity law. The details are available in S. Lang [10]. See also Parvathy Shastri [11], J. W. S. Cassels and A. Frölich [2]. For an exhaustive treatment of quadratic residues and quadratic congruences, see Don Redmond [12]. For problems and solutions relating to reciprocity laws, see J. Esmond and M. Ram Murthy [3].

6.8. A comment by W. C. Waterhouse

Gauss lemma given in theorem 41 states that if p is an odd prime and p does not divide a. $(a|b) = (-1)^\mu$ where μ denotes the number of elements in the set $\{1a, 2a, \ldots, (\frac{p-1}{2})a\}$ whose numerically least residues (mod p) are negative. The point is that

$$(-1)^\mu \equiv a^{\frac{p-1}{2}} \pmod{p}.$$

H. Hasse, W. J. Leveque and D. Shanks have pointed out that the factors $1, 2, \ldots, (\frac{p-1}{2})$ of a can be replaced by a 'half-system', any $\frac{p-1}{2}$ numbers not congruent to each other or each other's negatives. The first step towards a general understanding of the lemma is to notice that these are simply the coset representatives of the subgroup $\{1, -1\}$ which is the group of units in $\mathbb{Z}$. We take a look at the corresponding lemma for cubic residues as given by Eisenstein. See [3]. For a prime p congruent to 1 (mod 3) and an element ω of order 3 (mod p), we take coset representatives

$$R_1, R_2, \ldots, R_{\frac{p-1}{3}}$$

of $\{1, \omega, \omega^2\}$ (where $\omega = \exp(\frac{2\pi i}{3})$), and multiply by a not divisible by p. Let β be the number of these products of the form $\omega^2 R_i$ ($i = 1, 2, \ldots, \frac{p-1}{3}$). Then, Gauss lemma gives

$$a^{\frac{p-1}{3}} \equiv \omega^{2\beta + \gamma} \pmod{p} \quad (\gamma \geq 0)$$

So, in Gauss lemma, it is $(-1)^\mu$ rather than μ that matters. The expression in Gauss lemma is a product of 1's and -1's where μ of them are -1.

The general setting is as follows:

Let G be a group and H subgroup (of G) of finite order. Suppose that $a \in G$. Let $[H, H]$ denote the commutator subgroup of H. That is, $[H, H]$ is the subgroup of H generated by elements of the form $aba^{-1}b^{-1}$ where $a, b \in H$. Let b be a fixed element of G. We choose the coset representatives R_i of H and multiply to form bR_i. Rewriting bR_i as $h_i R_{\pi(i)}$ for $h_i \in H$ ($\pi(i)$ is the image of i under a permutation π of the set of suffixes). We form the product Πh_i.

Let $[G : H] = n$. h_i exists as $h_i^{-1} b = a_i$ (say) for some $h_i \in H$. Then,

$$bR_i = h_i a_i R_i = h_i R_{\pi(i)} \quad (i = 1, 2, \ldots, n)$$

As $[G : H] = n$, $\prod_{i=1}^n h_i R_{\pi(i)}$ can be associated with $\prod_{i=1}^n h_i$ where $h_i \in H$. A coset (mod H) is the sum of a finite number of cosets modulo $[H, H]$. Now, $[H, H]$ is a normal subgroup of H and $H/[H, H]$ is abelian. Also, $[G, G]$ is a normal subgroup of G with $G/[G, G]$ abelian.

Let $\phi : G \to H/[H, H]$ be given by

$$\phi(b) = \prod_{i=1}^n h_i R_{\pi(i)} = \text{a specified coset of } [H, H] \text{ in } H.$$

ϕ is a group homomorphism from G into $H/[H, H]$.

Let $\psi : H/[H,H] \to G/[G,G]$ be given by

$$\psi(\prod_{i=1}^{n} h_i R_{\pi(i)}) = \text{a coset of } [G.G] \text{ in } G \text{ determined by } b^n.$$

$(\psi \circ \phi)(b) = $ the coset determined by $b^{[G,H]}$ in $G/[G,G]$.
In other words, the composite map $G \to H/[H,H] \to G/[G,G]$ sends b (an element of G) to the class of $b^{[G:H]}$ in $G/[G,G]$.

In 1930, Emil Artin defined this general mapping from G to $H/[H,H]$ in a quite different number-theoretic set-up. He called it a 'transfer' and it served as a tool to prove the principal ideal theorem of class-field theory due to S. Iyanaga. See the reference in [13]. It turns out to be an extension of Gauss lemma to 'transfers'. This is related to group cohomology as pointed out by B. Eckmann. Generalizations here opened up the doors to deeper group-theoretic ideas.

We conclude by saying that the genesis of reciprocity laws lies in Gauss lemma as pointed out by W. C. Waterhouse in his 'tiny note' [13].

6.9. Worked-out examples

a) What are the odd primes p for which 3 is a quadratic residue ?
Answer: We have to find those odd primes p for which $(3|p) = 1$. We appeal to the quadratic reciprocity law (6.4.2). That is,

$$(3|p)(p|3) = (-1)^{(\frac{3-1}{2})(\frac{p-1}{2})},$$

or,

$$(3|p) = (-1)^{\frac{p-1}{2}}(p|3).$$

In order that $(3|p) = 1$, we must have

(6.9.1) $$\frac{p-1}{2} = 2k \text{ and } (p|3) = 1.$$

or

(6.9.2) $$\frac{p-1}{2} = 2l+1 \text{ and } (p|3) = -1$$

where $k, l \in \mathbb{Z}$.
(6.9.1) is satisfied when $p = 4k+1$ and $x^2 \equiv p\,(\text{mod } 3)$ is solvable. That is, p is of the form $p \equiv 1\,(\text{mod } 4)$ and $p \equiv 1\,(\text{mod } 3)$. Therefore, p has to be of the form $12k' + 1, k' \in \mathbb{Z}$.
(6.9.2) is satisfied of $p \equiv 3\,(\text{mod } 4)$ and $x^2 \equiv p\,(\text{mod } 3)$ is not solvable. That is, p is of the form $p \equiv 3\,(\text{mod } 4)$ and $p \equiv 2\,(\text{mod } 3)$. Therefore, p has to be of the form $12k' + 11, k' \in \mathbb{Z}$.

If p is of the form $12k+5$ or $12l+7$, 3 is a quadratic non-residue mod p. We conclude that 3 is a quadratic residue of p, when p is a prime of the form $12k \pm 1$.   □

b) What are the odd primes p for which 7 is a quadratic residue ?

Answer: We have to find those odd primes p for which $(7|p) = 1$. Appealing to quadratic reciprocity law (6.4.2),

$$(7|p) = (-1)^{\frac{p-1}{2}}(p|7).$$

So, $(7|p) = 1$ if, and only if,

(6.9.3) $\qquad \dfrac{p-1}{2} = 2k$ and $(p|7) = 1$

or

(6.9.4) $\qquad \dfrac{p-1}{2} = 2l+1$ and $(p|7) = -1$

where $k, l \in \mathbb{Z}$.

(6.9.3) holds if $p \equiv 1 \pmod 4$ and $p \equiv 1, 2, 4 \pmod 7$. Using Chinese Remainder Theorem, by forming pairs of congruences

$$\begin{array}{lll} p \equiv 1 \pmod 4 \\ p \equiv 1 \pmod 7 \end{array}\Big\} \quad \begin{array}{l} p \equiv 1 \pmod 4 \\ p \equiv 2 \pmod 7 \end{array}\Big\} \quad \begin{array}{l} p \equiv 1 \pmod 4 \\ p \equiv 4 \pmod 7 \end{array};$$

we note that p has to be one of the forms $p = 28k' + 1$, $p = 28k' + 9$ or $p = 28k' + 25$, where $k' \in \mathbb{Z}$.

(6.9.4) holds if, and only if, $p \equiv 3 \pmod 4$ and $p \equiv 3, 5, 6 \pmod 7$. This yields that p has to be one of the forms $p = 28k' + 3$, $p = 28k' + 19$, $p = 28k' + 27$.

Hence, an odd prime p is a quadratic residue modulo 7 if, and only if, p is one of the types $28k \pm 1$, $28k \pm 3$ or $28k \pm 9$. $\qquad \Box$

c) What are the primes π is $\mathbb{Z}[\omega]$, $(\omega = \exp(\frac{2\pi i}{3}))$ for which 3 is a cubic residue ?

Answer: Let π denote a prime in $\mathbb{Z}[\omega]$. $x^3 \equiv 2 \pmod \pi$ is solvable, if, and only if, $x^3 \equiv 2 \pmod{\pi'}$ where π' is an associate of π. π' can be taken as the primary prime associated with π. If π is a rational prime, say q, then $\chi(2, q) = 1$ and as 2 is a cubic residue of such primes.

Let $\pi = a + b\omega$ be a primary prime (which is complex). By cubic reciprocity law (see Theorem 47), $\chi(2, \pi) = \chi(\pi, 2)$. $N(2) = 4$. So,

$$\pi = \pi^{\frac{(4-1)}{3}} \equiv \chi(\pi, 2) \pmod 2.$$

So, $\chi(\pi, 2) = 1$ if, and only if, $\pi \equiv 1 \pmod 2$ in $\mathbb{Z}[\omega]$. Thus, $x^3 \equiv 2 \pmod \pi$ is solvable, if, and only if, $\pi \equiv 1 \pmod 2$, that is, if, and only if, $a \equiv 1 \pmod 2$ and $b \equiv 0 \pmod 2$. $\qquad \Box$

EXERCISES

1. *Mark the following statements true (T) or false (F) justifying your answer briefly.*

   a) *2 is a quadratic residue* (mod p), *if, and only if, p is a prime of the form $8k \pm 1$.*

   b) *Let p be a prime not dividing a. The number of solutions of $ax^2 + bx + c \equiv 0 \,(\mathrm{mod}\ p)$ is $1 + (b^2 - 4ac \mid p)$.*

   c) *It is correct to say that $(43 \mid 101) = 1$.*

   d) *$x^3 \equiv 2 \,(\mathrm{mod}\ p)$ if, and only if, p is a prime of the form $A^2 + 27B^2$, when $A, B \in \mathbb{Z}$.*

   e) *7 remains a prime in $D = \mathbb{Z}[\omega]$ where $\omega = \exp(\frac{2\pi i}{3})$.*

   f) *Let p be a rational prime. Then, $\chi(-3, p) = 1$.*

2. *Find all primes p for which 11 is a quadratic residue.*

3. *(Don Redmond) If p and q are odd primes and $q = 2p + 1$ show that*

   $$(p \mid q) = (-1 \mid p).$$

4. *Let $\alpha = 2 - 3\omega \in D = \mathbb{Z}[\omega]$, $\omega = \exp(\frac{2\pi i}{3})$. Prove that $x^3 \equiv \alpha \,(\mathrm{mod}\ 11)$ is not solvable in D.*

5. *Suppose $p \geq 5$ is a prime. Show that if $N(p)$ denotes the number of distinct nonzero cubic residues* (mod p), *then*

   $$N(p) = \begin{cases} p - 1, & \text{if } p \equiv 5 \text{ or } p \equiv 11 \,(\mathrm{mod}\ 12) \\ \dfrac{p-1}{3}, & \text{if } p \equiv 1 \text{ or } p \equiv 7 \,(\mathrm{mod}\ 12) \end{cases}$$

6. *Use Gauss lemma to determine $(-6 \mid 13)$.*

7. *Examine whether $x^2 \equiv -3 \,(\mathrm{mod}\ 53)$ is solvable.*

8. *Determine the primes p for which 23 is a quadratic residue.*

9. *Find the number of solutions of the congruence $x^2 \equiv 17 \,(\mathrm{mod}\ 21)$.*

10. *Solve $x^2 + 4x - 28 \equiv 0 \,(\mathrm{mod}\ 289)$.*

11. *If p is a prime of the form $12k + 1$, show that $(3 \mid p) = 1$.*

12. *If p and q are odd primes and $q = 2p + 1$, show that*

   $$(p \mid q) = (-1 \mid p).$$

13. *Let p be a prime > 3. Show that the sum of the quadratic residues of p is divisible by p.*

14. *(Don Redmond) Show that an integer is a square if, and only if, it is a quadratic residue of every prime.*

15. *Let π be a prime in $\mathbb{Z}[\omega]$. If $N(\pi) \neq 3$, prove that $\chi(\omega, \pi) = 1$, ω or ω^2, according as $N(\pi) \equiv 1, 4$ or $7 \,(\mathrm{mod}\ 9)$.*

16. *Find the primes π in $\mathbb{Z}[\omega]$ for which $x^3 \equiv 5 \,(\mathrm{mod}\ \pi)$ has a solution.*

17. *Find the primary prime which is an associate of $3 - \omega$ in $\mathbb{Z}[\omega]$.*

18. *[K. Ireland and M. I. Rosen] Let $\alpha \in \mathbb{Z}[\omega]$. Suppose that $x^3 \equiv \alpha \,(\mathrm{mod}\ \pi)$ is solvable for all but finitely many primes π in $\mathbb{Z}[\omega]$. Then, show that α is a cube in $\mathbb{Z}[\omega]$.*

REFERENCES

[1] Tom M. Apostol: Introduction to Analytic Number Theory, Undergraduate Texts in Mathematics Springer Verlag NY (1976) Narosa Publishing House, New Delhi, Reprint (1985) Chapters 6 and 9 pp 133–137 and 178–203.

[2] J. W. S. Cassels and A. Frölich: Algebraic Number Theory, Academic Press (1967).

[3] J. Esmonde and M. Ram Murthy: Problems in Algebraic Number Theory, GTM 190 Springer Verlag NY (1999) Chapters 2, 3, 7 and 9 pp 11–38, 81–96 and 115–124.

[4] J. B. Fraleigh: A first course in Algebra, Addison Wesley Pub Co., Reading Mass (1968) International Edition (3rd printing) 1994 chapters 35 and 38.

[5] C. F. Gauss: Disquisitiones Arithmetaicae (1801) Lipsiae English Translation: Arthur A Clarke, (reviewed by W. C. Waterhouse), Springer Verlag NY Inc. (1986).

[6] L. J. Goldstein: Density questions in algebraic number theory, Amer. Math. Monthly 78 (1971) 342–351.

[7] Emil Grosswald: Topics from the theory of numbers, Macmillan Company NY, Second Edn (1986) Chapter 5, pages 68–83.

[8] K. Ireland and M. I. Rosen: A classical Introduction to Modern number Theory. GTM No 84 Springer Verlag, NY (1990) Chapters 5 to 9 pp 55–122.

[9] Neal Koblitz: A course in number theory and cryptography, Undergraduate Texts in Mathematics. Springer Verlag NY (1987) Chapter 2 pp 29–52.

[10] S.Lang: Algebraic number Theory, Addison Pub Co, Reading, Mass. (1970).

[11] Parvathy Shastri: Reciprocity laws, Invited lecture: Symposium in Algebra and Number Theory. University of Poona, Pune, (Feb 1994) pp 1–13.

[12] Don Redmond: Number theory, Monographs and Textbooks in Pure and Applied Mathematics No:201, Marcel Dekker Inc., NY (1996) Chapters 2, 3 and 10 pp 53–94 and 693–721.

[13] W. C. Waterhouse: A tiny note on Gauss's lemma, J. Number Theory 30 (1988) 105–107.

ADDITIONAL REFERENCES

[A1] Sey Y. Kim: An elementary proof of the quadratic reciprocity law, Amer. Math. Monthly 111 (2004) 48–50.

[A2] F. Lemmermeyer: Reciprocity laws from Euler to Eisenstein, Springer-Verlag, Berlin (2000).

[A3] K. S. Williams: On Euler's criterion for cubic non-residues, Proc. Amer. Math. Soc., 49 (1975), 277–283.

[A4] B. F. Wyman: What is reciprocity law? Amer. Math. Monthly, 78 (1992), 571–586.

CHAPTER 7

Finite groups

Historical perspective

The notion of a group originated from ideas about transformations of geometrical objects. The study of groups of symmetry paved the way for the definition of an abstract group. By the end of 17th century, methods of solving quadratic, cubic and biquadratic equations were known. In the attempts to solve a quintic equation, permutation groups proved to be relevant. Lagrange got the idea that groups had something to do with equations. The permutation groups S_2, S_3 and S_4 were 'well-behaved' groups (in the sense of solvability) and were associated with a quadratic, cubic and bi-quadratic equation respectively. Lagrange knew that S_5 behaved 'differently'. It was Abel who showed that an equation of the fifth degree was not solvable by 'radicals'. During this period, Everiste Galois discovered a necessary and sufficient condition for an nth degree equation to be solvable by radicals. Galois showed that to each algebraic equation

$$f(x) = a_0 x^n + a_1 x^n + \cdots + a_n = 0 \quad (a_0 \neq 0, a_0, a_1, \ldots, a_n \in \mathbb{Q})$$

one could attach a group of permutations to the polynomial $f(x)$ of the equation. The equation is solvable by radicals if, and only if, the group associated with the polynomial is solvable. After Galois, Felix Klein (1849–1929) attempted to describe all geometries by their groups of symmetries. This, he called the Erlangen Programme. Since then, group theory has become a major tool in many branches of mathematics.

Arthur Cayley (1821–1895), Richard Dedekind and Kronecker gave a general definition of an abstract group. It was an example to make abstraction the general trend, just as the axiomatic or postulational development of mathematics laid the foundations of modern mathematics during the period 1880–1900. But, then, when did number theory interact with groups? It is to be emphasized that it was Augustin-Louis Cauchy (1789–1857) and the Norwegian high-school teacher Ludwig Sylow (1832–1918) who gave conditions for the existence of a subgroup of prime-power order in a finite group. The contributions of C. Jordan (1838–1922), Frobenius (a student of Karl Weierstrass), I. Schur (1875–1941) (a student of Frobenius) to the representation theory of groups are substantial. Indeed, the probing investigations of William Burnside (1852–1927) into group theory made it a well cut-out branch of algebra. Burnside's aim was to understand finite simple groups better. Burnside's conjecture: 'No simple group of odd order exists' was solved in the affirmative by Walter Feit and John Thompson in 1963. See [A2]. The classification problem of finite simple groups has also been settled by E. Zelmanov in 1992. Zelmanov received the Fields medal for his solution to 'Restricted Burnside Problem for groups' in 1994.

7.1. Introduction

Wherever there is a problem of enumeration or counting of objects, number theory lends a helping hand. While counting the number of conjugate classes of elements in the symmetric group S_n ($n \geq 2$), one gets a formula in terms of $p(n)$, the number of partitions of n. This aspect is narrated in Section 7.2. In chapter 5, the Rademacher formula for $N(n,r,s)$, the number of solutions of

$$x_1 + x_2 + \cdots + x_s \equiv n \,(\mathrm{mod}\ r)$$

under the restriction g.c.d $(x_i, r) = 1$, $i = 1, 2, \ldots, r$ was derived. See Section 5.6. A similar situation occurs in counting the number of distinguished representations of a group element of a finite group G. The main result in the context of groups is due to David Jacobson and K. S. Williams [8] and is shown in theorem 51. This is dealt with in Section 7.3. Section 7.4 is about the number of cyclic subgroups of a finite group. Burnside's lemma (see theorem 52) is applied to establish I. M. Richards' theorem (theorem 54) which says that if G is a finite group of order r, the number of cyclic subgroups of G is $d(r)$, the number of divisors of r if, and only if, G is cyclic. See [12]. Incidentally, P. Kesava Menon's identity [11]

$$\sum_{\substack{a\,(\mathrm{mod}\ r)\\ g.c.d\,(a,r)=1}} (a-1,r) = \phi(r)d(r)$$

is deduced from Burnside's lemma.

Next, a criterion for uniqueness of a cyclic group of order r due to Dieter Jungnickel [3] is proved in theorem 55. The criterion is that a group G of order r is a unique cyclic group if, and only if, g.c.d $(r, \phi(r)) = 1$.

7.2. Conjugate classes of elements in a group

We recall the notion of conjugacy in respect of elements of a group G.

Definition 7.2.1 : *For $a, b \in G$, b is called a conjugate of a in G, if there exists an element c in G such that $b = c^{-1}ac$.*

Conjugacy is an equivalence relation on the set G. For $a \in G$, we write

(7.2.1) $C(a) = \{x \in G : x$ is a conjugate of $a\}$

$C(a)$ denotes the equivalence class of a or the conjugate class of a (conjugacy class of a). If $N(a)$ denotes the set of elements of G which commute with a, $N(a)$ is a subgroup of G. $N(a)$ is known as the normalizer of a in G. We denote the number of elements in $C(a)$ by $|C(a)|$.

Fact 7.2.1 : If G is a finite group and $a \in G$

$$|C(a)| = \frac{|G|}{|N(a)|}$$

where $|G|$ and $|N(a)|$ denote the number of elements of G and $N(a)$ respectively.

For proof, see I. N. Herstein [7].

As G is a disjoint union of conjugate classes of elements, we deduce the class equation of G in the form

$$(7.2.2) \qquad\qquad |G| = \sum_a \frac{|G|}{|N(a)|}$$

where summation is over one element a in each conjugate class.

We apply (7.2.2) to connect the number of conjugate classes of elements in S_n, the permutation group on n symbols with partition function.

Definition 7.2.2 : *Let $T = \{a_1, a_2, \ldots, a_m, \ldots\}$ be a finite or infinite set of positive integers . If*

$$a_j^{(1)} + a_j^{(2)} + \cdots + a_j^{(r)} = n \text{ with } a_j^{(i)} \in T, (i = 1, 2, \ldots, r)$$

we regard $a_j^{(1)} + \cdots + a_j^{(r)}$ as a partition of n into r summands or parts belonging to T. If $n \in T$, n itself is counted as a partition of n. The summands do not have to be distinct.

Definition 7.2.3 : *Let T be a finite or infinite set of positive integers. The number of distinct partitions of n into summands (or parts) belonging to T is denoted by $p_T(n)$. p_T is the partition function relative to T.*

$p_T(n)$ is the number of unrestricted partitions of N into parts that belong to T. p_T is a function $p_T : \mathbb{N} \to \mathbb{N}$ where $\mathbb{N}$ denotes the set of positive integers. When $T = \mathbb{N}$, the partition function is denoted by p. In fact, $p(n)$ denotes the number of partitions of n. For instance, as 5=4+1 or 3+2 or 3+1+1 or 2+2+1 or 2+1+1+1 or 1+1+1+1+1. $p(5) = 7$ (as 5 = 5 is also to be counted for obtaining $p(5)$). By convention, we take $p(0) = 1$.

Definition 7.2.4 : *When $|x| < 1$, $F(x) = \sum_{n=0}^{\infty} p(n)x^n$ is called the generating function of $p(n)$.*

A result due to Euler says:

Fact 7.2.2 :

$$F(x) = \sum_{n=0}^{\infty} p(n)x^n = \prod_{k=1}^{\infty} (1 - x^k)^{-1} \quad (|x| < 1).$$

For proof, see Emil Grosswald [6].

Theorem 49 : *The number of conjugate classes in S_n is $p(n)$, the number of unrestricted partitions of n.*

Proof : It is known that every element α of S_n that is not a cycle by itself is expressible as a product of cycles. When we break a given permutation $\alpha \in S_n$ into a product of cycles, we obtain a partition of n. For, if the cycles appearing in α have lengths $n_1, n_2 \ldots, n_r$ respectively with $n_1 \leq n_2 \leq \cdots \leq n_r$, then

$$(7.2.3) \qquad\qquad n = n_1 + n_2 + \cdots + n_r.$$

A permutation $\alpha \in S_n$ has the cycle decomposition $\{n_1, n_2, \ldots, n_r\}$, if α is a product of disjoint cycles of lengths $n_1, n_2, \ldots n_r$.

Claim : Two permutations $\alpha, \beta \in S_n$ are conjugate if, and only if, they have the same cycle decomposition.

Suppose $\theta \in S_n$. Given $\alpha \in S_n$, we need to describe $\theta \alpha \theta^{-1}$. Suppose that α sends i to j and θ sends i to s and j to t. Then $\theta \alpha \theta^{-1}$ sends s to t. That is, to compute $\theta^{-1} \alpha \theta$, we replace every symbol in α by its image under θ.

Suppose

$$(7.2.4) \qquad \begin{aligned} \alpha &= (a_1 a_2 \cdots a_{n_1})(b_1 b_2 \cdots b_{n_2}) \cdots (x_1 x_2 \cdots x_{n_r}) \text{ and} \\ \beta &= (a'_1 a'_2 \cdots a'_{n_1})(b'_1 b'_2 \cdots b'_{n_2}) \cdots (x'_1 x'_2 \cdots x'_{n_r}) \end{aligned}$$

Then,

$$\beta = \theta \alpha \theta^{-1} \text{ where } \theta = \begin{pmatrix} a_1 a_2 \cdots a_{n_1} & b_1 b_2 \cdots b_{n_2} & \cdots & x_1 x_2 \cdots x_{n_r} \\ a'_1 a'_2 \cdots a'_{n_1} & b'_1 b'_2 \cdots b'_{n_2} & \cdots & x'_1 x'_2 \cdots x'_{n_r} \end{pmatrix}$$

It is clear that two conjugates have the same cycle decomposition. The rule for computing a conjugate of α is that we have to replace every element in a cycle by its image under the element $\theta \in S_n$ used for taking $\theta^{-1} \alpha \theta$. As each conjugate class gives a partition of n, $p(n)$ gives the number of conjugate classes in S_n. $\square$

Example 7.2.1 : In the case of S_5, as $p(5) = 7$, there are 7 conjugate classes of elements in S_5.

Remark 7.2.1 : The computation of $p(n)$ is possible, if $p(t)$ is known for all $t < n$. The well-known formula is

$$(7.2.5) \quad p(n) = p(n-1) + p(n-2) - p(n-5) - p(n-7) + \cdots + (-1)^{k+1} p(n-n_k) + \cdots$$

where $n_k = \frac{1}{2}k\,(3k \pm 1)$ are called pentagonal numbers.

For proof, see Emil Grosswald [6]. The series on the right side of (7.2.5) terminates when $\frac{k(3k+1)}{2} > n$.

7.3. Counting certain special representations of a group element

G denotes a finite group which is not necessarily abelian.
We call $a \in G$ a special element if a is characterised by a specific property P relating to G. For example, $a \in G$ is called a generator if there exists a subset T of

G such that $S = T \cup \{a\}$ generates G but the subgroup generated by T is not equal to G. The property P here is 'being a generator'.

Definition 7.3.1 : *The centre $Z(G)$ of G is defined by*

$$Z(G) = \{x \in G : xa = ax \text{ for all } a \in G\}.$$

In fact, the centre of G consists of those elements (of G) which commute with every element of G. $y \in G$ is called a central element if $y \in Z(G)$. We could take the property P of a special element to be 'being non-central'. Further, the property P could be such that the order of a special element is greatest.

In [8], David Jacobson and Kenneth S. Willams consider the number of representations of an element in a finite group G as a product of s special elements possessing a specified property P. More generally, one could consider a non-empty subset D of G and find the number $N(D,a,s)$ of solutions of the equation

(7.3.1) $x_1 x_2 \cdots x_s = a$,   where $x_i \in D$ $(i = 1, 2, \ldots s)$.

We observe that if a does not belong to the subgroup generated by D, then $N(D,a,s) = 0$ for all s.

For arbitrarily chosen D, the evaluation of $N(D,a,s)$ is not easy. If D is a subgroup of G,

(7.3.2) $N(D,a,s) = |D|^{s-1}$, for all $a \in D$.

Let D be a non-empty subset of G. We write $J(D) = J$ to denote the largest normal subgroup of G such that

$$xJ \subseteq D \text{ for all } x \in D.$$

Definition 7.3.2 : *The group G is said to be D-reduced, if $J = (e)$ where e denotes the identity element of G.*

G/J is denoted by $\bar{G}$ and elements of $\bar{G}$ (that is, cosets of J in G) are denoted by $\bar{a}$ for $a \in G$. In other words,

$$\bar{a} = aJ \text{ in } \bar{G}.$$

Lemma 7.3.1 : *Let G be a group. Suppose that J is a normal subgroup of G. If $x_1, x_2, \ldots, x_s$ are elements of G, the number of s-tuples $(y_1, y_2, \ldots y_s)$ such that*

$$x_1 x_2 \cdots x_s = y_1 y_2 \cdots y_s \text{ where } y_i \in x_i J \quad i = 1, 2, \ldots, s$$

is equal to $|J|^{s-1}$.

Proof : in two steps.

1. The case $s = 1$ is trivial as $x_1 e = x_1$ and $x_1 \in x_1 J$. So, the number of elements having this property equals $1 = |J|^0$.

2. Suppose that $s > 1$. Let $b_1, b_2, \ldots, b_{s-1}$ be arbitrary elements of J. Then,

$$b_1 x_2 b_2 x_3 \cdots b_{s-1} x_s = (x_2 x_3 \cdots x_s)(b_1 b_2 \cdots b_{s-1})$$

$$= (x_1 x_2 \cdots x_s)b \text{ where } b_1 b_2 \cdots b_{s-1} = b, b \in J$$

or

$$(b_1 x_2 b_2 x_3 \cdots b_{s-1} x_s) b^{-1} = x_1 x_2 \cdots x_s.$$

So, the number of representations such as $x_1 x_2 \cdots x_s = y_1 y_2 \cdots y_s$ where $y_i \in x_i J$ is the same as the number of $(s-1)$-tuples

$(b_1, b_2, \ldots, b_{s-1})$ where $b_i \in J$ $(i = 1, 2, \ldots, s-1)$.

Therefore, the number desired is $|J|^{s-1}$, as $\{b_1 b_2 \cdots b_{s-1}\}$ can be chosen from J in $|J|^{s-1}$ ways. $\square$

Theorem 50 : If $\bar{D} = \{\bar{x} : x \in D\}$, then $\bar{G}$ is $\bar{D}$-reduced and

(7.3.3) $N(D, a, s) = |J|^{s-1} N(\bar{D}, \bar{a}, s)$

Proof : $\bar{K} = K/J$ is a normal subgroup of $\bar{G} = G/J$, if K is a normal subgroup of G containing J. If $\bar{K}$ is the largest normal subgroup of $\bar{G}$ such that $\bar{x}\bar{K} \subseteq \bar{D}$ for all $\bar{x} \in \bar{D}$, then it is easy to check that $xK \subseteq D$ for all $x \in D$. As K is a normal subgroup of G, $K = J$ and so $\bar{G}$ is $\bar{D}$-reduced.

To find $N(D, a, s)$, we proceed as follows:

Let T denote the set of solutions $(x_1, x_2, \ldots, x_s)$ of (7.3.1), we introduce an equivalence relation on T by considering

$$(x_1, x_2, \ldots, x_s) \sim (y_1, y_2, \ldots, y_s)$$

if $y_i \in x_i J$ $(i = 1, 2, \ldots, s)$. With each equivalence class $C(x_1, x_2, \ldots, x_s)$ of the s-tuple $(x_1, x_2, \ldots x_s)$, we associate an s-tuple $(\bar{x}_1, \bar{x}_2, \ldots, \bar{x}_s)$ where

(7.3.4) $(\bar{x}_1 \bar{x}_2 \cdots \bar{x}_s) = \bar{a}$ in $\bar{G}$

Let E be the set equivalence classes of T, if $\bar{T}$ denotes the set of solutions of

(7.3.5) $\bar{x}_1 \bar{x}_2 \cdots \bar{x}_s = \bar{a}$ where $\bar{x}_1, \bar{x}_2 \ldots, \bar{x}_s \in \bar{D}$

we obtain a map $\psi : E \to \bar{T}$ defined by

(7.3.6) $\psi(C(x_1, x_2, \ldots x_s)) = (\bar{x}_1, \bar{x}_2, \ldots \bar{x}_s)$

Claim : ψ is a bijection.

For, if $(\bar{x}_1, \bar{x}_2, \ldots \bar{x}_s) = (\bar{t}_1, \bar{t}_2, \ldots \bar{t}_s)$, that is, if

$$\bar{x}_i = \bar{t}_i \quad i = 1, 2, \ldots s$$

then, $(x_1, x_2, \ldots x_s) \sim (t_1, t_2, \ldots, t_s)$ and

$$C(x_1, x_2, \ldots, x_s) = C(t_1, t_2, \ldots, t_s)$$

That is, ψ is one-one.

To show that ψ is onto, let $\bar{x}_1 \bar{x}_2 \cdots \bar{x}_s = \bar{a}$ where $\bar{x}_1, \bar{x}_2, \ldots, \bar{x}_s \in \bar{D}$.

Then, $x_1 x_2 \cdots x_s b = a$ where $x_1, x_2, \ldots, x_s \in D$ and $b \in J$.

Thus $(x_1, x_2, \ldots x_{s-1}, x_s b)$ is a solution of (7.3.1) and ψ maps $C(x_1, x_2, \ldots x_s b)$ onto $(\bar{x}_1, \bar{x}_2, \ldots, \bar{x}_s)$ in $\bar{T}$. So, the number of equivalence classes of T is equal to $N(\bar{D}, \bar{a}, s)$. By lemma 7.3.1, each equivalence class consists of $|J|^{s-1}$ elements and hence (7.3.3) follows. $\square$

Remark 7.3.1 : If $J = G \setminus D$ and J is a normal subgroup of G, $\bar{D} = \{\bar{a} : a \in D\}$ and $\bar{G}$ is $\bar{D}$-reduced. This shows that J consists of all the elements of $\bar{G}$ except the identity. For, $J \cap D = \emptyset$ and the coset eJ belonging to $\bar{G}$ is not in $\bar{D}$.

Next, we introduce a function $f : G \to \mathbb{Z}$ by

$$(7.3.7) \quad f(a) = \begin{cases} -1, & \text{if } a \in D \\ [G:H] - 1, a \notin D, & \text{where } H = G \setminus D \text{ is a subgroup of } G. \end{cases}$$

The formula for $N(D, a, s)$ is obtained in the following

Theorem 51 (D. Jacobson and K. S. Willams (1972)) : *Let G be a finite group. Suppose that D is a non-empty subset of G such that $H = G \setminus D$ is a subgroup of G. Then,*

$$(7.3.8) \quad N(D, a, s) = \frac{|D|^s}{|G|} \left\{ 1 + \frac{(-1)^s f(a)}{([G:H] - 1)^s} \right\}$$

where f is as defined in (7.3.7). $[G : H]$ denotes the index of H in G.

Proof : We, first, note that H is not taken as a normal subgroup. For $b \in G$, $N(D, b, s)$ denotes the number of representations of b as a product of s elements chosen from D. The number of ways of forming s-tuples from D is $|D|^s$. So,

$$(7.3.9) \quad \sum_{b \in G} N(D, b, s) = |D|^s$$

Now, all the solutions of

$$x_1 x_2 \cdots x_s x_{s+1} = a \quad \text{where } x_i \in D(i = 1, 2, \ldots, (s+1))$$

correspond to the solutions of the simultaneous system of equations

$$x_1 x_2 \cdots x_s = b, x_{s+1} = b^{-1} a, \quad b \in G, a \in G \text{ and } b^{-1} a \in D$$

Therefore, we deduce that

$$(7.3.10) \quad N(D, a, s+1) = \sum_{b^{-1} a \in D} N(D, b, s)$$

As $H = G \setminus D$, we obtain from (7.3.9) and (7.3.10) that

$$(7.3.11) \quad N(D, a, s+1) = |D|^s - \sum_{b^{-1} a \in H} N(D, b, s)$$

Now, $b^{-1} a \in H \Leftrightarrow a \in bH$. That is,

$$b^{-1} a \in H \Leftrightarrow b \in aH$$

Next, we claim that

$$(7.3.12) \quad N(D, a, s) = N(D, b, s), \text{ whenever } b \in aH.$$

For, suppose that $b = ac$ where $c \in H$.
If $a, b \in D$, $N(D, a, 1) = N(D, b, 1)$.

If $a \notin D$, then $a \in H$ and $b \in H$. So,

$$N(D,a,1) = N(D,b,1) = 0$$

So, (7.3.12) is true for $s = 1$.

Assume that $s > 1$. If $x_1 x_2 \cdots x_{s-1} x_s = a$ and each $x_i \in D$ $(i = 1, 2, \ldots, s)$ then,

$$x_1, x_2, \ldots, x_{s-1} (x_s c) = b \text{ where } c \in H$$

If $x_s c \notin D$, $x_s c \in H$. So, $x_s \in H$ and so $x_s \notin D$.

The contrapositive statement yields $x_s \in D \Rightarrow x_s c \in D$.

So, $N(D,a,s) = N(D,b,s)$, whenever $b \in aH$.

This establishes (7.3.12).

From (7.3.11), we obtain

(7.3.13) $N(D,a,s+1) = |D|^s - |H|\, N(D,a,s).$

This is a recurrence relation connecting $N(D,a,s)$ and $N(D,a,s+1)$. By repeated application of (7.3.13), one gets

(7.3.14) $N(D,a,s) = |D|^{s-1} - |H||D|^{s-2} + \cdots + (-1)^{s-2}|H|^{s-2}|D|$

$$+ (-1)^{s-1} N(D,a,1).$$

But,

(7.3.15) $N(D,a,1) = \begin{cases} 1, & \text{if } a \in D, \\ 0, & \text{if } a \notin D. \end{cases}$

So, the right side of (7.3.14) has either s or $(s-1)$ terms according as $a \in D$ or $a \notin D$. As $[G:H] = \frac{|G|}{|H|}$,

(7.3.16) $[G:H] - 1 = \frac{|D|}{|H|}$ and $1 + \frac{|H|}{|D|} = \frac{|G|}{|D|}.$

We note that we get the sum on the right side of (7.3.14) as that of a geometric progression with common ratio $-\frac{|H|}{|D|}$.

So,

(7.3.17) $N(D,a,s) = \begin{cases} |D|^{s-1} \left\{ \dfrac{1 - (-1)^s (\frac{|H|}{|D|})^s}{1 + \frac{|H|}{|D|}} \right\}, & \text{if } a \in D \\[4mm] |D|^{s-1} \left\{ \dfrac{1 - (-1)^{s-1}(\frac{|H|}{|D|})^{s-1}}{1 + \frac{|H|}{|D|}} \right\}, & \text{if } a \notin D \end{cases}$

So (7.3.17) reduces to

(7.3.18) $N(D,a,s) = \begin{cases} \dfrac{|D|^s}{|G|} \left(1 + \dfrac{(-1)(-1)^s}{([G:H]-1)^s} \right), & \text{if } a \in D \\[4mm] \dfrac{|D|^s}{|G|} \left(1 + \dfrac{(-1)^s}{([G:H]-1)^{s-1}} \right), & \text{if } a \notin D \end{cases}$

With f as given in (7.3.7), we obtain (7.3.8) from (7.3.18). $\qquad\square$

Remark 7.3.2 : Theorem 51 has been adapted from [8].

APPLICATIONS 7.3.1: G is a finite group. Suppose P is the property of an element being 'non-central'. The set of elements possessing the property P is the set of elements (of G) which are not in the centre $Z(G)$ of G. (See definition 7.3.1). So, D is the set of non central elements. Clearly, $G \setminus D = Z(G)$ is a subgroup of G. Therefore, theorem 51 is applicable to obtain $N(D, a, s)$.

We take, for instance the dihedral group D_4 giving the group of symmetries of a square. Its elements are

(7.3.19)

$$a_0 = \begin{pmatrix} 1 & 2 & 3 & 4 \\ 1 & 2 & 3 & 4 \end{pmatrix}, \quad b_1 = \begin{pmatrix} 1 & 2 & 3 & 4 \\ 2 & 1 & 4 & 3 \end{pmatrix},$$

$$a_1 = \begin{pmatrix} 1 & 2 & 3 & 4 \\ 2 & 3 & 4 & 1 \end{pmatrix}, \quad b_2 = \begin{pmatrix} 1 & 2 & 3 & 4 \\ 4 & 3 & 2 & 1 \end{pmatrix}$$

$$a_2 = \begin{pmatrix} 1 & 2 & 3 & 4 \\ 3 & 4 & 1 & 2 \end{pmatrix}, \quad c_1 = \begin{pmatrix} 1 & 2 & 3 & 4 \\ 3 & 2 & 1 & 4 \end{pmatrix},$$

$$a_3 = \begin{pmatrix} 1 & 2 & 3 & 4 \\ 4 & 1 & 2 & 3 \end{pmatrix}, \quad c_2 = \begin{pmatrix} 1 & 2 & 3 & 4 \\ 1 & 4 & 3 & 2 \end{pmatrix}.$$

a_i $(i = 0, 1, 2, 3)$ represent rotations, b_i $(i = 1, 2)$ present mirror images in perpendicular bisectors of sides and $c_i (i = 1, 2)$ represent diagonal flips.

$$Z(D_4) = \{a_0, a_2\}.$$

$$[D_4 : Z(D_4)] = 4.$$

D denotes the set of non-central elements. Taking $t \in D_4$, the expression for $N(D, t, 3)$ $(s = 3)$ is

(7.3.20)
$$N(D, t, 3) = \begin{cases} 28, & \text{if } t \in D \\ 24, & \text{if } t \notin D \end{cases}$$

We recall that a group G is said to be finitely generated if the set of elements which generate G is a finite set.

Definition 7.3.3 : *A group G satisfies the maximum condition for subgroups, if every non-empty chain of subgroups has a maximal element. That is, every ascending chain of subgroups becomes constant after a finite number of steps.*

The minimum condition is satisfied if every non-empty chain of subgroups has a minimal element.

Definition 7.3.4 : *a) Let G be a group with identity e. A subnormal series of G is a finite ascending chain of subgroups of G beginning with $H_0 = (e)$ as given below:*

$$H_0 = (e) \subset H_1 \subset H_2 \cdots \subset H_n = G$$

and H_i is a normal subgroup of $H_{i+1} (i = 0, 1, 2, \ldots, n-1)$. The factor groups H_{i+1/H_i} are called the factor of the subnormal series and the number of factors (> 1) is called the length of the subnormal series.
b) A subnormal series $\{K_j\}$ is a refinement of a subnormal series $\{H_i\}$ of a group G, if each H_i is a K_j.

c) Two subnormal series of the same group are called equivalent, if they have the same length and if there is a one-to-one correspondence between their factors where corresponding factors are isomorphic groups.

d) A composition series of a group G is a subnormal series whose factors are simple groups $\neq (e)$. That is, one has an ascending chain of subgroups:

$$H_0 = (e) \subset H_1 \subset \cdots \subset H_n = G$$

and each H_i is a maximal normal subgroup of $H_{i+1}(i = 0, 1, 2, \ldots, n-1)$.

Propositions 7.3.1 :

a) (Schreier's theorem) Any two subnormal series of the same group have isomorphic refinements.

b) (Jordan-Hölder theorem) Any two composition series of the same group are equivalent.

For proofs, see D. Robinson [13].

Fact 7.3.1 : A group satisfies the maximum condition if, and only if, each of its subgroups is finitely generated.

Fact 7.3.2 : A non-trivial finitely generated group has a maximal subgroup and also has a maximal normal subgroup (A maximal normal subgroup is a proper normal subgroup which is not a proper subgroup of a proper normal subgroup).

Fact 7.3.3 : If a group satisfies the minimum condition, then each of its elements has finite order.

Fact 7.3.4 : A group G is not finitely generated if it has an infinite ascending union $\cup_{n=1}^{\infty} A_n$ with each A_n properly contained in A_{n+1} where A_i, $(i = 1, 2, \ldots)$ are subgroups of G, or, if it is an infinite direct product $\prod_{n=1}^{\infty} B_n$ with each B_n a non-trivial subgroup of G.

For proofs of Facts 7.3.1 to 7.3.4, see Eugene Shenkman [14].

Definition 7.3.5 : *Let x be an element of a group G. x is called a non-generator if for every subset S (of G) such that $G = \langle S, x \rangle$ (the subgroup generated by S and x), one has $G = \langle S \rangle$ also.*

In the case of D_4, the maximal subgroups are

$$H_1 = \{a_0, a_2, b_1, b_2\}$$
(7.3.21)
$$H_2 = \{a_0, a_1, a_2, a_3\}$$
$$H_3 = \{a_0, a_2, c_1, c_2\}$$

Claim : It is easy to check that the set of non-generators of a group G is a subgroup of G. (It is non-empty, as it contains the identity element e). By convention, the empty set ϕ is a generating set for the trivial subgroup $\langle \phi \rangle = (e)$.

To substantiate the claim, let x, y be non-generators of G. Let S be a set of generators of G.

If $G = \langle xy^{-1}, S \rangle$, then $G = \langle x, y, S \rangle = \langle x, S \rangle = \langle S \rangle$

That is, xy^{-1} is a non-generator whenever x and y are non-generators. This establishes the claim.

Definition 7.3.6 : *The subgroup of non-generators of a group G is called the Frattini subgroup of G. (after Giovanni Frattini (1852–1925))*

Fact 7.3.5 : The Frattini subgroup of a group G is the intersection of G with the maximal subgroups of G.

For proof, see Eugene Shenkman [14].

In the case of D_4, from (7.3.21), we see that the Frattini subgroup of D_4 is given by K (say), where

$$K = H_1 \cap H_2 \cap H_3 = \{a_0, a_2\} = Z(D_4).$$

Remark 7.3.3 : If G is a finitely generated simple group, then the Frattini subgroup of G is (e) where e is the identity element in G.

Another illustration of theorem 51 is from the Frattini subgroup of a group G. If P is the property of an element being a generator of G, D is the set of generators. The complement of D in G is K, the Frattini subgroup of G. So the formula for $N(D, a, s)$ holds.

Taking the example of D_4 once again, the set of generators of D_4 is $\{a_1, a_3, b_1, b_2, c_1, c_2\}$ where a_i, b_i, c_i are as given in (7.3.19). Here, D is the set of generators and $|D| = 6$. Then, with $s = 4$

$$(7.3.22) \qquad N(D, t, 4) = \begin{cases} 2(3^4 - 1), & \text{if } t \in D; \\ 6(3^3 + 1), & \text{if } t \notin D. \end{cases}$$

Remark 7.3.4 : The dihedral group D_n is defined as a group of order $2n$ and having generators a, b such that

$$a^n = e, \, b^2 = e, \, ba = a^{-1}b$$

It is the group of symmetries of the regular n-gon. The elements of D_n [1, Proposition 3.6 page 165] are

$$\{e, a, a^2, \dots a^{n-1}; b, ab, a^2b, \dots, a^{n-1}b\} = \{a^i b^j : 0 \leq i < n, 0 \leq j < 2\}$$

It can be shown that if a group G is of order $2p$, where p is an odd prime, then G is either cyclic or D_p.

7.4. Number of cyclic subgroups of a finite group

It is known that if G is a finite cyclic group of order r, then for each divisor t of r, there is exactly one subgroup of order t and this subgroup is cyclic. This is established as follows:

Let r be equal to $p_1^{a_1} p_2^{a_2} \ldots p_k^{a_k}$ $a_i \geq 1$, $i = 1, 2, \ldots, k$. As $p_i^{b_i}$ $(1 \leq b_i \leq a_i)$ divides r, by Sylow's first theorem, there is a subgroup (of G) of order $p_i^{b_i}$ and it is cyclic. When m, n are relatively prime to one another the cyclic groups $C(m)$ and $C(n)$ of orders m and n respectively are such that $C(mn)$ is isomorphic to the direct product $C(m) \times C(n)$. It is also cyclic. So the subgroups of G are cyclic subgroups $C(t)$ for each divisor t of r. $C(t)$ is unique on account of the fact that $C(t)$ is isomorphic to the direct product of cyclic groups whose orders are relatively prime to one another. Therefore, if $d(r)$ denotes the number of divisors of r, G has exactly $d(r)$ cyclic subgroups. This statement can be sharpened by considering any finite group of order r.

We need the notion of a group action on a set X.

Definition 7.4.1 : *A group G is said to act on a set X, if for each $x \in X$, we associate the pair (g, x) denoted by $g(x)$ satisfying the following conditions:*

(a) $g(h(x)) = gh(x)$ for $g, h \in G$ and every $x \in X$
(b) $e(x) = x$ where e is the identity in G and $x \in X$.

For example, the symmetric group S_n is a group acting on the set $\{1, 2, \ldots, n\}$.

Definition 7.4.2 : *For $x, y \in X$, we define a congruence*

$$x \equiv y (\mathrm{mod}\ G)$$

if there exists $g \in G$ such that $g(x) = gx = y$.

$\equiv$ is an equivalence relation on X. The equivalence classes under $\equiv$ are called the orbits of G (in X).

Definition 7.4.3 : *For $x \in X$, we define*

$$G_x = \{g \in G : g(x) = x\}$$

G_x is called the stabilizer of x in G.

It is easy to check that G_x is a subgroup of G. It is also clear that if O_x denotes the orbit of G containing x, then, when $|G|$ is finite,

(7.4.1) $$|G| = |G_x||O_x|.$$

Theorem 52 (Burnside's lemma) : *Let G be a finite group. If G acts on a set X and if $\psi(g)$ denotes the number of elements of X which are left invariant by $g \in G$, the number $N(G)$ of orbits of G is given by*

(7.4.2) $$N(G) = \frac{1}{|G|} \sum_{g \in G} \psi(g)$$

Proof : We suppose that $S(G)$ denotes the set of orbits of G. We need to find out $N(G) = |S(G)|$. Let r denote the number of elements of the form $g(x)$ where $x \in X$, $g \in G$ and $g(x) = x$. For fixed g, the number of such elements is given by $\psi(g)$ (in the notation of the theorem). Then,

$$(7.4.3) \qquad\qquad r = \sum_{g \in G} \psi(g)$$

Now, for a fixed $x \in X$, the number of elements $g(x)$ for which $g(x) = x$, $g \in G$ is $|G_x|$ (see definition 7.4.3). If T denotes the union of orbits of G, we have

$$r = \sum_{y \in T} |G_y| = \sum_{O \in S(G)} \sum_{x \in O} |G_x| \quad \text{(where } O \text{ denotes an orbit).}$$

If x and y belong to the same orbit,

$$|G_x| = |G_y| = \frac{|G|}{|O_x|} = \frac{|G|}{|O_y|}, \quad \text{by (7.4.1).}$$

Thus,

$$r = \sum_{g \in G} \psi(g) = \sum_{O \in S(G)} \sum_{x \in O} \frac{|G|}{|O_x|},$$

$$= \sum_{O \in S(G)} |O_x| \frac{|G|}{|O_x|}$$

$$= |G| \sum_{O \in S(G)} 1.$$

Or,

$$(7.4.4) \qquad\qquad r = |G| N(G).$$

From (7.4.3) and (7.4.4), we arrive at the desired formula for $N(G)$ given in (7.4.2). $\qquad\qquad\qquad\qquad\qquad\qquad\qquad\qquad\qquad\qquad\qquad\qquad$ $\square$

Illustration 7.4.1 (Kesava Menon's identity [11]) :

$$(7.4.5) \qquad\qquad \sum_{\substack{a(\bmod r) \\ g.c.d\,(a,r)=1}} (a-1, r) = \phi(r) d(r)$$

where the summation on the left is over a reduced-residue system (mod r). $\phi(r)$ *and* $d(r)$ *are, respectively, the Euler ϕ-function and the divisor function.*

Proof : We consider a set X given by $\{1, 2, \ldots, r\}$. U_r denotes the multiplicative group of units in $\mathbb{Z}/r\mathbb{Z}$. U_r is of order $\phi(r)$. The action of U_r on X together with Burnside's lemma yields the desired identity.

$$\text{For} \quad \psi(g) = \#\{x \in X \ : \ g(x) = x \text{ where } g \in U_r\}$$

$\psi(g)$ is the number of solutions of the congruence

$$(7.4.6) \qquad\qquad gx \equiv x(\bmod r) \quad \text{where g.c.d } (g, r) = 1$$

(7.4.6) is written as

(7.4.7) $(g-1)x \equiv 0 \pmod{r}$

So, $\psi(g)$ is the number of solutions of the congruence (7.4.7). Therefore, $\psi(g) = $ g.c.d $(g-1,r)$. See Fact 5.2.1, chapter 5.

If $N(U_r)$ denotes the number of orbits of U_r in X, then, using Burnside's lemma (theorem 52), we obtain

(7.4.8) $$N(U_r) = \frac{1}{\phi(r)} \sum_{\substack{g \pmod{r} \\ g.c.d\,(g,r)=1}} (g-1,r)$$

Therefore, it will suffice to show that $N(U_r) = d(r)$, the number of divisors of r. Now, x, y ($\in X$) belong to the same orbit if there exists $g \in U_r$ such that $y = gx$, g.c.d $(g,r) = 1$.

If $y = gx$ for some $g \in U_r$, g.c.d $(y,r) = $ g.c.d (x,r). Those elements of x for which g.c.d $(x,r) = d$ fall into one orbit, where d is a divisor of r. In fact, if $d|r$, there are $\phi(\frac{r}{d})$ elements in an orbit O_d where $x \in O_d$ is such that g.c.d $(x,r) = d$. This is precisely a class-division of integers (mod r). It was first noticed by C. F. Gauss and later studied intensively by R. Vaidyanathaswamy. That the orbits of U_r in X are mutually disjoint follows from the fact that $\sum_{d|r} \phi(\frac{r}{d}) = r$. So, $N(U_r) = d(r)$. This proves (7.4.5).   □

Next, we give a characterization of a finite cyclic group. It is easy to note that a cyclic group of order r has $\phi(r)$ generators, ϕ being the Euler ϕ-function.

Lemma 7.4.1 : *If a group G contains an element of order s, then it contains at least $\phi(s)$ of them.*

Proof : Suppose that $g \in G$ has order s. Then the cyclic subgroup H (of G) generated by g has s elements. So, it has $\phi(s)$ generators all of which are elements of order s (in G). So, G has at least $\phi(s)$ elements of order s.   □

Corollary 7.4.1 : *A cyclic group G of order r has exactly $\phi(d)$ elements of order d for each divisor d of r.*

Proof : Let $\psi(d)$ denote the number of elements of order d in G, where $d|r$. Then, $\psi(d) \geq \phi(d)$. Since every element of G has order d for each divisor d of r,

$$r = \sum_{d|r} \psi(d) \geq \sum_{d|r} \phi(d) = r$$

So, equality can happen if, and only if, $\psi(d) = \phi(d)$ for each divisor d of r.   □

Lemma 7.4.2 : *Let G be a group of order r. If G has at most $\phi(d)$ elements of order d for each divisor d of r, then G is cyclic.*

Proof : If $\psi(d)$ denotes the number of elements of order d where $d|r$, we have, by hypothesis, $\psi(d) \leq \phi(d)$.
So, $r = \sum_{d|r} \psi(d) \leq \sum_{d|r} \phi(d) = r$.

Equality holds if, and only if, $\psi(d) = \phi(d)$ for each divisor d of r. In particular, $\psi(r) = \phi(r) \neq 0$ and so G has an element of order r. That is, G is cyclic. $\square$

Theorem 53 : *A finite group G of order r is cyclic if, and only if, it has $\phi(d)$ elements of order d for each divisor d of r.*

Proof : $:\Rightarrow G$ is cyclic.
So, for each divisor d of r, there is a cyclic subgroup H of order d and H has $\phi(d)$ generators.
$\Leftarrow$: Conversely, if G has $\phi(d)$ elements of order d for each divisor d of r, there is an element of order r or G is cyclic. $\square$

Lemma 7.4.3 (J. S. Jose) **:** *Let G be a group of order r. If, for each divisor d of r, the number of solutions of the equation $x^d = e$ (the identity in G) is less than or equal to d, then G is cyclic.*

Proof : [The abelian case] We give a proof of lemma 7.4.3 in the case where G is abelian. It is known [7] that G is a direct product of its Sylow p-subgroups. As a prime-power group is a direct product of cyclic subgroups, we consider two cyclic subgroups H and K (of G) of orders s and t respectively, where g.c.d $(s,t) = 1$.
   In $H \times K$, there are precisely st solutions for $x^{st} = e$.
If g.c.d $(s,t) > 1$, let g.c.d $(s,t) = q$.
Then $s = qs'$, $t = qt'$; g.c.d $(s',t') = 1$.
$x^q = e$ in $H \times K$ has $2q - 1$ solutions of the form $h^a k^b$ where (a,b) has values

$$(0,0), (s',0), (2s',0), \ldots, ((q-1)s',0); (0,t'), (0,2t'), \ldots (0,(q-1)t').$$

Then, when H and K are cyclic groups of orders s and t respectively and when $d \mid st$, then $x^d = e$ has exactly $d = $ g.c.d (d,st) solutions if, and only if, g.c.d $(s,t) = 1$. On the contrary, more than q solutions exist for $q = $ g.c.d $(s,t) > 1$. Further, $H \times K$ is cyclic if, and only if, g.c.d $(s,t) = 1$. So, for G, a group other than a cyclic group, when d divides $|G|$, $x^d = e$ has more than d solutions. When $x^d = e$ has less than or equal to d solutions, G has to be cyclic. $\square$

For the general case, see J. S. Jose [10].

Theorem 54 (I. M. Richards (1984)) **:** *Let G be a group of order r. If $d(r)$ denotes the number of divisors of r, then the number of cyclic subgroups of G is greater than or equal to $d(r)$. Further, the number of cyclic subgroups of G is $d(r)$ if, and only if, it is cyclic.*

Proof : U_r denotes the group of units in $\mathbb{Z}/r\mathbb{Z}$. As noted earlier, $|U_r| = \phi(r)$. We consider the action of U_r on the set G (stripped of its group structure). With each $s \in U_r$, we associate the permutation π_s defined by

(7.4.9) $\pi_s(g) = g^s$, for all $g \in G$,

Under this action, two elements belong to the same orbit of U_r if, and only if, they generate the same cyclic subgroup of G. Therefore, the number of orbits of U_r

denoted by $N(U_r)$ is equal to the number of cyclic subgroups of G. By Burnside's lemma (theorem 52), we get

(7.4.10) $$N(U_r) = \frac{1}{\phi(r)} \sum_{s \in U_r} \psi(s),$$

where $\psi(s)$ denotes the number of elements of G which are left invariant by $s \in U_r$. That is,

$$\pi_s(g) = g^s = g$$

Or, $g^{s-1} = e$ (the identity element).

In the case of a cyclic group $C(r)$ of order r, for $g \in C(r)$, $g^{s-1} = e$ $\Rightarrow t^{k(s-1)} = e$ where t is a generator of $C(r)$.
Reverting to a congruence modulo r, we have only to find the number of solutions of $(s-1)x \equiv 0 \pmod{r}$ to obtain the number of elements of order $(s-1)$ in $C(r)$. The number of solutions is given by g.c.d $(s-1,r)$. When G is an arbitrary group of order r, the centre $Z(G)$ of G may contain more than one element. Therefore, when products of elements of G belong to $Z(G)$, the number of elements of G of order $(s-1)$ is a multiple of g.c.d $(s-1,r)$. Thus,

(7.4.11) $$\psi(s) = k_s \text{ g.c.d } (s-1,r)$$

where k_s is an integer ≥ 1. (7.4.11) is due to Frobenius [5]. From (7.4.11), we get

(7.4.12) $$N(U_r) = \frac{1}{\phi(r)} \sum_{s \in U_r} k_s \text{ g.c.d } (s-1,r)$$

As $k_s \geq 1$,

(7.4.13) $$N(U_r) \geq \frac{1}{\phi(r)} \sum_{s \in U_r} \text{g.c.d } (s-1,r).$$

The case of equality in (7.4.13) occurs when G is a cyclic group of order r, as seen in illustration 7.4.1. Further, $N(U_r)$ is equal to the number of cyclic subgroups of G. From (7.4.13), $N(U_r) \geq d(r)$. This proves that the number of cyclic subgroups of a group G (of order r) $\geq d(r)$. This is the first part of the theorem.

Now, the number of cyclic subgroups of G is equal to $d(r)$ if, for each s with g.c.d $(s,r) = 1$ the number of solutions of the equation

(7.4.14) $$x^f = e \quad \text{in } G \quad (\text{where } f = \text{ g.c.d } (s-1,r))$$

is exactly g.c.d $(s-1,r)$.

Next, we set $T_r = \{\text{g.c.d } (s-1,r) : s \in U_r\}$. It follows that

(7.4.15) $$T_r = \begin{cases} \{d : d|r, d \text{ even}\}, & \text{if } r \text{ is even,} \\ \{d : d|r\}, & \text{if } r \text{ is odd.} \end{cases}$$

Therefore, if G is a group of odd order r such that the number of cyclic subgroups of G is $d(r)$, the number of divisors of r, then G is cyclic. This follows from the structure of T_r and lemma 7.4.3.

Next, assume that G is a group of even order r. Suppose that the number of cyclic subgroups of G is $d(r)$. As $T_r = \{t : t \,|\, r, t \text{ even}\}$, for each even divisor t of r, the number of solutions of $x^t = e$ in G is exactly t. In particular, the number of solutions of $x^2 = e$ is 2. Therefore a homomorphism $\psi : G \to G$ (onto G) defined by $\psi(x) = x^{-1}$ for all $x \in G$ is a unique involution in the sense that $\psi^2 = I$, the identity homomorphism. Those elements (of G) of the form y where $y^2 = e$ are fixed by ψ.

Let m be an odd divisor of r (An even integer other than 2^q ($q \geq 1$) can have non-trivial odd divisors). Suppose that $a \in G$ and $a^m = e$. Then, $a^{2m} = e$ and $\psi(a)^{2m} = (a^{-1})^{2m} = (a^{2m})^{-1} = e$. So, to each solution of $x^m = e$ in G, there correspond two solutions of the equation $x^{2m} = e$. So the number of solutions of the equation $x^{2m} = e$ is only $2m$. That is, the number of solutions of $x^m = e$ cannot exceed m. This is true of any divisor d of r. So, the number of solutions of $x^d = e$ (where d divides r) cannot exceed d. So, by lemma 7.4.3, G is cyclic.

This completes the proof of the second part of theorem 54. □

Remark 7.4.1 : The proof of theorem 54 has been adapted from I. M. Richards [12]. Theorem 52 and illustration 7.4.1 are also given in [15].

Remark 7.4.2 : In [A3], enumeration techniques giving a variation of Burnside's lemma have been pointed out. In particular, the proof of the following congruence:
Let a be an arbitrary integer. If μ denotes the Möbius function (definition 4.3.4, chapter 4), for $r \geq 1$,

$$(7.4.16) \qquad \sum_{d|r} \mu(\frac{r}{d}) a^d \equiv 0 (\bmod\ r)$$

is given using an analogue of Burnside's lemma. Incidentally, we mention that according to P. M. Neumann [A7], the orbit-counting lemma of Burnside is to be reckoned as Cauchy-Frobenius orbit-counting formula and this is endorsed in [A3]. Details of generalizations are not discussed here.

7.5. A criterion for the uniqueness of a cyclic group of order r

In [2], L. E. Dickson (1905) determined the positive integers r for which every group of order r is abelian. Naturally, one could ask for a criterion for a group of order r to be unique up to isomorphism. Of course, such a group must be cyclic. For, there is a cyclic group of order r for every positive integer $r \geq 2$. It is easily checked that there is a unique cyclic group of order p, a prime. For $r = pq$ where p and q are primes ($p < q$), there is a criterion for the existence of a cyclic group of order pq. It is given in the following

Lemma 7.5.1 : *(J. B. Fraleigh) Let $r = pq$ where p and q are primes and $p < q$. There exists a unique cyclic group of order r if, and only if, $q \not\equiv 1 \pmod{p}$, that is, if, and only if, p does not divide $(q-1)$.*

Proof : Let G be the group of order r. G has a Sylow q-group and the number of such subgroups is congruent to 1 (mod q) and divides pq by Sylow theorems. Since $p < q$, the only possibility is one. So, there is only one Sylow q-subgroup H of G and it is normal in G. Similarly, there is a Sylow p-subgroup K of G and the number of such subgroups is congruent to 1(mod p) and the number is either 1 or q. If q is not congruent to 1 (mod p), this number is also 1 and so K is normal in G. Since H and K are of orders q and p respectively, $H \cap K = (e)$, e being the identity, as every element a (other than e) in H is of order q and every element b (other than e) in K is of order p. So, $G \cong H \times K$ or $G \cong \mathbb{Z}/q\mathbb{Z} \times \mathbb{Z}/p\mathbb{Z}$. Clearly, G is cyclic.

Conversely, if G is cyclic and of order pq where p and q are primes, $G \cong \mathbb{Z}/p\mathbb{Z} \times \mathbb{Z}/q\mathbb{Z}$. If $p|(q-1)$, p is even and so $p = 2$ and there exists a Dihedral group of order $2p$ which is not abelian. (see remark 7.3.3). Then, G is not unique. So, if G is unique and cyclic, q is not congruent to 1 (mod p).   □

Remark 7.5.1 : Lemma 7.5.1 has been adapted from [4].

Theorem 55 (Dieter Jungnickel (1992)) **:** *Let r be a positive integer. The cyclic group $C(r)$ of order r is the only group of order r if, and only if, g.c.d $(r, \phi(r)) = 1$ where ϕ denotes Euler ϕ-function.*

Proof : We make the following

Observation 7.5.1 : *Suppose that $r = sp^a$ where p is a prime not dividing s and $a \geq 2$. Then, $\phi(r) = \phi(s)p^{a-1}(p-1)$, (by the multiplicativity of the ϕ-function and the fact that $\phi(p^a) = p^{a-1}(p-1)$).*

So, both r and $\phi(r)$ are divisible by p. Also, the direct product $C(s) \times C(p)^a$ is not isomorphic to $C(r)$. This implies that r has to be square-free. That is, r is a product of distinct primes. The following assumption is justified.

Assume that $r = p_1, p_2, \ldots, p_k$, a product of k primes. Then,

$$\phi(r) = \prod_{i=1}^{k} (p_i - 1)$$

Then, g.c.d $(r, \phi(r)) \neq 1$ implies that there exist primes p and q dividing r and $\phi(r)$. We write $r = pqt$ where $t > 1$. Also, p divides $(q-1)$. Then, there exists a non-abelian group H of order pq and so $H \times C(t)$ is a non-abelian group of order r. So, we could assume that g.c.d $(r, \phi(r)) = 1$.

Claim : There is only one cyclic group of order r, if g.c.d $(r, \phi(r)) = 1$.

Assume the contrary. That is, let r be the least positive integer for which a counterexample G exists. We will arrive at a contradiction.

Step 1. As r is square-free, $(t, \phi(t)) = 1$ for each divisor t of r.

Step 2. Every proper subgroup and every non-trivial quotient group of G are cyclic. This is clear, as r is minimal.

Step 3. The centre $Z(G)$ of G is trivial.

Otherwise, $G/Z(G)$ will be cyclic by step 2 and therefore G is abelian and hence cyclic.

Step 4. Let $x \neq e$ be an element of a maximal subgroup U of G. Then, U is the centralizer (normalizer) $C_G(x)$ of x in G. For, $C_G(x)$ is a proper subgroup of G by step 3 and U is cyclic and therefore, U is contained in $C_G(x)$ by step 2. The maximality of U shows that $U = C_G(x)$.

Step 5. If U and V are two maximal subgroups of G, then $U \cap V = (e)$ where e is the identity.

For, if $x \neq e$ and $x \in U \cap V$, by step 4, would give $U = V = C_G(x)$.

Step 6. Any maximal subgroup U equals its own normalizer

$$N_G(U) = \{ g \in G : g U_g^{-1} = U \}.$$

For, let $x \neq e$ be an element in $N_G(U)$. Then,

$\psi : N_G(U) \to N_G(U)$ given by $\psi(y) = x^{-1} y x$ gives an automorphism of the cyclic group U. If U has order s, then the order of ψ in the group of automorphisms of U has order $\phi(s)$ which divides $\phi(r)$, as r is square-free. Since x and hence ψ has order dividing r, order of ψ has to be 1. Then, x centralizes U and by step 3, belongs to U.

Step 7. Let U be a maximal subgroup of order u in G. Then, the conjugate subgroups of U contain exactly $r - \frac{r}{u}$ elements $\neq e$.

For, we recall that the number of conjugates of U is the index of the normalizer of U in G. By step 6, this index is $\frac{r}{u}$. By step 5, any two conjugates of U intersect trivially. Thus, the set of conjugates of U contain altogether $\frac{(u-1)r}{u}$ elements $\neq e$.

Step 8. With U as maximal subgroup of G of order u, choose an element x not contained in any of the conjugate subgroups of U. Let V be a maximal subgroup containing x and therefore not conjugate to U. Then, any conjugate of U and any conjugate of V intersect trivially by step 5. Applying step 7 also to V, we obtain $r - \frac{r}{v}$ elements $\neq e$ in the set of conjugates of V. But, there are only $r - 1$ elements $\neq e$ in G giving the inequality

$$r - \frac{r}{u} + r - \frac{r}{v} < r.$$

Or,

$$1 - (\frac{1}{u} + \frac{1}{v}) < 0 \text{ or } u + v > uv$$

which is a contradiction. $\qquad\square$

Remark 7.5.2 : The proof of theorem 55 has been adapted from Dieter Jungnickel [3]. For related results, see Jonathan Pakianathan and Krishnan Shankar [9].

For a detailed study of the theory of finite groups see M. A. Ashbacher [A1], T. Y. Lam [A4, A5] and Ronald Solomon [A9] given in additional references. Another interesting paper is that of Murthy and Murthy [A6].

7.6. Notes with illustrative examples

The theory of partitions is a well-developed branch of number theory. Euler introduced the notion of a generating function and obtained an identity connecting $p(n)$, the number of partitions of n with $\prod_{k=1}^{\infty}(1-x^k)^{-1}$ (see Fact 7.2.2). (Functions related to $p(n)$ are the so-called theta functions and modular functions which were extensively studied by Jacobi and others). We have seen that $p(n)$ is the number of conjugate classes of elements of S_n, the symmetric group on n symbols. It can be shown that for $n \geq 2$, the number of conjugate classes in S_n is also the number of different abelian groups of order p^n (up to isomorphism) where p is a prime.

The number of solutions of linear congruences with side conditions is made use of to count certain special elements of a finite group. See theorems 50 and 51. We observe that these theorems have their analogues in the context of finite rings.

Let R be a finite wing with unity. Suppose that D is a non-empty subset of R. The number of solutions of

$$(7.6.1) \qquad x_1 + x_2 + \cdots + x_s = a,\ a \in R,\ x_i \in D\ (i = 1, 2, \ldots, s)$$

is denoted by $N(D, a, s)$ with respect to the additive group $(R, +)$. The analogue of theorem 50 is valid when $J = J(D)$ is taken to be the largest ideal of R such that $x + J \subseteq D$ for all $x \in D$.

When $H = R \setminus D$ is an ideal of R, theorem 51 is applicable. We consider the ring $\mathbb{Z}/r\mathbb{Z}$ $(r \geq 1)$. If $N(n, r, s)$ denotes the number of solutions of

$$(7.6.2) \qquad x_1 + x_2 + \cdots + x_s \equiv n \pmod{r}$$

under the restriction g.c.d $(x_i, r) = 1$, $(i = 1, 2, \ldots, s)$ it is known [15] that $N(n, r, s)$ is multiplicative in r. See (5.5.6) in chapter 5. Therefore, we tackle the case where $r = p^m$ (p a prime, $m \geq 1$).

When $R = \mathbb{Z}/p^m\mathbb{Z}$, R is a quasi-local ring (that is, it has a unique maximal ideal). We take $D = U$, the group of units in R. Then,

$$(7.6.3) \qquad |U| = \phi(p^m) = p^{m-1}(p-1)$$

$H = R \setminus U$, H is the unique maximal ideal of R and
$|H| = |R \setminus U| = p^m - \phi(p^m) = p^{m-1}$.
So, $[R : H] = p$ and $[R : H] - 1 = (p - 1)$.
The formula for $N(U, p^m, s)$ using (7.3.8) is given by

$$(7.6.4) \qquad N(U, n, s) = \frac{(\phi(p^m))^s}{p^m}\left\{1 - \frac{(-1)^s}{(p-1)^s}\right\},\ \text{if } n \in U$$

and

$$(7.6.5) \qquad N(U, n, s) = \frac{(\phi(p^m))^s}{p^m}\left\{1 + \frac{(-1)^s(p-1)}{(p-1)^s}\right\},\ \text{if } n \notin U$$

(7.6.4) is the case when $p \nmid n$. (7.6.5) is the case when $p \mid n$. These two, together, yield

$$N(U,n,s) = \begin{cases} \frac{(p^{m-1}(p-1))^s}{p^m}\{1 - \frac{(-1)^s}{(p-1)^s}\} & \text{if } p \nmid n \\ \frac{(p^{m-l}(p-1))^s}{p^m}\{1 - \frac{(-1)^{s-1}}{(p-1)^{s-1}}\} & \text{if } p \mid n \end{cases}$$

In fact, we have derived the Rademacher formula for $r = p^m$ (see chapter 5, Corollary 5.6.1).

While studying the number of cyclic subgroups of a given group, we referred to Kesava Menon's identity. See (7.4.5). It is a remarkable example of an application of Burnside's lemma. Various generalizations of the identity are known.

Characterizations of finite cyclic groups are known. The uniqueness of a cyclic group of order r is tied up with $\phi(r)$, the Euler totient. Following Jonathan Pakianathan and Krishnan Shankar [9], we call a positive integer r a cyclic number if every group of order r is cyclic. The smallest non-prime cyclic number is 15. A positive integer is a cyclic number if, and only if, g.c.d $(r, \phi(r)) = 1$.

A group G is called a nilpotent group if, and only if, it is the internal direct product of its Sylow subgroups. See D. Robinson [13]. It is known that

Cyclic groups $\subset$ abelian groups $\subset$ nilpotent groups.

Based on these, one defines nilpotent and abelian numbers. Their characterizations are given in [9].

7.7. A worked-out example

Question: (T. Hungerford) Prove the fundamental theorem of arithmetic by applying Jordan-Hölder theorem to the group $(\mathbb{Z}/r\mathbb{Z}, \oplus)(r > 1)$.

Answer: We observe that $(\mathbb{Z}/r\mathbb{Z}, \oplus)$ is a finite cyclic group of order r. By theorem 54, it has $d(r)$ cyclic subgroups, where $d(r)$ denotes the number of divisors of r.

We recapture the notion of a nilpotent group. Let $N_0, N_1, N_2, \cdots$ be a sequence of normal subgroups of a group G such that $N_0 = (e)$,

$$N_1 = C(G) = \{a : ax = xa \text{ for all } x \in G\}, \text{ the centre of } G.$$

$C(G)$ is an abelian normal subgroup of G. N_2 is the inverse image of $C(G/N_1)$ under the canonical map $\psi : G \to G/N_1$. N_2 is normal in G and contains N_1. If N_i is the inverse image of $C(G/N_{i-1})$ $(i = 1, 2, \ldots)$, we obtain a sequence of normal subgroups of G, called the ascending central series of G, namely

(7.7.1) $$N_0 = (e) \subset N_1 \subset N_2 \subset \cdots$$

Definition 7.7.1 : *A group G is called nilpotent if $N_k = G$ for some k.*

An abelian group H is nilpotent, as $C(H) = H$. Even for a finite group G, ascending central series need not terminate in G. For instance, $C(S_3) = (e)$. So, S_3 is not nilpotent. However, S_3 is solvable. See exercise 10.

Lemma 7.7.1 : *A proper subgroup H of a nilpotent group G is properly contained in its normalizer*

$$N(H) = \{x \in G : xHx^{-1} = H\}$$

Proof : In the ascending central series for G, given by $(e) = N_0 \subset N_1 \subset \cdots$ and $H \subset N_k = G$, there exists $i \in \mathbb{N}$ such that $N_i \subset H$ and N_{i+1} is not contained in H. For, if $t_{i+1} \in N_{i+1}$ and $h \in H$,

$$t_{i+1}h^{-1}t_{i+1}^{-1}h \in N_i \subset H.$$

Therefore, $N_{i+1} \subset N(H)$ and $H \neq N(H)$.   □

Lemma 7.7.2 : *Every Sylow p-subgroup of a finite nilpotent group G is a normal subgroup of G.*

Proof : If p is a Sylow p-subgroup of G, $N(P)$ is given by

$$N(P) = \{x \in G : x^{-1}Px = P\}$$

By lemma 7.7.1, $P \subset N(P)$ and $P \neq N(P)$. As P is a maximal subgroup of G, $N(P) = G$. Hence P is a normal subgroup of G.   □

Remark 7.7.1 : A finite nilpotent group G possesses a subnormal series

$$N_0 = (e) \subset N_1 \subset \cdots \subset N_k = G$$

obtained from an ascending central series of G, N_i is a normal subgroup of N_{i+1} $(i = 0, 1, 2, \ldots (k-1))$. Its factors are abelian. G is said to be solvable. In fact, every finite cyclic group is nilpotent and has a composition series whose factors are of order a prime p dividing the order of the group. Every Sylow p-subgroup of a finite nilpotent group G is normal in G and the G is the internal direct product of its Sylow p-subgroups.

The problem is solved by noting that in a cyclic group G of order $r(> 1)$, if p_1 is a prime dividing r, there exists a Sylow p_1-subgroup of order $p_1^{a_1}$ which is normal in G. ($p_1^{a_1}$ is the highest power of p_1 dividing r). If $p_1, p_2, \ldots, p_n$ are the distinct prime divisors of r, G is the direct product of Sylow subgroups $P_1, P_2, \ldots, P_n$ corresponding to the prime factors $p_1, p_2, \ldots, p_n$ of r. Product of the orders of $P_1, P_2, \ldots, P_n$ is the order of G which is r. This accounts for the unique factorization of r into distinct prime factors.   □

7.8. An example from quadratic residues

r denotes an odd positive integer > 1. We consider the group $U(r)$ of units of $\mathbb{Z}/r\mathbb{Z}$. $U(r)$ is a group of order $\phi(r)$, the Euler ϕ- function.

We write

(7.8.1) $S(r) = \{[a] \in U(r) : \quad a = b^2, \ [b] \in U(r)\}.$

(where $[a]$ denotes the residue class of a modulo r).

It is verified that $S(r)$ is a subgroup of $U(r)$. To find the order of $S(r)$, we proceed as follows:

Given odd positive integer r, s, the Jacobi symbol $(r|s)$ (6.7.1) has the following properties

$$\text{(i) } (a|r)(b|r) = (ab|r)$$

(7.8.2) $$\text{(ii) } (a|rs) = (a|r)(a|s)$$

$$\text{(iii) } a \equiv b \,(\text{mod } r) \Rightarrow (a|r) = (b|r)$$

$$\text{(iv) } (r|s) = (-1)^{(\frac{r-1}{2})(\frac{s-1}{2})}(s|r).$$

Remark 7.8.1 : We consider quadratic residues modulo r both as integers and as elements of $\mathbb{Z}/r\mathbb{Z}$.

Lemma 7.8.1 : *Let r be an odd integer > 1, the set of quadratic residues modulo r corresponds to the subgroup $S(r)$ of $U(r)$. Consequently, the number of quadratic residues modulo r is equal to $\phi(r)/2^{\omega(r)}$, where $\omega(r)$ denotes the number of distinct prime factors of r.*

Proof : We note that if p is an odd prime and $e \geq 1$, the number of quadratic residues a modulo p^e with $0 \leq a < p^e$ is $\frac{\phi(p^e)}{2}$. For, $x^2 \equiv a \,(\text{mod } p^e)$ is solvable if, and only if, $x^2 \equiv a \,(\text{mod } p)$ is solvable.

The map $\psi : U(r) \to \{1, -1\}$ given by

$$\psi([a]) = (a|r), \text{ (the Jacobi Symbol)}$$

is a group homomorphism with

(7.8.3) $$\ker \psi = \{[a] \in U(r) : \psi([a]) = 1\}$$

Claim: If $S(r)$ is as given in (7.8.1), $S(r) = \ker \psi$.

For, $\ker \psi$ is a subgroup of $U(r)$ and $S(r) \subseteq \ker \psi$. If r is a perfect square, $\ker \psi = U(r)$. If r is not a perfect square, $[U(r) : \ker \psi] = 2$.

Let $[a] \in \ker \psi$. Then $\psi([a]) = 1$. This implies that a is a quadratic residue of r. If it is not, $[a] \notin S(r)$ and the image of $[a]$ under ψ would be -1. That is, $\ker \psi \subseteq S(r)$.

Thus, $S(r) = \ker \psi$.

Let $r = \prod_{i=1}^{k} p_i^{e_i}$ (p_i odd primes and $e_i \geq 1$, $i = 1, 2, \ldots, k$). As $S(r)$ is a direct product of groups $S_1, S_2, \ldots, S_k$ where $S_i (i = 1, 2, \ldots, k)$ is the group of quadratic residues modulo $p_i^{e_i}$,

$$|S(r)| = \prod_{i=1}^{k} |S_i| = \prod_{i=1}^{k} \frac{\phi(p_i^{e_i})}{2}.$$

($|X|$ denotes the number of elements of X). This completes the proof of lemma 7.8.1. $\square$

Remark 7.8.2 : Proof of lemma 7.8.1 has been adapted from Victor Shoup [A8].

Question: Let r be an odd positive integer > 1. Suppose that the group $S(r)$ of quadratic residues (mod r) acts on the set $\mathbb{Z}/r\mathbb{Z}$. If N denotes the number of orbits in $\mathbb{Z}/r\mathbb{Z}$ under the action of $S(r)$, show that

(7.8.4) $$N = \frac{2^{\omega(r)}}{\phi(r)} \sum_{[a^2] \in S(r)} \text{g.c.d } (a^2 - 1, r)$$

Answer: The identity (7.8.4) is yet another application of Burnside's lemma. (Theorem 52). $S(r)$ is of order $\frac{\phi(r)}{2^{\omega(r)}}$. We obtain the number of elements of $\mathbb{Z}/r\mathbb{Z}$ which are left invariant by the elements of $S(r)$. For $[a^2] \in S(r)$, let $t([a^2])$ denote the number of elements of $\mathbb{Z}/r\mathbb{Z}$ which are left invariant by $[a^2]$. Then, $t([a^2])$ is the number of solutions of the congruence

$$a^2 x \equiv x (\text{mod } r), \quad \text{g.c.d } (a^2, r) = 1.$$

That is, $(a^2 - 1)x \equiv 0 (\text{mod } r)$. It follows that $t([a^2])$ equals g.c.d $(a^2 - 1, r)$. By Theorem 52, we obtain

$$N = \frac{1}{|S(r)|} \sum_{[a^2] \in S(r)} \text{g.c.d } (a^2 - 1, r)$$

(7.8.4) is an immediate consequence.   □

Remark 7.8.3 : It is advisable to express N in terms of certain known arithmetic functions, as in illustration (7.4.1).

EXERCISES

1. *Mark the following statements true (T) or false (F) justifying your answer briefly.*
   a) *The number of conjugate classes in S_n is equal to the number of non-isomorphic abelian groups of order p^n, where p is a prime.*
   b) *The homomorphic image of a nilpotent group is nilpotent.*
   c) *There exist two or more non-isomorphic groups of order 75.*
   d) *The group S_4 of order 24 is nilpotent.*
   e) *An abelian group G has a composition series if, and only if, it is finite.*
   f) *A simple group of order 60 need not be isomorphic to A_5.*
2. *Prove that a finite group of order r is abelian if, and only if, it has r conjugate classes of elements.*
3. *Find the conjugate classes of elements of S_4 and S_5.*
4. *[I.M.Richards] Let $f(x) \in \mathbb{Z}[x]$. If U_r denotes the group of units in $\mathbb{Z}/r\mathbb{Z}$, show that*

$$\sum_{s \in U_r} (f(s), r) = \phi(r) \sum_{d|n} |\{t \in U_d : f(t) \equiv 0 (\text{mod } d)\}|$$

where $(f(s), r)$ denotes the g.c.d of $f(s)$ and r.
(We remark that this gives a generalization of Kesava Menon's identity (7.4.5)).

5. *Prove that a group of order 255 is cyclic.*

6. *If $r = pq$ where p, q are primes and $p < q$ with $q \equiv 1$ (mod p) show that there exists only one non-abelian group of order r up to isomorphism.*

7. *Let r, s be positive integers and $n = $ g.c.d (r, s). If we consider the additive groups*

$$G_1 = \mathbb{Z}/r\mathbb{Z} \text{ and } G_2 = s(\mathbb{Z}/r\mathbb{Z})$$

   Show that $G_1/G_2 \cong \mathbb{Z}/n\mathbb{Z}$.

8. *Let G be an abelian group. $T(G)$ denotes the set of elements of finite order in G. $T(G)$ is a subgroup of G called the torsion subgroup of G. If $T(G) = (e)$, the trivial subgroup, G is said to be torsion-free. If $T(G) = G$, G is called a torsion-group. When G is a abelian with torsion subgroup $T(G)$, show that $G/T(G)$ is torsion-free.*

9. *A group G is said to be indecomposable if it is not a direct product of proper subgroups. (See definition 5.3.3, chapter 5). Show that a finite cyclic group is indecomposable if, and only if, its order is p^m (p a prime, $m \geq 1$).*

10. *A composition series of a group G is a finite descending chain of subgroups:*

$$G = G_0 \supset G_1 \supset \cdots \supset G_r = (e)$$

   such that the factor groups G_i/G_{i+1} are simple groups $\neq (e)$ $(0 \leq i \leq r-1)$. G_i/G_{i+1} are called composition factors of G. A finite group G is said to solvable if, and only if, the composition factors of G are cyclic and of prime order.

   Show that a nilpotent group G is solvable.

REFERENCES

[1] M. Artin: Algebra, chapter 5, section 3 pp 162–166, Prentice Hall of India (P) Ltd, New Delhi (1994).

[2] L. E. Dickson: Definition of a group and a field by independent postulates. Trans. Amer. Math. Soc., 6 (1905) 198–204.

[3] Dieter Jungnickel: On the uniqueness of the cyclic group of order n, Amer. Math. Monthly, 99 (1992) 545–547.

[4] J. B. Fraleigh: A first course in Algebra, Addison Wesley International student Edition, 5th reprint (1999).

[5] G. Frobenius: Uber einen Fundamentalsatz der Grnppentheorie *I&II* Sitzungs. pre.Akademic Berlin (pp 987–991 and 1907 pp 428–431).

[6] Emil Grosswald: Topics from the theory of numbers, chapter 12 pp 216–232 Macmillan Pub. Co. NY (1985), 2nd Edition.

[7] I. N. Herstein: Topics in Algebra, chapter 2 section 11 pp 71–72, Wiley Eastern (P) Ltd New Delhi (1988) Reprint.

[8] D. Jacobson and K. S. Williams: On the number of distinguished representations of a group element, Duke Math. J., 39 (1972) 521–527.

[9] Jonathan Pakianathan and Krishnan Shankar: Nilpotent numbers, Amer. Math. Monthly, 107 (2000) 631–634.

[10] J. S. Jose: A course on group theory, Cambridge Univ. Press (1978) pp 215–216.

[11] P. Kesava Menon: On the sum $\sum(a-1,n)[(a,n)=1]$, J. Indian Math. Soc. 29 (1965) 155–163.

[12] I. M. Richards: A remark on the number of cyclic subgroups of a finite group, Amer. Math. Monthly, 91 (1984) 571–572.

[13] D. Robinson: A course in the theory of groups, GTM No 80, Springer Verlag NY (1993).

[14] Eugene Shenkman: Group theory, Affiliated East-West Press Pvt. Ltd, New Delhi, East-West student Edn. (1971).

[15] R. Sivaramakrishnan: Classical theory of arithmetic functions, Monographs in pure and applied Math. No 126, Marcel Dekker Inc. (1989).

ADDITIONAL REFERENCES

[A1] M. Ashbacher: Finite group theory: Cambridge studies in advanced Mathematics, 2nd Edition, Cambridge Univ. Press. Cambridge, UK, (2000).

[A2] Walter Feit and John Thompson: Groups of odd order are solvable Pac. J. Math, 13 (1963), 775–1029.

[A3] I. M. Issacs and M. R. Pournaki: Generalization of Fermat's Little Theorem via Group Theory, Amer. Math. Monthly, 112 (2005) 734–740.

[A4] T. Y. Lam: Representations of finite groups-A hundred years, Part I Notices of Amer. Math. Soc. 45 (1998) 361–372.

[A5] T. Y. Lam: Representations of finite groups-A hundred years, Part II Notices of Amer. Math. Soc. 45 (1998) 465–474.

[A6] M. R. Murthy and V. K. Murthy: On the number of groups of a given order, J. Number Theory 18 (1984) 178–191.

[A7] P. M. Neumann: A lemma that is not Burnside's, Math. Sci., 4 (1979) 133–141.

[A8] Victor Shoup: A computational introduction to Number Theory and Algebra, Cambridge University Press, Cambridge, UK (2005).

[A9] Ronald Solomon: A brief history of the classification of the finite simple groups. Bulletin (New series) of the Amer. Math. Soc., Vol 38. Number 3 (2001) 315–352.

Part II

THE RELEVANCE OF ALGEBRAIC STRUCTURES TO
NUMBER THEORY

CHAPTER 8

Ordered fields, fields with valuation and other algebraic structures

Historical perspective

The theory of fields is an extensive branch of algebra. It was in connection with the proof of the theorem:
'A fifth degree equation with coefficients from $\mathbb{Q}$, the field of rationals, is not always solvable by radicals', the notion of field extensions was laid on firm foundations.

The field $\mathbb{Q}$, the field $\mathbb{R}$ of real numbers and the field $\mathbb{C}$ of complex numbers pervade the whole of basic theory of fields. In these fields, 'order' and 'limit' concepts are fundamental. The evolution of the various forms of numbers is roughly as follows:

During the sixth century B.C., the Pythagorean school knew about rational and irrational numbers. It was the failure of repeated attempts to express $\sqrt{2}$ as a rational number that lead the followers of Pythagoras to the notion of an 'irrational number'. It is to be remarked that Babylonians conceived of and calculated square-roots of non-square integers, though they were satisfied with crude approximation methods. See K. von Fritz [7]. See also Nicholas Bourbaki [2]. Negative rationals and imaginary numbers were discovered during the period 1000–1500 A.D. They kept mathematicians puzzled as they were unbelievable, though accepted. Algebraic irrationals which cannot be written as radicals appeared through the Fundamental Theorem of Galois theory (1826). It also led to the study of lattices and other useful algebraic structures. George Boole (1815–1864) thought of an algebraic structure in which every element is an idempotent—for instance, the power set of a set X. By then, transcendental numbers were discovered. Transcendence of e, the exponential constant, was proved by C. Hermite (1822–1901) in 1873. The transcendence of π was shown by Lindemann (1852–1939) in 1882. It was only after this long sequence of events, classical algebra got unified with the trends in modern mathematics. The theory of continued fractions remained isolated for some time. But now, that also has been brought into the modern stream.

It is to be emphasized that a revelation of the nineteenth century was that there were fields which were different from the field of complex numbers and there were groups which made no reference to automorphisms of fields. This point of *view was brought out by Emil Artin (1898–1962). Indeed, thanks to the efforts of Emil Artin, an exposition of the fundamental Theorem of Galois Theory made it a landmark in the area of algebra where the classical theory of equations gets the correct interpretation via the splitting field of a polynomial $f(x)$ and the Galois group of the equation $f(x) = 0$. The role of a prime p is in the context of S_p. Abel gave the theorem: There exist equations of every degree ≥ 5 which are not solvable by radicals. He made use of the strong lemma: If f is an irreducible polynomial over $\mathbb{Q}$, of degree p (a prime) and if f has exactly $(p-2)$ real roots, then the Galois group of f is S_p. For details see Charles R. Hadlock [10]. This is just stated to give a sample of field theory in its classical form.*

The ideas of analysis and topology also had their share in the development of the theory of algebraic structures. Ordered fields, fields with valuation, archimedean and non-archimedean fields are some of the notions that proved to be useful in the study of fields during the first half of the twentieth century.

8.1. Introduction

The classical theorem: 'The field of real numbers is order complete' has a great impact on the development of algebra and analysis. Ordered fields generalize the concepts and properties of the field $\mathbb{Q}$ of rational numbers. It is also an important fact to note that -1 is not a sum of squares in $\mathbb{R}$, the field of reals. The formal 'real fields' are introduced. The idea of 'absolute values' has been exploited in the notion of 'fields with valuation'. The approximation theorem for fields with valuation has an interesting application to the proof of the Chinese Remainder Theorem, already done in chapter 5.

The notion of a normed division domain due to S. W. Golomb [9] generalised the idea of 'divisibility' in integral domains via weak partially ordered sets. Modular lattices and the Jordan-Hölder Theorem give rise to the idea of uniqueness of factorization of an element of a non-commutative integral domain into irreducibles. Theorem 68, shown, is due to P. M. Cohn [5]. While studying properties of Boolean algebras, an analogue of the fundamental theorem of arithmetic is pointed out.

This chapter is aimed at an understanding of abstract algebraic structures which are the outcome of generalizations of the familiar ideas in (i) divisibility (ii) absolute values and (iii) set-theoretic operations.

8.2. Ordered fields

In chapter 2, we considered ordered integral domains and showed that an integral domain in which the set of positive elements is well-ordered characterizes the ordered integral domain $\mathbb{Z}$. (See corollary 2.2.1)

Definition 8.2.1 : *A ring $(R, +, \cdot)$ is said to be ordered if there exists a non-empty subset P (of R) called the set of positive elements of R such that*

(a) whenever $a, b \in P$, $a+b \in P$ and $a \cdot b \in P$ and

(b) for each $a \in R$, only one of the following alternatives holds :

$$a \in P \ or \ -a \in P \ or \ a = 0 \quad (\textit{the law of Trichotomy}).$$

In fact, if R is an integral domain satisfying definition 8.2.1, R is called an ordered integral domain. $\mathbb{Z}$ and $\mathbb{R}$ (the field of real numbers) are ordered integral domains. (see definition 2.2.1 in chapter 2).

Lemma 8.2.1 : *In an ordered ring R, all squares of nonzero elements are positive.*

Proof : Let $a \in R$. By the law of trichotomy, either $a \in P$ or $-a \in P$ or $a \in 0$ (where P is the set of positive elements of R). Since P is closed under multiplication, $a^2 = (-a)(-a) = (-a)^2 \in P$.
Also $1 = 1^2 \in P$ and $(-1)^2 \in P$, though $-1 \notin P$.   □

The word 'ordered' is from the fact that one could define $a < b \Longleftrightarrow$ $(b-a) \in P$. We could write $b > a$ to mean $a < b$. The order $<$ satisfies the transitivity property:
Whenever $a < b$, $b < c$, then, one has $a < c$. One notes that it is not a partial order as reflexivity and antisymmetry of $\leq$ are not okay with $<$.

Definition 8.2.2 : *In an ordered ring R, the absolute value $|a|$ of $a \in R$ is defined as :*

$$|a| = \begin{cases} 0, & \textit{if } a = 0 \\ a, & \textit{if } a \in P, a \neq 0 \\ -a, & \textit{if } a \notin P, a \neq 0 \end{cases}$$

Fact 8.2.1 : For $a, b \in R$, an ordered ring in which absolute value of an element is defined as in definition 8.2.2,

(8.2.1) $|a \cdot b| = |a| \cdot |b|, \qquad |a+b| \leq |a| + |b|$

the inequality with $\leq$ (less than or equal to) shown in (8.2.1) is referred to as the triangle inequality.
Next, Q is the field of quotients of $\mathbb{Z}$. We can make $\mathbb{Q}$ into an ordered field:

Lemma 8.2.2 : *Let D be an ordered integral domain with K, the field of fractions of D. There is a unique way of defining a positive subset P_K of K in K so that the inclusion map $D \to K$ is an order-preserving homomorphism. This order is defined by*

$$\frac{a}{b} \in P_K \ \textit{if, and only if, } a \cdot b \in P \ \textit{in } D \ \textit{for } a, b \in D \ \textit{with } b \neq 0,$$

where P is the set of positive elements in D.

Proof : D can be embedded in K and K contains D as a subdomain. Every nonzero element of K is of the form $\frac{a}{b}$, $b \neq 0$ $(a, b \in D)$.
Suppose that K is ordered with P_K as the subset of positive elements. Then,

$$\frac{a}{b} = a \cdot b \cdot (\frac{1}{b})^2. \text{ Since } (\frac{1}{b})^2 \text{ is necessarily in } P_K, \frac{a}{b} \in P_K$$

if, and only if, $a \cdot b \in P_K$. As $i : D \to K$ is the inclusion map which is an order homomorphism, $a \cdot b \in P_K$ in $K \Longleftrightarrow a \cdot b \in P \subset D$. So, P_K must satisfy the property

$$\frac{a}{b} \in P_K \Longleftrightarrow a \cdot b \in P \text{ in } D.$$

Now, $\frac{a}{b} = \frac{c}{d}$ in $K \Rightarrow a \cdot d = b \cdot c$ in D.

Multiplication by $b \cdot d$ yields $(a \cdot d) \cdot (b \cdot d) = (b \cdot c) \cdot (b \cdot d)$ or $(a \cdot b) \cdot d^2 = (c \cdot d) \cdot b^2$. So, given $a \cdot b \in P$, we can define P_K by taking $\frac{a}{b} \in P_K \Longleftrightarrow a \cdot b \in P$ consistent with the condition stated in the lemma.

Next, defining $\frac{a}{b} \in P_K$ if, and only if, $a \cdot b \in P$, suppose $\frac{a}{b}$ and $\frac{c}{d}$ are in P_K. Their sum

$$\frac{a}{b} + \frac{c}{d} = \frac{a \cdot d + b \cdot c}{b \cdot d} \in P_K \text{ if, and only if, } (a \cdot d + b \cdot c) \cdot (b \cdot d) \in P;$$

that is, if, and only if, $(a \cdot b) \cdot d^2 + (b \cdot c) \cdot d^2 \in P$.

Since $a \cdot b \in P$ and $c \cdot d \in P$, $a \cdot b \cdot d^2 \in P$, $c \cdot d \cdot b^2 \in P$. So,

$$a \cdot b \cdot d^2 + c \cdot d \cdot b^2 = (a \cdot d + b \cdot c) \cdot b \cdot d \in P.$$

Or, $\frac{a}{b} + \frac{c}{d} \in P_K$ whenever $\frac{a}{b}$, $\frac{c}{d}$ are in P_K.

Similarly, $(\frac{a}{b}) \cdot (\frac{c}{d}) \in P_K$. Also P_K satisfies the trichotomy property. For $a \cdot b \in P$, $a \cdot b = 0_D$ and $(-a) \cdot b \in P$: only one of them holds. So with $\frac{a}{b} \in K$ either $\frac{a}{b} \in P_K$, or $\frac{-a}{b} \in K$ or $\frac{a}{b} = 0$.   □

We observe that the characteristic of an ordered integral domain D is zero. As $D \cong \mathbb{Z}$ (preserving the order), in any ordered field F, the prime subfield of F is isomorphic to $\mathbb{Q}$, the field of rational numbers. See R. Godement [8].

Observations 8.2.1 :

> (1) A sum of two or more squares in an ordered field is either positive or zero.
>
> (2) In any field K, a product of two sums of squares in K is again a sum of squares.
>
> > For if $a, b, c, d \in K$,
> > $$(a^2 + b^2)(c^2 + d^2) = (ad - bc)^2 + (ac + bd)^2.$$
>
> (3) If $a, b \in K$, $b \neq 0$, writing $\frac{a}{b} = a \cdot b(b^{-1})^2$, we see that if a and b are sums of squares, $\frac{a}{b}$ is also a sum of squares.
>
> (4) If K is a field of characteristic $\neq 2$ and -1 is a sum of squares in K, then every element a in K is a sum of squares. For, we could write

(8.2.2) $$4a = (1 + a)^2 + (-1)(1 - a)^2$$

> and so, $4a$ is a sum of squares. This implies that a is a sum of squares.
>
> (5) In the field $\mathbb{C}$ of complex numbers, $-1 = i^2 = i^2 + 0^2$ and so it follows that any complex number $a + bi$, $(a, b \in \mathbb{R}$, the field of real numbers) can be expressed as the sum of two squares. As in (8.2.2), we have

$$a + bi = \left(\frac{1 + (a + bi)}{2}\right)^2 + i^2\left(\frac{1 - (a + bi)}{2}\right)^2,$$

or,

$$(8.2.3) \qquad\qquad a+bi = x^2+y^2 \ where \ x,y \in \mathbb{C}.$$

We remark that $\mathbb{C}$ is not the only field in which -1 is a sum of two squares.

For instance, let $n \in \mathbb{N}$ and $n \equiv 3 (\mathrm{mod}\ 8)$. Let $\zeta = \exp(\frac{2\pi i}{n})$, an imaginary nth root of unity. We consider an extension of $\mathbb{Q}$ by adjoining ζ to $\mathbb{Q}$. We write

$$(8.2.4) \qquad\qquad K = \mathbb{Q}(\zeta),$$

where $[Q(\zeta) : \mathbb{Q}] = \phi(n)$.

We need a result on Gauss quadratic sums defined below:

Definition 8.2.3 : *For $n \in \mathbb{N}$,*

$$G(m,n) = \sum_{h=0}^{n-1} \exp\left(\frac{2\pi i h^2 m}{n}\right)$$

is called a Gauss quadratic sum.

In chapter 6, definition 6.6.4 gives a Gauss sum via characters.

Lemma 8.2.3 : *For $n \in \mathbb{N}$,*

$$(8.2.5) \qquad G(1,n) = \begin{cases} (1+i)\sqrt{n}, & \textit{if } n \equiv 0(\mathrm{mod}\ 4) \\ \sqrt{n}, & \textit{if } n \equiv 1(\mathrm{mod}\ 4) \\ 0, & \textit{if } n \equiv 2(\mathrm{mod}\ 4) \\ i\sqrt{n}, & \textit{if } n \equiv 3(\mathrm{mod}\ 4) \end{cases}$$

Proof : Let $e(x) = \exp(\frac{2\pi i x}{n})$. Then, when $n \equiv 2(\mathrm{mod}\ 4)$, $n = 4k+2$. So $\frac{n}{2} = 2k+1$ (say), $\frac{n-2}{2} = 2k$.

So $e((h+\frac{n}{2})^2) = e(h^2 + hn + \frac{n^2}{4}) = -e(h^2)$.

Then,

$$\sum_{h=0}^{n-1} \exp\left(\frac{2\pi i h^2}{n}\right) = \sum_{h=0}^{\frac{n-2}{2}} e(h^2) + \sum_{h=\frac{n}{2}}^{n-1} e(h^2),$$

or,

$$G(1,n) = \sum_{h=0}^{\frac{n-2}{2}} e(h^2) + \sum_{t=h-\frac{n}{2}}^{\frac{n-2}{2}} e((t+\frac{n}{2})^2)$$

$$= \sum_{h=0}^{\frac{n-2}{2}} e(h^2) - \sum_{t=0}^{\frac{n-2}{2}} e(t^2),$$

or,

$$(8.2.6) \qquad\qquad G(1,n) = 0, \ \text{when } n \equiv 2\,(\mathrm{mod}\ 4).$$

Now, $G(m,n)$ has the equivalent representation [17]

$$(8.2.7) \qquad G(m,n) = \prod_{t=1}^{\frac{n-1}{2}} 2c\sin\left(\frac{(4t-2)m\pi}{n}\right), \quad \text{when } n \text{ is odd.}$$

Taking $m = 1$, in (8.2.7), we obtain

$$(8.2.8) \qquad G(1,n) = \prod_{t=1}^{\frac{n-1}{2}} 2i\sin\left(\frac{(4t-2)\pi}{n}\right), \quad n \text{ odd.}$$

But,

$$(8.2.9) \qquad \prod_{t=1}^{\frac{n-1}{2}} 2\sin\frac{(4t-2)\pi}{n} = (-1)^{[\frac{n}{4}]}\sqrt{n} \quad (n \text{ odd }) [17]$$

where $[x]$ = the greatest integer not exceeding x.

Combining (8.2.8) and (8.2.9), we obtain

$$(8.2.10) \qquad G(1,n) = i^{\frac{n-1}{2}}(-1)^{[\frac{n}{4}]}\sqrt{n}.$$

When $n \equiv 1 \pmod 4$, $G(1,n) = i^{2k}(-1)^k\sqrt{n}$ (when $n = 4k+1$) $= (-1)^{2k}\sqrt{n}$, or we get $G(1,n)$ as in (8.2.5) if $n \equiv 1 \pmod 4$.

When $n \equiv 3 \pmod 4$, $G(1,n) = i\sqrt{n}$ (on simplification) which is as given in (8.2.5).

The case that remains is $n \equiv 0 \pmod 4$.

It is known [17] that $G(m,n)$ possesses the following multiplicative property: Whenever g.c.d $(m,n) = 1$,

$$(8.2.11) \qquad G(rm,n) \cdot G(rn,m) = G(r,mn), \text{ for any integer } r \geq 1.$$

Let n be odd. If $r = 1$, $m = 2^\beta$

$$G(1,2^\beta n) = G(n,2^\beta)G(2^\beta,n) \quad \text{(by (8.2.11))}.$$

If β is even,

$$G(2^\beta,n) = \sum_{h=0}^{n-1} e\left(\frac{2^\beta h^2}{n}\right) = \sum_{h=0}^{n-1} e\left(\frac{h^2}{n}\right) = G(1,n).$$

If β is odd,

$$G(2^\beta,n) = \sum_{h=0}^{n-1} e\left(\frac{2^\beta h^2}{n}\right) = \sum_{h=0}^{n-1} e\left(\frac{2h^2}{n}\right) = G(2,n).$$

Now,

$$G(2,n) = \prod_{k-1}^{\frac{n-1}{2}} 2\sin\left(\frac{8k-4}{n}\pi\right) = (-i)^{\frac{n-1}{2}}\sqrt{n}.$$

Also,

$$G(n,4) = 2(1+i^n)$$

and

$$G(n,8) = \sqrt{8}(1+i)i^{\frac{n-1}{2}}.$$

For $m = 2^\beta$, $\beta \geq 4$, we have

$$G(n,2^\beta) = \sum_{h=0}^{2^\beta-1} e\left(\frac{nh^2}{2^\beta}\right) = \sum_{h=0}^{2^{\beta-1}-1} e\left(\frac{n(2h+1)^2}{2^\beta}\right) + 2\sum_{h=0}^{2^{\beta-2}-1} e\left(\frac{nh^2}{2^{\beta-2}}\right).$$

Further simplification yields

$$G(n,2^\beta) = 2G(n,2^{\beta-2}), \text{ if } \beta > 3 \text{ and } n \text{ odd}$$

and

$$G(1,2^\beta n) = G(1,n)G(n,4)\sqrt{\frac{2^\beta}{4}} = (1+i)\sqrt{2^\beta n}.$$

If β is odd and ≥ 3,

$$G(1,2^\beta n) = G(2,n)G(n,8)\sqrt{\frac{2^\beta}{8}} = (1+i)\sqrt{2^\beta n}.$$

Therefore, if $s = 2^\beta n$ where n is odd and $\beta \geq 2$, $G(1,s) = (1+i)\sqrt{s}$ which is (8.2.5) when $s \equiv 0 \pmod 4$.

Fact 8.2.2 : A theorem of Legendre [19] says that if n is not of the form $4^a(8b+7)$ ($a \geq 0, b \geq 0$), then n can be written as a sum of three integral squares.

However, we need only a case where $n \equiv 3 \pmod 8$. As shown in [17], Gauss has proved that if $n \equiv 3 \pmod 8$, there exist rational integers x, y, z such that

(8.2.12) $$n = x^2 + y^2 + z^2$$

Theorem 56 (Paromita Chowla (1969)) **:** *Let n be a prime $\equiv 3 \pmod 8$. If $\zeta = \exp(\frac{2\pi i}{n})$ and $K = \mathbb{Q}(\zeta)$, then -1 is a sum of two squares: That is,*

(8.2.13) $$-1 = \alpha_1^2 + \alpha_2^2 \text{ where } \alpha_1, \alpha_2 \in K.$$

Proof : As $n \equiv 3 \pmod 8$, $n \equiv 3 \pmod 4$ as well. Using (8.2.5),

$$\sum_{h=0}^{n-1} \zeta^{h^2} = \sum_{h=0}^{n-1} \exp\left(\frac{2\pi i h^2}{n}\right) = G(1,n) = i\sqrt{n}, \text{ when } n \equiv 3 \pmod 8.$$

A basis for K as a vector space over $\mathbb{Q}$ is $\{1, \zeta, \zeta^2, \ldots, \zeta^t\}$. $t = n-2$. $\alpha \in K$ can be written as

(8.2.14) $$\alpha = a_0 + a_1\zeta + a_2\zeta^2 \cdots + a_t\zeta^t \quad a_i \in \mathbb{Q}, i = 0, 1, 2, \ldots, t.$$

α can be expressed in terms of $G(1,n)$ and so we write

(8.2.15) $$\alpha = a + ib\sqrt{n} \text{ where } a, b \in \mathbb{Q}.$$

We get through, if we show that

(8.2.16) $$-1 = (a + ib\sqrt{n})^2 + (c + id\sqrt{n})^2$$

where a, b, c, d are determinable rational numbers. Suppose that (8.2.16) is true. Then,

(8.2.17) $a^2 + c^2 - n(b^2 + d^2) = -1$

(8.2.18) $ab + cd = 0$

Taking $c = \frac{-ab}{d}$ (using (8.2.18))

$$a^2 + (\frac{-ab}{d})^2 - n(b^2 + d^2) = -1 \text{ (from (8.2.17))},$$

or,

(8.2.19) $(\frac{a^2}{d^2} - n)(b^2 + d^2) = -1$

As $n \equiv 3 \pmod 8$, from (8.2.12), we have

$$nd^2 = x^2 d^2 + y^2 d^2 + z^2 d^2.$$

Write $a = xd$. We get

$$n = \frac{a^2}{d^2} + y^2 + z^2$$

So, (8.2.19) becomes

$$(y^2 + z^2)(b^2 + d^2) = 1$$

We choose $b = \frac{y}{y^2 + z^2}$, $d = \frac{z}{y^2 + z^2}$.
Then $b^2 + d^2 = \frac{1}{y^2 + z^2}$.
   So, (8.2.19) is okay for the values of b, d chosen as above
with $a = \frac{xz}{y^2 + z^2}$. $c = \frac{-xy}{y^2 + z^2}$.
   We get

$$ab + cd = 0$$

and so with values of a, b, c, d thus determined, (8.2.16) is satisfied perfectly well. This proves (8.2.13).   □

Remark 8.2.1 : Theorem 56 has been adapted from [18].

   We are now justified in making the following

Definition 8.2.4 : *A field K is said to be real, if -1 is not a sum of squares in K. The field $\mathbb{R}$ of reals is one such.*

Theorem 57 (Serge Lang (1965)) : *Let K be a real field.*

   (i) *If $a \in K$, then $K(\sqrt{a})$ or $K(\sqrt{-a})$ is real. If a is a sum of squares in K, then $K(\sqrt{a})$ is real. If $K(\sqrt{a})$ is not real, then $-a$ is a sum of squares in K.*
   (ii) *If f is an irreducible polynomial of odd degree in $K[x]$, and α is a root of $f(x) = 0$, then $K(\alpha)$ is real.*

Proof : If $a \in K$ and a is a square say b^2 where $b \in K$, one has

$$K(\sqrt{a}) = K(b) = K$$

and K is real. Suppose that a is not a square in K. If $K(\sqrt{a})$ is not real, -1 is a sum of squares and so there exist elements $b_i, c_i (i = 1, 2, \ldots, r)$ such that

$$-1 = \sum_{i=1}^{r} (b_i + c_i \sqrt{a})^2 = \sum_{i=1}^{r} (b_i^2 + 2b_i c_i \sqrt{a} + c_i^2 a).$$

Since $\quad [K(\sqrt{a}) : K] = 2$, from

$$-1 = \sum_{i=1}^{r} (b_i^2 + c_i^2 a) + 2 \sum_{i=1}^{r} b_i c_i \sqrt{a}$$

equating the 'rational' parts,

$$-1 = \sum_{i=1}^{r} (b_1^2 + c_i^2 a).$$

If a is a sum of two or more squares in K, -1 is a sum of squares in K. This contradicts the fact that K is real.
However,

$$-a = \frac{1 + \sum_{i=1}^{r} b_i^2}{\sum_{i=1}^{r} c_i^2}.$$

So, $-a$ is a quotient of sums of squares. This would mean $-a$ is a sum of squares. Hence, $K(\sqrt{-a})$ is real.
This proves the first part of the theorem.

For the second part, we have to show that $K(a)$ is real, if a is a root of a polynomial equation $f(x) = 0$ where $f(x) \in K[x]$ and $deg\ f$ is odd. Suppose that $K(a)$ is not real. Then, we could write

$$-1 = \sum_{i=1}^{r} g_i(a)^2$$

where g_i are polynomials in $K[x]$ of degree $\leq n-1$, n being odd and $n = \deg f$ such that

$$-1 = \sum_{i=1}^{r} g_i(x)^2 + h(x) f(x).$$

$\sum_{i=1}^{r} g_i(x)^2$ has even degree > 0. Otherwise, -1 would be a sum of squares in K. It is clear that $\sum_{i=1}^{r} g_i^2(x)$ has degree $\leq 2(n-1)$. So, $\deg h(x) \leq n-2$. Since $\deg f$ is odd, $\deg h$ is also odd. If b is a root of $h(x) = 0$, then -1 will be a sum of squares in $K(b)$. Since $\deg h < \deg f$, it would mean that $K(b)$ is not real. Proceeding thus, by considering polynomials of lower odd degree, we will arrive at a polynomial of first degree say $x - \alpha$ and -1 will be a sum of squares in $K(\alpha) = K($ as $\alpha \in K)$ —a contradiction to the given data. Thus, K is real.
This completes the proof of the second part of theorem 57. $\qquad \square$

Remark 8.2.2 : For more results on real fields, see Serge Lang [14, Chapter XI].

8.3. Valuation rings

We start with a prime p which is arbitrary, but fixed. A subset of $\mathbb{Q}$, the field of rationals, is considered in the following

Definition 8.3.1 :

$$\mathbb{Z}_p = \{\frac{m}{n} \in \mathbb{Q} : \frac{m}{n} \text{ is in its lowest terms and } p \nmid n\}$$

$\mathbb{Z}_p$ consists of those rational numbers whose denominators do not contain p as a factor. If $p \neq 2$, $\frac{1}{2} \in \mathbb{Z}_p$, for example.

Theorem 58 : $\mathbb{Z}_p$ is a PID contained in $\mathbb{Q}$.

Proof : If $\frac{a}{b}$, $\frac{c}{d}$ are in $\mathbb{Z}_p$, by definition $p \nmid b$. $p \nmid d$. So $p \nmid bd$. Therefore $\frac{a}{b} + \frac{c}{d}$ and $\frac{ac}{bd}$ are in $\mathbb{Z}_p$. $0 \in \mathbb{Z}_p$. $(\mathbb{Z}_p, +)$ is an abelian group, multiplication is associative and distributive. Therefore, $\mathbb{Z}_p$ is a subring of $\mathbb{Q}$. $\mathbb{Z}_p$ is not an ideal, as $\mathbb{Q}$ is a field. When $\frac{a}{b}\frac{c}{d} = 0$, $ac = 0$. So either $a = 0$ or $c = 0$. So, $\mathbb{Z}_p$ is an integral domain properly contained in $\mathbb{Q}$.

Let I be an ideal of $\mathbb{Z}_p$.

An element $x \in I$ is of the form $\frac{a}{b} = p^m \frac{a'}{b'}$ where $p \nmid b'$ and $p \nmid a'$. Such an element is an element of $\mathbb{Z}_p$. Also $\frac{a'}{b'} \in \mathbb{Z}_p$ as $p \nmid b'$. So, every element $\frac{a}{b} \in I$ can be written as

$$\frac{a}{b} = p^m u \quad \text{where } u \in \mathbb{Z}_p.$$

Among the powers p^m, there is one which is the least, say p^n. So any element of I is of the form $p^n u$ where $u \in \mathbb{Z}_p$. So, I is generated by the unique positive integer p^n. Therefore, I is a principal ideal.

This being the case, $\mathbb{Z}_p$ is a PID.   □

Remark 8.3.1 : (1) If $x \in \mathbb{Q}$, then either $x \in \mathbb{Z}_p$ or $x^{-1} \in \mathbb{Z}_p$. For, if $x \in \mathbb{Q}$. $x = \frac{m}{n}$ (in its lowest terms). If $p \nmid n$, $x \in \mathbb{Z}_p$. If $p|n$, $n = p^a n'$ where $p \nmid n'$.

So $\frac{n}{m} = p^a(\frac{n'}{m})$ and $p \nmid m$. Therefore, $x^{-1} \in \mathbb{Z}_p$.

(2) If S is a subring of $\mathbb{Q}$ containing $\mathbb{Z}_p$, then either $S = \mathbb{Z}_p$ or $S = \mathbb{Q}$.

If $S = \mathbb{Z}_p$, there is nothing to prove. Suppose that S properly contains $\mathbb{Z}_p$. There exists $x \in S$ and $x \notin \mathbb{Z}_p$. As $x \in \mathbb{Q}$, when $x \notin \mathbb{Z}_p$, $x^{-1} \in \mathbb{Z}_p$. Then $x^{-1} \in \mathbb{Z}_p \subset S$ or $x^{-1} \in S$.

Therefore, if $x \in S$, $x \notin \mathbb{Z}_p$, $x^{-1} \in S$ also. So, for every $x \in S$ and $x \notin \mathbb{Z}_p$, $x^{-1} \in S$. Whenever x is not in $\mathbb{Z}_p$, both x and x^{-1} are in S, when one of them is in S. If $t \in \mathbb{Q}$ and $t \notin S$, then $t \notin \mathbb{Z}_P$ and so $t^{-1} \in \mathbb{Z}_p$. That is, $t^{-1} \in S$. But, then, $(t^{-1})^{-1} = t \in S$ or $\mathbb{Q} \subseteq S$. But, $S \subseteq \mathbb{Q}$. Therefore, $S = \mathbb{Q}$, whenever S contains $\mathbb{Z}_p$ properly.

(3) Every nonzero $x \in \mathbb{Q}$ is of the form $p^n u$, where u is a unit in $\mathbb{Z}_p$ and n is a unique integer depending on x. For, if $x \in \mathbb{Z}_p$, $x = \frac{r}{s}$ where $p \nmid s$. But, p may be a

factor of r. So,

$$x = p^n \frac{r'}{s}, \quad \text{where } p \nmid r'.$$

Then, $\frac{r'}{s}$ is a unit in $\mathbb{Z}_p$ (invertible) or $x = p^n u$, u being a unit in $\mathbb{Z}_p$.
If $x \notin \mathbb{Z}_p$, $x^{-1} \in \mathbb{Z}_p$ and so $x^{-1} = \frac{a}{b}$, where $p \nmid b$.
   Then, $x^{-1} = p^n \frac{a}{b}$, where $p \nmid a$. So, then, $x = p^{-n} \mathrm{v}$ where v is a unit in $\mathbb{Z}_p$.
   When $0 \neq x \in \mathbb{Q}$, there exists a unique integer n such that $x = p^n u$ where u is
a unit in $\mathbb{Z}_p$.
This fact is exploited to define what is called a valuation on $\mathbb{Q}$.

Definition 8.3.2 : *When $0 \neq x \in \mathbb{Q}$ and $x = p^n u$ where u is a unit in $\mathbb{Z}_p$, we define*
$\nu_p(x) = n$ *and use the convention that* $\nu_p(0) = +\infty$.

   We remark that $+\infty$ denotes an object which satisfies the 'rules of calculation'
given below:

$$n + (+\infty) = +\infty, \text{ for all } n \in \mathbb{Z}.$$

$$(+\infty) + (+\infty) = +\infty.$$

$$+\infty > n \text{ for all } n \in \mathbb{Z} \text{ and } +\infty \geq +\infty.$$

Lemma 8.3.1 : *If $\nu_p(x)$, $x \in \mathbb{Q}$ is as given in definition in 8.3.2, then, for $x, y \in \mathbb{Q}$,*
   (1) $\nu_p(xy) = \nu_p(x) + \nu_p(y)$ and
   (2) $\nu_p(x+y) \geq \min\{\nu_p(x), \nu_p(y)\}$.
Further, $\mathbb{Z}_p$ is the set of all $x \in \mathbb{Q}$ such that $\nu_p(x) \geq 0$.

Proof : When $0 \neq x \in \mathbb{Q}$, there exists a unique integer n such that

$$x = p^n u_1 \text{where } u_1 \text{ is a unit in } \mathbb{Z}_p.$$

Similarly, for $0 \neq y \in \mathbb{Q}$, there exists a unique integer m such that

$$y = p^m u_2 \text{ where } u_2 \text{ is a unit in } \mathbb{Z}_p.$$

As $xy = p^{n+m} u_1 u_2$, where $u_1 u_2$ is also a unit in $\mathbb{Z}_p$, (1) follows.

$$\text{Now, } x + y = p^n u_1 + p^m u_2 = p^n(u_1 + p^{m-n} u_2) \text{ if } n \leq m.$$

If $u_1 = \frac{a}{b}$, $u_2 = \frac{c}{d}$ are units in $\mathbb{Z}_p$, p does not divide any of a, b, c or d.
   Also, $u_1 + p^{m-n} u_2 = \frac{ad + p^{m-n} bc}{bd}$. Neither the numerator nor the denominator is
divisible by p. So, $u_1 + p^{m-2} u_2$ is a unit and (2) follows.
   In the situation where $x = 0$ or $y = 0$ or both are zero, the convention for rules
of calculation with $+\infty$ given after definition 8.3.2 makes it agree with the results
of (1) and (2).
   If $u \in \mathbb{Z}_p$ and u is a unit, then, $u = p^0 u$ and so $v_p(u) = 0$. If $x \in \mathbb{Z}_p$ and x is not
a unit, $x = \frac{a}{b}$ where p does not divide b. Let p^n be the highest power of p contained
in a. Then, $n = 0$ if, and only if, x is a unit. So, if x is not a unit in $\mathbb{Z}_p$, $n > 0$ and
thus $v_p(x) \geq 0$ whenever $x \in \mathbb{Z}_p$. $\square$

Definition 8.3.3 : *Let F be a field and R a subring of F. R is said to be a valuation ring of F, if $R \neq F$, and for each $x(\neq 0)$ in F either x or $x^{-1} \in R$.*

We remark that perhaps, the motivation for the definition of a valuation ring of a field comes from the example of $\mathbb{Z}_p$ contained in $\mathbb{Q}$.

Lemma 8.3.2 : *Let R be a valuation ring of a field F. Then, R is a quasi-local ring in the sense that R has a unique maximal ideal M.*

Proof : Let M be the set of non-units in R. If $a, b \in M$, then $a + b \in M$. For, if $a + b$ is a unit for any two elements $a, b \in M$, then $a + (-b)$ is also a unit. That is, $(a+b) + (a+(-b)) = a + a$ is a unit, or $(a+a) + (-a) = a$ is a unit which contradicts the assumption that a is a non-unit. For $r \in R$, $a \in M$, $r \cdot a \in M$. For, if $r \cdot a$ is a unit, $(r \cdot a)^{-1}$ exists. Therefore, $a^{-1} \cdot r^{-1} = r^{-1} \cdot a^{-1}$ exists, which implies that a^{-1} exists, a contradiction. So, $(M, +)$ is an abelian subgroup of $(R, +)$ and it absorbs products from right and left. That is, M is an ideal of R. As $R \setminus M$ is the set of units of R, M is the unique maximal ideal of R. Hence, R is a quasi-local ring. $\square$

Corollary 8.3.1 : *$\mathbb{Z}_p$ is a quasi-local ring contained in $\mathbb{Q}$.*

Definition 8.3.4 (Roger Godement [8]) : *A discrete valuation of a field F is a function $\nu : F \rightarrow \mathbb{Z} \cup \{+\infty\}$ such that*

(i) $\nu(0) = +\infty$, $\nu(x) \in \mathbb{Z}$, if $x \neq 0$;
(ii) $\nu(xy) = \nu(x) + \nu(y)$ for all $x, y \in F$;
(iii) $\nu(x+y) \geq \min\{\nu(x), \nu(y)\}$ for all $x, y \in F$.

Remark 8.3.2 : ν is said to be nontrivial, if $\nu(F)$ does not consist of 0 and $+\infty$ only.

It can be shown that the set R of elements $x \in F$ such that $\nu(x) \geq 0$ is a valuation ring of F. The maximal ideal M of R is the set of all $x \in F$ such that $\nu(x) > 0$. If we choose an element $t \in M$ such that $\nu(t)$ is a minimum, then M is generated by t. That is, $M = Rt$. Further, every ideal of R is of the form $(t^n) = Rt^n$ for some $n \geq 0$.

8.4. Fields with valuation

As remarked by Nathan Jacobson [12] 'valuation theory forms a solid link between number theory, algebra and analysis'. While discussing valuation rings (see Section 8.3), we considered $\mathbb{Z}_p$, the ring of rational numbers whose denominators do not contain the prime p (fixed) as a factor. $\mathbb{Z}_p$ is a valuation ring of $\mathbb{Q}$, the field of rational numbers. A non-trivial discrete valuation (see definition 8.3.4) of F is a real-valued valuation which is known by the name 'absolute-value', in view of 'triangle equality'.

Definition 8.4.1 : *Let $\tilde{\mathbb{R}}$ denote the set of non-negative real numbers. An absolute-value on a field F is a function $m : F \rightarrow \tilde{\mathbb{R}}$ given by*
(a) $m(a) \geq 0$ and $m(a) = 0$ if, and only if, $a = 0_F$

(b) $m(a \cdot b) = m(a)m(b)$

(c) $m(a+b) \leq m(a) + m(b)$ *(triangle inequality)*

For instance, if $F = \mathbb{C}$, the field of complex numbers, when $z = x + iy \in \mathbb{C}$, $|z| = \sqrt{x^2 + y^2}$, (i denotes $\sqrt{-1}$) is an absolute value on $\mathbb{C}$. The function $|\mathbb{C}| : \mathbb{C} \to \tilde{\mathbb{R}}$ reduces to the usual absolute value on the subfields $\mathbb{Q}$ and $\mathbb{R}$ of $\mathbb{C}$.

Recalling the definition of $\mathbb{Z}_p$, we obtain the p-adic absolute value of $\mathbb{Q}$ as follows. Suppose that p is an arbitrary but fixed prime.

Let $a \in \mathbb{Q}$, $a = (\frac{x}{y})p^k$ where $k \in \mathbb{Z}$ and x and y do not contain p as a factor. k is uniquely determined once a is given. If $p = 2$, and $a = \frac{5}{48} = \frac{5}{3} \cdot 2^{-4}$, $k = -4$.

Definition 8.4.2 : *For $a = (\frac{x}{y})p^k$, we define $\nu_p(a) = k$ with the convention that $\nu_p(0) = \infty$.*

We note that

(i) $\nu_p(a) = \infty$ if, and only if, $a = 0$

(ii) $\nu_p(a \cdot b) = \nu_p(a) + \nu_p(b)$

(iii) $\nu_p(a+b) \geq \min\{\nu_p(a), \nu_p(b)\}$

Let η be a real number with the property $0 < \eta < 1$.

Definition 8.4.3 : *The p-adic absolute value $m_p : \mathbb{Q} \to \tilde{\mathbb{R}}$ is defined by*

$$m_p(a) = \eta^{\nu_p(a)} \text{ where } \nu_p(a) \text{ is as given in definition } 8.4.2.$$

Taking $m_p(0) = 0$, we check that

(1) $m_p(a) = 0 \Leftrightarrow a = 0$, $m_p(a) \geq 0$ for all $a \in \mathbb{Q}$

(2) $m_p(a \cdot b) = m_p(a)m_p(b)$, $a, b \in \mathbb{Q}$ and

(3) $m_p(a+b) = \eta^{\nu_p(a+b)}$.

As $\nu_p(a+b) \geq \min\{\nu_p(a), \nu_p(b)\}$, $\eta^{\nu_p(a+b)} \leq \max\{\eta^{\nu_p(a)}, \eta^{\nu_p(b)}\}$. Or $m_p(a+b) \leq \max\{m_p(a), m_p(b)\} \leq m_p(a) + m_p(b)$.

So, $m_p : \mathbb{Q} \to \tilde{\mathbb{R}}$ is an absolute value on $\mathbb{Q}$ (See definition 8.4.1). We call m_p a p-adic valuation of $\mathbb{Q}$.

If we take $\eta = \frac{1}{p}$, $m_p(a) = |a|_p = p^{-n}$, where $v_p(a) = n$.

It is verified that 11_p is a p-adic valuation of $\mathbb{Q}$.

Remark 8.4.1 : For any field F, the trivial absolute value of F is given by

$$m(a) = \begin{cases} 0, & a = 0 \\ 1, & a \neq 0 \end{cases}$$

Let $F^* = F \setminus \{0\}$. If $\mathbb{R}^* = \tilde{\mathbb{R}} \setminus \{0\}$ is considered as a group under multiplication, $m : F^* \to \mathbb{R}^*$ where m is the absolute value map, is a group homomorphism. Also, $m(1_F) = 1$ and for $a \in F^*$, $m(a^{-1}) = \frac{1}{m(a)}$ and $m(-1_F) = 1$, as $-1_F \cdot -1_F = 1_F$.

If F is a finite field with $m : F \to \tilde{\mathbb{R}}$ as an absolute value map, as $m(1_F) = 1$, $m(\zeta^k) = 1$ where ζ is a kth root of 1_F contained in F. So, for a finite field, we get only the trivial absolute value.

Fact 8.4.1 : Let (X, d) be a metric space. Given $\epsilon > 0$, the set

$$B_d(a, \epsilon) = \{x \in X : d(a - x) < \epsilon\}$$

of all elements whose 'distance' from $a < \epsilon$ is called the ϵ-ball centred at a. The collection $\mathbb{B}$ of all ϵ-balls $B_d(a, \epsilon)$ for $a \in X$, $\epsilon > 0$, is a basis for a topology of X called the metric topology induced by d.

A topological space X is called Hausdorff if, for each pair a_1, a_2 of distinct points of X, there exist ϵ-balls U_1 and U_2 centred at a_1 and a_2 respectively such that $U_1 \cap U_2 = \emptyset$. The topology on X declares 'open subsets' of X. A subset A of X is called closed if $X \setminus A$ is open.

When $m : F \to \tilde{\mathbb{R}}$ is an absolute-value map, it is easily verified that $m(1_F) = 1$, $m(b) = 1$, if there exists $k \in \mathbb{N}$ such that $b^k = 1_F$.
$m(-a) = m(a)$ for all $a \in F$. If $a \neq 0_F$, $m(a^{-1}) = \frac{1}{m(a)}$ and

$$|m(a) - m(b)| \leq m(a - b)$$

Thus, an absolute-value on F defines a topology whose open sets are unions of ϵ-balls. Addition, subtraction and multiplication are continuous functions from $F \times F$ to F in the topology of F. We can talk about convergence of sequences and series in the usual manner. As m gives a metric topology on F, the topological space induced by m is Hausdorff.

Definition 8.4.4 : *If $\{a_n\}$ is a sequence with $a_i \in F (i = 1, 2, \ldots)$, we say that $\{a_n\}$ converges to $a \in F$ if, given $\epsilon > 0$, there exists $N = N(\epsilon) \in \mathbb{N}$ such that $m(a - a_n) < \epsilon$, for all $n \geq N$.*

Definition 8.4.5 : *Two absolute-value maps m_1 and m_2, defined on F, are said to be equivalent if they define the same topology on F.*

For example, if $m_{1,p}$ and $m_{2,p}$ are two p-adic valuations on $\mathbb{Q}$ given by

$$m_{1,p}(a) = \eta^{\nu_p(a)}, m_{2,p}(a) = \eta_2^{\nu_p(a)}, 0 < \eta_1, \eta_2, < 1$$

then, $m_{2,p}(a) = (m_{1,p}(a))^s$ where $s = \frac{\log \eta_2}{\log \eta_1} > 0$.
This implies that any ϵ-neighbourhood of a point defined by $m_{1,p}$ is an ϵ'-neighbourhood defined by $m_{2,p}$. So $m_{1,p}$ and $m_{2,p}$ define the same topology on F. This is no accident. This is the case for any field F and any two absolute-values m_1 and m_2 where $m_2(a) = m_1(a)^s$ with s a positive real number.

When the absolute value map is trivial, it defines the discrete topology on F.

If the absolute-value map is non-trivial, we have an $a \in F$ such that $0 < m(a) < 1$. Then, the sequence $\{a^n\}$ converges to 0_F as $\{m(a^n)\} \to 0$ and the set of points $\{a, a^2, \ldots, a^n, \ldots\}$ is denoted by A. By the closure $\bar{A}$ of A, we mean the intersection of all closed sets containing A. By the definition of $\bar{A}, 0_F \in \bar{A}$, as $m(a^n) < \epsilon$ for all $n \geq N(\epsilon)$ which is determinable. It is not necessary that 0_F should belong to A, as a subset S of F is closed if, and only if, $S \supseteq \bar{S}$. So, A is not closed in F. Therefore, the topology induced by m is not discrete. Or, if the topology induced by an absolute-value map is discrete, then it is trivial. In other words, the trivial absolute-value map is equivalent to itself. The equivalence of absolute-value maps is brought out in the following

Theorem 59 : *Suppose that m_1 and m_2 are absolute-value maps on a field F such that $m_1 : F \to \tilde{\mathbb{R}}$ is not trivial and $m_1(a) < 1$ for all $a \in F$ implies that $m_2(a) < 1$ for all $a \in F$. Then, there exists a positive real number s such that*

$$m_2(a) = m_1(a)^s \text{ for all } a \,(\neq 0) \in F$$

Proof : Let $a, b \in F$. If $m_1(a) < m_1(b)$, then $m_2(a) < m_2(b)$.
So, $m_1(a) > 1 = m_1(1_F)$ implies that $m_2(a) > 1$. Since $m_1 : F \to \tilde{\mathbb{R}}$ is non-trivial, we can choose $a_0 \in F$ such that $m_1(a_0) > 1$ and so $m_2(a_0) > 1$.
Suppose that $a \in F$ is such that $m_1(a) > 1$ and $m_2(a) > 1$. We write

$$(8.4.1) \qquad\qquad t = \frac{\log m_1(a)}{\log m_2(a_0)}$$

Then, $t > 0$ and $m_1(a) = m_1(a_0)^t$.
We have $m_2(a) = m_2(a_0)^{t'}$ with $t' > 0$. If $t' \neq t$, there exists a rational number r having the property $r > t$ and $r < t'$ or $r < t$ and $r > t'$.
If $r > t$ and $r < t'$, $m_2(a_0)^r < m_2(a)$. If $r < t$ and $r > t'$ $m_2(a_0)^r > m_2(a)$.

Claim: $t = t'$.
Let r be equal to $\frac{m}{n}(n \neq 0)$.
If $r > t$, $r > \frac{\log m_2(a)}{\log m_2(a_0)}$.
So, $m_2(a_0)^r > m_2(a)$.
If $r < t$, $m_2(a_0)^r < m_2(a)$.
With $r = \frac{m}{n}(m, n$ positive integers), if $r > t$,

$$m_1(a) < m_1(a_0)^{\frac{m}{n}}$$

and so, $m_1(a)^n < m_1(a_0)^m$.
Then, $m_2(a)^n < m_2(a_0)^m$ and so, $m_2(a) < m_2(a_0)^{\frac{m}{n}}$.
Similarly, if $r < t$, $m_2(a) > m_2(a_0)^{\frac{m}{n}}$.
The same argument works for r in relation to t'. Then, we are forced to conclude that $t = t'$. So,

$$t = \frac{\log m_2(a)}{\log m_2(a_0)} = \frac{\log m_1(a)}{\log m_1(a_0)}.$$

It follows that

$$\frac{\log m_2(a)}{\log m_1(a)} = \frac{\log m_2(a_0)}{\log m_1(a_0)}$$

or, $m_2(a) = m_1^s(a)$ where $s = \frac{\log m_2(a_0)}{\log m_1(a_0)} > 0$.
This holds for all a for which $m_1(a) > 1$. If $m_1(a) < 1$, we have

$$m_2(a^{-1}) = m_1(a^{-1})^s \text{ and so, } m_2(a) = (m_1(a))^s.$$

This completes the proof of theorem 59.   □

Corollary 8.4.1 : *The absolute-value maps* $m_1 : F \to \tilde{\mathbb{R}}$ *and* $m_2 : F \to \tilde{\mathbb{R}}$ *are equivalent as they define the same topology on* F.

Remark 8.4.2 : Theorem 59 has been adapted from Nathan Jacobson [12].

Lemma 8.4.1 : *Let* $m_1, m_2, \ldots, m_n$ *be inequivalent nontrivial absolute-value maps of* F. *There exists an element* $a \in F$ *such that*

$$m_1(a) > 1 \text{ and } m_2(a), m_3(a) \ldots, m_n(a) \text{ are all } < 1.$$

Proof : We prove the result by induction on n. If $n = 2$, we can find $b, c \in F$ such that

$$m_1(b) > 1, m_1(c) \leq 1 \text{ and } m_2(b) \leq 1, m_2(c) > 1.$$

This is possible since we are considering absolute-value maps which are not equivalent. See theorem 59.

We write $a = \frac{b}{c}$. Then,

$$m_1(a) = m_1(\frac{b}{c}) > 1, \text{ as } m_1(c^{-1}) = \frac{1}{m_1(c)} \geq 1$$

$$m_2(a) = m_2(\frac{b}{c}) < 1, \text{ as } m_2(c^{-1}) = \frac{1}{m_2(c)} < 1$$

So, the lemma is true for $n = 2$.

Suppose that the result holds for $n = r - 1 (> 2)$.

Let $b \in F$ and assume that

$$m_1(b) > 1, m_2(b) < 1, \ldots, m_{r-1}(b) < 1.$$

Let $c \in F$ such that

$$m_1(c) > 1, \text{ and } m_r(c) < 1.$$

If $m_r(b) \leq 1$, then $a = b^t c$, for sufficiently large t, is such that

$$m_1(a) > 1, m_2(a) < 1, \ldots, m_{r-1}(a) < 1$$

and $m_r(a) = m_r(b^t c) < (m_r(b))^t m_r(c) < 1$ is okay.

The case where $m_r(b) > 1$ is handled as follows:

If $m : F \to \tilde{\mathbb{R}}$ is an absolute value map, $m(b) < 1 \Longrightarrow b^t \to 0$ in the topology T_m induced by m. So $\dfrac{b^t}{1_F + b^t} \to 0$ in T_m.

So $m\left(\dfrac{b^t}{1_F + b^t}\right) \to 0$, as t grows large.

Therefore, $m(b) > 1$ gives $\frac{1}{b^t} \to 0$ in T_m.

$$\text{Therefore }, \frac{b^t}{1_F + b^t} = \frac{1_F}{1_F + \frac{1}{b^t}} \to 1_F \text{ in } T_m.$$

Thus, $m(\dfrac{b^t}{1_F+b^t}) \to 1$ for t growing large.

That is, for large t , $a = \dfrac{b^t}{1_F+b^t}c$ makes

$$m_1(a) > 1, m_2(a) < 1, \ldots, m_{r-1}(a) < 1$$

and $m_r(a) = m_r(\dfrac{b^t}{1_F+b^t}c) = m_r(\dfrac{b^t}{1_F+b^t})m_r(c) < 1.$

This proves the lemma 8.4.1.   □

Lemma 8.4.2 : Let $m_1, m_2, \ldots, m_r$ be inequivalent nontrivial absolute-value maps of F into $\widetilde{\mathbb{R}}$. Given $\epsilon > 0$, there exists $b \in F$ such that

$$m_1(1_F - b) < \epsilon, m_2(b) < \epsilon, \ldots, m_r(b) < \epsilon.$$

Proof : By lemma 8.4.1, we could choose $a \in F$ in such a way that

$$m_1(a) > 1, m_2(a) < 1, \ldots, m_r(a) < 1.$$

We write $b = \frac{a^t}{1_F+a^t}$ with t a real number to be found out.
For $j > 1$,

$m_j(b) = m_j\left(\frac{a^t}{1_F+a^t}\right) \to 1$ for the topology T_{mj} induced by m_j.

For $j = 1$,

$m_1(1_F - b) = m_1\left(\frac{1_F}{1_F+a^t}\right) = m_1\left((a^{-1})^t\right)m_1\left(\frac{1_F}{(a^{-1})^t+1_F}\right) \to 0$ for sufficiently large t.

The desired inequalities follow.   □

Theorem 60 (The approximation theorem (weak form)) :

Let $m_1, m_2, \ldots, m_r$ be inequivalent nontrivial absolute-value maps from a field F into $\widetilde{\mathbb{R}}$. Suppose that $a_1, a_2, \ldots, a_r$ are given elements of R. Given $\epsilon > 0$, there exists an element a in F such that

$$m_j(a - a_j) < \epsilon ; \quad j = 1, 2, \ldots r.$$

Proof : We write $\max_i\{m_i(a_j)\} = M$, $i, j = 1, 2, \ldots, r$.

By lemma 8.4.2, we note that there exist $b_1, b_2, \ldots, b_r \in F$ such that

$$m_i(1_F - b_i) < \frac{\epsilon}{rM} \quad i = 1, 2, \ldots, r ; \quad m_i(b_j) < \frac{\epsilon}{rM}, j \neq i.$$

As $m_i(a + b) \leq m_i(a) + m_i(b)$ for $a, b, \in F$; $i = 1, 2, \ldots r$;

when

$$a = a_1b_1 + a_2b_2, \cdots + a_rb_r,$$

$$m_i(a - a_i) = m_i(a_1b_1 + a_2b_2 + a_i(b_i - 1_F) + \cdots + a_rb_r)$$

$$\leq \sum_{\substack{j=1 \\ j \neq i}}^{r} m_i(a_jb_j) + m_i(a_i(b_i - 1_F))$$

$$\leq \frac{M\epsilon}{rM} + \frac{M\epsilon}{rM} + \cdots + \frac{M\epsilon}{rM}(r \text{ terms });$$

$$= \epsilon.$$

Or, $m_j(a - a_j) < \epsilon$ for $j = 1, 2, \ldots, r$; as required.   □

Remark 8.4.3 : If m_1, m_2 are two equivalent absolute value maps of F into $\widetilde{\mathbb{R}}$,

$$m_2(a) = m_1^s(a) \text{ for some } s > 0 \text{ and all } a \in F.$$

Or,

(8.4.2) $m_1^s(a) m_2(a)^{-1} = 1.$

The approximation theorem says that if $m_1, m_2, \ldots, m_r$ are inequivalent and nontrivial, there does not exist an r-tuple $(s_1, s_2, \ldots, s_r) \neq (0, 0, \ldots 0)$, $s_i \in \mathbb{R}$ $(i = 1, 2, \ldots r)$ such that

(8.4.3) $m_1(a)^{s_1} m_2(a)^{s_2} \cdots m_r(a)^{s_r} = 1$ for all $a \in F.$

For, suppose that $s_1 = 1$. Then, there exists an a_i such that $m_1(a_i) < \frac{1}{2^i}$ and $m_j(a_i - 1_F) < \frac{1}{2^i}$ for $j > 1$.

Then, $m_1(a_i) \to 0$ and $m_j(a_i) \to 1$. Therefore, (8.4.2) is not valid for all $a_i (i = 1, 2, \ldots r)$.

Next, we note that the p-adic absolute value on $\mathbb{Q}$ namely $m_p : \mathbb{Q} \to \widetilde{\mathbb{R}}$ given by $m_p(a) = \eta^{\nu_p(a)}$ (see definition 8.4.3) is such that

(8.4.4) $m_p(a + b) \leq \max\{m_p(a), m_p(b)\}.$

This motivates

Definition 8.4.6 : *An absolute value map $m : F \to \widetilde{\mathbb{R}}$ is called non-archimedean if*

(8.4.5) $m(a + b) \leq \max\{m(a), m(b)\}$ *for all $a, b \in F.$*

This condition is stronger than

$$m(a + b) \leq m(a) + m(b) \quad \text{for all } a, b \in F,$$

as $\max\{m(a), m(b)\} \leq m(a) + m(b)$ for all $a, b \in F$.
If (8.4.5) is not satisfied, we say that $m : F \to \widetilde{\mathbb{R}}$ is archimedean.

Theorem 61 : *An absolute-value map $m : F \to \widetilde{\mathbb{R}}$ is non-archimedean if, and only if, $m(n1_F) \leq 1$ for all $n \in \mathbb{Z}$.*

Proof : $\Rightarrow$ If $m : f \to \widetilde{\mathbb{R}}$ is non-archimedean, then for $a_i (i = 1, 2, \ldots n) \in F$.

$$m(a_1 + a_2 \cdots + a_n) \leq \max\{m(a_i)\}$$

So, $m(1_F + 1_F + \cdots + 1_F) \leq m(1_F) = 1$, where 1_F is added to itself $(n - 1)$ times.
So, $m(n1_F) \leq 1$ for all $n \in \mathbb{Z}$.
$\Leftarrow$: Suppose that $m(n1_F) \leq 1$ for all $n \in \mathbb{Z}$.

For any $n \in \mathbb{N}$, $a, b \in F$,

$$m\big((a+b)^n\big) = m\Big(a^n + \binom{n}{1}a^{n-1}b + \cdots + b^n\Big)$$

$$\leq m(a^n) + m(n1_F)m(a^{n-1})m(b) + \cdots + m(b^n).$$

Or,

$$m((a+b)^n) \leq (n+1)\max\{m(a^n), m(b^n)\}.$$

Or,

$$m(a+b) \leq (n+1)^{\frac{1}{n}}\max\{m(a), m(b)\}.$$

In $\mathbb{R}$, $\lim_{n\to\infty}(n+1)^{\frac{1}{n}} = 1$. Therefore,

$$m(a+b) \leq \max\{m(a), m(b)\}$$

□

Corollary 8.4.2 : *If F is a finite field, an absolute-value map $m : F \to \widetilde{\mathbb{R}}$ is non-archimedean.*

Proof : If $n1_F$ is in the prime-subfield of F and $n1_F \neq 0_F$,
then, $(n1_F)^{p-1} = 1_F$, where p is a prime which is the characteristic of F.
So $m(n1_F) = 1$, Also $m(0_F) = 0$.
So, $m : F \to \widetilde{\mathbb{R}}$ is non-archimedean.

□

Remark 8.4.4 : Theorem 61 has been adapted from Nathan Jacobson [12].

Remark 8.4.5 : The trivial absolute value map $m : F \to \widetilde{\mathbb{R}}$ is non-archimedean, since $m : F \to \widetilde{\mathbb{R}}$ is given by

$$m(a) = \begin{cases} 0, & a = 0_F \\ 1, & a \neq 0_F \end{cases}$$

So, for $a, b \in F$, $m(a+b) \leq \max\{m(a), m(b)\}$ holds.

Remark 8.4.6 : The absolute value map of $\mathbb{C}$ is archimedean. For, if $m : \mathbb{C} \to \widetilde{\mathbb{R}}$ is given by

$$m(x+yi) = \sqrt{x^2+y^2}, \quad x, y \in \mathbb{R};$$

for $a, b \in \mathbb{C}$, $m(a+b) = |a+b| \leq |a| + |b|$.
    $m(a+b) \leq \max\{m(a), m(b)\}$ if, and only if, a or b is zero. So, $m : \mathbb{C} \to \widetilde{\mathbb{R}}$ is archimedean.

Remark 8.4.7 : It can be shown that the absolute-value map $m : F \to \widetilde{\mathbb{R}}$ is non-archimedean if, and only if,

$$\{m(n1_F) : n \in \mathbb{Z}\}$$

is bounded above. See Edwin Weiss [23].

Observation 8.4.1 : *Valuation of a field (as in definition 8.3.4) and absolute-value map could be connected in the following manner.*

Let $v : F \to \mathbb{R} \cup \{\infty\}$ be a map given by

(i) $v(a) = \infty$ if, and only if, $a = 0$

(ii) $v(a \cdot b) = v(a)\, v(b)$

(iii) $v(a+b) \geqslant \min\{v(a), v(b)\}$.

  Then, F together with v forms a field with valuation. Suppose that $\alpha \in \mathbb{R}$ is such that $\alpha > 0$. We write $m(a) = |a|_\alpha = \alpha^{-v(a)}; a \in F$.

$|0|_\alpha = 0$. $m : F \to \mathbb{R}$ *satisfies*

(i) $m(a) \geqslant 0$ and $m(a) = 0$ if, and only if, $a = 0$

(ii) $m(a,b) = m(a)\,(b)$

(iii) $m(a+b) = |a+b|_\alpha = \alpha^{-v(a+b)} \leqslant \alpha^{-\min\{v(a),v(b)\}} = \max\{|a|_\alpha, |b|_\alpha\}$.

m is precisely an absolute value on F (compare definition 8.4.3).

  Given F together with v, we write

(8.4.6) $R_v = \{a \in F \;:\; v(a) \geqslant 0\}$.

R_v is a subring of F. In fact, F is the field of quotients of R_v.

  If $x \in F$, $x \notin R_v$, $v(x) < 0$ and so $v(x^{-1}) > 0$.

That is, if $x \notin R_v$, $x^{-1} \in R_v$. So, any $x \in F$ is such that either x or $x^{-1} \in R_v$. Consistent with definition 8.3.3, we note that R_v is a valuation ring of F.

  Thus, an integral domain D is said to be a valuation ring, if there exists a valuation v of the field F of quotients of D such that $D = R_v$ (8.4.6).

  Next, we observe that the equivalence of absolute-value maps of a field F gives an equivalence relation on F. (see definition 8.4.5). The equivalence classes with respect to this equivalence relation are called the 'prime divisors' of F. The prime divisor to which the trivial absolute value map belongs is known as the trivial prime divisor. All others are non-trivial prime divisors.

Fact 8.4.2 : Let P be a prime divisor of F. For any absolute-value map $m : F \to \widetilde{\mathbb{R}}$ belonging to P, we have

(8.4.7) $P = \{m^s : s > 0\}$

Definition 8.4.7 : *We call P an archimedean prime divisor when an absolute-value map which belongs to P is archimedean.*

  A similar terminology is used for a non-archimedean prime divisor. It may be verified that an archimedean absolute-value map cannot be equivalent to a non-archimedean absolute-value map.

  We now consider three disjoint subsets of a field F with respect to a non-archimedean prime divisor P of F. Let m be an absolute-value map belonging to P.

Definition 8.4.8 :

(1) The set
$$\mathcal{O} = \{a \in F : m(a) \leq 1\}$$
is called the valuation ring at P or the ring of integers of F at P.

(2) The Set
$$\mathscr{P} = \{a \in F : m(a) = 1\}$$
is a prime ideal of $\mathcal{O}$ and is called the prime ideal at P.

We check that

(8.4.8) $$\mathscr{U} = \{a \in F : m(a) = 1\}$$

is the multiplicative group of units of $\mathcal{O}$.

If $\mathcal{O}^{-1} = \{a^{-1} \in F : a \in \mathcal{O}, a \neq 0\}$

We have
$$\mathscr{U} = \mathcal{O} \cap \mathcal{O}^{-1}$$

Further,

(8.4.9) $$\mathcal{O} = \mathscr{P} \cup \mathscr{U} \text{ and } \mathscr{P} \cap \mathscr{U} = \emptyset$$

P consists of all the non-units of $\mathcal{O}$. Therefore, P is the unique maximal ideal of $\mathcal{O}$.

Definition 8.4.9 : *The quotient ring $\mathcal{O}/\mathscr{P}$ is a field and it is called the residue class field at P. It is denoted by $\overline{F}$.*

Remark 8.4.8 :

(1) The definitions given above are independent of the choice of the absolute-value map m belonging to P.

(2) If P is the trivial prime divisor, there is only one absolute value map $m : F \to \widetilde{\mathbb{R}}$ for which $m(a) = 1$ for all $a \, (\neq 0) \in F$. So, $\mathcal{O} = F$, $\mathscr{P} = (0_F)$ and $\mathscr{U} = F^* = F \setminus \{0_F\}$. Also $\overline{F} = F$.

Definition 8.4.10 : *Let F be a field. A function $\tau : F \to \mathbb{R} \cup \{\infty\}$ is called an exponential valuation of F, if*

(i) $\tau(a) = \infty \Leftrightarrow a = 0$;

(ii) $\tau(ab) = \tau(a) + \tau(b)$, for $a, b \in F$;

(iii) $\tau(a+b) \geq \min\{\tau(a), \tau(b)\}$.

It is easy to check that τ is related to a non-archimedean absolute-value map $m : F \to \widetilde{\mathbb{R}}$ by
$$\tau(a) = -\log m(a), \text{ for all } a \in F.$$

In fact, $m(a) = \exp\big(-\tau(a)\big)$ for all $a \in F$ and hence the name exponential valuation for τ.

All the properties of a non-archimedean absolute-value map can be carried over to τ.

We say that the exponential valuation τ and τ' are equivalent if, and only if, $\tau' = s\tau$ for some $s > 0$, $s \in \mathbb{R}$. Further, τ belongs to a prime divisor P of F whenever $e^{-\tau} = m \in P$. If $\tau : F \to \mathbb{R} \cup \{\infty\}$ belongs to P, one has

(8.4.10) $P = \{s\tau : s > 0\}$

Analogous to the definition 8.4.7 regarding subsets $\mathcal{O}$, $\mathcal{P}$ and $\mathcal{U}$, we have

(8.4.11) $\mathcal{O} = \{a \in F : \tau(a) \geq 0\}$

(8.4.12) $\mathcal{P} = \{a \in F : \tau(a) > 0\}$

(8.4.13) $\mathcal{U} = \{a \in F : \tau(a) = 0\}$

We observe that $\tau \in P$ determines a homomorphism $\tau : F^* \to (\mathbb{R}, +)$ the additive group of real numbers. $\tau(F^*)$ is a subgroup of $(\mathbb{R}, +)$. $\tau(F^*)$ is called the value-group of τ written $G(\tau)$. If $\tau' \in P$, then $\tau'(F^*) = s\tau(F^*)$ for some $s > 0$, by the property of equivalent absolute-value maps. Then, $G(\tau)$ and $G(\tau')$ are order isomorphic (as ordered abelian groups).

Fact 8.4.3 : In the case of subgroups of $(\mathbb{R}, +)$ two situations arise:
Let $B(0, \epsilon)$ denote the closed interval $\{-\epsilon, \epsilon\}$ (where $\epsilon > 0$) in $\mathbb{R}$. A subset X of $\mathbb{R}$ is called a bounded subset of $\mathbb{R}$ if $X \subseteq B(0, \epsilon)$ for some $\epsilon > 0$. A subset X of $\mathbb{R}$ is called a discrete subset of $\mathbb{R}$ if, and only if, X intersects every closed interval $B(0, \epsilon)$ of $\mathbb{R}$ in a finite set. A subgroup $(G, +)$ of $(\mathbb{R}, +)$ is either discrete or dense in $\mathbb{R}$. For instance, $(\mathbb{Z}, +)$ is a discrete subgroup of $(\mathbb{R}, +)$ whereas $(\mathbb{Q}, +)$ is a dense subgroup of $\mathbb{R}$. It is known [23] that only one of these happens in the case of a subgroup of $(\mathbb{R}, +)$.

Definition 8.4.11 : *The prime divisor P of F is called discrete or non-discrete according as $G(\tau) = \tau(F^*)$ for $\tau \in P$ is discrete or non-discrete in $(\mathbb{R}, +)$. This is independent of the choice of τ.*

Let P be a discrete prime divisor of F. If, for some $\tau \in P$, $G(\tau) = \tau(F^*) = (0)$, then $m = e^{-\tau}$ is the trivial absolute-value map, since $m(a)$ equals 1 for all $a \in F^*$ and so $G(\tau) = (0)$. If P is discrete and non-trivial, there exists a unique exponential valuation $(\tau : F \to \mathbb{R} \cup \{\infty\})$ belonging to P written τ_P such that $G(\tau_P) = \tau_P(F^*) = \mathbb{Z}$. Then, τ_P is referred to as the normalized exponential valuation belonging to P. From this, we get

$$\mathcal{O} = \{a \in F : \quad \tau_P(a) > -1\}$$
(8.4.14) $$\mathcal{P} = \{a \in F : \quad \tau_P(a) \geq 1\}$$
$$\mathcal{U} = \{a \in F : \quad \tau_P(a) = 0\}$$

With the notion of exponential valuation on hand, we go to the notion of an ordinary arithmetic field (OAF). See Edwin Weiss [23].

Definition 8.4.12 : *Let $\mathcal{S}$ be a non-empty collection of discrete prime divisors of a field F. The pair $\{F, \mathcal{S}\}$ is called an ordinary arithmetic field (OAF) if the following axioms are satisfied:*

(A1) For each $a \in F$, if τ_P is a normalized valuation of F, $\tau_P(a) \geq 0$ for a finite set of prime divisors P belonging to $\mathscr{S}$.

(A2) Given $P_1, P_2 \in \mathscr{S}$ with $P_1 \neq P_2$ and any integers k_1, k_2 there exists $a \in F$ such that

$$\tau_{P_1}(a - 1_F) \geq k_1$$
$$\tau_{P_2}(a) \geq k_2$$
$$\tau_P(a) \geq 0 \text{ for all other } P \in \mathscr{S}$$

Example 8.4.1 : For $\mathbb{Q}$ denoting the field of rational numbers, let $\mathscr{S}$ be the set of all non-archimedean prime divisors of $\mathbb{Q}$. Then $\{\mathbb{Q}, \mathscr{S}\}$ satisfies $(A1)$ and $(A2)$ of the definition of an OAF.

Definition 8.4.13 : *Let $\{F, \mathscr{S}\}$ be an OAF. We write*

$$R = \{a \in F \; : \; \tau_P(a) \geq 0 \text{ for all } P \in \mathscr{S}\}$$

R is a ring called the ring of integers of $\{F, \mathscr{S}\}$.
In the case of $\{\mathbb{Q}, \mathscr{S}\}$, the ring of integers is $\mathbb{Z}$. More generally, if D is a PID with field of quotients F, when $\mathscr{S}$ denotes the set of prime divisors of F determined by the primes in D, $\{F, \mathscr{S}\}$ is an OAF with ring of integers equal to D.

Theorem 62 (Strong approximation theorem) : *Let $\{F, \mathscr{S}\}$ be an OAF. If $T = \{P_1, P_2 \ldots, P_r\}$ is any finite subset of $\mathscr{S}$, then for arbitrary elements $a_1, a_2, \ldots a_r \in F$ and any integers $k_1, k_2, \ldots, k_r$, there exists an element $a \in F$ such that*

$$\tau_{P_i}(a - a_i) \geq k_i \quad i = 1, 2, \ldots r \text{ and } \tau_P(a) \geq 0 \text{ for } P \notin T \text{ and } P \in \mathscr{S}.$$

Proof : *Claim :* For each $i = 1, 2, \ldots, r$ there exists an element $b_i \in F$ such that

$$\tau_{P_i}(b_i - 1_F) \geq k_i, \quad i = 1, 2, \ldots, r$$
$$\tau_{P_i}(b_j) \geq k_j, \quad j \neq i$$
$$\tau_P(b_i) \geq 0, \quad P \notin T, P \in \mathscr{S}.$$

Step 1. Let us assume that $k_i \geq 1$ for $i = 1, 2, \ldots, r$
Fix i, take $j \neq i$. By $(A2)$ there exists an element $c_{ij} \in F$ such that

$$\tau_{P_i}(c_{ij} - 1_F) \geq k_i;$$
$$\tau_{P_i}(c_{ij}) \geq m_j$$
$$\tau_P(c_{ij}) \geq 0 \text{ for } P \in \mathscr{S}, P \neq P_i, P_j \, (i, j = 1, 2, \ldots r).$$

We set

$$b_i = \prod_{j \neq i} c_{ij}$$

we have

$$\mathbb{P}_i = \{a \in F \; : \; \tau_{P_i}(a) \geq 1\}$$

$c_{ij} - 1_F \in \mathbb{P}_i^{k_i}$. So, $c_{ij} \in 1_F + \mathbb{P}_i^{k_i}$.

So, $b_i \in 1_F + \mathbb{P}_i^{k_i}$. So $b_i - 1_F$ is such that

$$\tau_{P_i}(b_i - 1_F) \geq k_i.$$

Also,

$$\tau_{P_i}(b_j) \geq k_j, \ j \neq i$$

and

$$\tau_P(b_i) \geq 0 \text{ for } P \notin T, P \in \mathscr{S}.$$

This establishes the claim made.

Step 2. We enlarge T so as to include the finite set of all $P \in \mathscr{S}$ for which $\tau_P(a_i) < 0$. We denote the enlarged set by

$$T' = \{P_1, P_2, P_3, \ldots, P_r, P_{r+1}, \ldots, P_n\}$$

Suppose that $a_{r+1} = a_{r+2}, \ldots, a_n = 0_F$. For $P \in \mathscr{S}$, $\tau_P(0_F) = \infty$.
Let $k_{r+1} = k_{r+2} = \cdots = k_n = 0$.

Suppose that $k_i \geq 0$ for $i = 1, 2, \ldots, r$.

Now, $\tau_{P_i}(a_i) = \tau_{P_i}(a_i - 1_F + 1_F) \geq \min\{\tau_{P_i}(a_i - 1_F), \tau_{P_i}(1_F)\} = 0$. So, for each $i = 1, 2, \ldots, n$, there exists an element $b_i \in F$ such that

(8.4.15) $\qquad \begin{cases} \tau_{P_i}(b_i - 1_F) & \geq k_i - \tau_{P_i}(a_i), \quad i = 1, 2, \ldots r \\ \tau_{P_i}(b_j) & \geq k_j - \tau_{P_i}(a_j), \quad j \neq i, j \in \{1, 2, \ldots, n\} \\ \tau_P(b_i) & \geq 0, \quad P \notin T', P \in \mathscr{S}. \end{cases}$

Some of the inequalities given above may be deleted since their right sides may turn out to be $-\infty$.

We set

$$a = \sum_{i=1}^{n} b_i a_i = \sum_{i=1}^{r} b_i a_i$$

Then,

$$a - a_i = b_1 a_1 + \cdots + (b_i - 1_F) a_i + \cdots + b_r a_r$$

So,

$$\tau_{P_i}(a - a_i) \geq \min\{\min_j \tau_{P_i}(b_j a_j), \tau_{P_i}(a_i) + \tau_{P_i}(b_i - 1_F)\}$$

$$\geq \min\{\min_j \{\tau_{P_i}(b_j) + \tau_{P_i}(a_j)\}, \tau_{P_i}(a_i) + \tau_{P_i}(b_i - 1_F)\}$$

By step 1, $\tau_{P_i}(b_i - 1_F) \geq k_i$, $\tau_{P_i}(b_j) \geq k_j$.

For $j \neq i$, if $k_j \geq k_i$, we get through. If $k_j \leq k_i$, $\tau_{P_j}(a - a_j) \geq k_j$.
Further, $\tau_P(a) \geq 0$ for $P \notin T$, $P \in \mathscr{S}$.
So whatever be k_i ($i = 1, 2, \ldots, r$), the conclusion of theorem 62 holds. $\qquad \square$

Remark 8.4.9 : Theorem 62 is referred to as the strong approximation theorem, as it is valid in a more general setting when a_i are chosen from the completion of F with respect to P_i. For details, see Edwin Weiss [23].

Corollary 8.4.3 (The Chinese Remainder Theorem) : *If $a_1, a_2, \ldots a_r$ are integers and $m_1, m_2, \ldots, m_r$ are positive integers such that g.c.d $(m_i, m_j) = 1$, $i \neq j$, $i, j = 1, 2, \ldots, r$, the system of congruences*

$$x \equiv a_i \pmod{m_i}, i = 1, 2, \ldots, r$$

has a common solution modulo $m_1 m_2 \ldots m_r$.

For, one has to take $\{\mathbb{Q}, \mathscr{S}\}$ with $\mathscr{S}$ the set of all non-archimedean prime divisors of $\mathbb{Q}$. Then, there exists $a \in \mathbb{Z}$ such that

$$\tau_{P_i}(a - a_i) \geq m_i \quad (i = 1, 2, \ldots, r)$$

If p is a prime in $\mathbb{Z}$, P determines a discrete prime divisor $P(p) \in \mathscr{S}$. We define $\tau_{P(p)}$ by

$$\tau_{P(p)}(0) = \infty$$

and for any $a (\neq 0) \in \mathbb{Q}$, $\tau_{P(p)}(a) = $ the exponent to which p appears in the factorization of a. Since a has only a finite number of primes appearing in its denominator, $\tau_{P(p)}(a) \geq 0$ for a finite number of discrete prime divisors $P \in \mathscr{S}$. For distinct primes p_1, p_2 in $\mathbb{Z}$ there exist $x, y \in \mathbb{Z}$ such that $x p_1 + y p_2 = 1$. So $a = y p_2 \in F$ is such that

$$\tau_{P(p_1)}(a - 1) \geq k_1, \text{ say}$$
$$\tau_{P(p_2)}(a) \geq k_2$$
$$\text{and } \tau_P(a) \geq 0 \text{ for all other } P \in S.$$

$\{Q, \mathscr{S}\}$ is an OAF and $x \equiv a \pmod{m_1 m_2 \ldots m_r}$ gives a simultaneous solution. $\square$

Remark 8.4.10 : Theorems 60 and 62 have been adapted from Edwin Weiss [23]. The valuation-theoretic approach is due to Hensel (1861–1941) who was a student of Kronecker. The concepts of OAF, local and global fields which introduce Ade'les and Ide'les lead to abstract techniques in algebraic number theory. For the further study in this direction, see Edwin Weiss [23].

8.5. Normed division domains

In [9], Solomon W. Golomb gives an interesting account of the concepts of divisibility, primes and composites via a weak partially ordered set. One does not have to start with an integral domain. The underlying structure is a weak partially ordered set (to be defined below) in which a suitable multiplicative norm is considered. The approach is interesting and has applications to solutions of problems in graph theory. However, we confine ourselves to algebraic concepts.

Let X be a non-empty set. A relation R on X is a non-empty subset of $X \times X$. If $(a, b) \in R$, we write aRb to convey this information.

1. R is said to be reflexive if, aRa for all $a \in X$.
2. R is said to be transitive if aRb and bRc imply aRc for all $a, b, c \in R$.

Definition 8.5.1 :

(1) A relation R on a set X is called a weak partial order, if R is reflexive and transitive.

(2) A set X together with a weak partial order defined on it is called a weak partially ordered set. It is denoted by $(X, \triangle)$ where $\triangle$ denotes the weak partial order on X.

In the definition 8.5.1, 'anti-symmetry' of the relation R is not wanted. This is to accommodate units, if any, of an integral domain. In fact, when we consider a as a proper divisor of b, that is, $a|b$, $a \neq b$, 'divides' is a relation which is transitive. However, when $a|b$ and $a = b$, it is okay with $b|a$. That is, $a|b$ and $b|a \implies a = b$ (antisymmetry). But then, we do not provide for 'units' in the integral domain considered. Hence the motivation for a weak partially ordered set.

Definition 8.5.2 : *Let $(X, \triangle)$ be a weak partially ordered set. Suppose we associate a norm N with elements of X defined by*

$$N : X \to \mathbb{N}$$

such that for $a, b \in X$, $N(a)$ divides $N(b)$ whenever $a \triangle b$. Further if $u \in X$ is such that $N(u) = 1$ then $u \triangle y$ for all $y \in X$. The triple $(X, \triangle, N)$ is said to be a normed division domain, abbreviated as NDD.

Example 8.5.1 : The integral domain $\mathbb{Z}$ is an NDD. For, if the weak partial order in $\mathbb{Z}$ is 'divides' and the norm N is given by $N(a) = a^2$ for $a \in \mathbb{Z}$, then whenever a divides b, $N(a)$ divides $N(b)$. Further $N(1) = N(-1) = 1$ and $\{1, -1\}$ is the set of units in $\mathbb{Z}$. 1 or -1 divides a for any $a \in \mathbb{Z}$.

Notation 8.5.1 : Let us call the primes in $\mathbb{Z}$ to be rational primes (as mentioned in chapter 2).

The ring $\mathbb{Z}[i]$ of Gaussian integers is an NDD.

For, the weak partial order in $\mathbb{Z}[i]$ is 'divides' and the norm N is given by

$$N(a + bi) = a^2 + b^2 \text{ where } a + bi \in \mathbb{Z}[i].$$

$\{1, -1, i, -i\}$ is the set of units in $\mathbb{Z}[i]$. Clearly if u is a unit in $\mathbb{Z}[i]$, $N(u) = 1$ and u divides $(a + bi)$ for $a, b \in \mathbb{Z}$. So $\mathbb{Z}[i]$ is an NDD.

Definition 8.5.3 : *Let $(X, \triangle, N)$ be an NDD. $u \in X$ is called a unit, if $N(u) = 1$.*

The normed division domains that we consider are those for which there will be at least one unit u present in it. In $(X, \triangle, N)$, we read $a \triangle b$ as 'a divides b', where $a, b \in X$.

Definition 8.5.4 :

Let $D = (X, \triangle, N)$ be an NDD. An element π in X is called a prime of D if $N(\pi) > 1$ and there is no element a in X with $a \triangle \pi$ and $1 < N(a) < N(\pi)$.

Lemma 8.5.1 : *In the NDD $D = (X, \triangle, N)$, an element π in X is a prime of D if $N(\pi)$ is a rational prime.*

Proof : If $N(\pi)$ is a rational prime p, suppose that there exists $a \in D$ such that $a \triangle \pi$. Since $N(a)$ divides $N(\pi) = p$, $N(a)$ is either 1 or p. So, $1 < N(a) < N(\pi)$ is not possible when $a \triangle \pi$ in D. Thus, π is a prime in D. □

Theorem 63 : *Let $D = (X, \triangle, N)$ be an NDD. An element π in X with $N(\pi) > 1$ is a prime of D if, and only if, there is no element ρ in X which is a prime of D such that $\rho \triangle \pi$ and $N(\rho) < N(\pi)$.*

Proof : $:\Rightarrow$ If there a prime ρ of D with $\rho \triangle \pi$ and $N(\rho) < N(\pi)$ then clearly, π is not a prime of D. By contrapositive argument, the condition is sufficient.

$\Leftarrow$: If $N(\pi) > 1$ and π is not a prime of D, we must show that there exists a 'proper prime divisor' of π. For this we apply induction on $N(\pi)$.

If $N(\pi) = 2$, the smallest possible norm for a prime of D, then π is a prime by lemma 8.5.1. Assume that the result holds for all elements of norm $\leq m$ where $m > 2$ and consider an element π with $N(\pi) = m + 1$. By definition, either π is a prime or it has a divisor ρ with $2 \leq N(\rho) \leq m$. By induction hypothesis, either ρ is a prime of D in which case there is nothing more to prove or D has a prime σ which divides ρ with $2 \leq N(\sigma) < N(\rho) \leq m$. In this case, since $\sigma \triangle \rho$ and $\rho \triangle \pi$ and $\triangle$ is 'transitive', we conclude that $\sigma \triangle \pi$ and σ is a prime divisor of π with $1 < N(\sigma) < N(\pi)$. This contradicts the definition of a prime in D. So, if there is no element ρ in X which is a prime of D such that $\rho \triangle \pi$ and $N(\rho) < N(\pi)$, then, π with $N(\pi) > 1$ is a prime of D. □

Fact 8.5.1 : Let $D = (X, \triangle, N)$ be an NDD. Let Y be a subset of X. Suppose $D' = (Y, \triangle, N)$, the result of restricting $\triangle$ and N to Y. Then D' is an NDD and every prime of D which belongs to Y is a prime of D'.

Fact 8.5.2 : Let $D = (X, \triangle, N)$ be an NDD. Suppose that $\triangle'$ is a weak partial order on X such that $a \triangle' b$ whenever $a \triangle b$ but not in general conversely and such that $N(a)$ divides $N(b)$ whenever $a \triangle' b$, then $D^* = (X, \triangle', N)$ is an NDD. If π in X is a prime of D^*, then π is a prime of D.

Definition 8.5.5 : *Let $D = (X, \triangle, N)$ be an NDD. For $a \in X$, we define the principal ideal of a to be the set $\{y \in X : a \triangle y\}$ and denote it by (a).*
If $u \subset X$ with $N(u) = 1$, then $(u) = X$ is the unit ideal.

The analogy with $\mathbb{Z}$ is clear.

Definition 8.5.6 : *Let $D = (X, \triangle, N)$ be an NDD. Suppose that a, b are elements of X. We define the set $M = M(a,b)$ of common multiples of a and b to be $(a) \cap (b)$. If $M \neq \emptyset$, M has a non-empty subset $LCM(a,b)$ of elements of least norm, called the least common multiple of a and b.*

We remark that the least norm must be a common multiple of $N(a)$ and $N(b)$, but not necessarily their least common multiple.

For instance, we shall consider two elements $a = 1 + i$ and $b = 1 - i$ of $\mathbb{Z}[i]$.

$$(a) = \{(1+i)(c+di) : c, d \in \mathbb{Z}\}$$
$$(b) = \{(1-i)(c'+d'i) : c', d' \in \mathbb{Z}\}$$

As 2 is a common multiple of $(1+i)$ and $(1-i)$, an element t in $(a) \cap (b)$ of least norm is $t = 2$ with $N(t) = 4$.

But, $N(1+i) = 2$ and $N(1-i) = 2$. The least norm 4 is a multiple of 2. But, it is not the $l.c.m$ of $N(1+i)$ and $N(1-i)$.

Definition 8.5.7 : *Let $D = (X, \triangle, N)$ be an NDD. Suppose that $a \in X$, $b \in X$. We define the set $C = C(a,b)$ of common divisors of a and b as the set consisting of those elements $d \in X$ such that $a \in (d)$ and $b \in (d)$.*

If there exists $u \in X$ with $N(u) = 1$, then $u \in C(a,b)$ for all $a, b \in X$. If $C(a,b) \neq \emptyset$, $C(a,b)$ has a non-empty subset GCD (a,b) of elements of greatest norm, called the greatest common divisor of a and b.

We remark that the greatest norm will be a common divisor of $N(a)$ and $N(b)$, but not necessarily their greatest common divisor.

For instance, let us pick two elements a, b of $\mathbb{Z}(i)$ with $a = 2$, $b = 2i$

$$2i = (1+i)^2, \quad 2 = (1+i)(1-i)$$

$2i$ is an element of the principal ideal of $(1+i)$, 2 is an element of the principal ideal of $(1+i)$.

$$\text{GCD} \ (2, 2i) = \{1+i, -1+i, -1-i, 1-i\}.$$

$$N(1+i) = 2, N(2) = 4, N(2i) = 4$$

The greatest norm 4 is not the g.c.d of 2 and 4.

Definition 8.5.8 : *Let $D = (X, \triangle, N)$ be an NDD. Let X' be the set of equivalence classes of elements of X established by a norm-preserving equivalence relation. We define a new NDD by $D' = (X', \triangle, N)$ where whenever $\sigma \in X'$, $\tau \in X'$ then $\sigma \triangle \tau$ if, and only if, there exist elements s and t in X with $s \in \sigma$, $t \in \tau$ such that $s \triangle t$ in D.*

The elements of X belonging to the same equivalence class in X' are called associates relative to D'.

The equivalence classes or orbits of X arise from a group of transformations of the elements of X. In $\mathbb{Z}[i]$, it is the group U of units in $\mathbb{Z}[i]$ namely

$$U = \{1, -1, i, -i\}$$

It is clear that all the units are placed in a single orbit of X where each element of the orbit has norm equal to 1. If $z = a + bi$ is an element of $\mathbb{Z}[i]$, the orbit of z is

(8.5.1) $\{z, iz, -z, -iz, \}$

This corresponds to a transformation by the cyclic group C_4 of units in $\mathbb{Z}[i]$. If we take the group D_4 of symmetries of a square, a transformation of z would result in

(8.5.2) $\left\{ z, iz, -z, -iz, \bar{z}, i\bar{z}, -\bar{z}, -i\bar{z} \right\}$

D_4 consists of complex conjugation as well as multiplication by units. Each of the elements in (8.5.1) or (8.5.2) has norm equal to $N(z) = a^2 + b^2$. Associates of Gaussian integers could be defined via D_4 as well. Divisibility theory for $\mathbb{Z}[i]$ does not differ much from that of $\mathbb{Z}[i]$ in terms of C_4.

The purpose of introducing the notion of an NDD is to point out that when a norm function is suitably defined in respect of a weak partially ordered set, a lot of 'arithmetic' is worked out.

We conclude this section with an example from group theory. Let X denote the set of all finite groups. For $G \in X$, we define $N(G)$ to be $|G|$, the order of G. A weak partial order on X is given by $H \triangle_1 G$ if H is isomorphic to a subgroup of G whenever $H, G \in X$. By Lagrange's Theorem, $N(H)|N(G)$ whenever this happens. The resulting normed division domain $D_1 = (X, \triangle_1, N)$. D is such that the primes of D are cyclic groups of order a rational prime p. Those are the elements guaranteed to be primes of D by lemma 8.5.1.

We could introduce another weak partial order on X by stating that for $H, G \in X$, $H \triangle_2 G$ if H is isomorphic to a normal subgroup of G or by transitivity there is a normal series from H to G namely

$$H = H_0 \subset H_1 \subset H_2 \cdots \subset H_k = G$$

where H_i is a normal subgroup of H_{i+1} $(i = 0, 1, \ldots, k-1)$. Then $D_2 = (X, \triangle_2, N)$ satisfies the axioms of an NDD. The primes in D_2 are precisely finite simple groups. These will also include primes of D_1 as a subset. See Fact 8.5.1. For applications to graph theory, see S. W. Golomb [9].

8.6. Modular lattices and Jordan-Hölder theorem

While considering a weak partial order on a non-empty set X, we mentioned about reflexivity and transitivity. A relation R on X is said to be anti-symmetric if for $a, b \in X$, aRb and $bRa \Longrightarrow a = b$. As an example, the relation 'divides' on $\mathbb{N}$ (the set of positive integers) is anti-symmetric. However, it is not anti-symmetric on $\mathbb{Z}$, the set of rational integers.

Definition 8.6.1 : *A relation R on a non-empty set X is called a partial order if R is (i) reflexive (ii) antisymmetric and (iii) transitive. We call (X, R) a partially ordered set. R is normally expressed as '$\leq$'. By the symbol $(X, \leq)$, we mean a partially ordered set (or, briefly, a poset) wherein $\leq$ denotes the partial order relation.*

If $\mathbb{P}(X)$ denotes the power set of a non-empty set X, $(\mathbb{P}(X), \leq)$ is a poset under the partial order 'a subset of'. The relation 'divides' on $\mathbb{N}$ is a partial order in $\mathbb{N}$. Let G be a group. We define $\leq$ on the collection Y of subgroups of G by $H \leq K$ if H is a subgroup of K. Then $(Y, \leq)$ is a poset.

Let R be a commutative ring with unity. Let X denote the collection of ideals of R. If $\subseteq$ denotes 'set-inclusion', $(X, \subseteq)$ is a poset.

Definition 8.6.2 : *Let $(X, \leq)$ be a poset. For $a, b \in X$, we define $a < b$ to mean that $a \leq b$ and $a \neq b$. That is, a is strictly less than b. Examples of a relation of the type 'strictly less than' are 'a proper subset of', 'a proper divisor of' and 'a proper subgroup of'.*

Definition 8.6.3 : *Given $(X, \leq)$ let $a, b \in X$ be such that $a < b$. If, for any $c \in X$, $a < c < b$ is not true, we say that b covers a.*

When X is a finite set, the relation 'covers' can be used to represent the poset $(X, \leq)$ by means of a diagram. The diagram is known as 'the covering diagram' of the partial order on the given set X (also known as Hasse diagram). For instance, the covering diagram in the figure below shows the poset $(D(12), \leq)$ where $\leq$

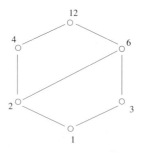

Figure 5

means 'divides'. $D(12)$ denotes the set of positive divisors of 12. When there are two or more elements such as a, b, c which are covered by different elements say a', b', c' respectively, it is customary to draw the edges vertical and parallel to one another, as in figure 5.

Definition 8.6.4 : *In a poset $(X, \leq)$, an element a in X is called a minimal element, if there does not exist an element b in X such that $b < a$.*

An element n in X is called a maximal element of $(X, \leq)$ if, there does not exist an element n in X such that $m < n$.

The poset $(\mathbb{Z}, \leq)$ with $\leq$ to mean 'less than or equal to' does not have either a minimal element or maximal element. In the case of the power set $\mathbb{P}(X)$ of X, $(\mathbb{P}(X), \subseteq)$, $\emptyset$ is the minimal element and X is the maximal element.

An element $0_X \in X$ is called a zero element of $(X, \leq)$ if $0_X \leq x$ for all $x \in X$. An element $1_X \in X$ is called unit element of $(X, \leq)$ if $x \leq 1_X$ for all $x \in X$. Obviously $\emptyset$ is the zero element of $(\mathbb{P}(X), \subseteq)$ and X is the unit element of $(\mathbb{P}(X), \subseteq)$.

A zero element is a minimal element but a minimal element need not be a zero element.

Definition 8.6.5 : *If* $(X, \leq)$ *has a zero element* 0_X, *the elements* p *in* X *that cover* 0_X *are called atoms.*

If $(X, \leq)$ *has a unit element* 1_X, *the elements* q *in* X *that are covered by* 1_X *are called coatoms of* X.

In $(\mathbb{P}(X), \subseteq)$, atoms are singletons $\{a\}, \{b\}$ etc. where $a, b, \ldots \in X$.

If $D(n)$ denotes the set of positive divisors of n, atoms are the prime divisors of n. In $(\mathbb{P}(X), \subseteq)$, the coatoms are subsets of the form $X \setminus \{a\}$ where $a \in X$. The coatoms of $(D(n), \leq)$ ($\leq$ means 'divides') are numbers of the form n/p_i where p_i ($i = 1, 2, \ldots k$) are the distinct prime divisors of n.

Definition 8.6.6 : *Suppose that* $(X, \leq)$ *and* $(Y, \leq')$ *are posets. A bijection* $f : X \to Y$ *is an isomorphism of posets if, wherever,* $x_1 \leq x_2$ *in* $(X, \leq)$, $f(x_1) \leq' f(x_2)$ *in* $(Y, \leq')$. *When* $f : X \to Y$ *is an isomorphism, the posets* $(X, \leq)$, $(Y, \leq')$ *are said to be order isomorphic.*

We consider the following finite partially ordered sets.

(8.6.1) $(D(24), \leq)$, the set of divisors of 24 with $\leq$ to mean 'divides'.

(8.6.2) $(C(54), \leq)$, the set of non-isomorphic subgroups of a cyclic group of order 54, $\leq$ means 'a subgroup of'

(8.6.3) $(S(40), \leq)$, the set of ideals of $\mathbb{Z}/40\mathbb{Z}$ with $\leq$ to mean 'a subset of'

We note that $D(24)$ has 8 elements. $C(54)$ has eight elements namely (e), $C(2), C(3), C(6), C(9), C(18), C(27)$ and $C(54)$ where (e) is the trivial subgroup of order 1. The number of subgroups of $C(54)$ is equal to the number of divisors of 54, by theorem 54 of chapter 7. The additive abelian group of $\mathbb{Z}/40\mathbb{Z}$ is of order $2^3 \cdot 5 = 40$. The ideals of $\mathbb{Z}/40Z$ are $I_1, I_2, \ldots, I_8$ where

$$I_1 = (0), |I_2| = 2, |I_3| = 4, |I_4| = 5, |I_5| = 8, |I_6| = 10,$$

$$|I_7| = 20, |I_8| = 40.$$

It is easy to check that any two of the partially ordered sets (8.6.1), (8.6.2), (8.6.3) are order isomorphic.

Definition 8.6.7 : *Let* $(X, \leq)$ *be a poset. Given* $a, b, \in X$, *an element* $g \in X$ *is called the greatest lower bound (g.l.b) of* a *and* b, *if*

(i) $g \leq a$ *and* $g \leq b$,

(ii) *whenever there exists* $h \in X$ *such that* $h \leq a$ *and* $h \leq b$ *then* $h \leq g$.

Similarly, an element $l \in X$ *is called the least upper bound (l.u.b) of* a *and* b, *if*

(i) $a \leq l$ *and* $b \leq l$ *and*

(ii) whenever there exists $m \in X$ such that $a \leq m$ and $b \leq m$, then $l \leq m$.

The *g.l.b* of a and b, if it exists, is denoted by $a \wedge b$ (read as a meet b). The *l.u.b* of a and b, if it exists, is denoted by $a \vee b$ (read as a join b). See [2] or [4].

The easiest illustration is from $(\mathbb{P}(X), \subseteq)$. For A, B subsets of X we have $A \wedge B = A \cap B$ and $A \vee B = A \cup B$. In the poset $(\mathbb{N}, \leq)$ where $\leq$ means 'divides', for $a, b, \in \mathbb{N}$, $a \wedge b$ is the g.c.d of a and b and $a \vee b$ is the l.c.m of a and b.

It is known that in a poset $(X, \leq)$ when $a, b \in X$ are given, if $a \wedge b$ or $a \vee b$ exist, then the *g.l.b* or *l.u.b* is unique. There are examples of posets $(X, \leq)$ where there are elements a, b in X and the pair $\{a, b\}$ does not have a *g.l.b* or *l.u.b*.

Definition 8.6.8 : *If $(X, \leq)$ is a poset in which any two elements have a g.l.b and l.u.b then $(X, \leq)$ is called a lattice.*

Remark 8.6.1 : Given a poset $(X, \leq)$, one could consider a subset S of X and define *l.u.b* and *g.l.b* with reference to the elements of S. We get *l.u.b* and *g.l.b* of S. Then, $(X, \leq)$ is said to be a lattice if every two-element subset of X has an *l.u.b* and a *g.l.b*.

Indeed, $(\mathbb{P}(X), \subseteq)$ is an example of a lattice, as also $(\mathbb{N}, \leq)$, where $\leq$ is to mean 'divides'.

The properties of a lattice $(L, \leq)$ are shown below:

(8.6.4)   For $a, b \in L, a \leq b$ if, and only if, $a \wedge b = a$,

(8.6.5)   For $a, b \in L, a \leq b$ if, and only if, $a \vee b = b$,

(8.6.6)   For $a \in L, a \wedge a = a$ and $a \vee a = a$ (idempotency law),

(8.6.7)   For $a, b \in L, a \wedge b = b \wedge a$ and $a \vee b = b \vee a$ (commutative law),

(8.6.8)   For $a, b, c \in L$, $\left. \begin{array}{l} a \wedge (b \wedge c) = (a \wedge b) \wedge c \\ a \vee (b \vee c) = (a \vee b) \vee c \end{array} \right\}$ (associative law) ,

(8.6.9)   For $a, b \in L$, $\left. \begin{array}{l} a \vee (a \wedge b) = a \\ a \wedge (a, b) = a \end{array} \right\}$ (absorption law) .

Definition 8.6.9 : *A lattice $(L, \leq)$ is said to be bounded, if L contains a zero element and a unit element.*

·   It is easy to check that any finite lattice is bounded. In the case of $(\mathbb{N}, \leq)$ where $\leq$ is to mean 'less than or equal to', it is not bounded, though it has a zero element.

If $B = \{0, 1\}$, B^n is the set of ordered n-tuples $(a_1, a_2, \ldots a_n)$ where $a_i = 0$ or 1 for $i = 1, 2, \ldots n$. For convenience, we write

$$a_1 a_2 \ldots a_n \text{ to denote } (a_1, a_2, \ldots, a_n)$$

For instance, 01100 is an element of B^5.

For $a_1 a_2 \ldots a_n, b_1 b_2 \ldots b_n \in B^n$, we define

$$a_1 a_2 \ldots a_n \leq b_1 b_2 \ldots b_n \text{ if } a_i \leq b_i \text{ for all } i = 1, 2 \ldots n.$$

In B^5, $10100 \leq 11101$

We set $a_1 a_2 \ldots a_n \wedge b_1 b_2 \ldots b_n = c_1 c_2 \ldots c_n$,

where $c_i = \min\{a_i, b_i\}$ $i = 1, 2, \ldots, n$ and

$$a_1 a_2 \ldots a_n \vee b_1 b_2 \ldots b_n = d_1 d_2 \ldots d_n,$$

where $d_i = \max\{a_i, b_i\}$ $(i = 1, 2, \ldots n)$.

It is easy to check that $(B^n, \leq)$ is a lattice. B^n has 2^n elements. $(B^3, \leq)$ can be looked upon as the set of vertices of a cube having an edge to be of unit length and having one vertex at $(0, 0, 0)$. B^n is referred to as a hyper cube.

It is possible to make a non-empty set L a lattice, once we introduce the operations of 'meet' and 'join' in L, provided the four laws from (8.6.6) to (8.6.9) hold for elements of L. When (8.6.6) to (8.6.9) are satisfied for a non-empty set L in which $\wedge$ and $\wedge$ are known; then, by defining a relation $\leq$ as

$$a \leq b \text{ if, and only if, } a \vee b = b,$$

we get $(L, \leq)$ as a lattice in which $l.u.b$ of a and b is $a \vee b$ and $g.l.b$ of a and b is $a \wedge b$. In fact, this alternative definition is analogous to that of rings where addition and multiplication are defined so as to satisfy certain axioms. Therefore, it is appropriate to recognize a lattice $(L, \leq)$ as a triple $(L, \vee, \wedge)$.

Definition 8.6.10 : *A subset L_1 of L is called a sublattice of $(L, \vee, \wedge)$, if $(L_1, \vee, \wedge)$ is itself a lattice obtained from the restrictions of $\vee$ and $\wedge$ to L_1.*

In other words, L_1 is a sublattice of L if the $l.u.b$ and $g.l.b$ of $a, b \in L_1$, are also in L_1, for all $a, b \in L_1$.

Definition 8.6.11 : *Let $(L, \vee, \wedge)$ be a lattice in which the partial order is denoted by $\leq$. $(L, \vee, \wedge)$ is called a modular lattice if for $a, b, c \in L$,*

$$a \leq c \Longrightarrow (a \vee b) \wedge c = a \vee (b \wedge c).$$

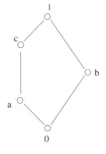

Figure 6

Not every lattice is modular. In the case of the pentagonal lattice given by Figure 6, one has $(a \vee b) \wedge c = 1 \wedge c = c$, whereas $a \vee (b \wedge c) = a \vee 0 = a$ and so the lattice shown is not modular.

Fact 8.6.1 : The normal subgroups of any group G form a modular lattice.

For proof see [5].

More generally, let M be an R-module. The lattice of R-submodules of M form a modular lattice.

If R is a non-commutative ring, we could consider the lattice of right or left-ideals of R. Fact 8.6.1 applied to rings says that the set of right or left ideals of R forms a modular lattice.

Definition 8.6.12 : *Given a lattice $(L, \vee, \wedge)$, a subset S of L is said to be a chain if, for $a, b \in S$ either $a \leq b$ or $b \leq a$. S is, in short, a totally ordered subset of L. If S is finite, S is said to be of finite length.*

Definition 8.6.13 : *Given a lattice $(L, \vee, \wedge)$, L is said to be of finite length, if*

$$\sup\{l : l \text{ is the length of a chain in } L\}$$

is finite. It is the l.u.b of lengths of chains in L.

A lattice $(L, \vee, \wedge)$ of finite length is defined up to isomorphism by its 'Covering relation': a covers b or $b < a$ if, and only if, a finite sequence $x_0, x_1, \ldots, x_n$ of elements of L exists such that

(8.6.10) $b = x_0$, $a = x_n$ and x_i covers x_{i-1} for $i = 1, 2, \ldots, n$.

A lattice $(L, \vee, \wedge)$ of finite length has a greatest and least element.

An alternate definition [5] of a modular lattice is as follows: We begin with the definition of an interval I in L.

Definition 8.6.14 : *Let $(L, \vee, \wedge)$ be a lattice. Suppose $a, b \in L$ are such that $a \leq b$. The subset*

$$[a, b] = \{x \in L : a \leq x \leq b\}$$

is called the interval defined by a and b.

Such an interval need not be a chain. But, $[a, b]$ is always a sublattice of L with greatest element b and least element a.

Lemma 8.6.1 : *[5] Let $I = [a, b]$ be an interval in a lattice $(L, \vee, \wedge)$. For all $x \in L$, the following inequality holds:*

(8.6.11) $(x \wedge b) \vee a \leq (x \vee a) \wedge b$

Proof : We define maps $\psi : L \to I$ and $\rho : L \to I$ by the equations

$$\left. \begin{array}{l} \psi(x) = (x \wedge b) \vee a \\ \rho(x) = (x \vee a) \wedge b \end{array} \right\}, x \in L$$

where $\psi(x)$, $\rho(x)$ belong to I.

When $x \in I$,

$$\psi(x) = (x \wedge b) \vee a = x \vee a = x, \quad \text{as } a \leq x \Longrightarrow a \vee x = x.$$
$$\rho(x) = (x \vee a) \wedge b = (a \vee x) \wedge b = x, \quad \text{since } x \wedge b = x, \text{ as } x \leq b.$$

Now, for $x \in L$,

$$(\psi \circ \rho)(x) = \psi(\rho(x)) = \psi(x), \text{ as } \rho(x) \in I.$$
$$(\psi \circ \psi)(x) = \psi(\psi(x)) = \psi(x) \text{ as } \psi(x) \in I.$$

We have

$$\psi \circ \rho = \psi^2 = \psi$$

and

$$\rho \circ \psi = \rho^2 = \rho.$$

Since $x \wedge b \leq (x \wedge b) \vee a = \psi(x) = x = \rho(x)$, $x \wedge b \leq \rho(x)$ and $a \leq \rho(x)$. It follows that

$$(x \wedge b) \vee a = \psi(x) \leq \rho(x)$$

Thus, (8.6.11) holds for all $x \in L$.   □

Remark 8.6.2 :
(i) If $(x \wedge b) \vee a = (x \vee a) \wedge b$ for all $x \in L$, $I = [a, b]$ is said to be modular.
(ii) Lemma 8.6.1 has been adapted from P. M. Cohn [5].

Definition 8.6.15 : *A lattice* $(L, \vee, \wedge)$ *in which all intervals are modular is called a modular lattice.*

In other words, $(L, \vee, \wedge)$ is modular if, and only if,

$$(c \vee a) \wedge b \leq (c \wedge b) \vee a \text{ for all } a, b, c \in L \text{ with } a \leq b.$$

In view of lemma 8.6.1, we conclude that a lattice $(L, \vee, \wedge)$ is modular if, and only if,

(8.6.12) $(c \vee a) \wedge b = (c \wedge b) \vee a$ for all $a, b, c \in L$ with $a \leq b$.

(8.6.12) is referred to as the 'modular law' for the lattice L.

Definition 8.6.16 : *In a lattice* $(L, \vee, \wedge)$ *with least element 0 and greatest element 1, two elements a and b are said to be complementary, if* $a \wedge b = 0$ *and* $a \vee b = 1$. *An element complementary to a is called the complement of a in L. Two elements which have a common complement b in L are called 'b-related' or simply related in L.*

Theorem 64 (P. M. Cohn (1965)) : *A lattice* $(L, \vee, \wedge)$ *is modular if, and only if, for each interval I of L, any two elements of I which are comparable and are related in I are equal.*

Proof : We know that $(L, \vee, \wedge)$ is modular if, and only if, definition 8.6.15 holds. So, $(L, \vee, \wedge)$ is not modular, if the inequality

(8.6.13) $(c \wedge b) \vee a \le (c \vee a) \wedge b$

is strict for at least one triple of values a, b, c such that $a \le b$. When $a = b$, the two sides of (8.6.13) are equal by absorption law: namely

$$a \wedge (a \vee c) = a \text{ and } a \vee (a \wedge c) = a.$$

Therefore let us take $a < b$.

$\Longleftarrow$: First assume that the inequality (8.6.13) is strict.
We write

$$a_1 = (c \wedge b) \vee a, \quad b_1 = (c \vee a) \wedge b$$

Then, $a \le a_1 < b_1 \le b$.

Now, $c \wedge b_1 = c \wedge ((c \vee a) \wedge b) = (c \wedge b) \wedge (c \vee a) \le (c \wedge b) \vee a = a_1$.

So, $c \wedge b_1 \le a_1$. Also $c \wedge b_1 \le c$.

That is,

(8.6.14) $c \wedge b_1 \le c \wedge a_1$.

Now, $a_1 < b_1$. So,

(8.6.15) $c \wedge a_1 \le c \wedge b_1$

From (8.6.14) and (8.6.15), we get

(8.6.16) $c \wedge a_1 = c \wedge b_1 = a_2 \text{ (say)}$

In the same manner, one has

(8.6.17) $c \vee a_1 = c \vee b_1 = b_2 \text{ (say)}$

So a_1 and b_1 are comparable and are related in $[a_2, b_2]$, although not equal. Thus, $(L, \vee, \wedge)$ is not modular $\Rightarrow$ inequality (8.6.13) is strict which, in turn, implies that any two elements of I which are comparable and are related in I are unequal.

So, the condition stated for $(L, \vee, \wedge)$ to be modular is sufficient.

Figure 7 shown below illustrates the proof.

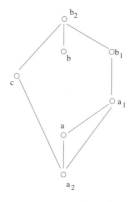

Figure 7

$:\Rightarrow$ Conversely, let a', b' be distinct elements of I which are comparable and related in $[a,b]$.
Suppose that

$$c \wedge a' = c \wedge b' = a, c \vee a' = c \vee b' = b$$

and $a \leq a' < b' \leq b$.
   Then, $(c \wedge b') \vee a' = a' < b' = (c \vee a') \wedge b'$
   That is,

$$(c \wedge b') \vee a' < (c \vee a') \wedge b'$$

So, the inequality (8.6.13) is strict and so, L is not modular.
So, the condition stated for $(L, \vee, \wedge)$ is necessary. $\square$

Remark 8.6.3 : The property stated in theorem 64 involves only five elements namely the end-points of the interval, the given element and its complements.

When we consider the sublattice formed by the five elements mentioned above, we obtain

Corollary 8.6.1 : *A lattice is modular if, and only if, it does not contain a sublattice isomorphic to the pentagonal lattice.*

Remark 8.6.4 : Theorem 64 has been adapted from P.M. Cohn [5].

Definition 8.6.17 : *A lattice $(L, \vee, \wedge)$ is said to be distributive, if it satisfies one of the following laws:*
   (i) $(a \vee b) \wedge c = (a \wedge c) \vee (b \wedge c)$, for all $a, b, c \in L$,
   (ii) $(a \wedge b) \vee c = (a \vee c) \wedge (b \vee c)$, for all $a, b, c \in L$.

Lemma 8.6.2 : *Every distributive lattice $(L, \vee, \wedge)$ is modular.*

Proof : Suppose $(L, \vee, \wedge)$ is distributive. Then, for $a, b, c \in L$,

(8.6.18) $(c \vee a) \wedge b = (c \wedge b) \vee (a \wedge b) \leq (c \wedge b) \vee a$

In particular, (8.6.18) is true whenever $a \leq b$ also. Thus, $(L, \vee, \wedge)$ is modular. $\square$

Fact 8.6.2 : A lattice $(L, \vee, \wedge)$ is distributive if, and only if, for each interval I of L, any two elements of I which are related in I are equal.

For proof, see [5, Proposition 4.5].

Lemma 8.6.3 : *Let $(L, \vee, \wedge)$ be a modular lattice and a, b any two elements in L. Suppose that $I = [a \wedge b, a]$, $J = [b, a \vee b]$; $a, b \in L$. Then, there exists a lattice-isomorphism between I and J.*

Proof : Given $x \in I$, $x \vee b \in J$.
   Also $a \wedge b \leq x \leq a$. So, $(x \vee b) \wedge a = x \vee (b \wedge a) = x$.

A dual statement is : if $y \in J$, $y \wedge a \in I$ and $(y \wedge a) \vee b = y$. So,

$$\xi : I \to J \text{ given by } \xi(x) = x \vee b$$

and

$$\eta : J \to I \text{ given by } \eta(y) = y \wedge a$$

are inverses of one another. Each is order-preserving and the lemma follows. $\square$

Definition 8.6.18 : *When maps ξ and η are defined as given above, the intervals I and J are said to be in perspective.*

More generally, two intervals, I and J are said to be projective, if there is a chain of perspectives from I to J. That is, there is a chain

$$I_0 = I, I_1, I_2, \ldots, I_n = J$$

of intervals such that I_{i-1}, and I_i are in perspective. $(i = 1, 2, \ldots, n)$. For instance, if a and b are related elements in a modular lattice $(L, \vee, \wedge)$ with 0 and 1, then the intervals $[0, a]$ and $[0, b]$ are projective. Since any two intervals in perspective are isomorphic, any two projective intervals are isomorphic. We are aiming at Jordan-Hölder theorem for lattices.

Definition 8.6.19 : *Let $(L, \vee, \wedge)$ be a lattice. Two chains in L given by*

(8.6.19) $e = a_0 \leq a_1 \leq \ldots \leq a_m = a$

(8.6.20) $e = b_0 \leq b_1 \leq \ldots \leq b_n = a$

between e and a are said to be isomorphic if $m = n$ and if there is a permutation π of $\{1, 2, \ldots, n\}$ such that the interval $[a_{i-1}, a_i]$ is isomorphic to $[b_{\pi(i-1)}, b_{\pi(i)}]$.

Any chain obtained from (8.6.19) by inserting more terms is called a refinement of (8.6.19).

Theorem 65 (P. M. Cohn(1965)) : *Let $(L, \vee, \wedge)$ be a modular lattice. Any two chains between the same two points in L have isomorphic refinements. More generally, the theorem holds in any lattice for two chains all of whose intervals are modular.*

Proof : Let

$$e = a_0 \leq a_1 \leq \cdots \leq a_m = a$$
$$e = b_0 \leq b_1 \leq \cdots \leq b_n = b$$

be two chains between the same elements e and $a(= b)$ of L.
We write

$$a_{i,j} = (a_i \wedge b_j) \vee b_{j-1}$$
$$b_{j,i} = (b_j \wedge a_i) \vee a_{i-1}$$

where $i = 1, 2, \ldots, m;\ j = 1, 2, \ldots, n$.
Then,

$$a_{i-1,j} \wedge (a_i \wedge b_j) = \left[(a_{i-1} \wedge b_j) \vee b_{j-1}\right] \wedge (a_i \wedge b_j)$$
$$= (a_{i-1} \vee b_{j-1}) \wedge b_j \wedge a_i$$

and

$$a_{i-1,j} \vee (a_i \wedge b_j) = \left[(a_{i-1} \wedge b_j) \vee b_{j-1}\right] \vee (a_i \wedge b_j)$$
$$= (a_i \wedge b_j) \vee b_{j-1}$$
$$= a_{i,j}$$

Now,

$$b_{j-1,i} \wedge (a_i \wedge b_j) = (a_{i-1} \vee b_{j-1}) \wedge b_j \wedge a_i$$

and

$$b_{j-1,i} \vee (a_i \wedge b_j) = b_{j,i}$$

We take $c = a_i \wedge b_j$, $d = a_{i-1,j}$. Then, by lemma 8.6.3, $[c \wedge d, c]$ and $J = [d, c \vee d]$ are isomorphic. But, then, $J = [a_{i-1,j}, a_{i,j}]$.
Taking $c = a_i \wedge b_j$, $d' = b_{j-1,i}$. We also have: $I' = [c \wedge d', c]$ and $J' = [d', c \vee d']$ are isomorphic. But, it is seen that $I = I'$. Thus, J and J' are isomorphic. That is, $[b_{j-1,i}, b_{j,i}]$ is isomorphic to $[a_{i-1,j}, a_{i,j}]$.
We get the chains

$$e = a_{0,1} \leq a_{1,1} \leq \cdots \leq a_{m,1} \leq a_{1,2} \leq \cdots \leq a_{m,2} \leq a_{1,3} \leq \cdots \leq a_{m,n} = a$$
$$e = b_{0,1} \leq b_{1,1} \leq \cdots \leq b_{n,1} \leq b_{1,2} \leq \cdots \leq b_{n,2} \leq b_{1,3} \leq \cdots \leq b_{n,m} = b$$

which refine (8.6.19) and (8.6.20) respectively.
They are isomorphic refinements of (8.6.19) and (8.6.20).   □

Theorem 66 (Jordan-Hölder theorem) : *If* $(L, \vee, \wedge)$ *is of finite length, any chain can be refined to a maximal chain and any two maximal chains between two given elements, say, a and b in L have the same length.*

Proof : By theorem 65, any two chains between a and b have isomorphic refinements and when a maximal chain is refined, its length is unaltered and so isomorphic maximal chains have the same length.   □

Remark 8.6.5 : Theorem 66 is the analogue of Jordan-Hölder theorem :

Any two composition series for a finite group have the same length k say, as shown below

$$G = G_0 \supset G_1 \supset G_2 \supset \cdots \supset G_k = (e)$$

and

$$G = G_0 \supset G_1' \supset G_2' \supset \cdots \supset G_k' = (e)$$

and the quotients

$$G/G_1, G_1/G_2, \cdots G_{k-1}/G_k$$
$$G/G_1', G_1'/G_2', \cdots, G_{k-1}'/G_k'$$

are factors which are simple groups and these are isomorphic in the sense that $G_i/G_{i+1} \cong G_j'/G_{j+1}'$ where $j = \pi(i)$ $(i, j = 1, 2, \ldots k)$ for a permutation π of the set $\{1, 2, \ldots, k\}$.

For proof see S. MacLane and G. Birkhoff [15, chapter XIV]. Also, see exercise 10, chapter 7.

8.7. Non-commutative rings

In the case of a commutative integral domain D, D is a UFD, if each irreducible element of D is a prime (see Proposition 2.4.2, chapter 2). An irreducible element in D is also referred to as an atom. D is called an atomic integral domain [6] if every nonzero non-unit of D is a finite product of atoms. It is easily checked that if D is an atomic integral domain, the ascending chain condition on principal ideals (ACCP) holds. Instead of primes, if one considers prime ideals, one has the following criterion for UFD property.

Fact 8.7.1 : An integral domain is a UFD if, and only if, every nonzero prime ideal in it contains a prime element.

For proof, see I. Kaplansky [13].

It follows that in a UFD, every minimal nonzero prime ideal is principal.

We will examine these in greater detail when we study Noetherian domains (see chapter 12).

We go to the non-commutative case.

Definition 8.7.1 : *An integral domain D which is not necessarily commutative is called an atomic integral domain if*

(1) D has no divisors of zero, either left or right.

(2) every nonzero non-unit $a \in D$ is such that there exists an irreducible element $q \in D$ dividing a either on the left or on the right.

In an atomic integral domain D, if $D^* = D \setminus \{0\}$, $a \in D^*$ has the representation.

(8.7.1) $a = a_1 a_2 \ldots a_r$ where a_i is an irreducible $(i = 1, 2, \ldots r)$

We consider a descending chain of right ideals from D to aD namely

(8.7.2) $D \supseteq a_1 D \supseteq a_1 a_2 D \supseteq \cdots \supseteq a_1 a_2 \cdots a_r D = aD.$

The corresponding quotient rings have the property:

$$a_1 D / a_1 a_2 D \cong D / a_2 D,$$
$$a_1 a_2 D / a_1 a_2 a_3 D \cong D / a_3 D,$$
$$\dots\dots\dots\dots\dots\dots\dots$$
$$a_1 a_2 \dots a_{r-1} D / a_1 a_2 \dots a_r D \cong D / a_r D.$$

So, we get a finite sequence of quotients

(8.7.3) $$D / a_1 D, D / a_2 D, \dots, D / a_r D.$$

If a has a second factorization into irreducibles of the form

$$a = b_1 b_2 \cdots b_s,$$

we could get a second sequence of quotients

(8.7.4) $$D / b_1 D, D / b_2 D, \dots, D / b_s D.$$

We say that (8.7.3) and (8.7.4) are isomorphic if $r = s$ and there is a permutation π of $\{1, 2, \dots, r\}$ such that $\pi(i) = j$, $1 \leq j \leq r$ and

$$D / a_i D \cong D / b_j D$$

Definition 8.7.2 : *An atomic integral domain D is called a general UFD if any two complete factorizations of a nonzero non-unit of D is a product of r atoms in two different ways and the corresponding quotients of right-ideals such as $a_i D$ and $b_j D$ are isomorphic.*

In the case of a commutative integral domain D, if a and $b \in D^*$ are such that $D / aD \cong D / bD$, then aD is the annihilator of D / aD and one will get $aD = bD$ and consequently a and b are associates. This is shown in

Theorem 67 : *Let D be a commutative integral domain in which a, b are nonzero non-units. Then a, b are associates if, and only if, $D / aD \cong D / bD$.*

Proof : $\Leftarrow$: Suppose that $D / aD \cong D / bD$.
There is an isomorphism $\psi : D / aD \to D / bD$ (which is onto) defined by $\psi(x + aD) = y + bD$; $x, y \in D$.
Then,

$$y_1 + bD = y_2 + bD \Longrightarrow y_1 - y_2 \in bD,$$
$$\Longrightarrow x_1 - x_2 \in aD,$$

where

$$\psi(x_1 + aD) = y_1 + bD,$$
$$\psi(x_2 + aD) = y_2 + bD.$$

In particular, there is a one-one correspondence between the elements of aD and bD. Since aD and bD have the same cardinality, $aD = bD$. This implies that a and b are associates.
$:\Rightarrow$ If a and b are associates, $aD = bD$ and so, $D / aD \cong D / bD$. $\square$

Now, considering left and right ideals of D say aD and Da one can talk about 'factorial duality' in D by saying that

(8.7.5) $$D/aD \cong D/bD \Leftrightarrow D/Da \cong D/Db.$$

We call (8.7.5) 'left-right symmetry'.

Analogous to the definition of associates in a commutative integral domain, we can introduce similarity between two nonzero divisors in a non-commutative integral domain.

Definition 8.7.3 : *Let a, a' be two nonzero divisors in a ring R. We say that a and a' are similar, if the matrices*

$$\begin{bmatrix} 1 & 0 \\ 0 & a \end{bmatrix} \text{ and } \begin{bmatrix} 1 & 0 \\ 0 & a' \end{bmatrix}$$

are similar in $M_2(R)$.

There are other equivalent statements possible. See P. M. Cohn [6].

An interesting example of a non-commutative UFD is the ring Q of integral quaternions. (A rational quaternion $a = a_0 + a_1 i + a_2 j + a_3 k$ is said to be integral if its coefficients a_0, a_1, a_2, a_3 are either integers or halves of odd integers).

Theorem 68 (P. M. Cohn) **:** *Let D be a non-commutative integral domain. D is a UFD whenever for each nonzero non-unit $a \in D$, the set $L(aD, D)$ of principal right ideals between D and aD forms a modular lattice of finite length, as a sublattice of the lattice of all right ideals of D.*

Proof : $L(D, aD)$ is a sublattice of the lattice of right ideals of D. $L(D, aD)$ is a modular lattice. When $L(D, aD)$ has a composition series, every chain in $L(D, aD)$ of intervals can be refined to a maximal chain and any two maximal chains between two given elements have the same length. The Jordan-Hölder theorem (theorem 66) says that when

$$a = a_1 a_2 \ldots a_r = b_1 b_2 \ldots b_s$$

are two different factorizations, $D/a_i D \cong D/b_j D$ and $i = s$ and so by definition (8.7.2) D is a UFD. $\square$

Remark 8.7.1 : We say that an integral domain D is a *PID*, if every right or left ideal of D is principal. In a PID, the principal right ideals between D and aD form a modular lattice. As every ideal is finitely generated, by a result on Noetherian rings, (see chapter 12) ascending chain condition and descending chain condition on principal ideals holds by the use of 'factorial duality'. This shows that a PID is a UFD in the non-commutative case as well. The ring of integral quaternions is a PID and so a UFD. More about unique factorization can be had from P. M. Cohn [6] and P. Samuel [21].

8.8. Boolean algebras

In Section 8.5, we have considered modular lattices and the Jordan-Hölder theorem. It is known that a distributive lattice is modular. For instance, if X is a non-empty set, $(\mathbb{P}(X), \subseteq)$ is a distributive lattice. $(\mathbb{N}, \leq)$ is a distributive lattice where $\leq$ means 'divides'. For $a, b \in \mathbb{N}$, $a \vee b$ is the l.c.m of a and b and $a \wedge b$ equals the g.c.d. of a and b.

Fact 8.8.1 : Let $(L, \vee, \wedge)$ be a distributive lattice and S an arbitrary non-empty set. The set L^S of functions $f : S \to L$ forms a distributive lattice.

Proof : Let $x \in S$. If $f : S \to L$, $f(x) \in L$. When $f, g \in L^S$, we define functions $J : S \to L$, $M : S \to L$ such that

$$J(x) = f(x) \vee g(x), \; M(x) = f(x) \wedge g(x), \text{ for all } x \in S.$$

Then, J, M defined by $J = f \vee g$ and $M = f \wedge g$ are in the poset $(L, \leq)$, where $\leq$ is obtained via $\vee$ and $\wedge$ in $(L, \vee, \wedge)$. So, L^S is a lattice. To prove that L^S is distributive, we proceed as follows:

For $f, g, h \in L^S$, we define

(8.8.1) $F = f \wedge (g \vee h)$ and $G = (f \wedge g) \vee (f \wedge h)$.

For $x \in S$, we have

$$F(x) = f(x) \wedge (g(x) \vee h(x))$$
$$= (f(x) \wedge g(x)) \vee (f(x) \wedge h(x)), \text{ by distributivity in } L,$$
$$= (f \wedge g)(x) \vee (f \wedge h)(x)$$
$$= G(x).$$

Thus, L^S is a distributive lattice.   □

Let $\tilde{\mathbb{Z}}$ denote the set of non-negative integers. For $r > 1$, suppose that $\Omega(r)$ denote the total number of prime factors of r. That is, if

$$r = \prod_{i=1}^{k} p_i^{a_i} (p_i \text{ primes such that } p_1 \leq p_2 \leq \cdots \leq p_k, a_i \geq 1, i = 1, 2, \ldots k).$$

$\Omega(r) = a_1 + a_2 + \cdots + a_k$. Further $\Omega(1) = 0$. It is easy to check that for $r, s \in \mathbb{N}$,

(8.8.2) $\Omega(rs) = \Omega(r) + \Omega(s)$.

$\Omega : \mathbb{N} \to \tilde{\mathbb{Z}}$ is a function onto $\tilde{\mathbb{Z}}$.

More generally, if

(8.8.3) $r = 2^{e(1)} \cdot 3^{e(2)} \cdot 5^{e(3)} \cdots p_k^{e(k)} \cdots (p_k, k^{\text{th}} \text{ prime})$

$e(j) \geq 0$ for $j = 1, 2, 3 \ldots$. The decomposition (8.8.3) assigns to each positive integer r a function $e : \mathbb{N} \to \tilde{\mathbb{Z}}$.

If $e : \mathbb{N} \to \tilde{\mathbb{Z}}$ and $f : \mathbb{N} \to \tilde{\mathbb{Z}}$ are two functions (e obtained from r and f obtained from $s = 2^{f(1)} \cdot 3^{f(2)} \ldots$), we take $e \wedge f$ to correspond with g.c.d (r, s) and $e \vee f$ to correspond with l.c.m (r, s). Clearly $e + f$ corresponds to $r \cdot s$. $\mathscr{L}$ denotes the set of functions $e : \mathbb{N} \to \tilde{\mathbb{Z}}$ which are such that $e(j) = 0$ except for a finite set of value of j. Let us agree that when r divides s, $e \le f$. Then, $(\mathscr{L}, \le)$ is a poset where '$\le$' in $\tilde{\mathbb{Z}}$ is induced by the partial order 'divides' in $\mathbb{N}$. $\tilde{\mathbb{Z}}$ is a distributive lattice. So, by Fact 8.8.1, $(\mathscr{L}, \lor, \land)$ is a distributive lattice.

Remark 8.8.1 : Fact 8.8.1 has been adapted from S. MacLane and G. Birkhoff [15].

Definition 8.8.1 : *Let $(L, \lor, \land)$ be a lattice having a least element 0 and a greatest element 1. By a complement of an element a in L, we mean an element y in L such that $a \land y = 0$ and $a \lor y = 1$. The lattice $(L, \lor, \land)$ is called a complemented lattice if all of its elements have complements.*

In $(\mathbb{P}(X), \subseteq)$, if A is in $\mathbb{P}(X)$, the complement A^c of A in X has the property $A \cap A^c = \emptyset$ and $A \cup A^c = X$. So $(\mathbb{P}(X), \cup, \cap)$ is a complemented lattice. We note that A^c is the only complement of A in $\mathbb{P}(X)$. The modular lattice of all subspaces of a finite dimensional vector space V (over a field F) is a complemented lattice. For, given a subspace W of V there exists an annihilator of W such that

$$\dim W + \dim(\operatorname{ann} W) = \dim V$$

and $W \cap \operatorname{ann} W = \{0\}$. See S. MacLane and G. Birkhoff [15].

Definition 8.8.2 : *A Boolean lattice is defined as a distributive complemented lattice.*

With the operations of join and meet suitably defined on a non-empty set L, one can make a partially ordered set $(L, \le)$ in which *l.u.b* $(a,b) = a \lor b$ and *g.l.b* $(a,b) = a \land b$, a lattice. The theory of a distributive complemented lattice was first applied by George Boole (1815–1864) to model structures in logic. For this reason, a lattice structure which is distributive and complemented is known as a Boolean algebra.

Definition 8.8.3 : *A Boolean algebra written $(B, \lor, \land)$ is such that one can intro-duce the operations '+' and '·' to replace '$\lor$' and '$\land$' symbols respectively. We write it as $(B, +, \cdot)$.*

Observation 8.8.1 : *The operations + and · in a Boolean algebra satisfy the following axioms.*

(i) $a+b = b+a$, and $a \cdot b = b \cdot a$, for a, $b \in B$,

(ii) $a+(b+c) = (a+b)+c$ and $a \cdot (b \cdot c) = (a \cdot b) \cdot c$ for, $a,b,c \in B$,

(iii) there exist elements 0 and 1 in B such that
$a+0 = a$, $a \cdot 0 = 0$; $a+1 = 1$ and $a \cdot 1 = a$, for all $a \in B$,

(iv) for $a \in B$, there exists a complement $\bar{a} \in B$ such that
$a+\bar{a} = 1$ and $a \cdot \bar{a} = 0$,

(v) for $a,b,c \in B$
$a \cdot (b+c) = (a \cdot b)+(a \cdot c)$; $a+b \cdot c = (a+b) \cdot (a+c)$,

(vi) for $a \in B$
$a+a = a$ and $a \cdot a = a$,

(vii) $a + a \cdot b = a(a+b) = a$, *for all* $a, b \in B$.

Remark 8.8.2 : A Boolean algebra $(B, +, \cdot)$ is not always a commutative ring. However, in an arbitrary ring $(R, +, \cdot)$ with unity, when every element of R is an idempotent, that is $a^2 = a$ for all $a \in R$, R is called a Boolean ring.

The 5-element lattice $(L, \vee, \wedge)$ shown below in figure 8, the element c has two complements. We note that

$$c \vee a = 1, \, c \wedge a = 0, \, c \vee b = 1, \, c \wedge b = 0.$$

$(L, \vee, \wedge)$ is a complemented lattice. But the complements of elements are not unique.

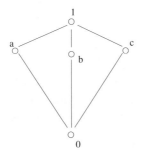

Figure 8

Fact 8.8.2 : In a distributive lattice $(L, \vee, \wedge)$, for $a, b, c \in L$, if

$$a \wedge b = a \wedge c \text{ and } a \vee b = a \vee c$$

then $b = c$. (law of cancellation).
For, as $(L, \vee, \wedge)$ is distributive,

$$b = b \wedge (a \vee b) = b \wedge (a \vee c) = (b \wedge a) \vee (b \wedge c) = (a \wedge b) \vee (b \wedge c)$$
$$= (a \wedge c) \vee (b \wedge c) = (a \vee b) \wedge c.$$

Or, $b = (a \vee b) \wedge c = (a \vee c) \wedge c$

$$= c.$$

On account of Fact 8.8.2, we see that in a Boolean algebra $(B, \vee, \wedge)$, the complement of an element is unique. That is, for each $a \in B$, $\bar{a}$ is unique. We recall that a bijection f between two lattices $(L_1, \cup, \cap)$ and $(L_2, \vee, \wedge)$ is an isomorphism, if it preserves least upper bound and greatest lower bound of a pair of elements either from L_1 or from L_2. Every lattice isomorphism is an isomorphism of the corresponding posets. It is known that if $(B, +, \cdot)$ is a finite Boolean algebra, it is a lattice isomorphic to $(\mathbb{P}(X), \cup, \cap)$ for an appropriate finite set X. It is also interesting to note that elements of a finite Boolean algebra can be written as a sum of atoms of B (elements that cover zero) in a unique way. In fact, we come across an analogue of the fundamental theorem of arithmetic in the context of a finite Boolean algebra. It is also possible to talk about uniqueness of factorization of elements of a lattice from a different point of view. See J. Martinez [16].

Lemma 8.8.1 : *Let $(B, +, \cdot)$ be a Boolean algebra. Suppose that a and b are distinct atoms of B. Then $a \cdot b = 0$.*

Proof : B is a poset under the corresponding partial order given by $a \leq b$ if, and only if, $a + b = b$, where $a, b \in B$.

Now, from axiom (vii) of a Boolean algebra (observation 8.8.1).

We have $a + a \cdot b = a$ and so, $a \cdot b \leq a$.

If $a \cdot b = a$, we have

$$a + b = a \cdot b + b = b + b \cdot a = b \text{ and so, } a \leq b.$$

Since a is an atom, $a > 0$. So, $b \geq a > 0$.

So $a = b$, a contradiction to the assumption that $a \neq b$.

So, $a \cdot b = a$ is impossible. So $a \cdot b < a$. As a is an atom, $a \cdot b = 0$. □

Theorem 69 : $(B, +, \cdot)$ *denotes a finite Boolean algebra. Every element a in B can be written as a sum of atoms. The atoms that appear in the representation are unique up to rearrangement.*

Proof : *Stage I:* We, first, show that a decomposition of $a \in B$, as a sum of atoms, exists.

If a is an atom, there is nothing to prove. If a is not an atom, there exists an atom $a_1 < a$ such that $a = a_1 + \bar{a}_1 \cdot a$.

For,

$$a_1 + \bar{a}_1 \cdot a = (a_1 + \bar{a}_1) \cdot (a_1 + a) = 1 \cdot (a_1 + a) = a_1 + a = a$$

If $\bar{a}_1 \cdot a$ is an atom, we are through. Otherwise, there exists an atom $a_2 \in B$ with $a_2 \neq a_1$ and $\bar{a}_1 \cdot a = a_2 + (\bar{a}_1 \cdot a) \cdot \bar{a}_2$. Then,

(8.8.4) $a = a_1 + a_2 + (\bar{a}_1 \cdot \bar{a}_2) \cdot a$

In (8.8.4) above, if $(\bar{a}_1 \cdot \bar{a}_2) \cdot a$ is an atom, the proof of the existence of such a decomposition is complete. Otherwise, we repeat the argument and get a decomposition of a in a finite number of steps as desired. Since B is finite, B has only a finite number of atoms. So, a decomposition of a as a sum of a finite number of atoms exists.

Stage II Uniqueness: Suppose a possesses two decompositions such as

(8.8.5) $a = a_1 + a_2 + \cdots + a_s = b_1 + b_2 + \cdots + b_t$

$(a_i, b_j$ are atoms, $i = 1$ to $s, j = 1$ to $t)$

We claim that $s = t$ and there exists a permutation σ of $\{1, 2, \ldots t\}$ such that $a_i = b_{\sigma(i)}$.

To prove that $t = s$, we proceed by induction on t.

If $t = 1$, $a = a_1 = b_1$.

Assume that the result is true for decompositions of a into a sum of less than t atoms. Since

$$b_1 \leq b_1 + b_2, \ldots + b_t = a = a_1 + a_2 + \cdots + a_s,$$

we have

$$b_1 = b_i \cdot a = b_1 \cdot (a_1 + a_2 \cdots + a_s) = b_1 \cdot a_1 + b_1 \cdot a_2 \cdots + b_1 \cdot a_s$$

Then, $b_1 \cdot a_i \leq b_1$ for $i = 1, 2, \ldots s$. By lemma 8.8.1, there exists a_k such that $b_1 \cdot a_k = b_1$ and $b_1 \cdot a_j = 0$ for all $j \neq k (j = 1, 2, \ldots s)$.

But, $b_1 \cdot a_k = b_1 \implies b_1 \leq a_k$ and so $b_1 = a_k$, since a_k is an atom.

Therefore,

$$a_k + (a_1 + a_2 + \cdots + a_{k-1} + a_{k+1} + \cdots + a_s) = b_1 + (b_2 + b_3 + \cdots + b_t)$$

Now, when $x + y = x + z$ and $x \cdot y = x \cdot z$ for $x, y, z \in (B, +, \cdot)$, then, $y = z$. This is true, as B is a distributive lattice (see Fact 8.8.2).

$$\text{Take } x = a_k = b_1, \quad y = (a_1 + a_2 \cdots + a_{k-1} + a_{k+1} + \cdots + a_s)$$
$$z = (b_2 + b_3 \cdots + b_t).$$

We have

$$x + y = x + z \text{ and } x \cdot y = a_k \cdot (a_1 + a_2 \cdots + a_{k-1} + \cdots + a_s)$$
$$x \cdot z = b_1 \cdot (b_2 + b_3 \cdots + b_t)$$

Since a_k and b_j are atoms

$$a_k \cdot (a_1 + \cdots + a_{k-1} + a_{k+1} + \cdots + a_s) = b_1 \cdot (b_2 + b_3 + \cdots + b_t) = 0$$

By cancellation,

$$a_1 + a_2 \cdots + a_{k-1} + a_{k+1} + \cdots + a_s = b_2 + b_3 + \cdots + b_t$$

By induction hypothesis $a_i = b_j$ for some i, j; $1 \leq i \leq k-1$, $k+1 \leq i \leq s, 2 \leq j \leq t$.

So the theorem holds for t whenever it holds for $(t-1)$, as $a_k = b_1$. So, the decomposition of a is unique up to rearrangement of terms. $\square$

Remark 8.8.3 : Theorem 69 has been adapted from David C. Buchthal and Douglas E. Cameron [3].

For more information about the theory of valuations see O. F. G. Schilling [A3], L. L. Dornhoff and F. E. Hohn [A1] for results on Lattices, Boolean algebras and their applications. See also Paulo Ribenboim [A2].

8.9. Notes with illustrative examples

There are methods of construction of real members starting from the field $\mathbb{Q}$ of rational members. An algebraic treatment due to Artin and Schreier [1] is via the notion of $\mathbb{Q}$-absolute values. This precedes the idea of a field with valuation. In the field $\mathbb{R}$ of real numbers, any finite sum of squares of nonzero elements is positive. That is, in $\mathbb{R}$, -1 is not a sum of squares. The notion of an ordered field specifies a subset P of positive elements. Thus, the idea of 'absolute values' is a natural outcome.

Let F be any field. Suppose $\mathbb{N}$ denotes the set of positive integers. The set of all maps from $\mathbb{N}$ into F is denoted by $F^{\mathbb{N}}$. Each map $a : \mathbb{N} \to F$ is a sequence with elements from F. We denote such a sequence by $\{a_n\}$, $a_n \in F$, $n \in \mathbb{N}$. By component-wise addition and multiplication, $F^{\mathbb{N}}$ is made a commutative ring. If $t \in F$, a scalar multiplication by $t \in F$ is possible to obtain $\{t a_n\}$ $(n = 1, 2, \ldots)$. $F^{\mathbb{N}}$ is thus a vector space over F. $F^{\mathbb{N}}$ has the structure of an 'F-algebra' which is commutative.

The valuation theory of fields originated from a number-theoretic source. It was Hensel who gave us p-adic numbers. In 1913, Kurshcha'k (1864–1933) introduced real-valued valuation of a field and showed that Hensel's field of p-adic numbers could be obtained as the 'completion' of $\mathbb{Q}$ relative to the p-adic valuation. (Von Neumann was one of his students). Analogous to the definition 8.4.1, we give below the definition of a $\mathbb{Q}$-valued field.

Definition 8.9.1 : *Let F be a field. A map $|| : F \to \mathbb{Q}$ is called a $\mathbb{Q}$-absolute value if, and only if, the following conditions hold:*

(i) $|a| \geq 0$ for all $a \in F$ with $|a| = 0$ when and only when $a = 0$,
(ii) $|a \cdot b| = |a||b|$ for all $a, b \in F$,
(iii) $|a+b| \leq |a| + |b|$ (triangle inequality) for all $a, b \in F$.

We call $(F, ||)$ a $\mathbb{Q}$-valued field.

If $|a| = 1$ for all $a \in F^*$ (the set of nonzero elements of F) and $|0_F| = 0$, then, $(F, ||)$ is a $\mathbb{Q}$-valued field and $||$ is the trivial $\mathbb{Q}$-absolute value on F. It is easy to check that as $||$ is multiplicative, when $0 \neq a \in F$ is such that $a^m = 1$ for some $m \in \mathbb{N}$, $|a| = 1$. That is, finite fields admit only the trivial $\mathbb{Q}$-absolute value. We can also talk about $\mathbb{Q}$-absolute values on the field $\mathbb{Q}$ itself. For instance, defining

$$|a|_\infty = \begin{cases} a & \text{if } a \text{ is positive}, a \in \mathbb{Q}, \\ -a & \text{if } a \text{ is negative}, a \in \mathbb{Q}, \\ 0, & \text{if } a = 0, \end{cases}$$

we see that $(\mathbb{Q}, ||_\infty)$ is a $\mathbb{Q}$-valued field.

Definition 8.9.2 : (a) *Let $(F, ||)$ be a $\mathbb{Q}$-valued field. If $\{a_n\} \in F^{\mathbb{N}}$ then $\{a_n\}$ is called a Cauchy sequence in F if, and only if, for each $n \in \mathbb{N}$, there exists $M(n) \in \mathbb{N}$ such that*

$$|a_i - a_j| < \frac{1}{n} \text{ whenever } i, j > M(n).$$

(b) $\{a_n\}$ *is called a null-sequence in F if, and only if, for each $n \in \mathbb{N}$, there exists $M(n) \in \mathbb{N}$ such that*

$$|a_i| < \frac{1}{n}, \text{ whenever } i > M(n)$$

The set of Cauchy sequences in $(F, ||)$ is denoted by $C(F)$. The set of null sequences in $(F, ||)$ is denoted by $N(F)$.

We remark that if $\{a_n\} \in C(F)$, there exists $B \in N$ such that $0 \leq |a_i| \leq B$ for all $i \in \mathbb{N}$. Further, if $(F, ||)$ is a $\mathbb{Q}$-valued field, then $\widetilde{F} = C(F)/N(F)$ is an extension of F under the natural identification of elements of F with the cosets determined by the constant sequences. The following equations are enough to justify the above statement:

(8.9.1) $(\{a_i\} + N(F))(\{b_i\} + N(F)) = \{a_i b_i\} + N(F).$

For multiplicative inverse, one has

(8.9.2) $(\{a\} + N(F))(\{b_i\} + N(F)) = \{1\} + N(F).$

$C(F)$ forms a ring under addition and component-wise multiplication. The set $N(F)$ of null sequences forms a maximal ideal of the ring of Cauchy sequences. One has the residue class field $\widetilde{F} = C(F)/N(F)$. $\widetilde{F}$ is a $\mathbb{Q}$-valued field under the extended $\mathbb{Q}$-absolute value. $\widetilde{F}$ is complete with respect to $\mathbb{Q}$-absolute value. That is, every Cauchy sequence in $(\widetilde{F}, ||)$ has a limit in $\widetilde{F}$.

We apply this to $(\mathbb{Q}, ||_\infty)$. Let $\{a_i\} \in C(\mathbb{Q})/N(\mathbb{Q})$. Then, there exist $B, b \in \mathbb{N}$ such that either for $i > B$, $a_i > \frac{1}{b}$, or, for $i > B$, $a_i < \frac{-1}{b}$. It means that for sufficiently large n either $a_n > \frac{1}{b}$ or $a_n < \frac{-1}{b}$. Only one of these is possible. This enables one to define positive and negative sequences in $C(\mathbb{Q})/N(\mathbb{Q})$, when $\mathbb{Q}$ is equipped with $||_\infty$ as the $\mathbb{Q}$-absolute value. That is,

$\{a_n\}$ is positive if, and only if, $\{-a_n\}$ is negative.
Constant sequences $\{a\}$ are positive if, and only if, $a > 0$ and $\{a\}$ is negative if, and only if, $a < 0$. We are now in a position to define the field $\mathbb{R}$ of real numbers.

Definition 8.9.3 : *Let $(\mathbb{Q}, ||_\infty)$ be the $\mathbb{Q}$-valued field considered under the absolute value $||_\infty$ given by*

$$|a|_\infty = a \text{ if } a \geq 0 \text{ and } |a|_\infty = -a, \text{ when } a < 0 \text{ for all } a \in \mathbb{Q}.$$

The extension $\widetilde{\mathbb{Q}} = C(\mathbb{Q})/N(\mathbb{Q})$ of $\mathbb{Q}$ is called the field of real numbers denoted by $\mathbb{R}$. An element $a \in \mathbb{R}$ is positive if, and only if, it is the coset of a positive sequence in $C(\mathbb{Q})$. $a \in \mathbb{R}$ is negative if, and only if, it is the coset of a negative sequence in $C(\mathbb{Q})$. Further, for $a, b \in \mathbb{R}$, we write $a > b$ to mean that $a - b > 0$. Also, '$<$' is a total order on $\mathbb{R}$ which extends the total order '$<$' in $\mathbb{Q}$.

$$\mathbb{R}^+ = \{a \in C(\mathbb{Q})/N(\mathbb{Q}) : a > 0\}$$

is the set of positive real numbers and it is closed under addition and multiplication.

Instead of $(\mathbb{Q}, ||\infty)$, we could consider the p-adic absolute value given in definition 8.4.3. If $m_p : \mathbb{Q} \to \tilde{\mathbb{R}}$ is replaced by $||_p : \mathbb{Q} \to \mathbb{Q}$, we get $(\mathbb{Q}.||_p)$ as a $\mathbb{Q}$-valued field. $\tilde{\mathbb{Q}}$ is obtained by considering Cauchy sequences and null sequences relative to $||_p$. The residue class field $\tilde{\mathbb{Q}}_p$ or simply $\mathbb{Q}_p$ is the field of (rational) p-adic numbers.

By lemma 8.3.1, $\mathbb{Z}_p$ is the set

$$\{x \in \mathbb{Q} : v_p(x) \geqslant 0\}.$$

It implies that $\mathbb{Z}_p = \{x \in \mathbb{Q} : 0 \leq |x|_p \leq 1\}$, where $|x|_p = p^{-v_p(x)}$. By definition 8.3.4, $v_p : \mathbb{R} \to \mathbb{Z} \cup \{\infty\}$ is a discrete valuation of $\mathbb{Q}$. Further, the only valuation rings (by definition 8.3.3) of the field $\mathbb{R}$ are the rings $\mathbb{Z}_p$ for every prime p. By theorem 58, $\mathbb{Z}$ is a PID and $\mathbb{Z}_p$ has exactly one nonzero prime ideal. $\mathbb{Z}_p$ is referred to as the ring of p-adic integers.

If $a \in \mathbb{Q}$, and $a \neq 0$, there are only finitely many primes p for which $|a|_p \neq 0$. As $\mathbb{R}$ is an *OAF*, for every prime p, the value–group of $||_p$ is $\mathbb{Z}$.

In the case of a field with a valuation map $m : F \Longrightarrow \tilde{\mathbb{R}}$, we can start with a prime divisor P of F. A sequence $\{a_n\}$ ($a_n \in F, n \in \mathbb{N}$) is called a P-Cauchy sequence in $\langle F, P \rangle$ if $\{a_n\}$ is Cauchy in the topology T_P specified by P. We say that F is P-complete, if every P-Cauchy sequence converges. The P-completion of $\langle F, P \rangle$ is denoted by $\langle \tilde{F}, \tilde{P} \rangle$. A field is always complete with respect to the trivial prime divisor. A theorem of Ostrowski (1893–1986) says that the only fields that are complete relative to a prime divisor P determined by an archimedean absolute-value map are $\mathbb{R}$ (the field of real numbers) and $\mathbb{C}$ (the field of complex numbers).

The idea of ordering of positive integers via 'divisibility' and the usual '$\leq$' relation happened to be in use, frequently, long back. However, the notion of a 'partial order relation' on a set came much later while introducing algebraic systems. Inclusion relations between sets led to the concepts of partial order relations and to lattices in which analogues of 'set-intersection' and 'set-union' are defined. The role of g.c.d and l.c.m of numbers also got exploited. Postulates for an algebraic system were declared or set down for specific structures such as groups, rings and fields. In the case of a Boolean algebra, there exist several independent sets of postulates. In his classic paper of 1904, E. V. Huntington [11] proclaimed an independent set of postulates defining a Boolean algebra. The standard definition of a Boolean algebra is along the lines suggested by Huntington. As is often remarked: 'Huntington list of postulates is mathematically elegant'. It will be an omission if we do not mention Marshal Stone's outstanding 75-page paper (1936) on the theory of representations of Boolean algebras [22]. The Stone Representation Theorem, as it is called, says that any Boolean ring R is isomorphic to a ring of subsets of some fixed set X. See D. M. Burton [4, Chapter 9].

Now, suppose that $(X, \leq)$ is a poset. A relation $\geq$ on X defined by $y \geq x$ if, and only if, $x \leq y$ $(x, y \in X)$ is called the converse of $\leq$ and $(X, \geq)$ is known as the dual of the poset $(X, \leq)$. That is, the converse of a partial ordering is a partial ordering. The atoms of $(X, \geq)$ are the coatoms of $(X, \leq)$. A result shown in respect of $(X, \leq)$ can be dualized by changing $\leq$ into $\geq$ to obtain a similar result for $(X, \geq)$. In the case of lattices, *l.u.b* and *g.l.b* are interchanged and $\leq$ replaced by $\geq$ to obtain dual statements. This phenomenon is called the principle of duality. The principle of duality holds in every Boolean algebra. The dual of theorem 69 is the following:

Any element in a finite Boolean algebra can be expressed uniquely as a product of coatoms. The proof is by the principle of duality.

8.10. Worked-out examples

a) In an ordered field F, the prime subfield is isomorphic to $\mathbb{Q}$, the field of rational numbers. Prove:

Answer: We recall that an ordered integral domain is of characteristic zero. So, an ordered field F is of characteristic zero. Any subfield of F must contain the multiplicative identity $1_F \in F$. Further, the prime subfield K of F is the intersection of all subfields of F. The prime subfield K is the subfield generated by 1_F.

The one-to-one order-preserving homomorphism $\psi : \mathbb{Z} \to F$ can be extended to a unique homomorphism $\psi' : \mathbb{Q} \to F$ of fields. The image of ψ' is the prime subfield of F.

This completes the proof. $\square$

8.10.1. ARCHIMEDEAN LAW.

Lemma 8.10.1 : *Given $a > 0$, $b > 0$ in $\mathbb{R}$, the field of real numbers. There exists $n \in \mathbb{N}$ such that $na > b$.*

Proof : Assume that the conclusion is false for two particular real numbers a and b so that for every n, $b \geq na$. Let

$$S = \{ x \in \mathbb{R} : x = na \text{ for all } n \in \mathbb{N} \}.$$

S has an upper bound b. So, S has a *l.u.b* say b'. Therefore, $b' \geq na$ for all $n \in \mathbb{N}$. It follows that

$$b' \geq (m+1)a, \text{ for all } m \in \mathbb{N}.$$

But then, $b' - a \geq ma$, for all $m \in \mathbb{N}$. This contradicts the minimality of b'. Lemma 8.10.1 follows. $\square$

Remark 8.10.1 : $\mathbb{R}$ is a complete ordered field in which archimedean law holds.

b) Give an example of an ordered field in which archimedean law does not hold.

Answer: Let $t > 0$ be an arbitrary, but fixed element of $\mathbb{R}$. We consider $\mathbb{Q}[[t]]$, the field of formal power series with coefficients from $\mathbb{Q}$. That is,

$$f(t) = a_0 + a_1 t + a_2 t^2 + \cdots \qquad \text{where } a_i \in \mathbb{Q}, i = 0, 1, 2, \ldots$$

belongs to $\mathbb{Q}[[t]]$ and is invertible when $f(t) \neq 0$. $f(t) \in \mathbb{Q}[[t]]$ is called positive whenever its first nonzero coefficient is positive. That is, $\sum_{i=0}^{\infty} a_i t^i > 0$ implies that for some k, $a_0 = a_1 = \cdots = a_{k-1} = 0$ and $a_k > 0$. So, $\mathbb{Q}[[t]]$ is an ordered field. For every $n \in \mathbb{N}$, $g(t) = 1 - nt > 0$.

$$1, 1/2, 1/3, \ldots \quad \text{are elements of } \mathbb{Q}[[t]].$$

$\lim_{n \to \infty} \frac{1}{n} \neq 0$, as $1 - nt > 0 \Rightarrow |\frac{1}{n}| > t$. So, there is no $n \in \mathbb{N}$ for which $|\frac{1}{n}| < t$, yet $t > 0$. Thus, $\mathbb{Q}[[t]]$ is not a complete ordered field. Moreover, given $t > 0$ and $1 > 0$, there does not exist $n \in \mathbb{N}$ such that $nt > 1$, since for every $n \in \mathbb{N}$, $nt < 1$. In other words, $\mathbb{Q}[[t]]$ is an incomplete ordered field in which archimedean law does not hold.

Remark 8.10.2 : The above example has been drawn from S. MacLane and G. Birkhoff [15].

EXERCISES

1. *Mark the following statements true (T) or false (F) justifying your answer briefly.*
   a) *Let D be an ordered integral domain with positive subset P. In D[x], we write*
   $$(a_0 + a_1 x + \cdots + a_n x^n) \in P^*, a_n \neq 0$$
   if, and only if, $a_n \in P$ in D. Then, D[x] with positive set P^* is an ordered integral domain.
   b) *$\mathbb{Z}[\sqrt{3}] = \{a + b\sqrt{3} : a, b \in \mathbb{Z}\}$ can be made an ordered integral domain, by defining the set P of positive elements.*
   c) *Let $n = p_1^{a_1}, p_2^{a_2}, \ldots, p_k^{a_k}(p_1 < p_2 \cdots < p_k)$ (p_i, primes, $i = 1, 2, \ldots, k$) $D(n)$, the set of divisors of n partially ordered by divisibility forms a distributive lattice $L_1 \times L_2 \times \cdots L_k$ where L_i is a chain of length $a_i (i = 1, 2, \ldots, k)$.*
   d) *There exists a modular lattice of 7 elements in which the complemented elements do not form a sublattice.*
   e) *Any Boolean algebra generated by n elements has more than 2^{2^n} elements.*
   f) *The modular lattice of all subspaces of a finite dimensional vector space V is also a complemented lattice.*
2. *An ordered integral domain D is said to be complete, if every non-empty set of positive elements of D has a g.l.b in D. Prove that the set $\mathbb{R}$ of real numbers is a complete ordered integral domain.*
3. *[S. Lang]*
   (a) *Starting from an absolute value map $m : \mathbb{Q} \to \tilde{\mathbb{R}}$, describe the archimedean and non-archimedean prime divisors of Q. (the field of rational numbers).*

(b) Let $\mathbb{Q}_p$ denote the field of p-adic numbers where p is a rational prime. Show that $\mathbb{Q}_p$ contains infinitely many fields of the type $\mathbb{Q}(\sqrt{-m})$, where m is a positive integer.

4. Prove that a lattice $(L, \vee, \wedge)$ is distributive, if the equations

$$x \vee y = x \vee z \text{ and } x \wedge y = x \wedge z \Longrightarrow y = z.$$

5. Suppose that we consider the lattice of normal subgroups of a group G. Show that it need not be a modular lattice. Check this for the lattice of normal subgroups of the Klein 4-group

$$V = \{e, a, b, c\} \text{ with } ab = bc = ca, \quad a^2 = b^2 = c^2 = e.$$

6. Let $F = \mathbb{Q}[x]/(x^2 + r)$, $r > 0$. Show that F can be ordered in two different ways. (One has to consider the monomorphisms of F into $\mathbb{C}$).

7. Let $(B, +, \cdot)$ be a ring containing 0 and 1 and for $a \in B$, the complement of a denoted by $\bar{a} = a + 1$ and

$$a + b = a \cdot \bar{b} + \bar{a} \cdot b, \quad a \cdot 1 = a, \quad a \cdot \bar{a} = a + a = 0$$

Check whether B is a Boolean algebra.

8. Let $(B, +, \cdot)$ be a Boolean algebra. For $a, b \in B$, show that $a = a \cdot b$ if, and only if, $a \cdot \bar{b} = 0$ where $\bar{b}$ is the complement of b in B.

9. Suppose that L_1 and L_2 are lattices which are isomorphic as posets. Show that L_1 and L_2 are isomorphic as lattices.

10. Give an example of a lattice $(L, \vee, \wedge)$ with a subset L_1 such that $(L_1, \vee, \wedge)$ is a lattice, but not a sublattice of L.

11. Prove that the positive divisors of a positive integer n form a complemented lattice if, and only if, n is square-free.

REFERENCES

[1] E. Artin and O. Schreier, Algebraische konstuktion reele Korper, Abhandhingen Math. Seminar Hamburgischen Universitat, 5, 85–99, (1927).

[2] Nicholas Bourbaki: Elements of the History of Mathematics (translated from French by John Meldrum) Springer Verlag, Berlin, Heidelberg (1994).

[3] David C. Buchthal and Douglas E. Cameron, Modern Abstract Algebra Chapter 23 pp 478–501, Prindle, Weber and Schmidt, Boston, USA, (1987).

[4] David M. Burton, A first course in rings and ideals, Chapter 9, Addison Wesley Pub Co., Reading Mass, 180–203, (1970).

[5] P. M. Cohn, Universal Algebra Chapter II, pp 63–73, Harper and Row, Publishers, Inc. New York, (1965).

[6] P. M. Cohn, Unique Factorization Domains, Amer. Math. Monthly, 80, 1–18, (1973).

[7] K. von Fritz: The discovery of incommensurability by Hyppasus of Metapontium, Annals of Math (2) Vol. XLXI (1945) pp. 244–264.

[8] Robert Godement, Algebra, Chapter 8, pp 160–167, Hermann Paris Houghton Mifflin Co., Boston USA, (1968).

[9] S. W. Golomb, Normed Division Domains, Amer. Math. Monthly, 88, (1981), 680–686.

[10] Charles R. Hadlock: Field Theory and its Classical Problems, Chapter 3, The Carus Math Monographs No.19, Math. Asso. of America, (1978), 123–179.

[11] E.V. Huntington, Sets of independent postulates for the algebra of logic, Trans. Amer. Math. Soc., 5, (1904), 288–309, See also do 35 (1933) 274–304, 557–558 & 571.

[12] Nathan Jacobson, Basic Algebra Vol 2, Chapter 9, pp 537–546, Hindustan Pub Corporation, Delhi, (1984).

[13] Kaplansky, Commutative rings, Allyn and Bacon Inc, Boston, 1970.

[14] S. Lang, Algebra, Chap XI, Addison Wesley Pub Co Inc, Reading Mass, USA, 271–282, (1965).

[15] S. Maclane and G. Birkhoff, Algebra Chapter XIV, The Macmillan Co., Collier Macmillan Ltd., London, (1967), 482–501.

[16] J. Martinez, Unique factorization in partially ordered sets, Proc. Amer. Math. Soc., 33, No.2, (1972), 218–220.

[17] T. Nagell, Introduction to Number Theory, Chapter V, Chelsea Pub. Co. Reprint, New York, (1981), 156–187.

[18] Paromita Chowla, On the representation of -1 as a sum of squares in a cyclotomic field, J. Number Theory, 1, (1969), 208–210.

[19] Don Redmond, Number Theory-An Introduction Chapter 6, Monographs and Textbooks in Pure and Applied Mathematics No:201, Marcel Dekker Inc, New York, (1996), 301–307.

[20] Chih-Han Sah, Abstract Algebra Chapter IX, 320–333, Academic Press, New York and London, (1967).

[21] P. Samuel, Unique factorization, Amer. Math. Monthly, 75, (1968), 945–952.

[22] M. H. Stone, The theory of representations of Boolean algebras, Trans. Amer. Math. Soc., 40, (1936), 37–111.

[23] Edwin Weiss, Algebraic Number Theory, Chelsea Pub. Co, New York, (1963), Chapter 1, pp 1–40 and Chapter 4 pp 119–122.

ADDITIONAL REFERENCES

[A1] L. L. Dornhoff and F. E. Hohn, Applied Modern Algebra, Chapter 6, 265–299, Macmillan Pub. Company and Collier Macmillan Publishers, London, (1978).

[A2] Paulo Ribenboim : The theory of classical valuations: Springer monographs in Mathematics Springer Verlag (1999) N. Y.

[A3] O. F. G. Schilling, The theory of valuations, Chapters 1 and 2 pp 1–59 Mathematical Surveys and Monographs, No:4, Amer Math Soc, Providence, RI, Fifth Printing, (1989).

CHAPTER 9

The role of the Möbius function— Abstract Möbius inversion

Historical perspective

Counting techniques and solving mathematical puzzles were part of the 'recreational hobby' of ancient mathematicians. Magic squares such as

4	9	2
3	5	7
8	1	6

(wherein numbers from 1 to 9 are used to get a total of 15, adding numbers horizontally, vertically or diagonally) were known to the Chinese (2000–1000 B.C.). The idea of permutations and combinations was also known in crude form. Significant developments in the study of combinatorial mathematics were noticed in the work of Isaac Newton. With the discovery of sets by George Cantor (1845– 1918), enumeration problems could be stated precisely and solutions found. 'Inversion' of a finite series proved to be useful. The classical inclusion-exclusion principle found a stronghold in the theory of probability and related topics. The natural 'ordering' of objects, for instance, 'divisibility ordering' of positive integers gave the clue to the formula of Möbius inversion via a function μ defined on the set of positive integers. The use of arithmetic functions like μ and the Euler totient ϕ found applications, ever since Euler codified the study of elementary number theory from the scribblings of Fermat. However, it was A. F. Möbius (1790–1868) who first investigated the properties of μ systematically. And that gave μ the name Möbius function. See Hardy and Wright [6]. The principle of inclusion-exclusion was investigated by several 19th century mathematicians and stated most clearly by Poincare' (1854–1912). It has been rediscovered many a time in different contexts with varying degrees of generalization. A good and complete account of the principle together with a history and development of classical applications in the theory of probability may be found in Frechet [4].

Formula for Möbuis inversion was first obtained by Weisner in [17]. It was also noticed by Philip Hall in [5] independently. Both of them were motivated by problems in finite groups. See also Morgan Ward [15] and Weigandt [18]. In 1964, Gian Carlo Rota [10] gave the theory of Möbius functions via locally finite partially ordered sets and emphasized its applications to combinatorial mathematics. This has had a great impact on the study of combinatorics, subsequently. The introduction of a Möbius function via a partially ordered set was also carried out independently by Harold Scheid [11] in 1968. It is no exaggeration to say that many results involving enumeration have a bearing on the theory of Möbius functions.

9.1. Introduction

This chapter introduces the notion of an 'incidence function' via a locally finite partially ordered set. See definitions (9.2.1) and (9.2.2) below. An incidence function is a generalized arithmetic function. The results of classical arithmetic functions are carried over to this general set-up. Naturally, the ζ-function and its inverse the Möbius function have their analogues among incidence functions. The Möbius function depends on the partial order considered for locally finite posets.

Section 9.2 is about the properties of the generalized Möbius function and abstract Möbius inversion given in theorem 70. The generalization to $n \times n$ matrices is also studied briefly. See Section 9.3. In Section 9.4, we consider an n-dimensional vector space $V_n(q)$ over a finite field $GF(q)$. The lattice $L(V_n(q))$ of subspaces of $V_n(q)$ with partial order 'a subspace of' is shown to be isomorphic to a 'geometric lattice' which is the lattice of projective subspaces of a projective space $P(V)$. See definition 9.4.13. The Möbius function of $L(V_n(q))$ is obtained in theorem 72.

9.2. Abstract Möbius inversion

In [10], G. C. Rota gives a detailed account of the theory of Möbius inversion via locally finite partially ordered sets. A good many applications have been found based on this generalization. See E. A. Bender and J. R. Goldman [2].

Let $(P, \leq)$ be a poset.

Definition 9.2.1 : *For, $x, y \in P$, by a segment $[x, y]$ in $(P, \leq)$ we mean*

$$[x, y] = \{t : x \leq t \text{ and } t \leq y\}.$$
$$Also, (x, y] = \{t : x < t \text{ and } t \leq y\},$$
$$[x, y) = \{t : x \leq t \text{ and } t < y\},$$
$$(x, y) = \{t : x < t < y\}.$$

A segment $[x, y]$ is not necessarily totally ordered. In $(\mathbb{N}, \leq)$ the segment $[2, 8] = \{t : 2 \leq t \leq 8\} = \{2, 3, 4, 5, 6, 7, 8\}$. If the partial order is 'divides',

$$[1, 6] = \{t : 1 | t \text{ and } t | 6\} = \{1, 2, 3, 6\}$$

2 does not divide 3 and so $[1, 6]$ is not totally ordered.

Definition 9.2.2 : *A poset $(P, \leq)$ is called a locally finite poset, if every segment $[x, y]$ of $(P, \leq)$ is finite.*

For instance, $(\mathbb{N}, \leq)$ where $\leq$ is to mean 'divides' is locally finite. In $(\mathbb{R}, \leq)$, where $\leq$ is to mean 'less than or equal to', a segment $[x, y]$ where $x \neq y$ is not finite.

Definition 9.2.3 : *Let R be a commutative ring with unity 1_R. A function $f : P \times P \to R$ is called an incidence function, if*

$$f(x, y) = 0, \ unless \ x \leq y.$$

For two incidence functions f, g, we define multiplication by

$$(9.2.1) \qquad (f \cdot g)(x, y) = \begin{cases} \sum_{x \leq t \leq y} f(x, t) g(t, y), & if \ x \leq y, \\ 0, & otherwise. \end{cases}$$

Clearly, $(f \cdot g)$ is an incidence function.

Let A_P be the set of incidence functions defined on $P \times P$.

For $f, g \in A_P$ their sum $f + g$ is given by

$$(9.2.2) \qquad (f + g)(x, y) = f(x, y) + g(x, y)$$

Lemma 9.2.1 : *The set A_P of incidence functions defined with respect to a locally finite poset P forms a commutative ring with unity element e_0 given by*

$$e_0(x, y) = \begin{cases} 1_R, & x = y \\ 0 & otherwise \end{cases}$$

Proof : Multiplication (9.2.1) is commutative and associative and addition (9.2.2) distributes multiplication. Further if $f \in A_P$,

$$f \cdot e_0 = e_0 \cdot f = f$$

So $(A_P, +, \cdot)$ is a commutative ring with unity e_0. □

Definition 9.2.4 : $(A_P, +, \cdot)$ *is called the incidence ring with unity e_0.*

Remark 9.2.1 : A_P can be considered as a vector space over $\mathbb{C}$, if

$$A_P = \{ f : f : P \times P \to \mathbb{C}, \text{ the field of complex numbers} \}.$$

Together with multiplication (9.2.1), A_P is a 'Dirichlet algebra' over $\mathbb{C}$.

Definition 9.2.5 : *The zeta-function of A_P is defined by*

$$(9.2.3) \qquad \zeta(x, y) = \begin{cases} 1_R, & if \ x \leq y \\ 0, & otherwise \end{cases}$$

Definition 9.2.6 : *The Möbius function μ of A_P is defined in two ways:*

$$(9.2.4) \qquad \mu(x, y) = \begin{cases} 1_R, & if \ x = y, \\ -\sum_{x \leq t < y} \mu(t, y), & if \ x < y, \\ 0_R, & otherwise. \end{cases}$$

$$(9.2.5) \qquad \mu_1(x,y) = \begin{cases} 1_R, & \text{if } x = y, \\ -\sum_{x < t \leq y} \mu(x,t), & \text{if } x < y, \\ 0_R, & \text{otherwise.} \end{cases}$$

The function μ_1 (9.2.5) is introduced only temporarily. We wish to show that $\mu_1 = \mu$.

Lemma 9.2.2 : *For ζ, μ and μ_1 as defined in (9.2.3), (9.2.4) and (9.2.5) respectively, the following equations hold*

$$\mu \cdot \zeta = e_0 \text{ and } \zeta \cdot \mu_1 = e_0$$

Proof : If x is not less than or equal to y, μ, ζ and e_0 have value zero.
For $x = y$,

$$(\mu \cdot \zeta)(x,x) = \mu(x,x)\zeta(x,x) = 1_R \cdot 1_R = e_0(x,x)$$

If $x < y$,

$$(\mu \cdot \zeta)(x,y) = \sum_{x \leq t \leq y} \mu(x,t)\zeta(t,y) = \sum_{x \leq t \leq y} \mu(x,t) = 0 = e_0(x,y).$$

So, $\mu \cdot \zeta = e_0$. Similarly, it is easy to note that $\zeta \cdot \mu_1 = e_0$. □

Lemma 9.2.3 : $\mu = \mu_1$

Proof :

$$\begin{aligned} \mu = \mu \cdot e_0 &= \mu \cdot (\zeta \cdot \mu_1), \text{ by } (9.2.2), \\ &= (\mu \cdot \zeta) \cdot \mu_1, \text{ by associativity,} \\ &= e_0 \cdot \mu_1 = \mu_1. \end{aligned}$$

This proves Lemma 9.2.3. □

Remark 9.2.2 : For $x \leq y$,

$$\sum_{x \leq t \leq y} \mu(x,t) = \sum_{x \leq t \leq y} \mu(t,y) = e_0(x,y)$$

Definition 9.2.7 : *$(P, \leq)$ is said to be left-finite if, for all $y \in P$, the set*

$$T = \{x : x \leq y\} \text{ is finite.}$$

As an example, $(\mathbb{N}, \leq)$ is left-finite where $\leq$ means 'less than or equal to' or 'divides':
We observe that if P has a least element, then it is left-finite.

Theorem 70 (Robert Spira (1972)) **:** *Let $f : P \to R$ and $g : P \to R$ be two functions. Suppose that P is left-finite. Then,*

$$(9.2.6) \qquad g(y) = \sum_{x \leq y} f(x) \Leftrightarrow f(y) = \sum_{x \leq y} g(x)\mu(x,y)$$

Proof : $:\Rightarrow$ Since P is left-finite, $\sum_{x \leq y} f(x)$ is a well-defined finite sum and so $g : P \to R$ is well-defined.

$$\sum_{x \leq y} g(x)\mu(x,y) = \sum_{x \leq y}(\sum_{t \leq x} f(t))\mu(x,y)$$

$$= \sum_{x \leq y}(\sum_{t \leq x} f(t)\zeta(t,x)\mu(x,y)$$

$$= \sum_{t \leq y} f(t) \sum_{x \leq y} \zeta(t,x)\mu(x,y)$$

$$= \sum_{t \leq y} f(t) \sum_{t \leq x \leq y} \zeta(t,x)\mu(x,y)$$

$$= \sum_{t \leq y} f(t)e_0(t,y), \text{ by lemmas 9.2.2 and 9.2.3.}$$

So, $\sum_{x \leq y} g(x)\mu(x,y) = f(y)$, as $e_0(t,y) = 0$ if $t \neq y$.

The reverse implication is proved easily and so, the details are omitted. $\qquad \square$

Remark 9.2.3 : Theorem 70 is adapted from [13].

Illustration 9.2.1 (Möbius inversion of number theory) : $(\mathbb{N}, \leq)$ *is a locally finite partially ordered set where $\leq$ means 'divides'. The Möbius function associated with $(\mathbb{N}, \leq)$ is given by*

(9.2.7) $$\mu(m,n) = \mu(1, \frac{m}{n})$$

For, let $n = km$. When $k = 1$, $\mu(m,m) = 1 = \mu(1,1)$.

We prove (9.2.7) by induction on k. Assume that it is true for $q \leq (k-1)$, where $k > 1$. Since m divides n properly,

$$\mu(m,n) = - \sum_{\substack{d|n \\ m|d, d \neq n}} \mu(m,d),$$

or,

$$\mu(m,n) = -\sum_{\substack{j|k \\ j \neq k}} \mu(m,jm)$$

Now, $\mu(m, jm) = \mu(1, j)$ by induction hypothesis for $j|k$, $j \neq k$.
     So,

$$\mu(m,jm) = -\sum_{\substack{j|k \\ j \neq k}} \mu(1,j) = \mu(1,k) = \mu(1, \frac{n}{m}) \text{ as in (9.2.7).}$$

By the definition of $\mu(1,n)$,

$$\mu(1,n) = \mu(n) = \begin{cases} 1, & \text{if } n = 1 \\ (-1)^r, & \text{if } n = p_1 p_2 \cdots p_r, \quad p_i \text{ distinct primes}, i = 1, 2, \ldots r, \\ 0, & \text{if } a^2 | n, a > 1. \end{cases}$$

As $(\mathbb{N}, \leq)$ is left-finite, theorem 70 yields

$$(9.2.8) \qquad g(n) = \sum_{d|n} f(d) \Leftrightarrow f(n) = \sum_{d|n} g(d)\mu(d,n) = \sum_{d|n} g(d)\mu(\frac{n}{d}),$$

as $\mu(d,n) = \mu(1, \frac{n}{d}) = \mu(\frac{n}{d})$ for d dividing n.
(9.2.8) is the classical Möbius inversion formula [6].
For more applications, see Robert Spira [13]. See, also, H. Scheid [10].

The generalization of μ via locally finite partially ordered sets tells us that there is a Möbius function associated with any locally finite poset. μ takes on different robes for different partial order relations.

Suppose $(P, \leq)$ is totally ordered and locally finite. That is, any two elements of P are comparable. That is, for $x, y \in P$; either $x \leq y$ or $y \leq x$.

If $x < y$, the segment $[x,y]$ is given by

$$\{x = t_0 < t_1 < t_2 < \ldots t_n = y\}$$

If y covers x there is no t_i $(i = 1, 2 \ldots n-1)$ such that $x < t_i < y$. If $x < y$ and y does not cover x, we get a chain as above. We could assume that between t_{i-1} and t_i, there is no element of P $(i = 1, 2, \ldots n)$. In other words, t_i covers t_{i-1}.
Then,

$$\mu(t_i, t_{i+1}) = - \sum_{t_i \leq t < t_{i+1}} \mu(t_i, t) = -\mu(t_i, t_i) = -1$$

For $n \geq 2$, one has

$$(9.2.9) \qquad \begin{cases} \mu(x,x) = 1 \\ \mu(x,t_1) = -1 \\ \mu(x,t_i) = 0, \quad i \geq 2. \end{cases}$$

Then, $\mu(x,y) = -\sum_{x \leq t < y} \mu(x,t) = -1 + 1 + 0 = 0$ for $2 \leq i \leq (n-2)$.

$$\mu(t_{n-1}, y) = -1,$$
$$\mu(t_i, y) = 0, \text{ for } 0 \leq i \leq (n-2).$$

The above method of computation could be applied in all situations, provided the poset is locally finite.

Definition 9.2.8 : *Let $\Sigma = (P, \leq)$ and $\Sigma' = (P', \leq')$ be two posets. We say that Σ is isomorphic to Σ' if, and only if, there is a bijection $\phi : P \to P'$ such that $\phi(x) \leq \phi(y)$ in $(P', \leq')$ whenever $x \leq y$ in $(P, \leq)$ and vice versa.*

Illustration 9.2.2 : *Let X be a finite set. We consider* $\Sigma = (\mathbb{P}(X), \subseteq)$.
Let $B^{(n)} = \{(a_1, a_2, \ldots a_n) : a_i = 0 \text{ or } 1, i = 1, 2, \ldots n\}$, $B = \{0, 1\}$. $\Sigma' = B^{(n)}$ *is a poset in the sense that*

$$(a_1, a_2, \ldots a_n) \le (b_1, b_2, \ldots b_n)$$

if, and only if, $a_i \le b_i$ *for each* $i = 1, 2, \ldots n$.
Let $X = \{x_1, x_2, \ldots, x_n\}$. $|X| = n$.
If T is a subset of X, $|T| \le n$.
We define a map $\varphi : (\mathbb{P}(X), \subseteq) \to B^{(n)}$ *by*

$$\phi(T) = (a_1, a_2, \ldots a_n) \text{ where}$$

$$a_i = \begin{cases} 1, & \text{if } a_i \in T, \\ 0, & \text{if } a_i \notin T. \end{cases}$$

To each subset T of X, there is an n-tuple $(a_1, a_2 \ldots a_n)$ *associated with T.*

$$T \subseteq T' \Leftrightarrow (a_1, a_2, \ldots a_n) \le (a_1', a_2' \ldots a_n')$$

where $(a_1', a_2', \ldots a_n')$ *is the image of T' under* ϕ.
ϕ *is an order-preserving isomorphism and so,* $\Sigma \cong \Sigma'$.

Definition 9.2.9 : *Let* $\Sigma = (P, \le)$, $\Sigma' = (P', \le')$ *be two partially ordered sets. Their direct product* $\Sigma = \Sigma \times \Sigma'$ *is a partially ordered set* $(S, \le)$ *where*

(i) $S = P \times P' = \{(a, a') : a \in P, a' \in P'\}$.
(ii) $s \le t$ *in S if, and only if,* $a \le b$ *in P and* $a' \le' b'$ *in P',*
where $s = (a, a')$ *and* $t = (b, b')$.

Theorem 71 (E. A. Bender and J. R. Goldman (1975)) : *If* $\Sigma = (P, \le)$ *has Möbius function* μ *and* $\Sigma' = (P', \le')$ *has Möbius function* μ', *then, the Möbius function of* $D = \Sigma \times \Sigma'$ *is given by*

$$(9.2.10) \qquad \mu((a, a'), (b, b')) = \mu(a, b)\mu'(a', b')$$

Proof : By definition 9.2.9, $\Sigma \times \Sigma'$ is a poset. The ζ-function of $\Sigma \times \Sigma' = D_1$ is given by

$$\zeta_D((a, a'), (b, b')) = \begin{cases} 1, & \text{if } (a, a') \le (b, b'), \\ 0, & \text{otherwise.} \end{cases}$$

Also, $(a, a') \le (b, b') \Leftrightarrow a \le b$ in Σ and $a' \le' b'$ in Σ'.
So, we get

$$(9.2.11) \qquad \zeta_D((a, a'), (b, b')) = \zeta(a, b)\zeta'(a', b'),$$

where ζ is the *zeta*–function of Σ and ζ' that of Σ' respectively.
Now,

$$\zeta \cdot \mu = e_0 \text{ in } \Sigma, \quad \zeta' \cdot \mu' = e_0' \text{ in } \Sigma'$$

For $D = \Sigma \times \Sigma'$,

$$e_{0,D}((a, a'), (b, b')) = \begin{cases} 1, & \text{if } (a, b) = (a', b') \\ 0, & \text{otherwise.} \end{cases}$$

So, $e_{0,D} = e_0 e_0'$.

So, if μ_D denotes the Möbius function of $D = \Sigma \times \Sigma'$, we have

$$\zeta_D \cdot \mu_D = e_{0,D} = e_0 e_0'.$$

Or,

$$\zeta\zeta' \cdot \mu_D = e_0 e_0' = (\zeta \cdot \mu)(\zeta' \cdot \mu')$$

Also,

$$(\zeta \cdot \mu)(\zeta' \cdot \mu')((a,a'),(b,b')) = \sum_{a \le t \le b} \mu(a,t) \sum_{a' \le t' \le b'} \mu'(a',t')$$

$$= \sum_{(a,a') \le (t,t') \le (b,b')} \lambda((a,a'),(t,t')) (\text{ say })$$

$$= e_{0,D}((a,a'),(b,b')).$$

It shows that $\lambda = \zeta_D^{-1} = \mu\mu'$. $\qquad\qquad\qquad\square$

Examples 9.2.1 :

1. We have noted that $(B^{(n)}, \le) \cong (\mathbb{P}(X), \subseteq)$ where $|X| = n$. We look at $B = \{0,1\}$. B is made a partially ordered set by the usual order relation:

$$0 \le 0, \quad 0 < 1, \quad 1 \le 1.$$

We denote this poset by $(B, \le)$. It is, obviously, locally finite and left-finite. The Möbius function of $(B, \le)$ is given by

(9.2.12) $$\mu(x,y) = \begin{cases} 1, & x = y, \\ -1, & x < y; \end{cases}$$

as $\sum_{x \le t \le y} \mu(x,t) = \mu(x,x) + \mu(x,y) = 0$ and $\mu(x,x) = 1$. We express this as

(9.2.13) $$\mu(x,y) = (-1)^{y-x},$$

since the only possibilities are $x = y$ or $x = 0, y = 1$.

Under the isomorphism $(P(X), \subseteq) \cong B^{(n)}$, if $A \mapsto (a_1, a_2, \ldots a_n)$ and $B \mapsto (b_1, b_2, \ldots b_n)$ where A, B are subsets of X,

$$\mu(A,B) = \mu((a_1, a_2, \ldots a_n), (b_1, b_2, \ldots b_n))$$

$$= \prod_{i=1}^{n} \mu(a_i, b_i)$$

and $\prod_{i=1}^{n} \mu(a_i, b_i) = (-1)^{\sum b_i - \sum a_i} = (-1)^{|B| - |A|}$
or, the Möbius function of $(\mathbb{P}(X), \subseteq)$ is

(9.2.14) $$\mu(A,B) = (-1)^{|B| - |A|}$$

where A and B are subsets of X and $|A|, |B|$ denote the number of elements of A and B respectively.

2. We take $(\mathbb{N}, \leq)$ where $\leq$ means 'divides'.
   Let $n \in \mathbb{N}$ be written as

$$n = \prod_{i=1}^{k} p_i^{a_i}, p_i \text{ are primes, } a_i \geq 1 (i = 1, 2, \ldots k)$$

Let $D(n)$ denote the set of divisors of n. By the uniqueness of factorization of n into prime powers,

$$D(n) = D(p_1^{a_1}) \times D(p_2^{a_2}) \times \ldots \times D(p_k^{a_k})$$

Let $\Sigma_i = (D(p_i^{a_i}), \leq)$ where $\leq$ means 'divides'.
Σ_i $(i = 1, 2, \ldots n)$ is a locally finite poset given by $\{1, p_i, p_i^2, \ldots p_i^{a_i}\}$. $D(p_i^{a_i})$ is totally ordered. It is a chain. Its Möbius function (9.2.9) is given by

$$\mu(p_i^{t_i}, p_i^{s_i}) = \begin{cases} 1, & \text{if } t_i = s_i \\ -1, & \text{if } s_i - t_i = 1 \\ 0, & \text{otherwise.} \end{cases}$$

By the product theorem (theorem 71), we arrive at

$$\mu(\pi_{i=0}^{k} p_i^{t_i}, \pi_{i=0}^{k} p_i^{s_i}) = \begin{cases} (-1)^{\sum(s_i - t_i)} & \text{if } s_i - t_i = 0 \text{ or } 1, \\ & \text{for } i = 0, 1, \ldots k \\ 0, & \text{if } s_i - t_i > 1 \text{ for some } i. \end{cases}$$

Then,

$$\mu(m, n) = \mu(1, \frac{n}{m}) = \mu(\frac{n}{m})$$

as obtained in (9.2.7).

Definition 9.2.10 : *Let* $(P, \leq)$ *be a locally finite poset. A function*
$f : P \times P \to \mathbb{C}$ *(the field of complex numbers)*
with the property $f(x, y) = 0$, *whenever* x *is not less than or equal to* y, *is referred to as an incidence function. (In fact, f is also called a generalized arithmetic function) (compare definition 9.2.3).*

Observation 9.2.1 : *Let* $A(P)$ *denote the set of incidence functions of* $(P, \leq)$. *As in lemma 9.2.1, we can make* $A(P)$ *an incidence ring by defining addition* $(+)$ *and multiplication* $(\cdot)$ *as follows:*

(9.2.15) $(f + g)(x, y) = f(x, y) + g(x, y)$, *for all* $x, y \in P$

(9.2.16) $(f \cdot g)(x, y) = \sum_{x \leq t \leq y} f(x, t) g(t, y)$, *for all* $x, y \in P$

The multiplicative identity of $(A(P), +, \cdot)$ *is given by*

(9.2.17) $e_0(x, y) = \begin{cases} 1, & x = y, \\ 0, & x \neq y. \end{cases}$

Definition 9.2.11 : *An incidence function f is said to have an inverse, say, g if $f \cdot g = g \cdot f = e_0$.*

If f has an inverse, it is unique and is denoted by f^{-1}.

Lemma 9.2.4 : *An incidence function $f \in A(P)$ has an inverse if, and only if, $f(x,x) \neq 0$ for all $x \in P$.*

Proof : Suppose that f^{-1} exists. Then for all $x \in P$,

$$(f \cdot f^{-1})(x,x) = f(x,x)f^{-1}(x,x) = e_0(x,x) = 1$$

so, $f(x,x) \neq 0$. Conversely, suppose that $f(x,x) \neq 0$ for all $x \in P$. We define a function $g \in A(P)$ such that g will serve as the inverse of f. Suppose $f \cdot g = e_0$ and $g(x,y) = 0$ whenever x is not $\leq y$. We are given that $(P, \leq)$ is locally finite. So the number of elements in $[x,y]$ where $x \leq y$ is finite. We denote the number of elements of $[x,y]$ by $\#[x,y]$.

If $y = x$ let $g(x,x) = \frac{1}{f(x,x)}$ which is valid as $f(x,x) \neq 0$. We prove the result by induction on $\#[x,y]$.

Let $x < y$. Suppose that $g(u,v)$ is known for all $u,v \in P$ such that $u \leq v$ and $\#[u,v] < \#[x,y]$.

If $x \leq t < y, \#[x,t] < \#[x,y]$. So $g[x,t]$ is known.
Let

(9.2.18)
$$g(x,y) = -\frac{1}{f(y,y)} \sum_{x \leq t < y} g(x,t)f(t,y)$$

As $g(x,x)f(x,x) = 1$ for all $x \in P$, when $x < y$, by (9.2.18)

$$\sum_{x \leq t \leq y} g(x,t)f(t,y) = 0$$

Thus, $g \cdot f = e_0$ for the interval $[x,y]$ we have chosen from P. As $[x,y]$ is arbitrary $g \cdot f = e_0$. We claim that $f \cdot g = e_0$ also holds. Since $g[x,x] \neq 0$, for all $x \in P$, there is a function $h \in A(P)$ such that $h.g = e_0$.
Then,

$$f \cdot g = e_0 \cdot (f \cdot g) = (h \cdot g) \cdot (f \cdot g)$$
$$= h \cdot (g \cdot f) \cdot g$$
$$= h \cdot e_0 \cdot g$$
$$= h \cdot g = e_0.$$

Or, $f \cdot g = e_0$. So, g is the unique inverse of f.   □

Remark 9.2.4 : The Möbius function $\mu \in A(P)$ is the inverse of the ζ-function defined as in ζ (9.2.3).

Now, we are in a set–up in which we could carry over the properties of arithmetic functions to incidence functions in the context of a locally finite partially ordered set.

Definition 9.2.12 : *A poset* $(P, \leq)$ *is called a local lattice, if every non-empty interval in P is a poset with respect to the partial ordering that it inherits from P. A local lattice* $(P, \vee, \wedge)$ *is called locally distributive, if every non-empty interval in P is a distributive lattice.*

One could consider incidence functions $f : P \times P \to \mathbb{C}$ where P is a locally distributive lattice. Multiplicativity and related concepts could be talked about in the set $A(P)$ of functions defined on $P \times P$. Details can be had from Paul J. McCarthy [8] and David A. Smith [12].

9.3. Incidence algebra of $n \times n$ matrices

$A(P)$ considered in observation (9.2.1) is also referred to as an incidence algebra. Yet another instance is provided by considering square matrices M with entries from an integral domain D with unity 1_D. The rows and columns of M are indexed by a locally finite partially ordered set P. That is, we write

(9.3.1) $M = [a_{ij}]$ where $i, j \in P$ and $a_{ij} = 0$ unless $i \leq j$

a_{ij} is the value of a function $f : P \times P \to \mathbb{C}$:

(9.3.2) $f(i, j) = \begin{cases} a_{i,j} & \text{if } i \leq j \\ 0 & \text{otherwise} \end{cases}$

We denote the set of square matrices M (9.3.1) by $\mathscr{M}(P)$. That is, we shall take the elements of $\mathscr{M}(P)$ to be $n \times n$ matrices with entries from D.

As usual, we have: when $M_1 = [a_{ij}], M_2 = [b_{ij}]$

$$M_1 + M_2 = [a_{ij} + b_{ij}]$$

$$M_1 M_2 = [c_{ij}] \text{ where } c_{ij} = \sum_{k=1}^{n} a_{ik} b_{kj}$$

We can consider $\mathscr{M}(P)$ as a D–algebra. If P is totally ordered, $\mathscr{M}(P)$ consists of $n \times n$ upper triangular matrices.

The analogue of lemma 9.2.4 is the following:

Lemma 9.3.1 : $M = [a_{ij}] \in \mathscr{M}(P)$ *is invertible if, and only if,* a_{ii} *is invertible for each* $i \in P$.

Proof : $:\Rightarrow$ If M is invertible, let $M^{-1} = [b_{ij}]$.

Then, $MM^{-1} = I$, the unit matrix, gives

(9.3.3) $a_{ii} b_{ii} = 1_D$

So, the condition is necessary.

$\Leftarrow$: Conversely, suppose that each a_{ii} is invertible in D. Then, for each a_{ii}, there exists $b_{ii} \in D$ such that (9.3.3) holds. For $i < j$, let b_{ik} be known for all k with $i \leq k < j$. Then, b_{ij} is uniquely determined from

(9.3.4)
$$\sum_{k=1}^{j} b_{ik}a_{kj} = 0$$

Then, the only unknown term in (9.3.4) is b_{ij} and it occurs with a_{jj} which is invertible by hypothesis. Thus, b_{ij} is determined for all $i \leq j$. Also, $b_{ij} = 0$ whenever i is not $\leq j$. So, then, $B = [b_{ij}]$ satisfies $BM = I$, the unit matrix, or M is invertible in $\mathcal{M}(P)$. $\square$

Illustration 9.3.1 : *The zeta matrix z of $\mathcal{M}(P)$ is given by*

$$z = [e_{ij}] = \begin{cases} 1_D, & \text{if } i \leq j \\ 0, & \text{otherwise.} \end{cases}$$

z is invertible. So, z has an inverse say $[\mu_{ij}]$. $[\mu_{ij}]$ is called the Möbius matrix.

If $(P, \leq)$ is replaced by $(\mathbb{N}, \leq)$ where $\leq$ means 'divides', $z = [e_{ij}]$ is such that

$$e_{ij} = \begin{cases} 1, & \text{if } i \text{ divides } j; \\ 0, & \text{otherwise.} \end{cases}$$

e_{ij} depends only on $\frac{j}{i}$. So, we write $e_{i,j} = e(\frac{j}{i})$ where $e(n) = 1$ for all $n \in \mathbb{N}$.

For $[\mu_{ij}]$, we have μ_{ij} as the Dirichlet inverse of $e(\frac{j}{i})$ in the set of arithmetic functions defined on $\mathbb{N}$.

So, then, $\mu_{ij} = \mu(\frac{j}{i})$ where μ is the classical Möbius function. That is,

$$\mu_{ij} = \begin{cases} \mu(\frac{j}{i}), & \text{if } i \text{ divides } j; \\ 0, & \text{otherwise.} \end{cases}$$

In the case of a 4×4 matrix,

$$z = \begin{bmatrix} 1 & 1 & 1 & 1 \\ 0 & 1 & 0 & 1 \\ 0 & 0 & 1 & 0 \\ 0 & 0 & 0 & 1 \end{bmatrix}$$

and

$$[\mu] = \begin{bmatrix} \mu(1) & \mu(2) & \mu(3) & \mu(4) \\ 0 & \mu(1) & 0 & \mu(2) \\ 0 & 0 & \mu(1) & 0 \\ 0 & 0 & 0 & \mu(1) \end{bmatrix} = \begin{bmatrix} 1 & -1 & -1 & 0 \\ 0 & 1 & 0 & -1 \\ 0 & 0 & 1 & 0 \\ 0 & 0 & 0 & 1 \end{bmatrix}$$

Illustration 9.3.2 : *Let $P = \{1, 2, \ldots n\}$ with partial order $\leq$. The zeta matrix is*

$$z = [e_{ij}], \text{ where } e_{ij} = \begin{cases} 1 & \text{if } i \leq j; \\ 0 & \text{otherwise.} \end{cases}$$

The Möbius matrix $[\mu_{ij}]$ is obtained as follows:

$$\mu_{ii} = 1, \quad i = 1, 2, \ldots$$
$$\mu_{ij} = 0, \ \textit{if } i \textit{ is not } \leq j.$$

Now,

$$\sum_{k=1}^{j} \mu_{ik} e_{kj} = 0, \ \textit{if } i \neq j.$$

$\mu_{11} = 1$, $\mu_{11}e_{12} + \mu_{12}e_{22} = 0$ or $1 + \mu_{12} = 0$. *That is,* $\mu_{12} = -1$

$$\mu_{11}e_{13} + \mu_{12}e_{23} + \mu_{13}e_{33} = 1 - 1 + \mu_{13} = 0, \ etc\ldots$$

It may be verified that

$$[\mu_{ij}] = I - V$$

where

$$V = [v_{ij}] \textit{ with } v_{ij} = \delta_{i,j} - 1 \textit{ and } \delta_{ij} = \begin{cases} 1, & i = j; \\ 0, & \textit{otherwise.} \end{cases}$$

In the case of 4×4 *matrices, we have*

$$z = \begin{bmatrix} 1 & 1 & 1 & 1 \\ 0 & 1 & 1 & 1 \\ 0 & 0 & 1 & 1 \\ 0 & 0 & 0 & 1 \end{bmatrix}, \quad [\mu_{ij}] = \begin{bmatrix} 1 & -1 & 0 & 0 \\ 0 & 1 & -1 & 0 \\ 0 & 0 & 1 & -1 \\ 0 & 0 & 0 & 1 \end{bmatrix}.$$

For more details, see P. M. Cohn [3].

9.4. Vector spaces over a finite field

We make use of Möbius inversion to study the number of k-dimensional subspaces of an n-dimensional vector space $V_n(q)$ over a finite field $GF(q)$, where $q = p^m$, p a prime and $m \geq 1$. The set of subspaces of $V_n(q)$ forms a poset under the partial order: 'a subspace of'. If J and K are subspaces of $V_n(q)$, we write $J \leq K$ if, and only if, J is a subspace of K. The resulting poset is denoted by $L(V_n(q))$. $L(V_n(q))$ is order isomorphic to the lattice of subspaces of a projective space. To make it clear, we need to know about the notion of a 'geometric lattice'. A slight diversion is worth attempting to understand the 'geometrical' terminology. See [7].

Definition 9.4.1 : *The real affine line $L = \mathbb{R}$ is defined as the set $\mathbb{R}$ of real numbers regarded as 'points' $l \in L$.*

A transformation $T_1 : L \longrightarrow L$ gives by $T_1(l) = kl$ where $k \in \mathbb{R}$ gives scalar multiplication. A transformation $T_2 : L \longrightarrow L$ given by $T_2(l) = l + a$ where $a \in \mathbb{R}$, gives a translation.

T_1 and T_2 are particular cases of the most general affine transformation $T : L \longrightarrow L$ given by

(9.4.1) $T(l) = kl + a; \quad k, a \in \mathbb{R},$

is a scalar multiplication followed by a translation. The word 'affine' literally means 'resemblance'. As T depends on the values of k and a, we shall denote T (9.4.1) by $T(k, a)$. Let $k', a' \in \mathbb{R}$. We consider the composition of $T(k, a)$ with $T(k', a')$ by writing

$$\left(T(k', a') \circ T(k, a) \right) l = T(k', a')(kl + a)$$
$$= k'(kl + a) + a'$$
$$= k'kl + (k'a + a')$$

It is also of the form in (9.4.1). $T(k, a)$ is a bijection or affine automorphism if, and only if, $k \neq 0$. The set of affine automorphisms forms a group under the composition of transformations.

Definition 9.4.2 : *The group of all affine automorphisms from L to L is called the one-dimensional real affine group denoted by A_1.*

Definition 9.4.3 : *Let M_1, M_2, M_3 be R-modules (see definitions 6.6.1 and 6.6.2, chapter 6), where R is a commutative ring with unity. A sequence $< f, g >$ of two module-homomorphisms*

$$M_1 \xrightarrow{f} M_2 \xrightarrow{g} M_3$$

of R-modules is said to be exact or exact at M_2 if im $f = \ker g$.

We note that $g \circ f$ is the zero-map as im $f \subseteq \ker g$. It is easy to check that $0 \longrightarrow M_1 \xrightarrow{f} M_2$ is exact at M_1 if f is a monomorphism. Further, $M_2 \xrightarrow{g} M_3 \longrightarrow 0$ is exact at M_3 if, and only if, g is an isomorphism.

Definition 9.4.4 : *A sequence of module-homomorphisms*

$$M_0 \xrightarrow{f_1} M_1 \xrightarrow{f_2} M_2 \longrightarrow \cdots \longrightarrow M_{n-1} \xrightarrow{f_n} M_n$$

is called an exact sequence when each sequence $< f_i, f_{i+1} >$ is exact at M_i for $i = 1, 2, \ldots, (n-1)$.

Definition 9.4.5 : *An exact sequence of the form*

$$0 \to M_1 \xrightarrow{f} M_2 \xrightarrow{g} M_3 \longrightarrow 0$$

with zero modules and so zero homomorphisms at the ends, is called a short exact sequence. In this case, f is a monomorphism and g is an epimorphism.

These ideas are applicable to other structures as well. In particular, we can consider sequences of group-homomorphisms. Let $\mathbb{R}^\times$ denote the multiplicative group of nonzero real numbers. With each affine automorphism $T(k, a)$, $k \neq 0$, we associate the nonzero real number k of $\mathbb{R}^\times$. This gives a homomorphism $\psi : A_1 \longrightarrow \mathbb{R}^\times$, where $\psi(T(k, a)) = k$ (see definition 9.4.2)

$$\ker \psi = \{ T(k, a) : \psi(T(k, a)) = 1 \} = \{ T(1, a) : a \in \mathbb{R} \}$$

We note that for $l \in L$

$$\big(T(1,a') \circ T(1,a)\big)l = T(1,a')(l+a)$$
$$= l+a+a'$$
$$= T(1,a+a')l.$$

We define $\xi : (\mathbb{R},+) \longrightarrow \ker\psi$ by $\xi(a) = T(1,a)$, $a \in \mathbb{R}$. ξ_1 defines a homomorphism from $(\mathbb{R},+)$ onto $\ker\psi$. So, the image of ξ in A_1 is $\ker\psi$ where $\psi : A_1 \longrightarrow \mathbb{R}^\times$. This enables us to form a short exact sequence of homomorphisms of groups in the form

(9.4.2) $$0 \longrightarrow (\mathbb{R},+) \xrightarrow{\xi} A_1 \xrightarrow{\psi} (\mathbb{R}^\times, \cdot) \longrightarrow 1$$

(9.4.2) is an illustration of definition (9.4.5).

Definition 9.4.6 : *A subset D of $\mathbb{R}^2$ is defined by*

$$D = \{(k,a) : k,a \in \mathbb{R} \text{ and } k \neq 0\}$$

D is made a group under the multiplication given by

(9.4.3) $$(k,a) \times (k',a') = (kk', k'a+a')$$

It is easy to check that $A_1 \cong D$.

Definition 9.4.7 : *A property of two or more points $\{l_1, l_2, \ldots\}$ of L is called an affine property if it is invariant under the action of the affine group A_1.*

Let w_1, w_2 be real numbers with $w_1 + w_2 = 1$.
The average of $l_1, l_2 \in L$ with weights w_1 and w_2 is given by

(9.4.4) $$l = w_1 l_1 + w_2 l_2$$

If $T(k,a)$ is given by the affine transformation (9.4.1), we obtain

$$T(k,a)(w_1 l_1 + w_2 l_2) = w_1 T(k,a)l_1 + w_2 T(k,a)l_2$$
$$= w_1(kl_1+a) + w_2(kl_2+a)$$
$$= kl+a, \text{ where } l \text{ is as given in (9.4.4),}$$
$$= T(k,a)l$$

So, an affine transformation preserves averages.

Conversely, if $A : L \longrightarrow L$ is any transformation which preserves averages, then A is an affine transformation.

Next, we go to the real affine plane. More generally, we give

Definition 9.4.8 : *Let F be a field of characteristic $\neq 2$. An affine space P over F is a non-empty set for which there exists a finite dimensional vector space V over F and a function $\Psi : V \times P \longrightarrow P$ denoting addition such that*

(a) for all vectors $u,v \in V$ and all points $t \in P$
   $0+t = t, (u+v)+t = u+(v+t),$

(b) for any two points $t, s \in P$, there is precisely one vector $v \in V$ satisfying

$$v + s = t.$$

Remark 9.4.1 : We take the dimension of P to be the dimension of V.

Observation 9.4.1 : *The operation $+$ is symbolic of the action of a vector u of V on a point $t \in P$.*

$$u \longmapsto u + t \text{ for } t \in P \text{ is a bijection of sets.}$$

The additive group of V acts on P. $t \longmapsto u + t$ is a translation of P. If $k, k' \in F$, and $u', v' \in V$

$$T(P) = \{ku' + k'v' : k, k' \in F, u, v' \in V\}$$

is such that $T(P)$ is a vector space isomorphic to V. The mapping is given by $\eta : V \longrightarrow T(P)$ where $\eta(u) = u' \in V$.

Then, for $k, k' \in F$, $ku \longmapsto ku'$ and $k'v \longmapsto k'v'$.

For $c \in F, u \in V$, $cu \longmapsto cu'$. $T(P)$ is called the space of translations of the affine space P.

Observation 9.4.2 : *The affine line L considered earlier is an affine space with the one-dimensional vector space $\mathbb{R}$ as its space of translations.*

Example 9.4.1 : V is a vector space (finite dimensional or not) over F. W is a subspace of V. $\dim W$ is finite. We write

$$P = \{W + x : x \in V \text{ giving distinct cosets of } W\}$$

P is an affine space over F. W makes the space of translations of P. That is , $T(P) = W$. We consider $w \in W$ as the bijection:

$$w + u \longmapsto v + (w + u), \text{ for all } w \in W.$$

It is the action of $v \in W$ on the coset $W + u$.

When $W = V$ and V is finite dimensional, let V_A stand for the set V. A point of V_A is a vector $u \in V$. We write

$$t = u \in V_A$$

If $v \in V$, we define $v + T$ as a point of V_A. The definition 9.4.8 holds. That is,

$$0 + t = t, \qquad (u + v) + t = u + (v + t)$$

Given $t, s \in V_A$, there exists a unique vector $v \in V$ such that $v + s = t$.

In other words, the vector space V has been converted into an affine space V_A. The space of translations $T(V_A)$ is V. Strictly, $T(V_A) \cong V$.

Summary: An affine space is defined in terms of a given vector space of translations.

$$\text{Let } w_1, w_2, \ldots w_n \in F \text{ such that } \sum_{i=1}^{n} w_i = 1_F.$$

For points $t_i (i = 1, 2, \ldots, n)$ elements of P considered as an affine space with associated finite dimensional vector space V, the scalar multiples $w_i t_i (i = 1, 2, \ldots, n)$ are such that $\sum_{i=1}^{n} w_i t_i$ is a point of P when $\sum_{i=1}^{n} w_i = 1_F$.

Definition 9.4.9 : *Let P, P' be a affine spaces over the same field F. An affine transformation $T : P \longrightarrow P'$ is such that*

$$T(w_1 t_1 + w_2 t_2 + \cdots + w_n t_n) = w_1 T(t_1) + w_2 T(t_2) + \cdots + w_n(t_n),$$

for all n and all points $t_i \in P$ with $\sum_{i=1}^{n} w_i = 1_F$.

It is verified that every translation is an affine transformation. For details and related results, see S. Maclane and G. Birkhoff [7].

We go to the notion of a projective plane.

Definition 9.4.10 : *A projective plane over a field is one in which*

(i) any two distinct points lie on a unique line.

(ii) any two distinct lines intersect in a unique point.

Definition 9.4.11 : *Given a three-dimensional vector space V over a field F, we define $P = P(V)$ by the statements:*

(i) A point $t \in P(V)$ is a one-dimensional subspace of V.

(ii) A line $L \in P(V)$ is a two-dimensional subspace of V.

A point t is on the line L (or incident with L) when the subspace t(of V) is contained in the subspace L(of V).

If t, s are points of $P(V)$ with $t \neq s$, they are different one-dimensional subspaces of V. So their sum is a two-dimensional subspace L of V and so is a line of $P(V)$ and that is the only line containing t and s. Property (i) of definition 9.4.11 holds. Property (ii) of definition 9.4.11 also holds on account of the fact that two distinct lines L_1 and L_2 of $P(V)$ are two-dimensional subspaces of V. So, their sum $L_1 + L_2$ must be the whole space V. By the theorem on the dimension of sum of two subspaces,

$$\dim(L_1 + L_2) = \dim L_1 + \dim L_2 - \dim(L_1 \cap L_2)$$

So,

$$\dim(L_1 \cap L_2) = 2 + 2 - 3 = 1 \text{ as claimed in } (ii) \text{ of definition 9.4.10.}$$

That is, $L_1 \cap L_2$ is a point t of $P(V)$ and t is the unique point in which L_1 and L_2 intersect. It is also clear that the points of $P(V)$ are the lines containing the origin of V and the lines of $P(V)$ are the planes containing the origin of V. In fact, the projective plane $P(V)$ has been constructed from a vector space of dimension 3. More generally, a projective space of dimension n is constructed from a vector space V of dimension $(n + 1)$ over a field.

Definition 9.4.12 : *A projective space is a set of points together with certain distinguished subsets called its projective subspaces (or hyperplanes).*

The construction of a projective space is shown below:

Let V be a vector space of dimension $(n+1)$. A projective space $P(V)$ of dimension n possesses the following properties:

Points t of $P(V)$ are the one-dimensional subspaces of V. For each subspace S of V, we take $P(S)$ to be the subset of all those points t of $P(V)$ (the one-dimensional subspace t of V) for which $t \subset S$. $P(V)$ has distinguished subsets as the sets $P(S)$ where S is a subspace of V. If S is a vector subspace of dimension $(k+1)$ contained in V, $P(S)$ is called a projective subspace of dimension k in $P(V)$. In particular, a two-dimensional vector subspace S is a projective line $P(S)$ in $P(V)$. A three-dimensional subspace W is a projective plane $P(W)$ in $P(V)$ and so on. For the trivial subspace (0) of V, $P((0)) = \emptyset$, the empty projective subspace of $P(V)$.

Fact 9.4.1 :

(a) If W is a vector subspace of another vector subspace S of V, that is, $W \subset S$, then $P(W) \subset P(S)$. For, each one-dimensional subspace of V contained in W is also contained in S. When $P(W) \subset P(S)$, we say that the projective subspaces $P(W)$ are 'incident' with $P(S)$.

(b) The projective subspaces of $P(V)$ are such that

$$P(S) \wedge P(W) = P(S \cap W),$$

corresponding to the intersection subspace $S \cap W$ in V and

$$P(S) \vee P(W) = P(S+W),$$

corresponding to the vector space sum $S+W$ in V.

(c) The correspondence $S \longmapsto P(S)$ gives a lattice isomorphism of the lattice $L(V)$ of subspaces of V with the lattice $L(P(V))$ of projective subspaces of $P(V)$.

Definition 9.4.13 : *The lattice $L(P(V))$ of projective subspaces of a projective space $P(V)$ is called a geometric lattice.*

$L(P(V))$ is isomorphic to the lattice $L(V)$ of subspaces of V. We remark that finite dimensional vector spaces have a role to play in the context of either an affine space or a projective space. To reach the stage where $L(V_n(q))$ is considered, we needed a narration up to the definition of a geometric lattice given in (9.4.13).

The study of subspaces of a finite dimensional vector space is analogous to the study of subsets of a finite set. Just as $\binom{n}{k}$ counts the k-element subsets of a set X having n elements, we can define $\binom{n}{k}_q$ to denote the number of k-dimensional subspaces of $V_n(q)$ where the ground field $GF(q)$ has $q = p^m$ elements (p a prime, $m \geq 1$). Let S_k denote a set of k linearly independent vectors belonging to $V_n(q)$. S_k forms a basis for a k-dimensional subspace of $V_n(q)$. S_k also belongs to $V_k(q)$, a k-dimensional vector space over $GF(q)$. Each k-dimensional subspace of $V_n(q)$ contains a collection $\mathbb{S}_k$ of k linearly independent vectors. Therefore, as $\binom{n}{k}_q$ counts the number of k-dimensional subspaces of $V_n(q)$, we have # sets of k linearly independent vectors in $V_k(q) \times \binom{n}{k}_q$ equals # sets of k linearly independent vectors in $V_n(q)$. The number of sets of k linearly independent vectors in $V_n(q)$ is computed in the following manner:

For a set of k linearly independent vectors of $V_n(q)$, the first vector, say v_1, can be chosen from $(q^n - 1)$ nonzero vectors of $V_n(q)$. The vectors v_1 generate q vectors namely, the vectors of the form αv_1 where $\alpha \in GF(q)$. Then, the second vector v_2 can be chosen in $(q^n - q)$ ways. v_1 and v_2 generate, by linear combinations, q^2 vectors. Therefore, the 3rd vector can be chosen in $(q^n - q^2)$ ways. Continuing this argument, we have:

sets of k linearly independent vectors in $V_n(q)$ is given by

$$(q^n - 1)(q^n - q)(q^n - q^2)\ldots(q^n - q^{k-1}).$$

In the same manner,

sets of k linearly independent vectors in $V_k(q)$

$$= (q^k - 1)(q^k - q)\cdots(q^k - q^{k-1})$$

So,

$$\binom{n}{k}_q = \frac{\text{\# sets of } k \text{ linearly independent vectors in } V_n(q)}{\text{\# sets of } k \text{ linearly independent vectors in } V_k(q)}.$$

Or,

(9.4.5) $$\binom{n}{k}_q = \frac{(q^n - 1)(q^n - q)\cdots(q^n - q^{k-1})}{(q^k - 1)(q^k - q)\cdots(q^k - q^{k-1})} = \frac{(q^n - 1)(q^{n-1} - 1)\cdots(q^{n-k+1} - 1)}{(q^k - 1)(q^{k-1} - 1)\cdots(q - 1)}$$

(9.4.5) gives a formula for the number of k-dimensional subspaces of $V_n(q)$.

Next, we aim at derivation of the Möbius function μ for the poset $L(V_n(q))$. As $V_n(q)$ is finite, $L(V_n(q))$ is a locally finite poset. It is also left-finite and the trivial subspace (0) is such that $(0) \subset W$ for any subspace W of $V_n(q)$. Further, let $\{u_1, u_2, \ldots u_t\}$ be a basis for a subspace W. If $\{u_1, u_2, \ldots, u_t, u_{t+1}, \ldots, u_s\}$ is a basis for S containing W, we write

$$f(u_i) = \begin{cases} u_i, & t+1 \leq i \leq s \\ 0 & i \leq t \end{cases}$$

Then, $[W, S]$ can be shown to be isomorphic to $[0, A]$ when A is a subspace of dimension $s - t$. In fact, we can take A to be the quotient space S/W. Since all k-dimensional spaces over $GF(q)$ are isomorphic, we have only to compute $\mu(0, V_n(q))$ for all n.

Theorem 72 (Bender and Goldman(1975)) : *The Möbius function of $L(V_n(q))$ $(n \geq 2)$ is given by*

(9.4.6) $$\mu(0, V_n(q)) = (-1)^n q^{\binom{n}{2}}.$$

Proof : We prove (9.4.6) by induction on n. The proof due to Bender and Goldman is by the method of undetermined coefficients. Let $G(q)$ be a vector space over $F(q)$ and having $|G(q)| = y$ vectors.

For a subspace $U \in L(V_n(q))$ let $N(U)$ be the number of linear transformations $\pi : V_n(q) \longrightarrow G(q)$ whose null-space is U. $N'(U)$ denotes the number of linear transformations $T' : V_n(q) \longrightarrow G(q)$ whose null space contains U. Then,

$$(9.4.7) \qquad N'(U) = \sum_{U \subseteq W} N(W) \Longleftrightarrow N(U) = \sum_{U \subseteq W} N'(W)\mu(U,W)$$

and with $U = (0)$,

$$(9.4.8) \qquad N\big((0)\big) = \sum_{W \in L(V_n(q))} \mu(0,W)N'(W)$$

By definition, $N((0))$ is the number of linear transformations whose null space is (0); that is, the number of one-one linear transformations. Such a transformation is identified by giving a list of n linearly independent vectors—the image of an ordered basis for $V_n(q)$. By the argument used to derive (9.4.5), we note that the number of one-one transformations from $V_n(q)$ into $G(q)$ is given by $(y-1)(y-2)\ldots(y-q^{n-1})$ where $y = |G(q)|$.

Next, we compute $N'(W)$. Let $\dim W = w$. A linear transformation has null space containing W if it maps W onto (0) and does anything at all with the rest of the vectors. So, if $\{v_1,v_2,\ldots v_n\}$ is a basis for $V_n(q)$ where $\{v_1,v_2,\ldots v_t\}$ is a basis for W, we must map $v_1,v_2,\ldots,v_t$ onto (0) and the remaining $(n-t)$ basis-vectors onto any vectors in $G(q)$. So,

$$(9.4.9) \qquad\qquad\qquad N'(W) = y^{n-t}$$

Substituting in (9.4.8) we get

$$(9.4.10) \qquad N\big((0)\big) = (y-1)(y-q)\ldots(y-q^{n-1}) = \sum_{W \in L(V_n(q))} \mu(0,W)y^{n-d(w)}.$$

Where $d(W)$ denotes the dimension of W.

(9.4.10) holds for all values of y. So, it is a polynomial identity. Equating the constant terms of both sides of (9.4.10) we get (when $W = V_n(q)$)
$$\mu(0,V_n(q)) = (-1)(-q)\ldots(-q^{n-1}) = (-1)^n q^b \text{ where } b = 1+2+\cdots+(n-1) = \binom{n}{2}.$$
This proves (9.4.6). $\qquad\qquad\qquad\qquad\qquad\qquad\qquad\qquad\qquad\qquad\square$

Remark 9.4.2 : From (9.4.10), as $\mu(0,V_n(q)) = (-1)^n q^{\binom{n}{2}}$, we see that

$$\mu(0,W) = (-1)^t q^{\binom{t}{2}} \text{ with } t = \dim W$$

That is,

$$(9.4.11) \qquad \prod_{i=0}^{n-1}(y-q^i) = \sum_{t=0}^{n} \binom{n}{t}_q (-1)^t q^{\binom{t}{2}} y^{n-t},$$

as there are $\binom{n}{t}_q$ t-dimensional subspaces W of $V_n(q)$. (9.4.11) is referred to as a q-identity. For more details and other illustrations of Möbius inversion, see Bender and Goldman [2].

9.5. Notes with illustrative examples

It was G.C. Rota's fundamental paper [9] of 1964 that sparked a lot of activity in the study of incidence functions and in the applications to combinatorial mathematics and the theory of graphs. The focal point in abstract Möbius inversion is that there is a Möbius function associated with every locally finite poset.

Let $\tilde{\mathbb{R}}$ be the set of real numbers which are ≥ 1. Following Niven and Zuckerman [9], we introduce a function $\beta : \tilde{\mathbb{R}} \longrightarrow \mathbb{C}$. We recall that an arithmetic function f has domain $\mathbb{N}$ or $\tilde{\mathbb{Z}}$ and range $\mathbb{C}$.

Definition 9.5.1 : *The 'Niven product' of f with β is given by*

$$(f\beta)(n) = \sum_{m=1}^{[n]} f(m)\beta(\frac{n}{m}). \quad n \in \tilde{\mathbb{R}},$$

where $[x]$ is the greatest integer not exceeding x.

$f\beta$ is a function from $\tilde{\mathbb{R}}$ into $\mathbb{C}$.
If g is another arithmetic function,

$$g(f\beta)(n) = \sum_{m=1}^{[n]} g(m)f\beta(\frac{n}{m})$$

$$= \sum_{m=1}^{[n]} \sum_{k=1}^{[\frac{n}{m}]} g(m)f(k)\beta(\frac{n}{mk})$$

If $mk = s$, $s = mk \leq m[\frac{n}{m}] \leq m\frac{n}{m} = n$.

If s is any positive integer $\leq n$, $s \leq [n]$ and if $s = mk$ is a factorization of s, we note that $1 \leq mk \leq [n]$ and

$$k = \frac{s}{m} \leq \frac{[n]}{m} = [\frac{n}{m}]$$

If $h(s) = \sum_{mk=s} g(m)f(k)$, h is the Dirichlet product of f and g and so

$$g(f\beta)(n) = \sum_{s=1}^{[n]} \sum_{m|s} g(m)f(\frac{s}{m})\beta(\frac{n}{s}) = \sum_{s=1}^{[n]} \left(\sum_{m|s} g(m)f(\frac{s}{m})\right)\beta(\frac{n}{s}).$$

Or,

(9.5.1) $$g(f\beta)(n) = \sum_{s=1}^{[n]} \left((g \cdot f)(s)\beta(\frac{n}{s})\right) = (g \cdot f)\beta(n)$$

Theorem 73 : *Suppose that β and γ are complex-valued functions defined on $\tilde{\mathbb{R}}$. Then for all $n \geq 1$,*

(9.5.2) $$\beta(n) = \sum_{m=1}^{[n]} \gamma(\frac{n}{m}) \Leftrightarrow \gamma(n) = \sum_{k=1}^{[n]} \mu(k)\beta(\frac{n}{k})$$

where μ is the classical Möbius function.

Proof : $:\Rightarrow$ By (9.5.1)

$$g(f\beta)(n) = (g \cdot f)\beta(n).$$

Take f to be the Möbius function μ. Then, $\gamma = \mu\beta$.

So,

$$\gamma(n) = (\mu\beta)(n) \text{ for all } n \geq 1.$$

Let $e(n) = 1$ for all $n \geq 1$. μ is the Dirichlet inverse of e. Or, $e \cdot \mu = e_0$, $(e_0(n) = [\frac{1}{n}])$.

Then,

$$e\gamma = e(\mu\beta) = (e \cdot \mu)\beta, \quad \text{by (9.5.1)}$$
$$= e_0\beta$$

and $e_0\beta(n) = \sum_{m=1}^{[n]} e_0(m)\beta(\frac{n}{m}) = \beta(n)$ or $e\gamma = \beta$.

So, $\gamma = \mu\beta \Longrightarrow e\gamma = \beta$. This proves (9.5.2) in one direction.

$\Leftarrow$: suppose that $\gamma = \mu\beta$.

$e\gamma = \beta \Longrightarrow \mu\beta = \mu(e\gamma) = (\mu \cdot e)(\gamma) = e_0(\gamma) = \gamma$.

Thus (9.5.2) is established. $\qquad\square$

We translate (9.5.2) into the set-up of partially ordered sets. We make $\tilde{\mathbb{R}}$ into a poset. For $x, y \in \tilde{\mathbb{R}}$, we say that x divides y or $x \leq y$ if $\frac{y}{x}$ is a positive integer. Then, $(\tilde{\mathbb{R}}, \leq)$ is a left-finite poset. It is also locally-finite.

$$x \leq y \Longrightarrow y \in \tilde{\mathbb{R}} \text{ and } x \in \{y, \frac{y}{2}, \frac{y}{3}, \dots, \frac{y}{[y]}\}.$$

We check that the associated Möbius function is $\mu(x, y)$ given by
$\mu(x, y) = \mu(1, \frac{y}{x}) = \mu(\frac{y}{x})$ as in illustration 9.2.1.

Then,

$$\sum_{m|n} f(m) = \sum_{m=1}^{[n]} f(\frac{n}{m}) = \sum_{j \leq n} f(\frac{n}{j})$$

Also,

$$\sum_{m|n} g(m)\mu(m, n) = \sum_{m=1}^{[n]} g(m)\mu(m, n) = \sum_{j \leq n} g(j)\mu(\frac{n}{j}),$$

or,

(9.5.3) $$\sum_{m=1}^{[n]} f(\frac{n}{m}) = g(n) \Leftrightarrow f(n) = \sum_{m=1}^{[n]} \mu(m)g(\frac{n}{m})$$

Corollary 9.5.1 : *If f and g are arithmetic functions, then*

(9.5.4) $$g(n) = \sum_{m=1}^{[n]} f([\frac{n}{m}]) \Leftrightarrow f(n) = \sum_{m=1}^{[n]} \mu(m)g([\frac{n}{m}])$$

Proof : (9.5.4) follows from (9.5.3), if we define $F : \tilde{\mathbb{R}} \longrightarrow \mathbb{C}$ as

$$F(n) = f([n]), \quad n \geq 1$$

and $G : \tilde{\mathbb{R}} \longrightarrow \mathbb{C}$ as $G(n) = g([n]), n \geq 1$. $\qquad\square$

For related results of classical number theory, see Ralph G. Archibald [1]. For a detailed study of the algebra of incidence functions defined on a locally finite poset, see Eugene Spiegel and Christofer J. O'Donnel [A1].

9.6. Worked-out examples

a) (Don Redmond) $\mathcal{A}$ denotes the commutative ring of arithmetic functions under the operations of addition and Dirichlet convolution. (See Section 4.3, chapter 4) Suppose that $U(\mathcal{A})$ denotes the group of units of $\mathcal{A}$. The logarithmic and von Mangoldt operators L and Λ (respectively) are given by

$$L : \mathcal{A} \to \mathcal{A} \quad \text{to mean} \quad L(f)(r) = f(r)\log r, \quad r \geq 1$$

$$\Lambda : U(\mathcal{A}) \to \mathcal{A} \quad \text{to mean} \quad \Lambda(f)(r) = (f^{-1} \cdot L(f))(r) = \sum_{t|r} f^{-1}(\frac{r}{t}) f(t) \log t$$

It is clear that if $f \in U(\mathcal{A})$, $L(f) = f \cdot \Lambda(f)$.

Let $e_0(r) = [\frac{1}{r}]$, where $[x]$ denotes the greatest integer not exceeding x. If μ stands for the Möbius function, evaluate (i) $\Lambda(e_0)$ (ii) $\Lambda(\mu)$ and (iii) $\Lambda(d)$ where $d(r)$ denotes the number of divisors of r.

Answer: (i) $\Lambda(e_0) = e_0^{-1} \cdot L(e_0)$. $e_0^{-1} = e_0$ and $L(e_0) = z$ where $z(r) = 0$, $r \geq 1$. So, $\Lambda(e_0) = e_0 \cdot z = z$.

(ii) Let $f \in U(\mathcal{A})$.

$$\Lambda(f^{-1}) = f \cdot L(f^{-1})$$

So, $\Lambda(f^{-1})(r) = \sum_{d|r} f(d) f^{-1}(\frac{r}{d}) \log(\frac{r}{d})$

$$= \log r \sum_{d|r} f(d) f^{-1}(\frac{r}{d}) - \sum_{d|r} f(d) f^{-1}(\frac{r}{d}) \log d$$

$$= L(e_0)(r) - (Lf \cdot f^{-1})(r).$$

Or,

(9.6.1) $\Lambda(f^{-1}) = -Lf \cdot f^{-1} \cdot = -\Lambda(f).$

when, $f = e$,

(9.6.2) $\Lambda(\mu) = -\Lambda(e) = -\sum_{t|r} \mu(\frac{r}{t}) \log t$

(iii)

$$\Lambda(d) = \Lambda(e \cdot e)$$

$$(e \cdot e) \cdot \Lambda(e \cdot e) = L(e \cdot e) = d \log$$

So,

$$\Lambda(e \cdot e) = (d \log) \cdot d^{-1} = (d \log) \cdot \mu \cdot \mu.$$

That is,

$$\Lambda(d)(r) = \sum_{t_1 t_2 t_3 = r} d(t_1)\log(t_1)\mu(t_2)\mu(t_3).$$

Remark 9.6.1 : von Mangoldt's function Λ is given by

$$(9.6.3) \qquad \Lambda(r) = \begin{cases} \log p, & \text{if } r \text{ is a prime-power } p^m, m \geq 1 \\ 0, & \text{otherwise.} \end{cases}$$

It is verified that $\sum_{t|r}\Lambda(t) = \log r$. By Möbius inversion,

$$\Lambda(r) = \sum_{t|r}\mu(t)\log(\frac{r}{t}) = -\sum_{t|r}\mu(t)\log t$$

From (9.6.2), we see that $\Lambda(\mu) = -\Lambda$.

It can be checked that $\Lambda(f \cdot g) = \Lambda(f) + \Lambda(g)$. As $\Lambda(e_0) = z$, $\Lambda(f) + \Lambda(f^{-1}) = \Lambda(e_0) = z$, where $f \in U(\mathcal{A})$. (9.6.1) is a consequence.

$\square$

b) (Nicol and Vandiver) Von Sterneck function $\Phi(n, r)$ is given by

$$\Phi(n, r) = \frac{\mu(\frac{r}{g})\phi(r)}{\phi(r/g)}, \quad g = \text{g.c.d } (n, r)$$

(see (5.5.4), chapter 5)

Let

$$g(n, r) = \sum_{d|r} f(d)\Phi(n, \frac{r}{d})$$

where f is any arithmetic function. Show that

$$(9.6.4) \qquad f(r) = \frac{1}{r}\sum_{k=1}^{r} g(k, r).$$

Answer: This inversion formula is deduced from the identity shown below: If $d|r$,

$$(9.6.5) \qquad \sum_{k=1}^{r}\Phi(k, d) = \begin{cases} r, & \text{if } d = 1 \\ 0, & \text{otherwise.} \end{cases}$$

For,

$$\sum_{k=1}^{r} g(k, r) = \sum_{k=1}^{r}\left(\sum_{t|r} f(d)\Phi(k, \frac{r}{t})\right)$$

$$= \sum_{t|r} f(t)\sum_{k=1}^{r}\Phi(k, \frac{r}{t})$$

By (9.6.5), the inner sum is r, if $t = r$ and is zero otherwise.
So,

$$\sum_{k=1}^{r} g(k,r) = r f(r), \text{ from which (9.6.4) follows.} \qquad \square$$

EXERCISES

1. **Mark the following statements true (T) or false (F) justifying your answer briefly.**

   a) *Let $\lambda : N \rightarrow \{-1,1\}$ be given by*

   $$\lambda(r) = \begin{cases} 1, & r = 1 \\ (-1)^{\Omega(r)}, & r > 1 \end{cases}$$

   where $\Omega(r)$ denotes the total number of prime factors of r, each counted according to its multiplicity. (λ is Liouville's function). We define

   $$M(r) = \sum_{j=1}^{r} \lambda(j)$$

   It is correct to say that $M(r) > 0$.

   b) *Let f, g be arithmetic functions satisfying*

   $$f(r) > 0, g(r) > 0 \,(r \geq 1) \text{ and } g(r) = \prod_{t \mid r} f(t).$$

   One obtains $\log f(r) = \sum_{t \mid r} \mu(\frac{r}{t}) \log g(t)$.

   c) *The lattice of subgroups of S_4 is a complemented lattice.*

   d) *The lattice of normal subgroups of a group G is a Boolean algebra.*

   e) *Let $(L_1, \vee, \wedge)$ and $(L_2, \cup, \cap)$ be two lattices. A bijection $\psi : L_1 \rightarrow L_2$ is a lattice isomorphism if*

   $$\psi(a \vee b) = \psi(a) \cup \psi(b),$$

   $$\text{and } (\psi \wedge b) = \psi(a) \cap \psi(b); \text{ for all } a, b \in L_1.$$

   If L_1 and L_2 are lattices that are isomorphic as posets, L_1 and L_2 are isomorphic as lattices.

   f) *For $a+bi, c+di \in \mathbb{C}$, we define*

   $$a+bi \otimes c+di, \text{ if } a \leq c \text{ and } b \leq d.$$

   It is correct to say that $(\mathbb{C}, \otimes)$ can be made a lattice.

2. Let x be real and $x \geq 1$. Suppose that $\varphi(x,n)$ denotes the number of positive integers $m \leq x$ such that $\gcd(m,n) = 1$. Show that

   $$\sum_{d \mid n} \varphi(\frac{x}{d}, \frac{n}{d}) = [x].$$

Deduce that

$$\varphi(x,n) = \sum_{d|n} \mu(d)[\frac{x}{d}]$$

where μ is the Möbius function and $[x]$ denotes the greatest integer not exceeding x.

3. Let $P = \{0,a,b,c,1\}$ with $0 < a < 1$, $0 < b < 1$, $0 < c < 1$. $a \wedge b = b \wedge c = c \wedge a = 0$, $a \vee b = b \vee c = c \vee a = 1$.
    Find the Möbius function of $(P, \vee, \wedge)$. (See Fig. 9)

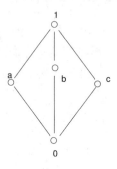

Figure 9

4. Let $P = \{a,b,c,d\}$ with $a < c < d$, $b < c < d$ as shown:
    Determine the Möbius function of $(P, \leq)$. (See Fig. 10)

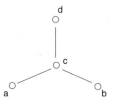

Figure 10

5. Let $(L_1, \vee, \wedge)$ and $(L_2, \vee, \wedge)$ be two lattices. Suppose that $L = L_1 \times L_2$, the product of the posets $(L_1, \leq)$, $(L_2, \leq')$.
    Show that L is a lattice. If L_1 and L_2 are modular, so is L: Prove.

6. Let $\gamma(n)$ denote the product of the distinct prime factors of n with $\gamma(1) = 1$. f and g are arithmetic functions given by

$$g(n) = \sum_{\substack{d|n \\ \gamma(d)=\gamma(n)}} f(d)$$

where d runs through those divisors of n such that d and n contain the same distinct prime factors.

Show that

$$f(n) = \sum_{\substack{d \mid n \\ \gamma(d) = \gamma(n)}} g(d) \mu(\frac{n}{d}).$$

7. *Let P, P' be isomorphic posets. That is, there exists a one-to-one map $\Psi : P \longrightarrow P'$ (onto) such that for $x, y \in P$,*

$$x \leq y \Longleftrightarrow \Psi(x) \leq' \Psi(y) \text{ in } P'.$$

   Show that

$$\mu'\big(\Psi(x), \Psi(y)\big) = \mu(x, y) \text{ for all } x, y \in P.$$

8. *Following R. P. Stanley [14], a locally finite poset $(P, \leq)$ is called a 'binominal poset' if*

   (i) *for every interval $[x, y]$ of $(P, \leq)$ all maximal chains in $[x, y]$ have the same length $n(x, y)$ or $[x, y]$ is an n-interval.*

   (ii) *any two n-intervals have the same number of maximal chains.*

   Show that $(\mathbb{N}, \leq)$ where $\leq$ means 'less than or equal to' is a binominal poset. Let $[a, b]$ be an n-interval in $(\mathbb{N}, \leq)$.

   If $\zeta(x, y) = 1$ for all $x, y \in P$ such that $1 \leq x \leq y$, show that

$$(\zeta \cdot \zeta)(a, b) = |[a, b]| = n + 1.$$

9. *A denotes the set of complex-valued functions on $\mathbb{N}$. $(\mathbb{N}, \leq)$ is a left-finite, locally finite poset where $\leq$ means 'divides'. Given $f \in A$, we define an incidence function $\bar{f}$ of $(\mathbb{N}, \leq)$ by*

$$\bar{f}(m, n) = \begin{cases} f(\frac{n}{m}), & \text{if } m \mid n; \\ 0 & \text{otherwise.} \end{cases}$$

   If $\bar{A}$ denotes the set of incidence functions of $(\mathbb{N}, \leq)$, show that the map $\theta : A \longrightarrow \bar{A}$ given by $\theta(f) = \bar{f}$ is one-to-one and preserves addition and Dirichlet convolution.

10. *[Paul J. McCarthy] Let L be a local lattice. An incidence function $\bar{f}$ of $(L, \vee, \wedge)$ is called factorable, if $\bar{f}$ has an inverse and if*

$$\bar{f}(a \vee b, c \vee d) = \bar{f}(a, c)\bar{f}(b, d)$$

   for all $a, b, c, d \in L$ such that a, b, c, d belong to the same interval in L and $a \leq c$, $b \leq d$ with $a \wedge b = c \wedge d$. If $\bar{f}$ is factorable, show that $\bar{f}(a, a) = 1$ for all $a \in L$. In particular, show that the ζ-function of L is factorable.

11. *[Paul J. McCarthy] Let L be a local lattice. If μ is factorable, show that $(L, \vee, \wedge)$ is locally distributive.*

REFERENCES

[1] Ralph G. Archibald, An introduction to the theory of numbers, Chap 4, Section 4.7, pp 86–91, Charles Merril Pub Co, Columbus Ohio (1970).

[2] E. A. Bender and J. R. Goldman, On applications of Möbius inversion in Combinatorial analysis, Amer. Math. Monthly, 82, (1975), 783–903.

[3] P. M. Cohn, Algebra Vol 2. Chapter 2 and Sec 2.3, pp 37–41, John Wiley and Sons, Brisbane, England, (1979).

[4] M. Frechet, Les Probabilities associee's a un systeme devenement compatible et dependants, Actualieis Sci.indust Paris, Hermann, 859 and 942, (1940 and 1943).

[5] Philip Hall, A contribution to the theory of groups of prime-power order, Proc. Lond. Math. Soc., 36, (1934), 24–80.

[6] G. H. Hardy and E. M. Wright, An Introduction to the theory of numbers Chapter XVI, pp 234–237, Oxford at the Clarendon Press, Reprint 1965.

[7] S. MacLane and G. Birkhoff, Algebra Chapter XII, pp 417–455, Macmillan Pub Co, NY, (1986).

[8] Paul J. McCarthy, Introduction to Arithmetical functions, Chapter 7, pp 293–332, Springer Verlag, Universitext, NY, (1986).

[9] I. Niven and H. S. Zuckerman, An introduction to the theory of numbers, Chap 4, pp 78–96, John Willey and Sons Inc., NY, (1972).

[10] G. C. Rota, On the foundations of Combinatorial theory I: Theory of Möbius functions, Z Wahr scheinlich Reitstheorie and Ver Gebiete, 2, (1964), 340–368.

[11] Harald Scheid, Arithmetische Funktionen üiber Halbord rung I and II, J. Reine Angew Math, 231, (1968), 192–214, 232, (1968), 207–220.

[12] David A. Smith, Generalized arithmetic function algebras, The theory arithmetic functions, Proc. Conf. at Western Michigan University, April 29–May 1, 1971. Lecture Notes in Mathematics No. 251 Antony A. Giora and Donald L. Goldsmith (Eds) Springer Verlag, Berlin, Heidelberg, New York (1972) pp 205–245.

[13] Robert Spira, Abstract Möbius Inversion, Math. Magazine, 45, (1972), 220–223.

[14] Richard P. Stanley, Generating functions (studies in Combinatorics) Studies in Math, Vol 17, G. C. Rota (Ed) Math. Association of America, (1978), 100–109.

[15] Morgan Ward, The algebra of lattice functions, Duke Math. J., 5, (1939), 357–371.

[16] L. Weisner, Abstract theory of inversion of finite series, Trans. Amer. Math. Soc., 38, (1935), 474–484.

[17] L. Weisner, Some properties of prime-power groups, Trans. Amer. Math. Soc., 38, (1935), 485–492.

[18] R. Wiegandt, On the general theory of Möbius inversion formula and Möbius product, Acta. Sci. Math. Szeged, 20, (1959), 164–180.

ADDITIONAL REFERENCE

[A1] Eugene Spiegel and Chistopher J. O'Donnel: Incidence Algebras, Monographs and Textbooks in Pure and Applied Mathematics No. 206, Marcel Dekker Inc., NY (1997).

CHAPTER 10

The role of generating functions

Historical perspective

It was Euler who gave us the idea of a generating function. He was interested in the theory of partitions of a positive integer n. We denote by p(n), the number of partitions of n. Through the introduction of generating functions and the progress in the theory of functions of a complex variable, the study of partitions became rigorous. The functions related to partitions and formulae connected with them were discovered as special cases of a more general set-up involving theta functions and modular functions. They were investigated thoroughly by Carl Gustav Jacob Jacobi and others. The results found a place in additive number theory legitimately. The role of generating functions in additive number theory is similar to the role of Dirichlet series of arithmetic functions in multiplicative number theory. In 1859, B. Riemann (1826–1866) undertook the study of $\pi(x)$, the number of primes less than or equal to x in establishing Gauss's conjecture $\pi(x) \sim \frac{x}{\log x}$ and connected this problem with the properties of $\zeta(s) = \sum_{n=1}^{\infty} n^{-s} (Re\ s > 1)$. Indeed, Riemann was one of the founders of the theory of functions of a complex variable and it was his interest in $\pi(x)$ that prompted him to pursue the general theory of functions of a complex variable. The Dirichlet series of an arithmetic function generalizes the Riemann ζ-function. In fact, $\zeta(s)$ is the generating function of the function e, where $e(n) = 1$, $n \geq 1$. The inverse of $\zeta(s)$ gives the Möbius function μ, where $1/\zeta(s) = \sum_{n=1}^{\infty} \mu(n) n^{-s}$, $Re\ s > 1$.

In combinatorial theory, one comes across the use of generating functions in solving enumeration problems. The development of the modern theory of generating functions is due to P. Doubilet, G. C. Rota and R. P. Stanley (1972). We look upon a generating function (representing a counting function) as an element of an algebra over $\mathbb{C}$, the field of complex numbers. The ring $\mathbb{C}[[x]]$ of formal power series in x with coefficients from $\mathbb{C}$ helps as a tool for enumeration and $\mathbb{C}[[x]]$ serves as the point of entry into algebra.

10.1. Introduction

The aim of this chapter is to go into the genesis of generating functions. As is well-known, Euler made use of generating functions in the study of partitions of a positive integer. In Section 10.2, we derive two classical theorems due to Euler for writing the generating function of $p(n)$, namely, $\sum p(n)x^n$ and its inverse. See

291

theorems 74 and 75. In Section 10.3, we discuss the notion of an elliptic function and introduce Weierstrass's $\mathbb{P}$-function. The purpose is to give an example of a modular function. Ramanujan's τ-function has the property that its generating function satisfies

$$\sum_{n=1}^{\infty} \tau(n)x^n = x \prod_{n=1}^{\infty} (1-x^n)^{24}, \quad |x| < 1$$

and it appears that certain generating functions could be expressed as a suitable infinite product. Theorem 76 in Section 10.3 gives an identity for τ. Section 10.4 is about properties of Stirling numbers and Bernoulli numbers. This is a context in which we exploit the nature of generating functions for deriving recurrence relations and from recurrence relations we get at generating functions. In Section 10.5, we introduce binomial posets which are originally due to Richard P. Stanley [19]. We work in the context of a locally finite partially ordered set. By defining an algebra of incidence functions, one is able to show that an algebra $B(P)$ of a special kind is isomorphic to the algebra $\mathbb{C}[[x]]$ of formal power series in x with coefficients from $\mathbb{C}$.

Dirichlet series are, in fact, generating functions of complex-valued arithmetic functions defined on $\mathbb{N}$. Dirichlet multiplication of arithmetic functions corresponds to ordinary multiplication of Dirichlet series. Very many interesting results follow. See theorem 80. See also Hardy & Wright [12]. Let F be a field. If K denotes the field of fractions of $F[[x]]$, it is shown that $F[[x]]$ is a valuation ring of K. (See theorem 81).

10.2. Euler's theorems on partitions of an integer

Perhaps, the first instance in the use of generating functions is from the theory of partitions. Let $\widetilde{\mathbb{Z}}$ denote the set of non-negative integers. A function $f : \widetilde{\mathbb{Z}} \to \mathbb{C}$ is an arithmetic function. The domain of f is also taken as $\mathbb{N}$, the set of positive integers, while doing multiplicative number theory.

Definition 10.2.1 : *Let f be an arithmetic function defined on $\widetilde{\mathbb{Z}}$. We call $F(x) = \sum_{n=0}^{\infty} f(n)x^n$, the generating function of f. x may be real or complex.*

The region of convergence of the series for $F(x)$ is specified when needed. Usually, $|x| < 1$ helps.

In [13] A. F. Horadam considers a sequence $\{w_n(a,b; p,q)\}$ where a, b, p, q are arbitrary complex numbers with $a \neq 0$, and

$$w_0 = a, w_1 = b \text{ and } w_n = pw_{n-1} - qw_{n-2} (n \geq 2)$$

$\{w_n\}$ is given by

(10.2.1) $w_n = A\alpha^n + b\beta^n$

where

$$A = \frac{b - a\beta}{\alpha - \beta}, \quad B = \frac{a\alpha - b}{\alpha - \beta}$$

and α, β are the zeros of the polynomial $x^2 - px + q$.

The generating function [13], [14] of $\{w_n\}$ is

$$(10.2.2) \qquad \frac{a+(b-pa)x}{1-px+qx^2} = \sum_{n=0}^{\infty} w_n x^n.$$

As illustrations, we note that the Fermat sequences $\{u_n\}$ and $\{v_n\}$ are such that $u_n = 2^n - 1$ and $v_n = 2^n + 1$. In fact,

$$(10.2.3) \qquad u_n = w_n(1,3; 3,2),$$

$$(10.2.4) \qquad v_n = w_n(2,3; 3,2).$$

Then,

$$(10.2.5) \qquad \frac{1}{1-3x+2x^2} = \sum_{n=0}^{\infty} u_n x^n, \quad |x| < 1/2.$$

$$(10.2.6) \qquad \frac{2-3x}{1-3x+2x^2} = \sum_{n=0}^{\infty} v_n x^n, \quad |x| < 1/2.$$

Next, in the case of Fibonacci sequence $\{F_n\}$ given by $F_0 = 1$, $F_1 = 1$ and $F_{n+1} = F_n + F_{n-1}$ ($n \geq 1$), the generating function of $\{F_n\}$ is given by

$$(10.2.7) \qquad \frac{1}{1-x-x^2} = \sum_{n=0}^{\infty} F_n x^n, \quad |x| < \min\{|\alpha|, |\beta|\},$$

where

$$(10.2.8) \qquad F_n = \frac{1}{\sqrt{5}} \{\alpha^{n+1} - \beta^{n+1}\}; \quad \alpha = \frac{1+\sqrt{5}}{2}, \beta = \frac{1-\sqrt{5}}{2}.$$

See A. F. Horadam [14] for more examples. A related reference is Pentti Haukkanen [11].

Definition 10.2.2 : *Let $A = \{a_1, a_2, \ldots, a_r, \ldots\}$ be a finite or infinite set of positive integers. If*

$$a_{i1} + a_{i2} + \cdots + a_{ir} = n \text{ with } a_{ij} \in A \ (j = 1, 2 \ldots, r),$$

we say that the sum $a_{i1} + a_{i2} + \cdots + a_{ir}$ is a partition of n into parts belonging to A.

If $n \in A$, n itself is counted as a partition of itself.

The summands or parts need not be distinct. Further, the order of the summands is immaterial. Every partition of n can be uniquely written as

$$(10.2.9) \qquad x = k_1 a_1 + k_2 a_2 \cdots + k_i a_i + \cdots$$

where a_i are distinct elements of A ($i = 1, 2, \ldots$) in an increasing order and $k_i \in \tilde{\mathbb{Z}}$. Only finitely many k_i are nonzero.

Definition 10.2.3 : $A = \{a_1, a_2, \ldots\}$ *is a finite or infinite set of positive integers. The number of distinct partitions of a positive integer n into parts belonging to A is denoted by $p_A(n)$. p_A is called the partition function relative to set A. $p_A : \mathbb{N} \to \tilde{\mathbb{Z}}$ is the desired function.*

If no restrictions are imposed, $p_A(n)$ gives the number of unrestricted partitions of n into parts belonging to A. If $A = \mathbb{N}$, $p_A(n)$ is written as $p(n)$. Various restrictions could be imposed to define $p_A(n)$. $p_A^{(0)}(n)$ stands for the number of partitions of n into an odd number of parts belonging to A. Likewise, $p^{(0)}(n)$ denotes the number of partitions of n into an odd number of parts. $p_A^{(e)}(n)(p^{(e)}(n))$ denotes the number of partitions of n into an even number of parts belonging to A (when $A = \mathbb{N}$). For instance, as

$$5 = 5$$
$$5 = 4 + 1$$
$$5 = 3 + 2$$
$$5 = 3 + 1 + 1$$
$$5 = 2 + 2 + 1$$
$$5 = 2 + 1 + 1 + 1$$
$$5 = 1 + 1 + 1 + 1 + 1$$

$p(5) = 7$. Also, $p^{(e)}(5) = 3$, $p^{(0)}(5) = 4$.

Next, some of the elementary ideas of complex analysis are described below for making the convergence of a series used for generating functions clear:

An open connected subset of the complex-plane is called a region. Let D be a region. A function $f : D \to \mathbb{C}$ is said to be differentiable at $z_0 \in D$, if $\lim_{z \to z_0} \frac{f(z) - f(z_0)}{z - z_0}$ exists and is independent of the path along which $z \to z_0$ in the complex-plane. The derivative is denoted by $f'(z_0)$.

$f : D \to \mathbb{C}$ is said to be analytic at $z_0 \in D$, if it is differentiable throughout some ϵ-neighbourhood of z_0. f is said to be analytic in D if it is analytic at every point of D. A region D is called a simply connected region, if D can be continuously deformed into a point without going outside D.

Let D be a simply connected region: Suppose that D contains an interval of the real-axis. Let $f : D \to \mathbb{C}$ be an analytic function. If $f(z)$ is real on the interval of the real axis contained in D, then $f(\bar{z}) = \overline{f(z)}$ where z or $\bar{z}$ belongs to D. As usual, $\bar{z}$ denotes the complex conjugate of z.

Let D_2 be a region containing a region D_1. Suppose that $f : D_1 \to \mathbb{C}$ is analytic in D_1. Let $g : D_2 \to \mathbb{C}$ be an analytic function. If $g(z) = f(z)$ at all points z of D_1, we say that g is an analytic continuation of f in the region D_2. For instance,

$$f(z) = 1 + z + z^2 + \cdots$$

is analytic at all points of the unit disc $|z| < 1$. The function $g : D_2 \to \mathbb{C}$ given by $g(z) = \frac{1}{1-z}$, $(z \neq 1)$ is analytic except at $z = 1$ of the complex plane. Further, $f(z) = g(z)$ for $|z| < 1$. We say that g is an analytic continuation of f in the region $D_2 = \mathbb{C} \setminus \{1\}$. There are several methods of analytic continuation of which the simplest is by power series.

Fact 10.2.1 : (Schwarz reflection principle) Let f be analytic in a region D of the upper half-plane H (of $\mathbb{C}$). Suppose that the boundary ∂D of D intersects the real axis in a line-segment L. Let f be continuous on $D \cup L$. Further, assume that $f(z)$ is real at all points z of L. If D^* is the reflection of D on the real axis, then, f can be continued analytically across L into D^* by taking

$$f(z) = \overline{f(\bar{z})} \quad (z \in D^*)$$

For proof, see L. V. Ahlfors [1].

We need to mention about infinite products. $\prod_{n=1}^{\infty}(1 - a_n)$ is said to converge, if $a_n \neq 1$ for all $n \geq N_0$ (specified) and if $\lim_{k \to \infty} \prod_{n=N_0}^{k}(1 - a_n)$ exists and is different from zero. If $\{a_n\}$ is real and $a_n \neq 1$ for all n and $\prod_{n=1}^{\infty}(1 - a_n)$ converges, then its value is nonzero.
See Tom Apostol [2].

Theorem 74 (Euler) : *For* $|x| < 1$,

$$(10.2.10) \qquad \sum_{n=0}^{\infty} p(n)x^n = \prod_{m=1}^{\infty}(1 - x^m)^{-1}, \quad where\ p(0) = 1.$$

Proof : $\prod_{m=1}^{\infty}(1 - x^m)^{-1} = (1-x)^{-1}(1-x^2)^{-1}(1-x^3)^{-1}\cdots$. One has

$$\prod_{m=0}^{\infty}(1 - x^m)^{-1} = (1+x+x^2+\cdots)(1+x^2+x^4+\cdots)(1+x^3+x^6+\cdots)$$

If we multiply the series on the right side, treating them as polynomials for the time being, we will get $1 + \sum_{k=1}^{\infty} a_k x^k$. It is our aim to show that $a_k = p(k)$. To get a term involving x^k, we can form

$$x^{k_1} x^{2k_2} x^{3k_3} \cdots x^{mk_m} = x^k$$

where $k = k_1 + 2k_2 + 3k_3 \cdots + mk_m$. In the partition of k, 1 occurs k_1 times, 2 occurs k_2 times and so on. The coefficient a_k of x^k is such that $a_k = p(k)$.

Now, suppose that $0 \leq x < 1$. We define

$$(10.2.11) \qquad F_m(x) = \prod_{k=1}^{m}(1 - x^k)^{-1}, \quad F(x) = \prod_{k=1}^{\infty}(1 - x^k)^{-1} = \lim_{m \to \infty} F_m(x)$$

$\prod_{k=1}^{\infty}(1 - x^k)$ converges absolutely, since $\sum_{k=0}^{\infty} x^k$ converges absolutely. So, $\prod_{k=1}^{\infty}(1 - x^k)^{-1}$ (the reciprocal of a convergent infinite product) converges absolutely for $0 \leq x < 1$.

Next, for a fixed x, $\{F_n(x)\}$ is an increasing sequence, as

$$F_{m+1}(x) = F_m(x)(1 - x^{m+1})^{-1} \geq F_m(x).$$

That is, $F_m(x) \leq F(x)$ for each x (when $0 \leq x < 1$) and $m \geq 1$. Also, $F_m(x)$ is the product of *a* finite number of absolutely convergent series. Therefore, it is an absolutely convergent series and we write it as

$$F_m(x) = 1 + \sum_{k=1}^{\infty} p_m(k)x^k$$

where $p_m(k)$ is the number of solutions of

$$k = k_1 + 2k_2 + \cdots + mk_m,$$

that is, the number of partitions of k into parts not exceeding m. If $m \geq k$, $p_m(k) = p(k)$. Also $p_m(k) \leq p(k)$ with equality when $m = k$. Thus,

(10.2.12) $$\lim_{m \to \infty} p_m(k) = p(k).$$

We have

$$F_m(x) = \sum_{k=0}^{m} p_m(k)x^k + \sum_{k=m+1}^{\infty} p_m(k)x^k$$

$$= \sum_{k=0}^{m} p_m(k)x^k + \sum_{k=m+1}^{\infty} p(k)x^k$$

For $x \geq 0$,

$$\sum_{k=0}^{m} p_m(k)x^k \leq F_m(x) \leq F(x).$$

For $t > m$,

$$\sum_{k=m+1}^{t} p(k)x^k < F_t(x) < F(x).$$

So, $\sum_{k=m+1}^{\infty} p(k)x^k$ converges. That is, $\sum_{k=0}^{\infty} p(k)x^k$ converges.

Also, $$\sum_{k=0}^{\infty} p_m(k)x^k \leq \sum_{k=0}^{\infty} p(k)x^k \leq F(x).$$

So, for each fixed x, $\sum_{k=0}^{\infty} p_m(k)x^k$ converges uniformly in m. Making $m \to \infty$, we have

$$F(x) = \lim_{m \to \infty} F_m(x) = \lim_{m \to \infty} \sum_{k=0}^{m} p_m(k)x^k = \sum_{k=0}^{\infty} \lim_{m \to \infty} p_m(k)x^k = \sum_{k=0}^{\infty} p(k)x^k.$$

This proves (10.2.10) for $0 \leq x < 1$. We can extend it by analytic continuation to the unit disc $|x| < 1$. $\qquad\square$

Definition 10.2.4 : *The pentagonal numbers are defined by the sums*

(10.2.13) $$w(n) = \sum_{k=0}^{n-1}(1 + 3k) = \frac{3n^2 - n}{2}.$$

For $n < 0$, $w(n)$ is given by $\frac{3n^2 + n}{2}$.

They are related to pentagonal-type graphs in which the number of vertices increases as shown:

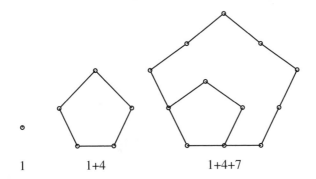

1   1+4   1+4+7

Figure 11

Theorem 75 (Euler's pentagonal number theorem) : *If* $|x| < 1$,

$$(10.2.14) \quad \prod_{m=1}^{\infty}(1 - x^m) = 1 + \sum_{n=1}^{\infty}(-1)^n\{x^{w(n)} + x^{w(-n)}\} = \sum_{n=-\infty}^{\infty}(-1)^n x^{w(n)}$$

where $w(n)$, $w(-n)$ *are pentagonal numbers* $(n = 1, 2, \ldots)$.

Proof : The method of proof is to show that for $0 \le x < 1$, if $P_n = \prod_{r=1}^{n}(1-x^r)$ and $S_n = 1 + \sum_{r=1}^{n}(-1)^r\{x^{w(r)} + x^{w(-r)}\}$ $|S_n - P_n| \le nx^{n+1}$. As $0 \le x < 1$, we will get

$$\lim_{n \to \infty} P_n = \lim_{n \to \infty} S_n.$$

We define $P_0 = S_0 = 1$.

We write $g(r) = \frac{r(r+1)}{2}$.

Let

$$(10.2.15) \qquad\qquad F_n = \sum_{r=0}^{n}(-1)^r\frac{P_n}{P_r}x^{rn+g(r)}$$

$F_1 = 1 - x - x^2 = S_1$.

$$F_n - F_{n-1} = \sum_{r=0}^{n}(-1)^r\frac{P_n}{P_r}x^{rn+g(r)} - \sum_{r=0}^{n-1}(-1)^r\frac{P_{n-1}}{P_r}x^{r(n-1)+g(r)}.$$

Now, $P_n = (1-x^n)P_{n-1}$. After simplification, we obtain

$$F_n - F_{n-1} = (-1)^n x^{n^2+g(n)} + (-1)^n x^{n^2+g(n-1)}$$

But, $n^2 + g(n) = n^2 + \frac{n(n+1)}{2} = w(-n)$ and $n^2 + g(n-1) = w(n)$.

So,

$$F_n - F_{n-1} = (-1)^n\{x^{w(n)} + x^{w(-n)}\} = S_n - S_{n-1}.$$

So, $F_n - S_n = F_{n-1} - S_{n-1} = F_{n-2} - S_{n-2} = \cdots = F_1 - S_1 = 0$
or $F_n = S_n$ for $n \geq 1$. From (10.2.15), we also have

$$(10.2.16) \qquad\qquad F_n = P_n + \sum_{r=1}^{n} (-1)^r \frac{P_n}{P_r} x^{rn+g(r)}.$$

Further, $0 < \frac{P_n}{P_r} \leq 1$ for $0 \leq x < 1$. $x^{rn+g(r)} \leq x^{n+1}$ for $r \geq 1$.
Therefore, the sum on the right side of (10.2.16) is bounded by nx^{n+1}. So, $|F_n - P_n| \leq nx^{n+1}$ and since $F_n = S_n$, given $\epsilon > 0$.

$$|S_n - P_n| < \epsilon \quad \text{for } n \geq N_0 \text{ (specified)}$$

So, (10.2.14) is okay for $0 \leq x < 1$. We extend it by analytic continuation to the disc $|x| < 1$. $\qquad\qquad\qquad\qquad\qquad\qquad\qquad\qquad\qquad \square$

Remark 10.2.1 : The proof of theorem 75 has been adapted from Tom Apostol [3]. See Emil Grosswald [9] also.

10.3. Elliptic functions

We begin by defining a lattice in the complex plane.

Let S be an arbitrary but fixed non-empty set. We recall definition 5.3.4, chapter 5, of a free abelian group. The definition holds good verbatim for any group F. We state

Fact 10.3.1 :

(a) If the group F together with the function $f : S \to F$ is a free group on S, then f is injective (one-one) and $f(S)$ generates F. We write the free group F on S by (F, f).
(b) (Existence theorem) For any set S, there always exists a free group on S.
(c) (Uniqueness theorem) If (F, f) and (F', f') are free groups on the same set S, there exists a unique isomorphism $j : F \to F'$ such that $j \circ f = f'$.

For proofs, See S. T. Hu [15].

If a free group is generated by a single element, then it is infinite cyclic. Infinite cyclic groups are also known as free cyclic groups.

We recall that an abelian group F together with a function $f : S \to F$ is a free abelian group on S if, given any function $g : S \to G$ where G is any abelian group, there exists a unique homomorphism $h : F \to G$ such that $h \circ f = g$. Further, f is injective and $f(S)$ generates F. Also, every set S of elements determines an essentially unique free abelian group (F, f). The abelian group F is called the free abelian group generated by the given set S.

Fact 10.3.2 : The direct sum of an arbitrarily indexed family

$$f : \{G_s : s \in S\}$$

of infinite cyclic groups G_s is isomorphic to the free abelian group generated by S.

For proof, see S. T. Hu [15].

Definition 10.3.1 : *A free abelian group F is said to be of rank n if, and only if, it is isomorphic to the direct sum of n infinite cyclic groups. The trivial group (e) is considered as a free abelian group of rank 0. r(F) denotes the rank of F.*

Definition 10.3.2 : *A subgroup $(L,+)$ of $(\mathbb{C},+)$ is called a lattice if $(L,+)$ is a free abelian group of rank 2.*

The lattice $(L,+)$ has a basis $\{\omega_1, \omega_2\}$. It is easy to see that

$$(10.3.1) \qquad L = \{z \in \mathbb{C} : z = m\omega_1 + n\omega_2; m, n \in \mathbb{Z}\}.$$

The set

$$(10.3.2) \qquad B = \{z \in \mathbb{C} : z = \alpha + t_1\omega_1 + t_2\omega_2; \alpha \in \mathbb{C}, 0 \le t_1 \le 1, 0 \le t_2 \le 1\}$$

is called a fundamental parallelogram with respect to the basis $\{\omega_1, \omega_2\}$ of $(L,+)$.

Definition 10.3.3 : *Let D be a region in C. A function $f : D \to \mathbb{C}$ is called a periodic function with period ω, if $f(z+\omega) = f(z)$, whenever $z, z+\omega \in D$.*

If ω is a period, so is $n\omega$ for every integer n. If ω_1 and ω_2 are periods, so is $m\omega_1 + n\omega_2$ for every choice of $m, n \in \mathbb{Z}$. We note that exp: $\mathbb{C} \to \mathbb{C}$ is periodic with period $2\pi i$.

Definition 10.3.4 : *Let D be a region of the complex plane containing ω_1 and ω_2, where $\frac{\omega_1}{\omega_2}$ is not real. $f : D \to \mathbb{C}$ is called a doubly-periodic function, if it has two periods ω_1 and ω_2.*

Definition 10.3.5 : *When f is a doubly-periodic function with periods ω_1 and ω_2 (with $\frac{\omega_1}{\omega_2}$ not real) the pair $\langle \omega_1, \omega_2 \rangle$ is called a fundamental pair if every period of f is of the form $m\omega_1 + n\omega_2$ for $m, n \in \mathbb{Z}$. We assume that $0 < \arg(\omega_1/\omega_2) < \pi$. We set $L(\omega_1, \omega_2) = \{z \in \mathbb{C} : z = m\omega_1 + n\omega_2, m, n \in \mathbb{Z}\}$. $L(\omega_1, \omega_2)$ is a lattice in the complex plane, given ω_1, ω_2 are complex numbers such that $\frac{\omega_1}{\omega_2}$ is not real.*

Next, let $\omega_1 = 1 + \sqrt{3}i$ and $\omega_2 = 1 - i$, so that $\frac{\omega_1}{\omega_2}$ is not real. $0 < \arg\omega_1 - \arg\omega_2 = 5\pi/12 < \pi$. The lattice generated by ω_1 and ω_2 is of the form given in (10.3.1). The fundamental parallelogram with respect to the basis $\{\omega_1, \omega_2\}$ is known. See figure 12 below:

Fact 10.3.3 : (i) Let α be a real number. Then,

$$\sum_{\substack{\omega \in L(\omega_1, \omega_2) \\ \omega \ne 0}} \omega^{-\alpha z} = \sum_{\substack{m,n \in \mathbb{Z} \\ (m,n) \ne (0,0)}} (m\omega_1 + n\omega_2)^{-\alpha}$$

is absolutely convergent if, and only if, $\alpha > 2$.

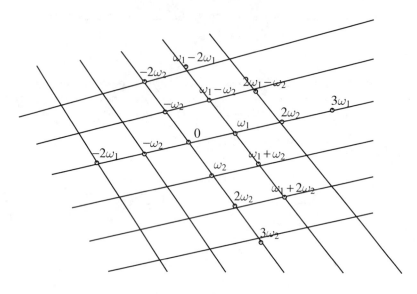

Figure 12

(ii) Let $\alpha > 2$ and A, a positive real number. Then, the series

$$\sum_{|\omega|>R} \frac{1}{(z-\omega)^\alpha} = \sum_{m} \sum_{\substack{n \\ |n\omega_1+n\omega_2|>R}} \frac{1}{(z-m\omega_1-n\omega_2)^\alpha}$$

is absolutely and uniformly convergent in the disc $|z| \le A$. For proofs, see Tom Apostol [2].

Next, let D be a simply connected region in the complex plane. Suppose that $f : D \to \mathbb{C}$ is analytic, except possibly, at $z_0 \in D$. Then z_0 is called an isolated singularity of f.

Definition 10.3.6 : *If $\lim_{z \to z_0}\{(z-z_0)f(z)\} = 0$, then $\lim_{z \to z_0} f(z)$ exists and is finite. By taking*

$$f(z_0) = \lim_{z \to z_0} f(z)$$

we can make f analytic throughout D and in that case, z_0 is called a removable singularity of f. In other words, we call z_0 an ordinary point of f.

Suppose that $\lim_{z \to z_0} f(z)(z-z_0) \neq 0$

Definition 10.3.7 : *If there exists $k > 0$ such that*

$$\lim_{z \to z_0} (z-z_0)^k f(z) = g(z_0)$$

and g is analytic at z_0, then z_0 is called a pole of f.

The smallest integer k for which $\lim_{z \to z_0}(z-z_0)^k f(z)$ exists and is finite, is called the order of the pole z_0 of f.

Further, if z_0 is neither a removable singularity nor a pole (of order k) of f, z_0 is said to be an essential singularity of f.

Let $f : D \to \mathbb{C}$ be single-valued and analytic in D. If $z_0 \in D$ and $f(z_0) = 0$, then, there exists a positive integer t such that

$$f(z) = (z-z_0)^t g(z)$$

with g analytic in D and $g(z_0) \neq 0$. z_0 is called a zero of order t, of f.

Let $f : D \to \mathbb{C}$ be single-valued in D and analytic except possibly at a finite number of points $z_1, z_2, \ldots, z_m$. Then, if $z_0 \in D$, f possesses a series expansion of the form

$$(10.3.3) \qquad f(z) = \sum_{n=-\infty}^{\infty} a_n(z-z_0)^n$$

which is convergent in some 'punctured disc': $0 < |z-z_0| < r$ having centre at z_0 and radius $r > 0$. If $z_0 \neq z_i$ $(i = 1, 2, \ldots, m)$ then, $a_n = 0$ for $n < 0$. (10.3.3) gives the Taylor series of f. The Taylor series converges at $z = z_0$ also (as $f(z_0) = a_0$). If $z_0 = z_i$ $(1 \leq i \leq m)$, a finite number or an infinite number of coefficients a_n $(n < 0)$ may be non-zero and the series expansion of f is termed a Laurent series.

Suppose $f(z) = \sum_{n=-k}^{\infty} a_n(z-z_0)^n$. Then $z = z_0$ is a pole of order k of f. If $a_n \neq 0$ for infinitely many negative values of n, then z_0 is an essential singularity of f.

Now, an entire function is one which is analytic everywhere in the complex plane. A meromorphic function is one whose only singularities are poles. These two kinds of functions are related. The reciprocal of an entire function is a meromorphic function. The reciprocal function has a pole at a point where the entire function has a zero. $f : \mathbb{C} \to \mathbb{C}$ given by

$$f(z) = \frac{1}{z} \exp(z)$$

is a meromorphic function in the complex plane. A rational function of the form $f = \frac{g}{h}$ where g and h are analytic in a region D of the complex plane, is meromorphic in D.

If we consider the set $\mathbb{A}$ of analytic functions defined on D (a simply connected region in $\mathbb{C}$), $\mathbb{A}$ can be made a commutative ring under the operations of ordinary addition and ordinary multiplication. $\mathbb{A}$ has no divisors of zero. That is, $\mathbb{A}$ is an integral domain. One can find the field of quotients of $\mathbb{A}$, say $\mathcal{F}$. Then, $\mathcal{F}$ is the field of meromorphic functions defined on D.

Definition 10.3.8 : *Let $f : \mathbb{C} \to \mathbb{C}$ be such that*

(i) f is doubly periodic with periods ω_1 and ω_2,

(ii) f is meromorphic on $\mathbb{C}$.

Then, f is called an elliptic function.

The function f defined by

(10.3.4)
$$f(z) = \sum_{\omega \in L(\omega_1, \omega_2)} \frac{1}{(z-\omega)^3}$$

is an elliptic function with periods ω_1 and ω_2 and it has a pole of order 3 at each period ω in $L(\omega_1, \omega_2)$.

For proof, see Tom Apostol [3].

An example of an elliptic function having a pole of order 2 at each period ω is Weierstrass's $\mathbb{P}$-function given by

(10.3.5)
$$\mathbb{P}(z) = \frac{1}{z^2} + \sum_{\substack{w \in L(\omega_1, \omega_2) \\ w \neq 0}} \left\{ \frac{1}{(z-\omega)^2} - \frac{1}{\omega^2} \right\}$$

For properties of $\mathbb{P}(z)$, see Tom Apostol [4].

Let H represent the upper-half of the complex plane. That is,

$$H = \{ \xi \in \mathbb{C} : \text{im}\,(\xi) > 0 \}$$

Let ω_1 and ω_2 be as in the definition of Weierstrass's $\mathbb{P}$ function (10.3.5). Let $\xi = \frac{\omega_2}{\omega_1}$ and im $\xi > 0$. Then, $\xi \in H$.

Definition 10.3.9 : $g_2(\xi)$, $g_3(\xi)$ are defined by

$$g_2(\xi) = 60 \sum_{\substack{(m,n) \in \mathbb{Z} \times \mathbb{Z} \\ (m,n) \neq (0,0)}} \frac{1}{(m+n\xi)^4}$$

$$g_3(\xi) = 140 \sum_{\substack{(m,n) \in \mathbb{Z} \times \mathbb{Z} \\ (m,n) \neq (0,0)}} \frac{1}{(m+n\xi)^6}$$

We observe that g_2 and g_3 occur in the differential equation satisfied by $\mathbb{P}$.

Definition 10.3.10 : *The discriminant* $\Delta(\xi)$ *is defined by*

$$\Delta(\xi) = g_2^3(\xi) - 27 g_3^2(\xi).$$

It is known [3] that g_2, g_3 and Δ are analytic in H. We write

(10.3.6)
$$J(\xi) = \frac{g_2^3(\xi)}{\Delta(\xi)}$$

J is also analytic in H.

Fact 10.3.4 : If $\xi \in H$, $\Delta(\xi)$ possesses the Fourier expansion

(10.3.7)
$$\Delta(\xi) = (2\pi)^{12} \sum_{n=1}^{\infty} \tau(n) e^{2\pi i n \xi}$$

where the coefficients $\tau(n)$ are integers. $\tau : \mathbb{N} \to \mathbb{Z}$ defined by (10.3.7) is known as Ramanujan's τ-function.

Further,

$$(10.3.8) \qquad 12^3 J(\xi) = e^{-2\pi i \xi} + 744 + \sum_{n=1}^{\infty} c(n) e^{2\pi i n \xi},$$

where $c(n)$ are integers.

For proofs, see Tom Apostol [4].

(10.3.6) gives J as an analytic function defined on H.

If

$$\xi' = \frac{a\xi + b}{c\xi + d}$$

where a, b, c, d are integers with $ad - bc = 1$, $\xi' \in H$ and $T : H \to H$ defined by $T(\xi) = \xi'$ is a unimodular transformation. It is known [4] that

$$(10.3.9) \qquad J(\xi') = J(\frac{a\xi + b}{c\xi + d}) = J(\xi),$$

where $a, b, c, d \in \mathbb{Z}$ and $ad - bc = 1$. (10.3.8) expresses $J(\xi)$ as an absolutely convergent Fourier series [4].

If the matrix of T is $A = \begin{bmatrix} a & b \\ c & d \end{bmatrix}$ with $ad - bc = 1$,

A and $-A$ represent the same transformation. The set Γ of all such transformations T forms a group under composition of transformations. Γ is called the modular group. Γ is generated by the transformations.

$T_1(\xi) = \xi + 1$ and $T_2(\xi) = -1/\xi$.

Let $k \in \mathbb{Z}$.

Definition 10.3.11 : *A function $f : \mathbb{C} \to \mathbb{C}$ is called an entire modular form of weight k, if the following conditions are satisfied:*

(i) f is analytic in the upper half-plane H,

(ii) $f(\frac{a\xi + b}{c\xi + d}) = (c\xi + d)^k f(\xi) \cdot (ad - bc = 1, a, b, c, d \in \mathbb{Z})$,

(iii) f possesses the fourier expansion

$$(10.3.10) \qquad f(\xi) = \sum_{n=0}^{\infty} q(n) \exp(2\pi i n \xi),$$

where $q(n)$ are integers.

A modular form of weight 0 is called a modular function. By (10.3.9), J is an example of a modular function. Further, J is meromorphic in H.

In the Fourier expansion (10.3.10) of f, $q(0)$ is the constant term and it is called the value of f at $i\infty$. If $q(0) = 0$, f is called a cusp form. For the discriminant Δ given in Fact 10.3.4, it is seen that Δ is a cusp form. Further, we observe that

$$(10.3.11) \qquad \Delta\left(\frac{a\xi + b}{c\xi + d}\right) = (c\xi + d)^{12} \Delta(\xi), \quad \begin{bmatrix} a & b \\ c & d \end{bmatrix} \in \Gamma.$$

(see [4, Theorem 3.2 of chapter 3].)

So, Δ is a cusp form of weight 12, by definition 10.3.11, as Δ is analytic in the upper half-plane and Δ has the Fourier expansion given by (10.3.7). Now, $J(\xi') = J(\xi)$ and J has a Fourier expansion exhibited in (10.3.8). It is known [4] that a non constant entire modular form exists only if $k \geq 4$ and k is even. For more details, see [4, chapter 6].

In 1877, R. Dedekind introduced an η-function as follows: for $\xi \in H$, where $\xi = \frac{\omega_2}{\omega_1}$ with Im $\xi > 0$.

$$(10.3.12) \qquad \eta(\xi) = e^{\pi i \xi / 12} \prod_{n=1}^{\infty} (1 - e^{2\pi i n \xi}).$$

The infinite product on the right side of (10.3.12) has the form $\prod_{n=1}^{\infty}(1 - x^n)$, where $x = e^{2\pi i \xi}$. If $\xi \in H$, $|x| < 1$. So the product in (10.3.12) converges absolutely and is nonzero. Further, it can be shown that the convergence of the product is uniform on compact subsets of H. Therefore, η is analytic in H.

Fact 10.3.5 : Let Δ and η be as given in Definition 10.3.10 and (10.3.9) respectively. If $\xi \in H$ and $x = e^{2\pi i \xi}$, then

$$(10.3.13) \qquad \Delta(\xi) = (2\pi)^{12} \eta^{24}(\xi) = (2\pi)^{12} x \prod_{n=1}^{\infty} (1 - x^n)^{24}$$

Consequently, from (10.3.7) one derives

$$(10.3.14) \qquad \sum_{n=1}^{\infty} \tau(n) x^n = x \prod_{n=1}^{\infty} (1 - x^n)^{24}, \text{ whenever } |x| < 1$$

where τ is Ramanujan's τ-function. For proof, see Tom M. Apostol [4].

We note that the left side of (10.3.14) is the generating function of τ. It is expressible as an infinite product. This happens, when one considers modular functions. There is a vast literature on the τ function. Ramanujan conjectured that τ is multiplicative. That is,

$$\tau(m)\tau(n) = \tau(mn) \quad \text{whenever g.c.d } (m,n) = 1$$

It was proved by L. J. Mordell [16] in 1920 using complex analysis. See Sivaramakrishnan [18].

A formula for $\tau(n)$ using the method of generating functions is due to John A. Ewell [8].

Let $k \in \mathbb{N}$. For each $n \in \mathbb{N}$, $\sigma_k(n)$ denotes the sum of the kth-powers of the positive divisors of n. That is,

$$\sigma_k(n) = \sum_{d|n} d^k$$

For instance, $\sigma_k(6) = 1^k + 2^k + 3^k + 6^k$. We write

$$(10.3.15) \qquad n = 2^{b(n)} M(n)$$

where $M(n)$ is the highest odd divisor of n. $2^{b(n)}$ is the highest power of 2 dividing n. For each integer n, we set

$$S_k(n) = \{(x_1, x_2, \ldots, x_k) \in \mathbb{Z}^k : n = x_1^2 + x_2^2 + \cdots + x_k^2\}$$

Let $r_k(n) = |S_k(n)|$ with $r_k(0) = 1$.
Next, we state four identities which are valid for complex x with $|x| < 1$.

(10.3.16)
$$\prod_{n=1}^{\infty}(1+x^n)(1-x^{2n-1}) = 1$$

(10.3.17)
$$\prod_{n=1}^{\infty}(1-x^n)(1-x^{2n-1}) = \sum_{n=-\infty}^{\infty}(-x)^{n^2}$$

(10.3.18)
$$\prod_{n=1}^{\infty}(1-x^{2n})(1+x^n) = \sum_{n=0}^{\infty}x^{\frac{n(n+1)}{2}}$$

(10.3.19)
$$x\left\{\sum_{n=0}^{\infty}x^{\frac{n(n+1)}{2}}\right\}^8 = \sum_{n=1}^{\infty}\frac{n^3 x^n}{1-x^{2n}}$$

Identities (10.3.16) to (10.3.18) are due to Euler and Gauss. See Hardy and Wright [12]. (10.3.19) is in Ramanujan's collected papers [17]. (10.3.18) can be restated as (using (10.3.16))

(10.3.20)
$$\prod_{n=1}^{\infty}(1-x^{2n-1})^{-2} = \sum_{n=0}^{\infty}x^{\frac{n(n+1)}{2}}$$

Theorem 76 (John A. Ewell (1984)) : *For each positive integer n,*

(10.3.21)
$$\tau(n) = \sum_{j=1}^{n}(-1)^{n-j}r_{16}(n-j)2^{3b(j)}\sigma_3(M(j))$$

where b and M are as defined in (10.3.15).

Proof : Using (10.3.17) and (10.3.18),

(10.3.22)
$$\prod_{n=1}^{\infty}(1-x^n)^{24} = \left(\sum_{n=0}^{\infty}x^{\frac{n(n+1)}{2}}\right)^8 \left(\sum_{-\infty}^{\infty}(-x)^{n^2}\right)^{16}$$

Multiplying both sides of (10.3.22) by x, we get, using (10.3.19),

(10.3.23)
$$x\prod_{n=1}^{\infty}(1-x^n)^{24} = \sum_{n=1}^{\infty}\tau(n)x^n = \sum_{n=1}^{\infty}\frac{n^3 x^n}{1-x^{2n}}\sum_{n=0}^{\infty}(-1)^n r_{16}(n)x^n$$

But,

$$\sum_{n=1}^{\infty} \frac{n^3 x^n}{1-x^{2n}} = \sum_{n=1}^{\infty} n^3 x^n \sum_{k=0}^{\infty} x^{2nk} = \sum_{n=1}^{\infty} \sum_{k=0}^{\infty} n^3 x^{n(2k+1)}$$

$$= \sum_{m=1}^{\infty} x^m \sum_{\substack{d|m \\ d \text{ odd}}} (\frac{m}{d})^3$$

$$= \sum_{m=1}^{\infty} 2^{3b(m)} \sigma_3(M(m)) x^m$$

From (10.3.23), we get

$$\sum_{n=1}^{\infty} \tau(n) x^n = \sum_{j=1}^{\infty} 2^{3b(j)} \sigma_3(M(j)) x^j \sum_{k=0}^{\infty} (-1)^k r_{16}(k) x^k.$$

Or,

$$\sum_{n=1}^{\infty} \tau(n) x^n = \sum_{n=1}^{\infty} x^n \sum_{j=1}^{n} (-1)^{n-j} r_{16}(n-j) 2^{3b(j)} \sigma_3(M(j)).$$

Comparing coefficients of x^n, we arrive at the identity (10.3.21). □

Remark 10.3.1 : Theorem 76 has been adapted from [8].

Before concluding this section, we mention that certain congruence properties of $\tau(n)$ could be deduced from (10.3.21). In particular, a remarkable congruence shown by Ramanujan [17] says that

(10.3.24) $\tau(n) \equiv \sigma_{11}(n) \, (\text{mod } 691).$

10.4. Stirling numbers and Bernoulli numbers:

In Section 10.2, the generating functions of certain sequences were mentioned. See (10.2.2) and (10.2.7). In the case of $\{a_n\}$, suppose that the following recurrence relation holds.

(10.4.1) $a_n = c_1 a_{n-1} + c_2 a_{n-2} + \cdots + c_k a_{n-k} \quad (n \geqslant k),$

where c_i ($i = 1$ to k) are real numbers.

Let

(10.4.2) $f(x) = \sum_{n=0}^{\infty} a_n x^n$

where a_n satisfies (10.4.1).

Claim : $f(x)$ is a quotient of two polynomials.

Assume that $C = \max\{|c_1|, |c_2| \cdots |c_k|\}$

We shall denote the sum

(10.4.3) $|a_0| + |a_1| + \cdots + |a_{k-1}| = S_k.$

Then, by (10.4.1)

(10.4.4) $|a_k| \leq C S_k$; as $|c_i| \leq C$ for $i = 1, 2, \cdots, k$.

We prove by induction on n that

(10.4.5) $|a_n| \leq C(C+1)^{n-k} S_k, n \geq k$.

Suppose that for $k \leq i \leq m$, $|a_i| \leq C(C+1)^{i-k} S_k$.
Then, as

$$\begin{aligned}
|a_{m+1}| &= |c_1 a_m + c_2 a_{m-2} + \cdots + c_k a_{(m+1)-k}| \\
&\leq C\{|a_m| + |a_{m-1}| + \ldots + |a_{m+1-k}|\} \\
&= C \sum_{i=1}^{k} |a_{m+1-i}|
\end{aligned}$$

By induction hypothesis,

$$\sum_{i=1}^{k} |a_{m+1-i}| \leq C S_k \sum_{i=i}^{k} (C+1)^{(m+1-i)-k}$$

So,

$$\begin{aligned}
\sum_{i=1}^{k} |a_{m+1-i}| &\leq C S_k (C+1)^{m-k} \sum_{i=1}^{k} (C+1)^{1-i} \\
&= C S_k (C+1)^{m-k} \left\{ \frac{1 - (C+1)^{-k}}{1 - (C+1)^{-1}} \right\} \\
&= S_k (C+1)^{m+1-k} (1 - (C+1)^{-k}).
\end{aligned}$$

So,

$$|a_{m+1}| \leq C S_k (C+1)^{m+1-k}.$$

So, if (10.4.5) holds for $k \leq i \leq m$, it also holds for $i = m+1$. But, by (10.4.4), it holds for $i = k$. Thus, (10.4.5) holds for all $n \geq k$.

Let

$$\lambda = \frac{1}{C+1}. \quad 0 \leq x < \lambda.$$

Suppose that

$$\begin{aligned}
g(x) &= \sum_{n=k}^{\infty} C S_k (C+1)^{n-k} |x|^n \\
&= \frac{C S_k}{(C+1)^k} \sum_{n=k}^{\infty} \frac{1}{\lambda^n} |x|^n.
\end{aligned}$$

(10.4.6) $$g(x) = \frac{C S_k}{(C+1)^k} \sum_{n=k}^{\infty} |\frac{x}{\lambda}|^n.$$

As $\left|\frac{x}{\lambda}\right| < 1$, the right side of (10.4.6) is convergent. So, the comparison test, (10.4.2) is valid for $|x| < \lambda$.

Let

(10.4.7) $\qquad h(x) = 1 - \sum_{i=1}^{k} c_i x^i.$

$$f(x)h(x) = \sum_{n=0}^{\infty} a_n x^n - \sum_{n=0}^{\infty} a_n \sum_{i=1}^{k} c_i x^{n+i}$$

$$= \sum_{n=0}^{\infty} a_n x^n - \sum_{i=1}^{k-1} \sum_{j=1}^{i} a_{i-j} c_j x^i - \sum_{j=k}^{\infty} \sum_{j=1}^{k} a_{i-j} c_j x^j,$$

where $i - j = n$ runs from 0 to $k-1$, first and then from k to infinity. So,

$$f(x)h(x) = a_0 + \sum_{i=1}^{k-1} (a_i - \sum_{j=1}^{i} a_{i-j} c_j) x^i + \sum_{i=k}^{\infty} (a_i - \sum_{j=1}^{i} a_{i-j} c_j) x^i.$$

By virtue of (10.4.1)

$$a_i = \sum_{j=1}^{i} a_{i-j} c_j; \quad i \geq k.$$

Therefore,

$$f(x)h(x) = a_0 + \sum_{i=1}^{k-1} (a_i - \sum_{j=1}^{i} a_{i-j} c_j) x^i$$

which is a polynomial, say $t(x)$, of degree at most $(k-1)$. As $h(x)$ is a nonzero polynomial,

(10.4.8) $\qquad f(x) = \dfrac{t(x)}{h(x)};$

where $h(x)$ is as given in (10.4.7). This proves that $f(x)$ is a quotient of polynomials, as claimed.

Illustration 10.4.1 :

(i) *In the case of* $w_n(a,b;p,q)$ *with* $w_0 = a, w_1 = b$ *and* $w_n = pw_{n-1} - qw_{n-2}(n \geq 2)$, *we have* $k = 2$. $h(x) = 1 - px - (-q)x^2 = 1 - px + qx^2.$

$$t(x) = w_0 + (w_1 - pw_0)x = a + (b - pa)x.$$

So, the generating function of $w_n(a,b;p,q)$ *is as given in* (10.2.2).

(ii) *For the Fibonacci* $\{Fn\}$ *with* $F_0 = 1, F_1 = 1$ *and*

$F_n = F_{n-1} + F_{n-2}(n \geq 2)$, *we have* $k = 2$.

$$h(x) = 1 - x - x^2,$$
$$t(x) = f_0 + (F_1 - F_0)x = 1$$

Thus, the generating function of $\{F_n\}$ *is as shown in* (10.2.7).

Remark 10.4.1 : The idea of proof of (10.4.8) is given in [A3].

Next, we note that there are many functions of an arithmetical nature which satisfy certain recurrence relations.

The Fibonacci and Lucas numbers satisfy the recurrence relation

$$(10.4.9) \qquad f(n+1) = f(n) + f(n-1), \quad n \geq 1$$

for all integral values of n. For Fibonacci numbers we have $f(0) = 1$ and $f(1) = 1$. For Lucas numbers, $f(0) = 2$, $f(1) = 1$.

In the case of the quantities $\binom{n}{r}$, we have

$$(10.4.10) \qquad \binom{n+1}{r} = \binom{n}{r} + \binom{n}{r-1}$$

with $\binom{n}{r} = 0$ for $r > n$ and $\binom{n}{n} = 1$.

$S(n,k)$, the Stirling numbers of the first kind are defined by the relation

$$(10.4.11) \qquad (x+1)(x+2)\cdots(x+n) = \sum_{k=0}^{n} S(n,k)\, x^{n-k}$$

where $S(n,0) = 1$ for $n \geq 1$. (James Stirling 1692–1770). By convention, we take $S(n,k) = 0$ for $k > n > 0$. $S(n,k)$ denotes the sum of the products of the first n natural numbers taken k at a time. (10.4.11) can be rewritten as

$$(10.4.12) \qquad (x+n)\{(x+1)(x+2)\cdots(x+n-1)\} = (x+n)\sum_{k=0}^{n-1} S(n-1,k)\, x^{n-k-1}$$

Comparing the coefficients of x^{n-k} on both sides of (10.4.12), we obtain a recurrence relation

$$(10.4.13) \qquad S(n,k) = S(n-1,k) + n\, S(n-1,k-1)$$

satisfied by Stirling numbers of the first kind.

For $x,y \in \widetilde{\mathbb{Z}}$, we consider numbers $G(x,y)$ satisfying

$$(10.4.14) \qquad G(x,y) = G(x-1,y) + x\, G(x-1,y-1)$$

(in analogy with $S(n,k)$).

For definiteness, let us write $G(j,0) = a_j$ and $G(0,k) = b_k$ for $j,k \in \widetilde{\mathbb{Z}}$ with $(j,k) \neq (0,0)$. For Stirling numbers, we have $a_j = 1$, $b_k = 0$ for $k \neq 0$. $G(-n,k)$ are called Stirling numbers of the second kind with $a_j = 1$, $b_k = 0$, $k \neq 0$.

It is verified that $G(n,k) = 0$ for $0 \leq n < k$, when $G(j,0) = a_j = 1$ and $G(0,k) = b_k = 0$ for $k \neq 0$.

For, suppose that $n < k$.

$$G(n,k) = G(n-1,k) + nG(n-1,k-1)$$
$$G(n-1,k) = G(n-2,k) + (n-1)G(n-2,k-2)$$
$$\cdots\cdots\cdots\cdots\cdots\cdots\cdots\cdots\cdots$$
$$G(1,k) = G(0,k) + 0\,G(0,k-2)$$

As $G(1,k) = 0$, it follows that $G(n,k) = 0$, whenever $n < k$.

Fact 10.4.1 :

(10.4.15) $$S(n,k) = \sum_{j=1}^{n} j\,S(j-1,k-1).$$

(10.4.16) $$S(n,1) = \frac{(n+1)(n)}{2} = \binom{n+1}{2}.$$

Next, we define positive integers $H(n,k)$ given recursively by the formula

(10.4.17) $$H(n,k) = (2n-k-1)H(n-1,k) + (n-k)H(n-1,k-1)$$

with

(10.4.18) $$\begin{cases} H(n-1,-1) & = 0 = H(n-1,n-1),\, n > 1 \\ \text{and } H(0,0) & = 1,\, H(0,-1) = 0. \end{cases}$$

Further, $H(1,0) = 1$.

We have from (10.4.17),

$$H(n,0) = (2n-1)H(n-1,0) = (2n-1)(2n-3)H(n-2,0) = \cdots,$$

or,

(10.4.19) $$H(n,0) = (2n-1)(2n-3)\cdots 5\cdot 3\cdot 1.$$

Also,

$$H(n,n-1) = nH(n-1,n-1) + H(n-1,n-2)$$

As $H(n-1,n-1) = 0$,

$$H(n,n-1) = H(n-1,n-2) = \cdots = H(1,0) = 1.$$

Fact 10.4.2 : [Morgan Ward]

(10.4.20) $$S(n,k) = \sum_{j=0}^{k-1} (-1)^j H(k,j) \binom{n+k-j}{2k-j}$$

Also, $S(n,k)$ is a polynomial in n of degree $2k$ with fractional coefficients, in general. For proof, see [20].

Stirling numbers of the second kind are $G(-n,k)$, $n \geq 1$. (See (10.4.6))

Fact 10.4.3 :

$$(10.4.21) \qquad n!\, G(-n-1,k) = \sum_{j=0}^{n-1} (-1)^j \binom{n}{j} (n-j)^{n+k}, \quad n \geq 1, k \geq 0$$

For proof, see Hansraj Gupta [10].
For instance, $G(-2,k) = 1$.
From (10.4.21), after cancelling n from both sides, we also have

$$(10.4.22) \qquad (n-1)!\, G(-n-1,k) = \sum_{j=0}^{n-1} (-1)^j \binom{n-1}{j} (n-j)^{n+k-1},$$

$(n \geq 1, k \geq 0)$.

Taking $k = 0$ in (10.4.22), we also get

$$(10.4.23) \quad (n-1)! = \binom{n-1}{0} n^{n-1} - \binom{n-1}{1}(n-1)^{n-1}$$
$$+ \binom{n-1}{2}(n-2)^{n-1} - \cdots + (-1)^{n-1}\binom{n-1}{n-1} 1^{n-1}.$$

Next, we look at the generating function of Bernoulli (Jacob Bernoulli (1654–1705)) numbers.

Definition 10.4.1 : *For $x \in \mathbb{C}$, the functions $B_n(x)$ are defined by*

$$\frac{ze^{xz}}{e^z - 1} = \sum_{n=0}^{\infty} \frac{B_n(x)}{n!} z^n, \quad \text{where } |z| < 2\pi$$

Taking $x = 0$ in the above equation, we have

$$\frac{z}{e^z - 1} = \sum_{n=0}^{\infty} \frac{B_n}{n!} z^n, \quad |z| < 2\pi$$

where $B_n = B_n(0)$, $n = 0, 1, \ldots$

$$B_0 = 1, B_1 = \frac{1}{2}, B_2 = \frac{1}{6}, B_3 = 0 \ldots$$

$B_n(x)$ are called Bernoulli polynomials.

Lemma 10.4.1 : *Bernoulli polynomials $B_n(x)$ are given by*

$$(10.4.24) \qquad B_n(x) = \sum_{j=0}^{n} \binom{n}{r} B_j x^{n-j}.$$

Proof : For,

$$\sum_{n=0}^{\infty} \frac{B_n(x)}{n!} z^n = \frac{z}{e^z - 1} e^{xz} = \left(\sum_{n=0}^{\infty} \frac{B_n}{n!} z^n\right)\left(\sum_{n=0}^{\infty} \frac{x^n}{n!} z^n\right)$$

Equating coefficients of z^n from both sides, we get

$$\frac{B_n(x)}{n!} = \sum_{j=0}^{n} \frac{B_j}{j!} \frac{x^{n-j}}{(n-j)!}$$

which gives (10.4.24).   □

Lemma 10.4.2 : *For $n \geq 1$,*

(10.4.25) $B_n(x+1) - B_n(x) = nx^{n-1}$.

Also, $B_n(1) = B_n(0)$, if $n \geq 2$.

Proof : We have

$$z\frac{e^{(x+1)z}}{e^z - 1} - z\frac{e^{xz}}{e^z - 1} = z\,e^{xz}$$

So,

(10.4.26) $\sum_{n=0}^{\infty} \frac{B_n(x+1) - B_n(x)}{n!} z^n = \sum_{n=0}^{\infty} \frac{x^n}{n!} z^{n+1}$.

Equating coefficients of z^n from both sides of (10.4.26), we obtain (10.4.25). For $n \geq 2$, $B_n(1) = B_n(0)$ follows from (10.4.25).   □

Corollary 10.4.1 : *For $n \geq 2$,*

$$B_n(0) = B_n(1) = \sum_{j=0}^{n} \binom{n}{j} B_j, \text{ by using (10.4.17) and so,}$$

(10.4.27) $B_n = \sum_{j=0}^{n} \binom{n}{j} B_j$,

which is a recurrence relation for computing Bernoulli numbers.

Fact 10.4.4 : Bernoulli numbers B_n are related to Stirling numbers of the second kind $G(-n,r)$ by the formula

(10.4.28) $B_n = G(-2,n) - \frac{1!}{2}G(-3,n-1) + \frac{2!}{3}G(-4,n-2)\cdots$

$$+(-1)^n \frac{n!}{n+1} G(-n-2,0),$$

where $m!G(-m-2,n) = \sum_{j=0}^{m}(-1)^j \binom{m}{j}(m-j+1)^{m+n}$ (which is the same as (10.4.22)). For proof, see Hansraj Gupta [10]. See, also, Morgan Ward [20].

10.5. Binomial posets and generating functions

Let Λ be an index set. We consider a family $\mathcal{S}$ of sets S_λ where each S_λ is a finite set.

$$S = \{S_\lambda : \lambda \in \Lambda\}.$$

To determine the cardinality of S_λ, we introduce a function $f : \Lambda \to \widetilde{\mathbb{Z}}$ where $f(\lambda) = |S_\lambda|$. In combinational problems, there will be a relationship between $\lambda \in \Lambda$ and S_λ. Let

$$X = \{1, 2, \ldots, n\}$$

Suppose that S_n denotes the set of subsets of X. Then, $f(n) = |S_n| = 2^n$. f is an example of a counting function. If we consider $\widetilde{\mathbb{Z}} \times \widetilde{\mathbb{Z}} = \{(k,n) : k, n \in \widetilde{\mathbb{Z}}\}$, we write $S_{(k,n)}$ to denote the set of all subsets of X ($|X| = n$) of cardinality k. Then, $|S_{(k,n)}| = f((k,n)) = \binom{n}{k}$. If D_n denotes the set of divisors of n, $f(n) = |D_n| = d(n)$, the number of divisors of n. Here, $S = \{D_1, D_2, \ldots, D_n, \ldots\}$.

Definition 10.5.1 : *A generating function is a representation of a counting function $f : S \to \widetilde{\mathbb{Z}}$ as an element $F(f)$ of some algebra $\mathcal{A}$. In the case of the partition function $p : \widetilde{\mathbb{Z}} \to \widetilde{\mathbb{Z}}$ one has $F(x) = \sum_{n=0}^\infty p(n)x^n$, $|x| < 1$. $F(x)$ can be considered as a formal power series in x with coefficients from $\mathbb{C}$. If $\mathbb{C}[[x]]$ denotes the ring of formal power series in x (with coefficients from $\mathbb{C}$), $F(x) \in \mathbb{C}[[x]]$. (See notation 4.5.1 in chapter 4).*

$\mathbb{C}[[x]]$ is a vector space over $\mathbb{C}$. If

$$A(x) = \sum_{n=0}^\infty a_n x^n, \qquad B(x) = \sum_{n=0}^\infty b_n x^n$$

$A(x)B(x) = \sum_{n=0}^\infty c_n x^n$ where $c_n = \sum_{k=0}^n a_k b_{n-k}$.

If $\alpha \in \mathbb{C}$, $\alpha A(x) \in \mathbb{C}[[x]]$. Multiplication distributes addition. $\mathbb{C}[[x]]$ satisfies the axioms of an algebra. Therefore, it is meaningful to look at a generating function as an element of an algebra. We write

$$(10.5.1) \qquad F(f,x) = \sum_{n=0}^\infty f(n)x^n$$

It is the ordinary generating function of f.

$$(10.5.2) \qquad E(f,x) = \sum_{n=0}^\infty \frac{f(n)}{n!} x^n$$

is referred to as the exponential generating function of f.

Let F be a finite field with q elements, where $q = p^m$ (p a prime, $m \geq 1$). The series $G(f,x)$, given by

$$(10.5.3) \qquad G(f,x) = \sum_{n=0}^\infty \frac{f(n)x^n}{(1+q)(1+q+q^2)\cdots(1+q+\cdots q^{n-1})}$$

is known as the Eulerian generating function of f.

The series $H(f,x)$ given by

(10.5.4)
$$H(f,x) = \sum_{n=0}^{\infty} \frac{f(n)x^n}{q^{\binom{n}{2}}n!}$$

is known as the chromatic generating function of f.

Our aim is to consider generating functions in a more general setting. While generalizing Möbius inversion, we looked upon $\mathbb{N}$ as a locally finite partially ordered set. Following Richard P. Stanley [19], we define a binomial poset as follows:

Definition 10.5.2 : *A poset* $(P, \leq)$ *is called a binomial poset if it satisfies the following three conditions:*

(a) *P is locally finite. That is, every interval $[x,y] = \{t : x \leq t \leq y\}$ is finite, and P contains arbitrarily large finite chains. (By a chain, we mean a totally ordered subset of P).*

(b) *For every interval $[x,y]$ of P, all maximal chains between x and y have the same length $n = n(x,y)$. We call $[x,y]$ an n-interval. We observe that the length of a chain is one less than its number of elements.*

(c) *For all $n \in \mathbb{N}$, any two n-intervals contain the same number $B(n)$ of maximal chains.*

Examples 10.5.1 :

(i) When $P = \mathbb{N}$, $(\mathbb{N}, \leq)$ is a binomial poset with $B(n) = 1$ for all $n \in \mathbb{N}$.

(ii) Suppose $(P, \leq)$ is the lattice of all finite subsets of $\mathbb{N}$, ordered by inclusion. Consider an n-element subset S of $\mathbb{N}$. A singleton from S can be chosen in n ways. Suppose $S_1 = \{a_1\}$ is one such. A subset containing two elements one of which is a_1 can be chosen in $(n-1)$ ways, call one such set $S_2 = \{a_1, a_2\}$. S_1 and S_2 together can be chosen in $n(n-1)$ ways. So a chain (of subsets) of length n obtained by taking the null set ϕ also, can be chosen in $n!$ ways. So, $B(n) = n!$.

(iii) Suppose $(P, \leq)$ is the lattice of all finite dimensional subspaces of a vector space of infinite dimension over $GF(q)$. $q = p^m$ (p a prime, $m \geq 1$) ordered by inclusion. As discussed in Section 9.4 of chapter 9, the number of k linearly independent vectors in a vector space $V_n(q)$ of dimension n is $(q^n - 1)(q^n - q) \cdots (q^n - q^{k-1})$.

As we need to take one-dimensional, two-dimensional, $\cdots$, $(n-1)$-dimensional subspaces of $V_n(q)$, the number of chains of length n from the zero subspace to $V_n(q)$ is obtained using the formula

(10.5.5)
$$\binom{n}{k}_q = \frac{(q^n - 1)(q^n - q) \cdots (q^n - q^{k-1})}{(q^k - 1)(q^k - q) \cdots (q^k - q^{k-1})}$$

for the number of k-dimensional subspaces of $V_n(q)$. See (9.4.5). In fact,

$$(10.5.6) \qquad B(n) = \binom{2}{1}_q \binom{3}{2}_q \cdots \binom{n}{n-1}_q.$$

$$
\begin{aligned}
\binom{n}{n-1}_q &= \frac{(q^n - 1)(q^n - q) \cdots (q^n - q^{n-2})}{(q^{n-1} - 1)(q^{n-1} - q) \cdots (q^{n-1} - q^{n-2})} \\
&= \frac{(q^n - 1)(q^{n-1} - 1)(q^{n-2} - 1) \cdots (q^2 - 1)}{(q^{n-1} - 1)(q^{n-2} - 1) \cdots (q^2 - 1)(q - 1)} \\
&= \frac{q^n - 1}{q - 1}.
\end{aligned}
$$

So,

$$(10.5.7) \quad B(n) = (\frac{q^2 - 1}{q - 1})(\frac{q^3 - 1}{q - 1}) \cdots (\frac{q^n - 1}{q - 1})$$
$$= (q + 1)(q^2 + q + 1) \cdots (q^{n-1} + q^{n-2} + \cdots + q + 1).$$

Now, we point out that there is a particular expression for $B(n)$ with respect to a binomial poset $(P, \leq)$ considered. In the case of $(P, \leq)$ where P is the lattice of finite subsets of $\mathbb{N}$, $B(n) = n!$. This accounts for the exponential generating function. So, with a binomial poset $(P, \leq)$, we could associate a generating function $F_P(f, x)$ defined by

$$(10.5.8) \qquad F_p(f, x) = \sum_{n=0}^{\infty} \frac{f(n)x^n}{B(n)}$$

To make the analogy stronger, we need to consider the incidence algebra of generalized arithmetic functions. See definition 9.2.10 of chapter 9.

Definition 10.5.3 : *Let $(P, \leq)$ be a locally finite poset. The incidence algebra $\mathcal{A}(P)$ of P over $\mathbb{C}$ is the vector space of all functions $f : P \times P \to \mathbb{C}$ given by*

$$f(x, y) = 0 \text{ whenever } x \text{ is not } \leq y$$

For $f, g \in \mathcal{A}(P)$, addition and multiplication (see definition 9.2.3) are

$$(10.5.9) \qquad (f + g)(x, y) = \begin{cases} f(x, y) + g(x, y), & x \leq y \\ 0, & \text{otherwise} \end{cases}$$

and

$$(10.5.10) \qquad (f \cdot g)(x, y) = \begin{cases} \sum_{x \leq t \leq y} f(x, t)g(t, y), & \text{if } x \leq y; \\ 0, & \text{otherwise.} \end{cases}$$

We take the particular case where $(P, \leq)$ is a binomial poset (see definition 10.5.2). In a binomial poset, any two n-intervals contain the same number $B(n)$ of maximal chains. Suppose that $f \in \mathcal{A}(P)$ is such that

$$f(x, y) = f(t, w)$$

whenever $[x, y]$ and $[t, w]$ are n-intervals having the same number $B(n)$ of maximal chains. In such a situation, f depends only on n. Let $\mathcal{B}(P)$ denote the set of incidence functions which are constants on n-intervals.

For $f, g, \in \mathcal{B}(P)$, $f + g$ and $f \cdot g$ are in $\mathcal{B}(P)$. The function e_0 given by

$$(10.5.11) \qquad\qquad e_0(x, y) = \begin{cases} 1, & x = y; \\ 0, & x \neq y. \end{cases}$$

is such that

$$(10.5.12) \qquad\qquad e_0(n) = \begin{cases} 1, & n = 0; \\ 0, & n \neq 0. \end{cases}$$

$e_0 \in \mathcal{B}(P)$. We write $f(n)$ for $f(x, y)$ when $[x, y]$ is an n-interval. It is verified that $\mathcal{B}(P)$ is a subalgebra of $A(P)$.

Let $[x, y]$ be an n-interval in $(P, \leq)$. For t such that $x \leq t \leq y$, suppose that $[x, t]$ is an i-interval. There are $B(i)$ maximal chains in $[x, t]$ and $B(n - i)$ maximal chains in $[t, y]$. Therefore, there are $B(i)B(n - i)$ maximal chains of $[x, y]$ passing through t. So the number of elements t in an n-interval $[x, y]$ such that $[x, t]$ is an i-interval is given by $\dfrac{B(n)}{B(i)B(n - i)}$.

Notation 10.5.1 : $\begin{bmatrix} n \\ i \end{bmatrix}$ denotes $\dfrac{B(n)}{B(i)B(n - i)}$.

$\begin{bmatrix} n \\ i \end{bmatrix}$ is the analogue of the familiar expression $\begin{pmatrix} n \\ k \end{pmatrix} = \dfrac{n!}{k!(n - k)!}$.

If $A(i) = \begin{bmatrix} i \\ 1 \end{bmatrix} = \dfrac{B(i)}{B(i-1)}$

$$(10.5.13) \qquad\qquad B(n) = A(n)A(n - 1) \cdots A(1)$$

which corresponds to the expression for $n!$. Further,

$$(10.5.14) \qquad (f \cdot g)(n) = \sum_{j=0}^{n} \begin{bmatrix} n \\ j \end{bmatrix} f(j)g(n - j), \quad \text{using } (10.5.10)$$

Theorem 77 (R. P. Stanley (1978)) : *If $(P, \leq)$ is a binomial poset and $\mathcal{B}(P)$ denotes the set of incidence functions which are constants on n-intervals, then, $\mathcal{B}(P) \cong \mathbb{C}[[x]]$.*

Proof : Let $f \in B(P)$. The generating function of f is given by

$$F_P(f, x) = \sum_{n=0}^{\infty} \frac{f(n)x^n}{B(n)}$$

where $f(n) = f(x, y)$ and $[x, y]$ is an n-interval. It is easy to check that

$$F_P(f + g, x) = F_P(f, x) + F_P(g, x)$$

Given $A(x) \in \mathbb{C}[[x]]$, we can write $A(x)$ as

$$A(x) = a_0 + a_1 x + a_2 x^2 + \cdots \quad \text{where } a_n \in \mathbb{C}$$

$\{a_n\}$ can be so chosen that $a_n = \dfrac{f(n)}{B(n)}$ for $f \in \mathcal{B}(P)$.

So, there exists a formal power series corresponding to an element $f \in \mathcal{B}(P)$ and vice-versa.

Let $\psi : \mathcal{B}(P) \to \mathbb{C}[[x]]$ be defined by

$$\psi(f) = F_P(f, x). \text{ Then, } \psi(f + g) = \psi(f) + \psi(g).$$

For $\alpha \in \mathbb{C}$, $\psi(\alpha f) = \alpha \psi(f)$.

$$\psi(f \cdot g) = \sum_{n=0}^{\infty} \frac{(f \cdot g)(n) x^n}{B(n)}.$$

But $(f \cdot g)(n) = \sum_{j=0}^{n} \begin{bmatrix} n \\ j \end{bmatrix} f(j) g(n - j).$

$$\text{Now, } F_P(f, x) F_P(g, x) = \left(\sum_{n=0}^{\infty} \frac{f(n) x^n}{B(n)} \right)\left(\sum_{n=0}^{\infty} \frac{g(n) x^n}{B(n)} \right)$$

$$= \sum_{n=0}^{\infty} \left(\sum_{j=0}^{n} \frac{f(j) g(n - j)}{B(j) B(n - j)} \right) x^n$$

$$= \sum_{n=0}^{\infty} \left(\sum_{j=0}^{n} \frac{B(n)}{B(j) B(n - j)} f(j) g(n - j) \right) \frac{x^n}{B(n)}$$

$$= \sum_{n=0}^{\infty} \frac{(f \cdot g)(n)}{B(n)} x^n.$$

So, $f \mapsto F_P(f, x)$ is an algebra homomorphism from $\mathcal{B}(P)$ into $\mathbb{C}[[x]]$.

$$\ker \psi = \{ f \in \mathcal{B}(P) : \psi(f) = 1(x) \}$$

where $1(x) \in \mathcal{C}[[x]]$ given by $1(x) = 1$.

So, $\sum_{n=0}^{\infty} \dfrac{f(n) x^n}{B(n)} = 1(x) \Rightarrow f(0) = 1$ and $f(n) = 0$ for $n \geq 1$.

Using (10.5.12), we note that $\ker \psi = e_0 \in \mathbb{B}(P)$, the unity element of the algebra $\mathbb{B}(P)$. So, ψ is one-one. It is also onto, as any element $\sum_{n=0}^{\infty} a_n x^n$ is the formal power series corresponding to a predetermined $f \in \mathcal{B}(P)$. Therefore, ψ is an isomorphism onto $\mathbb{C}[[x]]$ as desired. $\square$

Examples 10.5.2 : Let $\zeta \in \mathcal{B}(P)$ be defined by

(10.5.15) $\qquad\qquad \zeta(n) = 1 \quad \text{for all } n \in \bar{\mathbb{Z}}$

$$\zeta^2(n) = \sum_{t \in [x,y]} \zeta(x,t)\zeta(t,y)$$

$$= \sum_{t \in [x,y]} 1$$

= the cardinality of the set of elements in $[x,y]$.

As the cardinality of a chain of length n is $(n+1)$, we have

(10.5.16)
$$\sum_{n=0}^{\infty} \frac{(n+1)x^n}{B(n)} = \sum_{n=0}^{\infty} (n+1)x^n = (\sum_{n=0}^{\infty} x^n)^2 = (1-x)^{-2}$$

with $|x| < 1$.

In the case of the binomial poset of finite subsets of a set X, as $B(n) = n!$ (see example 10.5.1 (ii)), if $N(n)$ denotes the number of subsets of an n-element set, as $N(n) = \zeta^2(n)$,

$$\sum_{n=0}^{\infty} \frac{N(n)x^n}{n!} = (\sum_{n=0}^{\infty} \frac{x^n}{n!})^2 = e^{2x} = \sum_{n=0}^{\infty} \frac{2^n x^n}{n!}$$

and so, $N(n) = 2^n$.

10.6. Dirichlet series

Let f be an arithmetic function. Dirichlet introduced the series

(10.6.1)
$$\sum_{n=1}^{\infty} \frac{f(n)}{n^s} \quad (s \text{ complex}, \text{Re } s > a)$$

as the generating function of f where $f(n)$ is defined for $n \geq 1$. We call (10.6.1) the Dirichlet series of f.

A subset of $\mathbb{C}$ defined by $\{s \in \mathbb{C} : \text{Re } s > a\}$ is a half-plane.

To each Dirichlet series such as (10.6.1), there is a half-plane $\sigma > \sigma_a$ in which $\sum f(n)n^{-s}$ converges absolutely. ($\sigma = Re\, s$)

Fact 10.6.1 : Given f, suppose that $\sum |f(n)n^{-s}|$ does not converge for all s or does not diverge for all s. Then, there exists σ_a, a real number called the abscissa of absolute convergence such that $\sum f(n)n^{-s}$ converges absolutely for Re $s > \sigma_a$.

For proof see Tom M. Apostol [3].

Definition 10.6.1 : *When $\sum f(n)n^{-s}$ converges absolutely for Re $s > \sigma_a$, we write $F(s)$ to denote the sum function so that*

$$F(s) = \sum_{n=1}^{\infty} f(n)n^{-s}, \quad Re\, s > \sigma_a$$

Lemma 10.6.1 (Uniqueness theorem) **:** *If $F(s)$ and $G(s)$ represent two Dirichlet series:*

$$F(s) = \sum f(n)n^{-s}, \, G(s) = \sum g(n)n^{-s} \text{ where Re } s > \sigma_a$$

and both $\sum f(n)n^{-s}$ and $\sum g(n)n^{-s}$ have the same half-plane of absolute conver-
gence Re $s > \sigma_a$ and if $F(s) = G(s)$ for a sequence $\{s_k\}$ such that $\sigma_k = Re\ s_k \to \infty$
as $k \to \infty$, then, $f(n) = g(n)$ for all $n \geq 1$.

Proof : We write $h = f - g$ and $H(s) = F(s) - G(s)$, Re $s > \sigma_a$. By the hypothesis,
$H(s_k) = 0$ for $k \geq 1$ where Re $s_k \to \infty$ as $k \to \infty$.

Claim : $h(n) = 0$ for all $n \geq 1$.
On the contrary, suppose that $h(n) \neq 0$ for some n.
Let m be the smallest positive integer such that $h(m) \neq 0$.
Then,

$$H(s) = \sum_{n=m}^{\infty} \frac{h(n)}{n^s} = \frac{h(m)}{m^s} + \sum_{m+1}^{\infty} \frac{h(n)}{n^s}$$

Then,

$$h(m) = m^s H(s) - m^s \sum_{m+1}^{\infty} \frac{h(n)}{n^s}$$

Let $s = s_k$, then, $H(s_k) = 0$ or

$$h(m) = -m^{s_k} \sum_{m+1}^{\infty} h(n)n^{-s_k}$$

Choose k such that Re $s_k > c$ and $c > \sigma_a$. Then,

$$|h(m)| = |-m^{s_k} \sum_{m+1} h(n)n^{-s_k}|$$

$$\leq m^{\sigma_k}(m+1)^{-(\sigma_k - c)} \sum_{n=m+1}^{\infty} |h(n)|n^{-c} \text{ where } \sigma_k = Re\ s_k$$

$$= (\frac{m}{m+1})^{\sigma_k} M, \text{ where } M \text{ is independent of } k.$$

The quantity $(\frac{m}{m+1})^{\sigma_k} \to 0$ as $k \to \infty$.
So $h(m) = 0$, a contradiction. That is, $f = g$. $\square$

Remark 10.6.1 : Suppose that $F(s)$ given by $F(s) = \sum f(n)n^{-s}$ is such that
$F(s) \neq 0$ for some s with Re $s > \sigma_a$. Then, there is a half-plane $\sigma > c \geq \sigma_a$ in
which $F(s)$ is never zero.

For, if we take that no such half-plane exists, then for every $k \in \mathbb{N}$, there is
a complex number s_k with Re $s_k > k$ such that $F(s_k) = 0$. Since Re$(s_k) \to \infty$ as
$k \to \infty$, lemma 10.6.1 says that $f(n) = 0$ for all n, contradicting the hypothesis
that $F(s) \neq 0$ for some s.

Theorem 78 (Multiplication of Dirichlet series) : *Let $F(s)$ and $G(s)$ be the Dirichlet series of two arithmetic functions f and g, given by*

$$(10.6.2) \qquad F(s) = \sum_{n=1}^{\infty} f(n)n^{-s}, \quad Re\ s > a$$

$$(10.6.3) \qquad G(s) = \sum_{n=1}^{\infty} g(n)n^{-s} \quad Re\ s > b$$

If $h(n) = \sum_{d|n} f(d)g(\frac{n}{d})$, then,

$$(10.6.4) \qquad F(s)G(s) = \sum_{n=1}^{\infty} h(n)n^{-s} \quad Re\ s > \max\{a,b\}.$$

Proof : Suppose that $Re\ s > \max\{a,b\}$. Then,

$$F(s)G(s) = \sum_{n=1}^{\infty} f(n)n^{-s} \sum_{r=1}^{\infty} g(r)r^{-s} = \sum_{n=1}^{\infty} \sum_{r=1}^{\infty} f(n)g(r)(nr)^{-s}$$

The multiplication of series is valid and we can rearrange the terms in any manner without altering the sum.

Then,

$$(10.6.5) \qquad F(s)G(s) = \sum_{m=1}^{\infty} (\sum_{nr=m} f(n)f(r))m^{-s} = \sum_{m=1}^{\infty} h(m)m^{-s}$$

where $h = (f \cdot g)$. The abscissa of absolute convergence of the series on the extreme right of (10.6.5) determines the half-plane in which both (10.6.2) and (10.6.3) hold and so it is $Re\ s > \max\{a,b\}$. $\qquad \square$

Examples 10.6.1 : The Riemann ζ-function $\zeta(s)$ is given by

$$(10.6.6) \qquad \zeta(s) = \sum_{n=1}^{\infty} n^{-s}, Re\ s > 1$$

$\zeta(s)$ is the Dirichlet series of the arithmetic function e, given by $e(n) = 1$, $n \geq 1$. Further,

$$(10.6.7) \qquad \sum_{n=1}^{\infty} \mu(n)n^{-s} = \frac{1}{\zeta(s)}, \quad Re\ s > 1$$

as $e \cdot \mu = e_0$, where $e_0(n) = [\frac{1}{n}]$. ([x] denotes the greatest integer not exceeding x). It is easy to check the following assertions:

If ϕ denotes Euler's ϕ-function

$$(10.6.8) \qquad \sum_{n=1}^{\infty} \phi(n)n^{-s} = \frac{\zeta(s-1)}{\zeta(s)}, \quad Re\ s > 2.$$

If $\sigma_k(n)$ denotes the sum of the kth-powers of the divisors of n,

(10.6.9) $$\sum_{n=1}^{\infty} \sigma_k(n)n^{-s} = \zeta(s)\zeta(s-k), \quad \text{Re } s > \max\{1, k+1\}$$

Theorem 79 (Euler-product theorem) : *Let f be a multiplicative arithmetic function with the property that $\sum_{n=1}^{\infty} f(n)$ is absolutely convergent. Then, f satisfies*

(10.6.10) $$\sum_{n=1}^{\infty} f(n) = \prod_{p}\{1 + f(p) + f(p^2) + \ldots\}$$

where the product on the right extends over all primes.

Proof : We denote the product

$$\prod_{p \leq r}\{1 + f(p) + f(p^2) + \cdots\} \text{ by } P(r).$$

For each prime p, $1 + f(p) + f(p^2) + \cdots$ is absolutely convergent and so, $P(r)$ can be obtained by multiplying the series for primes $p \leq r$. A rearrangement of terms will give for a typical term:

$$f(p_1^{a_1})f(p_2^{a_2})\cdots f(p_k^{a_k}) = f(p_1^{a_1}p_2^{a_2}\cdots p_k^{a_k}),$$

as f is multiplicative. If S denotes the set of all $n \in \mathbb{N}$ for which each prime factor of n is less than r, we set

$$P(r) = \sum_{n \in S} f(n)$$

So,

$$\sum_{n=1}^{\infty} f(n) - P(r) = \sum_{n \in T} f(n)$$

where T consists of elements n for which at least one prime factor is bigger than r.

So,

$$|\sum_{n=1}^{\infty} f(n) - P(r)| \leq \sum_{n \in T} |f(n)| \leq \sum_{n > r} |f(n)|$$

For large r, $\sum_{n>r} |f(n)|$ can be made as small as we please, as $\sum_{n=1}^{\infty} |f(n)|$ is convergent. So,

(10.6.11) $$\lim_{r \to \infty} P(r) = \sum_{n=1}^{\infty} f(n)$$

Now, $\Pi(1+a_n)$ converges absolutely whenever the corresponding series $\sum a_n$ converges absolutely [1]. Here,

$$\sum_{p \leq r} |f(p) + f(p^2) + \cdots| \leq \sum_{p \leq r}(|f(p)| + |f(p^2)| + \cdots) \leq \sum_{n=1}^{\infty} |f(n)|$$

So, $\sum_{p \leq r} |f(p) + f(p^2) + \cdots|$ is bounded and hence
$\sum_p |f(p) + f(p^2) + \cdots|$ converges. Now, as the set of primes is countably infinite.
Hence, $\prod_p \{1 + f(p) + f(p^2) + \cdots\}$ converges absolutely and so, (10.6.10) holds.
$\square$

Corollary 10.6.1 : *If $F(s) = \sum_{n=1}^{\infty} f(n)n^{-s}$, Re $s > a$, where f is multiplicative, one has*

$$(10.6.12) \qquad \sum_{n=1}^{\infty} f(n)n^{-s} = \prod_p \{1 + f(p)p^{-s} + f(p^2)p^{-2s} + \cdots\}, \qquad Re\ s > a.$$

Further, if f is completely multiplicative,

$$(10.6.13) \qquad \sum_{n=1}^{\infty} f(n)n^{-s} = \prod_p (1 - f(p)p^{-s})^{-1}, \quad if\ Re\ s > a.$$

Remark 10.6.2 :
  (a) The Riemann ζ-function has the Euler-product form

$$(10.6.14) \qquad \zeta(s) = \prod_p (1 - p^{-s})^{-1}, \quad Re\ s > 1.$$

  (b) Proofs of lemma 10.6.1 and theorems 78, 79 have been adapted from [3].

We consider the set S of all sequences $(a_1, a_2, \ldots)$ of non-negative integers in which only finitely many a_i are nonzero.
Let $P = \{p_i : p_i$ is the ith-prime, $i \in \mathbb{N}\}$.
If $r = p_{i1}^{a_{i1}} p_{i2}^{a_{i2}} \cdots p_{ik}^{a_{ik}}$, p_{ij} is a prime in the set P having a predetermined place. $(a_{i1}, a_{i2}, \ldots)$ is a sequence $\in S$.
Let $\mathcal{A}$ denote the set of arithmetic functions $f : \mathbb{N} \to \mathbb{C}$.
For $f \in \mathcal{A}$, we associate

$$f(r) \text{ with } \sum_{r=1}^{\infty} f(r) x_{i1}^{a_{i1}} x_{i2}^{a_{i2}} \cdots x_{ik}^{a_{ik}}$$

So the power series $\sum_{r=1}^{\infty} f(r) x_{i1}^{a_{i1}} \cdots x_{ik}^{a_{ik}}$ corresponds to $f \in \mathcal{A}$.
Let $\mathbb{C}[[x_1, x_2, \ldots, x_n, \ldots]]$ denote the ring of formal power series in countably many indeterminates $x_1, x_2, \ldots, x_n, \ldots$. It was denoted by $\mathbb{C}_w$ (4.5.15) in chapter 4.

$$(10.6.15) \qquad \text{Then } F(f, x_1, x_2, \ldots) = \sum_{a_1, a_2, \ldots = 0}^{\infty} f(a_1, a_2, \ldots) x_1^{a_1} x_2^{a_2} \cdots$$

where $f(a_1, a_2, \ldots)$ corresponds to $f(r)$ with $r = p_1^{a_1} p_2^{a_2} \cdots$. (10.6.15) has been considered in definition 4.5.2 of chapter 4. As in theorem 77, $\mathcal{A}$ considered as a $\mathbb{C}$-algebra is isomorphic to $\mathbb{C}[[x_1, x_2, \ldots]]$. See corollary 4.5.1 in chapter 4.

Let $\mathcal{D}(\mathcal{A})$ denote the set of Dirichlet series of functions belonging to $\mathcal{A}$. Suppose for $f \in \mathcal{A}$,

$$(10.6.16) \qquad F(s) = \sum_{n=1}^{\infty} f(n)n^{-s}, \quad \text{Re } s > a \text{ (say)}$$

Theorem 80 : $\mathbb{C}[[x_1, x_2, \ldots]]$ *is isomorphic to* $\mathcal{D}(\mathcal{A})$.

Proof : If $\mathcal{A}$ is considered as a $\mathbb{C}$-algebra with Dirichlet multiplication as giving the product of arithmetic functions, by theorem 78, $\mathcal{A}$ is isomorphic to $\mathcal{D}(\mathcal{A})$ by the map $\psi : \mathcal{A} \to \mathcal{D}(\mathcal{A})$ where

$$\psi(f) = F(s) = \sum_{n=1}^{\infty} f(n)n^{-s} \quad (\text{Re } s > a).$$

If $G(s)$ is the Dirichlet series of g, $f + g$ has Dirichlet series

$$\sum_{n=1}^{\infty} (f(n) + g(n))n^{-s}, \quad \text{Re } s > s_0 \text{ (say)}.$$

Also, $\sum_{d|n} f(d)g(n/d)$ has Dirichlet series $F(s)G(s)$ for Re $s > s_1$, (say). So, the algebra $\mathcal{A}$ is isomorphic to $\mathcal{D}(\mathcal{A})$.

In $\mathbb{C}[[x_1, x_2, \ldots, x_i, \ldots]]$, we take x_i to correspond to p_i^{-s} where p_i is the ith-prime.

Then,

$$\sum_{n=1}^{\infty} f(n)n^{-s} \text{ corresponds to } \sum_{a_1,a_2,\ldots=0}^{\infty} f(a_i, a_2, \ldots) x_1^{a_1} x_2^{a_2} \cdots$$

For, if $n = p_{i1}^{a_{i1}} p_{i2}^{a_{i2}} \cdots p_{ik}^{a_{ik}}$,
$f(n)p_{i1}^{-a_{i1}s} p_{i2}^{-a_{i2}s} \cdots p_{ik}^{-a_{ik}s}$ corresponds to

$$f(a_{i1}, a_{i2}, \ldots) x_{i1}^{a_{i1}} x_{i2}^{a_{i2}} \cdots x_{ik}^{a_{ik}}.$$

In other words, the transformation is $p_{ij}^{-a_{ij}s} \mapsto x_{ij}^{a_{ij}}$.

Addition and multiplication are preserved and so

$$\mathcal{D}(\mathcal{A}) \cong \mathbb{C}[[x_1, x_2, \ldots, x_n, \ldots]].$$

That is, as $\mathbb{C}$-algebras, $\mathcal{D}(\mathcal{A})$ and $\mathbb{C}[[x_1, x_2, \ldots]]$ are isomorphic. $\qquad \square$

Observation 10.6.1 : *If* $\mathcal{A}^*$ *denotes the set of arithmetic functions* $f : \widetilde{Z} \to \mathbb{C}$, $\mathcal{A}^*$ *is a commutative ring with unity* c_0 *given by*

$$c_0(r) = \begin{cases} 1, & r = 0 \\ 0 & r \geq 1 \end{cases}$$

under the operations of ordinary addition and Cauchy multiplication given by

$$(f \cdot g)(r) = \sum_{i=0}^{r} f(i)g(r-i); \quad f, g \in \mathcal{A}^*$$

Any $f \in \mathcal{A}^*$ for which $f(0) \neq 0$ is a unit in $\mathcal{A}^*$. From the way Cauchy multiplication is defined, it is clear that $\mathcal{A}^* \cong \mathbb{C}[[x]]$. In fact, $\mathcal{A}^*$ and $\mathbb{C}[[x]]$ are two sides of the same coin as they have the structure of the set of complex-valued sequences $(a_0, a_1, a_2, \ldots)$. The important point that we observe is that any nonzero non-unit is of the form (x^r), $r \geq 1$ and $x = (0, 1, 0, 0, \ldots)$.

We take $\pi : \widetilde{\mathbb{Z}} \to \mathbb{C}$ as

$$\pi(r) = \begin{cases} 1, & r = 1 \\ 0, & r \neq 1 \end{cases}$$

π has the property: whenever $\pi | f \cdot g$, π either divides f or g. For

$$\pi^k(r) = \begin{cases} 1, & r = k \\ 0, & r \neq k. \end{cases}$$

If $f(k) = a_k \in \mathbb{C}$, $f(k) = a_k \pi^k$.
If $g(s) = b_s \in \mathbb{C}$, $g(s) = b_s \pi^s$.

In fact, any nonzero non-unit element of $\mathcal{A}^*$ is a multiple of π. That is, π serves as a prime in $\mathcal{A}^*$.

In other words, every nonzero non-unit in $\mathcal{A}^*$ is generated by a single prime. We, thus, have an example of a UFD in which there is only one prime, namely

$$\pi(r) = \begin{cases} 1, & r = 1, \\ 0, & otherwise. \end{cases}$$

Theorem 81 : *Let K be the field of quotients of $F[[x]]$ (F a field). Then, $F[[x]]$ is a valuation ring of K.*

Proof : In theorem 25 of chapter 4, we have shown that $F[[x]]$ is a P.I.D. Any nonzero non-unit of $F[[x]]$ is given by $A(x) = x^k B(x)$, where $k \geq 1$ and $B(x)$ is an invertible element of $F[[x]]$. That is, if $T(x)$ is in K, $T(x) = x^t U(x)$ where $U(x)$ is a unit in $F[[x]]$ and $t \in \mathbb{Z}$. If t is a positive integer $T(x) \in F[[x]]$. If t is a negative integer, $T^{-1}(x) = x^{-t} V(x)$ where $U(x)V(x) = (1_F, 0_F, 0_F, \ldots)$. As $V(x)$ is a unit in $F[[x]]$, either $T(x)$ or $T^{-1}(x) \in F[[x]]$.
So, $F[[x]]$ is a valuation ring of K. $\qquad\square$

Corollary 10.6.2 : *The valuation ring $F[[x]]$ is a PID having a unique nonzero prime ideal (x) where $x = (O_F, 1_F, O_F, \ldots)$.*

Remark 10.6.3 : For the field K of quotients of $F[[x]]$, given $A(x) \in K$, suppose that we define $v(A(x)) = v(x^t U(x)) = t$ where t is zero, positive or a negative integer, and $v(0) = \infty$. $U(x)$ is a unit in $F[[x]]$. One could verify that

$$v(A(x)B(x)) = v(A(x)) + v(B(x)), \quad \text{for all } A(x), B(x) \in K$$

and $v(A(x) + B(x)) \geq \min\{v(A(x)), v(B(x))\}$ for all $A(x), B(x) \in K$. Then, for $A(x) \in F[[x]]$, $v(A(x)) \geq 0$. The maximal ideal (x) of $F[[x]]$ is such that $v(A(x)) > 0$ for all $A(x) \in (x)$. $v : K \to \mathbb{Z}$ is a discrete valuation of K.

10.7. Notes with illustrative examples

The application of generating functions to the identity involving the partition function was shown in Theorem 74. Then came the generalization to arithmetical functions by means of Dirichlet series. The simplest Dirichlet series is that of $e, e(n) = 1, n \geq 1$.

$$\zeta(s) = \sum_{n=1}^{\infty} n^{-s} \quad \mathrm{Re}\, s > 1.$$

When $s = 2$, we get $\zeta(2) = \frac{\pi^2}{6}$. Further, $\zeta(2n)$ is a rational multiple of π^{2n} for all $n \in \mathbb{N}$. $\zeta(4) = \frac{\pi^4}{90}$ and

$$(10.7.1) \qquad \zeta(2n) = \frac{2^{2n-1} B_n}{(2n)!} \pi^{2n},$$

where B_n is the nth Bernoulli number defined in Section 10.4. A remarkable result of B. Riemann is the functional equation satisfied by $\zeta(s)$:

$$(10.7.2) \qquad \pi^{-s/2} \Gamma(\frac{s}{2}) \zeta(s) = \pi^{\frac{s-1}{2}} \Gamma(\frac{1-s}{2}) \zeta(1-s),$$

where $\Gamma(s)$ is given by

$$(10.7.3) \qquad \Gamma(s) = \int_0^{\infty} e^{-y} y^{s-1} dy, \quad \mathrm{Re}\, s > 0.$$

See Tom Apostol [3] or K. Chandrasekharan [6].

If $F(s) = \sum_{n=1}^{\infty} a_n n^{-s}$ ($\mathrm{Re}\, s > \alpha$) and $G(s) = \sum_{n=1}^{\infty} b_n n^{-s}$ ($\mathrm{Re}\, s > \beta$) then,

$$(10.7.4) \qquad \psi(x) = \sum_{n=1}^{\infty} \frac{a_n x^n}{1 - x^n} = \sum_{n=1}^{\infty} b_n x^n$$

if, and only if, $\zeta(s) F(s) = G(s)$ where $\mathrm{Re}\, s > \max\{1, \alpha, \beta\}$.
See Hardy and Wright [12].

As an example of (10.7.4) we take $F(s) = \sum_{n=1}^{\infty} \mu(n) n^{-s}$, $\mathrm{Re}\, s > 1$, μ being the Möbius function. Then, $F(s) = \zeta(s)^{-1}$ and so $G(s) = 1$.
Then,

$$(10.7.5) \qquad \sum_{n=1}^{\infty} \frac{\mu(n) x^n}{1 - x^n} = \sum_{n=1}^{\infty} b_n x^n = x. \quad (b_n = e_0(n) = [\frac{1}{n}]).$$

Let f be any arithmetic function. $\sum_{n=1}^{\infty} \frac{f(n) x^n}{1 - x^n}$ is called a Lambert series. Assuming the conditions for absolute convergence, we can arrive at

$$(10.7.6) \qquad \sum_{n=1}^{\infty} \frac{f(n) x^n}{1 - x^n} = \sum_{n=1}^{\infty} g(n) x^n,$$

whenever $g(n) = \sum_{d|n} f(d)$. (10.7.6) is a restatement of (10.7.4) without introducing $\zeta(s)$.

For $|x| < 1$, the following equations are easily verified :

If ϕ denotes Euler's totient,

(10.7.7)
$$\sum_{n=1}^{\infty} \frac{\phi(n)x^n}{1-x^n} = \frac{x}{(1-x)^2}.$$

If $\sigma_k(n)$ denotes the sum of the kth-powers of the divisors of n,

(10.7.8)
$$\sum_{n=1}^{\infty} \frac{n^k x^n}{1-x^n} = \sum_{n=1}^{\infty} \sigma_k(n)x^n$$

For more results of this type, see Hardy and Wright [12].

For an exhaustive study of generating functions, see Doubilet, Rota and Stanley [7].

Techniques of generating functionology have also been developed to go deeper into combinatorial aspects. See E. A. Bender and J. R. Goldman [A2] and M. Henle [A6] also. For a detailed account of results relating to Riemann ζ-function, see S. J. Petterson [A8].

Applications of Burnside's lemma (theorem 52) were discussed in chapter 7. One makes use of certain polynomials in a finite number of indeterminates for a counting technique. Though this is not a generating function technique, a particular value of a polynomial is needed for enumeration. The enumeration theorem is due to George Polya (1887–1985). See [A 5].

Let G be a subgroup of S_n. $\sigma \in G$ is a permutation of the set $X = \{1, 2, \ldots, m\}$ $(m \leq n)$. σ could be written as a product of disjoint cycles. If $l(\sigma)$ denotes the number of cycles in σ and if $\sigma \in G$ has $l_i(\sigma)$ cycles of length i,

$$l(\sigma) = l_1(\sigma) + l_2(\sigma) \ldots + l_m(\sigma).$$

Since, the union of disjoint cycles in σ is $\{1, 2, \ldots m\}$,

(10.7.9) $m = 1\,l_1(\sigma) + 2\,l_2(\sigma) + \cdots + m\,l_m(\sigma).$

Definition 10.7.1 : *If $x_1, x_2, \ldots x_m$ are interminates, the cycle index of $\sigma \in G$ is defined as*

(10.7.10) $x_1^{l_1(\sigma)} x_2^{l_2(\sigma)} \ldots x_m^{l_m(\sigma)}$

Definition 10.7.2 : *The cycle index polynomial of G is defined as*

$$P_G(x_1, x_2, \ldots x_m) = \frac{1}{|G|} \sum_{\sigma \in G} x_1^{l_1(\sigma)} x_2^{l_2(\sigma)} \ldots x_m^{l_m(\sigma)}$$

Let F be a set of r colours $f_1, f_2, \ldots f_r$. We write $X = \{1, 2, \ldots m\}$. $(m \geq r)$. F^X denotes the set of all functions from X into F. G is a permutation group on F^X given by

(10.7.11) $g(f_1, f_2, \ldots, f_r) = (f_{g(1)}, f_{g(2)}, \ldots, f_{g(m)})$

where g is a permutation of $\{1, 2, \ldots, m\}$ and $g : F^r \to F^X$ is as given in (10.7.11). In fact, G is a subgroup of S_{r^m} where $r^m = |F^X|$.

We state without proof

Proposition 10.7.1 (Polya's theorem) : *The number n of orbits of G on F^X is given by*

$$n = P_G(r, r, \ldots, r)$$

where $P_G(r, r, \ldots r)$ is the value of $P_G(x_1, x_2, \ldots x_m)$ at $x_1 = x_2 \ldots = x_m = r$.

For proof, see F. Harary and E. M. Palmer [A5]. See also Rudolf Lidl and Günter Pilz [A7].

Illustration 10.7.1 : *Suppose there are five beads and F = {red, yellow, blue} is a set of three colours. We take X = $\{1, 2, 3, 4, 5\}$. The group G of symmetries of a regular pentagon is the dihedral group D_5 of order 10 generated by a rotation (of order 5) and a reflection (of order 2). The cycle index polynomial of G is given by*

$$(10.7.12) \qquad P_G(x_1, x_2, x_3, x_4, x_5) = \frac{1}{10}(x_1^5 + 4x_5 + 5x_1 x_2^2).$$

If F = {r, y, b},

$$P_G(3, 3, 3, 3, 3) = \frac{1}{10}(3^5 + 4.3 + 5.3^3) = 39.$$

One can make 39 inequivalent bracelets with 3 colours and 5 beads.

10.8. Worked-out examples

a) Determine $\{a_n\}$ where $a_n = 2a_{n-1} - a_{n-2}$, $n \geq 2$ given that $a_0 = 1$, $a_1 = 3$.

Answer: a_n satisfies $a_n = c_1 a_{n-1} + c_2 a_{n-2}$ where $c_1 = 2$, $c_2 = -1$. Let $f(x) = \sum_{n=0}^{\infty} a_n x^n$ where $|x| < 1$.

As in (10.4.7), we take

$$h(x) = 1 - 2x + x^2 = (1-x)^2.$$

Then, by (10.4.8),

$$f(x) = \frac{t(x)}{h(x)}$$

where $t(x) = a_0 + (a_1 - a_0 c_1)x = 1 + (3-2)x = 1 + x$ and $f(x) = \frac{1+x}{(1-x)^2}$, $|x| < 1$. a_n is the coefficient of x^n in $(1+x)(1-x)^{-2}$ and so, $a_n = 2n + 1$. $\{a_n\}$ is the sequence of odd numbers. $\qquad\qquad \Box$

b) Solve the recurrence relation $a_n = -3a_{n-1} + n$, $n \geq 1$; given that $a_0 = 1$.

Answer: We wish to obtain a particular solution p_n of

$$a_n = -3a_{n-1} + n$$

by assuming that $p_n = a + bn$; where a and b are to be determined. Writing

$$a + bn = -3\{a + b(n-1)\} + n$$

we get

$$(10.8.1) \qquad a + bn = -3a + 3b + (1 - 3b)n.$$

So, $a = -3a + 3b$ and $b = 1 - 3b$. It follows that

(10.8.2) $$a = 3/16, \quad b = 1/4.$$

So, $p_n = \frac{3}{16} + \frac{n}{4}$ is a particular solution to the recurrence relation, if we ignore initial conditions.

From $a_n = -3a_{n-1}$ (without the term n), we see that $a_n = k(-3)^n$, where k is a constant.

We write $q_n = k(-3)^n$ as a solution of the homogeneous recurrence relation $a_n = -3a_{n-1}$. We have

$$p_n + q_n = \frac{3}{16} + \frac{n}{4} + k(-3)^n.$$

As $a_0 = 1$, $1 = \frac{3}{16} + k$ or $k = \frac{13}{16}$. The solution of the recurrence relation $a_n = -3a_{n-1} + n$ is given by

$$p_n + q_n = a_n = \frac{3}{16} + \frac{n}{4} + \frac{13}{16}(-3)^n.$$

$\square$

The result is based on

Fact 10.8.1 :   Let r, s be real number. If p_n is a particular solution of the recurrence

(10.8.3) $$a_n = ra_{n-1} + sa_{n-2} + f(n) \quad (n \geq 2)$$

where $f : \mathbb{N} \to \mathbb{R}$ is an arbitrary polynomial function, and if q_n is the solution of

(10.8.4) $$a_n = ra_{n-1} + sa_{n-2} \quad (n \geq 2)$$

(without imposing initial conditions), then $p_n + q_n$ is the solution to the recurrence relation (10.8.3). The initial conditions, if any, determine the constants in q_n. For details see E. E. Goodaire and M. M. Parmenter [A4].

10.9. Catalan numbers

10.9.1. EULER'S POLYGON DIVISION PROBLEM. We come across a sequence of integers in connection with an enumeration problem called Euler's polygon division problem described below:

In how many ways a regular polygon of n sides (written n-gon) can be divided into $n - 2$ triangles, if different orientations are counted as distinct?

If the required number is denoted by E_n, we note that by definition, $E_3 = 1$. For $n = 4$, one has 2 ways of dividing a regular quadrilateral into two triangles. (see figure 13)

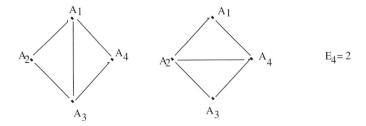

Figure 13

For $n = 5$, there are s ways of dividing a regular pentagon into 3 triangles. (see figure 14).

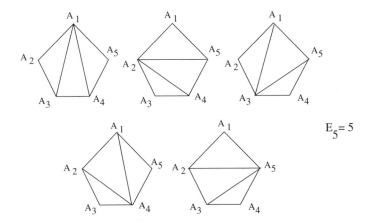

Figure 14

Lemma 10.9.1 : *(Segner (1758)). For $n \geq 4$,*

(10.9.1) $E_n = E_2 E_{n-1} + E_3 E_{n-2} + \cdots + E_{n-2} E_3 + E_{n-1} E_2.$

with $E_2 = 1$.

Proof : Let $A_k, A_{k+1}, \ldots, A_{k+i}$ denote the vertices of a regular i-gon, denoted by Δ_i. $(1 \leq k < i, 4 \leq i \leq (m+1))$ Δ_m has vertices $A_1, A_2, \ldots A_m$.
(10.9.1) holds for $n = 4$, as $E_4 = 2 = E_2 E_3 + E_3 E_2 = 1 + 1$. Assume that (10.9.1) holds for $n = m$ where $m > 4$. For convenience, we consider the vertices $A_2, A_3, \ldots,$ A_{m+1} of Δ_{m+1} as the vertices of an m-gon Δ_m. In Δ_{m+1}, we take $A_1 A_2 A_3$ as a triangle subdivision of Δ_{m+1} where $A_2 A_3$ is one side of Δ_m. E_{m+1} is evaluated by
(i) counting $E_2 = 1$ subdivision of $A_1 A_2$ and associating it with each of the E_m subdivisions of Δ_m formed by the vertices $A_2, A_3, \ldots, A_{m+1}$
(ii) counting $E_3 = 1$ subdivision of $\Delta_3 = A_1 A_2 A_3$ and associating it with each of the E_{m-1} subdivision of Δ_{m-1} having vertices $A_3, A_4, A_5, \ldots, A_{m+1}$

(iii) counting $E_4 = 2$ subdivisions of $A_1A_2A_3A_4$ and associating them with each of the E_{m-2} subdivisions of Δ_{m-3} having vertices $A_4, A_5, A_6, \ldots, A_{m+1}$ and so on. Finally, we count E_m subdivisions of $\Delta_m = A_1A_2 \ldots A_m$ and associate them with $E_2 = 1$ subdivision of $\Delta_2 = A_mA_{m+1}$.

Thus, we will obtain

$$E_{m+1} = E_2E_m + E_3E_{m-1} + \cdots + E_mE_2.$$

So (10.9.1) holds for $n = m+1$ and induction is complete.   □

Definition 10.9.1 : *A sequence of integers named after Catalan (1814–1894) is called a sequence of Catalan numbers C_n where $C_{n-2} = E_n (n \geq 2)$.*

The first eleven Catalan numbers are:
$C_0 = 1$, $C_1 = 1$, $C_2 = 2$, $C_3 = 5$, $C_4 = 14$, $C_5 = 42$, $C_6 = 132$, $C_7 = 429$, $C_8 = 1430$, $C_9 = 4862$, $C_{10} = 16796$.

Fact 10.9.1 :
(a) The only odd Catalan numbers are those of the form C_q where $q = 2^k - 1$. $k \geq 2$.
(b) The last digit of C_q is 5 for $q = 2^k - 1$ with $k = 9, 10, 11, 12, 13, 14$ and 15.
(c) The only prime Catalan numbers C_q with $q \leq 2^{15} - 1$ are those for which $q = 2$ and $q = 3$.

For proofs, see Alter R [A1].

Remark 10.9.1 : As $C_n = E_{n-2}$ $(n \geq 2)$, by virtue of lemma (10.9.1)

$$(10.9.2) \qquad C_n = C_0C_{n-1} + C_{n-2} + \ldots + C_{n-2}C_1 + C_{n-1}C_0.$$

Lemma 10.9.2 : *Catalan numbers C_n are given by*

$$(10.9.3) \qquad C_n = \frac{1}{n+1}\binom{2n}{n}, (n \geq 1)$$

with $C_0 = 1$.

Proof : Let $d_o = 1$. $d_n = \binom{2n}{n}, n \geq 1$.
For $0 < |x| < \frac{1}{4}$,

$$(10.9.4) \qquad (1-4x)^{-\frac{1}{2}} = \sum_{n=0}^{\infty} d_nx^n = D(x) \quad \text{(say)}$$

If C_n denotes the $(n+1)^{th}$ Catalan number, the generating function $G(x)$ of C_n is given by

$$(10.9.5) \qquad G(x) = \sum_{n=0}^{\infty} C_nx^n$$

(with appropriate interval of convergence). Our aim is to determine $G(x)$.

If $t_0 = t_1 = 1$ and $t_n = \sum_{j=0}^{n} C_j C_{n-j}$, by (10.9.2), $t_n = C_{n+1}$. So,

$$G^2(x) = C_1 + C_2 x + C_3 x^2 + \ldots.$$

Therefore,

(10.9.6) $1 + xG^2(x) = G(x).$

From (10.9.6), we note that

$$G(x) = \frac{1 \pm \sqrt{1-4x}}{2x}$$

The positive sign in $\pm\sqrt{1-4x}$ is not admissible. Therefore,

(10.9.7) $G(x) = \dfrac{1 - \sqrt{1-4x}}{2x}.$

Now,

$$(1-4x)^{1/2} = (1-4x)\,D(x) \text{ by } (10.9.4).$$

But, $(1-4x)^{1/2} = D^{-1}(x)$. So using (10.9.7), we get

$$D^{-1}(x) = 1 - 2xG(x).$$

Or,

(10.9.8) $(1-4x)\sum_{n=0}^{\infty} d_n x^n = 1 - 2x \sum_{n=0}^{\infty} C_n x^n.$

Comparing coefficients of like powers of x in (10.9.8), we have

$$d_{n+1} - 4d_n = -2C_n \ (n = 0, 1, 2, \ldots)$$

As $d_n = \binom{2n}{n}$, evaluation of C_n yields (10.9.3). $\square$

Remark 10.9.2 : An 'elementary' evaluation of C_n in the form

$$C_n = \frac{1}{2n+1} \binom{2n+1}{n}; \quad n = 1, 2, \ldots$$

is given in David Singmaster [A9]. See also Hansraj Gupta [10].

10.9.2. A WORKED-OUT EXAMPLE. Given a generating function G with $G(0) = 1$, let M be the infinite matrix $M = [m_{i,j}]$ with $(i,j)^{th}$ entry defined by

$$\sum_{i=0}^{\infty} m_{i,j} x^i = G(x)\,(xG(x))^j, (i, j \geq 0).$$

Note that M is lower triangular.

a) If $G(x) = \dfrac{1 - \sqrt{1-4x}}{2x}$ where $|x| < \frac{1}{4}$, show that

$$\sum_{k=j}^{\infty} m_{1,k} = m_{i+1, j+1} \text{ for all } i, j \geq 0.$$

b) Prove the converse of (a)

Answer: If $|x| < \frac{1}{4}$,

$$(10.9.9) \qquad \sum_{i=0}^{\infty}\left(\sum_{k=j}^{\infty} m_{i,k}\right) x^i = \sum_{k=j}^{\infty} G(x)\,(xG(x))^k = \frac{G(x)}{1-xG(x)}(xG(x))^j$$

As $G(x)$ is the generating function of $\{C_n\}$

$$1-xG(x) = \frac{1+\sqrt{1-4x}}{2} = \frac{1}{G(x)}.$$

So,

$$\frac{G(x)}{1-xG(x)}(xG(x))^j = \frac{G(x)}{x}(xG(x))^{j+1} = \frac{1}{x}\sum_{i=0}^{\infty} m_{i,j+1}\,x^i.$$

Hence,

$$\sum_{k=j}^{\infty} m_{i,k} = m_{i+1,j+1} \text{ for all } i,j \geq 0.$$

This proves (a).

Next, for the converse, we are given that

$$\sum_{k=j}^{\infty} m_{i,k} = m_{i+1,j+1}.$$

From (10.9.9),

$$(10.9.10) \qquad \sum_{i=0}^{\infty} m_{i+1,1} x^i = \sum_{i=0}^{\infty}\left(\sum_{k=0}^{\infty} m_{i,k}\right) x^i = \frac{G(x)}{1-xG(x)}.$$

From the definition of M and the fact that $m_{0,1} = 0$, we obtain

$$\sum_{i=0}^{\infty} m_{i+1,1} x^i = \frac{1}{x}G(x)\,(xG(x)).$$

From (10.9.10)

$$\frac{G(x)}{1-xG(x)} = \frac{1}{x}G(x)\,(xG(x))$$

It follows that

$$G(x) = \frac{1}{1-xG(x)}, |x| < \frac{1}{4}$$

which means that $G(x)$ is the generating function of $\{C_n\}$.   □

Remark 10.9.3 : The above worked-out example has been adapted from problem 10850 proposed by Wolfdieter Lang in Amer. Math. Monthly 109 (2002) 82–83. It is pointed out by Lin Tan that the matrix M is a recursive matrix in the sense that the generating functions of its columns are the powers of the generating function of first column. See L. Shapiro, S. Getu, W. Woan and I. Woodson: 'The Riordan group', Discrete Applied Math 34 (1991) 229–239.

EXERCISES

1. **Mark the following statements true (T) or false (F) justifying your answer briefly.**
   a) $\{a_n\}$ is defined by $a_1 = 1$, $a_{n+1} = (n+1)^2 - a_n$ ($n \geq 1$). Then, a_n is given by $a_n = \frac{n(n+1)}{2}$.
   b) For $\{a_n\}$ with $a_0 = 2$, $a_n = a_{n-1} + 2^n$. ($n \geq 1$). Using the method of generating functions or otherwise, one gets $a_n = 2^{n+1}$.
   c) For fixed a, we define

   $$m(a) = \inf_t |\zeta(a+it)|, \quad M(a) = \sup_t |\zeta(a+it)|$$

   where the infimum and supremum are taken over all real t. Then, for $a > 1$, one has

   $$M(a) = \zeta(a) \quad and \quad m(a) = \frac{\zeta(2a)}{\zeta(a)}.$$

   d) $\sum_{j=1}^{r} j^2 \binom{r}{j} = 2^{r-1} r.$ ($r \geq 1$)
   e) If $\tau(n)$ denotes Ramanujan's τ-function evaluated at n, it is correct to say that

   $$\tau(3n) \equiv -1 \pmod 3$$

   f) Let μ denote the Möbius function. Then,

   $$e^x = \prod_{n=1}^{\infty} (1 - x^n)^{\frac{-\mu(n)}{n}}, \quad whenever \ |x| < 1.$$

2. (*C. A. Nicol*) Let $F_r(x)$ denote the cyclotonic polynomial (whose zeros are the primitive r^{th}-roots of unity). If $C(n,r)$ denotes Ramanujan's sum, show that

   $$\sum_{a=1}^{r} C(a,r) x^{a-1} = (x^r - 1) \frac{F_r'(x)}{F_r(x)}$$

   where $F_r'(x)$ denotes the formal derivative of $F_r(x)$ with respect to x.

3. (*Ramanujan*) Let $C(n,r)$ denote Ramanujan's sum. Show that

   $$\sum_{a=1}^{\infty} \frac{C(a,r)}{a} = -\Lambda(r) \quad (r > 1),$$

   where

   $$\Lambda(r) = \begin{cases} \log p, & if \ r = p^m, m \geq 1; p \ a \ prime \\ 0, & otherwise. \end{cases}$$

4. Let n be a positive integer. Prove that the number of partitions of n into unequal parts is equal to the number of its partitions into odd parts. (*In Hardy and Wright [12], there is a proof of this result without using generating functions. See theorem 344 on page 277 of [12]*).

5. Let $p(n,k)$ denote the number of partitions of n into exactly k (≥ 0) summands with

$$p(n,k) = \begin{cases} 0, & 0 < n < k, \\ 1, & k = 0 = n. \end{cases}$$

Show that

$$kp(n,k) \geq \frac{1}{(k-1)!}\binom{n-1}{k-1}.$$

6. Prove Jacobi's identity

$$\sum_{j=0}^{\infty}(-1)^{j}(2j+1)x^{\frac{j(j+1)}{2}} = \prod_{n=1}^{\infty}(1-x^{n})^{3}.$$

7. Let $r_4(n)$ denote the number of solutions of

$$n = m_1^2 + m_2^2 + m_3^2 + m_4^2, \quad \text{where } m_1, m_2, m_3, m_4 \in \mathbb{Z}.$$

Two solutions differing only in sign or order of m_1, m_2, m_3, m_4 are reckoned as distinct. Show that $r_4(n)$ is the coefficient of x^n in

$$(1 + 2x + 2x^4 + 2x^9 + \ldots)^4 = (\sum_{m=-\infty}^{\infty} x^{m^2})^4.$$

Prove, further, that $r_4(n)$ is eight times the sum of the divisors (of n) which are not multiples of 4.

8. Let $m \in \mathbb{N}$ and m is a quadratic residue mod 5. If $R(5)$ denotes the set of quadratic residues (mod 5), we write $m \in R(5)$. Show that

$$\prod_{\substack{m \geq 1 \\ m \in R(5)}} (1 - x^m) = \sum_{n=-\infty}^{\infty} (-1)^n x^{\frac{n(5n+1)}{2}}.$$

9. Suppose that for $|x| < 1$,

$$\sum_{n=0}^{\infty} q(n)x^n = ((1-x)(1-x^2))^{-1}.$$

Give an arithmetical interpretation of $q(n)$.

10. [Ramanujan] Let $C(n,r)$ denote Ramanujan's sum (see (5.4.1), chapter 5). Assume that s is real and $s > 1$. If $\sigma_k(n)$ denotes the sum of the kth powers of the divisors of n, show that

$$\zeta(s) \sum_{r=1}^{\infty} \frac{C(n,r)}{r^s} = \frac{\sigma_{s-1}(n)}{n^{s-1}}, (n \geq 1).$$

(The above relation gives the generating function of $C(n,r)$).

11. Let $N(n)$ be an enumerating function satisfying the recurrence

$$N(n+3) = 5N(n+2) - 7N(n+1) + 4n, \quad n \geq 2$$

with $N(0) = 1$, $N(1) = 1$, $N(2) = 2$. Find the generating function of $N(n)$?

12. Let $P(n)$ be a polynomial in n of degree t. Show that

$$\sum_{n=0}^{\infty} P(n)x^n = \frac{F(x)}{(1-x)^{t+1}}$$

where $F(x)$ is a polynomial in x of degree at the most t and $F(1) \neq 0$ if, and only if,

$$\sum_{j=0}^{t+1} (-1)^j \binom{t+1}{j} P(n+j) = 0$$

and for some $n \geq 0$,

$$\sum_{j=0}^{t} (-1)^j \binom{t}{j} P(n+j) \neq 0.$$

(See Richard P. Stanley [19]).

13. [Richard P. Stanley] Let $\emptyset = s_0 \subset s_1 \subset s_2 \cdots \subset s_k = S$ be a chain of subsets of a finite set S with $|S| = n$. Suppose, further, that $|S_{i+1} - S_i| \geq 2$ for $0 \leq i < k$. If $M(n)$ denotes the number of chains

$$\phi = S_0 \subset S_1 \cdots \subset S_k = S \text{ and } |S_{i+1} - S_i| \geq 2, \quad 0 \leq i < k$$

in $\mathbb{P}(S)$, show that

$$\sum_{n=0}^{\infty} M(n)\frac{x^n}{n!} = (2+x-e^x)^{-1}.$$

It is assumed that the series on the left converges in an appropriate interval for x.

14. The Legendre polynomial $P_n(x)$ is defined as a solution of the differential equation:

$$(1-x^2)y'' - 2xy' + n(n+1)y = 0$$

It is known that

$$P_n(x) = \sum_{k=0}^{N} \frac{(-1)^k(2n-2k)!}{2^n k!(n-k)!(n-2k)!} x^{n-2k}$$

where

$$N = \begin{cases} \frac{n}{2}, & \text{if } n \text{ is even} \\ \frac{n-1}{2}, & \text{if } n \text{ is odd.} \end{cases}$$

Show that the generating function of $P_n(x)$ is given by

$$(1-2xt+t^2)^{-1/2} = \sum_{n=0}^{\infty} P_n(x)t^n, \quad |t| < 1.$$

Deduce that

$$(n+1)P_{n+1}(x) - (2n+1)xP_n(x) + nP_{n-1}(x) = 0 \quad (n \geq 1)$$

15. *Bessel polynomials $y_n(x)$ are defined by*

$$y_n(x) = \sum_{k=0}^{n} \frac{(n+k)!}{(n-k)!k!} \left(\frac{x}{2}\right)^k,$$

which satisfy the differential equation

$$x^2 y'' + (2x+2)y' - n(n+1)y = 0.$$

Show that $y_n(x)$ satisfies the recurrence relation

(10.9.11) $y_n(x) - (2n-1)x\, y_{n-1}(x) - y_{n-2}(x) = 0$ $(n \geq 2)$,

with $y_0(x) = 1$, $y_1(x) = 1+x$.
Obtain the generating function of $y_n(x)$. (In (10.9.11), the coefficients are not constant real numbers).

Following L. Carlitz, we write $p_n(x) = x^n y_{n-1}(\frac{1}{x})$.
     Show that

$$\sum_{k=0}^{\infty} \frac{p_k(x)}{k!} t^k = e^{x(1-\sqrt{1-2t})}. \ (0 < t < 1/2)$$

(Ref: H. L. Krall and O. Frink: A new class of orthogonal polynomials : the Bessel polynomials. Trans. Amer. Math. Soc. 65 (1949) 100–115)

REFERENCES

[1] L. V. Ahlfors: Complex analysis Chapter 4, pp 172–174 McGrew Hill Book Co. International Edn., Singapore (1977).

[2] Tom M. Apostol: Mathematical Analysis Chapter 8, pp. 206–210, Narosa Pub. House, New Delhi, 12nd Edn., (1993).

[3] Tom M. Apostol: Introduction to Analytic Number Theory, Chapters 11 and 14 pp. 225–234 and 304–315, Narosa Pub. House, New Delhi (Reprint).

[4] Tom M. Apostol: Modular functions and Dirichlet series, Chapters 1–3, pp. 1–52 GTM No 41, Springer Verlag NY (1976).

[5] D. M. Burton: A first course in rings and ideals, Chapter 7, pp 112–118, Addison Wesley Pub. Co., Reading Mass. (1970).

[6] K. Chandrasekharan: Arithmetical functions, Chapter 2 pp. 28–33, Springer Verlag, Berlin, Heidelberg (1970).

[7] P. Doubilet, G. C. Rota and R. Stanley: On the foundations of combinational theory VI: The idea of a generating function, Sixth Berkeley symposium on mathematical sciences and probability, Vol II Probability Theory, Univ. of California (1972) 267–318.

[8] John A. Ewell: A formula for Ramanujan's tau function, Proc. Amer. Math. Soc., Vol. 91 (1984) 37–40.

[9] Emil Grosswald: Topics from the theory of numbers Chapter 12, pp. 216–235, MacMillan Co. NY, Second Edn. (1985).

[10] Hansraj Gupta: Selected topics in number theory Chapters 5 and 7 pp. 137–154 and 188–231, Abacus Press (1980), Ringwood, Hampshire, UK.

[11] Pentti Haukkanen: A note on a curious polynomial identity, Nieuw Archief voor Wiskunde, New series 13 (1995) 181–186.

[12] G. H. Hardy and E. M. Wright: An Introduction to the theory of numbers, Chapter XVII pp. 244–259, Oxford at the Clarendon Press 4th Edn. (1965).

[13] A. F. Horadam: Basic properties of certain generalized sequence of numbers, Fibonacci Quarterly 3 (1965) 161–177.

[14] A. F. Horadam: Special properties of sequence $w_n(a,b;b,q)$ Fibonacci Quarterly, 5 (1967) 424–434.

[15] S. T. Hu: Elements of Modern algebra, Chapter 3 pp. 65–68 and 80–84, Holden Day Inc., San Francisco (1965).

[16] L. J. Mordell: On Mr Ramanujan's empirical expansion of modular functions, Proc. Cambridge Phil. Soc., 19 (1920) 117–124.

[17] S. Ramanujan: Collected Papers. Chelsea Pub Co NY, Reprint (1962) pp. 232–238.

[18] R. Sivaramakrishnan: Classical theory of arithmetic functions, Monographs and Textbooks in Mathematics, No. 126, Chapter XII pp. 248–265, Marcel Dekker (1989).

[19] Richard P. Stanley: Generating functions, Studies in combinatorics, Vol. 17 pp. 100–141, Math Association of America (1978).

[20] Morgan Ward: Stirling numbers and polynomials, Amer. J. Math., 56 (1934) 87–95.

ADDITIONAL REFERENCES

[A1] Alter. R: Some remarks and results on catalan numbers: Proc. 2nd Louisiana Cof. Comb. Graph Th. and Comput. (1971) 109 –132.

[A2] E. A. Bender and J. R. Goldman: Enumerative uses of generating functions, Indiana Univ. Math. J. 20 (1971), 753–765.

[A3] Joseph B. Dence and Thomas P. Dence: Elements of the theory of numbers, Harcourt/Academic Press, (2001), Burlington, MA 01803.

[A4] E. E. Goodaire and M. M. Parmenter: Discrete Mathematics with Graph Theory, chapter 4, 194–210, Prentice Hall International, Inc (1998), Upper Saddle River, NJ 07458.

[A5] Frank Harary and Edgar M. Palmer: Graphical Enumeration, Academic Press, NY and London (1973).

[A6] M. Henle: Dissection of generating functions, Studies in Appl. Math, 51, (1972) 397–410.

[A7] Rudolf Lidl and Günter Pilz: Applied Abstract Algebra, Second Edn, Springer Verlag, NY (1998).

[A8] S. J. Petterson: An introduction to the theory of the Riemann Zeta-function, Cambridge University Press (1988).

[A9] David Singmaster: An elementary evaluation of the catalan numbers. Amer. Math. Monthly 85 (1978) 366–368.

CHAPTER 11

Semigroups and certain convolution algebras

Historical perspective

Perhaps, multiplicative systems would have come about from various sources. One such is the set $\mathbb{N}$ of positive integers where ordinary multiplication gives rise to a monoid $(\mathbb{N}, \cdot)$ with 1 as the unity element. George Cantor, the father of set theory, gave the example of all maps: Map (A, B) from a non-empty set A into a non-empty set B. The composition of maps defines a 'multiplication' which is associative. For composition of Map (A, B) and Map (B, C) where A, B, C are non-empty sets one gets Map (A, C). In particular, Map (A, A) satisfies the associative law of composition and the identity map $i_A : A \rightarrow A$ (onto A) serves as the multiplicative identity. As is well-known, a semigroup is a multiplicative system in which the associative law for multiplication holds. The evolution of groups has its roots in geometry and in analysis, and group-theoretic ideas were used around 1800. As $(\mathbb{N}, \cdot)$ is a semigroup, it is natural to consider functions from a semigroup, G into the field $\mathbb{C}$ of complex numbers. In [11], J. Knopfmacher gives an interesting exposition of abstract analytic number theory wherein many of the results of classical number theory are generalized in a suitable context of semigroups satisfying certain axioms. The treatment is all the more remarkable when one gets an abstract analogue of the Prime Number Theorem.

We have had groups, rings, fields. Structures such as algebras become useful when we look at the underlying vector space structure of $\mathbb{C}$ as a 2-dimensional algebra over $\mathbb{R}$, the field of real numbers. In fact, we realize $\mathbb{C}$ as a 2-dimensional algebra over $\mathbb{R}$, the field of reals. One might ask the question : who gave us $\mathbb{C} = \mathbb{R}^2$? As C. F. Gauss gave the fundamental theorem of algebra, there is reason to believe that Gauss was aware of $\mathbb{C} = \mathbb{R}^2$. However, the identification $\mathbb{C} = \mathbb{R}^2$ was first done by Casper Wessel (1745–1818) in 1798. But his paper came to the noticed or became well-known only by the end of 19th century. It was Jean Robert Argand (1768–1822) who made a greater effect by the well-known 'Argand plane'. W. R. Hamilton (1805–1865) gave the definition of complex numbers with suitable rules for addition and multiplication in 1837. We remark that G. Cantor defined real numbers in terms of rationals in 1871. R. Dedekind gave the definition independently in 1872.

In a field $(F, +, \cdot)$ one has the additive group $(F, +)$ as well as multiplicative group $(F^*, \cdot)$ where $F^* = F \setminus \{0_F\}$. Permutation groups play a role in Galois theory of equations when one considers a Galois group associated with a polynomial *equation. In linear algebra, it was J. W. Gibbs (1839–1903) who created a vector algebra and it was important to note that a vector in 3-dimensions could take the form*

$$\vec{r} = x\vec{i} + y\vec{j} + z\vec{k}$$

where $\vec{i}, \vec{j}, \vec{k}$ form a rectangular triad of unit vectors and $x, y, z \in \mathbb{R}$. $\vec{r}$ is thought of as a directed line segment emanating from the origin $(0,0,0)$ to the point (x,y,z). In a more general setting, one talks about a finite dimensional vector space over a field F. This paved the way for the enrichment of linear algebra. A ring which differs from a field only 'slightly' is called a division ring. More specifically, a ring which satisfies all the properties of a field except commutative law of multiplication is a division ring. If the ring is also an algebra over a field F, it is called a division algebra. In 1843, W. R. Hamilton discovered a division algebra by considering real quaternions. The set

$$Q = \{a + bi + cj + dk : a, b, c, d \in \mathbb{R}\}$$

in which $i^2 = j^2 = k^2 = -1$, $ij = k = -ji$, $jk = i = -kj$ and $ki = j = -ik$, serves as a division algebra having a basis $\{1, i, j, k\}$. Q is called the real quaternion algebra. In 1844, Grassman (1809–1877) obtained a most general algebra of which Q is a special case. Such algebras are called 'Grassman algebras' in the literature. Very quickly, algebras found application in many branches of Mathematics and Physics. Modern mathematics got a face-lift via the notion of algebras. A powerful theorem of J. H. M. Wedderburn (1882–1948) says that a finite division ring is a field. See T. Hungerford [10]. It is also to be emphasized that A. Cayley's (1821–1895) contributions to matrix algebra are significant.

How does number theory enter into the realm of finite or infinite-dimensional algebras? Perhaps, a valid answer is that it happened when an analyst and a number-theorist collaborated. In 1955, E. Hewitt and H. S. Zuckerman [9] have studied in detail finite dimensional convolution algebras. They have established that a convolution of functions is just a technique for 'multiplication' of linear functionals in a 'well-set' background. The operations of Dirichlet convolution (see chapter 4), Cauchy convolution (see observation 10.6.1, chapter 10) and such other convolutions can be brought under one roof when one considers the vector space of complex-valued functions defined on a semigroup.

The study of rings of arithmetic functions was begun by Eckford Cohen [4], [5] during 1952–54. It was followed up by Leonard Carlitz [2], [3] in a slightly different situation which he called an 'unusual setting'.

11.1. Introduction

In this chapter, properties of semigroups are studied. The definition of a semicharacter of a semigroup G is given. The idea is to use the notion of a character of a group sitting inside the semigroup G. The vector space of complex valued-functions defined on a semigroup is considered. Convolution algebras are defined as given in Hewitt and Zuckerman [9]. This fits well into the concepts of convolutions of arithmetic functions studied in number theory. Finite dimensional convolution algebras arise in a natural way from the vector space of functions defined on a finite semigroup. An interesting theorem borrowed from [9] says that any finite dimensional algebra over a field is, indeed, a convolution algebra.

Following M. Tainiter [14], a finite dimensional algebra of complex-valued functions defined on a finite commutative semigroup of idempotents is shown to provide an analogue of Dirichlet convolution of arithmetic functions. The functions considered are called abstract arithmetical functions. The analogue of product theorem for Dirichlet series is proved. See theorem 86. Some examples are given. Some new convolutions of functions are introduced in Section 11.6. Finally, the notion of a functional-theoretic algebra is introduced. A class of linear functionals defined on the vector space $\mathcal{A}$ of arithmetic functions is used to define a new product of arithmetic functions.

11.2. Semigroups

By a semigroup, we mean a non-empty set G together with a binary relation, dot ($\cdot$) on G, called multiplication which is associative. A semigroup G is called commutative if $x \cdot y = y \cdot x$ for all $x, y \in G$. An element $u \in G$ is called an idempotent if $u \cdot u = u^2 = u$. We denote a semigroup by $(G, \cdot)$.

Definition 11.2.1 : *Let $(G, \cdot)$ be a semigroup. $x \in G$ is said to divide an element y in G (or y is divisible by x) if there exists $t \in G$ such that $y = x \cdot t$.*

Definition 11.2.2 : *An element $x \in G$ is said to be an irreducible element, if the equation $x = y \cdot t$ implies that either $y = x$ or $t = x$.*

Definition 11.2.3 : *A semigroup $(G, \cdot)$ is said to have a zero, if it has a unique element 0 which is divisible by every element x in G. That is, $x \cdot 0 = 0$, $0 \cdot x = 0$ for all $x \in G$.*

Fact 11.2.1 :

(i) Let $(G, \cdot)$ be a semigroup. Suppose that z is not an element in G. Then, $G \cup \{z\}$, written G_z, is made a semigroup under the multiplication rules:

$$x \cdot y = x \cdot y \text{ as in } (G, \cdot), \text{ for all } x, y \in G,$$

$$x \cdot z = z \cdot x = z, \text{ for all } x \in G_z.$$

In fact, $(G_z, \cdot)$ is obtained from $(G, \cdot)$ by adjoining zero.

(ii) Let $(G, \cdot)$ be a semigroup and suppose that e is not an element of G. Then $G \cup \{e\}$, written G_e, is made a semigroup under the multiplication rules:

$$x \cdot y = x \cdot y \text{ as in } (G, \cdot), \text{ for all } x, y \in G,$$
$$e \cdot x = x \cdot e = x, \text{ for all } x \in G_e.$$

$(G_e, \cdot)$ is obtained from $(G, \cdot)$ by adjoining a unit.

(iii) Let $(G, \cdot)$ be a semigroup containing an idempotent u. Let a be an object not contained in G. Then $G \cup \{a\}$, written G_a, is made a semigroup under the multiplication rules:

$$x \cdot y = x \cdot y, \text{ as in } (G, \cdot), \text{ for all } x, y \in G,$$
$$x \cdot a = x \cdot u, \text{ for all } x \in G$$
$$a \cdot x = u \cdot x, \text{ for all } x \in G$$
$$a \cdot a = a^2 = a.$$

Then $(G_a, \cdot)$ is a semigroup obtained from $(G, \cdot)$ by idempotent adjunction. See Hewitt and Zuckerman [9].

Definition 11.2.4 : *Let $(G, \cdot)$ be a semigroup and $x \in G$. x is said to be of finite order, if there exist two integers $k, l \in \mathbb{N}$ with $k \geq l$ and $l \geq 1$ such that*

$$x^{k+l} = x^l.$$

Clearly, idempotent elements in a semigroup $(G, \cdot)$ are of finite order. In fact, all the elements of a finite semigroup are of finite order. In $(\mathbb{N}, \cdot)$, no element other than 1 is of finite order.

Notation 11.2.1 : [9] Given a semigroup $(G, \cdot)$, let $x \in G$ be such that x is of finite order. Then, the sequence

$$x, x^2, x^3, \ldots$$

contains at the most $k + l - 1$ distinct elements. For, one has, by definition, the elements

$$x^l, x^{l+1}, \ldots, x^{l+k-1}, x^{l+k} = x^l$$

If r is the smallest integer such that $x^r = x^s$, $1 \leq s < r$, we write

(11.2.1) $l_x = s \text{ and } k_x = r - s$

Then, $x^p = x^q$, $p > q \Leftrightarrow q \geq l_x$ and $p = q + j k_x$ for some integer j.

Remark 11.2.1 : Let $(G, \cdot)$ be a semigroup.

(a) If $x \in G$ is of finite order, then,

$$(x^m)^2 = x^m \quad (m \geq 1) \text{ if, and only if, the condition}$$

$m = j k_x \geq l_x$ holds for some positive integer j.

(b) If $(x^m)^2 = x^m$ for some $m \geq 1$, x is of finite order.

(c) If $(x^m)^2 = x^m$ and $(x^r)^2 = x^r$, then $(x^m)^r = x^m = (x^r)^m = x^r$.

(d) If $x \in G$ is of finite order and $l_x = 1$, $(x^{k_x})^2 = x^{k_x}$.

(e) If $x, y \in G$ are such that $l_x = l_y = 1$, then,

$$(x \cdot y)^{k_x k_y + 1} = (x^{k_x})^{k_y} \cdot (y^{k_y})^{k_x} \cdot x \cdot y$$
$$= x^{k_x} \cdot y^{k_y} \cdot x \cdot y$$
$$= x^{k_x+1} \cdot y^{k_y+1}$$
$$= x \cdot y, \text{ as } k_x + 1 = r \text{ (least)},$$

where $x^r = x$ and $k_y + 1 = t$ (least). Further $x^t = x$ (and $l_x = l_y = 1$).
So, $l_{x \cdot y} = 1$.

Fact 11.2.2 : Let $(G, \cdot)$ be a finite or infinite commutative semigroup all of whose elements are of finite order. Suppose that $l_x = 1$ for all $x \in G$. Then,

(a) G is a disjoint union of groups.
(b) The set H of idempotents of G forms a semigroup $(H, \cdot)$.

Proof :
(a) For each idempotent $a \in G$, we write

$$S_a = \{x \in G : x^m = a \text{ for some } m \geq 1\}$$

every $x \in G$ is in some S_a, as all elements of G are of finite order. For $x \in S_a$, using remark 11.2.1(a), we see that

$$x^{k_x} = a$$

Similarly, $y \in S_a \Rightarrow y^{k_y} = a$.
So, $(x \cdot y)^{k_x k_y} = (x^{k_x})^{k_y} \cdot (y^{k_y})^{k_x} = a \cdot a = a$.
Therefore, $x \cdot y \in S_a$ whenever x, y are in S_a.
Also, $a \cdot x = x^{k_x} \cdot x = x^{k_x+1} = x^r = x$ as $l_x = 1$.
Further, $x \cdot x^{2k_x - 1} = x^{2k_x} = a^2 = a$. Thus, a serves as the identity element in $(S_a, \cdot)$ and $x^{2k_x - 1}$ is the inverse of x in S_a. So, $(S_a, \cdot)$ is a group. As $S_a \cap S_b = \emptyset$ for $a \neq b$, $(G, \cdot)$ is a disjoint union of groups.
(b) Suppose that $x \in S_a$ and $y \in S_b$ $(a \neq b)$.
Then,

$$(x \cdot y)^{k_x k_y} = (x^{k_x})^{k_y} \cdot (y^{k_y})^{k_x} = x^{k_x} \cdot y^{k_y} = a \cdot b$$

So, $x \cdot y \in S_{a \cdot b}$. Associativity follows.
Thus, $(H, \cdot)$ is a semigroup.

$\square$

Remark 11.2.2 : Fact 11.2.2 has been adapted from [9].

Example 11.2.1 : In $(\mathbb{P}(x), \cap)$ every element is an idempotent. $(P(x), \cap)$ is a semigroup of idempotents. $\mathbb{P}(x)$ is a union of groups of order 1, trivially.

Next, we go to the particular case where $(G, \cdot)$ is a finite commutative semigroup all of whose elements are of finite order.

Notation 11.2.2 : We denote the set $\{x : x \in G, l_x = 1\}$ by G^0.
   If $x, y \in G^0$, then

$$(x \cdot y)^{k_x k_y + 1} = (x^{k_x})^{k_y} \cdot (y^{k_y})^{k_x} \cdot (x \cdot y) = x^{k_x} \cdot y^{k_y} \cdot (x \cdot y)$$
$$= x^{k_x + 1} \cdot y^{k_y + 1},$$

or,

(11.2.2) $(x \cdot y)^{k_x k_y + 1} = x \cdot y.$

For, if $l_x = 1$, $k_x = r - 1$, where r is the least positive integer such that $x^r = x^s$.
   From (11.2.2), we see that $l_{x \cdot y} = 1$ and so, G^0 is a semigroup. By Fact 11.2.1(b), if H denotes the set of idempotents of G^0, H is a semigroup contained in G^0.

Definition 11.2.5 : *For each $a \in H$, we write*

$$T_a = \{x : x \in G, x^m = a, \text{ for some } m \geq 1\}$$

As in the proof of Fact 11.2.1(a), T_a are pairwise disjoint and every $x \in G$ is in some T_a. Each T_a is a semigroup and $T_a \cap G^0$ is a group. It is simply the group S_a considered in the proof of Fact 11.2.2(a).
   For, if $x \in T_a$, $y \in T_b$, then, $x^m = a$, $y^{m'} = b$.
   So, $(x \cdot y)^{mm'} = (x^m)^{m'} \cdot (y^{m'})^m = a^{m'} \cdot b^m = a \cdot b$ as $a, b \in H$.
   Further, T_a is a semigroup containing the group $T_a \cap G^0$.

Observation 11.2.1: If $x \in T_a$, $x^m = a$, for some $m \geq 1$. So,

$$(a \cdot x)^{m+1} = a \cdot x^{m+1} = a \cdot a \cdot x = a \cdot x.$$

So, $l_{a \cdot x} = 1$. Therefore,

(11.2.3) $aT_a = T_a \cap G^0.$

11.3. Semicharacters

We introduce the notion of semicharacters in the same way as is done for characters of a group.

Definition 11.3.1 : *Let $(G, \cdot)$ be a semigroup. A function $\chi : G \to \mathbb{C}$ is called a semicharacter of G, if $\chi(a) \neq 0$ for some $a \in G$ and*

$$\chi(a)\chi(b) = \chi(a \cdot b) \text{ for all } a, b \in G.$$

Observations 11.3.1:

(a) Let $(G, \cdot)$ be a semigroup. If $x \in G$ is of finite order and χ is a semicharacter of G, then $\chi(x)$ is either zero or a root of unity.

(b) If $x \in G$ is an idempotent and χ is a semicharacter of G, then $\chi(x)$ is either 0 or 1.

(c) If $(G, \cdot)$ is a monoid (containing the unity element e) and if χ is a semicharacter of G, then $\chi(e) = 1$.

(d) If $(G, \cdot)$ contains a zero z and χ is a semicharacter not identically equal to 1, then $\chi(z) = 0$.

(e) A semicharacter of a group is a character in the normal sense.

Fact 11.3.1 : Suppose $(G, \cdot)$ is a finite semigroup such that for every pair of distinct elements $a, b \in G$, there exists a semicharacter χ not assuming the value zero (for any $x \in G$) and $\chi(a) \neq \chi(b)$, then G is an abelian group.

For, by hypothesis, any semicharacter $\chi : G \to \mathbb{C}$ is a one–one map. So, as G is finite, $\chi(a^n) = \chi(a)$ for $n \geq 1$. Similarly, $\chi(b^m) = \chi(b)$ for $m \geq 1$. Thus, $\chi(a^{n-1}) = \chi(b^{m-1}) = 1$ or $a^{n-1} = b^{m-1}$ serves as the multiplicative identity. There is a unique identity element and any $a \in G$ has a unique inverse. So, $(G, \cdot)$ is a group. As $\chi(a \cdot b) = \chi(b \cdot a)$, $a \cdot b = b \cdot a$ and so $(G, \cdot)$ is abelian.

Theorem 82 (Hewitt and Zuckerman (1955)) **:** *Let $(G, \cdot)$ be a commutative semigroup. Suppose that χ is a semicharacter of G. Let H denote the semigroup of idempotents of G. For each $a \in H$, we write*

(11.3.1) $$T_a = \{x \in G : x^m = a, \text{ for some } m \geq 1\}$$

and

(11.3.2) $$G^0 = \{x \in G : l_x = 1\}$$

Then, there exists $a_0 \in H$ and a character χ_{a_0} of the group $T_{a_0} \cap G^0$ such that

(11.3.3) $$\chi(y) = \begin{cases} 0, & \text{if } a_0 \cdot a \neq a_0, \text{ for the element } a \text{ of G such} \\ & \text{that } y \in T_a; \\ \chi_{a_0}(a_0 \cdot y), & \text{if } a_0 \cdot a = a_0, \text{ for the element } a \\ & \text{such that } y \in T_a. \end{cases}$$

Proof : For some $y \in G$, there exists a semicharacter χ of G such that $\chi(y) \neq 0$. Then, for some $a \in H$, $\chi(a) \neq 0$. So, $\chi(a) = 1$, as a is an idempotent.

We let

(11.3.4) $$a_0 = \prod_{\substack{a \in H \\ \chi(a)=1}} a$$

It is clear that $\chi(a_0) = 1$ and for $a \in H$,

$$\chi(a) = \begin{cases} 1, & \text{if } a_0 \cdot a = a_0; \\ 0, & \text{if } a_0 \cdot a \neq a_0. \end{cases}$$

For all $y \in G$, we have $\chi(y) = \chi(a_0)\chi(y) = \chi(a_0 y)$.
If $y \in T_a$ and $a_0 \cdot a = a_0$, then $a_0 \cdot y \in T_{a_0 \cdot a} = T_{a_0}$.
So,

$$a_0 \cdot y = a_0 \cdot (a_0 \cdot y) \in a_0 T_{a_0} = T_{a_0} \cap G^0 \qquad \text{(by (11.2.3))}$$

Further, if $y \in T_a$ and $a_0 \cdot a \neq a_0$, then $y^m = a$ for some $m \geq 1$ and so,

$$(\chi(y))^m = \chi(a) = 0, \quad \text{for } a \in H \text{ and } \chi(a) \neq 1.$$

Therefore, $\chi(y) = 0$.

Now, $T_{a_0} \cap G^0$ is a group and $\chi(a_0) \neq 0$. When the domain of χ is restricted to the group $T_{a_0} \cap G^0$, χ is a character of the group $T_a \cap G^0$ and we denote the restriction of χ to $T_{a_0} \cap G^0$ by χ_{a_0}. This proves theorem 82. $\square$

Theorem 82 can be reset for any arbitrary commutative semigroup in which every element is of finite order and in which H is finite. We have

Theorem 83 (Hewitt and Zuckerman (1955)) : *Let G be a finite or infinite commutative semigroup all of whose elements have finite order. If $a_0 \in H$ and χ_{a_0} is a character of the group $T_{a_0} \cap G^0$, then, the function $\chi : G \to \mathbb{C}$ defined by*

$$\chi(y) = \begin{cases} 0, & \text{if } a_0 \cdot a \neq a_0, \text{ for some } a \text{ such that } y \in T_a \\ \chi_{a_0}(a_0 \cdot y), & \text{if } a_0 \cdot a = a_0, \text{ for the element } a \text{ for which } y \in T_a \, ; \end{cases}$$

is a semicharacter of G.

Proof : $\chi(a_0) = \chi_{a_0}(a_0)$, as $a_0 \in H$ and $a_0 \in T_{a_0}$.
So, $\chi(a_0) \neq 0$. We have only to show that

$$\chi(a \cdot b) = \chi(a)\chi(b), \quad \text{for } a, b \in G.$$

If $y \in T_a$ and $t \in T_b$, then $y \cdot t \in T_{ab}$.
If $a_0 \cdot (y \cdot t) = a_0$, then,

$$a_0 \cdot y = a_0 \cdot y \cdot t \cdot y = a_0 \cdot (y \cdot t) = a_0$$

and $a_0 \cdot t = a_0 \cdot (y \cdot t) \cdot t = a_0 \cdot y \cdot t = a_0$.

If $a_0 \cdot y = a_0 \cdot t = a_0$, then, $a_0 \cdot y \cdot t = a_0 \cdot y \cdot a_0 \cdot t = a_0 \cdot (a_0 \cdot y \cdot t) = a_0 \cdot a_0 = a_0$.
Therefore, $\chi(y \cdot t) \neq 0$ if, and only if, $\chi(y) \neq 0$ and $\chi(t) \neq 0$.

If $\chi(y) \neq 0$, $\chi(t) \neq 0$, then,

$$\chi(y)\chi(t) = \chi_{a_0}(a_0 \cdot y)\chi_{a_0}(a_0 \cdot t)$$
$$= \chi_{a_0}(a_0 \cdot y \cdot a_0 \cdot t)$$
$$= \chi_{a_0}(a_0 \cdot y \cdot t)$$
$$= \chi(y \cdot t), \text{ by hypothesis.}$$

Thus, χ is a semicharacter on G. $\square$

Corollary 11.3.1 : *Let G be a finite commutative semigroup. The semicharacters of G form a linearly independent set of functions.*

Proof : By Fact 6.6.2, chapter 6, for a finite abelian group A, the group char(A) of characters of A is isomorphic to A. If H denotes the semigroup of idempotents of G, the semicharacter χ of G evolves from the group characters of the finite group $T_a \cap G^0$, where $a \in H$. It follows that the number m of distinct semicharacters of G is such that $m \leq |G|$.

Now, let $\chi_1, \chi_2, \ldots \chi_m$ be the semicharacters of G. Suppose that $\alpha_i \in \mathbb{C}$ $(i = 1, 2, \ldots, m)$ and $\sum_{i=1}^{m} \alpha_i \chi_i = 0$. We consider an idempotent $a \in G$ and the group $T_a \cap G^0$. On this group, every semicharacter χ_j is either identically zero or is a character of the group. Since the characters of a finite group are linearly independent, we get $\alpha_j = 0$ for each j such that $\chi_j(a) \neq 0$. Since a is arbitrary, it follows that $\alpha_j = 0$, for $j = 1, 2, \ldots, m$. $\quad\square$

Theorem 84 (Hewitt and Zuckerman (1955)) : *Let G be a semigroup all of whose elements are of finite order and having the property that for all $a, b \in G$ such that $a \neq b$, there is a semicharacter χ satisfying $\chi(a) \neq \chi(b)$. Then, G is commutative and $l_x = 1$, for all $x \in G$.*
Conversely, suppose that G is a commutative semigroup in which every element is of finite order and $l_x = 1$ for all $x \in G$. Then, for $a, b \in G$ such that $a \neq b$, there is a semicharacter χ for which $\chi(a) \neq \chi(b)$.

Proof : $:\Rightarrow$ For all semicharacters χ, we have

$$\chi(a \cdot b) = \chi(a)\chi(b) = \chi(b)\chi(a) = \chi(b \cdot a)$$

By the property of χ, $a \cdot b = b \cdot a$ for all $a, b \in G$. So G is commutative. Also,

$$(\chi(a))^{k_a + l_a} = \chi(a^{k_a + l_a}) = \chi(a^{l_a}) = (\chi(a))^{l_a}$$

(by the definition of k_a and l_a). So, $(\chi(a))^{k_a + 1} = \chi(a)$, as $(\chi(a))^{k_a} = 1$. So, $a^{k_a + 1} = a$. That is, $l_x = 1$ for all $x \in G$.
$\Leftarrow$: If $\chi(a) = \chi(b)$ for all semicharacters χ given in theorem 83, we take $a_0 = a^{k_a}$ and obtain

$$\chi(a) = \chi_{a_0}(a_0 \cdot a) = \chi_{a_0}(a^{k_a} \cdot a) = \chi_{a_0}(a).$$

Since $\chi_{a_0}(a) \neq a_0$, we have $\chi(a) = \chi(b) \neq 0$ and

(11.3.5) $\qquad a_0 \cdot b^{k_b} = a_0, \quad \chi(b) = \chi_{a_0}(a^{k_a} \cdot b).$

Thus, we have

(11.3.6) $\qquad \chi_{a_0}(a) = \chi_{a_0}(a^{k_a} \cdot b)$

But, χ_{a_0} can be any character of the abelian group $T_{a_0} \cap G^0$.
If $a \in T_{a_0}$ and $a \notin G^0$, $a_0 \cdot a \in a_0 \cdot T_{a_0} = T_{a_0} \cap G^0$ (by (11.2.3)). So, $a_0 \cdot a \in G$ and $a^m = a_0$ for some $m \geq 1$. As we have assumed that $a_0 = a^{k_a}$, $a_0 \cdot a = a^{k_a + 1} = a \in G^0$.
So, $T_{a_0} \subseteq G^0$ or $T_{a_0} \cap G^0 = T_{a_0}$.
So, from (11.3.6),

(11.3.7) $\qquad a = a^{k_a} \cdot b$

(It is on account of the fact that distinct elements of an abelian group can be separated by characters. That is, $a \neq b \Rightarrow \chi_{a_0}(a) \neq \chi_{a_0}(b)$).
Also, we have

$$a^{k_a} \cdot b^{k_b} = a^{k_a} \quad \text{since } a_0 \cdot b^{k_b} = a_0, \text{ by (11.3.5)}.$$

Since a and b are interchangeable, we get, from (11.3.7),

(11.3.8) $b = b^{k_b} \cdot a.$

So, $b^{k_b} \cdot a^{k_a} = b^{k_b}$.
So,

$$a^{k_a} = a^{k_a} \cdot b^{k_b} \text{ and } a = a^{k_a} \cdot b = b^{k_b} \cdot b = b.$$

That is, $\chi(a) = \chi(b) \Rightarrow a = b$ and this completes the proof of the converse. $\square$

Remark 11.3.1 : In the case of a finite commutative semigroup G, all semicharacters of G could be obtained in terms of a character of the group $T_{a_0} \cap G^0$ where $a_0 \in H$. Taking G^0 as a semigroup, each semicharacter of G^0 can be extended to a semicharacter of G in one and only one way.

Remark 11.3.2 : Given a semigroup $(G, \cdot)$, let $\widehat{G}$ denote the set of semicharacters of G. If χ, θ are in $\widehat{G}$, the product $\chi\theta$ is defined by

$$\chi\theta(a) = \chi(a)\theta(a) \quad \text{for all } a \in G.$$

So, $\widehat{G}$ can be given an algebraic structure.

Fact 11.3.2 : The semicharacters of a semigroup either form a semigroup by themselves or they form a semigroup, if an additional element zero is supplied. Further,

(a) If G is a semigroup, $\widehat{G}$ has a unity element.
(b) If G is a semigroup with unity element, then $\widehat{G}$ is a semigroup.

For details, see Hewitt and Zuckerman [9].

Definition 11.3.2 : *Let H be a finite commutative semigroup of idempotents. $a \in H$ is called a prime element of H, if the equality $a = b \cdot c$ $(b, c \in H) \Rightarrow b = c = a$.*

It follows that $a \in H$ is a prime element, if, and only if, $a = a \cdot b$ where $b \in H \Rightarrow b = a$.

Lemma 11.3.1 (Hewitt and Zuckerman) **:** *Every finite commutative semigroup of idempotents contains at least one prime element.*

Proof : We choose an arbitrary element $a_1 \in H$. If a_1 is not a prime element, then there exists an element $a_2 \in H$ such that $a_1 = a_1 \cdot a_2$ with $a_1 \neq a_2$. Repeating this procedure, we obtain a finite sequence $a_1, a_2, \ldots, a_m$ where $a_i = a_i \cdot a_{i+1}$ and $a_{i+1} \neq a_i$ $(i = 1, 2, 3, \ldots (m-1))$. If $h \leq j \leq m$, we have

$$a_h = a_h \cdot a_{h+1} \cdot a_{h+2} \cdots a_j.$$

Therefore, $a_h \cdot a_j = a_h$ as $a_j^2 = a_j$.
For, $a_{j-1} \cdot a_j = a_{j-1}$,
$a_{j-2}a_{j-1}a_j = a_{j-2}a_{j-1} = a_{j-2}$.
Or, $a_h \cdot a_{h+1} \cdot a_j = a_h$.

That is, $a_n \cdot a_{n+1} \cdots a_j^2 = a_h \cdot a_j$,

or, $a_h = a_h \cdot a_j$.

If $h < j \leq m$ and $a_h = a_j$, then,

$$a_h = a_h \cdot a_{h+1} = a_j \cdot a_{h+1} = a_{h+i} \cdot a_j = a_{h+1}.$$

This is a contradiction. As H is finite, $a_1, a_2, \ldots, a_m$ are such that the sequence $a_1, a_2, \ldots, a_m$ will eventually end with a prime element a_m. This proves the lemma.
□

Corollary 11.3.2 : *Corresponding to every element $a_j \in H$, there is a prime element a_m (of H) such that $a_j = a_j \cdot a_m$.*

Corollary 11.3.3 : *If H has just one prime element a_m and $b = b \cdot a_m$ for all $b \in H$, then a_m is a unity element of H (an identity element). Further, if H has just one unity element e, then H has just one prime element e.*

Proofs of corollaries 11.3.2 and 11.3.3 are omitted.

Next, we examine more closely, a commutative semigroup of idempotents. We recall that if $(G, \cdot)$ is a semigroup, for $x, y \in G$, x divides y if there exists $t \in G$ such that $y = x \cdot t$. If G is a commutative semigroup of idempotents, both x and y divide $x \cdot y$. For, $y \cdot (x \cdot y) = x \cdot (x \cdot y) = x \cdot y$.

A prime element $a \in G$ is such that $a = y \cdot t \Leftrightarrow y = t = a$. So, a prime element is an irreducible (see definition 11.2.2). A prime element has no divisors other than itself. By lemma 11.3.1, a finite commutative semigroup H of idempotents has at least one prime element. Further, a zero element z in a commutative semigroup, if it exists, has the property that z is divisible by every $x \in H$. (see definition 11.2.3).

Lemma 11.3.2 : *A finite commutative semigroup H of idempotents has a zero element.*

For, the semigroup is finite and so we can find an element z such that $z \cdot x = z$ for all $x \in H$.

Definition 11.3.3 : *Let H be a commutative semigroup of idempotents. An element x in H is said to cover an element y in H if x divides y and x does not divide any other divisor of y.*

For any non-empty set S, the power set $(\mathbb{P}(S), \cap)$ is a commutative semigroup of idempotents. For $A \subset S$ and $B \subset S$ such that $A \setminus \{a\} = B$, where $a \in A$, then B covers A. In the set of divisors of a positive integer $n = p^a m$ where p is a prime and m and p are relatively prime, $p^a | n$ and p^a does not divide any other divisor of n. p^a covers n.

Definition 11.3.4 : *Let $(P, \leq)$ be a poset. $(P, \leq)$ is called a lower semilattice if every pair of elements of P has a greatest lower bound. That is, given $a, b \in P$, there exists an element $g \in P$ such that $g \leq a$ and $g \leq b$ and if $c \in P$ is such that $c \leq a$ and $c \leq b$, then $c \leq g$. We write $g = a \wedge b$.*

Example 11.3.1 : Let X be any non-empty set. The power set $\mathbb{P}(X)$ of X ordered by inclusion is a lower semilattice. The g.l.b. of two elements A, B belonging to $\mathbb{P}(X)$ is $A \cap B$. In fact, any collection of subsets of X (which is closed under intersection) is a lower semilattice when partially ordered by set-inclusion.

It is easy to check that the subgroups of a group, subrings of a ring, ideals of a ring, subspaces of a vector space and submodules of a module all form lower semilattices.

Lemma 11.3.3 : *A commutative semigroup of idempotents is a lower semilattice up to isomorphism.*

Proof : Let $(H, \cdot)$ be a commutative semigroup of idempotents. For $x, y \in H$, we write $x \leq y$ if x divides y (and $y \leq x$ if y divides x). x and y are said to be isolated if neither divides the other. As $x^2 = x$, $x \leq x$ for all $x \in H$ (reflexivity). Suppose that for $x, y \in H$, $x \leq y$ and $y \leq x$. Then,

$$y = x \cdot t \text{ and } x = y \cdot s \text{ for } t, s \in H.$$

So,

(11.3.9) $$y = x \cdot t = (y \cdot s) \cdot t = y \cdot (s \cdot t)$$

Therefore, $x = y \cdot s = y \cdot (s \cdot t) \cdot s$ (by (11.3.9)) or,

$$x = y \cdot t \cdot s^2 = y \cdot (t \cdot s) = y \cdot (s \cdot t) = y$$

That is, $x \leq y$ and $y \leq x \Rightarrow x = y$ (antisymmetry).
When $x \leq y$ and $y \leq t$, it is easily checked that $x \leq t$ (transitivity). Thus $(H, \leq)$ is a poset.

Next, for $x, y \in H$, as $x \cdot (x \cdot y) = y \cdot (x \cdot y) = x \cdot y$, one has

$$x \leq x \cdot y \text{ and } y \leq x \cdot y$$

Also, if $t \in H$ and $t \leq x$, $t \leq y$, then $t \leq x \cdot y$ by transitivity. Therefore, the g.l.b of x and y is given by $x \cdot y$. This makes $(G, \cdot)$ a lower semilattice. $\square$

Next, in the case of a finite semigroup H of idempotents, for $a \in H$,

$$T_a = \{x \in G : x^m = a, m \geq 1\} = \{a\}$$

and $G^0 = \{x \in G : l_x = 1\} = G$. Further, $aT_a = \{a\} = T_a \cap G^0$ is a group of order 1. If $\chi : G \to \mathbb{C}$ is a semicharacter, for $t \in G$,

$$\chi(t^2) = \chi(t) \text{ and so } \chi(t) = 0 \text{ or } 1.$$

If $a \in H$, any semicharacter $\chi_a : H \to \mathbb{C}$ is given by

(11.3.10) $$\chi_a(y) = \begin{cases} 1, & \text{if } y | a, \\ 0, & \text{otherwise.} \end{cases}$$

(11.3.10) follows from the proof of theorem 82. Again, by corollary 11.3.1, the semicharacters of H form a linearly independent set. We deduce below

Theorem 85 : *Given a finite commutative semigroup H of idempotents,*

(11.3.11) $$B = \{\chi_a : a \in H\}$$

the set of semicharacters of H spans a vector space $V(H)$ of dimension $|H|$ over $\mathbb{C}$. Further, any function $f : H \to \mathbb{C}$ is an element of $V(H)$.

11.4. Finite dimensional convolution algebras

Let F be a field. By an F-algebra (or a linear algebra over F) we mean a ring $\mathcal{A}_F$ which is also a vector space over F. The scalar multiplication, that is, the map $F \times \mathcal{A}_F \to \mathcal{A}_F$ satisfies

$$\alpha(a+b) = \alpha a + \alpha b \text{ for all } \alpha \in F, a, b \in \mathcal{A}_F,$$
$$(\alpha + \beta)a = \alpha a + \beta a \text{ for all } \alpha, \beta \in F, a \in \mathcal{A}_F,$$
$$(\alpha\beta)a = \alpha(\beta a) \text{ for all } \alpha, \beta \in F, a \in \mathcal{A}_F,$$
$$1_F a = a \text{ for all } a \in \mathcal{A}_F,$$

and $\alpha(ab) = (\alpha a)b = a(\alpha b)$ for all $\alpha \in F, a, b \in \mathcal{A}_F$.

If there exists $e \in \mathcal{A}_F$ such that $ea = ae = a$ for all $a \in \mathcal{A}_F$, $\mathcal{A}_F$ is an F algebra with a unity element e. When $\mathcal{A}_F$, as a vector space over F, is finite dimensional, $\mathcal{A}_F$ is called a finite dimensional F-algebra. When the ground field F is understood, $\mathcal{A}_F$ is referred to as an algebra.

We examine a finite dimensional algebra $\mathcal{A}_F$. The multiplication in $\mathcal{A}_F$ is completely determined by the products of the basis elements. If $\{u_i\}$ is a basis for $\mathcal{A}_F$, we have

$$u_i \cdot u_j = \sum_k \gamma_{ijk} u_k$$

where γ_{ijk} are called the multiplication constants of the algebra. When $\mathcal{A}_F$ is infinite dimensional, $\sum_k \gamma_{ijk} u_k$ is such that only finitely many of γ_{ijk} are non-zero for any choice of i, j. If $a = \sum \alpha_i u_i$, $b = \sum \beta_j u_j$, then,

$$a \cdot b = \sum_k \alpha_i \beta_j \gamma_{ijk} u_k.$$

Since the multiplication in $\mathcal{A}_F$ is associative, the multiplication constants satisfy the equations arising from $(u_i \cdot u_j) \cdot u_k = u_i \cdot (u_j \cdot u_k)$. That is,

$$\sum_l \gamma_{ijl}\gamma_{lkm} = \sum_l \gamma_{jkl}\gamma_{ilm}$$

Further, when $\mathcal{A}_F$ is commutative, it is true that $\gamma_{ijk} = \gamma_{jik}$.

We begin with an arbitrary semigroup G and consider complex-valued functions having domain G.

Let $\mathcal{F}$ denote the vector space of functions $f : G \to \mathbb{C}$ with the usual definitions of addition and scalar multiplication.

For $x \in G$, $f \in \mathcal{F}$, we define $_x f$ by

(11.4.1) $$_x f(y) = f(xy)$$

$_xf$ is an element of $\mathcal{F}$ (for fixed $x \in G$).

Let $\mathcal{L}$ denote the vector space of linear functionals defined on $\mathcal{F}$. For $L \in \mathcal{L}$, $f \in \mathcal{F}$ and $x \in G$ we write

(11.4.2) $L_y(f(xy))$ to denote $L(_xf)$

For all $L \in \mathcal{L}$ and $f \in \mathcal{F}$, $L_y(f(xy))$ is such that the value of $L_y(f(xy))$ at x, is an element of $\mathbb{C}$.

(11.4.3) $L_y(f(xy)) = L(_xf)$ is an element of $\mathcal{F}$ when x is taken as

an element of the domain G.
For all $L, M \in \mathcal{L}$, we define

(11.4.4) $N(f) = M_x(L_y(f(xy)))$, as an element of $\mathcal{L}$, for all $f \in \mathcal{F}$.

Definition 11.4.1 : *[9] Using the relations* (11.4.1) *to* (11.4.4), *N is called the convolution of M and L and we call $\mathcal{L}$ a convolution algebra. The notation for convolution of M and L is $M * L$.*

Examples 11.4.1 :

(a) A CONVOLUTION ALGEBRA:

Let G be a non-empty set. Take $a \in G$ to be a fixed element of G. For $x, y \in G$, we define multiplication $(\cdot)$ by $x \cdot y = a$. $(G, \cdot)$ is a semigroup which is commutative. Let $\mathcal{F}$ be the vector space of complex-valued functions defined on G. The function $e : G \to \mathbb{C}$ given by $e(x) = 1$ for all $x \in G$ is an element of $\mathcal{F}$.

If $\mathcal{L}$ denotes the vector space of linear functionals on $\mathcal{F}$, $\lambda_a \in \mathcal{L}$ is defined by

(11.4.5) $\lambda_a(f) = f(a) \ (\in \mathbb{C})$, for all $f \in \mathcal{F}$.

Then, for $x, y \in G$, $_xf(y) = f(xy) = f(a) = \lambda_a(f)$.
For any $x \in G$, $_xf(y)$ gives $f(a)$.

$$L_y(f(xy)) = L_y(f(a)e) = f(a)L_y(e) = \lambda_a L_y(e)$$
$$(M * L)(f) = M_x(L_y(f(xy))) = M_x(\lambda_a L_y(e)e) = M_x(e)L_y(e)\lambda_a(f)$$

Convolution of M and L is given by $M(e)L(e)\lambda_a$ for all $M, L \in \mathcal{L}$. $\mathcal{L}$ is a convolution algebra.

(b) CAUCHY CONVOLUTION:

$\widetilde{\mathbb{Z}}$ denotes the set of non-negative integers. Let $\mathcal{A}^*$ denote the set of functions $f : \widetilde{\mathbb{Z}} \to \mathbb{C}$. f is, in fact, a sequence of complex numbers, say, $\{a_n\}$. If $g = \{b_n\}$,

(11.4.6) $f + g = \{a_n + b_n\}$

and

(11.4.7) $f * g = \{c_n\}$ where $c_n = \displaystyle\sum_{k=0}^{n} a_k b_{n-k}$

$(\mathcal{A}^*, +, *)$ is an algebra over $\mathbb{C}$. We claim that $(\mathcal{A}^*, +, *)$ is a convolution algebra in the sense of definition 11.4.1.

Let $\widetilde{\mathcal{F}}$ be the space of all functions defined on $\widetilde{\mathbb{Z}}$ which vanish except on finite subsets of $\widetilde{\mathbb{Z}}$. We know that $(\widetilde{\mathbb{Z}}, +)$ is a semigroup. Let $\mathcal{L}$ be the space of all functionals of $\widetilde{\mathcal{F}}$. A linear functional $A : \mathcal{F} \to \mathbb{C}$ is such that there is a unique sequence $\{a_n\} \in \mathcal{A}^*$ for which

$$(11.4.8) \qquad A(f) = \sum_{n=0}^{\infty} a_n f(n) \text{ for all } f \in \widetilde{\mathcal{F}}$$

The right side of (11.4.8) is a finite sum as f vanishes except for a finite number of elements of $\widetilde{\mathbb{Z}}$. Conversely, every sequence $\{a_n\}$ defines a linear functional on $\widetilde{\mathcal{F}}$. For $f \in \widetilde{\mathcal{F}}$, $_m f(n) = f(mn)$ determines $_m f$ as an element of $\widetilde{\mathcal{F}}$.

Let $f \not\equiv 0$. When $f \in \widetilde{\mathcal{F}}$, let $(t-1)$ be the greatest integer such that $f(t-1) \neq 0$. Then,

$$(11.4.9) \qquad f(m+n) = 0 \quad \text{ for all } n \in \widetilde{\mathbb{Z}} \text{ and } m \geq t.$$

Therefore,

$$A_n(f(m+n)) = \sum_{n=0}^{\infty} a_n f(m+n) = 0 \text{ for all } m \geq t.$$

So $A_n(_m f) \in \widetilde{\mathcal{F}}$. (11.4.1), (11.4.2) and (11.4.3) hold.

Now, $\mathcal{L}$ consists of all linear functionals defined on $\widetilde{\mathcal{F}}$. Let e_n be defined by

$$(11.4.10) \qquad e_n(m) = \begin{cases} 1, & n = m \\ 0, & \text{otherwise} \end{cases}$$

Then, $\{e_n\}$ $(n \in \widetilde{\mathbb{Z}})$ forms a basis for $\widetilde{\mathcal{F}}$ and $A(e_n) = a_n$ (by (11.4.8)). For $A, B \in \mathcal{L}$, we obtain

$$(11.4.11) \qquad (A * B)(e_n) = A_k(B_l(e_n(k+l))) = \sum_{k=0}^{\infty} \sum_{l=0}^{\infty} a_k b_l e_n(k+l).$$

Or,

$$(11.4.12) \qquad (A * B)(e_n) = \sum_{k=0}^{n} a_k l_{n-k}$$

So, multiplication given in (11.4.7) for elements of $\mathcal{A}^*$ is a convolution in the sense of definition 11.4.1.

(c) DIRICHLET CONVOLUTION:

$\mathbb{N}$, the set of positive integers is a semigroup under multiplication $f : \mathbb{N} \to \mathbb{C}$ is an arithmetic function. $\{f(1), f(2), \ldots\}$ is a complex-valued sequence. We write

(11.4.13) $$f = \{f(1), f(2), f(3), \ldots\}$$

Let

(11.4.14) $$g = \{g(1), g(2), g(3), \ldots\}$$

The Dirichlet product of f and g is given by

(11.4.15) $$h = (f \cdot g) = \{h(1), h(2), \ldots\}$$

where

(11.4.16) $$h(r) = \sum_{d|r} f(d) g\left(\frac{r}{d}\right), \quad \text{(see definition 4.3.1 of chapter 4)}$$

(the summation on the right is over the positive divisors d of r). The set $\mathcal{A}$ of arithmetic functions is a vector space over $\mathbb{C}$. The function e_0 given by

(11.4.17) $$e_0(r) = \left[\frac{1}{r}\right],$$

where $[x]$ denotes the greatest integer not exceeding x serves as the identity for Dirichlet multiplication (11.4.16) see (4.3.4) of chapter 4. As in Cauchy multiplication, we look at a subspace $\mathcal{A}_0$ of $\mathcal{A}$ defined by

(11.4.18) $$\mathcal{A}_0 = \{f \in \mathcal{A} : f \text{ vanishes except on finite subsets of } \mathbb{N}\}$$

Let $\mathcal{L}$ denote the vector space of linear functionals defined on $\mathcal{A}_0$. For all $A \in \mathcal{L}$, there is a unique $\{a_n\}$ ($n \in \mathbb{N}$) such that $A(f) = \sum_{n=1}^{\infty} a_n f(n)$ for all $f \in \mathcal{A}_0$. $A(f)$ is a finite sum. For $r \in \mathbb{N}$, we write

$$_r f(n) = f(rn), \quad n \in \mathbb{N}$$

Then, $_r f \in \mathcal{A}_0$. When $f \in \mathcal{A}_0$, let $(s - 1)$ be the greatest integer such that $f(s - 1) \neq 0$. Then,

$$f(rn) = 0 \text{ for } n \in \mathbb{N} \text{ and } r \geq s$$

Further,

(11.4.19) $$A_n f(rn) = \sum_{n=1}^{\infty} a_n f(rn) = 0, \quad \text{for } r \geq s$$

$A_n(_r f) \in \mathcal{A}_0$. If B is a linear functional on $\mathcal{A}_0$ and $f \in \mathcal{A}_0$, $B_r(A_n(f(rn)))$ is a linear functional. We note that B_r is defined the same way as A_n is given in (11.4.19). That is,

$$B_r(f(rn)) = \sum_{r=1}^{\infty} b_r f(rn).$$

The functions e_r given by

(11.4.20) $$e_r(n) = \begin{cases} 1 & n = r \\ 0 & \text{otherwise} \end{cases}$$

form the basis for $\mathcal{A}_0$, as $f \in \mathcal{A}_0$ can be written uniquely as

(11.4.21)
$$f(n) = \sum_{r=1}^{\infty} f(r)e_r(n)$$

That is,

(11.4.22)
$$f = \sum_{r=1}^{\infty} f(r)e_r$$

We check Dirichlet convolution for the basis elements $e_r, r = 1, 2, \ldots$

$$B * A(e_r) = B_k(A_l(e_r(kl)))$$
$$= \sum_{k=1}^{\infty} \sum_{l=1}^{\infty} b_k a_l e_r(kl)$$

Therefore,

(11.4.23)
$$B * A(e_r) = \sum_{kl=r} b_k a_l$$

which is the gist of Dirichlet multiplication. That is, Dirichlet multiplication is also brought under convolution of linear functionals as in the earlier example. In other words, $\mathcal{L}$, the space of linear functionals on $\mathbb{A}_0$ is indeed a convolution algebra.

Definition 11.4.2 : *Let G be a group. The group algebra FG of G over F is defined as the F-algebra obtained by taking G as a basis for the vector space FG over F. Elements of FG are finite sums*

$$\sum_{g \in G} \alpha_g g \text{ where } \alpha_g \in F.$$

Multiplication in FG is defined by

(11.4.24)
$$(\sum_{g \in G} \alpha_g g)(\sum_{h \in G} B_h h) = \sum_{g \in G} \sum_{h \in G} \alpha_g \beta_h gh, \quad \alpha_g, \beta_h \in F.$$

Addition is defined by

(11.4.25)
$$\sum_{g \in G} \alpha_g g + \sum_{g \in G} \beta_g g = \sum_{g \in G} (\alpha_g + \beta_g)g; \quad \alpha_g, \beta_g \in F$$

and for $\beta \in F$,

(11.4.26)
$$\beta \sum_{g \in G} \alpha_g g = \sum_{g \in G} (\beta \alpha_g)g$$

The multiplication in FG is given by the multiplication table of G.

	e	a	b
e	e	a	b
a	a	b	e
b	b	e	a

Figure 15

Let C_3 denote a cyclic group of order 3. $C_3 = \langle a \rangle$, say. Taking C_3 as a basis for a vector space $\mathbb{C}C_3$, $\mathbb{C}C_3$ is made an algebra with the multiplication table given in the Figure 15 shown above.

Any element x of the group algebra $\mathbb{C}C_3$ is of the form $\alpha_1 e + \alpha_2 a + \alpha_3 b$, where $\alpha_1, \alpha_2, \alpha_3 \in \mathbb{C}$. If $y \in \mathbb{C}C_3$ is given by $\beta_1 e + \beta_2 a + \beta_3 b$,

$$x \cdot y = (\alpha_1 e + \alpha_2 a + \alpha_3 b) \cdot (\beta_1 e + \beta_2 a + \beta_3 b)$$

or

$$x \cdot y = (\alpha_1 \beta_1 + \alpha_2 \beta_3 + \alpha_3 \beta_2)e + (\alpha_1 \beta_2 + \alpha_2 \beta_1 + \alpha_3 \beta_3)a + (\alpha_1 \beta_3 + \alpha_3 \beta_1 + \alpha_2 \beta_2)b$$

(using the multiplication table in Figure 15).

$\mathbb{C}C_3$ is a group algebra of dimension 3. In the same manner, we can talk about a semigroup algebra FG when G is a semigroup.

If $M_n(F)$ denotes the set of $n \times n$ matrices with entries from F ($n \geq 1$), $M_n(F)$ forms a ring and it is also a vector space of dimension n^2 over F. If

$$E_{ij} = [e_{ij}]$$

where E_{ij} is the matrix whose entry at i, i is 1_F and having all other entries O_F, $\{E_{ij}\}$ ($i = 1, 2, \ldots n$, $j = 1, 2, \ldots n$) is a basis for $M_n(f)$. The multiplication is given by

$$E_{ij}E_{kl} = \delta_{jk}E_{il} \quad \text{where } \delta_{jk} \text{ is Kronecker delta, given by}$$

(11.4.27) $$\delta_{jk} = \begin{cases} 1, & j = k \\ 0 & \text{otherwise.} \end{cases}$$

$M_n(f)$ is the matrix algebra of dimension n^2 over F.

Returning to the case of an arbitrary commutative semigroup $G = \{x_1, x_2, \ldots, x_n\}$ with $x_i \cdot x_j = x_{i*j}$ where $i * j$ is the suffix which will make x_{i*j} the product of x_i and x_j, we consider another algebra $\mathcal{L}_1(G)$ described below:

Suppose that $\mathcal{F}_1(G)$ denotes the vector space of functions $f : G \to \mathbb{C}$. The space of linear functionals on $\mathcal{F}_1(G)$ is denoted as $\mathcal{L}_1(G)$.

Let $\phi_i \in \mathcal{F}_1(G)$ be given by

(11.4.28) $$\phi_i(x_j) = \delta_{ij} \quad \text{(Kronecker delta; } i, j = 1, 2, \ldots n)$$

If $\lambda_i \in \mathcal{L}_1(G)$ is given by

(11.4.29) $$\lambda_i(\phi_j) = \delta_{ij} \quad (i, j = 1, 2, \ldots, n)$$

it is verified that $\mathbb{B}_1 = \{\phi_1, \phi_2, \ldots \phi_n\}$ forms a basis for $\mathcal{F}_1(G)$ and any element $f \in \mathcal{F}_1(G)$ can be written as

$$(11.4.30) \qquad f = \sum_{i=1}^{n} f(x_i)\phi_i$$

Also, the functionals λ_i $(i = 1, 2, \ldots n)$ form a basis for $\mathcal{L}_1(G)$.
   If $L \in \mathcal{L}_1(G)$,

$$(11.4.31) \qquad L = \sum_{i=1}^{n} L(\phi_i)\lambda_i$$

and $\lambda_i(f) = \sum_{j=1}^{n} f(x_j)\lambda_i(\phi_j) = f(x_i)$ $(i = 1, 2, \ldots n)$, for all $f \in \mathcal{F}_1(G)$.

Lemma 11.4.1 (Hewitt and Zuckerman (1955)) : $\mathcal{L}_1(G)$ *is a convolution algebra and is isomorphic to the semigroup algebra* $\mathbb{C}G$.

Proof : We have observed that $\{\lambda_1, \lambda_2, \ldots, \lambda_n\}$ forms a basis for $\mathcal{L}_1(G)$ where λ_i is as given in (11.4.29).
   Now, we use the identity:

$$(11.4.32) \qquad f(x \cdot y) = \sum_{k,l=1}^{n} f(x_k \cdot x_l)\phi_k(x)\phi_l(y)$$

which is valid for all $f \in \mathcal{F}_1(G)$ and $x, y \in G$. Then, we need to find the product of the linear functionals λ_i and λ_j where $\lambda_i, \lambda_j \in \mathcal{L}_1(G)$.
   When $f \in \mathcal{F}_1(G)$, $\lambda_i(f) = f(x_i)$ $(i = 1, 2, \ldots n)$. Fix the suffixes r, s of x_r and x_s for the time being. We get

$$(11.4.33) \qquad f(x_r \cdot x_s) = \sum_{k,l=1}^{n} f(x_k \cdot x_l)\phi_k(x_r)\phi_l(x_s).$$

We write

$$\lambda_{i,x_r}(f(y)) = \sum_{k,l=1}^{n} f(x_k \cdot x_l)\phi_k(x_r)\lambda_i(\phi_l).$$

Then,

$$\lambda_{i,x_r}(\lambda_{j,x_s}(f(x_r \cdot x_s))) = \lambda_{i,x_r}\left(\sum_{k,l=1}^{n} f(x_k \cdot x_l)\phi_k(x_r)\lambda_j(\phi_l)\right)$$

$$= \sum_{k,l=1}^{n} f(x_k \cdot x_l)\lambda_i(\phi_k)\lambda_j(\phi_l)$$

$$= \sum_{k,l=1}^{n} f(x_k \cdot x_l)\delta_{ik}\delta_{jl}$$

$$= f(x_i \cdot x_j)$$

or,

$$\lambda_i \otimes \lambda_j(f) = \lambda_{i*j}(f), \text{ where } i*j \text{ is such that } x_i \cdot x_j = x_{i*j} \, (1 \leq i, j \leq n).$$

That is, $\lambda_i \otimes \lambda_j = \lambda_{i*j}$. Hence, $\mathcal{L}_1(G)$ is a convolution algebra.

The semigroup algebra $\mathbb{C}G$ consists of complex linear combinations $\sum_{x \in G} \alpha_x x$ and

$$(\sum_{x \in G} \alpha_x x) + (\sum_{x \in G} \beta_x x) = \sum_{x \in G}(\alpha_x + \beta_x)x.$$

For $r \in \mathbb{C}$, $r \sum_{x \in G} \alpha_x x = \sum_{x \in G}(r\alpha_x)x$. And

$$(\sum_{x \in G} \alpha_x x)(\sum_{y \in G} \beta_y y) = \sum_{x \in G} \sum_{y \in G} \alpha_x \beta_y (x \cdot y).$$

Let $\psi : \mathcal{L}_1(G) \rightarrow \mathbb{C}G$ be defined by $\psi(\sum_{i=1}^{n} \alpha_i \lambda_i) = \sum_{i=1}^{n} \alpha_i x_i$.
It is easily verified that ψ is an isomorphism of $\mathcal{L}_1(G)$ onto $\mathbb{C}G$. In other words, $\mathcal{L}_1(G) \cong \mathbb{C}G$.   □

Fact 11.4.1 :

(i) A finite dimensional algebra $\mathcal{A}_F$ is isomorphic to an algebra $\mathcal{L}_1(G)$ for an appropriate finite semigroup G if, and only if, $\mathcal{A}_F$ has a basis which is closed under multiplication.

(ii) Every finite dimensional algebra $\mathcal{A}_F$ is a convolution algebra.
For proofs of (i) and (ii), see [9].

(iii) Let $\mathcal{A}_F$ be an n-dimensional algebra over a field F. Then, $\mathcal{A}_F$ is isomorphic to a subalgebra of the matrix algebra $M_n(f)$.
For proof, see P. M. Cohn [7].

11.5. Abstract arithmetical functions

When H is a finite commutative semigroup of idempotents, we form the vector space $V(H)$ of functions $f : H \rightarrow \mathbb{C}$. By theorem 85, the set $B = \{\chi_a : a \in H\}$ of semicharacters of H spans $V(H)$. We recall that χ_a are functions defined in (11.3.10). We wish to make $V(H)$ an algebra. Elements of $V(H)$ are abstract arithmetical functions. $V(H)$ is finite dimensional. It is a convolution algebra, when multiplication of functions is defined as below:

For $f, g \in V(H)$

(11.5.1) $(f \cdot g)(x) = h(x) = \displaystyle\sum_{yz=x} f(y)g(z),$ for all $x \in H$.

(11.5.1) is an analogue of Dirichlet convolution of arithmetic functions considered in definition 4.3.1, chapter 4. There, it was the infinite semigroup $\mathbb{N}$ with multiplication as the binary operation. Here, $(\mathbb{N}, \cdot)$ is replaced by a finite semigroup $(H, \cdot)$ of idempotents. (11.5.1) is meaningful as H is finite. The convolution operator $\cdot$ is a map $\psi : V(H) \times V(H) \rightarrow V(H)$ which gives (11.5.1). The n-fold convolution of functions $f_1, f_2, \cdots f_n$ could be defined in the same manner. In the case of an arithmetic function f, we have considered the Dirichlet series $\sum f(n)n^{-s}$ (Re $s > \sigma_a$) of f (see definition 10.6.1, chapter 10). An analogous definition for abstract arithmetical functions is given in

Definition 11.5.1 (M. Tainiter) **:** *For $f \in V(H)$, we write*

$$(11.5.2) \qquad F(a) = \sum_{y \in H} f(y)\chi_a(y) = \sum_{y \leq a} f(y), \quad \text{for all } a \in H;$$

where χ_a is a semicharacter of H specified at $a \in H$.
In (11.5.2), summation $\sum$ is over all divisors y of $a \in H$. F can be viewed as a linear operator on $V(H)$. For more details see M. Tainiter [15]. F is called the generating function of f.

Theorem 86 (M. Tainiter (1968)) **:** *Let $f, g, t \in V(H)$. If F, G, T are the generating functions of f, g and t respectively, then*

$$(11.5.3) \qquad t = f \cdot g \Leftrightarrow T(a) = F(a)G(a), \quad \text{for all } a \in H.$$

Proof : $:\Rightarrow$ Suppose that $t = f \cdot g$.
By definition, for $a \in G$

$$F(a) = \sum_{y \leq a} f(y), \qquad G(a) = \sum_{y \leq a} g(y).$$

Then,

$$F(a)G(a) = (\sum_{y \leq a} f(y))(\sum_{z \leq a} g(z))$$

$$= \sum_{y \leq a} \sum_{z \leq a} f(y)g(z)$$

$$= \sum_{y \wedge z \leq a} f(y)g(z), \text{ as } H \text{ is a lower semilattice.}$$

As $y \wedge z = yz$, we get

$$(11.5.4) \qquad F(a)G(a) = \sum_{yz \leq a} f(y)g(z)$$

Now,

$$(11.5.5) \qquad T(a) = \sum_{x \leq a} t(x) = \sum_{x \leq a} \sum_{yz=x} f(y)g(z) = \sum_{yz \leq a} f(y)g(z)$$

From (11.5.4) and (11.5.5), $T(a) = F(a)G(a)$. When $T(a) = F(a)G(a)$, $a \in H$ we conclude that $t = f \cdot g$, as the generating function of an abstract arithmetical function is unique. $\qquad \square$

Corollary 11.5.1 : *As $(H, \cdot)$ is a poset (more precisely a lower semilattice) there is a Möbius function μ associated with $(H, \leq)$ and*

$$F(a) = \sum_{y \leq a} f(a) \Leftrightarrow f(x) = \sum_{a \leq x} F(a)\mu(a,x)$$

Examples 11.5.1 :: The function $e \in V(H)$ is given by $e(x) = 1$ for all $x \in H$. Then,

$$(11.5.6) \qquad E(a) = \sum_{y \in H} e(y)\chi_a(y) = \sum_{y \leq a} 1 = \text{ the number of divisors of } a.$$

Let

$$(11.5.7) \qquad f(x) = \begin{cases} 1, & \text{if } x \text{ is an irreducible;} \\ 0, & \text{otherwise.} \end{cases}$$

Then, $F(a) = \sum_{y \leq a} f(y) = $ the number of irreducible divisors of a.
The analogue of the ζ-function ζ_H of $(H, \leq)$ is given by

$$(11.5.8) \qquad \zeta_H(x,y) = \begin{cases} 1, & x \leq y; \\ 0, & \text{otherwise.} \end{cases}$$

Its generating function $Z(b,a)$ is expressed as

$$Z(b,a) = \sum_{y \leq a} \chi_b(y)\chi_a(y) = \sum_{\substack{y \leq a \\ y \leq b}} 1.$$

$Z(b,a)$ is the number of common divisors of a and b.
For more related results, see M. Tainiter [15].

Next, we consider the power set $\mathbb{P}(S)$ of a finite set S having n elements. $|\mathbb{P}(S)| = 2^n (n \geq 1)$. $\mathbb{P}(S)$ forms a semigroup of idempotents under set-intersection. We define

$$(11.5.9) \qquad \chi_A(B) = \begin{cases} 1, & \text{if } B \text{ is a subset of } A; \\ 0, & \text{otherwise.} \end{cases}$$

Möbius function of $(\mathbb{P}(S), \subseteq)$ is given by

$$(11.5.10) \qquad \mu(B,A) = \begin{cases} (-1)^{|A|-|B|}, & \text{if } B \text{ is a subset of } A; \\ 0, & \text{otherwise.} \end{cases}$$

Theorem 87 (Tainiter (1968)) : *Let A be a subset of S with $|S| = n$, such that $|A| = m$, $(m \leq n)$. If $h_k(A)$ denotes the number of ways of expressing A as the intersection of k subsets of S, then*

$$(11.5.11) \qquad h_k(A) = (2^k - 1)^{n-m}.$$

Proof : As $(\mathbb{P}(S), \cap)$ is a semigroup, the problem is to find the number of ways of writing A as a product of k elements belonging to $\mathbb{P}(S)$. Let

$$e(A) = 1, \quad \text{for all } A \in \mathbb{P}(S).$$

Then,

$$h_k(A) = \sum_{A_1 A_2 \cdots A_k = A} e(A_1)e(A_2)\cdots e(A_k).$$

As semigroup multiplication corresponds to set-intersection, we write

(11.5.12) $$E(A) = \sum_{B \supseteq A} e(B).$$

$E(A)$ gives the number of subsets of S which contain A. The generating function of $h_k(A)$ is given by

(11.5.13) $$H_k(A) = \sum_{B \in \mathbb{P}(S)} h_k(B)\chi_A(B) = \sum_{B \supseteq A} h_k(B).$$

By Möbius inversion,

(11.5.14) $$h_k(A) = \sum_{B \supseteq A} H_k(B)\mu(A,B).$$

However, using (11.5.12) and (11.5.13), we note that

$$H_k(A) = d^k(A),$$

where $d(A)$ = the number of subsets of S which contain A.
If we count the number of subsets of $S \setminus A$, and add to each subset the elements of A, we get the number of subsets of S which contain A. That is, $d(A) = 2^{n-|A|}$. So, (11.5.14) is rewritten as

$$h_k(A) = \sum_{B \supseteq A} 2^{k(n-|B|)}(-1)^{|B|-|A|}$$

As $|A| = m$, $|B| - |A|$ decreases from $(n-m)$ to 0. So, we obtain

$$h_k(A) = \sum_{j=0}^{n-m} 2^{k(n-(m+j))}(-1)^j \binom{n-m}{j},$$

$$= 2^{k(n-m)} \sum_{j=0}^{n-m} (-1)^j 2^{-kj} \binom{n-m}{j}.$$

Or, $h_k(A) = 2^{k(n-m)}(1 - \dfrac{1}{2^k})^{n-m} = 2^{k(n-m)} \dfrac{(2^k-1)^{n-m}}{2^{k(n-m)}} = (2^k - 1)^{n-m}$

This proves (11.5.11). $\square$

11.6. Convolutions in general

There are several instances for obtaining generalisations of arithmetical convolutions. One is the case of functions defined on the semigroup X of finite abelian groups G with respect to the direct product.
Let $f : X \to \mathbb{C}$, $g : X \to \mathbb{C}$ be functions defined on X. The direct convolution [6] of f and g is defined by

(11.6.1) $$(f \cdot g)(G) = \sum_{D \times E = G} f(D)g(E),$$

where the summation is over all pairs of groups D, E such that $D \times E = G$. The group of order 1 (the identity group) is denoted by G_0. If T denotes the sub-semigroup (of X) consisting of all finite cyclic groups, then T is isomorphic to the multiplicative semigroup $(\mathbb{N}, \cdot)$ of positive integers under the correspondence associating with each integer $r \in \mathbb{N}$, the unique finite cyclic group of order r in X. The elementary functions defined on X are

$$E(G) = 1, \quad G \in X$$

(11.6.2) $\qquad E_0(G) = \begin{cases} 1, & \text{if } G = G_0 \\ 0, & \text{if } G \neq G_0 \end{cases}$

$$d(G) = \text{ the number of direct factors of } G \text{ in } X.$$

The theory of arithmetic functions could be developed parallely in the context of the semigroup X. When X is replaced by T, Dirichlet convolution results from (11.6.1). A generalization of the Prime Number Theorem [1] is obtained by Eckford Cohen in the generalized set-up. See [6] for details.

Next, we consider the set

(11.6.3) $\qquad\qquad L = \{1, 2, \ldots, r\}, \quad r \geq 1$

For $x, y \in L$, the product $x \cdot y$ is defined by

$$x \cdot y = \max\{x, y\}$$

Then, $(L, \cdot)$ is a finite commutative semigroup of idempotents. We note that y divides x if, and only if, $y \leq x$.
That is,

$$y \leq x \Leftrightarrow y \text{ is less than or equal to } x \text{ and}$$

$$y \leq x \Leftrightarrow \max\{x, y\} = x.$$

The element r in L serves as the zero element in $(L, \cdot)$. $(L, \leq)$ is a lower semilattice. The set $\{\chi_a : a \in L\}$ of semicharacters of L consists of the elements χ_a given by

(11.6.4) $\qquad\qquad \chi_a(y) = \begin{cases} 1, & \text{if } y \leq a; \\ 0, & \text{otherwise.} \end{cases}$

The vector space $V(L)$ spanned by the semicharacters χ_a (11.6.4) is finite dimensional. $\dim V(L) = r$. A function $f : L \to \mathbb{C}$ belongs to $V(L)$.

Definition 11.6.1 : *For $f, g \in V(L)$, the MAX convolution of f and g written $h = (f \circ g)$ is given by*

(11.6.5) $\qquad h(x) = (f \circ g)(x) = \sum_{\max\{y, z\} = x} f(y) g(z), \quad \text{for all } x \in L.$

When $x = r$, one has

(11.6.6) $\qquad h(r) = \sum_{\max\{y, z\} = r} f(y) g(z) = g(r) \sum_{i=1}^{r} f(i) + f(r) \sum_{i=1}^{r} g(i) - f(r) g(r)$

In terms of χ_a *(11.6.4), the* ζ*-function related to* $(L, \leq)$ *is given by*

(11.6.7) $$\zeta(a,x) = \chi_a(x).$$

The generating function F of $f \in V(L)$ *is defined by*

$$F(a) = \sum_{y \in L} f(y)\chi_a(y) = \sum_{y \leq a} f(y), \quad \text{for all } a \in L.$$

Theorem 88 (Haukkanen and Sivaramakrishnan (1998)) : *For* $f, g \in V(L)$, *let* $h = (f \circ g)$, *the MAX product of f and g. If F, G and H are the generating functions of* f, g *and h respectively, then,*

$$H(a) = F(a)G(a), \quad a \in L.$$

Proof follows on lines similar to the proof of theorem 86.

When L is as defined in 11.6.3, $(L, \leq)$ is a locally finite partially ordered set. The Möbius function μ of $(L, \leq)$ exists and is given by

(11.6.8) $$\mu(x,y) = \begin{cases} 1, & \text{if } x = y \\ -1, & \text{if } y - x = 1 \ (x \leq y) \\ 0, & \text{otherwise} \end{cases}$$

Next, we consider the set

(11.6.9) $$D = \{d \ : \ 1 \leq d \leq r \text{ and } d | r\}$$

D is the set of divisors of r. For $x, y \in D$, we define their product $x * y = [x, y]$, the l.c.m of x and y. Then $(D, *)$ is a finite commutative semigroup of idempotents. $|D| = d(r)$, the number of divisors of r. We write $y|x$, if y divides x in the number-theoretic sense. The partial order $\leq$ in D is given by

$$y \leq x \Leftrightarrow y | x \quad \text{or} \quad y \leq x \Leftrightarrow [x, y] = x$$

r serves as the zero element in $(D, *)$. Clearly $(D, \leq)$ is a lower semilattice. The semicharacters of D are defined as in (11.6.4). We denote the semicharacters by χ_a $(a \in D)$. The set of semicharacters spans a vector space $V(D)$ of dimension $d(r)$.

Definition 11.6.2 : *For* $f, g \in V(D)$, *the LCM convolution of f and g, written* $h = f \otimes g$, *is defined by*

(11.6.10) $$h(x) = (f \otimes g)(x) = \sum_{[y,z]=x} f(y)g(z), \quad x \in D,$$

where the summation is over all $y, z \in D$ *such that their l.c.m. is equal to x.*

The ζ *-function related to* $(D, \leq)$ *is given by*

$$\zeta(y,x) = \begin{cases} 1, & \text{if } x \leq y; \\ 0, & \text{otherwise.} \end{cases}$$

We note that $\zeta(a,x) = \chi_a(x)$. *The generating function F of* $f \in V(D)$ *is given by*

$$F(a) = \sum_{y \in D} f(y)\chi_a(y) = \sum_{y \leq a} f(y), \quad \text{for all } a \in D,$$

That is,

(11.6.11) $$F(a) = \sum_{[y,a]=a} f(y)$$

Theorem 89 : *For* $f, g \in V(D)$, *let* $h = (f \otimes g)$. *If* F, G, H *are the generating functions of* f, g *and* h *respectively*

$$H(a) = F(a)G(a), \quad a \in D.$$

Proof is omitted as it is similar to those of theorems 85 and 87.

We remark that the Möbius function of $(D, \leq)$ is

(11.6.12) $$\mu(y,x) = \begin{cases} 1, & y = x, \\ (-1)^k, & \text{if } \frac{x}{y} = t \text{ is a product of } k \text{ distinct primes,} \\ 0, & \text{if } \frac{x}{y} = a^2, a > 1. \end{cases}$$

More results of this nature are given in Haukkanen and Sivaramakrishnan [8].

11.7. A functional-theoretic algebra

Various types of arithmetical convolutions have been studied by many authors. See M. V. Subbarao [14]. When the set $\mathcal{A}$ of arithmetical functions is considered as a vector space over $\mathbb{C}$, the field of complex numbers, we can associate a Dirichlet series $\sum_{n=1}^{\infty} f(n)n^{-s}$ (Re $s > a$) with each $f \in \mathcal{A}$ and it gives rise to a correspondence between $f \in \mathcal{A}$ and $F(s) = \sum_{n=1}^{\infty} f(n)n^{-s}$ (Re $s > \sigma_a$). Multiplication of Dirichlet series corresponds to Dirichlet convolution of arithmetic functions.

It is known [1] that when $\sum_{n=1}^{\infty} |f(n)n^{-s}|$ is not either convergent or divergent for all s, there exists an abscissa of absolute convergence σ_a (say) such that $\sum_{n=1}^{\infty} f(n)n^{-s}$ converges absolutely for all Re $s > \sigma_a$. See Fact 10.6.1. We consider a subset $\mathcal{A}(k)$ of the vector space of Dirichlet series as follows:

Let k be a real number > 1. $\mathcal{A}(k)$ consists of all arithmetic functions $f \in \mathcal{A}$ such that their Dirichlet series $\sum_{n=1}^{\infty} f(n)n^{-s}$ converge absolutely for all Re $s > k - \delta$ (δ positive). Then, it follows that if $f \in \mathcal{A}(k)$, $\sum_{n=1}^{\infty} f(n)n^{-k}$ is a complex constant which we denote by $F(f,k)$. Let $f, g \in \mathcal{A}(k)$. Then, $F(f,k)$ and $F(g,k)$ are elements of $\mathbb{C}$. Further, for $\alpha, \beta \in \mathbb{C}$

(11.7.1) $$\sum_{n=1}^{\infty} (\alpha f(n) + \beta g(n))n^{-k} = \alpha F(f,k) + \beta F(g,k)$$

So, $\alpha f + \beta g \in \mathcal{A}(k)$. In other words, $\mathcal{A}(k)$ is a subspace of $\mathcal{A}$.
Let $L : \mathcal{A}(k) \to \mathbb{C}$ be given by

(11.7.2) $$L(f) = F(f,k), \quad f \in \mathcal{A}(k).$$

Then, L is a linear functional on $\mathcal{A}(k)$. Further, if $e_0 \in A$ is given by

(11.7.3)
$$e_0(r) = \begin{cases} 1, & r = 1, \\ 0, & r > 1. \end{cases}$$

$e_0 \in \mathcal{A}(k)$ and $L(e_0) = 1$. That is $F(e_0, k) = 1$.

We introduce a multiplication of elements of $\mathcal{A}(k)$ in terms of the linear functional L (11.7.2).

Definition 11.7.1 : *For $f, g \in \mathcal{A}(k)$, the product of f and g denoted by $f \times g$ is defined as*

(11.7.4)
$$f \times g = F(g, k)f + F(f, k)g - F(f, k)F(g, k)e_0$$

We verify that

(i) $f \times g = g \times f$ for all $f, g \in \mathcal{A}(k)$
(ii) $f \times e_0 = e_0 \times f$ for all $f \in \mathcal{A}(k)$
(iii) $L(f \times g) = L(f)L(g)$ for all $f, g \in \mathcal{A}(k)$.

Further, if $f, g, h \in \mathcal{A}(k)$ we have

(iv) $f \times (g \times h) = (f \times g) \times h$ (associativity)
(v) $f \times (g + h) = f \times g + f \times h$ (distributivity)

We verify that $f \times g \in \mathcal{A}(k)$, whenever f, g are $\mathcal{A}(k)$. Further,

(11.7.5)
$$L(f \times g) = F(g, k)F(f, k) = L(f)L(g).$$

So, $(\mathcal{A}(k), +, \times)$ is a commutative algebra. We refer to $(\mathcal{A}(k), +, \times)$ as a functional-theoretic algebra, briefly written as F-T.A.

As an example, consider the functions ϕ (Euler totient) and $e \equiv 1$. ϕ and e are elements of $\mathcal{A}(3)$.

We recall that

$$\sum_{n=1}^{\infty} \frac{\phi(n)}{n^s} = \frac{\zeta(s-1)}{\zeta(s)}, \quad \mathrm{Re}\, s > 2$$

and

$$\sum_{n=1}^{\infty} \frac{e(n)}{n^s} = \zeta(s), \quad \mathrm{Re}\, s > 1.$$

From (11.7.4), we see that when $\mathrm{Re}\, s > 2$

(11.7.6)
$$\phi \times e = \zeta(s)\phi + \frac{\zeta(s-1)}{\zeta(s)}e - \zeta(s-1)e_0$$

and

(11.7.7)
$$L(\phi \times e) = L(\phi)L(e) = \zeta(s-1), \mathrm{Re}\, s > 2.$$

(11.7.7) follows from (11.7.6).

Now, (11.7.4) could be considered in a more general set-up.

Definition 11.7.2 : *Let $\mathcal{A}_F$ be a vector space over a field F. Suppose that L is a linear functional on $\mathcal{A}_F$ such that $L(e_0) = 1$ for some $e_0 \in \mathcal{A}_F$. For $x, y \in \mathcal{A}_F$, the product $x \cdot y$ is defined by*

(11.7.8) $x \cdot y = L(y)x + L(x)y - L(x)L(y)e_0$

Then, $(\mathcal{A}_F, +, \cdot)$ is called a functional-theoretic algebra (F-T.A).

A further generalization of (11.7.8) is as follows:

Definition 11.7.3 : *[12] Let $\mathcal{A}_F$ be a vector space over a field F. Given two linear functionals L_1 and L_2 on $\mathcal{A}_F$ with the property $L_1(e_0) = L_2(e_0) = 1_F$ for some $e_0 \in \mathcal{A}_F$, we define multiplication of two elements $x, y \in \mathcal{A}_F$ by*

(11.7.9) $x \cdot y = L_1(x)y + L_2(y)x - L_1(x)L_2(y)e_0.$

It is easy to check that multiplication (11.7.9) is associative and e_0 serves as the unity element in $(\mathcal{A}_F, +, \cdot)$. $(\mathcal{A}_F, +, \cdot)$ is called a functional-theoretic algebra. It is denoted by $\mathcal{A}_F(L_1, L_2)$.

There are interesting properties of $(\mathcal{A}_F, +, \cdot)$ when $\mathcal{A}_F$ is a finite dimensional vector space over F. These have been considered by Sebastian Vattamattam and Sivaramakrishnan in [12] and [13]. It is shown in [13] that certain functional-theoretic algebras are indeed convolution algebras in the sense of Hewitt and Zuckerman [9]. It is also known that there exist convolution algebras which are not functional-theoretic.

Postscript : To provide an essential background to the study of algebras, it is desirable to study 2-dimensional real algebras. See Steven C. Althoen and Lawrence D. Kugler [A1].

11.8. Notes with illustrative examples

The set $\widetilde{\mathbb{Z}}$ of non-negative integers is the simplest example of a semigroup under addition. Let R be any commutative ring with unity 1_R. The set $\mathcal{A}(R)$ of all functions $f : \widetilde{\mathbb{Z}} \to R$ is an R-module. We can define a multiplication (Cauchy) in $\mathcal{A}(R)$ by writing

(11.8.1) $(f \cdot g)(n) = \sum_{i=0}^{n} f(i)g(n-i), \quad f, g \in \mathcal{A}(R)$

$f \cdot g$ belongs to $\mathcal{A}(R)$. Multiplication is associative and commutative. It distributes addition. $\mathcal{A}(R)$ is an algebra over R. Suppose that we define $x : \widetilde{\mathbb{Z}} \to R$ by

(11.8.2) $x(n) = \begin{cases} 1, & n = 1, \\ 0, & n \neq 1, \end{cases}$

Then,

$$x \cdot x(n) = x^2(n) = \begin{cases} 1, & \text{if } n = 2, \\ 0, & n \neq 2. \end{cases}$$

In general, for $k \geq 1$

$$(11.8.3) \qquad x \cdot x \cdot x(n) \ (k \text{ factors}) = \begin{cases} 1, & \text{if } n = k. \\ 0, & \text{if } n \neq k. \end{cases}$$

We express the left side of (11.8.3) by x^k. Any $f \in \mathcal{A}(R)$ can be symbolically shown as a power series:

$$(11.8.4) \qquad f = f(0) + f(1)x + f(2)x^2 \cdots + f(k)x^k + \cdots$$

On account of (11.8.4), $\mathcal{A}(R)$ is called the algebra of formal power series with coefficients from R. It is written as $R[[x]]$ (See Section 4.5, chapter 4). The subset $\mathcal{B}(R)$ of $\mathcal{A}(R)$ which consists of all functions $f : \widetilde{\mathbb{Z}} \to R$ such that $f(n) = 0$ except for at most a finite number of integers $n \in \widetilde{\mathbb{Z}}$ forms a subalgebra of $\mathcal{A}(R)$. $\mathcal{B}(R)$ is the ring $R[x]$ of polynomials with coefficients from R. $R[x]$ is commutative and has the unity element 1_R, as an element of $R[x]$. $R[x]$ is called the polynomial algebra over R. As we have $\mathbb{N}$ (set of positive integers), as a semigroup under multiplication, functions defined on an arbitrary semigroup G are considered. The role of semicharacters is in obtaining a basis for the vector space of complex valued functions defined on G. The familiar arithmetical convolutions are convolutions in the general set-up. So, there is sufficient justification in introducing finite dimensional convolution algebras. A specific example is that of abstract arithmetical functions defined on a commutative semigroup of idempotents [15].

11.9. Worked-out examples

a) $\mathcal{A}$ denotes the ring of arithmetic functions with addition and Dirichlet convolution as the ring operations. (By Corollary 4.5.1 of chapter 4, $\mathcal{A}$ is a UFD). Suppose that M denotes the set of functions $f : \mathbb{R}^+ \to \mathbb{C}$ where $\mathbb{R}^+$ denotes the set of positive real numbers $f(x)$ is assumed to be zero for $0 < x < 1$. $(M, +)$ is an abelian group. For $\alpha \in \mathcal{A}$, $f \in M$, we define

$$(11.9.1) \qquad (\alpha \circ f)(x) = \sum_{n \leq x} \alpha(n) f(\frac{x}{n}).$$

Show that M is a left $\mathcal{A}$-module using (11.9.1) for scalar multiplication.

Answer: From (11.9.1), it is clear that for $\alpha \in \mathcal{A}$, $f \in M$, $\alpha \circ f \in M$. $\alpha \circ f$ also vanishes for x in $(0, 1)$. The operator $\circ$ is neither commutative nor associative. However, we have

Lemma 11.9.1 : *(Apostol) For α, $\beta \in A$ and $f \in M$,*

$$\alpha \circ (\beta \circ f) = (\alpha \cdot \beta) \circ f$$

where $(\alpha \cdot \beta)$ is the Dirichlet product of α and β.

Proof : Let $x > 0$, Then,

$$(\alpha \circ (\beta \circ f)) = \sum_{n \leq x} \alpha(n) \left(\sum_{m \leq \frac{x}{n}} \beta(m) f(\frac{x}{mn}) \right)$$

$$= \sum_{mn \leq x} \alpha(n)\beta(m) f(\frac{x}{mn})$$

$$= \sum_{r \leq x} \left(\sum_{n|r} \alpha(n)\beta(\frac{r}{n}) \right) f(\frac{x}{r}), \text{ where } r \text{ is chosen as } mn.$$

$$= \sum_{r \leq x} (\alpha.\beta)(r) f(\frac{x}{r})$$

$$= ((\alpha \cdot \beta) \circ f)(x).$$

$\square$

We deduce from lemma (11.9.1) that as $(M, +)$ is an abelian group and the function $\psi : \mathcal{A} \times M \to M$ defined by $\psi(\alpha, f) = \alpha \circ g$ satisfies

$$\alpha \circ (f + g) = \alpha \circ f + \alpha \circ g,$$
(11.9.2) $\qquad (\alpha + \beta) \circ f = \alpha \circ f + \beta \circ f,$
$$(\alpha \cdot \beta) \circ f = \alpha \circ (\beta \circ f), \text{ by lemma } 11.9.1$$
$$\text{and } e_0 \circ f = f,$$

where e_0 is the multiplicative identity in $\mathcal{A}$. Thus, M is a unital left $\mathcal{A}$- module.

$\square$

Remark 11.9.1 : Apostol [1] calls multiplication (11.9.2) a generalized Dirichlet convolution. Therefore, it is appropriate to call M a convolution module. Generalized Dirichlet convolution given above has been considered as 'Niven product' mentioned in Section 9.5, chapter 9. Generalized Möbius inversion was illustrated in theorem 73, in that context.

b) Describe the structure of a 2-dimensional algebra over a field F.

Answer: Let $\{e, a\}$ be a basis for an algebra A (over a field F). We assume that $a \notin Fe$, so that A is 2-dimensional. Multiplication is uniquely determined by $a \cdot a = a^2$.

Suppose that $a^2 = me + na$; $m, n \in F$. We consider $h(x) = x^2 - nx - m \in F[x]$. As $a^2 - na - me = 0_A$, a is a 'zero' of $h(x)$. The following cases arise.

Case (i): $h(x) = 0$ has two distinct roots α, β with $\alpha \neq \beta$. α, β are determined from the equation $\alpha + \beta = n$, $\alpha\beta = m$. $(x - \alpha)(x - \beta) = 0$ yields

$$(a - \alpha e)(a - \beta e) = 0.$$

As $\alpha \neq \beta$, we write $b = \frac{a - \alpha e}{\beta - \alpha}$. $\beta \notin Fe$. So, $\{e, b\}$ forms a basis for A. It is verified that $b^2 = b$.

Case (ii): $h(x) = 0$ has a unique repeated root in F. That is, $h(x) = (x - \alpha)^2$, where $\alpha \in F$.

We write $b = a - \alpha e$. $\{e, b\}$ is a basis for A and $b^2 = (a - \alpha e)^2 = h(a) = 0$.

Case (iii): $h(x) = 0$ has no roots in F. Then, $h(x)$ is an irreducible polynomial over F.

Claim : A is a field.

Here, we write $b = se + ta$, $s, t \in F$, $b \in A$, $b \neq 0_A$.

$$h(x) = (tx + s)q(x) + r(x),$$

where either $r(x) = 0$ or $\deg r(x) = 0$. That is $r(x) = r\,(say)$ and $r \in F$. Writing $q(x) = t'x + s'$, we note that

$$h(a) = (ta + se)(t'a + s'e) + re.$$

Or,

$$h(a) - re = (ta + se)(t'a + s'e)$$

By the choice of a, $a^2 - na - me = 0_A$.

So,

(11.9.3) $$e = (ta + se)(\frac{-(t'a + s'e)}{r}) = bb' \text{ (say)}.$$

It follows that b' serves as the inverse of $b(\neq 0)$. Hence, A is a field. □

Remark 11.9.2 : A 2-dimensional algebra A over a field F is either a field or possesses a basis $\{e, b\}$ wherein either $b^2 = d$ or $b^2 = 0_A$. If F is algebraically closed (meaning that any polynomial $f(x) \in F[x]$ has all its zeros in F), case (iii) in the above solution does not arise.

EXERCISES

1. *Mark the following statements true (T) or false (F) justifying your answer briefly.*

   a) *Let A be an algebra over the field $\mathbb{C}$ of complex numbers. Suppose that $\{e, i, j, k\}$ be a basis for A (e, the identity) $i^2 = j^2 = k^2 = -e$, $ij = -ji = k$, $jk = -kj = i$, $ki = -ik = j$. If $M_2(\mathbb{C})$ stands for the algebra of 2×2 matrices with entries from $\mathbb{C}$, then $A \cong M_2(\mathbb{C})$.*

   b) *Consider the algebra A over $\mathbb{Q}$ (the field of rational numbers) given by a basis $\{1, i, j, k\}$ and having the multiplication table*

	i	j	k
i	-1	k	$-j$
j	$-k$	-2	$2i$
k	j	$-2i$	-2

It happens that the four-dimensional algebra A given above has divisors of zero.

c) *(Alexander Abian) Let A be a 2-dimensional algebra over $\mathbb{R}$, the field of real numbers and having a basis: $\{e,b\}$ where $b^2 = 0_A$. It is correct to say that any product $\alpha \cdot \beta = 0_A$ for $\alpha, \beta \in A$.*

d) *(Alexander Abian) Let A be a 2-dimensional algebra over $\mathbb{Q}$, the field of rational numbers. If $\{e,b\}$ is a basis for A such that $e^2 = b$, $e \cdot b = b \cdot e = b^2 = 0_A$, then, $\alpha \cdot \beta \cdot \gamma = 0_A$ for all $\alpha, \beta, \gamma \in A$.*

e) *Let S be an m-dimensional division subalgebra of an n-dimensional division $\mathcal{A}$ with unity e_A. If the unity element of S is also e_A, then, m divides n.*

f) *Let A be a finite dimensional algebra over a field F. A subspace of A which is at the same time a right and left ideal of A is called an ideal of A. If $M_n(F)$ denotes the algebra of $n \times n$ matrices with entries from F, $M_n(F)$ has no ideals other than the zero ideal and $M_n(F)$.*

2. *Let $(G, \cdot)$ be an arbitrary commutative semigroup. Suppose that x_i $(i = 1, 2, \ldots, n)$ belongs to G. If $\{\pi(1), \pi(2) \ldots \pi(n)\}$ is a permutation of the set $\{1, 2, \ldots, n\}$, show that*

$$x_1 \cdot x_2, \ldots x_n = x_{\pi(1)} \cdot x_{\pi(2)} \cdots x_{\pi(n)}$$

3. *The semigroup $(\mathbb{N}, \cdot)$ has the unity element 1 and so is a monoid. Let $P = \{2, 3, 5, \ldots\}$. P denotes the set of primes in $\mathbb{N}$. Show that P generates $\mathbb{N}$ and that P is contained in every set of generators of $\mathbb{N}$.*

4. *Let $GL_n(\mathbb{C})$ denote the general linear group of order n over $\mathbb{C}$. That is, $GL_n(\mathbb{C})$ stands for the group of invertible $n \times n$ matrices with entries from $\mathbb{C}$. An n dimensional matrix representation of a semigroup G is a homomorphism $\rho : G \to GL_n(\mathbb{C})$. If P_g is the image of $g \in G$ under ρ, $\rho_{gh} = \rho_g \rho_h$. ρ_g and ρ_h are invertible matrices belonging to $GL_n(\mathbb{C})$. ρ is said to be faithful if ρ is an isomorphism. That is, ρ maps G isomorphically onto its image. Show that every semigroup of order n, commutative, or not, has a faithful representation by matrices of order not exceeding $(n+1)$.*

5. *Describe all semigroups of order 2 up to isomorphism. (There are four of them)*

6. *Let G be a finite semigroup having 3 elements. There are 9 algebras $\mathcal{L}_1(G)$ of dimension 3. Obtain one of them.*

7. *[Fröbenius] Let $\mathcal{A}$ be a finite dimensional division algebra over $\mathbb{R}$, the field of reals. If $\mathcal{A}$ is commutative, show that either $\mathcal{A} = \mathbb{R}$, the field of reals or $\mathcal{A} = \mathbb{C}$, the field of complex numbers. If $\mathcal{A}$ is not commutative, show that $\mathcal{A}$ is the four-dimensional algebra H of real quaternions.*

8. *Let H be a finite commutative semigroup of idempotents. An element e in H is called a unit for multiplication, if $e \cdot x = x \cdot e = x$ for all $x \in H$. If H has a unit e, show that H has just one prime element e.*

9. Let $\mathcal{A}_F$ be a vector space over a field F. Suppose that $L : \mathcal{A}_F \to F$ is a linear functional such that $L(e) = 1_F$ for some $e \in \mathcal{A}_F$. For $x, y \in \mathcal{A}_F$, we define

$$x \cdot y = L(x)y + L(y)x - L(x)L(y)e$$

Then, $(\mathcal{A}_F, +, \cdot)$ is an algebra over F. (in fact, a functional-theoretic algebra). Show that $(\mathcal{A}_F, +, \cdot)$ is a quasi-local algebra. If $(\mathcal{A}_F, +, \cdot)$ is a commutative algebra, show that it has an idempotent other than e.

10. Give an example of a 3-dimensional convolution algebra which is not commutative.

11. For arithmetic functions f, g; one defines unitary convolution of f and g by

$$(f \oplus g)(r) = \sum_{\substack{d \mid r \\ g.c.d\,(d, \frac{r}{d})=1}} f(d)g(\frac{r}{d})$$

where the summation is over those divisors d (of r) for which g.c.d $(d, \frac{r}{d}) = 1$. Examine whether unitary convolution can be obtained via a vector space of complex-valued functions defined on an appropriate semigroup.

REFERENCES

[1] Tom M. Apstol: Introduction to Analytic Number Theory, Chapter 11 pp 224–248 UTM, Springer Verlag (1976).

[2] L. Carlitz: Arithmetical functions in an unusual setting, Amer. Math. Monthly, 73 (1966), 582–590.

[3] L. Carlitz: Arithmetical functions in an unusual setting II, Duke Math. J., 34 (1967), 757–759.

[4] Eckford Cohen: Rings of arithmetic functions, Duke Math. J., 19 (1952), 115–129.

[5] Eckford Cohen: Rings of arithmetic functions II. The number of solutions of quadratic congruences, Duke Math. J., 21 (1954), 9–28.

[6] Eckford Cohen: Arithmetical functions of finite abelian groups, Math. Annalen 142 (1961), 165–182.

[7] P. M. Cohn: An Introduction to Ring Theory, Chap 2 pp 53–58, SUMS, Springer Verlag, Berlin-Heidelberg, 2nd printing 2001.

[8] Pentti Haukkanen and R. Sivaramakrishnan: On semigroups of idempotents and certain arithmetical convolutions (unpublished manuscript) (1998).

[9] E. Hewitt and H. S. Zuckerman: Finite dimensional convolution algebras, Acta Mathematica, 93 (1955), 67–119.

[10] T. W. Hungerford: Algebra GTM No:73 Springer Verlag. (1986) Chapter VII, pp. 327–370.

[11] J. Knopfmacher: Abstract analytic Number Theory, North Holland Pub. Co., Amsterdam (1975).

[12] Sebastian Vattamattam and R. Sivaramakrishnan: Associative algebras via linear functionals, Proc. Annual Conference of Kerala Math. Association and International seminar on Mathematical Tradition of Kerala: Jan 17–19, 2000. pp. 81–89.

[13] Sebastian Vattamattam and R. Sivaramakrishnan: A note on convolution algebras: Paper read at the International Conf. on 'Recent Trends in Analysis' held at St. Joseph's college, Irinjalakuda (Kerala), December 16–18, 2000.

[14] M.V. Subbarao: On some arithmetical convolutions, Lecture Notes in Math. # 251, Springer Verlag (1972), 247–271.

[15] M. Tainiter: Generating functions on idempotent semigroups with applications to Combinatorial Analysis, J. Comb. Theory, 5 (1968), 273–288.

ADDITIONAL REFERENCE

[A1] Steven C. Althoen and Lawrence D. Kugler: When is $\mathbb{R}^2$ a division algebra?, Amer. Math. Monthly, 90 (1983), 625–635.

Part III

A GLIMPSE OF ALGEBRAIC NUMBER THEORY

CHAPTER 12

Noetherian and Dedekind domains

Historical perspective

The birth of algebraic number theory took place while attempts were made to prove Fermat's Last Theorem during the 18th century and early 19th century. The creation of ideals had a bearing on the growth of ring theory. Ideals were used long before homomorphisms reached the scene. There was a shift of emphasis from ideals to homomorphisms. This was brought about by a German mathematician Emmy Noether (1882–1935). Her view of ring theory has had a tremendous influence on the growth of ring theory. Van der Waerden's (1903–1996) Modern Algebra published in 1931 gave the modern view of algebra as he drew the theorems and their applications from the courses presented by E. Artin and Emmy Noether, which he had attended at Gottingen. The class of rings which Emmy Noether had investigated is known as the class of Noetherian rings in honour of the great woman algebraist Emmy Noether. Among Noetherian rings are the ring of integers and rings of polynomials.

The study of rings of algebraic integers strengthened the development of algebraic number theory. The algebraic analogue of the fundamental theorem of arithmetic is found in unique factorization domains. But, the uniqueness of factorization of an element of a ring into a product of irreducibles is not true in certain rings such as $\mathbb{Z} [\sqrt{-5}]$. However, if one introduces the notion of 'ideal' numbers, uniqueness of factorization is retrieved. That is, ideals of a certain ring called a Dedekind Domain, are expressible uniquely as a product of prime ideals. This was a major contribution during early 20th century.

12.1. Introduction

This chapter is about the study of Noetherian rings, Artinian rings and Dedekind domains. These are rings with certain 'finiteness' conditions: A Noetherian ring R possesses three equivalent properties:

(i) R satisfies the ascending chain condition for ideals.
(ii) The maximum condition for ideals holds in R.
(iii) Every ideal of R is finitely generated.

This is discussed from the point of view of the ring of integers of quadratic number fields. Fermat's two-square theorem is revisited. The Jacobson radical plays an important role in the study of primary ideals of a ring. The Lasker-Noether decomposition theorem for Noetherian rings is proved. (See theorem 93). Artinian rings are studied. The integral closure of a ring is introduced. It is shown that any PID is integrally closed in its field of quotients. Properties of Dedekind domains are given. A substitute for the fundamental theorem of arithmetic is shown for Dedekind domains. It is about the unique factorization of a nonzero ideal of a Dedekind domain into a product of prime ideals. (See theorem 101). The ring of algebraic integers (a number ring corresponding to an algebraic number field) is shown to be a Dedekind domain. Introducing congruences modulo an ideal of a Dedekind domain, an analogue of the Chinese Remainder theorem is proved in Section 12.7. Section 12.8 deals with integral domains having finite norm property.

12.2. Noetherian rings

The ring $\mathbb{Z}$ of integers being a PID has the property that every ideal of $\mathbb{Z}$ is generated by a single element. If $d_1 = p$ a prime, $d_2, d_3, \ldots, d_t = r$ are the divisors of r, we can find a chain of divisors:

$$d_1 | d_{s_1}, d_{s_1} | d_{s_2}, \ldots, d_{s_{k-1}} | d_{s_k} = r$$

This gives an ascending chain of principal ideals.

(12.2.1) $(r) \subset (d_{s_{k-1}}) \subset \cdots \subset (d_{s_2}) \subset (d_{s_1}) \subset (p)$

As (p) is a maximal ideal, the ascending chain of principal ideals (12.2.1) terminates. In the case of a ring which is not commutative, one can look at ascending chains of right ideals and left ideals separately. If each such ascending chain of left and right ideals terminates, the ring is called a Noetherian ring. This phenomenon of 'ascending chain condition' (briefly written as a.c.c) was noticed by Emmy Noether in 1917. In what follows, we consider R to be a commutative ring with unity.

Definition 12.2.1 : *A ring R is said to satisfy a.c.c for ideals if, given any sequence* $\{I_n\}$ *of ideals of R with*

$$I_1 \subseteq I_2 \subseteq \cdots \subseteq I_n \subseteq \cdots,$$

there exists an integer m depending on $\{I_n\}$ *such that* $I_n = I_m$ *for all* $n \geq m$.

A field F satisfies a.c.c for ideals trivially, as its only ideals are (0) and F. In the case of $\mathbb{Z}$, if $m, n \in \mathbb{Z}$ are such that m divides n, then, $(n) \subseteq (m)$ and $(n) \subseteq (m) \Rightarrow m$ divides n. As a nonzero integer $n \in \mathbb{Z}$ has only a finite number of distinct (non-associated) divisors, $\mathbb{Z}$ satisfies a.c.c for ideals which are principal ideals, though. We referred to this as ACCP when we considered GCD domains in Section 2.5 (see definition 2.5.1), chapter 2.

Not all rings satisfy a.c.c for ideals. Take for instance, the set R of all finite subsets of $\mathbb{Z}^+$ with symmetric difference $\triangle$ as addition and intersection as multiplication. $(R, \triangle, \cap)$ is a commutative ring without unity. If $I_n = \{1, 2, 3, \ldots n\}$, the power set $\mathscr{P}(I_n)$ of I_n is an ideal of $(R, \triangle, \cap)$. Further,

$$\mathscr{P}(I_1) \subset \mathscr{P}(I_2) \subset \cdots \subset \mathscr{P}(I_n) \subset \cdots$$

is a strictly ascending chain of ideals of R which does not terminate, as n goes on increasing.

Definition 12.2.2 : *A ring R is said to possess the property of maximum condition for ideals, if every non-empty set $\mathscr{M}$ of ideals of R partially ordered under set-inclusion has at least one maximal member, that is, an ideal which is not properly contained in any other ideal belonging to $\mathscr{M}$.*

Definition 12.2.3 : *An ideal I of a ring R is said to be finitely generated if there exists a finite set S of elements of I such that S generates I. In other words, $(S) = I$.*

Lemma 12.2.1 : *The following statements about the ideals of a ring R are equivalent:*

(i) R satisfies a.c.c for ideals.
(ii) The maximum condition for ideals holds in R.
(iii) Every ideal of R is finitely generated.

Proof : (i)$\Rightarrow$(ii)

Let $\mathscr{M}$ be a non-empty collection of ideals of R. Suppose that $\mathscr{M}$ has no maximal element. We will arrive at a contradiction to the data in (i).

Since $\mathscr{M}$ is non-empty, we can pick an ideal $I_1 \in \mathscr{M}$. By our assumption, I_1 is not a maximal element of $\mathscr{M}$. So I_1 is contained in a proper ideal I_2 belonging to $\mathscr{M}$. Likewise, as I_2 is not a maximal element in $\mathscr{M}$, there exists an ideal I_3 belonging to $\mathscr{M}$ such that $I_2 \subseteq I_3$. Proceeding thus, we obtain an infinite ascending chain of ideals of R, namely,

$$I_1 \subset I_2 \subset I_3 \subset \cdots$$

all of whose inclusions are proper. This violates a.c.c for ideals and hence $\mathscr{M}$ has a maximal element. Or, (i) $\Rightarrow$ (ii).

(ii) :$\Rightarrow$ (iii)

Suppose that the maximum condition for ideals holds in R. Let I be an ideal of R. If $I = (0_R)$, it is finitely generated (generated by 0_R). If $I \neq (0_R)$, we pick a nonzero element $a_1 \in I$. If the principal ideal $(a_1) = I$, we are through. Otherwise, there exists an element $a_2 \in I$ which does not belong to (a_1). Then,

$$(a_1) \subset (a_1, a_2) \subseteq I.$$

If $(a_1, a_2) \neq I$, there exists $a_3 \in I$, $a_3 \notin (a_1, a_2)$ such that

$$(a_1, a_2) \subset (a_1, a_2, a_3) \subseteq I$$

Proceeding thus, we obtain an ascending chain of ideals of R in the form

(12.2.2) $\qquad (a_1) \subset (a_1, a_2) \subset (a_1, a_2, a_3) \subset \cdots \subset (a_1, a_2, \ldots, a_n) \subset \cdots$

The maximum condition for ideals ensures that the collection $\mathscr{M}$ of ideals (a_1), $(a_1, a_2), \ldots, (a_1, a_2, a_3, \ldots, a_n), \ldots$ has a maximal element say $(a_1, a_2, \ldots, a_m)$ for some $m \in \mathbb{N}$. If $I \neq (a_1, a_2, \ldots, a_m)$, we could pick an element $a \in I$ and $a \notin (a_1, \ldots, a_m)$. Then, the ideal $(a_1, a_2, a_3, \ldots, a_m, a)$ contains $(a_1, a_2, \ldots, a_m)$ properly and so spoils the maximality of $(a_1, a_2, \ldots, a_m)$ in the collection $\mathscr{M}$ of ideals contained in I. This contradiction shows that the ascending chain of ideals in (12.2.2) terminates and so $I = (a_1, a_2, \ldots, a_m)$. Thus, I is finitely generated. This proves (ii) $\Rightarrow$ (iii).

(iii) $\Rightarrow$ (i)

We choose an ascending chain of ideals (of R) of the form

$$(12.2.3) \qquad I_1 \subseteq I_2 \subseteq I_3 \subseteq \cdots \subseteq I_n \subseteq \cdots$$

Let $I = \cup_n I_n$. Then, I is an ideal of R. By (iii), I is finitely generated.

Suppose that $I = (a_1, a_2, \ldots, a_t)$.

Each element a_k $(k = 1, 2, \ldots, t)$ of the set of generators is an element of some ideal I_{i_k} of the chain in (12.2.3). Choosing m to be the largest of the suffixes i_k we note that all the elements a_k $(k = 1, \ldots t)$ are contained in I_m. However, for $n \geq m$,

$$I = (a_1, a_2, \ldots, a_t) \subseteq I_m \subseteq I_n \subseteq I$$

Therefore, $I_n = I_m$ for $n \geq m$ and so the chain of ideals in (12.2.3) terminates. This is true for any arbitrary chain of ideals. That is, R satisfies a.c.c for ideals. Thus, (iii) $\Rightarrow$ (i).   □

Definition 12.2.4 : *A ring R which satisfies any one of the equivalent conditions of lemma 12.2.1 is called a Noetherian ring (in honour of Emmy Noether who initiated the study of such rings). An integral domain which is Noetherian is said to be Noetherian domain.*

We remark that it is advantageous to utilize the fact that in a Noetherian ring, every ideal is finitely generated. $\mathbb{Z}$ is, indeed, the simplest non-trivial example of a Noetherian ring. If $n\mathbb{Z}$ is an ideal of $\mathbb{Z}$, $\mathbb{Z}/n\mathbb{Z}$ is also Noetherian, as $\mathbb{Z}/n\mathbb{Z}$ is finite.

Definition 12.2.5 : *A ring R' is called a homomorphic image of a ring R, if there exists a homomorphism $\psi : R \to R'$ which is onto.*

Fact 12.2.1 :

(i) Any homomorphic image of a Noetherian ring is Noetherian.
(ii) If I denotes an ideal of a Noetherian ring R, then R/I is Noetherian.

For proof, see D. M. Burton [3].

12.3. More about ideals

We confine ourselves to a commutative ring R with unity 1_R.

Let R, R' be commutative rings. If $\psi : R \to R'$ is a ring homomorphism, when I is an ideal of R, it is not necessary that $\psi(I)$ is an ideal of R'. For instance, if (a) denotes a nonzero ideal of $\mathbb{Z}$, when $\psi : \mathbb{Z} \to \mathbb{Q}$ (the field of rationals) is an embedding of $\mathbb{Z}$ in $\mathbb{Q}$, $\psi : (a) = \frac{a}{1} \in \mathbb{Q}$. But, if $p/q \in \mathbb{Q}$, $\frac{p}{q}\psi(ka) = \frac{p}{q}\frac{ka}{1} \notin \psi((a))$. However, when J is an ideal of R', $\psi^{-1}(J)$ is an ideal of R.

Definition 12.3.1 : *Let $\psi : R \to R'$ be a ring homomorphism. If J is an ideal of R', $\psi^{-1}(J) = J^c$ is called the contraction of J via ψ.*

We observe that if J is a prime ideal of R', J^c is also a prime ideal of R.

Definition 12.3.2 : *Let $\psi : R \to R'$ be a ring homomorphism. Suppose that I denotes an ideal of R. The extension I^e of I via ψ is defined as the ideal $\psi(I)R'$ which is generated by $\psi(I)$ in R'.*

To be precise, I^e is given by the set

(12.3.1) $$\{\sum_{\text{finite}} b_i\psi(a_i) : a_i \in I, b_i \in R'\}.$$

When I is a prime ideal of R, I^e need not be a prime ideal of R'. For, if $\psi : \mathbb{Z} \to \mathbb{Q}$ is the embedding of $\mathbb{Z}$ in $\mathbb{Q}$ and $I = (p)$, where p is a prime in $\mathbb{Z}$, I^e is a non-trivial ideal of $\mathbb{Q}$ implies that $I^e = \mathbb{Q}$ (as $\mathbb{Q}$ is a field). So, I^e is not a prime ideal of $\mathbb{Q}$.

We look at the diagram of homomorphisms

(12.3.2) $$R \xrightarrow{\psi} \psi(R) \xrightarrow{\psi'} R'.$$

$\psi(R)$ is a subring of R'. ψ is onto $\psi(R)$. $\psi' : \psi(R) \to R'$ is one-one. It is known [3] that there is a one-one correspondence between the ideals of $\psi(R)$ and the ideals of R which contain ker ψ. Also, the prime ideals of $\psi(R)$ correspond to the prime ideals of R which contain ker ψ. Nothing specific can be said about ψ'. We wish to examine the situation in $\mathbb{Z}[i]$, the ring of Gaussian integers.

Let $\mathbb{Q}(\sqrt{m})$ be the quadratic extension of $\mathbb{Q}$ by $\sqrt{m}$ where m is a square-free integer. $R(m)$ is the ring of algebraic integers of $\mathbb{Q}(\sqrt{m})$. The structure of $R(m)$ is given in theorem 14, chapter 3.
If $\alpha \in \mathbb{Q}(\sqrt{m})$, the norm $N(\alpha)$ (see definition 3.3.2, chapter 3) is given by

$$N(\alpha) = a^2 - b^2m, \text{ if } \alpha = a + b\sqrt{m}$$

Let $S(m)$ denote the set of norms of nonzero algebraic integers of $\mathbb{Q}(\sqrt{m})$. Since norms of algebraic integers are rational integers, $S(m) \subseteq \mathbb{Z}$. As $1 \in R(m)$, $1 = N(1) \in S(m)$. Now, $a + b\sqrt{m} \in R(m)$ is such that $a, b \in \mathbb{Z}$, when $m \not\equiv 1 \pmod 4$.

Now, if $m \not\equiv 1 \pmod 4$, $n \in \mathbb{Z} \Rightarrow n \in S(m)$ if, and only if, the Diophantine equation $x^2 - my^2 = n$ has a solution. Suppose that $\alpha \in R(m)$. Then, $N(\alpha) = n$ if, and only if, $\alpha = a + b\sqrt{m}$ has the property that $a, b \in \mathbb{Z}$. This happens only when $m \not\equiv 1 \pmod 4$.

Lemma 12.3.1 : *Suppose that $\alpha \in R(m)$ and $N(\alpha)$ is an irreducible in $S(m)$. Then, α is an irreducible in $R(m)$.*

Proof : Suppose that $\beta \in R(m)$ and $\beta|\alpha$. Then, $N(\beta)|N(\alpha)$. As α is an irreducible in $R(m)$, $N(\beta) = \pm 1$ in which case β is unit or $N(\beta) = \pm N(\alpha)$. In the latter case, α and β are associates. So, α is an irreducible in $R(m)$. $\square$

Corollary 12.3.1 : *If $N(\alpha)$ is a prime in $\mathbb{Z}$, then, α is a prime in $R(m)$.*

For, an element of $S(m)$ which is a prime in $\mathbb{Z}$ is, of course, a prime in $S(m)$ (with stronger reason, as $S(m) \subseteq \mathbb{Z}$). So α is a prime in $R(m)$.

Let $R(m)^* = R(m) \setminus \{0\}$. A non-unit α in $R(m)^*$ is a prime if, and only if, its only divisors are its associates and the units. Since the conjugation map $\bar\psi : R(m) \to R(m)$ given by $\bar\psi(\alpha) = \bar\alpha$ is an automorphism of the ring $R(m)$, α is a prime in $R(m)$ if, and only if, $\bar\alpha$ is a prime in $R(m)$.

The essence of lemma 12.3.1 is that elements α of $R(m)$ with norm $N(\alpha)$, a prime, in $S(m)$, are irreducible in $R(m)$.

Let p be a rational prime in $\mathbb{Z}$. Then, p may or may not be an element of $S(m)$. If neither p nor $-p$ is in $S(m)$, then $p^2 = N(p) \in S(m)$ is a prime in $S(m)$. By lemma 12.3.1, p is an irreducible in $R(m)$. We say that p stays prime in $R(m)$ or p is inertial. If p is not an irreducible in $R(m)$, $\pm p$ is a norm $\in S(m)$, such that for some $\alpha \in R(m)$,

$$N(\alpha) = \alpha\bar\alpha = \pm p \in S(m)$$

Then, α and $\bar\alpha$ are irreducible in $R(m)$, by lemma 12.3.1. There are occasions where α and $\bar\alpha$ are associates. In such cases, we say that p ramifies. Where α and $\bar\alpha$ are not associates, we say that p splits in $R(m)$ (as the word suggests).

Lemma 12.3.2 : *If $R(m)$ is a UFD and α is a prime in $R(m)$, then, $N(\alpha) = \pm p$ or $\pm p^2$ for some rational prime p.*

Proof : If $|N(\alpha)|$ denotes the absolute value of $N(\alpha)$, we note that α divides $|N(\alpha)|$. So, there is a least positive integer t such that $\alpha|t$ in $R(m)$. Let $t = r \cdot s$ in $\mathbb{Z}$. Then, $\alpha|r$ or $\alpha|s$ in $R(m)$. The minimality of t forces $t = r$ or $t = s$. So, t is a prime in $\mathbb{Z}$. But, then,

$$\alpha\bar\alpha = N(\alpha)|N(t) \text{ or } N(\alpha)|t^2.$$

So, $N(\alpha) = \pm t$ or $\pm t^2$ and t is a prime. $\qquad\square$

Next, we consider odd primes p in $\mathbb{Z}$ and the case where $R(m)$ is a UFD.

Theorem 90 (Ethan D. Bolker [1970]) **:** *Suppose that $R(m)$ is a UFD. If p is an odd prime, the nature of p in $R(m)$ is determined in the following manner: Let $(m|p)$ denote the Legendre symbol (see $(6.0.1)$ in chapter 6).*

(i) p ramifies in $R(m)$, if $(m|p) = 0$.
(ii) p splits in $R(m)$, if $(m|p) = 1$.
(iii) p is inertial in $R(m)$, if $(m|p) = -1$.

Proof : We begin with (iii). If $(m|p) = -1$, m is not a quadratic residue (mod p). So, the Diophantine equation

$$(12.3.3) \qquad\qquad x^2 - my^2 = \pm p$$

has no solution. For, if $x^2 - my^2 = kp$ has a solution, where $p \nmid k$, $p \nmid m$, g.c.d $(p,x) = 1$ and g.c.d $(p,y) = 1$. Then,

$$x^2 \equiv my^2 (\text{mod } p) \text{ implies } (xy')^2 \equiv m(\text{mod } p),$$

where $yy' \equiv 1 \pmod{p}$. So, then, m is a quadratic residue mod p—a contradiction. Therefore, $\pm p \notin S(m)$. So, p is inertial in $R(m)$ or p stays prime in $R(m)$. That is, p is inertial in $R(m) \Leftrightarrow (m|p) = -1$, as $p^2 = N(\pm p) \in S(m)$ is a prime in $S(m)$ (as $\pm p \notin S(m)$).

Next, we go to (i) and (ii). If $p|m$, then, p divides $p^2 - m$. Now, $x^2 \equiv m(\text{mod } p)$ has a solution, if $(m|p) = 1$. In either case,

$$p|(x^2 - m) = (x + \sqrt{m})(x - \sqrt{m}), \text{ for some } x \in \mathbb{Z}.$$

If p is a prime in $R(m)$ (assumed to be a UFD) p divides either $x + \sqrt{m}$ or $x - \sqrt{m}$. Since $p \neq 2$, neither $\frac{x+\sqrt{m}}{p}$ nor $\frac{x-\sqrt{m}}{p}$ is an algebraic integer. So, p is not a prime in $R(m)$.

Next, take $\pm p \in S(m)$: Therefore, there are integers $a, b \in \mathbb{Z}$ such that $\alpha = a + b\sqrt{m} \in R(m)$ and $\alpha\bar{\alpha} = N(\alpha) = \pm p = a^2 - mb^2$, where $m \equiv 1 \pmod{4}$. Also,

$$r = \frac{a + b\sqrt{m}}{2} \in R(m)$$

and

$$4r\bar{r} = 4N(r) = \pm 4p = a^2 - mb^2, \text{ if } m \equiv 1(\text{mod } 4).$$

So, for each value of m (either $\equiv 1 \pmod{4}$ or $\not\equiv 1 \pmod{4}$)

$$p|a \Rightarrow p|m$$

When $\dfrac{\alpha}{\bar{\alpha}}$ is a unit in $R(m)$, α ramifies in $R(m)$. But,

$$\frac{\alpha}{\bar{\alpha}} = \frac{\alpha^2}{N(\alpha)} = \frac{\alpha^2}{\pm p}$$

So, p ramifies in $R(m)$ if, and only if, $p|\alpha^2$. Now,

$$(a^2 + mb^2) + 2ab\sqrt{m} = \begin{cases} \alpha^2, & \text{if } m \not\equiv 1(\text{mod } 4); \\ 4\alpha^2, & \text{if } m \equiv 1(\text{mod } 4). \end{cases}$$

So $p|a$ implies $p|\alpha^2$. Conversely, if $p|\alpha^2$, $p|2ab$. So, p (being odd) divides a or b. If $p|a$, we are done. If $p|b$, we will have $p|a$, as $p|(a^2 + mb^2)$. So, $p|\alpha^2 \Rightarrow p|a$. So, p ramifies in $R(m) \Leftrightarrow p|m \Leftrightarrow (m|p) = 0$.

p splits in $R(m) \Leftrightarrow x^2 \equiv m \pmod{p}$ has a solution $\Leftrightarrow (\frac{m}{p}) = 1$. These are the statements in (i) and (ii). $\square$

Remark 12.3.1 : Theorem 90 has been adapted from Ethan D. Bolker [2].

Corollary 12.3.2 : *In the case of* $\mathbb{Z}[i]$, *a prime of the form* $4k+1$ *splits in* $\mathbb{Z}[i]$ *as* $(-1|p) = 1$. *A prime of the form* $4k+3$ *is inertial in* $\mathbb{Z}(i)$, *as* $(-1|p) = -1$.

Remark 12.3.2 : We give another proof of Fermat's Two-square theorem: Any prime of the form $4k+1$ can be expressed as a sum of two squares. (See theorem 4 of chapter 1).

Proof : We consider the diagram of homomorphisms

$$\mathbb{Z} \xrightarrow{\psi} \mathbb{Z}[i] \xrightarrow{\psi'} \mathbb{C}$$

For $a \in \mathbb{Z}$, $\psi(a) = (a,0)$ and ψ' is the inclusion map.

By an inclusion map we mean the following:

Let $B \subset A$. If $1_A : A \to A$ is the identity map, $1_A : B \to A$ is called the inclusion map of B into A.

A prime ideal (p) in $\mathbb{Z}$ may or may not stay as a prime ideal in $\mathbb{Z}[i]$. If $p \equiv 3$ (mod 4), p is inertial in $\mathbb{Z}[i]$ and so stays a prime in $\mathbb{Z}[i]$.

If $p \equiv 1$ (mod 4), p splits in $\mathbb{Z}[i]$. So, if (p) is an ideal of $\mathbb{Z}$, $(p)^e$ is a product of two prime ideals in $\mathbb{Z}[i]$. That is, $(p)^e = IJ$ where $I = (a+bi)$ and $J = (a-bi)$. I, J are prime ideals of $\mathbb{Z}[i]$, as $a+bi$ and $a-bi$ are non-associated primes in $\mathbb{Z}[i]$. So, p can be written as $a^2 + b^2$. $\qquad\square$

Illustration 12.3.1 :

(a) The ideal $(2)^e$ in $\mathbb{Z}[i]$ is generated by $(1+i)^2$.

(b) The ideal $(5)^e$ in $\mathbb{Z}[i]$ is the product of the prime ideals $(2+i)$ and $(2-i)$ in $\mathbb{Z}[i]$.

Definition 12.3.3 : *Let I be an ideal of a ring R. The nilradical of I, written $\sqrt{I}$, is defined by*

$$\sqrt{I} = \{r \in R \ : \ r^n \in I, \text{ for some } n \text{ (depending on } r) \in \mathbb{N}\}$$

Definition 12.3.4 : *$x \in R$ is called a nilpotent element (or simply nilpotent), if there exists a positive integer n such that $x^n = 0_R$.*

We observe that the nilradical of I may be considered as the set of those elements $r \in R$ such that the coset $r+I$ in R/I is nilpotent (as an element of R/I).

Definition 12.3.5 : *The nilradical of the zero ideal (of R) is referred to as the nilradical of R.*

We observe that the nilradical of R consists of all the nilpotent elements of R.

Lemma 12.3.3 : *Given an ideal I of R, $\sqrt{I}$ is an ideal for which $I \subseteq \sqrt{I}$ holds.*

Proof : It is obvious from the definition of $\sqrt{I}$ that $I \subseteq \sqrt{I}$, as $x \in I \Rightarrow x^1 = x \in I$. Now, R is a commutative ring with unity. The binomial expansion for $(a-b)^n$ holds for all $n \geq 2$. Also, every term in the expansion of $(a-b)^{n+m}$ contains either a^n or b^m as a factor. So, $(a-b)^{n+m} \in I$. This shows that when $a, b \in \sqrt{I}$, $(a-b)$ is also an element of $\sqrt{I}$, when $a^n \in I$, $b^m \in I$ respectively. If $r \in R$ and $a \in \sqrt{I}$, as $a^n \in I$ for some $n \in \mathbb{N}$, $(ra)^n = r^n a^n \in I$. So, $ra \in \sqrt{I}$. Thus, $\sqrt{I}$ is an ideal of R. $\qquad\square$

Theorem 91 : *Let $n = p_1^{a_1} p_2^{a_2} \cdots p_k^{a_k}$ (p_i primes, $a_i \geq 1$; $i = 1, 2, \ldots, k$). If $I = (n)$, the ideal generated by n in $\mathbb{Z}$, then the nilradical of I is given by*

$$\sqrt{I} = (p_1 p_2 \cdots p_k)$$

(the ideal generated by the product of the prime factors of n).

Proof : Let $q = p_1 p_2 \cdots p_k$. If $a = \max\{a_1, a_2, \ldots, a_k\}$, then $q^a \in (n) = I$. By definition, $q \in \sqrt{I}$. So $(q) \subseteq \sqrt{I}$.

Next, let $m \in \mathbb{N}$. If m^b is divisible by n for some $b \geq 1$, then, $m \in \sqrt{I}$. Then, m is divisible by each of the prime factors $p_1, p_2, \ldots p_k$ of n. So,

$$m \in (p_1) \cap (p_2) \cdots \cap (p_k) = (p_1 p_2 \cdots p_k) = (q)$$

Therefore, $\sqrt{I} \subseteq (q)$. Thus, $\sqrt{I} = (q)$. $\qquad\qquad\square$

Definition 12.3.6 : *An ideal I of a ring R is called a primary ideal, if the conditions $ab \in I$ and $a \notin I$ together imply that $b^n \in I$ for some $n \in \mathbb{N}$.*

Clearly, a prime ideal is a primary ideal.

The motivation for considering primary ideals is from the fact that any integer $n > 1$ contains a prime factor p occurring to a power $m \geq 1$. If $I = (p^m)$, given $ab \in I$, $a \notin I$, b is divisible by an sth-power of p ($s \geq 1$).

Fact 12.3.1 : *An ideal I (of R) is primary, if, and only if, whenever $ab \in I$, $a \notin I$, $b \in \sqrt{I}$.*

Proof follows from definition 12.3.6.

In $\mathbb{Z}$, primary ideals are precisely the principal ideals generated by the prime powers p^m ($m \geq 1$), together with the improper ideals (0) and $\mathbb{Z}$.

Lemma 12.3.4 : *If Q is a primary ideal of a ring R, its nilradical $\sqrt{Q}$ is a prime ideal.*

Proof : If $ab \in \sqrt{Q}$, $a \notin \sqrt{Q}$, then there exists a positive integer n such that $(ab)^n = a^n b^n \in Q$ and $a^n \notin Q$. (For, otherwise, a will be in $\sqrt{Q}$). As Q is a primary ideal, there exists a positive integer m such that $(b^n)^m = b^{nm} \in Q$. That is, $b \in \sqrt{Q}$. So, $\sqrt{Q}$ is a prime ideal. $\qquad\qquad\square$

Note : $\sqrt{Q}$ is called the prime ideal associated with the primary ideal Q.

Remark 12.3.3 :

(1) In the case of $\mathbb{Z}$, we have seen that if $I = (n)$, where $n = p_1^{a_1} p_2^{a_2} .. p_k^{a_k}$ (p_i primes, $a_i \geq 1$; $i = 1, 2, \ldots, k$), $\sqrt{I} = (q)$ where $q = p_1 p_2 \cdots p_k$. So, different principal ideals may have the same nilradical (q). In the same manner, different primary ideals may have the same associated prime ideal as their nilradical.

(2) It can be shown that if Q is a primary ideal (of R), $\sqrt{Q}$ is the smallest prime ideal containing Q.

(3) If Q is a primary ideal of R, R/Q will have divisors of zero.

When Q is a primary ideal, the zero divisors of R/Q have a special property.

Fact 12.3.2 : Let I be an ideal of R. Then I is primary $\Leftrightarrow$ every zero divisor $a+I$ of R/I is nilpotent.

For, if I is a primary ideal of R , we consider $a+I$ of R/I. Then, there exists $b+I, b \notin I$ such that

$$(a+I)(b+I) = ab+I = I$$

This shows that $ab \in I$ and as $b \notin I$, there exists $n \in \mathbb{N}$ such that $a^n \in I$, since I is a primary ideal of R. So, $(a+I)^n = a^n+I = I$. Therefore $a+I$ is nilpotent as an element of R/I.

Next, suppose that for an ideal I of R, every zero divisor of R/I is nilpotent. Let $ab \in I$ with $b \notin I$. Then,

$$ab+I = (a+I)(b+I) = I \text{ with } b+I \neq I$$

If $a+I \neq I$, $a+I$ is a zero divisor of R/I. $a+I$ is given to be nilpotent. So, $(a+I)^n = I$ for some $n \in \mathbb{N}$. This means that $a^n \in I$. Therefore, whenever $ab \in I, b \notin I, a^n \in I$, for some $n \in \mathbb{N}$. That is, I is a primary ideal of R.

12.4. Jacobson radical

The structure of a ring is studied from various points of view. In the case of an integral domain, we check whether it belongs to any of the types (i) a PID (ii) a Euclidean domain (iii) a GCD domain or (iv) a Noetherian domain. $\mathbb{Z}$ is all of these.

The notion of the Jacobson radical of a ring helps to understand the structure of a ring in a different way. We come across the idea of semisimplicity which is yet another property possessed by $\mathbb{Z}$. Henceforth, R denotes a commutative ring with unity 1_R.

Definition 12.4.1 : *The Jacobson radical of a ring R, denoted by J(R), is the set*

$$(12.4.1) \qquad J(R) = \cap \{M : M \text{ is a maximal ideal of } R\}$$

Definition 12.4.2 : *If $J(R) = (0_R)$, R is said to be a semisimple ring.*

If F is a field, its only ideals are (0_F) and F. (0_F) is a maximal ideal of F and so any field is a semisimple ring. A theorem of Krull-Zorn [3] says that in a commutative ring with unity, every proper ideal is contained in a maximal ideal. Therefore, the Jacobson radical exists, as every commutative ring with unity has at least one maximal ideal.

We observe that $\mathbb{Z}$ has an infinite number of maximal ideals of the form (p) where p is a prime. We claim that $\mathbb{Z}$ is semisimple. That is, $J(\mathbb{Z}) = (0)$. For, if $J(\mathbb{Z}) \neq (0)$, $J(\mathbb{Z})$ is a principal ideal generated by an integer n (say). As n has only a finite number of distinct prime divisors, (n) cannot belong to an infinite number of maximal ideals of $\mathbb{Z}$. Therefore,

$$\cap_{i=1}^{\infty} (p_i) = (0),$$

where the intersection is over all the prime ideals (p_i) (which are maximal) of $\mathbb{Z}$. See Theorem 92.

Lemma 12.4.1 : *Let I be an ideal of R. Then, $I \subseteq J(R)$ if, and only if, each element of the coset $1_R + I$ is a unit in R.*

Proof : $\Leftarrow$: Suppose that each member of $1_R + I$ is a unit in R and I is not contained in $J(R)$. There exists a maximal ideal M of R such that I is not contained in M.

Let $a \in I$ and $a \notin M$. The maximality of M implies that the ideal generated by M and a is R. We express this by writing $(M,a) = R$. From this, we have $1_R = m + ra$ where $m \in M$, $r \in R$.

Then, $m = 1_R - ra \in 1_R + I$, as $a \in I$. Therefore, m is a unit in R. This is impossible, since no proper ideal of R contains a unit. So, $I \subseteq J(R)$.

$\Rightarrow$: Conversely, suppose that $I \subseteq J(R)$ and that there is some element $a \in I$ for which $1_R + a$ is not a unit in R. This will lead to a contradiction. As $1_R + a$ is not a unit, $1_R + a$ belongs to some maximal ideal M of R (as every ideal is contained in a maximal ideal). Since $a \in I$, and $I \subseteq J(R)$, $a \in J(R)$. Thus, $a \in M$.

So, $(1_R + a) - a = 1_R \in M$, a contradiction. Therefore, if $I \subseteq J(R)$, every element of $1_R + I$ is a unit in R. $\qquad\square$

Corollary 12.4.1 : *In a commutative ring with unity 1_R, an element a (of R) belongs to $J(R)$ if, and only if, $1_R - ra$ is a unit for each $r \in R$.*

Lemma 12.4.2 : *Let R be a commutative ring with unity 1_R. Then, $R/J(R)$ is a semisimple ring.*

Proof : As $J(R)$ is an ideal of R, it makes sense to talk about the quotient ring $R/J(R)$. We consider a coset $a + J(R)$ in $J(R/J(R))$.
As $a + J(R) \in J(R/J(R))$, by corollary 12.4.1,

$$(1_R + J(R)) - (r + J(R))(a + J(R)) = 1_R - ra + J(R)$$

is a unit in $R/J(R)$ for each choice of $r \in R$ giving distinct cosets in $R/J(R)$. So, there exists $b + J(R) \in R/J(R)$ such that

$$(1_R - ra + J(R))(b + J(R)) = 1_R + J(R),$$

That is, $b - rab + J(R) \in 1_R + J(R)$ or $1_R - (b - rab) \in J(R)$.
Now, $b - rab = 1_R - (1_R - b + rab)$ is a unit in R as $1_R - b + rab \in J(R)$. So,

$$b(1_R - ra)c = 1_R, \text{ for } c, \text{ a unit in } R.$$

So, $1_R - ra$ is a unit in R for each $r \in R$. By corollary 12.4.1 $a \in J(R)$. Therefore, $a + J(R) = J(R)$ which is an intersection of maximal ideals of R.

There is a one-one correspondence between the maximal ideals of $R/J(R)$ and the maximal ideals of R containing $J(R)$.
So,

$$a + J(R) \in J(R/J(R)) \Rightarrow a \in J(R),$$

or, $J(R/J(R)) = $ the zero ideal $(J(R))$ in $R/J(R)$.
Thus, $R/J(R)$ is semisimple. $\qquad\square$

Remark 12.4.1 : Given a ring R, $J(R)$ is the smallest ideal of R such that $R/J(R)$ is semisimple.

For, suppose that I is an ideal of R such that R/I is semisimple.
We consider the natural homomorphism $\nu : R \to R/I$ which is onto R/I.

$$J(R/I) = \cap\{B : B \text{ is a maximal ideal of } R/I\}$$

If B is a maximal ideal of R/I, its preimage A under ν is given by

$$A = \{x \in R : \nu(x) \in B\}$$

A is a maximal ideal of R containing I. So,

$$J(R/I) = \cap\{\nu(A) : A \text{ is a maximal ideal of } R \text{ containing } I\}.$$

Now, it is easily checked that $\nu(A) = A/I$. So,

$$\nu(I + J(R)) = (I + J(R))/I.$$

Since R/I is semisimple, $J(R/I) = (I)$.
But, $\cap\{\nu(A) : A$ is a maximal ideal of R containing $I\}$ contains
$\nu(I + J(R)) = (I + J(R))/I$.
So, $J(R/I) = (I) \supseteq (I + J(R))/I$.
As (I) is the zero element in R/I, $(I) \subseteq (I + J(R))/I$.
So, $(I + J(R))/I = (I)$.
This can happen only when $J(R) \subseteq I$.
This establishes Remark 12.4.1. Next, we obtain an analogue of Euclid's theorem on infinitude of primes using the fact that $\mathbb{Z}$ is a semisimple ring.

Theorem 92 (D. M. Burton (1970)) **:** *Let R be a PID. Then, R is semisimple if, and only if, R is a field or R has an infinite number of maximal ideals.*

Proof : $\Leftarrow$: Given R is a PID. Suppose $\{p_i\}$ denotes the set of primes in R. The maximal ideals in R are precisely the principal ideals (p_i). Let $a \in J(R)$. If $a \neq 0_R$ then, $a \in J(R)$ if, and only if, a is divisible by each prime p_i. If R has an infinite number of maximal ideals, then, a has to be 0_R. So, $J(R) = (0_R)$. Thus, R is semisimple.

$:\Rightarrow$ Suppose that R is semisimple. If R contains only a finite number of primes, say, $p_1, p_2, \ldots, p_k$; then

$$J(R) = \cap_{i=1}^{k}(p_i) = (p_1 p_2 \ldots p_k) \neq (0_R)$$

So, then, R cannot be semisimple. So, R is semisimple $\Rightarrow R$ has an infinite number of maximal ideals.

When the set $\{p_i\}$ of primes in R is empty, each nonzero element of R is a unit. Since a nonzero non-unit in R is divisible by a finite number of primes by the uniqueness of factorization of a non-unit into a product of primes, $\{p_i\} = \emptyset$ implies that R is a field in which case $J(R) = (0_R)$. $\qquad\square$

Corollary 12.4.2 : *As $\mathbb{Z}$ is a PID and $\mathbb{Z}$ is semisimple, $\mathbb{Z}$ has an infinite number of maximal ideals which in turn implies that the number of primes in $\mathbb{N}$ is infinite.*

12.5. The Lasker–Noether decomposition theorem

In $\mathbb{Z}$, a nonzero non-unit has the factorization

$$n = \pm p_1^{a_1} p_2^{a_2} \cdots p_k^{a_k} \quad (p_i \text{ are primes}, a_i \geq 1, i = 1, 2, \ldots k).$$

When expressed in terms of ideals, one gets

$$(n) = (p_1^{a_1}) \cap (p_2^{a_2}) \cap \cdots \cap (p_k^{a_k})$$

where each $(p_i^{a_i})$ is a primary ideal of $\mathbb{Z}$. This sort of representation of an ideal is valid in a Noetherian ring. In this connection, it is appropriate to quote O. Zariski and P. Samuel [16]:

"The theorem we are going to prove states that in a Noetherian ring, every ideal is an intersection of primary ideals. In many aspects, this theorem reduces the study of arbitrary ideals to that of primary ideals. The theorem does not extend, however to non-Noetherian rings, even if infinite intersections are allowed. The theorem was first proved in the case of polynomial rings by the chess-master Emanuel Lasker who introduced the notion of a primary ideal. His proof was involved and computational. To Emmy Noether is due the recognition that the theorem is a consequence of a.c.c and the proof given here is essentially hers!".

We consider, as usual, commutative rings with unity and describe the proof of Lasker-Noether theorem (see theorem 93).

Definition 12.5.1 : *A ideal I of a ring R is said to be irreducible if it is not the intersection of ideals (of R) properly containing I. If it is, otherwise, I is called reducible.*

An example of an irreducible ideal is a prime ideal P of R. For, suppose that there exist ideals I and J of R such that

$$P = I \cap J, \quad P \subset I, \quad P \subset J.$$

We pick elements $a, b, \in R$ such that $a \in I \setminus P$, $b \in J \setminus P$. Then, ab is in both I and J. Therefore $ab \in I \cap J = P$. As P is a prime ideal, either $a \in P$ or $b \in P$ —a contradiction to the choice of a, b. So, $P = I \cap J$ with $P \subset I$, $P \subset J$ is unacceptable. That is, P is an irreducible ideal of R. However, there exist primary ideals Q (of R) which are reducible.

Lemma 12.5.1 : *Every ideal of a Noetherian ring R is a finite intersection of irreducible ideals.*

Proof : Let $\mathcal{F}$ be the family of all ideals (of R) which are not finite intersections of irreducible ideals. If $\mathcal{F} \neq \emptyset$, $\mathcal{F}$ has a maximal element J, as R is Noetherian. Any ideal (of R) containing J must be a finite intersection of irreducible ideals. Since $J \notin \mathcal{F}$, J is not irreducible, we write $J = I \cap K$ where I, K are ideals strictly containing J. The maximality of J implies that I and K are finite intersections of irreducible ideals. Then, J is also so. This contradicts the fact that $J \in \mathcal{F}$. Thus, $\mathcal{F} \neq \emptyset$ is incorrect and so $\mathcal{F}$ is empty. This proves the lemma. □

Lemma 12.5.2 : *In a Noetherian ring, every irreducible ideal is a primary ideal.*

Proof : We establish a statement which is the contrapositive of the assertion in the statement of lemma 12.5.2.

Let I be an ideal (of R) which is not primary. There exists a pair $a, b \in R$ such that $ab \in I$, $b \notin I$ and $a^m \notin I$ for all $m \geq 1$.

Now,

$$I : (a) \subseteq I : (a^2) \subseteq \cdots \subseteq I : (a^n) \subseteq \cdots$$

forms an ascending chain of ideals of R. By definition 2.3.6 of chapter 2,

$$I : (a) = \{x \in R : x(a) \subseteq I\}.$$

When $xa \in I$, $xa^2 \in I, \cdots$. So, $x \in I : (a) \Rightarrow x \in I : (a^2)$.
That is, if $xa^n \in I$, $xa^{n+1} \in I$ $(n \geq 1)$.
Since R is Noetherian, there exists an integer $m \in \mathbb{N}$ such that $I : (a^m) = I : (a^{m+1})$.

Claim :

(12.5.1) $I = I : (a^m) \cap (I, b),$

where (I, b) denotes the ideal generated by I and b. (We have assumed that $b \notin I$).
Now, $I \subseteq I : (a^m)$ and also $I \subseteq (I, b)$. So,

(12.5.2) $I \subseteq I : (a^m) \cap (I, b)$

To obtain the reverse inclusion, we proceed as follows:

Let $r \in I : (a^m) \cap (I, b)$.
Then, r can be written as $r = s + ta^m = x + yb$, where $s, x \in I$, $t, y \in R$.

$$ta^{m+1} = (x + yb)a - sa$$
$$= (x - s)a + yab \in I.$$

So, $t \in I : (a^{m+1}) = I : (a^m)$, so $ta^m \in I$.
Therefore, $r = s + ta^m \in I$. This shows that

(12.5.3) $I : (a^m) \cap (I, b) \subseteq I.$

From (12.5.2) and (12.5.3), (12.5.1) follows.

Therefore, $I \subset I : (a^m)$ (inclusion is strict as $a^m \notin I$).
Also, $b \notin I$. So $I \subset (I, b)$. From (12.5.1), we note that I is reducible.

So, an irreducible ideal of R is a primary ideal, by contrapositive argument. □

Lemma 12.5.3 : *Let P, Q be ideal of R such that $Q \subset P \subset \sqrt{Q}$. Suppose that for $a, b \in R$. $ab \in Q$ with $b \notin P$, one has $a \in Q$. Then, Q is a primary ideal of R with $P = \sqrt{Q}$.*

Proof : *Stage 1:* Q is a primary ideal.
Let $ab \in Q$ with $b \notin Q$. Then $a \in P \subseteq \sqrt{Q}$. So, $a^n \in Q$ for some $n \in \mathbb{N}$. So Q is primary.
Stage 2: $P = \sqrt{Q}$.
It suffices to show that $\sqrt{Q} \subseteq P$. Let $b \in \sqrt{Q}$. Then, there exists $m \in \mathbb{N}$ such that $b^m \in Q$. Suppose that m is chosen in such a way that m is the least positive integer

which serves to make $b^m \in Q$.

If $m = 1$, $b \in Q \subseteq P$. So, $b \in \sqrt{Q} \Rightarrow b \in P$ or $\sqrt{Q} \subseteq P$.

If $m > 1$, we have $b^m = b^{m-1} \cdot b \in Q$ with $b^{m-1} \notin Q$.

Hence $b \in Q$ or $b \in P$. So $\sqrt{Q} \subseteq P$. As P is given to be contained in $\sqrt{Q}$, $P = \sqrt{Q}$. $\square$

Note : $\sqrt{Q}$ is a primary ideal (of R) associated with Q.

Definition 12.5.2 : *For an ideal I of R, $I = \cap_{i=1}^{n} Q_i$ is called an irredundant primary decomposition of I, if each Q_i ($i = 1, 2, \ldots n$) is a primary ideal (of R) and if*

(i) no Q_i contains the intersection of other primary components, that is,

(12.5.4) $\qquad \cap_{i \neq j} Q_i \neq \cap_{i=1}^{n} Q_i$, *for any $j = 1, 2, \ldots, n$.*

(ii) $\sqrt{Q_i} \neq \sqrt{Q_j}$ for $i \neq j$.

The purpose of the above definition is the following: If an ideal I admits a finite primary decomposition say $I = \cap_{i=1}^{n} Q_i$, some of the Q_i's may be omitted to yield an *irredundant primary decomposition*. For, suppose that Q_i is the intersection of all those primary components which have the same associated prime ideal say I_P. In other words,

$$I_P = \sqrt{Q_{i_1}} = \sqrt{Q_{i_2}} = \cdots = \sqrt{Q_{i_k}}.$$

We take $Q_i' = Q_{i_1} \cap Q_{i_2} \cap \cdots \cap Q_{i_k}$. Then, Q_i' is primary and its associated prime ideal is $I_p = \sqrt{Q_i'}$. Further,

(12.5.5) $\qquad\qquad\qquad I = \cap_i Q_i'$

which is an irredundant primary decomposition. It follows that every ideal in a Noetherian ring has an irredundant primary decomposition.

Theorem 93 (Lasker-Noether) : *Every ideal in a Noetherian ring can be represented as a finite intersection of primary ideals.*

Proof : Follows from the fact that every ideal in a Noetherian ring R is a finite intersection of irreducible ideals and each irreducible ideal is primary. (see lemmas 12.5.1 and 12.5.2). $\square$

Corollary 12.5.1 : *Every ideal of a Noetherian ring has an irredundant primary decomposition.*

For, a finite intersection of primary ideals can be converted into an irredundant primary decomposition.

Example 12.5.1 : Let n be equal to 600. $600 = 2^3 \times 3 \times 5^2$.

In $\mathbb{Z}$, $(600) = (2^3) \cap (3) \cap (5^2)$ is an irredundant primary decomposition of the ideal generated by 600.

Fact 12.5.1 : (A characteristic property of primary decomposition) Let R be the Noetherian ring. Suppose that an ideal I of R has a finite irredundant primary decomposition:

$$I = \cap_{i=1}^{n} Q_i,$$

where Q_i are primary ideals with associated prime ideals $\sqrt{Q_i}$. If P is a prime ideal of R, then, $P = \sqrt{Q_i}$ for some i if, and only if, there exists an element $a \notin I$ such that $P = \sqrt{I : (a)}$.

For proof, see D. M. Burton [3, chapter 12, theorem 12.4, pp 238–239]. The above result gives the associated prime ideals in a primary decomposition via the ideal and an element $a \notin I$.

Fact 12.5.2 : A consequence of the result in Fact 12.5.1 is the following: Suppose that I denotes an ideal of a Noetherian ring R and that I possesses two finite irredundant primary decompositions, namely,

$$(12.5.6) \qquad I = Q_1 \cap Q_2 \cap \cdots Q_n = Q_1' \cap Q_2' \cap \cdots \cap Q_m'.$$

Then, $n = m$ and the associated prime ideals of these two decompositions are equal. That is, under a reordering of suffixes,

$$\sqrt{Q_i} = \sqrt{Q_i'} \quad (1 \le i \le n)$$

This provides a characterisation of the irredundant primary decomposition of an ideal I of a Noetherian ring R which connects the associated prime ideals with I. Further, it is the number n of primary components for I in (12.5.6) that is unique.

See T. W. Hungerford [9, chapter VIII, sections viii.1 to viii.4] and D. M. Burton [3, chapter 12, pp 234–240].

Definition 12.5.3 : *A ring R is said to satisfy the descending chain condition for ideals (d. c. c) if, given any descending chain of ideals*

$$I_1 \supset I_2 \supset \cdots \supset I_n \supset \cdots$$

there exists an integer m such that $I_m = I_{m+1} = I_{m+2} = \cdots$

Fact 12.5.3 : A ring R satisfies the descending chain condition on ideals if, and only if, every non-empty set of ideals of R, partially ordered by set-inclusion contains a minimal element.

In such a situation, we say that if R satisfies d.c.c on ideals, the minimum condition on ideals holds and conversely. Proof is omitted.

Definition 12.5.4 : *A ring R is said to be an Artinian ring, if it satisfies either of the conditions stated in Fact 12.5.3.*

We observe that the ring $\mathbb{Z}$ of integers is Noetherian, but not Artinian. For, we can give an infinite descending chain of ideals

$$(m_1) \supset (m_2) \supset (m_3) \supset \cdots$$

where m_1 divides m_2, m_2 divides $m_3 \cdots, m_t$ divides m_{t+1} and so on.

Examples 12.5.2 :

(1) Any field is both Artinian and Noetherian. For $n > 1$, as $\mathbb{Z}/n\mathbb{Z}$ is finite. $\mathbb{Z}/n\mathbb{Z}$ is both Artinian and Noetherian.

(2) Let $\mathbb{R}$ denote the field of real numbers. Suppose that Map $\mathbb{R}$ denotes the set of functions $f : \mathbb{R} \to \mathbb{R}$. We define

$$I_k = \{f \in \text{Map } \mathbb{R} : f(x) = 0 \text{ whenever } -k \leq x \leq k\}.$$

We note that $I_1 \supset I_2 \supset I_3 \supset \cdots$ and

$$I_1 \subset I_{1/2} \subset I_{1/3} \subset \cdots$$

Map $\mathbb{R}$ is a commutative ring under the operations of addition and ordinary multiplication. I_k is an ideal of Map $\mathbb{R}$. Map $\mathbb{R}$ contains ascending and descending chains that do not terminate. Map $\mathbb{R}$ is neither Artinian nor Noetherian.

(3) Let p be an arbitrary, but fixed prime. We define

$$\mathbb{Z}(p^\infty) = \{\frac{m}{p^n} : m, n \in \tilde{\mathbb{Z}} \text{ and } 0 \leq m < p^n\}$$

$\mathbb{Z}(p^\infty)$ contains rational numbers r of the form $\frac{m}{p^n}$ such that $0 \leq r < 1$. $\mathbb{Z}(p^\infty)$ is an abelian group under addition modulo 1. It is a ring without identity, if we introduce multiplication by defining the product $a \cdot b = 0$ for $a, b \in \mathbb{Z}(p^\infty)$.

Let I be a nontrivial ideal of $\mathbb{Z}(p^\infty)$. We choose k to be the smallest positive integer such that for some $a \in \tilde{\mathbb{Z}}$, $\frac{a}{p^k} \notin I$. We need to choose a as relatively prime to p. Then,

$$I = \{\frac{s}{p^{k-1}} : 0 \leq s \leq p^{k-1} - 1\}.$$

To specify k, we write $I = I_{k-1}$. The only ideals of $\mathbb{Z}(p^\infty)$ (which are subgroups of the additive group $\mathbb{Z}(p^\infty)$ under addition modulo 1) are of the form $\{I_k\}$ $(k \geq 0)$ and

$$(0) \subset I_1 \subset I_2 \subset \cdots \subset I_k \subset \cdots \subset \mathbb{Z}(p^\infty)$$

$\mathbb{Z}(p^\infty)$ has a strictly ascending chain of ideals which does not terminate. So, $\mathbb{Z}(p^\infty)$ is not Noetherian. However, any descending chain is of finite length. So, $(\mathbb{Z}p^\infty)$ is Artinian.

(4) The polynomial ring $F[x_1, x_2, \cdots]$ satisfies neither a.c.c nor d.c.c on ideals. For instance, one has

$$(x_1) \subset (x_1, x_2) \subset (x_1, x_2, x_3) \cdots$$

and

$$(x_i) \supset (x_i^2) \supset (x_i^3) \supset \cdots.$$

((2) and (3) above have been adapted from D. M. Burton [3]).

Fact 12.5.4 :

(i) Let I be an ideal of a Noetherian ring R. Then R/I is Noetherian. This is true for Artinian rings as well.

(ii) Let I be an ideal of a ring R. If I, R/I are Noetherian, so is R.

Likewise, if I and R/I are Artinian, so is R.

For proofs, see D. M. Burton [3] [Chapter 11, theorems 11.6 and 11.7, pp. 225–226].

The a.c.c and d.c.c on ideals are meant to study the structure of rings and modules. In a more general set-up, the following situation characterizes a.c.c.

Fact 12.5.5 : Let $(P, \leq)$ be a partially ordered set. Let $\{x_n\}$ be any sequence of elements of P. Then, the increasing sequence

$$x_1 \leq x_2 \leq x_3 \leq \cdots \leq x_n \leq \cdots$$

is stationary (that is, there exists $m \in \mathbb{N}$ such that $x_m = x_{m+1} = \cdots$) if, and only if, every non-empty subset T of P has a maximal element. For, if there is a non-empty subset T of P which has no maximal element, then a.c.c on $\{x_n\}$ would fail. Conversely, when the set $\{x_n, n \in \mathbb{N}\}$ has a maximal element, $\{x_n\}$ satisfies a.c.c.

For instance, if $(P, \leq)$ is the poset of submodules of an R-module M ordered by set-inclusion, a.c.c on submodules holds $\Leftrightarrow$ every non-empty collection of submodules of M has a maximal element. An R-module possessing this property is called a Noetherian module.

The dual of Fact 12.5.5 is

Fact 12.5.6 : Let $(P, \geq)$ be a poset. Any sequence $\{x_n\}$ of elements of P satisfies d.c.c, that is,

$$x_1 \geq x_2 \geq x_3 \geq \cdots$$

implies that there exists $m \in \mathbb{N}$ such that $x_m = x_{m+1} = \cdots$

if, and only if, every non-empty subset of P has a minimal element.

If $(P, \leq)$ is the poset of submodules of an R-module M partially ordered by set-inclusion, then $(P, \leq)$ satisfies d.c.c on submodules $\Leftrightarrow$ every non-empty collection of submodules of $(P, \leq)$ has a minimal element. An R-module M possessing this property is called an Artinian module.

Now, we note that an R-module M is Noetherian $\Leftrightarrow$ every R-submodule of M is finitely generated. Proof follows on lines similar to that of lemma 12.2.1.

Next, we consider a finite sequence $\{M_k\}$ $(0 \leq k \leq n)$ of R-submodules of an R-module M. If

(12.5.7) $$M = M_0 \supset M_1 \supset M_2 \cdots \supset M_n = (0)$$

where $\supset$ is strict containment, (12.5.7) gives a chain of length n. A composition series is a maximal chain such as (12.5.7) wherein no finite number of extra submodules could be inserted. This is equivalent to saying that M_{k-1}/M_k $(1 \leq k \leq n)$ is a simple module.

Fact 12.5.7 : Let M be an R-module. Suppose that M has a composition series of finite length n. Then, every composition series of M has length n and every chain in M can be extended to a composition series.

For proof, see M. F. Atiyah and I. G. McDonald [1, chapter 6, p 77].

Suppose that an R-module M has a composition series. Then, all chains in M are of finite length and so both a.c.c, d.c.c hold for submodules of M. Conversely, if both a.c.c and d.c.c hold for submodules of M, we can construct a strictly descending chain $M = M_0 \supset M_1 \supset \ldots$ which terminates. So, by Fact 12.5.7, M has a composition series.

Definition 12.5.5 : *An R-module M satisfying both a.c.c and d.c.c on submodules is called a module of finite length.*

We remark that all composition series of M have the same length $l(M)$ (say). Further, Jordan-Hölder theorem [5] (see theorem 66, chapter 8) applies to modules of finite length. In the case of a finite dimensional vector space V over a field F, if $\dim V = n$, a maximal chain in V is of length n (starting from zero subspace and reaching V: If K is a subspace of dimension k, it is contained in a subspace of dimension $(k+1)$, $0 \le k \le n-1$). So, V satisfies a.c.c on subspaces if, and only if, it satisfies d.c.c. on subspaces.

Theorem 94 (Atiyah and MacDonald (1969)) **:** *Let R be a ring in which the zero ideal is a product of a finite number of maximal ideals (not necessarily distinct). Then, R is Noetherian $\Leftrightarrow$ R is Artinian.*

Proof : Let $M_1, M_2, \ldots, M_n$ be maximal ideals of R with the property

$$(0_R) = M_1 M_2 \cdots M_n.$$

One has a descending chain of ideals

(12.5.8) $\qquad R \supset M_1 \supseteq M_1 M_2 \supseteq M_1 M_2 M_3 \supseteq \cdots \supseteq M_1 M_2 \cdots M_n = (0_R)$

As M_i is a maximal ideal ($1 \le i \le n$), R/M_i is a field. $M_1/M_1 M_2$ is a vector space over R/M_2. For, if $r + M_2 \in R/M_2$ and $x + (M_1 M_2)$ is an element of $M_1/M_1 M_2$.

$$(r + M_2)(x + M_1 M_2) = rx + xM_2 + rM_1 M_2$$
$$= rx + M_1 M_2, \text{ as } M_1 M_2 \subseteq M_1 \text{ and } x \in M_1.$$

As $M_1/M_1 M_2$ is an abelian group and scalar multiplication by elements of R/M_2 is taken care of, $M_1/M_1 M_2$ is a vector space over R/M_2.

In general, $(M_1 M_2 \cdots M_{k-1})/(M_1 M_2 \cdots M_k)$ ($2 \le k \le n$) is a vector space over R/M_k. Each $V_{k-1} = (M_1 M_2 \cdots M_{k-1})/(M_1 M_2 \cdots M_k)$ is finite-dimensional ($1 \le k \le n$). Therefore, for each V_{k-1}, a.c.c holds $\Leftrightarrow$ d.c.c holds.

Let M be any maximal ideal of R. We consider an ascending chain $J_1 \subseteq J_2 \subseteq \cdots$ of ideals of R. Then, $J_1 \cap M \subseteq J_2 \cap M \subseteq \cdots$ is a chain of ideals of M.

Claim: $J_{k-1} \cap M = J_k \cap M$. When $J_{k-1} \subseteq J_k$, $J_{k-1} \cap M \subseteq J_k \cap M$. If $x \in J_k$, there exists $y \in J_{k-1}$ such that $x, y \in R$ and the cosets $x + M$ and $y + M$ are equal (considering R/M). So, $x + M = y + M$ and so, $y - x \in M$.

As $J_{k-1} \subseteq J_k$, when $y \in J_{k-1}$, $y \in J_k$. So, $y - x \in J_k$. That is, $y - x \in J_k \cap M$. We get through if we show that $y - x \in J_{k-1} \cap M$.

If $y - x \notin J_{k-1} \cap M$, $y - x \notin J_k \cap M$—a contradiction.

Therefore, $y - x \in J_{k-1}$ as well as M.

Or, if $y-x \in J_k \cap M$, $y-x \in J_{k-1} \cap M$. Thus, $J_{k-1} \cap M = J_k \cap M$. So, $j_{k-1} = J_k$. Therefore, M is Noetherian, as also R. In the same manner M is Artinian $\Rightarrow R$ is Artinian. As a.c.c $\Leftrightarrow$ d.c.c for each V_{k-1} ($1 \le k \le n$), R is Noetherian $\Leftrightarrow R$ is Artinian. $\square$

Remark 12.5.1 : Technique employed is by using Fact 12.5.4 (ii). Theorem 94 has been adapted from [1].

12.6. Dedekind domains

Before we go to the definition of a Dedekind domain, we will give the meaning of 'integral closure'. The characterization of integral closure is tied up with the definition of a finitely generated R-module.

Let $(M,+)$ be an abelian group. Suppose that R is a commutative ring which acts on M linearly. Then, M gets the structure of an R-module.

Definition 12.6.1 : *A subset S of an R-module M is said to generate M, if M is the smallest R-submodule that contains S. In other words, there is no proper submodule of M that contains S.*

When S generates M, every element a of M can be written as

$$(12.6.1) \qquad a = n_1 s_1 + n_2 s_2 \cdots + n_k s_k + r_1 s_1 + r_2 s_2 \cdots + r_k s_k$$

where $n_i \in \mathbb{Z}$, $r_i \in R$ and $s_i \in S$ ($i = 1, 2, \ldots, k$).

When S is finite, we say that M is finitely generated. When S is a singleton, we say that M is a cyclic R-module.

Examples 12.6.1 :

(i) Let R be a commutative ring with unity 1_R. Then $R[x]$, the polynomial ring is an R-module when we take 'scalar multiplication' as multiplication of a polynomial $f(x)$ by $r \in R$ given by
$r f(x) \in R[x]$.

(ii) Let M, N be R-modules. If $\psi : M \to N$ is an R-isomorphism onto N, then,

$$\ker \psi = \{a \in M \; : \; \psi(a) = 0\}$$

is an R-submodule of M. Further, $M/\ker\psi$ and N are R-isomorphic.

(iii) Let I be a nonzero ideal of a PID R. Then, R/I considered as an R-module is both Noetherian and Artinian.

Definition 12.6.2 : *Given a commutative ring T with unity 1_T, let R be a subring of T containing 1_T. Then T is called an extension ring of R.*

For example, if R is a commutative ring with unity 1_R, the polynomial ring $R[x]$ is an extension ring of R. However, though $2\mathbb{Z}$ is a subring of $\mathbb{Z}$, as $2\mathbb{Z}$ does not contain 1, $\mathbb{Z}$ is not considered as an extension ring of $2\mathbb{Z}$.

Definition 12.6.3 : *Suppose that T is an extension ring of R and $t \in T$. If there exists a monic polynomial $f(x) \in R[x]$ such that t is a zero of $f(x)$, t is said to be integral over R.*
If every element of T is integral over R, then, T is called an integral extension of R.

We consider the set
$$\mathbb{Z} \times \{0\} = \{(a,0) : a \in \mathbb{Z}\}.$$
For $(a,0), (b,0) \in \mathbb{Z} \times \{0\}$, we define
$$(a,0) + (b,0) = (a+b,0)$$
and $(a,0) \cdot (b,0) = (ab,0)$.

Then, $\mathbb{Z} \times \{0\}$ is a ring with unity $(1,0)$. If i denotes $\sqrt{-1}$, the ring $\mathbb{Z}[i]$ of Gaussian integers is easily seen to be an extension ring of $\mathbb{Z} \times \{0\}$.

Definition 12.6.4 : *Let T be an extension ring of a ring R. Suppose that Y is a non-empty subset of T. The subring of T generated by Y over R is the intersection of all subrings of T which contain $Y \cup R$. It is denoted by $R[Y]$.*

If $Y = \{t\}$, a singleton, the subring generated by t over R is $R[t]$, as each element of $R[t]$ is a polynomial in t with coefficients from R. If $f(t) \in R[t]$, and f is monic, it may happen that $f(t) = 0_R$. It is the situation where t is integral over R. For this reason, $R[t]$ is not isomorphic to $R[x]$, the polynomial ring in the indeterminate x.

We remark that if $f(x) \in R[x]$, $f(x) \neq 0_R$ even when $f(t) = 0_R$.

When $t \in T$ is integral over R, the module structure of $R[t]$ is 'neat'. This is brought out in

Theorem 95 : *Let T be an extension ring of R. Given $t \in T$, t is integral over $R \Leftrightarrow$ the ring $R[t]$ is a finitely generated R-module.*

Proof : $\Rightarrow$ Given t is integral over R. Therefore, t satisfies a monic polynomial equation with coefficients from R. That is, t is a root of

$$(12.6.2) \qquad x^n + a_{n-1}x^{n-1} + \cdots + a_1 x + a_0 = 0_R, \quad a_i \in R(i = 0, 1, \ldots n).$$

Denoting the left side of (12.6.2) by $f(x)$, we see that $f(t) = 0$. Now, every element of $R[t]$ is of the form $g(t)$ where $g(x) \in R[x]$. As $R[x]$ is a Euclidean ring, (see Remark 3.2.1(ii), chapter 3) we can apply division algorithm to $g(x)$ with $f(x)$ for 'division' to get

$$(12.6.3) \qquad g(x) = q(x)f(x) + r(x), \text{ where either } r(x) = 0 \text{ or } \deg r(x) < \deg f(x).$$

Replacing x by t in (12.6.3) and noting that $f(t) = 0_R$, we get $g(t) = r(t)$. So, $g(t)$ is a polynomial of degree $< n$. Also, $g(t)$ is a linear combination of $1_R, t, t^2, \ldots, t^m$ with $m = \deg r(x) < \deg f(x) = n$. So, $R[t]$ is finitely generated as an R-module.

$\Leftarrow$: $t \in T$ is such that $R[t]$ is finitely generated R-module.
Let $\{y_1, y_2, \ldots, y_n\}$ be a set of generators of $R[t]$. $t \in R[t]$.
As $R[t]$ is a subring of T, $ty_i \in R[t]$ for $i = 1, 2, \ldots, n$.

Using the generating set $\{y_1, y_2, \ldots, y_n\}$, we obtain

$$ty_1 = a_{11}y_1 + a_{12}y_2 + \cdots + a_{1n}y_n,$$
$$ty_2 = a_{21}y_1 + a_{22}y_2 + \cdots + a_{2n}y_n,$$
$$\ldots\ldots\ldots\ldots\ldots\ldots\ldots\ldots\ldots$$
$$ty_n = a_{n1}y_1 + a_{n2}y_2 + \cdots + a_{nn}y_n,$$

where a_{ij} $(i, j = 1, 2, \ldots, n)$ are elements of R.
So,

$$(t - a_{11})y_1 - a_{12}y_2 \cdots - a_{1n}y_n = O_R,$$
$$-a_{21}y_1 + (t - a_{22})y_2 - \cdots - a_{2n}y_n = O_R,$$
$$\ldots\ldots\ldots\ldots\ldots$$
$$-a_{n1}y_1 - a_{n2}y_2 \cdots + (t - a_{nn})y_n = O_R.$$

or,

(12.6.4)
$$\begin{bmatrix} t - a_{11} & -a_{12} & \cdots & -a_{1n} \\ -a_{21} & t - a_{22} & \cdots & -a_{2n} \\ \cdots & \cdots & \cdots & \cdots \\ \cdots & \cdots & \cdots & \cdots \\ -a_{n1} & -a_{n2} & \cdots & (t - a_{nn}) \end{bmatrix} \begin{bmatrix} y_1 \\ y_2 \\ \vdots \\ y_n \end{bmatrix} = 0_R$$

This gives a system of n homogeneous linear equations in $y_1, y_2 \ldots y_n$.
Let $B = \det[b_{ij}]$ where $[b_{ij}]$ is the coefficient matrix given by

$$b_{ij} = \begin{cases} t - a_{ii}, & \text{if } j = i \\ -a_{ij}, & \text{otherwise.} \end{cases}$$

Using Cramer's rule, we see that $By_i = 0_R$ for each $i = 1, 2, \ldots, n$. So, $Bq = 0_R$ for every $q \in R[t]$. As $1_R \in R[t]$, we get $B1_R = 0_R$ from which it follows that $B = 0_R$. But B, is a monic polynomial in t of degree n, on account of (12.6.4). Thus, t is integral over R. $\square$

Corollary 12.6.1 : *Let T be a ring extension of R. If T is a finitely generated R-module, then T is an integral extension of R.*

Proof : Let $t \in T$. We claim that t is integral over R. Let S be a subring of T containing 1_R and $R[t]$. Then, S is a finitely generated R-module, as we need only take $S = R[t]$. Then, t is integral over R and so, every element of T is integral over R which proves that T is an integral extension of R. $\square$

Next, let $Y = \{y_1, y_2, \ldots, y_n\}$. We have seen that the subring of T generated by Y over R is the smallest subring of T containing $R \cup Y$. $R[Y]$ consists of elements

$$f(y_1, y_2, \ldots, y_n), \quad n \geq 1$$

such that $f(x_1, x_2, \ldots x_n) \in R[x_1, x_2, \ldots, x_n]$. For any set $y_1, y_2, \ldots, y_n$ of elements of T, an element of $R[y_1, y_2, \ldots y_n]$ is a polynomial in $y_1, y_2, \ldots, y_n$. As noted earlier, in the case of $R[t]$ and $R[x]$, $R[y_1, y_2, \ldots, y_n]$ need not be isomorphic to

$R[x_1, x_2, \ldots, x_n]$, where $x_1, x_2, \ldots, x_n$ are indeterminates. It can be checked that for $1 \leq i \leq n$,

$$R[y_1, y_2, \ldots, y_{i-1}][y_i] = R[y_1, y_2, \ldots, y_i].$$

Since $R[y_1, y_2, \ldots y_n]$ is a ring containing R, $R[y_1, y_2, \ldots, y_n]$ is indeed an R-module.

Theorem 96 : *Let T be an extension ring of R. Suppose that $y_1, y_2, \ldots y_n \in T$ are integral over R. Then, $R[y_1, y_2, \ldots, y_n]$ is a finitely-generated R-module and is an integral extension of R.*

Proof : We consider R-modules $R[y_1]$, $R[y_1, y_2], \cdots, R[y_1, y_2, \ldots, y_n]$. We can form a strict ascending chain of extension rings:

$$R \subset R[y_1] \subset R[y_1, y_2] \subset \cdots \subset R[y_1, y_2, \ldots, y_n].$$

For each i, y_i is integral over R. So, y_i is integral over $R[y_1, y_2, \ldots, y_{i-1}]$. Since $R[y_1, y_2, \ldots, y_{i-1}][y_i] = R[y_1, y_2, \ldots y_i], R[y_1, y_2, \ldots, y_i]$ is a finitely-generated $R[y_1, y_2, \ldots y_{i-1}]$-module, by iteration, $R[y_1, y_2, \ldots y_n]$ is a finitely-generated R-module. Each element of $R[y_1, y_2 \ldots y_n]$ is integral over R. So, $R[y_1, y_2, \ldots, y_n]$ is an integral extension of R. □

Corollary 12.6.2 : *Let T be an extension ring of R. If S denotes the set of all elements of T which are integral over R, then, S is an integral extension of R which contains every subring of T that is integral over R.*

Proof : Let $s, t \in S$. Then, s and t are elements of $R[s, t]$. So, $s - t \in R[s, t]$. Also $st \in R[s, t]$. Since s and t are integral over R, $R[s, t]$ is an integral extension ring of R (by theorem 96). Also, $s - t \in S$ and $st \in S$. So, S is a subring of T. $R \subset S$, as every element of R is trivially integral over R. By the definition of S, S is an integral extension of R and contains all subrings of T that are integral extensions of R, as $R[s, t] \subseteq S$, whenever $s, t \in S$. □

Definition 12.6.5 : *If T is an extension ring of R, the ring S containing all elements of T which are integral over R is called the integral closure of R in T. If $S = R$, R is said to be integrally closed in T.*

Given an integral domain D, its field F of quotients D is an extension ring of D up to isomorphism. Further, F is the smallest field containing D. See T. W. Hungerford [9].

Definition 12.6.6 : *An integral domain D is said to be integrally closed, if it is integrally closed in its field of quotients .*

Theorem 97 : *A PID is integrally closed.*

Proof : Let D be an integral domain with field of quotients F. We pick an element $\frac{a}{b} \in F$ where $a, b \, (\neq 0) \in D$.
Suppose that $\frac{a}{b}$ is integral over D. Then $\frac{a}{b}$ satisfies a monic polynomial equation

$$(12.6.5) \qquad x^n + a_{n-1}x^{n-1} + \cdots + a_1 x + a_0 = 0_D \quad (a_i \in D, i = 0, 1, \ldots n)$$

Suppose that a and b are relatively prime to one another. That is, a and b are in their lowest terms in $\dfrac{a}{b}$.

Substituting $\dfrac{a}{b}$ for x in (12.6.5), we see that

$$a^n + b(a_{n-1})a^{n-1} + \cdots + a_0 b^{n-1} = 0_D$$

Therefore,

$$b(a_{n-1}a^{n-1} + \cdots + a_1 b^{n-2} + a_0 b^{n-1}) = -a^n$$

So, b divides a^n. But b does not divide a. As $a^n = a^{n-1} \cdot a$, b divides a^{n-1}. Proceeding thus, we arrive at the conclusion: b divides a, after a finite number of steps. This contradicts the fact that a and b are relatively prime to one another. Consequently, 'b divides a' can happen only when b is a unit. In such a situation, we note that $\frac{a}{b} \in D$. So, every element of F which is integral over D belongs to D. Hence, D is integrally closed. $\qquad\square$

Arguing on the same lines, one can show that every unique factorization domain is integrally closed. We deduce from theorem 97 that $\mathbb{Z}$ is integrally closed.

Fact 12.6.1 : A GCD domain is integrally closed.

For proof, see G. Karpilovsky [10].

Observation 12.6.1 :

(i) *The only integral domains that satisfy d.c.c on ideals are fields. In other words, any Artinian integral domain is a field. For, let R be an integral domain and $0_R \neq a \in R$. We consider the strictly descending chain of ideals (of R)*

$$Ra \supset Ra^2 \supset Ra^3 \supset \cdots \supset Ra^n \supset \cdots$$

Then $Ra^m = Ra^{m+1}$ for some integer m. So, there exists $r \in R$ such that $a^m = ra^{m+1}$ or $a^n(1_R - ra) = 0_R$. As R is an integral domain, we note that $ra = 1_R$. That is, every nonzero element of R has a multiplicative inverse and so, R has to be a field.

(ii) *For a commutative ring without unity, maximal ideals need not be prime. We have only to check this in the ring $2\mathbb{Z}$ of even integers.*

(iii) *In commutative rings with unity, prime ideals need not be maximal. For example, in $\mathbb{Z}$, the ideal (0) is a prime ideal which is not a maximal ideal of $\mathbb{Z}$. In $R = \mathbb{Z} \times \mathbb{Z}$, $\mathbb{Z} \times \{0\}$ is a prime ideal of R; for, $R/\mathbb{Z} \times (0) \cong \mathbb{Z}$ an integral domain. If $\mathbb{Z}_e = 2\mathbb{Z}$, $\mathbb{Z} \times \mathbb{Z}_e$ is an ideal of R. Further, $\mathbb{Z} \times \{0\} \subset \mathbb{Z} \times \mathbb{Z}_e \subset R$. So, $\mathbb{Z} \times \{0\}$ is not a maximal ideal of R.*

(iv) *Let R be a commutative ring with unity. Suppose that R is Artinian. We claim that every proper prime ideal of R is maximal:*

When P is a prime ideal of R, R/P is an integral domain. Further, the ideals of R/P have the form I/P where I is an ideal of R containing P. We consider a descending chain

$$I_1/P \supseteq I_2/P \supseteq \cdots$$

of ideals of R/P. Then, $I_1 \supseteq I_2 \supseteq \cdots$ is a descending chain of ideals (of R) containing P. As R is Artinian, the descending chain terminates. That is, one has $I_m = I_{m+1} = \cdots$ for some integer m. Then, $I_m/P = I_{m+1}/P = \cdots$ and so R/P is Artinian. Therefore, R/P being an integral domain which is Artinian turns out to be a field (by observation 12.6.1 (i) above) from which we deduce that P is a maximal ideal of R.

(v) An Artinian ring R has only a finite number of prime (hence maximal) ideals. For, suppose that there exists an infinite sequence $\{P_n\}$ of distinct prime ideals of R. We form a descending chain of ideals

$$P_1 \supseteq P_1 P_2 \supseteq P_1 P_2 P_3 \supseteq \cdots .$$

Since R is Artinian, there exists a positive integer m for which

$$P_1 P_2 \cdots P_m = P_1 P_2 \cdots P_{m+1}.$$

It follows that $P_1 P_2 \cdots P_m \subseteq P_{m+1}$. So $P_s \subseteq P_{m+1}$ for some $s \leq (m+1)$. But, P_s is a maximal ideal of R. So, we must have $P_s = P_{m+1}$. This contradicts the hypothesis that the prime ideals P_n ($n \geq 1$) are distinct. So, R has only a finite number of maximal ideals.

Remark 12.6.1 : Observation 12.6.1(iii) has been adapted from D. M. Burton [3].

Theorem 98 : *Let R be a commutative ring with unity. Then,*

$$R \text{ is Artinian } \Rightarrow R \text{ is Noetherian}$$

Proof : When R is a field, R is both Artinian and Noetherian. So, let us assume that R is not a field.

Claim : When R is Artinian, the zero ideal is a product of a finite number of maximal ideals.

As R has only a finite number of maximal ideals, the Jacobson radical $J(R)$ of R takes the form

(12.6.6) $$J(R) = M_1 \cap M_2 \cdots \cap M_n,$$

where $M_i (i = 1, 2, \ldots n)$ is a maximal ideal of R. $J(R)$ has the property that a positive power of $J(R)$ is the zero ideal. In other words, $J(R)$ is a nilpotent ideal. For, if we consider the descending chain

$$J(R) \supseteq (J(R))^2 \supseteq (J(R))^3 \supseteq \cdots ,$$

there exists a positive integer t such that $(J(R))^t = (J(R))^{t+1} = \cdots$. If $I = (J(R))^t$, $I \subseteq J(R)$ and $I^2 = I$.
We have to show that $I = (0)$. Suppose the contrary. That is, assume that $I \neq (0)$. Let $\mathcal{F}$ denote the family of all ideals A (of R) such that
(i) $A \subseteq I$ and   (ii) $AI \neq (0)$.
Then, $\mathcal{F}$ is non-empty as I belongs to $\mathcal{F}$. As R is Artinian, $\mathcal{F}$ has a minimal element, say B. By (ii) $BI \neq (0)$. So, there exists $b \in B$ such that $bI \neq (0)$.

But then, $(bI)I = bI^2 = bI \neq (0)$.

Also, $bI \subseteq B \subseteq I$ (by (i)).

By the minimality of B, $bI = B$.

Therefore, there exists $a \in I$ such that $ba = b$. But, $I \subseteq J(R)$. So, $a \in J(R)$. As $J(R)$ is the radical of R, it is clear that $1_R - a$ is a unit in R. (See Corollary 12.4.1). So, there exists $c \in R$ such that

$$(1_R - a)c = 1_R$$

Then, $b = b1_R = b(1_R - a)c = (b - ba)c = 0_R$, as $ba = b$. This contradicts the fact that $bI \neq (0)$.

So, $I = (0)$ and so $J(R)$ is nilpotent.

Next, if $J(R)$ is as given in (12.6.6), M_i and M_j are comaximal whenever $i \neq j$ $(i, j = 1, 2, \ldots n)$. That is, $M_i + M_j = R$ $(i \neq j)$.

For, as M_i is a maximal ideal of R, for every $r \notin M_i$, there exists $x \in R$ such that

(12.6.7) $1_R - rx \in M_i$

(It is a consequence of the fact that when $r \notin M_i$, the ideal (M_i, r) (generated by M_i and r) equals R).

Take $r \in M_j \setminus M_i$. Then, there exists $x \in R$ such that (12.6.7) holds. But, as $r \in M_j$, $rx \in M_j$ and therefore,

$$(1_R - rx) + rx = 1_R \in M_i + M_j.$$

In (12.6.6), M_i, M_j are pairwise comaximal whenever $i \neq j$. By a familiar argument in the study of ideals, one gets

(12.6.8) $M_1 M_2 \cdots M_n = M_1 \cap M_2 \cap \cdots \cap M_n$

Therefore,

$$(0_R) = (J(R))^t = M_1^t M_2^t \cdots M_n^t$$

Thus, (0_R) is a product of maximal ideals with repetitions allowed. By theorem 94, the desired conclusion follows.   □

Theorem 99 (I. S. Cohen (1950)) : *Let R be a commutative ring with unity. R is Noetherian if, and only if, every prime ideal of R is finitely generated.*

Proof : $:\Rightarrow$ It is true that if R is Noetherian, every prime ideal (of R) is finitely generated (as it is the case with all proper ideals).

$\Leftarrow$: Let R be a commutative ring with unity in which every prime ideal is finitely generated. Suppose that R is not Noetherian. This means that the collection $\mathcal{F}$ of ideals (of R) which are not finitely generated is non-empty. By Zorn's lemma, $\mathcal{F}$ has a maximal element, say I. By our hypothesis, I cannot be a prime ideal of R. Therefore, there exist elements a, b of R which are not in I such that $ab \in I$. Both the ideals (I, b) and $I : (b)$ contain I properly. So, $a \in I : (b)$ (say), as $a \in I$. By the maximality of I in $\mathcal{F}$, these ideals are finitely generated. Let

(12.6.9) $(I, b) = (x_1, x_2, \ldots x_n)$

(12.6.10) $I : (b) = (y_1, y_2, \ldots y_m)$

Then, $x_i = a_i + b r_i$ where $a_i \in I$, $r_i \in R$ $(i = 1, 2, \ldots n)$. We rewrite (I, b) as

$$(I, b) = (a_1, a_2, \ldots a_n, b)$$

We define the ideal J by

(12.6.11) $J = (a_1, a_2, \ldots, a_n, by_1, by_2, \ldots, by_m)$.

Since $by_j \in I$ for every j, we get $J \subseteq I$.

Claim : $I \subseteq J$.

Let $t \in I$. As $t \in (I, b)$,

$$t = a_1 s_1 + \ldots + a_n s_n + bs \quad (s_i, s \in R; i = 1, 2, \ldots, n)$$

As $a_i \in I$ $(i = 1, \ldots, n)$ $s \in I : (b)$.

So, we find elements $c_i \in R$ such that

$$s = y_1 c_1 + y_2 c_2 \cdots + y_m c_m \ \text{(by (12.6.10))}$$

It follows that $t = a_1 s_1 + \cdots + a_n s_n + (by_1) c_1 + \cdots + (by_m) c_m$, and $t \in J$ by (12.6.11). So, the claim is okay. Therefore, I itself is finitely generated which is impossible, as $I \in \mathcal{F}$.

This contradicts the assumption that $\mathcal{F}$ is non-empty. So, $\mathcal{F} = \emptyset$. Hence R is Noetherian.   □

Remark 12.6.2 : The original version of proof of theorem 99 is in I. S. Cohen [6].

Definition 12.6.7 : *An integral domain D is said to be a Dedekind domain if*

  (i) *D is Noetherian,*
  (ii) *D is integrally closed and*
  (iii) *every nonzero prime ideal of D is a maximal ideal.*

From theorem 97 and by virtue of the fact that a PID is Noetherian and that a nonzero prime ideal of a PID is maximal, we conclude that a PID is a Dedekind domain. In particular, $\mathbb{Z}$ is a Dedekind domain.

Definition 12.6.8 : *Let D be an integral domain with field of quotients F. A fractional ideal L of D is one for which L is a nonzero D-submodule of F such that $aL \subset D$ for some nonzero element a in D.*

This means that elements of L have a 'common denominator' $a \in D$. The ideals of D are fractional ideals for which $a = 1_D$.

Lemma 12.6.1 : *A nonzero finitely generated D-submodule M of F is a fractional ideal of D.*

Proof : For, if $\{b_1, b_2, \ldots b_k\}$ generates M, then,

$$M = Db_1 + Db_2 + \cdots + Db_k, \quad (b_i \in F, i = 1, 2, \ldots, k).$$

Let $b_i = \frac{c_i}{a_i}$, $a_i \neq 0$ and $a_i, c_i \in D$. If $a = a_1 a_2 \cdots a_k$, then,

$$aM = Da_2 a_3 \cdots a_k c_1 + Da_1 a_3 \cdots a_k c_2 + \cdots + Da_1 a_2 \cdots a_{k-1} c_k \subset D.$$

So, there exists $a \in D$ such that a $M \subset D$. This completes the proof of lemma 12.6.1. $\qquad\square$

Remark 12.6.3 : If L is a fractional ideal of D with $aL \subset D$ for $a \neq 0$, $a \in D$, then, aL is an ideal of D and $\psi : L \to aL$ defined by $\psi(x) = ax$, $x \in L$, is a D-module homomorphism.

Definition 12.6.9 : *Given two fractional ideals L_1 and L_2 of D, their product L_1L_2 is defined by*

$$L_1L_2 = \{\sum_{i=1}^{n} a_ib_i \ : \ a_i \in L_i, b_i \in L_2; n \in \mathbb{N}\}$$

Definition 12.6.10 : *A fractional ideal L of D is said to be invertible if there exists a fractional ideal M of D such that $LM = D$.*

We remark that fractional ideals of D are of the form $a^{-1}I$ where $a \in F$ and I denotes an ideal of D. $a^{-1}I$ is a D-submodule of F. In the case of $\mathbb{Z}$ with field of quotients Q, the field of rational numbers, the fractional ideals of $\mathbb{Z}$ are of the form $q\mathbb{Z}$ where $q \in Q$. So, if $L_1 = q_1\mathbb{Z}$ and $L_2 = q_2\mathbb{Z}$, $L_1L_2 = q_1q_2\mathbb{Z}$. Obviously, if $q_1 \neq 0$, $L_1^{-1} = \dfrac{1}{q_1}\mathbb{Z}$.

Lemma 12.6.2 : *If a factional ideal L of D possesses an inverse denoted by L^{-1} then,*

$$L^{-1} = \{a \in F : aL \subset D\}$$

Further, L^{-1} is unique.

Proof : Suppose that $M = \{a \in F : aL \subset D\}$.
As $ML \subset D$, M is a fractional ideal of D. (Any $t \in L$ is such that $tM \subset D$). Further $LM = ML \subset D$.

If L is invertible, $LM = ML = D$ and $M \subseteq L^{-1}$.
Conversely, since L^{-1} and M are D-submodules of F,

$$L^{-1} = DL^{-1} = (ML)L^{-1} = M(LL^{-1}) \subset MD = DM = M$$

So, $L^{-1} = M$.
To prove the uniqueness of L^{-1}, we proceed as follows:
Suppose that L, M, M' are fractional ideals of D such that $LM = LM'$ and L is invertible. Then, $M = DM = (L^{-1}L)M = L^{-1}(LM) = L^{-1}(LM') = (L^{-1}L)M' = DM' = M'$. $\qquad\square$

Remark 12.6.4 :

(i) Let I be an ideal of D. If I is invertible, then $D \subset I^{-1}$. For,

$$I^{-1} = \{a \in F : aI \subset D\}.$$

If $y \in D$, $yI = I \subset D$. So, $y \in I^{-1}$ and $D \subset I^{-1}$.

(ii) Every nonzero principal ideal of an integral domain D is invertible. For, let $I = (a)$, a principal ideal generated by $a \in D$ $(a \neq 0_D)$. We take $q = 1_D/a \in F$, the field of quotients of D. Then, $L = qD \subset F$ and L is a fractional ideal (of D) having the property $IL = D$. So, I is invertible.

We are now ready to prove the structure theorem (for Dedekind domains) which states that every ideal of a Dedekind domain D is a product of prime ideals and the representation is unique. This will serve as a substitute for unique factorization of elements in D when the uniqueness of factorization of an element into irreducibles fails. An example is that of $\mathbb{Z}[\sqrt{-5}]$ which is not a UFD. However, any ideal of $\mathbb{Z}[\sqrt{-5}]$ can be expressed uniquely as a product of prime ideals. $\mathbb{Z}[\sqrt{-5}]$ will be shown to be a Dedekind domain in the subsequent narration of this chapter.

Lemma 12.6.3 : *In a Dedekind domain D, every ideal contains a product of prime ideals.*

Proof : Assume the contrary. Then, the set S of ideals which do not contain products of prime ideals is non-empty. As D is Noetherian, S has a maximal member, say, J. J is not a prime ideal, since it does not contain a product of prime ideals. We pick two elements $a, b \in D \setminus J$ such that $ab \in J$. The ideals $J + (a)$ and $J + (b)$ are strictly bigger than J. So, by hypothesis, they contain products of prime ideals. This property is shared by the ideal $(J + (a))(J + (b))$ also. As $ab \in J$,

$$(J + (a))(J + (b)) \subset J.$$

This contradicts the assumption that J does not contain products of prime ideals. That is, $S = \emptyset$. □

Lemma 12.6.4 : *Let J be a proper ideal of a Dedekind domain having field of quotients F. Then, we can pick an element $q \in F \setminus D$ such that $qJ \subset D$.*

Proof : We take a nonzero element a in J. By lemma 12.6.3, the principal ideal (a) contains a product of prime ideals. Suppose that $P_1, P_2 \ldots P_r$ are prime ideals having the property

$$P_1 P_2 \cdots P_r \subset (a).$$

Let r be the least such suffix. $r \geq 2$. We know that every proper ideal is contained in a maximal ideal P, by Krull-Zorn theorem, see [3]. P is a prime ideal. So, $P_1 P_2 \ldots P_r \subset P$. It is verified that P contains some P_i. The argument is as follows:
Suppose that P does not contain any of the P_i. Let $a_i \in P_i \setminus P$, $i = 1, 2, \ldots r$. As $P \supset P_1 P_2 \cdots P_r$, $a_1 a_2 \cdots a_r \in P$. So, some a_i $(i = 1, 2, \ldots r)$ belongs to P. This contradicts the assumption that $a_i \notin P$ for each i. So, P contains some P_i. Without loss of generality, we take $P_1 \subseteq P$. As D is a Dedekind domain, any proper prime ideal is maximal. So, $P_1 = P$. Since (a) cannot contain a product of fewer than r prime ideals, there exists $b \in P_2 \cdots P_r \setminus (a)$. We write $q = \frac{b}{a} \in F \setminus D$.

Claim : $qJ \subset D$.

As $P_1 P_2 \ldots P_r \subset (a)$ and $P_1 = P$, we have $bP \subseteq (a)$. Now $\frac{1_D}{a} D \in F$ and $J' = \frac{1_D}{a} D$ is a fractional ideal which is such that (a) $J' = D$. Since P is a maximal ideal, P does not contain any unit $\in D$. So, there exist nonzero element $y \in F$ such that $yP \subseteq D$. Therefore,

$$T = \{x \in F \; : \; xP \subseteq D\}$$

is non-empty and so, by lemma 12.6.2 $P^{-1} = T$ (say) exists. That is, P is invertible. Further, when $(a) \subset P$, $D \subseteq P^{-1} \subseteq (a)^{-1}$. From the fact that $b \in P_2 P_3 \cdots P_r \setminus (a)$ and $bP \subseteq (a)$, we see that for $q = \frac{b}{a}$,

$$qP = \frac{b}{a} P \subseteq D \text{ and } \frac{b}{a} \in P^{-1}$$

But $b \notin (a)$. So, $q = \frac{b}{a}$ is such that $qJ \subset qP \subseteq D$ and so, $qJ \subset D$, as claimed.

$\square$

Theorem 100 : *Let D be a Dedekind domain. Given I, an ideal of D, there exists an ideal J (of D) such that IJ is a principal ideal.*

Proof : We pick a nonzero element a from I. We define

(12.6.12) $J = \{x \in D \; : \; xI \subset (a)\}$,

where (a) denotes the principal ideal generated by a.

As $a \in I$, $aI \subset (a)$ and so $a \in J$ and J is a nonzero ideal of D. Further,

$$JI = IJ \subset (a).$$

So, we get through, if we show that $IJ \supset (a)$.

We set $L = \frac{1_D}{a} IJ$.

If $y = \frac{1_D}{a} \sum_{\text{finite}} x_i y_i$ where $x_i \in I$, $y_i \in J$; $y \in L$. Also, $y \in L \Rightarrow y \in D$. Therefore, $L \subseteq D$. L is also an ideal of D.

If $L = D$, then $IJ = (a)$, showing that IJ is a principal ideal.

Claim : The assertion: 'L is is a proper ideal of D' is false.

Suppose that L is a proper ideal of D. F denotes the field of quotients of D. By lemma 12.6.4, we can pick an element $q \in F \setminus D$ such that $qL \subset D$.

Since D is a Dedekind domain, D is integrally closed.

Since $J = \frac{1_D}{a} aJ$ and $a \in I$, $J \subset L$. So,

$$qJ \subset qL \subset D$$

As q is a unit in $F \setminus D$, $qJ = \{qx : x \in J\}$

(12.6.13) $qxI = xqI = xI \subset (a)$

Let $qJ \subset J$. Let $\{b_1 b_2 \ldots b_m\}$ be a set of generators of I.

$$qb_1 = a_{11} b_1 + a_{12} b_2 \cdots + a_{1m} b_m,$$

$$\ldots\ldots\ldots\ldots\ldots\ldots\ldots\ldots\ldots\ldots$$

$$qb_m = a_{m1} b_1 + a_{m2} b_2 \cdots + a_{mm} b_m.$$

Or,

$$q \begin{bmatrix} b_1 \\ b_2 \\ \vdots \\ b_m \end{bmatrix} = M \begin{bmatrix} b_1 \\ b_2 \\ \vdots \\ b_m \end{bmatrix}$$

where $M = [a_{ij}]$; $a_{ij} \in D$, $i, j = 1, 2, \ldots m$. Or,

(12.6.14) $(qI_m) \begin{bmatrix} b_1 \\ b_2 \\ \vdots \\ b_m \end{bmatrix} = 0_D$ (I_m is the $m \times m$ unit matrix).

As $\{b_1, b_2, \ldots, b_m\}$ is a set of nonzero elements of J, we note from (12.6.14) that q satisfies a monic polynomial with coefficients from D. Then, $q \in D$ — a contradiction. So, the claim is established and $L = D$ is the only possibility, or, $IJ = (a)$ as desired.   □

Corollary 12.6.3 : *Given ideals L, M, N of a Dedekind domain, $LM = LN$ implies $M = N$.*

For, by theorem 100, given L an ideal of D, there exists an ideal J of D such that $LJ = (a)$, a principal ideal. This shows that $aM = aN$ and so, $M = N$.

Corollary 12.6.4 : *If I and J are ideals of a Dedekind domain D, I divides J if, and only if, $I \supset J$.*

Proof : If I divides J, it is true that $I \supset J$. Conversely, assume that $I \supset J$. Fix an ideal L of D such that IL is principal, say (a). Writing $M = \frac{1}{a} LJ$, we see that M is an ideal contained in D. Thus, $IM = (\frac{1}{a}) ILJ = J$. So, I divides J.   □

Theorem 101 : *Every ideal of a Dedekind domain D is expressible as a product of prime ideals and this representation is unique.*

Proof : Suppose that the set of ideals (of D) which are not expressible as a product of prime ideals (of D) is not empty. As D is Noetherian, this set has a maximal member say $M \neq D$, considering proper ideals only. Then, M is contained in a prime ideal P (which is a maximal ideal). $M \subset P \Rightarrow P$ divides M, by corollary 12.6.4. So, we rewrite M as PJ for some ideal J of D. Then, J contains M strictly, by cancellation property (See corollary 12.6.3). For, if $J = M$, $DM = PM \Rightarrow D = P$, which is false. So, J is strictly bigger than M. Also, J is a product of prime ideals. As $M = PJ$, M is also a product of prime ideals, a contradiction to the assumption that the set of ideals not expressible in this manner is non-empty. Therefore, the set of ideals which are not expressible as a product of prime ideals is empty.

Thus, any proper ideal I of D is expressible in the form

(12.6.15) $I = P_1 P_2 \ldots P_r$ (P_i not necessarily distinct)

where P_i are prime ideals. $(i = 1, 2, \ldots r)$

As for uniqueness, suppose that I has also the representation

(12.6.16) $I = Q_1 Q_2 \cdots Q_s$ (Q_j not necessarily distinct)

where Q_j are prime ideals ($j = 1, 2, \ldots s$). Then,

$$P_1 P_2 \cdots P_r = Q Q_2 \ldots Q_s \Rightarrow P_1 \supset Q_1 Q_2 \cdots Q_s$$

So, P_1 contains some Q_j. Rearranging Q_j, if needed, we get $P_1 \supset Q_1$. As the prime ideals are maximal, $P_1 = Q_1$. Using the cancellation property, we get

$$P_2 P_3 \cdots P_r = Q_2 Q_3 \cdots Q_s$$

Proceeding as before, we note that $P_i = Q_i$ for all i and $r = s$. This completes the proof of theorem 101. $\square$

Next, we examine whether a Dedekind domain could be a UFD under certain restrictions. Though $\mathbb{Z}$ is a UFD, not all Dedekind domains have the property of a UFD.

Theorem 102 (D. A. Marcus (1977)) : *A Dedekind domain is a UFD, if, and only if, it is a PID.*

Proof : $\Leftarrow$: The 'if' part is a consequence of theorem 97, since a PID is Noetherian, integrally closed and every nonzero prime ideal is maximal. So, a Dedekind domain which is a PID is a UFD.

$:\Rightarrow$ Let D be a Dedekind domain which is a UFD. We show that D is a PID. Assume the contrary. That is, we suppose that D is not a PID. Let P be a non-principal prime ideal. Such an ideal exists, since, otherwise, all ideals would be principal ideals. Since D is a Dedekind domain, by theorem 100, there exists an ideal J of D such that PJ is a principal ideal. We consider the set Σ of ideals I such that PI is principal. This set is non-empty as $J \in \Sigma$. As D is Noetherian, Σ has a maximal member, say M. Suppose that $PM = (a)$, $a \in D$. We claim that a is an irreducible element in D. For, if not, one has $a = bc$ where b, c are nonzero non-units. Then, either (b) or (c) will be of the form PL for some L dividing M. As M is a maximal element in Σ, L dividing $M \Rightarrow L = M$. But, then, b or c is a unit.

As P divides (a), $P \supset (a)$. Similarly $M \supset (a)$. We pick two elements x, y such that $x \in P \setminus (a)$, $y \in M \setminus (a)$. Now, $xy \in (a)$, as $PM = (a)$. So, a divides xy. But, a does not divide x, a does not divide y. This is not allowed in a UFD. This contradiction makes us accept the fact that there does not exist a non-principal prime ideal of D. As every proper ideal of D is a unique product of prime ideals and as every proper prime ideal is principal, every proper ideal of D is principal. Further, D is generated by the multiplicative identity 1_D. So, D is a PID. $\square$

Remark 12.6.5 : Theorem 102 has been adapted from [12].

Next, we consider the ring of integers of an algebraic number field. It may be recalled that by an algebraic number field, we mean a subfield K of $\mathbb{C}$ such that

K has finite dimension when it is considered as a vector space over Q. Further, a complex number α is an algebraic integer if, and only if, α is a zero of a monic polynomial with coefficients from $\mathbb{Z}$.

Lemma 12.6.5 : $\alpha \in \mathbb{C}$ *is an algebraic integer if, and only if, $\mathbb{Z}[\alpha]$ is a finitely generated $\mathbb{Z}$-module.*

Proof : $:\Rightarrow$ Given α, an algebraic integer, there exists a monic polynomial of degree n (≥ 1) with coefficients from $\mathbb{Z}$ and having α as a zero. That is, one gets

$$\alpha^n + a_{n-1}\alpha^{n-1} + \cdots a_1\alpha + a_0 = 0 \; ; \; a_i \in \mathbb{Z}, \; i = 1, 2, \ldots, n.$$ This means that the $\mathbb{Z}$-module $\mathbb{Z}[\alpha]$ is generated by $\{1, \alpha, \alpha^2, \ldots, \alpha^{n-1}\}$. Therefore, $\mathbb{Z}[\alpha]$ is finitely generated.

$\Leftarrow$: suppose that $\alpha \in \mathbb{C}$ is such that $\mathbb{Z}[\alpha]$ is a finitely generated $\mathbb{Z}$-module. We denote $\mathbb{Z}[\alpha]$ by M. As $\alpha \in M$, $\alpha M \subseteq M$. Let $\{e_1, e_2, \ldots, e_m\}$ generate M.

$$\alpha e_1 = a_{11}e_1 + a_{12}e_2 \ldots + a_{1m}e_m,$$
$$\alpha e_2 = a_{21}e_1 + a_{22}e_2 + \cdots + a_{2m}e_m,$$
$$\ldots\ldots\ldots\ldots$$
$$\ldots\ldots\ldots\ldots$$
$$\alpha e_m = a_{m1}e_1 + a_{m2}e_2 \ldots + a_{mm}e_m,$$

where $a_{ij} \in \mathbb{Z}$ ($i, j = 1, 2, \ldots, m$). Or,

$$(\alpha I_m - A) \begin{bmatrix} e_1 \\ e_2 \\ \vdots \\ e_m \end{bmatrix} = 0; \; (I_m, \text{ the } m \times m \text{ unit matrix and } A = [a_{ij}]).$$

This shows that $\det(\alpha I - A) = 0$. That is, α satisfies a monic polynomial

$$x^m + b_{m-1}x^{m-1} + \cdots + b_1 x + b_0 = 0,$$

where b_j ($j = 0, 1, 2, \ldots, (m-1)$) are integers. Hence α is an algebraic integer. $\square$

Corollary 12.6.5 : *If α, β are algebraic integers, so are $\alpha + \beta$ and $\alpha\beta$.*

Proof : By lemma 12.6.5, $\mathbb{Z}[\alpha]$ and $\mathbb{Z}[\beta]$ are finitely generated $\mathbb{Z}$-modules. Let $\{a_1, a_2, \ldots a_m\}$ generate $\mathbb{Z}[\alpha]$ and $\{b_1, b_2, \ldots b_n\}$ generate $\mathbb{Z}[\beta]$. Then, the mn products $a_i b_j$ ($i = 1, \ldots m; j = 1, 2, \ldots n$) generate a $\mathbb{Z}$-module which we shall denote by $\mathbb{Z}[\alpha, \beta]$. $\mathbb{Z}[\alpha, \beta]$ is a finitely generated $\mathbb{Z}$-module. It is verified that $\alpha + \beta$ and $\alpha\beta$ are elements of $\mathbb{Z}[\alpha, \beta]$. So, by lemma 12.6.5, $\alpha + \beta$ and $\alpha\beta$ are algebraic integers. $\square$

From lemma 12.6.5, we deduce that the set of algebraic integers (which are elements of $\mathbb{C}$) forms a ring denoted by $\mathscr{A}$. If K is any algebraic number field, $\mathscr{A} \cap K$ is a subring of K.

Definition 12.6.11 : *Given an algebraic number field K, we call $\mathscr{A} \cap K$, the number ring corresponding to K.*

For instance, if $K = Q(\sqrt{m})$, where Q is the field of rational numbers, m square free, the number ring corresponding to K is given by

$$\mathscr{A} \cap K = \begin{cases} \mathbb{Z}[\sqrt{m}], & \text{if } m \not\equiv 1 \ (\text{mod } 4); \\ \mathbb{Z}[\zeta], & \text{where } \zeta = \frac{1+\sqrt{m}}{2}, \text{ if } m \equiv 1 \ (\text{mod } 4). \end{cases}$$

(See theorem 14, chapter 3).

Definition 12.6.12 : *If $\omega = \exp(2\pi i/m)$, $(i = \sqrt{-1})$, $m \in \mathbb{N}$; one has*

$$Q(\omega) = \{a_0 + a_1\omega + \cdots + a_{m-2}\omega^{m-2} : a_i \in Q, (i = 0, 1, 2, \ldots (m-2))\}$$

is an algebraic number field, called the mth cyclotomic field.

It could be verified that the number ring corresponding to $Q(\omega)$ is given by

$$(12.6.17) \qquad\qquad \mathscr{A} \cap Q(\omega) = \mathbb{Z}[\omega]$$

In particular, for $m = 4$, $\exp(\frac{2\pi i}{4}) = i$. (See [12]). So, $\mathbb{Z}[i]$, the ring Gaussian integers is the number ring corresponding to the algebraic number field $Q(i)$. We recall that $Q[\omega]$ is a finite extension of Q of degree $\phi(m)$ where ϕ denotes the Euler ϕ-function.

Next, we examine the additive structure of $\mathscr{A} \cap K$ where $\mathscr{A}$ is the ring of algebraic integers and K an algebraic number field.

Recalling the definition 10.3.6 of a free abelian group of rank n, we give for ready reference

Definition 12.6.13 : *A free abelian group G of rank n (n finite) is a group which is a direct sum of n subgroups each of which is isomorphic to $\mathbb{Z}$.*

In other words, G is isomorphic to the group $(\mathbb{Z}^n, +)$ where $\mathbb{Z}^n$ is the set of lattice points in $\mathbb{R}^n$. The rank of a free abelian group is well-defined, as $\mathbb{Z}^m$ and $\mathbb{Z}^n$ are non-isomorphic for $m \neq n$ (see [9]). It is also verified that a subgroup of a free abelian group of rank n is also a free abelian group of rank $\leq n$.

Let α be an algebraic number. α is the zero of a polynomial $f(x)$ where $f(x) \in Q[x]$, $\deg f = n$, say. That is, the equation $f(x) = 0$ is

$$(12.6.18) \qquad a_0\alpha^n + a_1\alpha^{n-1} + \cdots + a_n = 0, \quad a_0 \neq 0, \quad a_i \in Q, i = 0, 1, \ldots n$$

can be rewritten as

$$\alpha^n + b_1\alpha^{n-1} + \cdots + b^n = 0, \quad \text{where } b_i = \frac{a_i}{a_0}, i = 0, 1, 2 \ldots n.$$

As $b_1, b_2 \ldots b_n$ are elements of Q, we take m to be the l.c.m. of denominators of $b_1, b_2 \ldots b_n$, $m \in \mathbb{Z}$. We obtain

$$m^n\alpha^n + b_i m^n \alpha^{n-1} \cdots + b_n m^n = 0.$$

Writing $\theta = m\alpha$, we get a monic polynomial equation of the form

$$\theta^n + c_1\theta^{n-1} + \cdots + c_n = 0 \text{ where } c_i = b_i m^i, i = 1, 2, \ldots n$$

We note that $c_1, c_2 \ldots, c_n$ are in $\mathbb{Z}$. So, θ is an algebraic integer. Thus, given an algebraic number α, we can find $m \in \mathbb{Z}$ such that $m\alpha$ is an algebraic integer.

Fact 12.6.2 : If K denotes an algebraic number field and $\mathscr{A}$ the ring of algebraic integers, $\mathscr{A} \cap K$, the number ring corresponding to K is a free abelian group of rank n, where $n = [K : Q]$.

For proof, see D. A. Marcus [12, corollary to theorem 9].

Next, if α is an algebraic number, α is a zero of an irreducible polynomial $f(x) \in Q[x]$. If $\deg f = n$, $[Q[\alpha] : Q] = n$. We write $K = Q[\alpha]$. The zeros of $f(x)$ all lie in $\mathbb{C}$. The zeros of $f(x)$ are called conjugates of α. An embedding of K in $\mathbb{C}$ is a ring homomorphism of K into $\mathbb{C}$. There are n embeddings of K in $\mathbb{C}$, obtained by mapping α into any one of its conjugates. Each conjugate β (of α) determines a unique embedding $\psi_\beta : K \to \mathbb{C}$ given by $g(\alpha) \mapsto g(\beta)$ for every $g \in Q[x]$. Every embedding arises in this way, since in each ring homomorphism α goes to one of its conjugates. These, in fact, give n monomorphisms of K, say, $\sigma_1, \sigma_2, \ldots \sigma_n$. For $\theta \in K$, we write

(12.6.19) $$N(\theta) = \sigma_1(\theta)\sigma_2(\theta) \cdots \sigma_n(\theta).$$

$N(\theta)$ depends on K as well as θ. Clearly, $N(\theta) \in Q$, as $N(\theta)$ is the product of the conjugates of θ and θ satisfies an irreducible polynomial with coefficients from Q. $N(\theta)$ is called the norm of θ.

Theorem 103 : *The number ring corresponding to an algebraic number field K is a Dedekind domain.*

Proof : First, we observe that if $R = \mathscr{A} \cap K$ where K is the given number field and $\mathscr{A}$ is the ring of algebraic integers, R is an integral domain. Further, R is a free abelian group of finite rank, say n. If I denotes an ideal of R, $(I, +)$ is also a free abelian group of rank $\leq n$ and so I is finitely generated.
That is,

(1) R is a Noetherian domain.
We claim that
(2) R is integrally closed in K.
Reason: For $\alpha, \beta \in R$, if we take $\theta = \frac{\alpha}{\beta}$, θ is an algebraic integer. For, if θ satisfies a monic polynomial

$$x^n + a_{n-1}x^{n-1} + \cdots + a_1 x + a_0$$

where a_i $(i = 0, 1, \ldots (n-1))$ are algebraic integers, $\mathbb{Z}[a_0, a_1, \ldots a_{n-1}, \theta]$ is finitely generated as a $\mathbb{Z}$-module. So $\theta \in R$. So, R is integrally closed in K.
Next, we have to show that
(3) every nonzero prime ideal of R is maximal. It suffices to show that if P is a nonzero prime ideal of R, R/P is a finite integral domain. Let I be a nonzero ideal of R.

Claim : R/I is finite.

Let α be a nonzero element of I. Let $m = N(\alpha)$, the norm of α (in terms of automorphisms of K) see (12.6.19). As $\alpha \in R$, m is an integer, $m \neq 0$. From the definition of norm, $m = \alpha\beta$, where β is a product of conjugates of α (other than α). These conjugates need not be in R. $\beta \in R$, since $\beta = \frac{m}{\alpha} \in K$ and it is verified that $\beta \in \mathscr{A}$. So, $m \in I$, as $\alpha \in I$, $\beta \in R$. So, I contains the nonzero integer m. We consider $R/(m)$. Now, if G is a free abelian group of rank n, for $m \in \mathbb{Z}$, $G/m\mathbb{Z}$ is a direct sum of n cyclic groups of order m (in additive notation). So, $G/m\mathbb{Z}$ is finite and has m^n elements. So, $R/(m)$ has order m^n. As $(m) \subset I$, $R/I \subset R/(m)$ and so R/I is finite. R/I has order t where $t \mid m^n$. So, every nonzero prime ideal of R is maximal, since R/P is a field.

Thus, $R = \mathscr{A} \cap K$ is a Dedekind domain.

$\square$

Remark 12.6.6 : A number ring R corresponding to an algebraic number field K has the following characteristic properties:

  (i) R is a Dedekind domain and so any ideal I of R is expressible as a finite product of prime ideals uniquely, which is the analogue of the uniqueness of factorization of an element as a product of primes in a UFD.

 (ii) Every nonzero ideal I of R is such that R/I is finite.

12.7. The Chinese remainder theorem revisited

Let D be an integral domain with field of quotients F. From Definition (12.6.8), we recall that if L is a fractional ideal of D, L is a nonzero D submodule of F such that $aL \subset D$ for some nonzero element $a \in D$. If L possesses an inverse denoted by L^{-1}, then

$$L^{-1} = \{a \in F : aL \subset D\}$$

and $LL^{-1} = D$.

Definition 12.7.1 : *For a fractional ideal L of D, we define*

$$D_L = \{a \in F : aL \subset L\}$$

We notice that D_L is an extension ring of D. For, if $t \in D$, $tL = L$ and so $t \in D_L$. As D_L is a subring of F, D_L is an extension ring of D.

Lemma 12.7.1 : *If D is a Noetherian domain which is integrally closed and L is a proper ideal of D, then $D_L = D$.*

Proof : A D is Noetherian, L is finitely generated. Suppose that $\{a_1, \ldots a_n\}$ generates L. Let $b \in D_L$. Then $bL \subset L$.
So,

$$ba_1 = c_{11}a_1 + \cdots\cdots + c_{1n}a_n,$$

$$ba_2 = c_{21}a_1 + \cdots\cdots + c_{2n}a_n,$$

$$\cdots\cdots\cdots\cdots$$

$$ba_n = c_{n1}a_1 + \cdots\cdots + c_{nn}a_n$$

where $c_{ij} \in D$ $(i, j = 1, 2, \ldots n)$. Or,

$$(bI_n - M) \begin{bmatrix} a_1 \\ a_2 \\ \vdots \\ a_n \end{bmatrix} = 0_D,$$

where I_n is the $n \times n$ unit matrix and $M = [c_{ij}]$ with $c_{ij} \in D$; $i, j = 1, 2, \ldots, n$. So, as $a_i \neq 0$ for $i = 1, 2, \ldots n$, $\det(bI_n - M) = 0$. This gives a monic polynomial equation of the nth degree in b with coefficients from D. As D is integrally closed, $b \in D$. So, $D_L = D$. $\qquad \square$

Theorem 104 : *Let D be a Dedekind domain which is not a field. If M denotes a nonzero maximal ideal of D, then M is an invertible prime ideal of D.*

Proof : As M is a maximal ideal, M is prime. Moreover, every nonzero prime ideal of D is maximal. If M possesses an inverse, say M^{-1}, then,

$$M^{-1} = \{a \in F : aM \subset D\}, \quad \text{where } F \text{ is the field of quotients of } D.$$

M is Noetherian and integrally closed. So, by lemma 12.7.1, $D_M = D$. As $M^{-1} \supset D_M, MM^{-1} \supset MD_M = D_M = D$. So, $MM^{-1} \supseteq D$. Now, $M \subset D$. So, $MM^{-1} \subset DM^{-1}$. If $t \in D$, and $q \in M^{-1}$, as $qM \subseteq D$, $tqM \subset D$. Or, $MM^{-1} \subseteq D$. This proves that $MM^{-1} = D$. That is, M is an invertible prime ideal of D. $\qquad \square$

Corollary 12.7.1 : *Every proper prime ideal of a Dedekind domain D is invertible, as every maximal ideal of D is invertible.*

Next, we observe that while proving theorem 103, we came across the fact that given a nonzero ideal I of a number ring $\mathcal{A} \cap K$ corresponding to an algebraic number field K, $A \cap K / I$ is finite. As $\mathcal{A} \cap K$ is a Dedekind domain, we have plenty of examples of Dedekind domains D for which whenever I is a nonzero ideal of D, D/I is finite.

Definition 12.7.2 : *Let D be a Dedekind domain. Assume that for every nonzero proper ideal I of D, D/I is finite. The number of cosets of I in D is called the norm of I and is denoted by $N(I)$.*

In the case of $\mathbb{Z}$, the integral domain of integers, if $I = n\mathbb{Z}$, where n is a positive integer > 1, $N(I) = |\mathbb{Z}/n\mathbb{Z}| = n$.

Now, if D is as given in the definition 12.7.2,

$$(12.7.1) \qquad N(I)D = \{N(I)x : x \in D\}$$

Claim : $N(I)D \subset I$.

For let $\psi : N(I)D \rightarrow D/I$ be given by

$$\psi(N(I)x) = N(I)x + I, \text{ the coset of } I \text{ determined by } N(I)x \in D.$$

Then,

$$\psi(N(I)1_D) = N(I)1_D + I$$
$$= (1_D + I) \oplus (1_D + I) \oplus \cdots \oplus (1_D + I)(N(I)\text{times}).$$

As $N(I)$ is the order of the additive group $(D/I, \oplus)$, where $\oplus$ denotes addition modulo I,

$$\psi(N(I)1_D) = [0_D] \text{ in } D/I.$$

Therefore, if $x \in D$

$$\psi(N(I)x) = (N(I)1_D + I)(x + I) = [0_D]. \text{ Or, } x \in I.$$

It means that $N(I)D \subset I$.

Using the notation 'divides' for ideals, we have $I \mid N(I)D$.

Remark 12.7.1 : If P is a proper prime ideal of D, P is a maximal ideal of D, as D is a Dedekind domain. Therefore, D/P, when finite, has $q = p^m$ elements, where p is a prime and $m \geq 1$. (D/P is a finite field).

Lemma 12.7.2 : *Let D be a Dedekind domain. Given a prime ideal P of D and any positive integer m, the quotient rings D/P (a field) and P^m/P^{m+1} have isomorphic additive groups.*

Proof : We choose an element a in $P^m \setminus P^{m+1}$. Let θ be a map given by $\theta : (D, +) \rightarrow (P^m, +)$ with $\theta(x) = ax$ for all $x \in D$. As $a \in P^m \setminus P^{m+1}$, $ax \in P^m$ for any $x \in D$.

If y is an element of P^m, $\theta(y) = ay \in P^{m+1}$. So $\theta(P) \subset P^{m+1}$. So, θ induces a homomorphism $\bar{\theta} : (D/P, \oplus) \rightarrow (P^m/P^{m+1}, \oplus)$.

If $\bar{x} \in \ker \bar{\theta}$ and x belongs to the coset $\bar{x}$ of P in D, $ax \in P^{m+1}$.

That is, $x \in P$. So $\bar{x} = 0$ and so, $\bar{\theta}$ is an isomorphism. It is also onto. For, if y belongs to the coset $\bar{y}$ of P in D and if we take $\bar{y} \in P^m/P^{m+1}$, taking $b \in aD + P^m$, we see that $b \in P^{m+1}$ and $b = ax + y$ for some $x \in D$. So, the element $\bar{x} \in D/P$ has the property:

$$\theta(\bar{x}) = \bar{y} \in P^m/P^{m+1}.$$

So, $\bar{\theta}$ is a surjective homomorphism. That is, D/P and P^m/P^{m+1} have isomorphic additive groups.   □

Lemma 12.7.3 : *Let D be a Dedekind domain. Any nonzero ideal of D is divisible by only finitely many ideals.*

For, if P is a prime ideal which contains a given ideal I of D, then $P|I$. Suppose that for an integer m, $P^m|I$, but $P^{m+1} \nmid I$. Then, the number of ideals which serve as divisors of I will be finite, as I is expressible uniquely as a finite product of prime ideals.

Theorem 105 : *Let D be a Dedekind domain such that for any proper ideal I (of D) the quotient ring D/I is finite. Writing $N(I) = |D/I|$, the number of ideals J*

such that $N(J) \leq r$ (a given positive integer) is finite. Further, for ideals I, J in D with norms $N(I)$ and $N(J)$ respectively,

$$(12.7.2) \qquad N(IJ) = N(I)N(J).$$

Proof : We assume that $a_i \in D$, $i = 1, 2, \ldots, k$ and $k > (r+1)$.

Let $S = \{a_1, a_2, \ldots a_k\}$, $a_i \neq a_j$ for $i \neq j$.

S has k distinct elements belonging to D. For an ideal J with $N(J) \leq r$, we can select two elements a_s, a_t from S such that $a_s - a_t \in J$.

This is expressed by writing the congruence $a_s \equiv a_t \pmod{J}$. The set of elements of the form $a_s - a_t$ is finite as, $|D/J|$ is finite and $\leq r$. If $a_s - a_t \in I$ (say), then, $J \subset I$ or $I | J$. As the number of ideals of D dividing J is finite (by lemma 12.7.3) and since the set of elements $a_s - a_t$ is finite, there are only a finite number of ideals J for which $N(J) \leq r$.

For the second part, we choose a prime ideal P of D. It is easy to check that for additive groups D/P, D/P^2 and P/P^2

$$\frac{D/P^2}{D/P} \cong P/P^2,$$

or,

$$(12.7.3) \qquad |D/P| |P/P^2| = |D/P^2|$$

That is, $N(P)^2 = N(P^2)$, by lemma 12.7.2.

Further,

$$\frac{D/P^{m+1}}{D/P^m} \cong P^m/P^{m+1}$$

So, $N(P^{m+1}) = N(P^m)N(P)$, as $D/P \cong P^m/P^{m+1}$.

So, by induction on m, $N(P^m) = (N(P))^m$, $m \geq 1$.

By theorem 101, as D is a Dedekind domain, any nonzero ideal I of D has the unique representation

$$(12.7.4) \qquad I = P_1^{a(P_1)} P_2^{a(P_2)} \cdots P_k^{a(P_k)}$$

where P_i $(i = 1, 2, \ldots k)$ are distinct prime ideals and $a(p_i) \in \mathbb{N}$ $(i = 1, 2, \ldots k)$. So, it follows from (12.7.4) that $N(IJ) = N(I)N(J)$, as described. □

Lemma 12.7.4 (Fermat's little theorem) : *Let D be a Dedekind domain in which D/I is finite for any nonzero ideal I of D. If P is a nonzero prime ideal of D, for all $x \in D$,*

$$(12.7.5) \qquad x^{N(P)} \equiv x \pmod{P}$$

where $N(P) = |D/P|$ and $N(P)$ is the least exponent for which (12.7.5) holds.

Proof : If $x \in P$, (12.7.5) holds trivially. Suppose that $x \notin P$. Since D/P is a finite field, $\bar{x} = x + P$ belongs to a cyclic group of order $N(P) - 1$ and so,

$$(\bar{x})^{N(P)-1} = [1_D], \quad \text{the congruence class of } 1_D.$$

Or, $x^{N(P)-1} \equiv 1_D \pmod{P}$ and so (12.7.5) holds for all $x \in D$. Further, $N(P)$ is the least positive integer satisfying (12.7.5), as the multiplicative group $(D^*/P, \otimes)$, where $D^*/P = D/P \setminus [0_D]$ has order $N(P) - 1$.   □

Theorem 106 (Euler ϕ-function for a class of Dedekind domains) : *Let D be a Dedekind domain in which D/I is finite for any nonzero ideal I of D. We write $N(I) = |D/I|$. If $\phi(I)$ denotes the number of invertible elements in D/I,*

$$(12.7.6) \qquad \phi(I) = N(I) \prod_P (1 - 1/N(P))$$

where $\prod_P$ means that the product is over all the prime ideals (of D) dividing I. Moreover, if $x \in D$ and the greatest common divisor of xD and I is D, then

$$(12.7.7) \qquad x^{\phi(I)} \equiv 1_D \pmod{I}.$$

Proof : We note that if $I = \prod_P P^{a(P)}$ is the factorization of I into a product of prime ideals, it is clear that

$$(12.7.8) \qquad D/I \cong \oplus \sum D/P^{a(P)}.$$

((12.7.8) is obtained along the lines of proof of exercise 10 in chapter 5). Therefore, it suffices to prove the theorem for $I = P^m$, $m \geq 1$ where P is a prime ideal. When $m = 1$, it is the case of the number of nonzero elements of D/P. That is,

$$\phi(P) = N(P) - 1.$$

For $m > 1$, $\phi(P^m) = $ # invertible elements of D/P^m or

$$\phi(P^m) = N(P^m) - N(P^{m-1}) = N(P)^m \{1 - 1/N(P)\}.$$

So, from (12.7.8), (12.7.6) follows.

The second part of the theorem gives the analogue of Eulers' theorem. Given that x belongs to a unit in the residue class ring $(D/I, \oplus, \otimes)$, (12.7.7) follows naturally, as $[1_D]$ is the multiplicative identity.   □

Remark 12.7.2 : Theorem 106 has been adapted from W. Narkeiwicz [13].

Next, we go to the analogue of the Chinese Remainder Theorem in the context of a Dedekind domain. Let D be a Dedekind domain. Every proper nonzero ideal of D is a product of prime ideals. If J is a fractional ideal of D, we can choose a nonzero element b from D such that $bJ \subset D$. bJ is an ideal in D. So,

$$(12.7.9) \qquad J = (bD)^{-1} bJ.$$

Therefore, we can express J as a product of prime ideals P with exponents $a(P)$ (say) which may be a positive or a negative integer. That is, the prime ideals of D generate the abelian group of all fractional ideals of D with identity D. Further, a nonzero ideal of D is divisible by only a finite number of prime ideals of D. The idea is that any fractional ideal J of D has a unique representation

$$(12.7.10) \qquad J = \prod_P P^{a(P)}$$

where the product extends over all prime ideals of D where $a(P)$ are integers (positive or negative or zero) and only a finite number of them is nonzero.

Definition 12.7.3 : *Let I, J be ideals of a Dedekind domain D. If*

$$I = \prod_P P^{a(P)}, \qquad J = \prod_P P^{b(P)};$$

we say that I divides J if, and only if, $a(P) \leq b(P)$ for distinct primes P occurring in the factorizations of I and J.

It follows that g.c.d of I and J written g.c.d (I, J) is given by

$$(12.7.11) \qquad \gcd(I, J) = \prod_P P^{c(P)} \text{ where } c(P) = \min\{a(P), b(P)\}.$$

The l.c.m. of I and J written $[I, J]$ is given by

$$(12.7.12) \qquad [I, J] = \prod_P P^{d(P)} \text{ where } d(P) = \max\{a(P), b(P)\}.$$

As in elementary number theory, one has

$$(12.7.13) \qquad \gcd(I, J)[I, J] = IJ.$$

It can be checked that g.c.d $(I, J) = I + J$. I and J are said to be relatively prime, if g.c.d $(I, J) = $ the ideal generated by 1_D that is, D. I and J are relatively prime to one another. So, I and J are relatively prime to one another if, and only if, $I + J = D$. We know that two such ideals are comaximal (when I, J are proper ideals of D).

We can discuss solutions of linear congruences the way it is done in elementary number theory. We consider linear congruences modulo an ideal I in a Dedekind domain.

Lemma 12.7.5 : *Let D be a Dedekind domain and let I be a proper nonzero ideal of D. For $a, b \in D$, the congruence*

$$ax \equiv b \pmod{I}$$

has a solution in D if, and only if, b belongs to the ideal $I + aD$.

Proof : $\Leftarrow$: If $b \in I + aD$, b can be written as

$$b = y + at$$

where $y \in I$, $t \in D$. So, $b - at \in I$. So, t is such that $at \equiv b \pmod{I}$. So, t is a solution of $ax \equiv b \pmod{I}$.

$:\Rightarrow$ If $ax \equiv b \pmod{I}$ has a solution, say, $t \in D$, then for suitable $y \in I$, $b = at + y \in aD + I$. So, b belongs to the ideal $aD + I$. $\square$

Corollary 12.7.2 : *If P is a prime ideal of a Dedekind domain D and $a \in D \setminus P$, then, the congruence*

$$ax \equiv b \pmod{P^n}$$

has a solution in D for any $b \in D$ and $n \geq 1$.

For, we observe that $a \notin P^n$ and so, $P^n + aD = D$ yields the result, as P is a maximal ideal in D.

Corollary 12.7.3 : *If $P_1, P_2, \ldots P_m$ are distinct prime ideals in a Dedekind domain D, then, for $a_1, a_2, \ldots a_m \in D$ and $n \geq 1$, there exists a common solution to the system of congruences*

$$x \equiv a_1 \,(\mathrm{mod}\ P_1^n),$$

$$x \equiv a_2 (\mathrm{mod}\ P_2^n),$$

$$\ldots\ldots\ldots\ldots$$

$$\ldots\ldots\ldots\ldots$$

$$x \equiv a_m (\mathrm{mod}\ P_m^n).$$

Proof : We choose $b_i \in J_i$ where $J_i = (P_1 \cdot P_2 \cdot \cdots \cdot P_{i-1} \cdot P_{i+1} \cdot \cdots \cdot P_m)^n \setminus P_i$, $(i = 1, 2, \ldots m)$. Let t_i be a solution of

$$b_i x \equiv a_i \,(\mathrm{mod}\ P_i^n).$$

t_i exists, as $a_i \in b_i D + P_i^n$ $(i = 1, 2, \ldots m)$. Then, $t = b_1 t_1 + b_2 t_2 + \cdots + b_m t_m$ has the desired property.   □

Theorem 107 (The Chinese Remainder Theorem) : *Let $I_1, I_2 \ldots I_m$ be pairwise comaximal ideals of a Dedekind domain D. If $a_1, a_2 \ldots a_m$ are elements of D, then, there exists a common solution to the system of congruences:*

$$x \equiv a_1 (\mathrm{mod}\ I_1),$$

$$x \equiv a_2 (\mathrm{mod}\ I_2),$$

$$\ldots\ldots\ldots\ldots$$

$$\ldots\ldots\ldots\ldots$$

$$x \equiv a_m (\mathrm{mod}\ I_m).$$

Proof : We observe that any congruence of the form

$$x \equiv a \,(\mathrm{mod}\ I)$$

is equivalent to a system of congruences

$$x = a_i (\mathrm{mod}\ P_i^{b_i})\, i = 1, 2, \ldots t$$

where $I = \prod_{i=1}^{t} P_i^{b_i}$, $b_i \geq 1\,(i = 1, 2, \ldots t)$.

The corollary 12.7.3 of lemma 12.7.5 gives a common solution to given congruences, as the m congruences $x \equiv a_i$ (mod I_i) $(i = 1, 2, \ldots, m)$ give rise to km $(k \geq 1)$ congruences in terms of the prime-power moduli. So, they have a common solution which is unique modulo $\prod P^{a(P)}$, the l.c.m of $I_1, I_2, \ldots, I_m$.   □

As in the case of rings (see theorem 32, chapter 5), another version of the Chinese Remainder Theorem for Dedekind domains is the following

Theorem 108 (Frölich and Taylor) : *Let P_i $(i = 1, \ldots n)$ denote distinct prime ideals of a Dedekind domain D. Suppose that a_i $(i = 1, 2, \ldots n)$ denote positive integers. Let $\psi : D \to \prod_{i=1}^n D/P_i^{a_i}$ be given by*

$$\psi(x) = (x + P_1^{a_1}, x + P_2^{a_2} \ldots, x + P_n^{a_n}); \quad x \in D.$$

Then, ψ is a surjective homomorphism and $D/\prod_{i=1}^n P_i^{a_i}$ and $\prod_{i=1}^n D/P_i^{a_i}$ are isomorphic.

Proof : It is clear that ψ is a homomorphism.

$$\ker \psi = P_1^{a_1} \cap P_2^{a_2} \cap \ldots P_n^{a_n}$$

Now, $P_n^{a_n}$ and $\prod_{i=1}^{n-1} P_i^{a_i}$ are relatively prime to one another and when I and J are comaximal, $I + J = D$ and $I \cap J = IJ$.

So,

$$(\prod_{i=1}^{n-1} P_1^{a_1}) \cap P_n^{a_n} = \prod_{i=1}^n P_i^{a_i}$$

Therefore, $\ker \psi = P_1^{a_1} P_2^{a_2} \cdots P_n^{a_n}$.

To show that ψ is surjective, we proceed as follows:

Let x_i $(i = 1, 2, \ldots n)$ be specified elements of D. For each k, $1 \leq k \leq n$, the ideals $P_k^{a_k}$ and $J_k = \prod_{\substack{i=1 \\ i \neq k}}^n P_i^{a_i}$ are relatively prime to one another.

So, we can find $r_k \in J_k$; $s_k \in P_k^{a_k}$ such that $r_k + s_k = 1_D$. Then, $x_k \equiv x_k r_k \in P_i^{a_i}$, $i \neq k$. However, $x_k s_k = x_k(1_D - r_k) \in P_k^{a_k}$. Therefore,

$$x_k \equiv x_k r_k \, (\text{mod } P_k^{a_k})$$

Therefore, $x = x_1 r_1 + x_2 r_2 + \cdots + x_k r_k$ is such that

$$x \equiv x_k (\text{mod } P_k^{a_k}) \text{ for each } k.$$

So, ψ is surjective. By the fundamental homomorphism theorem, the desired isomorphism of rings $D/\Pi P_i^{a_i}$ and $\Pi_{i=1}^n D/P_i^{a_i}$ is established. $\square$

Remark 12.7.3 : What we have shown is that given $x_1, x_2 \ldots, x_n \in D$ there exists $x \in D$ such that

$$x \equiv x_1 (\text{mod } P_1^{a_1}),$$

$$\ldots \ldots \ldots \ldots$$

$$\ldots \ldots \ldots \ldots$$

$$x \equiv x_n (\text{mod } P_n^{a_n})$$

and the common solution determines the class of $x (\text{mod } \prod_{i=1}^n P_i^{a_i})$ uniquely.

Example 12.7.1 : Let $I = P_1^2 P_2 P_3$ $(P_1, P_2, P_3$ distinct prime ideals)

Let $x_1 \in P_1 \setminus P_1^2$, $x_2 = 1_D$, $x_3 = 1_D$.

Then, we can find $y_1 \in D$ such that $y_1 \equiv x_1 \pmod{P_1^2}$, $y_1 \notin P_2$, $y_1 \notin P_3$. Similarly, we can find $y_2 \in P_2 \setminus P_2^2$, $y_2 \notin P_1$, $y_2 \notin P_3$ and $y_3 \in P_3 \setminus P_3^2$, $y_3 \notin P_1$, $y_3 \notin P_2$.

For the system of congruences :

$$x \equiv x_1 \pmod{P_1^2}; \quad x \equiv x_2 \pmod{P_2}; \quad x \equiv x_3 \pmod{P_3}$$

a unique solution is $x \equiv y_1 \pmod{I}$.

For the system of congruences :

$$x \equiv x_1 \pmod{P_1} \quad x \equiv x_2 \pmod{P_2^2} \quad x \equiv x_3 \pmod{P_3}$$

a unique solution is $x \equiv y_2 \pmod{J}$ where $J = P_1 P_2^2 P_3$.

For the system of congruences :

$$x \equiv x_1 \pmod{P_1} \quad x \equiv x_2 \pmod{P_2} \quad x \equiv x_3 \pmod{P_3^2}$$

one has $x \equiv y_3 \pmod{J'}$ where $J' = P_1 P_2 P_3^2$.

When $\{P_1, P_2, P_3\}$ is the only set of prime ideals of D, P_1 will contain only multiples of y_1 or P_1 is the principal ideal (y_1). Similarly, P_2 is the principal ideal (y_2) and P_3 is the principal ideal (y_3).

Remark 12.7.4 : Example 12.7.1 has been adapted from [8]. We have a generalization: A Dedekind domain D having only a finite number of prime ideals reduces to a PID. See P.M. Cohn [7].

For more properties of Dedekind domains and applications, see A. Frölich and M. J. Taylor [8], W. Narkeiwicz [13], P. Samuel [15].

12.8. Integral domains having finite norm property

Ideals of rings have played a major role in the development of algebraic number theory. If K denotes an algebraic number field, we consider the ring R of integers corresponding to K. Every proper ideal of R is either a prime ideal or a unique product of prime ideals (except order). Further, the residue class ring of R modulo an ideal I is finite for every proper ideal I of R. The fact that R is a Dedekind domain is exploited to know more about R. Using the unique factorization theorem (see theorem 101), it is shown that the norm of I denoted by $N(I)$ is multiplicative (see theorem 105). We wish to establish that the multiplicative property of the norm implies the unique (prime ideal) factorization theorem in an integral domain with unity in which every proper ideal has finite norm. Definition 12.5.1 is recast as follows:

Definition 12.8.1 : *Let D be an integral domain with unity. An ideal I of D is called irreducible if I is proper and $I = AB$ (where A, B are ideals of D) implies that either $A = D$ or $B = D$. In other words, A or B is the unit ideal.*

In a general integral domain, an irreducible ideal may not be a prime ideal and a prime ideal may not be an irreducible. (However, a 'prime' element in an integral domain is irreducible).

Definition 12.8.2 : *An integral domain D with unity is said to have the finite norm property, if for every proper ideal I of D, the quotient ring D/I is finite. When D/I is finite, we denote the number of elements of D/I by $N(I)$. $N(I)$ is called the norm of I.*

Examples 12.8.1 : Dedekind domains having finite norm property [5] are

i The ring D_K of algebraic integers of a number field K.

ii The ring $\mathbb{F}_q[x]$ of polynomials over a finite field $\mathbb{F}_q$.

iii Let p be a prime. We denote by $\mathbb{Z}_p$ (see definition 8.3.1), the set of rational numbers of the form $\frac{a}{b}$ where a and b are integers with $p \nmid b$. By Theorem 58, $\mathbb{Z}_p$ is a PID contained in $\mathbb{Q}$. By Corollary 8.3.1, $\mathbb{Z}_p$ is a quasilocal ring having a unique maximal ideal P which is generated by p. Every proper ideal of $\mathbb{Z}_p$ is a power of P. $\mathbb{Z}_p$ is Noetherian and is integrally closed. $\mathbb{Z}_p$ is a Dedekind domain having the finite norm property. $\mathbb{Z}_p$ is a special primary ring (see definition 12.10.1).

Lemma 12.8.1 : *If I and J are proper ideals of the D (having finite norm property) such that $I \subset J$, then, there exists a positive integer k such that $N(I) = kN(J)$. Further, I is a proper subset of J if, and only if, $k \geq 2$.*

Proof : Since $(I, +)$ is a subgroup of $(J, +)$, there exist elements $t_i \in J$ $(i = 1, 2, \ldots, k)$ such that the cosets $I + t_i$ are disjoint for $i = 1, 2, \ldots, k$. Also,

$$(12.8.1) \qquad J = \cup_{i=1}^{k} \{I + t_i\}$$

It is clear that if $J = I$, then, $k = 1$ and if $J \neq I$, then, $k > 1$. If

$$(12.8.2) \qquad D = \cup_{j=1}^{n} \{J + s_j\}$$

is the (finite) coset decomposition of D modulo J, then

$$(12.8.3) \qquad D = \cup_{\substack{1 \leq j \leq n \\ 1 \leq i \leq k}} \{I + t_i + s_j\}$$

is the coset decomposition of D modulo I. It follows that $N(I) = kN(J)$. $\square$

Lemma 12.8.2 : *If every proper ideal of D has finite norm, then every proper ideal is finitely generated.*

Proof : Let I be a proper ideal of D and $a_1 \neq 0_D$ be an element of I. If $I = (a_1)$, then I is finitely generated. If $I \neq (a_1)$, let $a_2 \in I$ and $a_2 \notin (a_1)$. If $I = (a_1, a_2)$ then, I is finitely generated. If $I \neq (a_1, a_2)$, then, there exists $a_3 \in I$ such $a_3 \notin (a_1, a_2)$. If this process is continued, we obtain a chain of ideals

$$(12.8.4) \qquad (a_1) \subset (a_1 a_2) \subset (a_1, a_2, a_3) \subset \ldots$$

By lemma 12.8.1, we have

$$N(a_1) > N((a_1, a_2)) > N((a_1, a_2, a_3)) > \ldots$$

Now, the norm of a proper ideal is a positive integer and so the chain (12.8.4) of ideals terminates in a finite number of steps. Therefore, we arrive at

$$I = (a_1, a_2, \ldots a_k).$$

Hence I is finitely generated. $\square$

Theorem 109 (H. S. Butts and L. I. Wade (1966)) : *Let D be an integral domain having finite norm property. Assume that $N(IJ) = N(I)N(J)$ for every pair of ideals I and J in D. If I, J are ideals in D such that $I \subset J$, there exists an ideal A in D such that $I = JA$.*

Proof : *Case (i)* $J = I + (t), t \notin I$.

If, either J or $J + (t)$ is not a proper ideal or if $t \in J$, then it is clear that there is an ideal A such that $J = IA$. Suppose that $I = J + (t)$ and J are proper ideals such that $t \notin I$ and

$$A = \{x \in D : xt \in J\}$$

Then, A is an ideal of D and $IA \subset J$. In order to show that $J = IA$, it is enough if we prove that $N(IA) = N(J)$, using lemma 12.8.1.

Since $IA \subset J$, by lemma 12.8.1, $N(IA) = N(I)N(A) \geq N(J)$. To show that $N(IA) = N(J)$, we make a

Claim : If $\frac{N(I)}{N(J)} = k$, a positive integer, $N(A) \leq k$.

Let $J = \cup_{i=1}^{k} \{I + t_i\}$ be a coset decomposition of J modulo I. Suppose that $D = \cup_{i=1}^{l} \{A + s_i\}$ be a coset decomposition of D modulo A. We write

$$T = \{I + t_1, I + t_2, \ldots, I + t_k\}$$
$$S = \{A + s_1, A + s_2, \ldots, A + s_l\}$$

Suppose that $J = I + (t)$. Let $\Psi : T \to S$ be defined by $\Psi(I + t_i) = A + s_j$, where t_{s_j} occurs in $I + t_i$. If $A + s_m = A + s_n$, $ts_m - ts_n \in I$. So, $t(s_m - s_n) \in I$. As $I \subset J$, $t(s_m - s_n) \in J$. It follows that ts_m and ts_n belong to the same coset of J modulo I. So, Ψ is independent of the coset representative. Since each element of J belongs to one and only one coset of I, Ψ is single-valued. Suppose that $A + s_m$ and $A + s_n$ come from the same coset $I + t_q$ ($1 \leq q \leq k$). Then, $t(s_m - s_n) \in I$ and so, $s_m - s_n \in A$ or $A + s_m = A + s_n$. So, Ψ is one-one. That is, $N(A) = l \leq k$. Further, when $J = I + (t)$, $J = IA$.

Case (ii) $J \neq I + (t)$ where $t \notin I$.

Since $I \subset J$, by lemma 12.8.2, J is finitely generated and consequently, there exist elements $t_1, t_2, \ldots t_h$ such that

$$I \subset I + (t_1) \subset I + (t_1) + (t_2) \subset \ldots \subset I + (t_1) + (t_2) \ldots + (t_h) = J$$

I is finitely generated and so, J is expressible as

(12.8.5) $J = I + (t_1) + (t_2) \ldots + (t_h)$; $t_i \in D$, $(i = 1, 2, \ldots, h)$.

As in the argument used for case (i), we obtain an ideal A_1 such that $I = (I + (t_1))A_1$. Similarly, there is an ideal A_2 such that $I + t_1 = (I + (t_1) + (t_2))A_2$. So, $I = (I + (t_1) + (t_2))A_1 A_2$. By induction, there is an ideal A such that $I = JA$. $\square$

Next, we observe that the definition (2.4.3) of a prime ideal P (see chapter 2) may be recast in the following manner:

If D is an integral domain and A, B are ideals of D such that $AB \subset P \Rightarrow$ either $A \subset P$ or $B \subset P$, then, P is a prime ideal of D. Further, if D is an integral domain with unity 1_D, the following cancellation property holds:

If $0_D \neq a \in D$ and B, C are ideals of D,

$$(12.8.6) \qquad (a)B = (a)C \Rightarrow B = C.$$

For, if $b \in B$, then $ab \in (a)B = (a)C$. Therefore

$$ab = \sum_{i=1}^{n} (r_i a) c_i = \sum_{i=1}^{n} a(r_i c_i) = \sum_{i=1}^{n} a d_i$$

where $r_i \in D$, $c_i \in C$, $d_i \in C$; $i = 1, 2, \ldots, n$.

So,

$$ab = a \sum_{i=1}^{n} d_i \text{ or, } b = \sum_{i=1}^{n} d_i \in C$$

Thus, $B \subseteq C$. Similarly, $C \subseteq B$ or $B = C$.

Remark 12.8.1 : In the case of a Dedekind domain D, we have shown in Corollary 12.6.3 that given nonzero ideals L, M, N of D, $LM = LN \Rightarrow M = N$.

Theorem 110 (H. S. Butts and L.I. Wade (1966)) **:** *Let D be an integral domain having finite norm property. If the norm of a proper ideal of D is multiplicative, then every proper ideal of D is either a prime ideal or a unique product of prime ideals, except for order.*

Proof : Let I be a proper ideal of D. Let I be irreducible (or nonfactorable). By theorem 109, if $I \subset J$, there exists an ideal A such that $I = JA$. It follows that either J or A is D. If J is proper, $A = D$. So $I = J$. That is, I is a maximal ideal of D. So, I is a prime ideal.

If I is reducible, then, $I = I_1 I_2$ where I_1 and I_2 are proper ideals. So, $N(I) = N(I_1)N(I_2)$. Since the norm of a proper ideal ≥ 2, it follows that I can be expressed as

$$(12.8.7) \qquad I = I_1 I_2 \ldots I_m$$

where $I_j (j = 1, 2, \ldots m)$ is a proper irreducible (nonfactorable) ideal.

Since proper irreducible ideals are maximal, it follows that I is a (finite) product of maximal ideals and hence a product of prime ideals.

UNIQUENESS: Suppose that I admits two different factorizations in the form

$$(12.8.8) \qquad I = I_1 I_2 \ldots I_m = J_1 J_2 \ldots J_n$$

where $I_j(j = 1, 2, \ldots m)$ are maximal ideals and $J_i(i = 1, 2, \ldots n)$ are proper prime ideals. By the property of a prime ideal, $I_j \subset J_i$ for some j, say $j = 1$. Then, $I_1 = J_i$. Let $x(\neq 0_D)$ be an element of I_1. Since $(x) \subset I_1$, there exists an ideal Q such that $(x) = I_1 Q$ (by theorem 109). We get

$$QI_1 I_2 \ldots I_m = QJ_1 \ldots J_{i-1} I_1 J_{i+1} \ldots J_n$$

or

$$(x)I_2 \ldots I_m = (x)J_1 J_2 \ldots J_{i-1} J_{i+1} \ldots J_n$$

By the cancellation property (12.8.6)

$$I_2 \ldots I_m = J_1 J_2 \ldots J_{i-1} J_{i+1} \ldots J_n$$

By the standard argument used in such situations, we get $m = n$ and any two factorizations differ only in the order of the factors which are prime ideals.   □

Remark 12.8.2 : The converse of theorem 110 holds for integral domains (with unity element 1_D) having finite norm property. However, an integral domain (with unity element 1_D) in which unique prime ideal factorization theorem holds, need not have the finite norm property. For example, the ring $F[x]$ of polynomials over an infinite field F (of characteristic zero or not) is a Dedekind domain which does not have the finite norm property, since the quotient ring $F[x]/(x)$ is isomorphic to F. We add that one can construct an infinite field of characteristic p (a prime), by considering the ring $\mathbb{F}_q[x]$ where $\mathbb{F}_q$ is a finite field with $q = p^m$ (p a prime). $\mathbb{F}_q[x]$ is an integral domain. Its field of quotients $\mathbb{F}_q(x)$ is an infinite field having characteristic p.

Definition 12.8.3 : *Let D be an integral domain with unity element 1_D. Suppose that I denotes a proper ideal of D. We denote by S, the set of all positive integers n obtained by taking the lengths of all possible finite chains of ideals*

(12.8.9) $I \subset I_1 \subset I_2 \ldots \subset I_n = D, \quad I \neq I_1 \text{ and } I_j \neq I_{j+1}$

($j = 1, 2, \ldots, (n-1)$). I is said to be of finite length, if S is finite and in this case, the largest integer in S is called the length of I, denoted by $L(I)$.

Theorem 111 (H. S. Butts and L. I. Wade (1966)) : *Let D be an integral domain. Suppose that every proper ideal I of D is of finite length, say, $L(I)$. If $L(AB) = L(A) + L(B)$ for proper ideals A, B of D and if, for ideals I, J of D, one has $I \subset J$, then, there exists an ideal A (of D) such that $I = JA$.*

Proof : As in the proof of theorem 109, it is enough if we prove the theorem for the case where $J = I + (t), t \notin I$. We define the set A by

$$A = \{x \in D : xt \in I\}$$

A is an ideal of D and $JA \subset I$. So,

$$L(JA) = L(J) + L(A) \geq L(I).$$

To show that $JA = I$, it suffices to prove that

(12.8.10) $L(I) \geq L(JA) = L(J) + L(A)$.

We consider a strictly ascending chain of ideals from A to D as follows:

(12.8.11) $A \subset A + (t_1) \subset A + (t_1) + (t_2) \subset \ldots \subset A + (t_1) + \ldots + (t_n) = D$.

Since $J = I + (t)$, we take

(12.8.12) $A \subset A + (tt_1) \subset A + (tt_1) + (tt_2) \subset A + (tt_1) + \ldots + (tt_n)$

(12.8.12) is a strictly increasing chain of ideals each contained in J. For, if $I = I + (tt_i), i = 1, 2, \ldots n$, then $tt_i \in I$. By definition $t_i \in A$. Then, $A = A + (t_i)$ and then, (12.8.12) is not a strictly increasing chain. So, from (12.8.12), it follows that by induction, $A + (tt_1) + \ldots + (tt_n) \subset J$. Hence, (12.8.10) holds and $I = JA$. $\square$

Lemma 12.8.3 : *Let D be an integral domain and $a \in D$. If J is an ideal in D such that $J \subset (a)$, then, there exists an ideal Q in D such that $(a)Q = J$.*

Proof : Let $Q = \{x \in D : ax \in J\}$. It is an ideal of D and $(a)Q = J$. $\square$

Lemma 12.8.4 : *Let D be an integral domain.*

(1) *Suppose that every proper ideal J in D is either maximal or a product of maximal ideals. If M is a maximal ideal in D, then, there exists an element $a \in D$ and an ideal $Q \neq (0)$ (in D) such that $MQ = (a)$.*

(2) *Suppose that every proper ideal in D is either maximal or a unique product of maximal ideals (except for order). If I and J are proper ideals of D such that $I \subset J$, then there exists an ideal A in D such that $I = JA$.*

Proof : If $M = 0_D$, we take $Q = D$ and $a = 0_D$. If $M \neq (0_D)$, let $0_D \neq a$ be an element of M. There exist maximal ideals $M_1, M_2, \ldots M_n$ in D such that

(12.8.13) $(a) = M_1 M_2 \ldots M_n$

since $(a) \subset M$, it follows that as M is a prime ideal, $M = M_i$ for some i, say $i = 1$. Then,

$(a) = MM_2 \ldots M_n$ and we have only to take $M_2 M_3 \ldots M_n = Q$

This proves the first part of lemma 12.8.4.

By the result in (1), there exists an ideal $Q \neq (0)$ in D and an element $t \in D$ such that $(t) = JQ$ and $IQ \subset JQ = (t)$. So, by lemma 12.8.3, there exists an ideal A in D such that $(t)A = IQ$. Therefore, $(t)A = JQA = IQ$. We do the factorizations into products of maximal ideals for I, J, A and Q and by using the fact that the factorization is unique except for order, we obtain $I = JA$.

This proves the second part of lemma 12.8.4. $\square$

Remark 12.8.3 : By theorem 101, we have seen that a Dedekind domain D has the property: Every proper ideal I of D is either a prime ideal or a unique product of prime ideals except for order. Lemma 12.8.4 shows that if D is a Dedekind domain, for proper ideals I, J of D such that $I \subset J$, there exists an ideal A of D such that $I = JA$. This condition is also sufficient to make D a Dedekind domain. See W. Krull [11]. See also H. S. Butts and L. I. Wade [5]. So, then, theorem 111 says that an integral domain satisfying the conditions stated therein makes it a Dedekind domain. The following theorem gives yet another sufficient condition for an integral domain D (with unity element 1_D) to be a Dedekind domain.

Theorem 112 (H. S. Butts (1964)) : *If D is an integral domain with unity element 1_D such that every proper ideal of D is either an irreducible ideal or can be factorized uniquely into a product of irreducible ideals of D, then irreducible ideals are prime ideals.*

Proof : Let J be a proper prime ideal of D. Suppose that $0_D \neq a$ is an element of J. There exist irreducible ideals $P_1, P_2, \ldots P_n$ such that

$$(12.8.14) \qquad (a) = P_1 P_2 \ldots P_n.$$

We observe that (a) is a fractional ideal and so each $P_i (i = 1, 2, \ldots n)$ is a fractional ideal. (See Remark 12.6.4 (ii)). Also, by the application of lemma 12.8.2, each P_i is finitely generated.

Let $x \in D$ and $x \notin P_i$ (i arbitrary, but fixed). Since P_i is finitely generated, $P_i + (x)$ is finitely generated. As uniqueness of factorization of an ideal into irreducible ideals holds in D, cancellation law for ideals holds in D. So, finitely generated ideals of D are irreducible. See Heinz Prüfer (1896–1934) [14]. Therefore, $P_i + (x)$ is invertible. By hypothesis, every ideal of D has finite norm. So, by theorem 109, as $P_i \subset P_i + (x)$, there exists an ideal A in D such that $P_i = (P_i + (x))A$. Now, P_i is an irreducible ideal. So, $P_i + (x) = D$. It implies that P_i is a maximal ideal of D. We began with a proper prime ideal J of D. As $P_1 P_2 \ldots P_n \subseteq J$, it follows that $J = P_i$ for some i. So, J is invertible. A converse of theorem 104 is due to I. S. Cohen [6]. It says that if every proper prime ideal of D is invertible, then D is a Dedekind domain. (See Exercise 12). Hence, D satisfying the hypothesis of the theorem is a Dedekind domain. Clearly, irreducible ideals are maximal ideals and so prime ideals. $\qquad \square$

Remark 12.8.4 : Theorem 112 has been adapted from H. S. Butts [4].

Theorem 113 (H. S. Butts and L. I. Wade (1966)) : *D denotes an integral domain with unity 1_D. Then, every proper ideal of D is either a maximal ideal or a unique product of maximal ideals if, and only if, every proper ideal A has finite length $L(A)$ and for proper ideals I, J of D,*

$$(12.8.15) \qquad L(IJ) = L(I) + L(J).$$

Proof : $\Leftarrow$: Let I be a proper ideal of D. If I is irreducible, $I = AB$ implies either A or B is the unit ideal and so I is a maximal ideal. If I is reducible, we write $I = I_1 I_2$

where I_1, I_2 are proper ideals such that $L(I) = L(I_1) + L(I_2)$, by (12.8.15). So, since $I_1 I_2 \subseteq I_1 \cap I_2$ and $L(I_1 I_2) = L(I_1) + L(I_2)$, I is a product of irreducible ideals. That is, I is a product of maximal ideals. The uniqueness of factorization of a proper ideal into a product of maximal ideals is shown as in the proof of theorem 108.

$:\Rightarrow$ Let I be a proper ideal of D. We are given that either I is maximal or a unique product of maximal ideals (except for order).

Claim : Every proper ideal is of finite length. Further, if I, J are proper ideals of D, the lengths $L(I)$ and $L(J)$ of I and J respectively, satisfy the relation (12.8.15). Let

(12.8.16) $I = M_1 M_2 \ldots M_n$

and

(12.8.17) $J = P_1 P_2 \ldots P_m,$

where $M_i (i = 1, 2, \ldots n)$ and $P_j (j = 1, 2, \ldots m)$ are maximal ideals and the representations (12.8.16) and (12.8.17) of I and J (respectively) are unique. By appealing to lemma 12.8.4(2), we obtain a strict ascending chain of ideals

(12.8.18) $I \subset M_2 M_3 \ldots M_n \subset M_3 M_4 \ldots M_n \subset \ldots M_{n-1} \subset M_n \subset D$

(where D may be considered as M_{n+1}). By the uniqueness of factorization, the ascending chain in ((12.8.18)) is the longest. So, $L(I) = n$. Similarly, $L(J) = m$. We get through, if we show that $L(IJ) = n + m$.

If a maximal ideal P occurs in I as well as J, one gets a factor P^2 in IJ. Also, there is no ideal between P^2 and P when P is a maximal ideal. See I.S. Cohen [6]. Noting that $L(P^2) = 2$, we could calculate $L(IJ)$. Suppose that q of the maximal ideals occurring in ((12.8.16)) and ((12.8.17)) are common. Then

$$L(IJ) = (n - q) + 2q + (m - q) = n + m.$$

This completes the proof of theorem 113. $\square$

Remark 12.8.5 : Theorems 109–111 and 113 have been adapted from H. S. Butts and L. I. Wade [5].

We, now, give an example of an integral domain which has finite norm property and which is not a Dedekind domain. Let $\mathbb{F}_q[x]$ be the ring of polynomials with coefficients from a finite field $\mathbb{F}_q$ where $q = p^m$ (p a prime, $m \geq 1$). We define

(12.8.19) $D = \{a_0 + a_2 x^2 + \ldots + a_n x^n : a_0, a_2, \ldots, a_n \in \mathbb{F}_q, n \geq 2\}.$

D consists of polynomials chosen from $\mathbb{F}_q[x]$ such that the x-coefficient in each such polynomial is zero. D is an integral domain. In fact, D is a subring of $\mathbb{F}_q[x]$. As $\mathbb{F}_q$ is finite, proper ideals of D are of finite length. They have also finite norm. Norm of a proper ideal is not multiplicative. Every ideal in D is generated by either one or two elements. Any proper prime ideal is maximal. Every proper ideal is not a unique product of prime ideals. See I.S. Cohen [6]. For proper ideals I, J of D, it is not true that $L(IJ) = L(I) + L(J)$, in general. By theorem 112, D is not a Dedekind domain. It is also obvious, otherwise.

For supplementary reading in rings and modules one may refer to A. J. Berrick and M. E. Keating [A1].

12.9. Notes with illustrative examples

As $\mathbb{Z}$ is a Noetherian domain, so is $\mathbb{Z}[x]$, by Hilbert basis theorem (see (a) in Section 12.10). If $n\mathbb{Z}$ denotes the ideal generated by n (≥ 2), $\mathbb{Z}/n\mathbb{Z}$ is a finite ring and so satisfies both a.c.c and d.c.c. In general, if I denotes a nonzero proper ideal of a PID, say R, the quotient ring R/I satisfies both chain conditions.

In the case of $\mathbb{Z}[x]$, the ideal

$$(x) = \left\{ a_1 x + a_2 x^2 + \ldots a_n x^n : a_k \in \mathbb{Z}, 1 \leq k \leq n \right\}$$

is a principal ideal of $\mathbb{Z}[x]$ generated x. $\mathbb{Z}[x]/(x) \cong \mathbb{Z}$, an integral domain. So (x) is a prime ideal of $\mathbb{Z}[x]$. Let $J = (x, 2)$ the ideal generated by x and 2 in $\mathbb{Z}[x]$. J is a maximal ideal (of $\mathbb{Z}[x]$) containing polynomials of the form

$$a_0 + a_1 x + a_2 x^2 + \cdots + a_n x^n ; \; a_i \in \mathbb{Z}, 0 \leq i \leq n$$

where a_0 is even. As $\mathbb{Z}[x]/J$ is a field having two elements, J is a maximal ideal of $\mathbb{Z}[x]$ containing the prime ideal (x). So, the prime ideal (x) is not a maximal ideal of $\mathbb{Z}[x]$. Thus, $\mathbb{Z}[x]$ is not a Dedekind domain. $\mathbb{Z}[x]$ is not a PID also, as a principal ideal domain has to be a Dedekind domain. This conclusion is also obvious from the fact that J is not a principal ideal of $\mathbb{Z}[x]$. Further, the ideal $Q = (x, 4)$ generated by x and 4 in $\mathbb{Z}[x]$ is a primary ideal of $\mathbb{Z}[x]$. It is not the power of any prime ideal of $\mathbb{Z}[x]$.

In the case of $\mathbb{Z}[x, y]$, the ideals (x), (x, y) and $(2, x, y)$ are prime ideals in $\mathbb{Z}[x, y]$. However, $(2, x, y)$ is the only maximal ideal among these ideals. Further, the ideal (x^k, xy, y^k) ($k \geq 1$) is a primary ideal of $\mathbb{Z}[x, y]$. In the case of $I = (x^2, xy)$, $\sqrt{I}$ is the prime ideal (x). But I is not a primary ideal of $\mathbb{Z}[x, y]$. If F is any field, one can show that $F[x, y]/(x + y) \cong F[x]$. Now, $F[x, y]$ is Noetherian. The ideal $J = (x^2, 2xy)$ has an irredundant primary decomposition:

$$(12.9.1) \qquad J = (x^2, xy, y^2) \cap (x) \cap (x^2, 2x, 4).$$

Next, we have seen that any PID is integrally closed. Let us consider $D = \mathbb{Z}[\sqrt{-3}]$, D is a ring of algebraic integers, not fully corresponding to the number field $Q(\sqrt{-3})$. An imaginary cube-root of unity $\omega = \frac{-1 + \sqrt{-3}}{2}$ is integral over D as it satisfies the monic polynomial $x^2 + x + 1$. As $\omega \notin \mathbb{Z}(\sqrt{-3})$, $\mathbb{Z}(\sqrt{-3})$ is not integrally closed in its field of quotients.

A Dedekind domain D has the property that any nonzero ideal of D is generated, at the most, by two elements. We prove this in the following manner: First we need a

Lemma 12.9.1 : *Let I, J be nonzero ideals of a Dedekind domain D. Then, there exists $a \in I$ such that $aI^{-1} + J = D$.*

Proof : We consider the principal ideal generated by a. As $a \in I$, I divides (a). So, $(a) = II'$ or, $I' = (a)I^{-1} = aI^{-1}$ is an ideal of D. So, $aI^{-1} + J$ is the g.c.d of aI^{-1} and J.

Let $J = P_1 P_2 \cdots P_k$ where P_i $(i = 1, 2, \ldots k)$ is a prime ideal of D. So, it is sufficient to show that

$$(12.9.2) \qquad aI^{-1} + P_i = D \quad (i = 1, 2, \ldots k)$$

Since P_i is a maximal ideal of D, (12.9.2) holds, if $aI^{-1} \neq P_i$. So, we have only to choose $a \in I \setminus IP_i$ for all $i = 1, 2, \ldots k$.

If $k = 1$, the unique factorization of ideals implies that $I \neq IP_1$. So, (12.9.2) holds for $k = 1$. If $k > 1$, let

$$I_i = IP_1 P_2 \cdots P_{i-1} P_{i+1} \cdots P_k$$

By the case $k = 1$, we can choose $a_i \in I_i \setminus I_i P_i$. We define

$$a = a_1 + \cdots + a_k$$

Then, each $a_i \in I_i \subseteq I$. So, $a \in I$.

If, by chance, $a \in IP_j$ and for $j \neq i$, $a_j \in I_j \subset IP_i$, we will get

$$a_i = a - a_1 - \cdots - a_{i-1} - a_{i+1} - \cdots - a_k \in IP_i \ .$$

This contradicts the choice of a_i. So, $a \notin IP_i$ $(i = 1, 2, \ldots k)$ and (12.9.2) is true. So, aI^{-1} and J are comaximal in D. $\qquad\square$

Theorem 114 : *Let I be a nonzero ideal of a Dedekind domain D. If $0 \neq b \in I$, there exists $a \in I$ such that I is generated by a and b. That is, $I = (a, b)$.*

Proof : Let $J = bI^{-1}$. Then, there exists $a \in I$ such that

$$aI^{-1} + J = D. \text{ That is, } aI^{-1} + bI^{-1} = D$$

This means that $(a)I^{-1} + (b)I^{-1} = D$.

That is, $((a) + (b))I^{-1} = D$.

But, $(I)(I^{-1}) = D$. So, $I = (a) + (b) =$ the ideal generated by a and b. That is, $I = (a, b)$. $\qquad\square$

Remark 12.9.1 : The above theorem exhibits the structure of a nonzero ideal of a Dedekind domain and, in particular, of the number ring corresponding to an algebraic number field.

Sometimes, a power of a prime ideal becomes a principal ideal. For instance, if $P = (2, 1 + \sqrt{-5})$ in $D = \mathbb{Z}[\sqrt{-5}]$, D/P is a field isomorphic $\mathbb{Z}/2\mathbb{Z}$ and so P is a prime ideal which is maximal. Any element of P is of the form $a = 2a_1 + (1 + \sqrt{-5})b_1, a_1, b_1 \in \mathbb{Z}$.

$$a^2 = 4a_1^2 + b_1^2(1 - 5 + 2\sqrt{-5}) + 4a_1 b_1(1 + \sqrt{-5})$$

$$= 2(c + d\sqrt{-5}) \quad \text{where } c, d \in \mathbb{Z}.$$

So, $P^2 = (2)$, the ideal generated by 2 in $\mathbb{Z}(\sqrt{-5})$.

It is appropriate to mention that a study of the properties of Noetherian rings and Dedekind domains opens the door to the vast arena of the theory of algebraic numbers. Further, looking at $\mathbb{Z}$ and $\mathbb{Z}[x]$, we note the following:

1) $\mathbb{Z}$ is a UFD : $\mathbb{Z}[x]$ is a UFD.
2) $\mathbb{Z}$ is a Noetherian domain : $\mathbb{Z}[x]$ is Noetherian.
3) $\mathbb{Z}$ is a *PID*: $\mathbb{Z}[x]$ is not a PID.
4) $\mathbb{Z}$ is a Euclidean domain: $\mathbb{Z}[x]$ is not a Euclidean domain.
5) $\mathbb{Z}$ is a Dedekind domain : $\mathbb{Z}[x]$ is not a Dedekind domain.
6) $\mathbb{Z}$ is semisimple: $\mathbb{Z}[x]$ is semisimple.*
* This is deduced from Lemma 16.3.2, chapter 16.

12.10. Worked-out examples

a) (Hilbert Basis theorem): If R is a Noetherian ring with identity 1_R, and if x is transcendental over R, then show that $R[x]$ is a Noetherian ring.

Answer: We have to show that every ideal in $R[x]$ is finitely generated.

Let $i = 0, 1, 2, \ldots$. $\{B_i\}$ denotes a family of subsets of R such that there exists a polynomial $t(x)$ of degree i given by

(12.10.1) $t(x) = b_i x^i + b_{i-1} x^{i-1} + \cdots + b_0.$

$t(x) \in R[x]$ and $t(x)$ belongs to an ideal A of $R[x]$. Further, b_i is the leading coefficient of $t(x)$.

Now, $b_1, b_2 \in B_i \Rightarrow b_1 - b_2 \in B_i$ and since $b \in B_i$, $r \in R$ implies $r \cdot b = b \cdot r \in B_i$, B_i is an ideal of R.

When $b \in B_i$ and b is the leading coefficient of a polynomial $t(x)$ as in (12.10.1), $xt(x) = bx^{i+1} + \cdots + b_0 x \in A$ (an ideal of $R[x]$). This shows that $b \in B_{i+1}$. Or,

(12.10.2) $B_i \subseteq B_{i+1} \quad (i = 0, 1, 2, \ldots)$

As R is given to be Noetherian, the ascending chain

(12.10.3) $B_0 \subseteq B_1 \subseteq B_2 \subseteq \cdots \subseteq B_n \subseteq \cdots$

terminates, say at $n = N$. So,

$$B_N = B_{N+1} = \cdots$$

As each B_i is finitely generated, we write

$$B_i = (b_{i1}, b_{i2}, \ldots, b_{ik_i}), b_{ij} \in R(j = 1, 2, \ldots, k_i).$$

Let $c_{ij}(x)$ be a polynomial in A of degree i and with leading coefficient b_{ij}, $1 \le j \le k_i$.

Since $b_{ij} \in B_i$, there exists at least one polynomial like $c_{ij}(x)$ in A.

The number of such polynomials $c_{ij}(x)$ $(1 \le j \le k_i)$ is $k_0 + k_1 + \cdots + k_N$. The set of polynomials $c_{ij}(x)$ is finite. These polynomials $c_{ij}(x)$ generate A.

For, all polynomials of degree 0 in A form the set $B_0 \subset R$. B_0 is an ideal of R. So, it is finitely generated.

We assume that all polynomials in A of degrees $< r$ are generated by $c_{ij}(x)$. Let $s(x) \in A$ be of degree r.
Then,

$$s(x) = c_r x^r + \cdots + c_0, \quad \text{where } r \le N \text{ and } c_r \in B_r.$$

This yields

$$c_r = \sum_{j=1}^{k_r} a_{rj} b_{rj}, \quad a_{rj} \in R.$$

Now,

$$c_{rj}(x) = b_{rj} x^r + \cdots + b_{rk_r}.$$

So,

$$q(x) = s(x) - \sum_{j=1}^{k_r} a_{rj} c_{rj}(x) \in A.$$

This polynomial has degree $< r$. By induction hypothesis, $q(x)$ is generated by $c_{ij}(x)$. Hence, $s(x)$ is generated by $c_{ij}(x)$.

If $r > N$, $c_r \in B_r = B_N$. So,

$$c_r = \sum_{j=1}^{k_N} a_{Nj} b_{Nj}, \quad a_{Nj} \in R (j = 1, 2, \ldots, k_N).$$

Therefore, $s(x) - \sum_{j=1}^{k_N} a_{Nj} x^{r-N} c_{Nj}(x)$ has degree $< r$ and so $s(x)$ is generated by $c_{ij}(x)$.

Thus, all polynomials of degree r are generated by $c_{ij}(x)$. Hence, by the induction argument, this is true of all polynomials in A. So, A is finitely generated (generated by a finite number of polynomials $c_{ij}(x)$). Since A is an arbitrary ideal of $R[x]$, $R[x]$ is Noetherian, whenever R is so. $\qquad\square$

Remark 12.10.1 : If $x_1, x_2, \ldots, x_n$ are independent transcendental elements over a Noetherian ring R, then $R[x_1, x_2, \ldots, x_n]$ is Noetherian.

Proof follows by induction on n, the number of transcendental elements.

Now, we note that any principal ideal ring is Noetherian. As $R[x]$ is Noetherian whenever R is, there are plenty of Noetherian rings.

Example 12.10.1 : Let F be a field. The polynomial ring $F[x, y]$ is Noetherian. This follows from Hilbert Basis Theorem. (x, y) is a maximal ideal of $F[x, y]$. (x^2, xy, y^2) is an ideal contained in (x^2, y) which is a primary ideal. For, $\sqrt{(x^2, y)}$ is (x, y) which is maximal (we have only to appeal to lemma 12.5.3).

If $I = (x, y)$, $I^2 = (x^2, xy, y^2)$. So,

$$(x^2, xy) = (x^2, xy, y^2) \cap (x) \quad \text{(an irredundant primary decomposition)}.$$

(x^2, xy) can also be written as $(x^2, xy) = (x^2, y) \cap (x)$.
We remark that (x^2, xy, y^2) and (x^2, y) are both primary ideals. (x) is a prime ideal in $F[x, y]$ and so primary.

Next, we recall that if R is a ring with unity 1_R, R is called quasi-local if R has a unique maximal ideal. A local ring is a Noetherian quasi-local ring.

A ring S is said to be a primary ring if S contains at least one proper prime ideal. A prime ideal is said to be a minimal prime ideal, if it is minimal in the set of prime ideals. When S is a commutative ring with unity 1_S, a minimal prime ideal is necessarily a proper ideal.

Fact 12.10.1 : A ring S is a primary ring if, and only if, S has a minimal prime ideal which contains all the zero divisors in S.

For proof, see worked-out example (b) in chapter 14.

Definition 12.10.1 : *A special primary ring R is a local ring with maximal ideal M such that each proper ideal of R is a power of M. (R is assumed to be commutative and has 1_R)*

It means that a special primary ring is a principal ideal ring (PIR) with only finitely many ideals, by virtue of Noetherian property.

Definition 12.10.2 : *Given R is a commutative ring, R is said to be a ZPI- ring (for Zerlegung Primideal) if each nonzero ideal of R is uniquely expressible as a product of prime ideals of R. (If R has identity 1_R, we exclude factors of 1_R in considering uniqueness of factorization). R is said to be a general ZPI ring, if each ideal of R can be expressed as a finite product of prime ideals of R.*

Fact 12.10.2 : Let R be a commutative ring with identity 1_R. The following statements are equivalent:
(i) R is a general ZPI-ring.
(ii) R is Noetherian and for each maximal ideal M of R, there are no ideals between M and M^2.
(iii) R is a finite direct sum of Dedekind domains and special primary rings. That is,

(12.10.4) $R = D_1 \oplus D_2 \oplus \cdots \oplus D_n \oplus S_1 \oplus S_2 \cdots \oplus S_m$

where $D_i(i = 1$ to $n)$ is a Dedekind ring and $S_j(j = 1$ to $m)$ is a special primary ring.

For proof, see R. Gilmer [A2].

b) (Paul Arne Storer) Suppose that R is a Noetherian ring in which all maximal ideals are principal. Show that R is a principal ideal ring.

Answer: Let M be a maximal ideal of R. If $M = (m)$ (say), M/M^2 is a vector space (over the field R/M) of dimension at most 1. Therefore, there are no ideals of R properly squeezed in between M and M^2. Using Fact 12.10.2, one notes that R has the structure given in (12.10.4).

Now, $D_i(i = 1, 2, \ldots, n)$ inherits from R the property that each of its maximal ideals is principal. So, every nonzero prime ideal of D_i is principal. As D_i

is a Dedekind ring, when all its maximal ideals are principal, D_i is a PIR. As each $S_j (j = 1, 2, \ldots, m)$ is a special primary ring, each S_j is a PIR, by definition 12.10.1.
This shows that R is a PIR. ☐

Remark 12.10.2 : Example (b) has been adapted from Problem 10534, Amer. Math. Monthly 106 (1999) p 265.

Remark 12.10.3 : An editorial comment (in Amer. Math. Monthly) on problem 10534 is that R. Gilmer and W. Heinzer have given the following theorem (generalizing the content of (b)):
    If R satisfies ACCP and each maximal ideal of R is principal, then R is a PIR.
See R. Gilmer, W. Heinzer: Principal ideal rings and a condition of Kummer, J. Algebra 83 (1983) 285–292.

Remark 12.10.4 : Example (b) gives a condition for making a Noetherian domain a UFD. By remark 12.10.3, one gets a condition for an integral domain in which ACCP is satisfied to become a UFD. Unfortunately the ring of arithmetic functions $\mathcal{A}$ (which is shown to be a UFD, by using theorem 27 of Cashwell and Everett) does not fall under Example (b), as $\mathcal{A}$ is not Noetherian. However, one knows that $\mathcal{A}$ is a quasi-local ring in which ACCP holds. What additional requirement is needed for $\mathcal{A}$ to possess UFD property, is a question that remains an open problem (of interest to algebraists, but not to number-theorists, as remarked by Paul J. McCarthy years ago!).

EXERCISES

1. *Mark the following statements true (T) or false (F) justifying your answer briefly.*
    a) *Let F be a field, $\alpha \in F$. The set M_α of all polynomials $f(x) \in F[x]$, having α as a zero is given by*

    $$M_\alpha = \{ f(x) \in F[x] : f(\alpha) = 0_F \}$$

    It is correct to say that M_α is a maximal ideal $F[x]$ and $F[x]/M_\alpha \cong F$.
    b) *Let $n(> 1)$ be a square-free integer. $(x^2 - n)$ denotes the ideal generated by $x^2 - n$ in $\mathbb{Q}[x]$, ($\mathbb{Q}$ being the field of rational numbers). Then, $\mathbb{Q}[x]/(x^2 - n)$ need not be a field.*
    c) *Let R be a quasi-local ring (commutative and has multiplicative identity 1_R). Suppose that the maximal ideal M (of R) is a principal ideal, say (p). Assume that $\cap_{n=1}^\infty M^n = (0_R)$. If I denotes a proper ideal of R, then, $I = M^k \quad (k \in \mathbb{N})$.*

d) Let $\mathbb{C}\{z\}$ denote the ring of rational functions of the complex variable z and having no pole on the unit circle $|z| = 1$. $\mathbb{C}\{z\}$ is not Noetherian.

e) Let R be an integral domain. Then, R is a Dedekind domain if, and only if, every nonzero fractional ideal of R is invertible.

f) Let D be a Dedekind domain. Suppose that S is a multiplicatively closed subset of D. Then $S^{-1}D$ is either a Dedekind domain or the field of quotients of D.

2. (D. M. Burton) Let I be a proper ideal of a Noetherian ring R. Show that

$$\cap_{n=1}^{\infty} I^n = \{x \in R : (1_R - a)x = 0 \text{ for some } a \in I\}$$

3. Prove that a prime ideal in a commutative ring is irreducible.

4. Show that in $\mathbb{Z}[x]$, the ideal $J = (9, 3x+3) = (3) \cap (9, x+1)$.

5. We shall call an ideal I of a commutative ring R (with unity 1_R) a semiprime ideal, if I is an intersection of prime ideals of R. Exhibit the semiprime ideals of $\mathbb{Z}$. Give an equivalent definition of a semiprime ideal I in terms of $\sqrt{I}$.

6. Let R be a commutative ring with unity 1_R. Show that a product of principal ideals of R is again a principal ideal.

7. Let I be an ideal of a Noetherian ring R in which every non-trival prime ideal is maximal. Show that I is expressible as a product of primary ideals.

8. Let R be a commutative ring with unity 1_R. $e \in R$ is called an idempotent, if $e^2 = e$. If each maximal ideal of R has an idempotent generator $e \neq 1_R$ or 0_R, show that every primary ideal of R is maximal.

   Use Cohen's theorem (Theorem 99) to deduce that R is Noetherian.

9. (Dennis Spellman) Let R be a principal ideal ring with unity 1_R in which two elements a, b are multiples of one another. Show that they are unit multiples of one another. That is, show that there is an invertible element $u \in R$ such that $a = ub$.

   [Ref. Problem 10495 Amer. Math. Monthly 105 No.1 (Jan 1998) 70]

10. An integral domain D is called a Prüfer domain if every finitely-generated ideal of D is invertible. Prove that a Prüfer domain is a Dedekind domain if, and only if, it is Noetherian.

11. Prove that $\mathbb{Z}[\sqrt{10}] = \{a + b\sqrt{10} : a, b \in \mathbb{Z}\}$ is a Dedekind domain, but not a PID.

12. Let D be an integral domain D. If every proper prime ideal of D is invertible, show that D is a Dedekind domain.

    (Conversely) Let D be a Dedekind domain. Show that every proper prime ideal of D is invertible. (See corollary 12.7.1)

13. Illustrate theorem 94 with a non-trivial example.

14. Let $\mathbb{Q}_p$ denote the completion of $\mathbb{Q}$ with respect to a p-adic valuation $||_p$. (See Section 8.4, chapter 8) Find the integral closure of $\mathbb{Z}$ in $\mathbb{Q}_p$.

REFERENCES

[1] M. F. Atiyah and I. G. Macdonald : Introduction to commutative algebra, (Addison-Wesley Pub Co Reading, Mass (1969)) Chapters 4 and 6 to 9, pp. 36–49 and 74–99.

[2] Ethan D. Bolker : Elementary Number Theory, Chapter 6, W. A. Benjamin Inc. (1970) N. Y. pp. 82–113.

[3] D. M. Burton : A first course in rings and ideals, Addison-Wesley Pub. Co. Reading Mass (1970) Chapters 5, 11, and 12 pp. 71–89 and 217–261.

[4] H. S. Butts: Unique factorization of ideals into nonfactorable ideals, Proc. Amer. Math. Soc. 15 (1964) p 21.

[5] H. S. Butts and L. I. Wade: Two criteria for Dedekind domains, Amer. Math. Monthly, 73, (1966) 14–21.

[6] I. S. Cohen : Commutative rings with restricted minimum condition, Duke Math. J. 17 (1950) 27–42.

[7] P. M. Cohn : An introduction to ring theory, Springer Undergraduate Math. Series, Springer-Verlag London (2000), Chapters 2 and 3, pp 53–133.

[8] A Frölich and M. J. Taylor : Algebraic Number Theory, Cambridge Studies in Advanced Mathematics, No. 27 Chapters 1 and 2, Cambridge University Press (1993) pp. 8–70.

[9] T. W. Hungerford : Algebra, GTM No 73, Springer Verlag NY (1988), Chapter 8, pp 371–409.

[10] G. Karpilovsky : Commutative group algebras, Monographs and Textbooks in Pure and Applied Math No 78, Marcel Dekker Inc. NY (1983), Chapters 3 and 4, pp 44–108.

[11] W. Krull: Idealtheorie, Chelsea Pub Co. New York (1948).

[12] D. A. Marcus : Number fields, Univestitext Springer Verlag, NY (1977), Chapters 2 and 3, pp 12–97.

[13] W. Narkeiwicz : Elementary and Analytic Theory of Algebraic Numbers, Springer Verlag PWN Warzawa (1990), Chapter 1, pp 1–41.

[14] Heinz Prüfer: Untersuchungen über Teilbarkeitseigenshaften in Korpern, J. Reine Angew. Math. 168 (1932) 1–36.

[15] P. Samuel : Algebraic theory of numbers, Houghton Mifflin Co. Boston Hermann Paris (1970), Chapters 2 and 3, pp. 27–30 and 46–52.

[16] O. Zariski and P. Samuel: Commutative algebra Vol I, Springer Verlag, GTM No: 28 (1982).

ADDITIONAL REFERENCES

[A1] A. J. Berrick and M. A. Keating: An introduction to rings and modules with
     K- theory in view, Cambridge Studies in Advanced Mathematics No. 65,
     Cambridge University Press, The Edinburgh building, Cambridge CB2 2RU,
     UK (2000).

[A2] Robert Gilmer: Multiplicative ideal theory, Pure and Applied Mathematics
     12, a series of monographs and textbooks, Marcel Dekker, Inc. NY (1972)
     (Theorem 392).

CHAPTER 13

Algebraic number fields

Historical perspective

The origin of algebraic number theory is in the 'attempted-proofs' of Fermat's Last Theorem: The motivation is to be found in the generalizations of the integral domain $\mathbb{Z}$ (of rational integers) giving rise to the notion of algebraic integers: (See 'Euclidean domains', chapter 3). Many of the results of number theory are tackled in a more general set-up in algebraic number theory. For instance, Fermat's two-square theorem (theorem 4, chapter 1) is proved by considering the Euclidean domain $\mathbb{Z}[i]$ of Gaussian integers. We recall that an algebraic integer is a zero of a monic polynomial in $\mathbb{Z}[x]$. Very often, the study of a suitable ring of algebraic integers helps in the solution of a problem which is initially stated in terms of ordinary (rational) integers. For example, we have the context of solutions of the Pell equation $x^2 - my^2 = 1$. The consideration of ideals instead of elements of a ring was indeed a breakthrough for purposes of factorization. This is achieved in Dedekind domains considered in chapter 12.

The contributions of Kummer, Dirichlet, Kronecker and Dedekind to the development of the theory of rings and ideals have been indicated earlier, (see chapters 2 and 3). It was L. J. Mordell (1888–1972) who gave a series of lectures on 'Fermat's Last Theorem' which were recorded in print and that perhaps is the first authentic account of a survey of the beginnings of Algebraic Number Theory.

13.1. Introduction

In this chapter, we make a brief study of the number ring R corresponding to a number field K. By introducing equivalence classes of ideals in R, namely, the ideals I, J of R are equivalent if, and only if, there exist elements α, β in R such that $\alpha I = \beta J$. The equivalence classes form a group under multiplication. It is called the class group of R. It is shown that the class-group is finite. The order of the class group is referred to as the class-number of K. The ring corresponding to the cyclotomic field $Q[\omega]$ where $\omega = \exp(\frac{2\pi i}{m})$ is obtained. The 'Carlitz theorem' on the characterization of a number field of class-number 2 is proved in theorem 118. This leads to the notion of half-factorial domains. See Section 13.4.

The Pell equation $x^2 - my^2 = 1$ (m square-free) is studied in detail. It is related to the quadratic number field $Q(\sqrt{m})$. The existence of a nontrivial solution to the Pell equation is shown in theorem 120. The structure of the group of units in a number ring corresponding to an algebraic number field K is brought out in Dirichlet's unit theorem (see theorem 124). Some examples are pointed out.

Bhaskara's (Bhaskaracharya (1114–1185)) *Cakravala method* of solving the Pell equation is shown in Section 13.6.

13.2. The ideal class group

K denotes an algebraic number field, that is, a finite extension of the field $\mathbb{Q}$ of rational numbers. Let R denote the ring $\mathscr{A} \cap K$ where $\mathscr{A}$ stands for the ring of algebraic integers.

Definition 13.2.1 : *For ideals I, J of R, we say that $I \sim J$ if, and only if, there exist elements $\alpha, \beta \in R$ such that $\alpha I = \beta J$.*

Clearly, $\sim$ is reflexive and symmetric. It is also transitive, as whenever $I \sim J$ and $J \sim L$ where I, J, L are ideals of R, one has $\alpha I = \beta J$ for $\alpha, \beta \in R$. Also $\gamma J = \delta L$ for $\gamma, \delta \in R$. So,

$$\beta\gamma J = \gamma\beta J = \gamma\alpha I = \beta\delta L \text{ and so } I \sim L.$$

So, $\sim$ determines an equivalence relation on the set of ideals of R. Our aim is to show that there are only finitely many equivalence classes of ideals under $\sim$. When R is a principal ideal domain, every ideal of R is a principal ideal and so any two ideals are equivalent and therefore the set of ideals of $\mathbb{R}$ forms one class only. In the case of $\mathbb{Z}[\sqrt{-5}]$, it can be shown that all non-principal ideals of $\mathbb{Z}[\sqrt{-5}]$ belong to one class under $\sim$ and so one will conclude that there are only two classes of ideals in the set of ideals of $\mathbb{Z}[\sqrt{-5}]$, namely

(i) The class c_1 of principal ideals of $\mathbb{Z}[\sqrt{-5}]$ and
(ii) The class c_2 of non-principal ideals of $\mathbb{Z}[\sqrt{-5}]$.

Definition 13.2.2 : *In R, the classes induced by $\sim$ are known as ideal classes.*

Next, we consider two algebraic number fields K and F such that $K \subset F$. Let $R = \mathscr{A} \cap K$ and $S = \mathscr{A} \cap F$. ($\mathscr{A}$ denotes the ring of algebraic integers).

Lemma 13.2.1 : *Let P, Q be prime ideals of R and S respectively. If $Q \supset P$, $Q \cap R = P$ and $Q \cap R = P \Rightarrow Q \cap K = P$.*

Proof : As $Q \supset P$, $Q \cap R$ contains P and $Q \cap R$ is an ideal of R. As P is a prime ideal of R and R is a Dedekind domain, P is a maximal ideal of R. So $Q \cap R = P$ or $Q \cap R = R$. If $Q \cap R = R$, $1_R \in Q$ and so $Q = S$, a contradiction. So, $Q \cap R = P$. Further, $Q \cap R = P \Rightarrow Q \cap K = P$, as $Q \subset \mathscr{A}$.
For, $R \subset \mathscr{A} \cap F$ and $Q \cap R = P \subset \mathscr{A} \cap K = R$. So, $Q \cap K = Q \cap R = P$. $\square$

Definition 13.2.3 : *Let P, Q be prime ideals of $R = \mathscr{A} \cap K$ and $S = \mathscr{A} \cap F$ respectively with $K \subset F$. We say that Q lies over P, if $Q \cap R = P$. It also means that P lies under Q.*

Lemma 13.2.2 : *Every prime ideal Q of S lies over a unique prime ideal P of R. Further, every prime ideal P of R lies under at least one prime ideal Q of S.*

Proof : The first part of the lemma wants that $Q \cap R = P$ is unique. It is so, as P is a maximal ideal of R.

For the second part, we observe that the prime ideals lying over P are prime ideals of S. Let Q be a prime ideal of S. We consider the ideal PS of S. As Q is an ideal of S, $PS \subset Q \Rightarrow P \subset Q$. Therefore, Q divides PS. So, the prime ideals of S lying over P are the prime divisors of PS. Now, $PS \neq S$. For, assume $PS = S$, $1_s \notin P$. Further, by lemma 12.6.4 of chapter 12, there exists $t = \frac{b}{a} \in K \setminus R$ such that $tP \subset R$. Then, $tPS \subset RS = S$. If $1_s \in PS$, then $t \in S$. Then, t is an algebraic integer contradicting the fact that $t \in K \setminus R$.

Therefore, the prime ideals of S which lie over P are the prime ideals which occur in the prime decomposition of PS. That is, every prime ideal of R lies under at least one prime ideal Q of S and Q happens to be a prime divisor of PS. $\square$

Remark 13.2.1 : Lemma 13.2.2 has been adapted from [10].

Definition 13.2.4 : *If P is a prime ideal of R which lies under Q, a prime ideal of S, Q divides PS and the exponent e to which $\mathbb{Q}$ occurs in the prime-power decomposition of PS is called the ramification index of Q, written $e(Q|P)$.*

Example 13.2.1 : We take $K = \mathbb{Q}$. $F = \mathbb{Q}[i]$, $R = \mathbb{Z}$ and $S = \mathbb{Z}[i]$, $i = \sqrt{-1}$. Let $\alpha = 1 - i$. The principal ideal $Q = (1 - i)$ of $\mathbb{Z}[i]$ is such that $Q \cap \mathbb{Z} = (2)$. Q is a prime ideal, since $\mathbb{Z}[i]/(1 - i) \cong \mathbb{Z}/5\mathbb{Z}$ which is a field. Now, Q lies over $P = (2)$. For, $2S = 2\mathbb{Z}[i] = Q^2$. So, $e(Q|(2)) = 2$.

If p is an odd prime in $\mathbb{Z}$ and a prime ideal Q of $\mathbb{Z}[i]$ lies over $P = (p)$, $PS = p\mathbb{Z}[i] = Q$ and so, $e(Q|P) = 1$.

Next, as $R = \mathscr{A} \cap K$, R is a Dedekind domain and it is true that for any nonzero proper ideal I of R, R/I is finite. (See Remark 12.6.6 (ii) of chapter 12)

Definition 13.2.5 : *Let $R = \mathscr{A} \cap K$ be the number ring corresponding to the algebraic number field K. Let I be a nonzero ideal of R. The norm (or index) of I is defined as $||I|| = |R/I|$, the cardinality of the quotient ring R/I which is finite.*

We used the notation $N(I)$ for $||I||$ in the context of a Dedekind domain (see definition 12.7.2).

We recall that the norm $N(\alpha)$ of an element $\alpha \in R$ is given by

$$(13.2.1) \qquad N(\alpha) = \prod_{i=1}^{n} \sigma_i(\alpha)$$

where $\sigma_1, \sigma_2 \ldots \sigma_n$ are the $n (= [K : Q])$ embeddings of K in $\mathbb{C}$ which fix $\mathbb{Q}$.

Fact 13.2.1 :

(a) For ideals I, J in R, $||IJ|| = ||I|| \, ||J||$

(b) Let I be an ideal of R. For the S-ideal IS of $S = \mathscr{A} \cap F$ where $K \subset F$,

$$||IS|| = ||I||^m \quad \text{where } m = [F : K]$$

(c) Let $\alpha \in R$ and $\alpha \neq 0$. If (α) denotes the principal ideal generated by α,

$$||(\alpha)|| = |N(\alpha)| \quad \text{where } N(\alpha) \text{ is as given in (13.2.1)}.$$

For proof see D. A. Marcus [10, chapter 3, theorem 22].

Theorem 115 : *K denotes a number field. R is the number ring corresponding to K. For any ideal I of R, if $0 \neq \alpha \in I$, there exists a positive real number j (depending on K) such that*

(13.2.2) $|N(\alpha)| \leq j||I||,$

where $N(\alpha)$ is as given in (13.2.1). Further, j is independent of $||I||$.

Proof : As R is a finitely generated $\mathbb{Z}$-module, R has an 'integral basis'. That is, if $[K : Q] = n$, we can choose a set of n elements $\alpha_1, \alpha_2 \ldots \alpha_n$ of R such that any element $\beta \in R$ can be written as

$$\beta = \sum_{i=1}^{n} a_i \alpha_i, \quad a_i \in \mathbb{Z}, i = 1, 2, \ldots n.$$

As $||I||$ is finite, we can find a unique positive integer t such that

$$t^n \leq ||I|| < (t+1)^n$$

We consider $\sum_{s=1}^{n} t_s \alpha_s$, $t_s \in \mathbb{Z}$, $0 \leq t_s \leq t$.
Each t_s $(s = 0, 1, 2, \ldots t)$ takes $(t+1)$ values. So, there are $(t+1)^n$ elements $\sum_{s=1}^{n} t_s \alpha_s$ belonging to R. As $||I|| < (t+1)^n$, two of the elements say $\delta_1 = \sum_{s=1}^{n} t_s^{(1)} \alpha_s$ and $\delta_2 = \sum_{s=1}^{n} t_s^{(2)} \alpha_s$ are such that $\delta_1 - \delta_2 \in I$ where $0 \neq \delta_1 - \delta_2$. Taking $\delta_1 - \delta_2 = \alpha$, we obtain

$$\alpha = \sum_{s=1}^{n} m_s \alpha_s, \quad m_s \in \mathbb{Z} \text{ and } |m_s| \leq t, s = 1, 2, \ldots, n.$$

Now,

(13.2.3) $$|N(\alpha)| = \prod_{i=1}^{n} |\sigma_i(\alpha)| = \prod_{i=1}^{n} \left| \sum_{s=1}^{n} m_s \sigma_i(\alpha_s) \right|.$$

As $|m_s| \leq t$, we have

$$\sum_{s=1}^{n} m_s |\sigma_i(\alpha_s)| \leq t \sum_{s=1}^{n} |\sigma_i(\alpha_s)|.$$

We write

(13.2.4) $$\prod_{i=1}^{n} \sum_{s=1}^{n} |\sigma_i(\alpha_s)| = j.$$

From (13.2.3) and (13.2.4) we see that

$$|N(\alpha)| \leq t^n j. \quad \text{But, } t^n \leq ||I||.$$

So, $|N(\alpha)| \leq ||I||j$ which is (13.2.2).

j is defined only in terms of $\sigma_i(\alpha_s)$ ($s = 1, 2, \ldots n$) and so comes from the embeddings of K in $\mathbb{C}$ which fix $\mathbb{Q}$. That is, j is independent of $||I||$. $\square$

If R is the number ring corresponding to an algebraic number field K, the set $\mathscr{S}$ of ideals of R could be partitioned into mutually disjoint classes under the equivalence relation $I \sim J$, if, and only if, there exist elements $\alpha, \beta \in R$ such that $\alpha I = \beta J$ where I, J belong to $\mathscr{S}$. It is verified that two ideals I_1, I_2 of R are isomorphic ring-theoretically, if, and only if, they belong to the same ideal class.

Let I be an ideal of R. If αI is principal for some $\alpha \in R$, then I is principal. For if $\alpha I = (\beta)$ (say), there exists $\gamma \in I$ such that $\alpha\gamma = \beta$. Let $t \in I$. Then $\alpha t = \beta\delta$ for some $\delta \in R$. So, $\alpha t = \alpha\gamma\delta$. As R is an integral domain, cancellation of α gives $t = \gamma\delta$. So any element of I is a multiple of γ or I is principal. Further, the principal ideals form an ideal class. In the case of R, by theorem 100, when I is a given ideal of R, we can select an ideal J of R such that IJ is principal. If $C(I)$ and $C(J)$ are the ideal classes containing I and J respectively, then $C(IJ)$ is the class containing principal ideals (of R). Therefore, if we define multiplication of ideal classes by

(13.2.5) $C(I_1)C(I_2) = C(I_1 I_2)$ for $I_1, I_2 \in \mathscr{S}$,

(13.2.5) is well-defined on the set of ideal classes. By virtue of the fact that $C(IJ)$ is the class of principal ideals, the ideal classes form an abelian group, the so-called ideal class group C_K. In the case of a PID, $C_K = \{1\}$. We refer to the ideal classes $C(I)$ as the ideal classes of R.

Theorem 116 : *Let R be the number ring corresponding to an algebraic number field K.*

  (i) *Every ideal class of R contains an ideal I such that $||I|| \leq j$ where j is as given in theorem 115.*

  (ii) *There are only finitely many ideal classes of R.*

Proof :

  (i) Let C be an ideal class of R. Suppose that an ideal I belongs to C. By theorem 100, there exists an ideal J of R such that IJ is principal. That is, J belongs to the class C^{-1}. Let $\alpha \in J$ be such that

(13.2.6) $|N(\alpha)| \leq j||J||$

  J contains the principal ideal (α). So, $(\alpha) = IJ$ where I belongs to C. By Fact 13.2.1 (c), $|N(\alpha)| = ||(\alpha)|| = ||I|| \, ||J||$. From (13.2.6), we see that $||I|| \, ||J|| \leq j||J||$.

  So, I belonging to C is such that $||I|| \leq j$. This proves the first part.

  (ii) We observe that if an ideal $I = \prod_{i=1}^{n} P_i^{n(P_i)}$ where P_i are distinct prime ideals dividing I,

$$||I|| = \prod_{i=1}^{n} ||P_i||^{n(P_i)}$$

Each ideal of R has only a finite number of prime divisors. So, given j as in theorem 115, using (i) of this theorem, only finitely many ideals I satisfy $||I|| \leq j$. Every ideal class contains an ideal I with $||I|| \leq j$. If there are an infinite number of ideal classes, there would occur an infinite number of ideals J with $||J|| \leq j$ — a contradiction. Therefore, the number of ideal classes is finite.

$\square$

Corollary 13.2.1 : *Given an algebraic number field K, the ideal class group C_K is finite.*

Proof is immediate from (ii) of theorem 116.

Definition 13.2.6 : *The order of the ideal class group C_K of an algebraic number field K is called the class number of the number ring R of K and is denoted by $h(K)$.*

$h(K)$ is also referred to as the class-number of the field K.

Illustration 13.2.1 :

(i) *The number field $\mathbb{Q}(\sqrt{2})$.*

The number ring corresponding to $\mathbb{Q}[\sqrt{2}]$ is $\mathbb{Z}[\sqrt{2}]$. It has an integral basis $\{1, \sqrt{2}\}$. For any ideal I of $\mathbb{Z}[\sqrt{2}]$, one has

$$||I|| \leq 5.$$

For, $j = \prod_{i=1}^{2}(|\sigma_i(1)| + |\sigma_i(\sqrt{2})|)$
$= (1+\sqrt{2})(1+\sqrt{2}) = (1+\sqrt{2})^2 > 5.$

The prime divisors of I are among the prime ideals lying over (2), (3) and (5) of $\mathbb{Z}$.

$$2\mathbb{Z}[\sqrt{2}] = (\sqrt{2})^2$$

$3\mathbb{Z}[\sqrt{2}]$ and $5\mathbb{Z}[\sqrt{2}]$ are prime ideals of $\mathbb{Z}[\sqrt{2}]$. It is checked that the ideals I for which $||I|| \leq 5$ are $\mathbb{Z}[\sqrt{2}]$, $(\sqrt{2})$ and $2\mathbb{Z}[\sqrt{2}]$. Each one of these is a principal ideal and so the class number of $\mathbb{Z}[\sqrt{2}]$ is 1. That is, $\mathbb{Z}[\sqrt{2}]$ is a PID.

(ii) *The number field $Q[\sqrt{-5}]$.*

The number ring corresponding to $Q[\sqrt{-5}]$ is $\mathbb{Z}[\sqrt{-5}]$. Every ideal class of $\mathbb{Z}[\sqrt{-5}]$ contains an ideal I with $||I|| \leq 10$. We consider the prime ideals of $\mathbb{Z}[\sqrt{-5}]$ which lie over (2), (3), (5) and (7). It is verified that

$$2\mathbb{Z}[\sqrt{-5}] = (2, 1+\sqrt{-5})^2,$$

$$3\mathbb{Z}[\sqrt{-5}] = (3, 1+\sqrt{-5})(3, 1-\sqrt{-5}),$$

$$5\mathbb{Z}[\sqrt{-5}] = (\sqrt{-5})^2,$$

$$7\mathbb{Z}[\sqrt{-5}] = (7, 3+\sqrt{-5})(7, 3-\sqrt{-5}).$$

$(2, 1+\sqrt{-5})$ is not a principal ideal of $\mathbb{Z}[\sqrt{-5}]$. Similarly, the prime divisors of the ideals $3\mathbb{Z}[\sqrt{-5}]$ and $7\mathbb{Z}[\sqrt{-5}]$ are not principal ideals. The class C_2

containing $(2, 1 + \sqrt{-5})$ *is an element of order 2 in the ideal class group. All non-principal ideals are in* C_2. *The ideal class group is* $C_K = \{C_1, C_2\}$, *when* $K = Q[\sqrt{-5}]$. C_1 *is the class of principal ideals. Thus, the class number of* $\mathbb{Z}[\sqrt{-5}]$ *is 2.*

(iii) *The number field* $Q(\sqrt{-6})$.

   The number ring corresponding to $K = Q[\sqrt{-6}]$ *is* $\mathbb{Z}[\sqrt{-6}]$. *It has an integral basis* $\{1, \sqrt{-6}\}$

$$j = \prod_{i=1}^{2} \{|\sigma_i(1)| + |\sigma_i(\sqrt{-6})|\} = (1 + \sqrt{6})^2 \geq 7.$$

For any ideal I of $\mathbb{Z}[\sqrt{-6}]$, $\|I\| \leq 3$. *So the class number* ≤ 3.
It is checked that if $P_1 = (2, \sqrt{-6})$, $P_1^2 = 2\mathbb{Z}[\sqrt{-6}]$. *If* $P_2 = (3, \sqrt{-6})$,
$P_2^2 = 3\mathbb{Z}[\sqrt{-6}]$, C_2 *containing* P_1 *and* P_2 *is an element of order 2 in the ideal class group. Further,* P_1 *and* P_2 *are the only prime ideals of norm 2 and 3 respectively. Thus, if* $K = Q[\sqrt{-6}]$, C_K *is of order 2 and* $h(K) = 2$.

13.3. Cyclotomic fields

Let m be a positive integer. We write $\omega = \exp(\frac{2\pi i}{m})$, i denotes $\sqrt{-1}$. When $m = 1, 2$, $\omega = \pm 1$, an extension of $\mathbb{Q}$ by adjoining ω to $\mathbb{Q}$ is $\mathbb{Q}$ itself. So, we take $m > 2$. The extension of $\mathbb{Q}$ by adjoining ω to $\mathbb{Q}$ is called a cyclotomic extension of Q. It is denoted by $\mathbb{Q}(\omega)$. $\mathbb{Q}(\omega)$ is referred to as the mth cyclotomic field.

If $\omega = \exp(\frac{2\pi i}{6})$, $\omega = -\omega^4 = -(\omega^2)^2$ and we see that $\mathbb{Q}(\omega) = \mathbb{Q}(\omega^2)$.

Observation 13.3.1 :

   (1) *Let* m *be an odd integer. Writing* $\omega = \exp(\frac{2\pi i}{2m})$ *we see that* $\omega = -\omega^{m+1}$ *and* $\omega \in \mathbb{Q}(\omega^2)$. *That is, the mth cyclotomic field and 2mth cyclotomic field are one and the same.*

   (2) *When* m *is even, the cyclotomic fields* $\mathbb{Q}(\omega)$ *are all distinct since* $[\mathbb{Q}(\omega) : \mathbb{Q}]$, *(the degree of the extension* $\mathbb{Q}(\omega)$ *over* $\mathbb{Q}$) *is given by* $\phi(m)$, *the Euler* ϕ-*function at m.*

   (3) *There are an infinite number of cyclotomic fields. The other infinite set of algebraic number fields that we have considered is the class of quadratic number fields* $Q(\sqrt{m})$ *where m is square-free and* $m \in \mathbb{Z}$.

We wish to concentrate on the pth cyclotomic field $\mathbb{Q}(\omega)$ where $\omega = \exp(\frac{2\pi i}{p})$, p being an odd prime. ω is a complex p^{th} root of unity. Also, $\omega^2, \omega^3, \ldots \omega^{p-1}$ are complex p^{th}-roots of unity. It is known [5] that

(13.3.1) $$f(x) = x^{p-1} + x^{p_2} + \cdots + x + 1$$

is irreducible over $\mathbb{Z}$ (by Eisenstein criterion) and $[\mathbb{Q}(\omega) : \mathbb{Q}] = (p-1)$. $f(x)$ is expressible as

(13.3.2) $$f(x) = (x - \omega)(x - \omega^2) - (x - \omega^{p-1})$$

$\omega, \omega^2, \ldots \omega^{p-1}$ are the 'conjugates' of ω in $f(x) = 0$. The embeddings of $\mathbb{Q}(\omega)$ in $\mathbb{C}$ are monomorphisms

$$\sigma_i : \mathbb{Q}(\omega) \to \mathbb{C}$$

given by $\sigma_i(\omega) = \omega^i$ $(1 \le i \le p-1)$.

In fact, $\{1, \omega, \omega^2, \ldots \omega^{p-2}\}$ is a basis for the vector space $\mathbb{Q}(\omega)$ over $\mathbb{Q}$. (It may be recalled that $\{1, \omega, \ldots \omega^{p-1}\}$ is a linearly dependent set, as $1 + \omega + \omega^2 + \ldots + \omega^{p-1} = 0$).

$\alpha \in \mathbb{Q}(\omega)$ is of the form

(13.3.3) $\alpha = a_0 + a_1\omega + \cdots + a_{p-2}\omega^{p-2}$ $(a_i \in \mathbb{Q}, 0 \le i \le p-2)$

The monomorphism $\sigma_i : Q(\omega) \to \mathbb{C}$ is given by

(13.3.4) $\sigma_i(\alpha) = a_0 + a_1\omega^i + \cdots + a_{p-2}\omega^{(p-2)i}$

The norm $N(\alpha)$ and trace $T(\alpha)$ of α are defined by

(13.3.5) $$N(\alpha) = \prod_{i=1}^{p-1} \sigma_i(\alpha)$$

(13.3.6) $$T(\alpha) = \sum_{i=1}^{p-1} \sigma_i(\alpha)$$

It is easy to check that

(13.3.7) $T(\omega^i) = -1$ $(1 \le i \le p-1)$

(13.3.8) $N(\omega^i) = 1, \quad i \in \mathbb{Z};$

and

(13.3.9) $$N(1-\omega) = \prod_{i=1}^{p-1}(1 - \omega^i) = p$$

Theorem 117 : *The number ring corresponding to $\mathbb{Q}(\omega)$ is $\mathbb{Z}[\omega]$.*

Proof : Let D be the number ring corresponding to $Q(\omega)$.

Claim : D has an integral basis $\{1, \omega, \omega^2, \ldots \omega^{p-2}\}$. That is, if $\alpha \in D$, one has

(13.3.10) $\alpha = a_0 + a_1\omega + \cdots + a_{p-2}\omega^{p-2}$ $a_i \in \mathbb{Z}, (0 \le i \le p-2)$

Let $\beta = a_0 + a_1\omega \cdots + a_{p-2}\omega^{p-2}$ where $a_i \in \mathbb{Q}, (0 \le i \le p-2)$. If $\beta \in D$, we have to show that each $a_i \in \mathbb{Z}, (0 \le i \le p-2)$. Now,

$$\beta(1-\omega) = a_0(1-\omega) + a_1(\omega - \omega^2) + \cdots + a_{p-2}(\omega^{p-2} - \omega^{p-1})$$

Also, $T(\omega) = -1$ and $T(1) = (p-1)$. Further,

(13.3.11) $T(1-\omega) = T(1-\omega^2) = \cdots = T(1-\omega^{p-1}) = p.$

So,

$$T(\beta(1-\omega)) = a_0 T(1-\omega) + a_1 T(\omega - \omega^2) \cdots + a_{p-2} T(\omega^{p-2} - \omega^{p-1})$$

$$T(\omega^{j-1} - \omega^j) = \sum_{i=1}^{p-1} \sigma_i(\omega^{j-1} - \omega^j)$$

$$= \sum_{i=1}^{p-1} \sigma_i(\omega^{(j-1)}) \sigma_i(1-\omega)$$

$$= \sum_{i=1}^{p-1} \omega^{i(j-1)}(1-\omega^i)$$

$$= \sum_{i=1}^{p-1} \omega^{i(j-1)} - \sum_{i=1}^{p-1} \omega^{ij}.$$

Or,

$$T(\omega^{j-1} - \omega^j) = 0 \; ; \quad 2 \leq j \leq (p-1).$$

That is,

$$T(\beta(1-\omega)) = a_0 T(1-\omega) = a_0 p, \quad \text{by (13.3.11)}.$$

As $(1-\omega)(1-\omega^2)\ldots(1-\omega^{p-1}) = N(1-\omega) = p$, by (13.3.9),

$p \in (1-\omega)D$. So, $(1-\omega)D \cap \mathbb{Z}$ contains $p\mathbb{Z}$. As $p\mathbb{Z}$ is a maximal ideal of $\mathbb{Z}$, $(1-\omega)D \cap \mathbb{Z} \neq p\mathbb{Z} \Rightarrow (1-\omega)D \cap \mathbb{Z} = \mathbb{Z}$. That is, $1-\omega$ is a unit in D. So, the conjugates $1-\omega^i$ $(1 \leq i \leq p-1)$ of $1-\omega$ are also units. As $N(1-\omega) = p$, p is a unit in $D \cap \mathbb{Z}$. It will mean that p is a unit in $\mathbb{Z}$ which is false. So,

$$(13.3.12) \qquad\qquad\qquad (1-\omega)D \cap \mathbb{Z} = p\mathbb{Z}$$

Now, $T(\beta(1-\omega)) \in p\mathbb{Z}$. So, as $pa_0 \in p\mathbb{Z}$. Or, $a_0 \in \mathbb{Z}$.
Since $\omega^{-1} = \omega^{p-1} \in D$, (as ω^{p-1} has a representation $-1 - \omega - \omega^2 \cdots - \omega^{p-2}$ as in (13.3.10),

$$(\beta - a_0)\omega^{-1} = a_1 + a_2\omega + \cdots + a_{p-2}\omega^{p-2} \in D$$

So, arguing as before, $a_1 \in \mathbb{Z}$. In the same manner, we arrive at $a_i \in \mathbb{Z}$, $2 \leq i \leq (p-2)$. This shows that $D \subseteq \mathbb{Z}[\omega]$. But, $\mathbb{Z}[\omega] \subseteq D$. Thus, $D = \mathbb{Z}[\omega]$. $\square$

13.4. Half-factorial domains

We look at a characterization of algebraic number fields K having class number $h(K) = 2$. We know that the number ring corresponding to an algebraic number field K is a PID if, and only if, $h(K) = 1$. Further, $h(K) = 1$ implies that the number ring corresponding to K is a UFD.

Theorem 118 (L. Carlitz (1960)) : *An algebraic number field K has class-number ≤ 2 if, and only if, every nonzero element $\alpha \in R$, the number ring corresponding to K is such that whenever*

$$\alpha = \pi_1 \pi_2 \cdots \pi_s = \tau_1, \tau_2, \ldots \tau_t$$

(i) $s = t$ and π_i and τ_j are associated $i, j = 1, 2 \ldots s$, provided, $h(K) = 1$.

(ii) $s = t$ and π_i and τ_j $(i \neq j)$ need not be associated, provided $h(K) = 2$.

Proof :

(i) When $h(K) = 1$, every ideal of R is principal and so R is a PID. Then, it is essential that $s = t$ and π_i, τ_j are associates $(i, j = 1, 2, \ldots s)$, as R is a UFD.

(ii) Let us take $h(K) = 2$. As R is a Dedekind domain, every nonzero proper ideal of R is a product of distinct prime ideals and factorization into prime ideals is unique. (See theorem 101, chapter 12). For $\alpha \in R$, one has

$$(13.4.1) \qquad (\alpha) = \pi_1 \pi_2 \ldots \pi_m \cdot \tau_1 \tau_2 \ldots \tau_n,$$

where $\pi_1, \pi_2 \ldots \pi_m$ are principal ideals and $\tau_1, \tau_2, \ldots \tau_n$ are prime ideals which are not principal ideals. Then,

$$(13.4.2) \qquad \pi_i = (p_i) \quad (i = 1, 2, \ldots m).$$

As τ_i, τ_j are non-principal ideals and as $h(K) = 2$, τ_i, τ_j belong to the class C_2 of non-principal ideals where the class group $H = \{C_1, C_2\}$ with $C_2^2 = C_1$. Therefore, there exist elements β, γ in R such that

$$\beta \tau_i = \gamma \tau_j.$$

It implies that $\beta \tau_i \tau_j = \gamma \tau_j^2 = (\lambda_{i,j})$ where $\lambda_{i,j} \in R$. This shows that

$$(13.4.3) \qquad \tau_i \tau_j = (\rho_{i,j}) \quad (i, j = 1, 2, \ldots n).$$

The nonprincipal ideals τ_i, τ_j are so paired as to make $\tau_i \tau_j$ a principal ideal. From (13.4.1) and (13.4.2), we also note that n has to be even, say $2u$.

Every factorization of $\alpha \in R$ has a unique representation in the form

$$(13.4.4) \qquad \alpha = \epsilon \pi_1 \pi_2 \cdots \pi_m \rho_{1,2} \rho_{3,4} \cdots \rho_{n-1,n}$$

where ϵ is a unit and α has $m + u$ prime factors.

An important result about ideal classes is that every class of ideals contains at least one prime ideal. This is given in connection with more stronger theorems in E. Hecke [7]. We examine the case $h(K) > 2$. Assume that there exists an ideal class C of order $t > 2$. Let P be a prime ideal in C and P' a prime ideal in C^{-1}. Then, we have

$$(13.4.5) \qquad P^t = (\pi), \ (P')^t = (\pi') \quad \text{and} \ PP' = (\pi_1),$$

where π, π', π_1 are primes. For, $C^t = R$, $(C^{-1})^t = R$ and $CC^{-1} = R$. (13.4.5) implies that

$$(13.4.6) \qquad \pi_1^t = \epsilon \pi \pi'$$

where ϵ is a unit.

Next, we assume that there exist two classes C_1 and C_2 such that each is of order 2 and $C_3 = C_1 C_2$ is not principal. We choose prime ideals $P_j \in C_j$

($j = 1, 2, 3$). Then,

(13.4.7) $P_j^2 = (\pi_j), \quad (j = 1, 2, 3)$

(13.4.8) $P_1 P_2 P_3 = (\pi)$

where π_1, π_2, π_3 and π are primes. From (13.4.8) we get

(13.4.9) $\pi^2 = \pi_1 \pi_2 \pi_3.$

Let $\alpha \in R$ be given by

(13.4.10) $\alpha = \pi_1 \pi_2 \cdots \pi_s.$

From (13.4.7), (13.4.8) and (13.4.9), we see that when $h(K) > 2$, the number s of primes in (13.4.10) is not independent of the factorization. So, when $h(K) = 2$, one has

(13.4.11) $\alpha = \pi_1 \pi_2 \ldots \pi_s = \tau_1 \tau_2 \ldots \tau_t$

with $s = t$ and π_i and τ_j need not be associates $i, j = 1, 2, \ldots s$, $i \neq j$. $s = t$ follows from the factorization of α as a product of primes, given in (13.4.4).

$\square$

Remark 13.4.1 : Theorem 118 has been adapted from [2].

Example 13.4.1 : It was observed in illustration (13.2.1) (ii) that $\mathbb{Z}[\sqrt{-5}]$ is the number ring of $Q[\sqrt{-5}]$ and that the class number of $Q[\sqrt{-5}]$ is 2. A representative set of ideals is $[\mathbb{Z}[\sqrt{-5}]]$ belonging to the class of principal ideals and $(2, 1 + \sqrt{-5})$ belonging to the class of non-principal ideals. Further,

(13.4.12) $(2, 1 + \sqrt{-5})^2 = (2)$

The two different factorizations of 6 are

$$6 = 2 \cdot 3 = (1 + \sqrt{-5})(1 - \sqrt{-5})$$

Let $P_1 = (2, 1 + \sqrt{-5})$, $P_2 = (3, 1 + \sqrt{-5})$, $P_3 = (3, 1 - \sqrt{-5})$. It is verified that $P_1^2 = (2)$ as in (13.4.12), $P_2 P_3 = (3)$, $P_1 P_2 = (1 + \sqrt{-5})$, $P_1 P_3 = (1 - \sqrt{-5})$. So, in terms of ideal classes,

$$[P_1]^2 = [\mathbb{Z}[\sqrt{-5}]], \ [P_1][P_2] = [\mathbb{Z}[\sqrt{-5}]], \ [P_1][P_3] = [\mathbb{Z}[\sqrt{-5}]]$$

So, P_1, P_2 and P_3 are in the class of non-principal ideals. Every element of $\mathbb{Z}[\sqrt{-5}]$ which is not a power of a prime can be expressed as a product of primes in just two ways.

Observation 13.4.1 : *Carlitz's theorem (Theorem 118) motivates the definition of a half-factorial domain.*

Definition 13.4.1 : *Let D be an integral domain. D is called a half-factorial domain if, given any two factorizations of an element $a \in D$,*

$$a = \pi_1 \pi_2 \cdots \pi_s = \tau_1 \tau_2 \cdots \tau_t$$

with each π_i, τ_j $(i = 1, 2, \ldots s)$, $(j = 1, 2, \ldots t)$ is an irreducible, one has $s = t$.

We do not insist that π_i and τ_j should be associated irreducibles, which is what is required of a UFD. A UFD is, of course, a half-factorial domain (HFD).

A number ring corresponding to an algebraic number field K is a HFD if, and only if, its class-number is less than or equal to 2. Number rings fall into two classes: those which are UFD's (class number 1 case) and those which are non-UFD half-factorial domains (class number 2 case). For a detailed discussion of properties of half-factorial domains in the context of number rings corresponding to quadratic number fields $Q(\sqrt{m})$ (m square-free), see Jim Coykendall [4]. In [4], Jim Coykendall investigates half-factorial domains from the point of view of a 'norm' associated with $\mathbb{Z} + n\mathbb{Z}[\sqrt{m}]$ given by

$$f(x, y) = x^2 - mn^2 y^2.$$

See also A. Zaks [18], [19]. Number fields having class-number 2 have also been studied by E. Hecke [7].

The breakdown of unique factorization in an integral domain could be seen through a semigroup ring. See R. Gilmer [A3].

Let R be a commutative ring and $(s, +)$ an abelian monoid. $\tilde{\mathbb{Z}}$ denotes the set of non-negative integers.

Definition 13.4.2 : *The set*

$$R[x; S] = \{ \sum_{i=0}^{n} r_0 x^{s_i} : n \in \tilde{\mathbb{Z}}, 0 \leq i \leq n, r_i \in R \text{ and } s_i \in S \}$$

endowed with polynomial type addition and multiplication is called the semigroup ring of R over S.

When $S \subset \tilde{\mathbb{Z}}$, $R[x; S]$ may be looked upon as a subring of $R[x]$ (the polynomial ring with x as an indeterminate).

When

(13.4.13) $S = \{ 2m + 3n : m, n \in \tilde{\mathbb{Z}} \}$

S is a submonoid of $\tilde{\mathbb{Z}}$, generated by 2 and 3.

Let k be any field. If S is as given in (13.4.13), we write

(13.4.14) $K[x; S] = \{ \sum_{i=0}^{n} a_i x^{s_i} : n \in \tilde{\mathbb{Z}}, a_i \in K, s_i \in S, 0 \leq i \leq n \text{ and } a_1 = 0 \}$

For instance,

$$a_0 + a_2 x^{s_2} + a_3 x^{s_3} + a_4 x^{s_4} = a_0 + a_2 x^5 + a_3 x^7 + a_4 x^8$$

When $s_2 = 2+3 = 5$, $s_3 = 4+3 = 7$, $s_4 = 2+6 = 8$.

$$K[x;S] \subset K[x^2,x^3]$$

$K[x;S]$ is an integral domain. The units in $K[x;S]$ are the non zero elements of K. x^2, x^3 are irreducibles. One has

(13.4.15) $$x^6 = x^2 \cdot x^2 \cdot x^2 = x^3 \cdot x^3.$$

In (13.4.15), x^6 is a product of two irreducibles. It is also a product of 3 irreducibles. So, $K[x;S]$ is neither a UFD nor a HFD.

Remark 13.4.2 : The above example has been adapted from Scott T. Chapman [A2].

Next, let $\tilde{\mathbb{Q}}$ denote the set of non-negative rational numbers. $\mathbb{C}$ denotes the field of complex numbers.

$(\tilde{\mathbb{Q}},+)$ is an abelian monoid. As in (13.4.14), we write the semigroup ring

(13.4.16) $$\mathbb{C}[x,\tilde{\mathbb{Q}}] = \{\sum_{i=0}^{n} c_i x^{q_i} : n \in \tilde{\mathbb{Z}} \text{ and for } 0 \leq i \leq n, c_i \in \mathbb{C} \text{ and } q_i \in \tilde{Q}\}$$

$\mathbb{C}[x,\tilde{\mathbb{Q}}]$ is an integral domain. Every element of $\mathbb{C}[x,\tilde{\mathbb{Q}}]$ can be factorized into elements which are polynomial like, but having rational numbers for 'degrees'. So, $\mathbb{C}[x,\tilde{\mathbb{Q}}]$ has no irreducible elements.

Another example is from a subring of $\mathbb{C}[x]$. Let $\mathbb{R}$ denote the field of real numbers. We consider

(13.4.17) $$\mathbb{R}+x\mathbb{C}[x] = \{\sum_{i=0}^{n} c_i x^i \in \mathbb{C}[x] : c_i \in \mathbb{C}, \ 0 \leq i \leq n \text{ and } c_0 \in \mathbb{R}\}.$$

It is verified that $\mathbb{R}+x\mathbb{C}[x]$ is a subring of $\mathbb{C}[x]$. $\mathbb{R}+x\mathbb{C}[x]$ is an integral domain. It is a HFD, but not a UFD. See Paulo Ribenboim [A5].

13.5. The Pell equation

An equation of the form $x^2 - my^2 = 1$ where m is square-free is called the Pell equation. The problem is to solve the equation in positive integers for a given m. The British mathematician John Pell (1610–1685) had nothing to do with the equation. Euler had erroneously credited a method of solution to Pell, though the method was found out by another British mathematician William Brounckner (1620–1684) in response to a challenge by Fermat. It is reported that Fermat did have solutions for the cases $m = 109, 149$ and 433 which require large values of x, y. Owing to the reference made by Euler to Pell, the equation continues to be known by Pell's name.

In the case of a linear indeterminate equation of the first degree, say $ax + by = c$, a method of solution in x and y is the same as finding the g.c.d of a and b and is treated by an Indian who called it Kuttaka or Kuttākara (Pulverization). It is referred to in the classical work in Sanskrit: 'Aryabhatiya' written by the Indian mathematician Aryabhata II (920–1000). The equation $x^2 - my^2 = k$ was tackled by the Indian mathematician Brahmagupta (620 A.D.) in his treatise: Brahma sphuta Siddhanta. Brahmagupta considered the identity

$$(13.5.1) \qquad (x^2 - my^2)(z^2 - mt^2) = (xz \pm myt)^2 - m(xt \pm yz)^2$$

In modern terminology, $x^2 - my^2$ is called a binary quadratic form with discriminant m and (13.5.1) says that two binary quadratic forms can be 'composed' to yield another quadratic form with discriminant m in the variables $X = xz \pm myt$. $Y = xt \pm yz$. During the 19th century, Gauss and Dirichlet spoke about composition of quadratic forms and Jacobi used quadratic forms to study determinants arising in the theory of invariants.

Brahmagupta made use of (13.5.1) in the following manner:
Suppose that there exist integers k, k' such that

$$(13.5.2) \qquad x^2 - my^2 = k, \quad z^2 - mt^2 = k'$$

are both solvable with $x = \alpha$, $y = \beta$; $z = \alpha'$, $t = \beta'$ and $\beta\beta' \neq 0$. Then, there exists a solution of the equation

$$(13.5.3) \qquad x^2 - my^2 = kk'.$$

In fact, $x = \alpha\alpha' \pm \beta\beta'$, $y = \alpha\beta' \pm \alpha'\beta$ is also a solution.
When $k' = k$, one has $x^2 - my^2 = k^2$ and it is solvable, provided $x^2 - my^2 = k$ is solvable. That is, if $\alpha^2 - m\beta^2 = k$, we write $\lambda = \alpha^2 + m\beta^2$ $\mu = 2\alpha\beta$. Then,

$$\lambda^2 - m\mu^2 = k^2.$$

Thus, one gets

$$(13.5.4) \qquad (\frac{\lambda}{k})^2 - m(\frac{\mu}{k})^2 = 1$$

That is, the equation $x^2 - my^2 = 1$ is solvable in rational numbers. A solution is $\langle \frac{\alpha^2 + m\beta^2}{k}, \frac{2\alpha\beta}{k} \rangle$. Further, if there is one solution, there are infinitely many. Precisely, if $\langle \alpha, \beta \rangle$ is a solution, $\langle \alpha^2 + m\beta^2, 2\alpha\beta \rangle$ is also a solution. Therefore, by iteration, there are an infinity of solutions. The gist of the argument is that if $x^2 - my^2 = 1$ has a solution, it has infinitely many. However, Brahmagupta did not solve the equation $x^2 - my^2 = 1$ for all m. He solved it for the case $m = 92$ and left the general case open. In Bijaganitha (1150 A.D.), Bhaskara (Bhaskaracharya) mentions an algorithm called 'Cakravala' to solve $x^2 - my^2 = 1$. Both the works of Brahmagupta and Bhaskara were translated by the Englishman H. T. Colebrooke in 1817. His text is entitled 'Algebra with arithmetic and mensuration from the Sanskrit of Brahmagupta and Bhaskara'.

13.6. The Cakravala method

To solve the equation $x^2 - my^2 = 1$, we start with an auxiliary equation $x^2 - my^2 = k$ which can be solved for x and y, say, $x = a$ and $y = b$ so that $a^2 - mb^2 = k$. In other words, we chose k for which $a^2 - mb^2 = k$ holds where $a, b \in \mathbb{N}$. We also assume that a and b are relatively prime to one another.

With $a = l$, $b = 1$ one has

(13.6.1) $$x^2 - my^2 = l^2 - m,$$

where l is so chosen that l^2 is close to m.
Analogous to (13.6.1), we have

(13.6.2) $$(al + mb)^2 - m(bl + a)^2 = k(l^2 - m).$$

Dividing by k^2, we have

(13.6.3) $$\left(\frac{ab + mb}{k}\right)^2 - m\left(\frac{bl + a}{k}\right)^2 = \frac{l^2 - m}{k}.$$

If $\dfrac{l^2 - m}{k}$, $\dfrac{al + mb}{k}$ and $\dfrac{bl + a}{k}$ are integers, we get an integral solution of

(13.6.4) $$x^2 - my^2 = \frac{l^2 - m}{k}$$

and l^2 is close to m. We choose λ such that $bl - k\lambda = -a$. We have solved the indeterminate equation

$$bx - ky = -a$$

by Kuttākara. If $\dfrac{bl + a}{k}$ is an integer,

$$(l^2 - m)b^2 = l^2 b^2 + k - a^2$$
$$= k\left(\frac{bl + a}{k}(bl - a) + 1\right).$$

As $\dfrac{bl + a}{k} \in \mathbb{N}$, k divides $(l^2 - m)b^2$. But k and b are so chosen that g.c.d $(k.b) = 1$ and k divides $l^2 - m$. Then, $\frac{l^2 - m}{k}$ is also an integer. It follows from (13.6.3) that $\dfrac{la + mb}{k}$ is an integer. Therefore, we arrive at

(13.6.5) $$a_1^2 - mb_1^2 = k_1,$$

where $a_1 = \dfrac{la + mb}{k}$, $b_1 = \dfrac{bl + a}{k}$ and $k_1 = \dfrac{l^2 - m}{k}$.

That is, $x^2 - my^2 = k$, has a solution. Repeating the process of transforming the equation $x^2 - my^2 = k$ to $x^2 - my^2 = k_1$, we obtain (a_2, b_2, k_2), (a_3, b_3, k_3) and so on. The claim of the Chakravala method is that this process will eventually lead to (a_r, b_r, k_r) where $k_r = \pm 1, \pm 2, \pm 4$ which could be handled the way Brahmagupta treated such cases.

As R. Sridharan [14] has remarked, Chakravala refers to a cycle in a deeper sense which corresponds to the periodicity of the continued fraction for $\sqrt{m}$. We point out that Fermat had probably a method of solution of the Pell equation for every m. See A. Weil [17].

The Pell equation may be rewritten as

(13.6.6) $$(x + y\sqrt{m})(x - y\sqrt{m}) = 1$$

Therefore, finding a solution of the Pell equation is equivalent to finding a non-trivial unit (whose norm equals 1) in the ring $\mathbb{Z}[\sqrt{m}]$. This leads to the fact that if we know a solution of the Pell equation, we can find infinitely many. That is, if the solutions are ordered by magnitude, the nth solution $\langle x_n, y_n \rangle$ can be obtained from the first one $\langle x_1, y_1 \rangle$ by writing

(13.6.7) $$x_n + y_n \sqrt{m} = (x_1 + y_1 \sqrt{m})^n$$

provided the group of units in $\mathbb{Z}[\sqrt{m}]$ is infinite.

Accordingly, the first solution $\langle x_1, y_1 \rangle$ is called the fundamental solution to the Pell equation. Solving the Pell equation amounts to finding $\langle x_1, y_1 \rangle$ for a given m.

Next, if $R(m)$ denotes the number ring corresponding to an algebraic number field $\mathbb{Q}(\sqrt{m})$, we denote by $N(\alpha)$, the norm of α ($\alpha \in \mathbb{Q}(\sqrt{m})$) (see definition 3.3.2, chapter 3). That is,

(13.6.8) $$\alpha \bar{\alpha} = N(\alpha)$$

Fact 13.6.1 : Let $\alpha, \beta \in R(m)$.

(i) α is a unit if, and only if, $N(\alpha) = \pm 1$.

(ii) If α and β are associates in $R(m)$, $N(\alpha) = \pm N(\beta)$.

(iii) If $N(\alpha)$ is a rational prime, α is irreducible in $R(m)$.

(i) and (ii) are easy to observe. And (iii) has already been noted in chapter 12. See Lemma 12.3.1 and its corollary 12.3.1.

Lemma 13.6.1 : *Let $G(m)$ denote the group of units of $R(m)$. If $m < 0$*

(i) $G(m) = \{\pm 1, \pm i\}$ for $m = -1$.

(ii) $G(m) = \{\pm 1, \pm \omega, \pm \omega^2\}$ where $\omega = \exp(\frac{2\pi i}{3})$ for $m = -3$.

(iii) $G(m) = \{\pm 1\}$ for $m < -3$.

Proof : Suppose that $\alpha \in R(m)$ is a unit. Then, there exists $\beta \in R(m)$ such that $\alpha\beta = 1$. As the norm is multiplicative,

$$N(\alpha)N(\beta) = N(\alpha\beta) = N(1) = 1$$

If $\alpha = a + b\sqrt{m}$, $a, b \in \mathbb{Z}$, $N(\alpha) = a^2 - mb^2$ and $N(\alpha)$ is positive for $m < 0$. So, $N(\alpha) = 1$ if, and only if, $m < 0$

(a) For $m = -1$, one has $N(\alpha) = a^2 + b^2 = 1$.

So, $a = \pm 1$, $b = 0$ and $a = 0$, $b = \pm 1$ are the possible solutions of $a^2 + b^2 = 1$. So, $G(-1) = \{\pm 1, \pm i\}$.

(b) For $m < -3$, $a^2 - mb^2 = 1$ would imply $a = \pm 1$, $b = 0$.

So, $G(m) = \{\pm 1\}$ for $m < 3$.

(c) We are left with the case $m = -3$. As $-3 \equiv 1 \pmod 4$

$$R(-3) = \mathbb{Z}(\zeta) \text{ where } \zeta = \frac{1 + \sqrt{-3}}{2}; \text{ by Remark 3.3.4, chapter 3.}$$

For $\alpha \in R(-3)$, $N(\alpha) = 1 \Rightarrow x^2 + 3y^2 = 1$, where x, y are halfs of rational odd integers. If $x = \frac{s}{2}$, $y = \frac{t}{2}$ where s, t are odd, we get

$$s^2 + 3t^2 = 4$$

The solutions are $\langle \pm 1, \pm 1 \rangle$.
When $s = -1$, $t = 1$, we have $\alpha \in R(-3)$ expressed as

$$\alpha = \frac{s + t\sqrt{-3}}{2} = \frac{-1 + \sqrt{-3}}{2} = \omega \ (\text{say})$$

ω is an imaginary cube-root of unity. The other cases are $\langle 1, -1 \rangle$, $\langle -1, -1 \rangle$ and $\langle 1, 1 \rangle$. These give $\alpha = -\omega, \omega^2, -\omega^2$ respectively. So, $G(-3) = \{\pm 1, \pm \omega, \pm \omega^2\}$, where $\omega = \exp(\frac{2\pi i}{3})$.

All this put together says that we have found out the group of units of $R(m)$ for all $m < 0$, as described. □

Definition 13.6.1 : *A unit δ of $R(m)$ is called a proper unit, if $N(\delta) = 1$. δ is called an improper unit, if $N(\delta) = -1$.*

We remark that there are no improper units in $R(m)$ when $m < 0$. However, for $m > 0$, as in $R(2)$, there exist improper units of $R(m)$. $1 + \sqrt{2}$ is an improper unit of $R(2)$, the number ring corresponding to $Q(\sqrt{2})$.

When m is a square-free positive integer, $Q(\sqrt{m})$ is a subfield of the field $\mathbb{R}$ of real numbers.

Fact 13.6.2 :

(a) When $m \not\equiv 1 \pmod 4$, solving the equation

$$x^2 - my^2 = \pm 1$$

is equivalent to finding the units in the number ring $R(m)$ corresponding to the algebraic number field $Q(\sqrt{m})$.

(b) When $m \equiv 1 \pmod 4$, solving the equation

$$x^2 - my^2 = \pm 4$$

is equivalent to finding the units in the number ring $R(m)$ corresponding to the algebraic number field $(Q\sqrt{m})$.

While considering real quadratic number fields $Q(\sqrt{m})$, $(m > 0)$, we will need the structure of the number ring $R(m)$ corresponding to $Q(\sqrt{m})$. We write

(13.6.9)
$$\zeta = \begin{cases} \sqrt{m} & \text{if } m \not\equiv 1 \pmod 4 \\ \frac{-1 + \sqrt{m}}{2} & \text{if } m \equiv 1 \pmod 4 \end{cases}$$

Then,

(13.6.10)
$$R(m) = \mathbb{Z} + \mathbb{Z}[\zeta].$$

(See Remark 3.3.4, chapter 3. It is immaterial whether we take $\zeta = \frac{1+\sqrt{m}}{2}$ or $\frac{-1+\sqrt{m}}{2}$). For $\alpha = a+b\zeta \in R(m)$,

$$(13.6.11) \qquad \bar{\alpha} = \begin{cases} a-b\zeta & \text{, if } m \not\equiv 1 \,(\text{mod } 4) \\ (a-1)-b\zeta & \text{, if } m \equiv 1 \,(\text{mod } 4) \end{cases}$$

The real-axis is split into four open intervals:

$$I_1 = (-\infty,-1), I_2 = (-1,0), I_3 = (0,1) \text{ and } I_4 = (1,\infty).$$

Let $\delta = a+b\zeta$ $(a,b \in \mathbb{Z})$ be a unit other than ± 1, belonging to $R(m)$. For $m = 2$, $\delta = 1+\sqrt{2}$ is a non-trivial unit in $R(2)$. As $1+\sqrt{2} > 1$, $\delta \in I_4$. $\bar{\delta} = 1-\sqrt{2}$ lies in I_2. Further, $-\delta = -1-\sqrt{2}$ lies in I_1 and $-\bar{\delta} \in I_3$. That is, I_1, I_2, I_3 and I_4 contain exactly one and only one of the units $\pm\delta, \pm\bar{\delta}$. This is true of $R(m)$, m a square-free positive integer.

Fact 13.6.3 : (i) Let $\delta = a+b\zeta$ be a unit other than ± 1 in $R(m)$. Then, just one of the four units $\pm\delta, \pm\bar{\delta}$ lies in each of the intervals I_1, I_2, I_3, I_4. Further, $\delta > 1$ if, and only if, $a > 0$, $b > 0$.
(ii) Let $M \in \mathbb{R}$ be such that $M > 1$. The interval $(1,M]$ contains only a finite number of units of $R(m)$.

For proof, see Ethan D. Bolker [1, chapter 6, Lemma 31.1 and corollary 31.2].

Theorem 119 (Dirichlet's inequalities) :

(a) Let t be a positive irrational number. Given M, a positive integer, there exist integers y,x ; $0 \le y \le M$ and $x \ge 0$ such that

$$(13.6.12) \qquad |x-yt| < \frac{1}{M}.$$

(b) There are infinitely many pairs $\langle x,y \rangle$ (with $y \ne 0$) of integers such that

$$(13.6.13) \qquad |\frac{x}{y}-t| < \frac{1}{y^2}.$$

Proof : (a) Let $[s]$ denote the greatest integer not exceeding s. We have

$$0 \le s-[s] < 1.$$

We divide the interval $[0,1)$ into M subintervals:

$$[1,\frac{1}{M}),[\frac{1}{M},\frac{2}{M}),\ldots,[\frac{M-1}{M},1).$$

Each of the subintervals shown above has length $1/M$. By the pigeon-hole principle, two of the $M+1$ numbers

$$t-[t], 2t-[2t],\ldots,Mt-[Mt], (M+1)t-[(M+1)t]$$

must lie in the same interval. So, there exist integers j, k such that $0 < j < k \le M+1$ and

$$|[kt]-[jt]-(k-j)t| < \frac{1}{M}.$$

We take $x = [kt] - [jt]$ and $y = k - j$ $(< M)$. Then,

$$|x - yt| < \frac{1}{M} \quad \text{as desired.}$$

(b) By the result in (a), there exists one pair $\langle x, y \rangle$ of integers satisfying (13.6.12) namely $x = 0$, $y = M$ giving $|t| < 1/M^2$.

Suppose that we have found n pairs $\langle x_j, y_j \rangle$ $(j = 1, 2, \ldots n)$ such that (13.6.13) holds. We consider the absolute differences

$$d_j = |x_j - ty_j| \quad (j = 1, 2, \ldots n).$$

Let d_q be the least among d_j. Since t is irrational, $d_q > 0$. Suppose M denotes an integer $> \frac{1}{d_q}$. Then, by (a), we can find x and $y \neq 0$ such that

$$|x - yt| < \frac{1}{M}.$$

Since $\frac{1}{M} < d_q$, $\langle x, y \rangle$ is not one of the pairs $\langle x_j, y_j \rangle$ $(j = 1, 2, \ldots n)$. But, $0 < y \leq M$ gives

$$|\frac{x}{y} - t| < \frac{1}{My} \leq \frac{1}{y^2}.$$

So $\langle x, y \rangle$ is the $(n+1)$th pair satisfying (13.6.13). Hence, by induction on n, the number of pairs $\langle x, y \rangle$ for which (13.6.13) holds is infinite. $\qquad \square$

Theorem 120 (Existence of a nontrivial solution of the Pell equation) :
There exist integers $y \neq 0$ and x such that

(13.6.14) $\qquad x^2 - my^2 = 1$, *where m is a square-free integer ≥ 2.*

Proof : By theorem 119(b), we can find integers x, y $(\neq 0)$ satisfying

(13.6.15) $\qquad |\frac{x}{y} - \sqrt{m}| < \frac{1}{y^2}.$

Let $\alpha = x + y\sqrt{m}$, $\alpha \in \mathbb{Z} + \mathbb{Z}[\sqrt{m}] \subseteq R(m)$, the number ring corresponding to the algebraic number field $\mathbb{Q}(\sqrt{m})$.

$$|N(\alpha)| = |(x - y\sqrt{m})(x + y\sqrt{m})|$$
$$= y^2 |\frac{x}{y} - \sqrt{m}| |\frac{x}{y} + \sqrt{m}|$$
$$< |\frac{x}{y} + \sqrt{m}|, \text{ by (13.6.15).}$$

But,

$$|\frac{x}{y} + \sqrt{m}| = |\frac{x}{y} - \sqrt{m} + 2\sqrt{m}|$$
$$\leq |\frac{x}{y} - \sqrt{m}| + 2\sqrt{m}$$
$$< \frac{1}{y^2} + 2\sqrt{m}.$$

Therefore, $|N(x+y\sqrt{m})| < 1+2\sqrt{m}$, as $y \geq 1$.

So, $N(\alpha)$ is an integer lying between $-1-2\sqrt{m}$ and $1+2\sqrt{m}$ for infinitely many pairs $\langle x,y \rangle$, $y \neq 0$. This is deduced from theorem 119(b). So, there exists a rational number γ with the property: $|\gamma| < 1+2\sqrt{m}$ and

(13.6.16) $N(\alpha) = N(x+y\sqrt{m}) = \gamma$,

for infinitely many $\alpha \in \mathbb{Z}+\mathbb{Z}[\sqrt{m}]$.

Suppose that $\beta \in \mathbb{Z}+\mathbb{Z}[\sqrt{m}]$ satisfies (13.6.16) and $\alpha \neq \pm\beta$. Then $\alpha\beta^{-1} \neq \pm 1$ and $N(\alpha\beta^{-1}) = \gamma\gamma^{-1} = 1$. Therefore, $\alpha\beta^{-1}$ is a nontrivial unit in $R(m)$, if $\alpha\beta^{-1} \in R(m)$. But, then,

$$\alpha\beta^{-1} = \frac{\alpha\bar{\beta}}{N(\beta)} = \frac{\alpha\bar{\beta}}{\gamma}.$$

Thus, $\alpha\beta^{-1} \in R(m)$, provided γ divides $\alpha\bar{\beta}$ in $R(m)$. We will show that this is, indeed, the case.

If we consider residue classes modulo $|\gamma|$, there are $|\gamma|$ possible values for x and $|\gamma|$ possible values for y and so there are γ^2 possible values for pairs $\langle x,y \rangle$ modulo $|\gamma|$. Among the infinitely many α satisfying (13.6.16), there are two, namely, $\pm\beta \neq \alpha$ such that γ divides $\alpha-\beta$ in $R(m)$. Then,

$$\gamma \text{ divides } (\alpha-\beta)\bar{\beta} = \alpha\bar{\beta} - \beta\bar{\beta} = \alpha\bar{\beta} - \gamma.$$

So, γ divides $\alpha\bar{\beta}$ and hence $\alpha\beta^{-1}$ is a proper unit in $R(m)$ with $N(\alpha\beta^{-1}) = 1$. It means that there exists a pair $\langle x_0,y_0 \rangle$ with $y_0 \neq 0$, such that $x_0^2 - my_0^2 = 1$. That is, the Pell equation $x^2 - my^2 = 1$ has a non-trivial solution.   □

Theorem 121 : *There exists a unit $\eta \in R(m)$, $\eta \neq \pm 1$ such that every unit in $R(m)$ is of the form $\pm\eta^n$, $n \in \mathbb{Z}$.*

Proof : The statement of the theorem presumes that there exists a unit $\eta \neq \pm 1$ belonging to $R(m)$. This is precisely the same as saying that the Pell equation $x^2 - my^2 = 1$ has a non-trivial solution. In other words, $R(m)$ has a proper unit other than ± 1. This has been established in theorem 120 for square-free $m \geq 2$. The following step is meaningful:

There exists a unit $\delta \neq \pm 1$ in $R(m)$.

$\delta = a+b\zeta$ is an algebraic integer. For $m > 0$, $\mathbb{Q}(\sqrt{m})$ is a real quadratic extension of $\mathbb{Q}$. Any element $\alpha = c+d\sqrt{m}$ belonging to $\mathbb{Q}(\sqrt{m})$ belongs to $\mathbb{R}$, the field of real numbers. $\mathbb{R}$ is a complete ordered field under the partial order $\leq$ and so an order structure in $(Q(\sqrt{m}), \leq)$ is inherited from $(\mathbb{R}, \leq)$. We can compare $a+b\zeta$ and $c+d\zeta$; $a,b,c,d \in \mathbb{Z}$. That is, when $\delta \neq \pm 1$ in $R(m)$ is a unit, we can assume that $\delta > 1$. By Fact 13.6.3(ii), there are only a finite number of units in the interval $(1, \delta]$. Let η be the least among them.

For $\alpha, \beta \in R(m)$, $|N(\alpha)| = |N(\beta)|$ implies that α and β are associates. So, when δ is a unit, $\pm\delta^n$ is also a unit, $n \in \mathbb{Z}$.

We claim that every non-trivial unit in $R(m)$ is of the form $\pm\eta^n$. Suppose that $\beta > 1$ is a unit. There exists $n \in \mathbb{Z}$ such that

(13.6.17) $0 < \eta^{n-1} < \beta \leq \eta^n$.

Multiplying throughout by η^{1-n}, we get

(13.6.18) $$0 < 1 < \beta\eta^{1-n} \leq \eta.$$

But, then, $\beta\eta^{1-n}$ is a unit. As η is the least among the units in $[1, \delta)$,

$$\beta\eta^{1-n} = \eta. \quad \text{Therefore, } \beta = \eta^n.$$

This completes the proof. $\qquad\qquad\qquad\qquad\qquad\qquad\qquad\qquad\qquad\qquad\qquad\qquad$ □

Remark 13.6.1 : Theorem 121 stated differently means that the group of units of $R(m) = \{\pm\eta^n : n \in \mathbb{Z}\}$. If $G(m)$ denotes the group of units of $R(m)$,

$$G(m) \cong \mathbb{Z}/2\mathbb{Z} \times \mathbb{Z}.$$

It is a special case of Dirichlet's unit theorem which gives the structure of $G(m)$ for all square-free m (see Section 13.7).

Definition 13.6.2 : η *given in theorem 121 is called the fundamental unit of $R(m)$.*

As examples, we take $R(2)$ and $R(3)$.
In the case of $R(2)$, $1 + \sqrt{2}$ is the fundamental unit and $1 + \sqrt{2}$ is the least algebraic integer > 1. Any unit of $R(2)$ is either ± 1 or $\pm(1 + \sqrt{2})^n$, $n \in \mathbb{Z}$.
In the case of $R(3)$, $1 + \sqrt{3}$ is not a unit. But, $2 + \sqrt{3}$ is the least algebraic integer > 1 which is also a unit. So, $2 + \sqrt{3}$ is the fundamental unit in $R(3)$.

Observation 13.6.1 : *The fundamental solution $\langle x_0, y_0 \rangle$ to the Pell equation $x^2 - my^2 = 1$ is the one which minimizes $x + y\sqrt{m}$ for positive integers x and y.*

Case (i) $m \not\equiv 1$ (mod 4):
    $x_0 + y_0\sqrt{m}$ *is either the fundamental unit η of $R(m)$ if η is a proper unit or $x_0 + y_0\sqrt{m}$ is equal to η^2 if η is an improper unit of $R(m)$.*

Case (ii) $m \equiv 1$ (mod 4):
    If η is the fundamental unit of $R(m)$, $x_0 + y_0\sqrt{m} = \eta^k$ for a suitable positive integer k.

Examples 13.6.1 : In the case of $R(2)$, the fundamental unit is $1 + \sqrt{2}$. It is an improper unit. $\eta^2 = 3 + 2\sqrt{2}$. $\langle x_0, y_0 \rangle = \langle 3, 2 \rangle$ gives the fundamental solution of $x^2 - 2y^2 = 1$.
In the case of $R(3)$ the fundamental unit is $2 + \sqrt{3}$ which is proper. So, $\langle x_0, y_0 \rangle = \langle 2, 1 \rangle$ gives the fundamental solution of $x^2 - 3y^2 = 1$.
In the case of $R(5)$, the fundamental unit is $2 + \sqrt{5}$ which is improper. Let $\eta = 2 + \sqrt{5}$.

$$(2 + \sqrt{5})^6 = 2889 + 1292\sqrt{5}$$

$\langle x_0, y_0 \rangle = \langle 2889, 1292 \rangle$ is the fundamental solution of $x^2 - 5y^2 = 1$ as, $2889^2 - 5 \times 1292^2 = 1$.
In the case of $R(6)$, the fundamental unit is $5 + 2\sqrt{6}$ which is proper. So, $\langle x_0, y_0 \rangle = \langle 5, 2 \rangle$ is the fundamental solution to $x^2 - 6y^2 = 1$.

In the case of $R(14)$, the fundamental unit is $15 + 4\sqrt{14}$ which is proper. $\langle x_0, y_0 \rangle = \langle 15, 4 \rangle$ is the fundamental solution to $x^2 - 14y^2 = 1$.

In the case of $R(23)$, the fundamental unit is $24 + 5\sqrt{23}$ which is proper. $\langle x_0, y_0 \rangle = \langle 24, 5 \rangle$ is the fundamental solution to $x^2 - 23y^2 = 1$.

In the case of the prime 17, $4 + \sqrt{17}$ is an improper unit of $R(17)$. Also, $(4 + \sqrt{17})^2 = 33 + 8\sqrt{17}$ is such that $\langle x_0, y_0 \rangle = \langle 33, 8 \rangle$ is the fundamental solution of $x^2 - 17y^2 = 1$. The square-root of $33 + 8\sqrt{17}$ is a unit in $R(17)$. More generally, it can be shown that if p is a prime of the form $4k + 1$, the number ring $R(p)$ corresponding to the number field $\mathbb{Q}(\sqrt{p})$ has an improper unit whose square yields the fundamental solution to the Pell equation $x^2 - py^2 = 1$. See Ethan D. Bolker [1, chapter 6, p.134] J. Esmonde and M. Ram Murthy give related results and problems in [5, chapter 8].

13.7. Dirichlet's unit theorem

We make a closer look at the multiplicative group U of units in a number ring R corresponding to an algebraic number field K. Dirichlet's unit theorem says that U is the direct product of a finite cyclic group consisting of the roots of unity in R and a free abelian group. It is known that $K = \mathbb{Q}(\theta)$ where θ is an algebraic integer. We assume that $[K : Q] = n$. Let $\sigma_1, \sigma_2, \ldots \sigma_n$ be the embeddings of K in $\mathbb{C}$. In fact, there are exactly n distinct monomorphisms of K into $\mathbb{C}$. They are given by $\sigma_i : K \to \mathbb{C}$ where $\sigma_i(\theta) = \theta_i$ (say) ($i = 1, 2, \ldots n$). The minimum polynomial of θ over $\mathbb{Q}$ is of degree n and θ_i ($i = 1, 2, \ldots, n$) are the distinct zeros of the minimum polynomial of θ in $\mathbb{C}$. If $\sigma_i(\theta) \in \mathbb{R}$, we say that σ_i is real. Otherwise, σ_i is called complex ($i = 1, 2, \ldots, n$). Suppose that $\overline{\sigma_i}(\alpha) = \overline{\sigma_i(\alpha)}$. Since complex conjugation is an automorphism of $\mathbb{C}$, $\bar{\sigma}_i$ is also a embedding of K in $\mathbb{C}$. Complex embeddings occur in conjugate pairs. We write $n = r + 2s$ where r denotes the number of real embeddings of K in $\mathbb{C}$ and $2s$ denotes the number of complex embeddings of K in $\mathbb{C}$. Hereafter, $n = r + 2s$ will mean that r gives the component for real embeddings of K in $\mathbb{C}$ and $2s$ gives the component for non-real embeddings of K in C. The following notation, therefore, stands:

(13.7.1) $n = r + 2s$.

Next, we note that R is a Noetherian domain with field of quotients F such that $K \subseteq F$. Let M be an R-module. We recall that $x \in M$ is called a torsion element if $ax = 0_M$ for some nonzero element $a \in R$. If M is a finitely-generated R-module, the torsion elements of M form a finitely-generated R-submodule, denoted by T_M, where

(13.7.2) $T_M = \{ x \in M : ax = 0_M, \text{ for some } a \in R \}$

The annihilator of M is written as

(13.7.3) $ann M = \{ a \in R : ay = 0_M \text{ for all } y \in N \}$.

$annM$ is an ideal of R. M is called torsion-free, if $T_M = (0_M)$ implying that 0_M is the only torsion element of M. We recall that an R-module M' is said to be a free

R-module on $S = \{g_1, g_2, \ldots g_k\}$ of generators if, and only if, every element m in M' can be uniquely expressed as

$$(13.7.4) \qquad m = \sum_{i=1}^{k} a_i g_i, \quad a_i \in R, (i = 1, 2, \ldots k).$$

Lemma 13.7.1 : *Let M be a nonzero finitely-generated R-module. M is torsion-free if, and only if, one of the following conditions hold:*

(a) M is isomorphic to a R-submodule of a free R-module of finite rank.

(b) M is isomorphic to an R-submodule of a finite dimensional K-vector space V, where R is contained in K.

Proof : Suppose that (a) holds. Let N be a free R-module of finite rank. N has a basis which is a set of linearly independent generators of N. M is isomorphic to a R-submodule of N implies that the R-submodule N' of N for which $M \cong N'$ is such that N' has a basis of linearly independent generators and so, N' is torsion-free. That is, M is torsion-free.

Conversely, if M is finitely-generated and M is torsion-free, then M has a basis of linearly independent generators and so, M is isomorphic to a R-submodule of a free R-module of finite rank.

Next, suppose that (b) holds. Let $\{e_j : j = 1, 2, \ldots, k\}$ be a K-basis for the vector space V, where K is a field containing R. Assume that $\{g_1, g_2 \ldots g_m\}$ denotes a generating set of M over R. If $b \neq 0_R$ and $b \in R$, then bg_i $(i = 1, 2, \ldots, m)$ is an element of M. By embedding M in V, bg_i can be taken as an element of V. So,

$$(13.7.5) \qquad bg_i = \sum_{j=1}^{k} a_{ij} e_j, \quad a_{ij} \in K (i = 1, 2, \ldots, m).$$

Considering b^{-1} as an element of K, $\{b^{-1} e_j : j = 1, \ldots, k\}$ forms a basis for an R-module N containing M. That is, M is contained in a free R-module of finite rank.

So, by (a), M is torsion-free. Conversely, if M is finitely generated and torsion-free, taking K as the field of quotients of R, M can be embedded in a finite dimensional K-vector space V which yields (b). $\qquad \square$

Next, we need some geometrical ideas.

Let V be a k-dimensional vector space over $\mathbb{R}$ (the field of real numbers). Let $\{e_1, e_2 \ldots, e_k\}$ be a basis for V.

Let $\Omega = \{v \in V : v = \sum_{i=1}^{k} a_i e_i, a_i \in \mathbb{Z}\}$.

$(\Omega, +)$ is a subgroup of $(V, +)$, generated by $\{e_1, e_2 \ldots e_k\}$.

For $x, y \in V$, using the usual 'distance' scalar $d(x, y) = \{\sum_{i=1}^{k} (x_i - y_i)^2\}^{1/2}$ (see Fact 8.4.1, chapter 8) where $x = \sum x_i e_i$, $y = \sum y_i e_i$. $d(x, y)$ defines a metric on V and we can consider a metric topology on V.

Definition 13.7.1 : *A subset S of V is said to be bounded, if S is contained in $B_r(0) = \{x \in V : d(x, 0) \leq r\}$, (a spherical k-ball of radius r, $r > 0$) for some $r > 0$.*

Definition 13.7.2 : *An additive subgroup Λ of V is said to be discrete, if the set $\Lambda \cap S$ is finite for all bounded subsets S of V.*

For $t \leq k$,

$$(13.7.6) \qquad \Omega_t = \{x \in V : \sum_{i=1}^{t} a_i e_i = x, a_i \in \mathbb{Z}\}$$

is a discrete subgroup of V, since it has finite intersection with any bounded subset of V. (see Fact 8.4.1, chapter 8).

Definition 13.7.3 : $\Omega_t = \{x \in V : \sum_{i=1}^{t} a_i e_i = x, a_i \in \mathbb{Z}\}$ *is called a lattice of dimension t ($t \leq n$) generated by $\{e_1, e_2, \ldots e_t\}$.*

Theorem 122 : *An additive subgroup Λ of V is a discrete subgroup of V if, and only if, Λ is a lattice of dimension t ($\leq k$).*

Proof : $\Leftarrow$: Suppose that Λ is a lattice of dimension t, contained in V. Then, $(\Lambda, +)$ is a free abelian group of finite rank t. So, Λ is generated by an $\mathbb{R}$-linearly independent set of vectors $\{v_1, v_2, \ldots v_t\}$. We enlarge this to an $\mathbb{R}$ basis of V, say, $\{v_1, v_2 \ldots v_t, v_{t+1}, \ldots v_k\}$. If S denotes a bounded subset of V, suppose that $x \in S$ is of the form $x = \sum_{i=1}^{k} x_i v_i$. It follows that the coordinates x_i are all bounded and so $\Lambda \cap S$ is finite. That is, Λ is a discrete subgroup of V.

$:\Rightarrow$ Suppose that Λ is a discrete subgroup of V. We have to show that Λ is a lattice of dimension t (say) ($t \leq k$).

Let $\mathbb{B} = \{v_1, v_2, \ldots v_t\}$ denote a maximal subset of Λ such that $\mathbb{B}$ is a linearly independent set in V. We define

$$(13.7.7) \qquad T_{\mathbb{B}} = \{\sum_{i=1}^{t} a_i v_i : 0 \leq a_i \leq 1\}$$

$T_{\mathbb{B}}$ is a subset of V. By the maximality condition imposed on $\mathbb{B}$, given $x \in \Lambda$, we can write

$$(13.7.8) \qquad x = \sum_{i=1}^{t} x_i v_i, \quad x_i \in \mathbb{R}, \text{set of real numbers } (i = 1, 2, \ldots k).$$

If $[y]$ denotes the greatest integer not greater than y,

$$(13.7.9) \qquad x = \sum_{i=1}^{t} (x_i - [x_i]) v_i + \sum_{i=1}^{t} [x_i] v_i.$$

$[x_i] \in \mathbb{Z}$ and $[x_i] \leq x_i < [x_i] + 1$, $i = 1, 2, \ldots k$.
As $0 \leq x_i - [x_i] < 1$ $(i = 1, 2, \ldots, k)$,

$$\sum_{i=1}^{t} (x_i - [x_i]) v_i \text{ is in } T_{\mathbb{B}} \text{ as well as } \Lambda.$$

From (13.7.9), we note that Λ is generated over $\mathbb{Z}$ by $T_B \cap \Lambda$ and $\{v_1, v_2, \ldots v_t\}$. Since $T_{\mathbb{B}} \cap \Lambda$ is finite, Λ is finitely generated as a $\mathbb{Z}$-module. Since V has no torsion elements other than $\vec{0}$, Λ is torsion-free. By definition (13.7.3), Λ is a free $\mathbb{Z}$-module, being isomorphic to a $\mathbb{Z}$-submodule of the finite dimensional vector space V over $\mathbb{R}$.

Let $x \in T_\mathbb{B} \cap \Lambda$. Take $x = \sum_{i=1}^{t} x_i v_i$. For each positive integer j we introduce

$$(13.7.10) \qquad y_j = \sum_{i=1}^{t} (x_{ij} - [x_{ij}]) v_i.$$

Then, $y_j \in T_\mathbb{B} \cap \Lambda$. Therefore, we can find distinct integers l, m such that $y_l = y_m$. That is,

$$\sum_{i=1}^{t} x_i (l-m) v_i = \sum_{i=1}^{k} ([x_i l] - [x_i m]) v_i$$

Since $\{v_1, \ldots v_t\}$ is a linearly independent set in V, we may equate the coefficients to obtain

$$x_i(l-m) = [x_i l] - [x_i m]$$

This shows that x_i is a rational number ($i = 1, 2 \ldots t$). Since $T_B \cap \Lambda$ is finite, $\sum_{i=1}^{t} a_i v_i$ with $a_i \in \mathbb{Z}$ is such that

$$\Lambda' = \{ y \in V : \sum_{i=1}^{t} a_i y_i = y \text{ with } a_i \in \mathbb{Z} \}$$

is a subgroup of Λ such that the quotient group Λ/Λ' is finite. Now, as $\mathbb{B} = \{v_1, v_2, \ldots v_t\}$ is a linearly independent set in V, by transforming $\mathbb{B}$ into a $\mathbb{Z}$-basis $\{v_1', v_2', \ldots v_t'\} = \mathbb{B}'$, we can take $\mathbb{B}'$ as a $\mathbb{Z}$-basis for Λ. The matrix of the transformation is non-singular over $\mathbb{Q}$, the field of rationals. Thus, the matrix of transformation is non-singular over $\mathbb{R}$. So $\mathbb{B}' = \{v_1', v_2', \ldots v_t'\}$ is a linearly independent set over $\mathbb{R}$. So, Λ is a free $\mathbb{Z}$-module of finite rank and is contained in V. Hence, Λ is a lattice of dimension t. $\qquad \square$

Starting again with number field K and having R as its ring of integers, we consider U_K, the group of units of K. Let μ_K denote the group of roots of unity in K. $\zeta \in \mu_K$ is an algebraic integer, as ζ satisfies a polynomial equation $x^m - 1 = 0$ for some $m \geq 1$. As $\zeta \cdot \zeta^{m-1} = \zeta^m = 1$, we note that a root of unity say ζ is indeed a unit. So, $\zeta \in U_K$. Let T_K denote the $\mathbb{Z}$-torsion submodule of the $\mathbb{Z}$-module U_K.

Clearly, $\mu_K \subset T_K$. Now, if $u \in T_K$, there exists a positive integer m such that $u^m = 1$. So, $u \in \mu_K$ or

$$(13.7.11) \qquad \mu_K = T_K$$

We are interested in studying the torsion-free part of U_K. Let $F_K = U_K / \mu_K$. If $n = r + 2s$ in the notation of (13.7.1), we claim that $F_K \cong \mathbb{Z}^{(r+s-1)}$. Then,

$$(13.7.12) \qquad U_K \cong \mu_K \times \mathbb{Z}^{(r+s-1)}$$

which implies that U_K is a finitely-generated $\mathbb{Z}$-module.
(13.7.12) is, in fact, the essence of Dirichlet's unit theorem. $(r+s-1)$ is called the Dirichlet rank of U_K.

Let $\sigma : K \to \mathbb{R}^r \times \mathbb{C}^s$ be defined by

$$(13.7.13) \qquad \sigma(\alpha) = (\sigma_1(\alpha), \ldots \sigma_r(\alpha); \sigma_{r+1}(\alpha), \ldots, \sigma_{r+s}(\alpha))$$

where $\sigma_1, \sigma_2, \ldots \sigma_{r+s}$ are the embeddings of K in $\mathbb{C}$. We observe that when K is considered as a vector space over $\mathbb{Q}$, Q-linearly independent elements of K map into $\mathbb{R}$-linearly independent elements of $\mathbb{R}^r \times \mathbb{C}^s$. If G is a finitely-generated $\mathbb{Z}$-module contained in K which is $\mathbb{Z}$-free of rank m and having a basis $\{\alpha_1, \alpha_2 \ldots \alpha_m\}$, the image of G in $\mathbb{R}^r \times \mathbb{C}^s$ under σ is a lattice of dimension m in $\mathbb{R}^r \times \mathbb{C}^s$ with the generating set $\{\sigma(\alpha_1), \sigma(\alpha_2), \ldots, \sigma(\alpha_m)\}$.

Next, we define a map ψ :

$$(13.7.14) \qquad \psi : \mathbb{R}^r \times \mathbb{C}^s \to \mathbb{R}^{r+s}$$

in terms of $\psi_i(i = 1, 2, \ldots, r : r+1, \ldots, r+s)$, where

$$(13.7.15) \qquad \psi_i(x) = \begin{cases} \log|x_i|, & i = 1, 2, \ldots r; \\ \log|x_i|^2, & i = r+1, \ldots, r+s; \end{cases}$$

where $x = (x_1, x_2, \ldots x_r; x_{r+1}, \ldots x_{r+s}) \in \mathbb{R}^r \times \mathbb{C}^s$.
Then, $\psi(x) = (\psi_1(x), \psi_2(x), \ldots, \psi_{r+s}(x))$.
It is easy to check that when $x \cdot y = (x_1 y_1, x_2 y_2, \ldots, x_{r+s} y_{r+s}) : x, y \in \mathbb{R}^r \times \mathbb{C}^s$.

$$\psi(x \cdot y) = \psi(x) + \psi(y), \quad \text{for } x, y \in \mathbb{R}^r \times \mathbb{C}^s$$

So, we have maps $\sigma : K \to \mathbb{R}^r \times \mathbb{C}^s$ and $\psi : \mathbb{R}^r \times \mathbb{C}^s \to \mathbb{R}^{r+s}$.
If $\sigma' = \psi \circ \sigma$, $\sigma' : K \to \mathbb{R}^{r+s}$ is such that for $\alpha \in K$, we have

$$(13.7.16) \qquad \sigma'(\alpha) = \psi(\sigma(\alpha)).$$

That is,

$$\sigma'(\alpha) = (\log|\sigma_1(\alpha)|, \ldots, \log|\sigma_r(\alpha)|; \log|\sigma_{r+1}(\alpha)|^2, \ldots \log|\sigma_{r+s}(\alpha)|^2).$$

$\sigma' : K \to \mathbb{R}^{r+s}$ is known as the logarithmic representation of K. $\mathbb{R}^{r+s}$ is called the logarithmic space. It is verified that

$$(13.7.17) \qquad \sigma'(\alpha\beta) = \sigma'(\alpha) + \sigma'(\beta) \quad \text{for } \alpha, \beta \in K$$

Let $K^* = K \setminus \{0\}$, $\sigma' : K^* \to \mathbb{R}^{r+s}$ is a homomorphism from the multiplicative group K^* into the additive group $(\mathbb{R}^{r+s}, +)$.

We restrict σ' to U_K, the group of units of R, the number ring corresponding to the number field K. We get a homomorphism $\sigma_R' : U_K \to \mathbb{R}^{r+s}$. In what follows, we will see that σ_R' does the job.

Let H denote the kernel of σ_R'.

$$H = \{u \in U_K : \sigma_R'(u) = 0\}$$

$\sigma_R'(u) = 0 \Rightarrow |\sigma_i(u)| = 1$ for $i = 1$ to $r+s$.
u is an algebraic integer. The minimum polynomial of u given by

$$f(t) = \prod_{i=1}^{r+s} (t - \sigma_i(u))$$

has integer coefficients. It is known that if $p(t) \in \mathbb{Z}[t]$ is a monic polynomial all of whose zeros are in $\mathbb{C}$ and have absolute value 1, then, every zero of $p(t)$ is a root of unity. Therefore, all the $\sigma_i(u)$ are roots of unity. In particular, u itself is a root of unity. The image of R in $\mathbb{R}^r \times \mathbb{C}^s$ under σ is a lattice and so is a discrete subgroup of $\mathbb{R}^r \times \mathbb{C}^s = \mathbb{R}^{r+2s}$. So, H contains only finitely many roots of unity. But any finite subgroup of K^* is cyclic. H contains -1 which is a unit of order 2. So, $\pm u$ is a unit of finite order when it belongs to H. So, H is a finite cyclic group of even order. We have denoted H by μ_K.

Let J be the image of U_K under σ'_R.

Claim :

(13.7.18) J is a lattice of dimension $\leq r + s - 1$

We note that the norm of a unit is ± 1.
From (13.7.15), we see that if $x = (x_1, x_2, \ldots x_r; x_{r+1} \ldots x_{r+s})$.

$$\sum_{i=1}^{r+s} \psi_i(x) = \sum_{i=1}^{r} \log|x_i| + \sum_{i=r+1}^{r+s} \log|x_i|^2$$

If $u \in U_K$,

$$\sum_{i=1}^{r+s} \psi_i(u) = \log|N(u)| = \log 1 = 0.$$

So, if $y \in J$, $y = (y_1, y_2, \ldots, y_{r+s})$ is such that

(13.7.19) $y_1 + y_2 \ldots + y_{r+s} = 0$

If $V = \{y \in \mathbb{R}^{r+s} : y_1 + y_2 \ldots + y_{r+s} = 0\}$, V is a subspace of $\mathbb{R}^{r+s}$ and $J \subseteq V$. Clearly, V has dimension $r + s - 1$, because of the restriction (13.7.19) on the coordinates of an element $y \in V$.

Let $|| \ ||$ denote the usual norm on $\mathbb{R}^{r+s}$. For $t > 0, t \in \mathbb{R}$ and $u \in U_K$, $||\psi(u)|| < t \Rightarrow |\psi_i(u)| \leq ||\psi(u)|| < t$.

So,

$$\log|\psi_i(u)| < t \quad i = 1, 2, \ldots, r;$$
$$\log|\psi_i(u)|^2 < t \quad i = r+1, \ldots, r+s;$$
$$\text{or, } |\psi_i(u)| < e^t \quad i = 1, 2, \ldots, r;$$
$$|\psi_i(u)|^2 < e^t \quad i = r+1, \ldots, r+s.$$

So, the set of points $\psi(u)$ in $\mathbb{R}^{r+s}$ corresponding to the units u with $||\psi(u)|| < t$ is bounded and finite. So, J intersects a closed ball $B_t(0)$ in $\mathbb{R}^{r+s}$ in a finite set and so J is a discrete subgroup of $\mathbb{R}^{r+s}$. Therefore, J is a lattice of dimension $\leq r + s - 1$, by theorem 122, as claimed in (13.7.18).

Lemma 13.7.2 : *In $\mathbb{R}^m$, there exists a lattice L of dimension m and having a basis $\{e_1, e_2 \ldots e_m\}$ which is also a basis for $\mathbb{R}^m$.*

Proof : We define the set

$$D = \{\sum_{i=1}^{m} a_i e_i : 0 \le a_i < 1\}$$

as a fundamental domain for the lattice L. For $x \in L$,

$$D + x = \{\sum_{i=1}^{m} a_i e_i + x : 0 \le a_i < 1\}$$

is a subset of $\mathbb{R}^m$ whose elements are of the form $d + x$ where $d \in D$. It is easy to check that each element v of $\mathbb{R}^m$ lies in one of the subsets $D + x$, $x \in L$. For, if $v \in D + x_1$ as well as $D + x_2$ for $x_1, x_2 \in L$ where $x_1 \ne x_2$ there exist $d_1, d_2 \in D$, $d_1 \ne d_2$ such that

$$v = d_1 + x_1$$
$$v = d_2 + x_2$$

It follows that $x_1 - x_2 = d_2 - d_1$. But $x_1 - x_2 \in L$ and $d_2 - d_1 \notin L$ when $d_1 \ne d_2$. Further, L has dimension m if, and only if, there exists a bounded subset B of $\mathbb{R}^m$ such that

(13.7.20) $$\mathbb{R}^m = B + \bigcup_{x \in L} x$$

For, if L has dimension m, we have only to take B to be a fundamental domain for L. It means that every vector in $\mathbb{R}^m$ is in exactly one of the sets $B + l$ for $l \in L$. To prove the converse, suppose that L has dimension $m' < m$. We could write $\mathbb{R}^m = W \oplus W'$ where W has dimension m' and W' has dimension $m - m'$. Now, (13.7.20) is the same as

$$\mathbb{R}^m = B + \bigcup_{v \in W} v$$

Then, the image of B under the projection $\pi : \mathbb{R}^m \to W'$ is W'. As π preserves distances, W' will have to be bounded—a contradiction to the fact that W' is unbounded. That is, L has dimension m. $\qquad\square$

Definition 13.7.4 : *A subset E of $\mathbb{R}^m$ is called a convex set if, whenever $x, y \in E$, the element $\lambda x + (1 - \lambda)y$ also belongs to E, for all λ with $0 \le \lambda \le 1$.*

Definition 13.7.5 : *A subset E of $\mathbb{R}^m$ is said to be symmetric, if, whenever $x \in E$, $-x$ is also an element of E.*

Let $\{v_1, v_2, \ldots v_m\}$ be a basis for a lattice of dimension m in $\mathbb{R}^m$. If D denotes the fundamental domain of L, the volume of D is given by

(13.7.21) $$\nu(D) = \text{ absolute value of } \det[a_{ij}],$$

where $v_i = a_{i1}e_1 + a_{i2}e_2 + \cdots + a_{im}e_m$ with $\{e_1, e_2, \ldots e_m\}$ forming a standard basis of $\mathbb{R}^m$. The volume of a subset S of $\mathbb{R}^m$ is countably additive and determinable.

Theorem 123 (Minkowski's theorem) : *Let L be an m-dimensional lattice in* $\mathbb{R}^m$. *Suppose that D denotes the fundamental domain of L. Let E be a bounded symmetric convex subset of* $\mathbb{R}^m$. *If*

$$\nu(E) > 2^m \nu(D),$$

then E contains a nonzero element of L.

Proof : We enlarge L into a lattice $2L$ with fundamental domain $2D$ and having volume $2^m\nu(D)$. We consider the m-dimensional torus $\mathbb{R}^m/2L$. We denote it by T^m. Then,

$$\nu(T^m) = \nu(2D) = 2^m\nu(D).$$

Let $\xi : \mathbb{R}^m \to T^m$ be defined by $\xi(x) = x + 2L$, $x \in \mathbb{R}^m$. Since $\nu(E) > 2^m\nu(D)$, ξ cannot preserve the volume of E. But, ξ is onto T^m. So $\xi(E)$ is contained in T^m. So, $\xi(\nu(E)) \leq \nu(T^m) = 2^m\nu(D) < \nu(E)$.

So, the restriction of ξ to E is not injective. Therefore, there exist $x_1, x_2 \in E$ with $x_1 \neq x_2$ such that

$$\xi(x_1) = \xi(x_2)$$

It means that $x_1 - x_2 \in 2L$. As $x_2 \in L$, $-x_2 \in L$, by symmetry. By convexity, $\frac{1}{2}x_1 + \frac{1}{2}(-x_2) \in E$ or $\frac{1}{2}(x_1 - x_2) \in E$. But, as $x_1 - x_2 \in 2L$ $\frac{1}{2}(x_1 - x_2) \in L$. So, there exists a nonzero element $\frac{1}{2}(x_1 - x_2) \in E \cap L$. □

Lemma 13.7.3 (Stewart and Tall) : *Let L be a lattice of dimension $r + 2s$ in $\mathbb{R}^r \times \mathbb{C}^s$. Suppose D denotes the fundamental domain of L and volume of D is $\nu(D)$. If $c_1, c_2 \ldots c_{r+s}$ are positive real numbers whose product*

$$c_1 c_2 \ldots c_{r+s} > (\frac{4}{\pi})^s \nu(D),$$

then, there exists in L a nonzero element $y = (y_1, y_2, \ldots y_r; y_{r+1} \ldots y_{r+s})$ such that

$$(13.7.22) \qquad \begin{cases} |y_1| < c_1, \ |y_2| < c_2, \ \ldots, \ |y_r| < c_r; \\ |y_{r+1}|^2 < c_{r+1}, \ \ldots, \ |y_{r+s}|^2 < c_{r+s}. \end{cases}$$

Proof : We denote by Y the set of points $y \in \mathbb{R}^r \times \mathbb{C}^s$ such that the coordinates of y satisfy the conditions stated in (13.7.22). We compute the volume of Y.

$$\nu(Y) = \int_{-c_1}^{c_1} dy_1 \cdots \int_{-c_r}^{c_r} dy_r \times \iint_{x_1^2 + z_1^2 < c_{s+1}} dx_1 dz_1 \times \cdots \times \iint_{x_s^2 + z_s^2 \leq c_{r+s}} dx_s dz_s$$

$$= 2c_1 \cdot 2c_2 \cdots 2c_r \cdot \pi c_{r+1} \cdots \pi c_{r+s}$$

$$= 2^r \pi^s c_1 c_2 \cdots c_{r+s}$$

Y is a cartesian product of line segments and circular discs. So, Y is bounded, symmetric and convex. By theorem 123, if

$$(13.7.23) \qquad \nu(Y) > 2^{r+2s}\nu(D),$$

Y has a nonzero element in common with L. (13.7.23) is satisfied, if

$$2^r \pi^s c_1 c_2 \cdots c_{r+s} > 2^{r+2s}\nu(D),$$

that is, if $c_1 c_2 \cdots c_{r+s} > (\frac{4}{\pi})^s \nu(D)$ and this is the stated hypothesis of the lemma. Y contains a nonzero element of L.

Therefore, there exists $y = (y_1, y_2 \cdot; y_{r+1}, \ldots, y_{r+s})$ in L satisfying (13.7.22).

This completes the proof of the lemma.   □

In (13.7.18), we made the claim that the image J of U_K under σ'_R is a lattice of dimension $\le (r + s - 1)$. The claim was shown to be correct by using theorem 122. What is more important is that the next lemma specifies the dimension of J.

Lemma 13.7.4 : *The image J of U_K in $\mathbb{R}^{r+s}$ under σ'_R is a lattice of dimension equal to $(r + s - 1)$.*

Proof : The subset V defined by

$$(13.7.24) \qquad V = \{(y_1, y_2, \ldots, y_{r+s}) \in \mathbb{R}^{r+s} : y_1 + y_2 \cdots + y_{r+s} = 0\}$$

is a subspace (of $\mathbb{R}^{r+s}$) of dimension $r + s - 1$. Every element of J is the image of some element in $\mathbb{R}^r \times \mathbb{C}^s$ under ψ (13.7.14). Further, when $x \in \mathbb{R}^r \times \mathbb{C}^s$, $\psi(x) \in V$ if, and only if, $|N(x)| = 1$.

We recall that if $x = (x_1, x_2, \ldots, x_r; x_{r+1}, x_{r+2}, \ldots, x_{r+s})$

$$(13.7.25) \qquad N(x) = x_1 x_2 \ldots x_r |x_{r+1}|^2 \cdots |x_{r+s}|^2.$$

Let

$$(13.7.26) \qquad S = \{x \in \mathbb{R}^r \times \mathbb{C}^s : |N(x)| = 1\}$$

Then, $\psi(S) = V$. If $T \subseteq S$ and T is bounded, then $\psi(T)$ is also bounded. Further, as the norm is multiplicative, when $x \in S$, $xT \subseteq S$, if $T \subseteq S$. $\sigma : K \to \mathbb{R}^r \times \mathbb{C}^s$ is given. When u is a unit in K, $\sigma(u)T \subseteq S$. For, using (13.7.25), as $\sigma_i(u) = \pm 1$, $i = 1, 2, \ldots (r+s)$, $|N(\sigma(u))| = 1$ and thus, $\sigma(u) \in S$.

Therefore,

$$(13.7.27) \qquad \sigma(u)T \subseteq S, \text{ where } T \subseteq S \text{ and } T \text{ is bounded.}$$

We now proceed to find a suitable subset T of S satisfying (13.7.27) shown after some steps that follow. $\mathbb{R}^r \times \mathbb{C}^s$ has a standard basis given by

$$e_1 = (1, 0 \ldots 0; 0, 0, \ldots 0),$$

$$e_2 = (0, 1 \ldots 0; 0, 0, \ldots 0),$$

$$\cdots \cdots \cdots$$

$$e_r = (0, 0 \ldots, 1; 0, 0, \ldots 0),$$

$$e_{r+1} = (0, \ldots 0; 1, 0, \ldots 0),$$

$$e'_{r+1} = (0, 0 \ldots 0; i, 0, \ldots 0),$$

$$\cdots \cdots \cdots$$

$$e_{r+s} = (0, 0 \ldots 0; 0, 0, \ldots 0, 1),$$

$$e'_{r+s} = (0, 0 \ldots 0; 0, 0, \ldots 0, i).$$

When R is the number ring corresponding to the number field K, the image of R under σ written $\sigma(R)$ is a lattice M of dimension m (say).

The product of two vectors in $\mathbb{R}^r \times \mathbb{C}^s$ is obtained by component-wise multiplication. Also, $\sigma : K \to \mathbb{R}^r \times \mathbb{C}^s$ is as given in (13.7.13).

σ is both a ring homomorphism as well as an $\mathbb{R}$-algebra homomorphism. Let y be an arbitrary but fixed element of $\mathbb{R}^r \times \mathbb{C}^s$. We define the map

$$(13.7.28) \qquad \lambda_y : \mathbb{R}^r \times \mathbb{C}^s \to \mathbb{R}^r \times \mathbb{C}^s$$

by $\lambda_y(x) = yx$ for all $x \in \mathbb{R}^r \times \mathbb{C}^s$. λ_y is a linear operator on $\mathbb{R}^r \times \mathbb{C}^s$.

Claim : $\det(\lambda_y) = N(y)$.

We fix y as $(y_1, y_2, \ldots y_r; x_1 + iz_1, x_2 + iz_2, \ldots x_s + iz_s) \in \mathbb{R}^r \times \mathbb{C}^s$
Now,

$$\lambda_y(e_{r+1}) = ye_{r+1} = (0, \ldots, 0; x_1 + iz_1, 0, \ldots, 0).$$

Or,

$$\lambda_y(e'_{r+1}) = 0e_1 + \cdots + 0e_r + (x_1 e_{r+1} + z_1 e'_{r+1}) + \cdots + 0e_{r+s} + 0e'_{r+s},$$

$$\lambda_y(e'_{r+1}) = ye'_{r+1} = (0, 0, \ldots 0; -z_1 + ix_1, 0, \ldots, 0),$$

or,

$$\lambda_y(e'_{r+1}) = 0e_1 + \cdots + 0e_r - z_1 e_{r+1} + x_1 e'_{r+1} + \cdots + 0e_{r+s} + 0e'_{r+s},$$

$$\ldots \ldots \ldots \ldots$$

$$\lambda_y(e_{r+s}) = (0, 0, \ldots 0; 0, 0, \ldots, x_s + iz_s),$$

and so,

$$\lambda_y(e_{r+s}) = 0e_1 + \cdots + 0e_r + 0e_{r+1} + \cdots + x_s e_{r+s} + z_s e'_{r+s}.$$

Also,

$$\lambda_y(e'_{r+s}) = (0, 0 \cdots 0; 0; \cdots, 0, -z_s + ix_s),$$

or,

$$\lambda_y(e'_{r+s}) = (0e_1 + 0e_2 \cdots + 0e_r + 0e_{r+1} + 0e'_{r+1} + \cdots - z_s e_{r+s} + x_s e'_{r+s}.$$

Therefore,

$$(13.7.29) \quad \det \lambda_y = \det \begin{bmatrix} y_1 & 0 & \cdots & 0 & \vdots & 0 & 0 & \vdots & \cdots & \vdots & 0 & 0 \\ 0 & y_2 & \cdots & 0 & \vdots & 0 & 0 & \vdots & & \vdots & 0 & 0 \\ \cdots & \cdots & \cdots & \cdots & \vdots & \cdots & \cdots & \vdots & \cdots & \vdots & \cdots & \cdots \\ 0 & 0 & \cdots & y_r & \vdots & 0 & 0 & \vdots & \cdots & \vdots & 0 & 0 \\ \hdashline 0 & 0 & \cdots & 0 & \vdots & x_1 & -z_1 & \vdots & & \vdots & 0 & 0 \\ 0 & 0 & \cdots & 0 & \vdots & z_1 & x_1 & \vdots & & \vdots & 0 & 0 \\ \hdashline \cdots & \cdots & \cdots & \cdots & \vdots & \cdots & \cdots & \vdots & \cdots & \vdots & \cdots & \cdots \\ \hdashline 0 & 0 & \cdots & 0 & \vdots & 0 & 0 & \vdots & & \vdots & x_s & -z_s \\ 0 & 0 & \cdots & 0 & \vdots & 0 & 0 & \vdots & & \vdots & z_s & x_s. \end{bmatrix}$$

Or,

$$(13.7.30) \qquad \det(\lambda_y) = y_1 y_2 \cdots y_r (x_1^2 + z_1^2) \cdots (x_s^2 + z_s^2) = N(y).$$

Matrix of λ_y has the block form

$$\begin{bmatrix} A & 0 & 0 & 0 & \cdots \\ 0 & B_1 & & & \\ 0 & & B_2 & & \\ & & & \ddots & \\ 0 & 0 & \cdots & 0 & B_s \end{bmatrix} \quad (r+2s \times r+2s \text{ matrix}),$$

where

$$A = \begin{bmatrix} y_1 & 0 & \cdots & 0 \\ & y_2 & \cdots & \\ 0 & \cdots & & y_r \end{bmatrix} \quad (r \times r \text{ matrix}),$$

$$B_i = \begin{bmatrix} x_i & -z_i \\ z_i & x_i \end{bmatrix} \quad (2 \times 2 \ matrix) \qquad (i = 1, 2, \ldots s).$$

We remark that (13.7.29) gives $\det \lambda_y$ as the determinant of a $r+2s \times r+2s$ matrix. That is, $\mathbb{R}^r \times \mathbb{C}^s$ is considered as a vector space of dimension $(r+2s)$ over $\mathbb{R}$.

Now, $N(y) = \pm 1$, if $y \in S$. So, $\det(\lambda_y)$ is unimodular, if $y \in S$. It implies that the fundamental domain for the lattice $\lambda_y(M)$ has the same volume as a fundamental domain for M. We denote it by $\nu(D)$. We choose real numbers $c_i > 0$ $(i = 1, 2, \ldots, r+s)$ such that

$$(13.7.31) \qquad c_1 . c_2 \ldots c_{r+s} > (\frac{4}{\pi})^s \nu(D).$$

Let E be the set of elements $x \in \mathbb{R}^r \times \mathbb{C}^s$ for which

(13.7.32)
$$|x_i| < c_i \, (i = 1, 2, \ldots r)$$
$$|x_{r+j}|^2 < c_{r+j} \, (j = 1, 2, \ldots s).$$

Then, there exists in $\lambda_y(M)$ a nonzero element $x \in E$ such that (13.7.32) holds. (See lemma 13.7.3). Now, $x \in \lambda_y(M)$ is such that

(13.7.33) $x = y\sigma(\alpha), \quad (0 \neq \alpha \in R).$

Since $N(x) = N(y)N(\sigma(\alpha)) = N(y)N(\alpha) = \pm N(\alpha)$, as $N(\alpha) = \prod_{i=1}^{r+s} \sigma_i(\alpha)$, we get

(13.7.34) $|N(\alpha)| < c_1 c_2 \ldots c_r c_{r+1} \ldots c_{r+s}$ using (13.7.32).

We denote the right side of (13.7.34) by A. In R, there exist only finitely many ideals I having norm $||I|| < A$.

Reason : For, given a positive rational integer m equal to the norm of an ideal I contained in R, by definition

$$m = ||I|| = |R/I|.$$

For $x \in R$, $m(x+I) = I$ as $(R/I, \oplus)$ is an abelian group of order m. If $m = 1$, $x \in I$. For $m \geq 2$,

$$mx + mI = mx + (m-1)I + I = I$$

gives $mx + (m-1)I$ which could be reduced to the form $y + I$ equals to I. Or, $mx + (m-1)I \in I$. Proceeding thus, we obtain $mx + I = I$ or $mx \in I$, for $x \in R$. As this is true for any $x \in R$, taking $x = 1$, we get $m \in I$. Therefore,

$$(m) \subseteq I,$$

or, $I|(m)$. As R is a Dedekind domain, we could write $(m) = P_1^{a_1} P_2^{a_2}, \ldots, P_k^{a_k}$ where $P_1, P_2, \ldots P_k$ are distinct prime ideals of R and $a_i \geq 1$, $i = 1, 2, \ldots, k$.

Since I divides $P_1^{a_1} P_2^{a_2} \cdots P_k^{a_k}$, I has only a finite number of prime divisors. Further, the factorization of I into a product of prime ideals is unique. I has the form $P_1^{b_1} P_2^{b_2} \ldots P_k^{b_k}$, $0 \leq b_i \leq a_i (i = 1, 2, \ldots, k)$. Therefore, there are only a finite number of choices for I.

The generators of these ideals (which are finitely generated) are known and are unique up to multiplication by units. So, there exist only a finite number of nonassociated elements say $\beta_1, \beta_2 \ldots \beta_k$ whose norms are less than A in absolute value. As $|N(\alpha)| < A$, $\alpha u = \beta_i$ for some $i = 1, 2, \ldots k$, where u is a unit. As $x = y\sigma(\alpha)$, $x = y\sigma(\beta_i u^{-1}) = y\sigma(\beta_i)\sigma(v)$, v a unit and

(13.7.35) $y\sigma(v) = x\sigma(\beta_i^{-1})$

We define

(13.7.36) $T' = S \cap (\cup_{i=1}^k \sigma(\beta_i^{-1})E),$

where E is a bounded subset of $\mathbb{R}^r \times \mathbb{C}^s$. So, $\sigma(\beta_i^{-1})E$ is bounded. Therefore, a finite union intersected with S is bounded. That is, T' is bounded. We remark that T' does not depend on the choice of $y \in S$. Since y and $\sigma(v)$ are in S, $x\sigma(\beta_i^{-1})$ is in

S $(i = 1, 2 \ldots k)$ (by (13.7.35)). Therefore, $y \in \sigma(v)T'$. As y is an arbitrary element of S, we have

$$(13.7.37) \qquad \bigcup_{v \in U_k} \sigma(v)T' = S.$$

Thus, we have succeeded in finding a bounded subset T' of S such that (13.7.37) holds. That is, we have succeeded in finding in V a bounded subset T' of V such that $\psi(T') = B$ (say) and

$$V = B + \bigcup_{x \in J} x$$

But then, from (13.7.24), J is a lattice of dimension $(r+s-1)$ by the statement leading to (13.7.20). This proves the lemma. $\qquad \square$

Next, using the lemma above and noting that $\ker(\sigma'_R) = H$ is a finite cyclic group of even order denoted by μ_K (see the steps before (13.6.18)), we see that

$$(13.7.38) \qquad U_K / \mu_K \cong \mathbb{Z}^{r+s-1}$$

which is the celebrated Dirichlet's unit theorem stated below:

Theorem 124 : *The group U_K of units in R, the number ring corresponding to the number field K is the direct product $\mu_K \times F_K$ where μ_K is a finite cyclic group consisting of roots of unity in K which is of even order and F_K is a free abelian group of rank $r + s - 1$ (isomorphic to $\mathbb{Z}^{r+s-1}$) where $n = r + 2s = [K : Q]$.*
F_K consists of elements

$$(13.7.39) \qquad u = u_1^{t_1} u_2^{t_2} \cdots u_{r+s-1}^{t_{r+s-1}}, \quad t_i \in \mathbb{Z}, \quad i = 1, 2, \ldots (r+s-1);$$

where $u_1, u_2, \ldots u_{r+s-1}$ are units in R. Such a set $\{u_1, u_2, \ldots u_{r+s-1}\}$ is called a 'fundamental system of units' in K. The exponents $t_1, t_2 \ldots t_{r+s-1}$ are uniquely determined for a given member $u \in F_K$.

Remark 13.7.1 : Theorems 122 and 123 have been adapted from I. N. Stewart and D. O. Tall [15].

Examples 13.7.1 :

(i) **Units in quadratic number fields** : Let $m > 0$ be square-free. $K = \mathbb{Q}(\sqrt{m})$ is a real quadratic field extension of $\mathbb{Q}$. Since $\mathbb{Q}(\sqrt{m}) \subset \mathbb{R}$, the only roots of unity in K are ± 1. So, $\mu_K = \{\pm 1\}$. Now, $[K : \mathbb{Q}] = 2$. So, $r + 2s = 2$ giving $r = 2, s = 0$. There are two real embeddings of K in $\mathbb{C}$ and no complex embeddings in $\mathbb{C}$, $r + s - 1 = 1$. So,

$$U_K \cong \mu_K \times F_k = \mathbb{Z}/2\mathbb{Z} \times \mathbb{Z}.$$

If $K = \mathbb{Q}(\sqrt{-m})$,

$$(13.7.40) \qquad U_K = \begin{cases} \mathbb{Z}/4\mathbb{Z}, & \text{if } m = 1; \\ \mathbb{Z}/6\mathbb{Z}, & \text{if } m = 3; \\ \mathbb{Z}/2\mathbb{Z}, & \text{otherwise.} \end{cases}$$

For, when $m = 1$, $K = \mathbb{Q}(i)$, i denoting $\sqrt{-1}$. $R = \mathbb{Z}[i]$. The group of units is $\{\pm 1, \pm i\}$. So, $G(-1) \cong \mathbb{Z}/4\mathbb{Z}$, see lemma 13.6.1. $r + 2s = 2 \Rightarrow r = 0, s = 1$.

$$r + s - 1 = 0, \quad U_K = \mu_K \cong \mathbb{Z}/4\mathbb{Z}.$$

When $m = 3$, $R = \mathbb{Z}[\zeta]$, where $\zeta = \frac{-1 + \sqrt{-3}}{2}$, as $-3 \equiv 1 \pmod 4$. $r + 2s = 2 \Rightarrow r = 0, s = 1$. $U_K = \mu_K$.
$\alpha \in U_K \Rightarrow \alpha = a + b\,\zeta$ is such that $a^2 + 3b^2 = 4$.
Then, $(a, b) \in \{(\pm 2, 0), (\pm 1, \pm 1)\}$.
$U_K = \{\pm 1, \pm \omega, \pm \omega^2\}$ where $\omega = \exp(\frac{2\pi i}{3})$ (already discussed) or, $U_K \cong \mathbb{Z}/6\mathbb{Z}$ when $m = 3$.

For other negative values of m, if $m \equiv 1 \pmod 4$, $a^2 + bm^2 = 1 \Rightarrow \pm 1$ is the only root of unity in K, or $U_K \cong \mathbb{Z}/2\mathbb{Z}$.
If $m \equiv 1 \pmod 4$, $a^2 + mb^2 = 4 \Rightarrow (a, b) = (\pm 2, 0)$ for $m > 4$.
This shows that $U_K \cong \mathbb{Z}/2\mathbb{Z}$.

(ii) **Units in cyclotomic fields :**

Let $\omega(m) = \exp(\frac{2\pi i}{m})$. We consider $K = \mathbb{Q}(\omega(m))$.
The number ring corresponding to K is $\mathbb{Z}(\omega(m))$.

(a) If m is even, the only roots of unity in K are the mth-roots of unity. We have $[K : \mathbb{Q}] = \phi(m)$, the Euler ϕ-function.
Let $\theta \in \mathbb{Q}(\omega(m))$ be a primitive k^{th} root of unity, k not dividing m. Then, if $r = l.c.m(k, m)$, there exists $\alpha \in \mathbb{Q}(\omega(m))$ such that $\alpha^r = 1$ and r is the least positive integer with this property.
So, $\mathbb{Q}(\omega(r)) \subseteq \mathbb{Q}(\omega(m))$, or,

$$(13.7.41) \qquad \phi(r) = [\mathbb{Q}(\omega(r)) : \mathbb{Q}] \leq [\mathbb{Q}(\omega(m)) : \mathbb{Q}] = \phi(m)$$

Now, m is even and $r > m$.
But, m is even and properly divides r. This implies that $\phi(m)$ divides $\phi(r)$ property or $\phi(m) < \phi(r)$, a contradiction to (13.7.41).
So, $\mathbb{Q}(\omega(m))$ contains only m^{th} roots of unity as roots of unity. So, $\mu_k \cong \mathbb{Z}/m\mathbb{Z}$.

(b) If m is odd, the roots of unity in $\mathbb{Q}(\omega(m))$ are the $2m^{\text{th}}$ roots of unity. For, if m is odd,

$$\omega(2m) = \exp(\frac{2\pi i}{2m}) = \exp(\frac{\pi i}{m}) = -\{\omega(m)\}^{\frac{m+1}{2}}.$$

So, $\mathbb{Q}(\omega(m)) = \mathbb{Q}(\omega(2m))$. So, $\mu_k \cong \mathbb{Z}/2m\mathbb{Z}$.

(c) When $K = \mathbb{Q}(\omega(5))$.
$U_K = \{\pm \zeta^h (1 + \zeta)^k : 0 \leq h \leq 4, k \in \mathbb{Z}, \zeta = \exp(\frac{2\pi i}{5})\}$.
For, if $\omega(m) = \exp(\frac{2\pi i}{m})$, writing $\omega(m) = \omega$, we see that

$1+\omega+\omega^2+\cdots\omega^{k-1}$ is a unit in $\mathbb{Z}[\omega(m)]$, if $g.c.d(k,m)=1$. Its inverse is $\frac{\omega-1}{\omega^k-1}$. Also, as $g.c.d(k,m)=1$, there exist integers x,y such that $xk+ym=1$. So $\omega^{kx}=1$, for some $x\in\mathbb{Z}$.

Let $\beta=\zeta+1/\zeta$, $\zeta=\exp(\frac{2\pi i}{5})$. Then, $\beta^2+\beta-1=0$.

Since $\beta=2\cos(\frac{2\pi}{5})>0$, $\beta=\frac{-1+\sqrt{5}}{2}$. So, $\mathbb{Q}(\beta)=\mathbb{Q}(\sqrt{5})$.

Also, $[\mathbb{Q}(\omega(5)):\mathbb{Q}]=\phi(5)=4$. When $r+2s=4$, $r=0$, $s=2$, $r+s-1=1$. The roots of unity in $\mathbb{Z}[\omega(5)]$ are the 10th roots of unity contained in $\mathbb{Z}[\omega(5)]$. So,

$$\mu_K\cong\{\pm\zeta^h:0\le h\le 4,\ \zeta=\exp(\frac{2\pi i}{5})\}$$

As $r+s-1=1$, $F_K\cong\mathbb{Z}$. We have only to obtain a unit in $\mathbb{Z}[\omega(5)]$ other than a root of unity.

If $\alpha=1+\beta=1+\zeta+1/\zeta=\frac{1+\zeta+\zeta^2}{\zeta}$, α is a unit.

$\alpha=\frac{1+\sqrt{5}}{2}$. So, $F_K\cong\alpha^k$, $k\in\mathbb{Z}$. Thus,

$$U_K\cong\{\pm\zeta^h\alpha^k:0\le h\le 4,k\in\mathbb{Z}\}.$$

For more results and worked-out problems, see J. Esmonde and M. Ram Murthy [5, chapter 8]. For an interesting survey of the theory behind the Pell equation, one may refer to H. W. Lenstra Jr. [9]. He discusses the method of algorithms for solving the Pell equation.

13.8. Notes with illustrative examples

When m is a square-free positive integer ≥ 2, the fundamental unit in the number ring $R(m)$ corresponding to the number field $\mathbb{Q}(\sqrt{m})$ is the least algebraic integer greater than 1 and having positive coefficients. In $\mathbb{Z}[\zeta]$, where $\zeta=\exp(\frac{2\pi i}{5})$, it is $\frac{1+\sqrt{5}}{2}$. The fundamental unit is large enough for certain other values of m. In $\mathbb{Z}[\sqrt{31}]$, it is $1520+273\sqrt{31}$. In $\mathbb{Z}[\sqrt{94}]$, it is $2143295+221064\sqrt{94}$. See D.A. Marcus [10]. For $\mathbb{Z}[\sqrt{95}]$, it is $39+4\sqrt{95}$. There are algorithms for determining fundamental units. One method is by the use of continued fractions.

In the case of cubic fields, that is, K with $[K:\mathbb{Q}]=3$, there is only one real embedding of K in $\mathbb{C}$ and $3=r+2s$ gives $r=s=1$. So, $r+s-1=1$. The group U_K of units of the number ring R corresponding to K has the form

(13.8.1) $U_K=\{\pm u^k:k\in\mathbb{Z}\}$,

where u is a unit and ± 1 is the only roots of unity in K. For fields K of degree 4 over $\mathbb{Q}$ and having no real embedding (in $\mathbb{R}$) we have $4=r+2s=0+2.2$. $r+s-1=1$. If θ is a root of unity in K, the group U_K of units is given by

(13.8.2) $U_K=\{\theta u^k:k\in\mathbb{Z},\ \theta$ is a root of unity$\}$.

For the 5th cyclotomic field $\mathbb{Q}[\zeta]$ where $\zeta=\exp(\frac{2\pi i}{5})$, we have noted that a unit is $1+\zeta$.

Next, let K be a number field of degree n over $\mathbb{Q}$. Let $\sigma_1, \sigma_2, \ldots \sigma_n$ be the embeddings of K in $\mathbb{C}$. Suppose that $\langle \alpha_1, \alpha_2 \ldots \alpha_n \rangle$ be an n-tuple of elements of K.

Definition 13.8.1 : *The discriminant of* $\{\alpha_1, \alpha_2, \ldots \alpha_n\}$ *is defined as*

$$disc\,(\alpha_1, \alpha_2, \ldots, \alpha_n) = (\det[\sigma_i(\alpha_j)])^2,$$

where $[\sigma_i(\alpha_j)]$ *is the* $n \times n$ *matrix whose ith row-jth column element is* $\sigma_i(\alpha_j)$.

The role of the discriminant is to check whether the n-tuple $\langle \alpha_1, \alpha_2 \ldots \alpha_n \rangle$, forms a linearly independent set of elements or not.

Proposition 13.8.1 : *$disc(\alpha_1, \alpha_2, \ldots, \alpha_n) = 0$ if, and only if,* $\{\alpha_1, \alpha_2, \ldots, \alpha_n\}$ *is a linearly dependent set over* $\mathbb{Q}$.
For proof, see D.A. Marcus [10].

Let $\{\alpha_1, \alpha_2, \ldots, \alpha_n\}$ be a basis for K over $\mathbb{Q}$ consisting of algebraic integers. Suppose that $\Delta = disc(\alpha_1, \alpha_2, \ldots \alpha_n)$.

Proposition 13.8.2 : *If* $\alpha \in R$, *the number ring corresponding to the number field* K, *then*

(13.8.3) $$\alpha = \frac{1}{\Delta} \sum_{j=1}^{n} a_j \alpha_j \text{ where, } a_j \in \mathbb{Z}; j = 1, 2, \ldots n.$$

For proof, see D. A. Marcus [10].

This enables one to obtain a basis for R over $\mathbb{Z}$. We see that R contains a free abelian group A of rank n. $\beta \in A$ is given by

$$\beta = \sum_{j=1}^{n} m_j \alpha_j \ldots \qquad m_j \in \mathbb{Z}; \quad j = 1, 2, \ldots n.$$

That is, $A = \mathbb{Z}\alpha_1 \oplus \mathbb{Z}\alpha_2 \oplus \cdots \oplus \mathbb{Z}\alpha_n$.
R is contained in the free abelian group $\frac{1}{\Delta} A$ given by

$$\frac{1}{\Delta} A = \mathbb{Z} \frac{\alpha_1}{\Delta} \oplus \cdots \oplus \mathbb{Z} \frac{\alpha_n}{\Delta}.$$

As R is sandwiched between two free abelian groups of rank n, R, itself, is a free abelian group of rank n.

We remark that the discriminant is well-defined and any two integral bases for K give rise to the same discriminant. We call this the discriminant of K. If Δ denotes the discriminant of K where $[K : \mathbb{Q}] = n$, then, $\Delta \equiv 0$ or $1 \pmod 4$. This is referred to as Stickelberger's criterion. For proof, see Esmonde and Murthy [5].

Next, we make a few observations about $\mathbb{Q}(\omega(m))$ where $\omega(m) = \exp(\frac{2\pi i}{m})$. $\mathbb{Q}(\omega(m))$ is a cyclotomic extension of $\mathbb{Q}$, that is an extension obtained by adjoining $\omega(m)$ to $\mathbb{Q}$. It is known that the square-root of a positive integer is always contained in $\mathbb{Q}(\omega(m))$ for a suitable choice of m.

By a genuine square root of an integer a, we mean $a^{1/2}$ and there exists $y \in \mathbb{R}$ such that $y^2 = a$.

Definition 13.8.2 : *If a is an integer greater than 1, then the real number $a^{1/n}$ is said to be a genuine nth-root of a, if it cannot be written in the form $b^{1/m}$ for some integer b and $m < n$.*

A genuine nth root of a (for $a > 1$) is an irrational number. For $a > 1$, the real square root $a^{1/2}$ of a is contained in a cyclotomic field.

Theorem 125 (Rajat Tandon (2001)) **:** *Let a be an integer > 1. If $a^{1/n}$ is a genuine nth-root of a with $n > 2$, then, $a^{1/n}$ is not contained in a cyclotomic field.*

An outline of the proof is along the following lines:
The group $G(\mathbb{Q}[\omega(m))$ of automorphisms of $\mathbb{Q}(\omega(m))$ keeping $\mathbb{Q}$ fixed is isomorphic to the group of units of $\mathbb{Z}/m\mathbb{Z}$ for $m > 2$ and $G(\mathbb{Q}[\zeta_m])$ is abelian. Further, if $\mathbb{Q} \subseteq F \subseteq K$ where F and K are extensions of $\mathbb{Q}$ obtained by adjoining zeros of polynomials in $\mathbb{Q}[x]$, the group $G(K)$ of automorphisms of K leaving $\mathbb{Q}$ fixed and the group $G(F)$ of automorphisms of F leaving $\mathbb{Q}$ fixed are related and

$$G(F) \cong G(K)/G(K/F),$$

where $G(K/F)$ denotes the group of automorphisms of K leaving F fixed. The point is that if $G(K)$ is abelian, so is $G(F)$.

If $\mathbb{Q}(a^{1/n}) \subseteq \mathbb{Q}(\omega(n))$, then $\mathbb{Q}(a^{1/n}, \omega(n)) \subseteq \mathbb{Q}(\omega(m), \omega(n))$ where $m < n$. But, if $[m,n]$ denotes the l.c.m of m and n,

$$\mathbb{Q}(\omega(m), \omega(n)) = \mathbb{Q}(\omega([m,n])).$$

If p is a prime > 2 and if $\mathbb{Q}[a^{1/p}, \omega(p)] \subseteq \mathbb{Q}(\omega(m))$ for some m, the group of automorphisms of $\mathbb{Q}(a^{1/p}, \omega(p))$ leaving $\mathbb{Q}$ fixed would be abelian. However, Rajat Tandon [16] proves

Proposition 13.8.3 : *If p is an odd prime or 4 and if $a^{1/p}$ is genuine with $a > 1$ then $G(a^{1/p}, \omega(p))$ is not abelian.*

This proves theorem 125. For details see Rajat Tandon [16].
Rajat Tandon's theorem speaks about an analogue of Fermat's Last theorem : $x^n + y^n = z^n$ has non-trivial solutions, for $n = 2$. But, for $n > 2$, $x^n + y^n = z^n$ has no non-trivial solutions. In the same manner, one notes that the square-root of an integer $a > 1$ is contained in a cyclotomic extension of $\mathbb{Q}$. But, for $n > 2$, a genuine nth root of an integer $a > 1$ is not contained in a cyclotomic extension of $\mathbb{Q}$.

In Section 8.5, chapter 8, we have given the definition of a normed division domain. (See definition 8.5.2). It is a weak partially ordered set $(X, \triangle)$ endowed with a norm $N : X \to \mathbb{N}$ such that for $a, b \in X$, $N(a)$ divides $N(b)$ whenever $a\triangle b$. Further, if $n\triangle x$ is such that $N(n) = 1$, then $n\triangle y$, for all $y \in X$. ($\triangle$ is reflexive and transitive)

The number ring R corresponding to a number field K is such that $R^* = R \setminus \{0\}$ is a multiplicative normed division domain where $N(\alpha) = |\alpha\alpha_2 \ldots \alpha_n|$; $(\alpha_1\alpha_2 \ldots \alpha_n$ are the conjugates of α in K) where $\alpha \in R^*$. The norm is multiplicative and hence the name multiplicative normed division domain. See Rajendran Valiaveetil [12].

If p is a prime $\equiv 2 \pmod 3$, $D^* = \mathbb{Z}[\sqrt{-p}\,] \setminus \{0\}$ is a multiplicative normed division domain. We consider

$$M_2 = \{a + b\sqrt{-p} : a - b \equiv 0 \pmod 2\}$$
$$M_3 = \{a + b\sqrt{-p} : a - b \equiv 0 \pmod 3\}$$

M_2 and M_3 are maximal ideals of $\mathbb{Z}[\sqrt{-p}\,]$

$$M_2 \cap M_3 = \{a + b\sqrt{-p} : a - b \equiv 0 \pmod 6\}$$

Let J denote $M_2 \cap M_3$. J is an ideal of $\mathbb{Z}[\sqrt{-p}\,]$ and is such that whenever $\alpha\beta \in J$ with $\alpha \notin J, (\alpha, \beta \in \mathbb{Z}[\sqrt{-p}\,])$ either $\beta \in J, 2\beta \in J$ or $3\beta \in J$. For proof, see [12]. J is referred to as a quasi-prime ideal. The details and a generalisation may be found in [12].

We wish to point out that only the basic ideas about algebraic number theory have been brought out in this chapter. Indeed, algebraic number theory is very vast. Many outstanding books on the subject are available. For instance, one may consult P. Samuel [13], A. Fröhlic and M.J. Taylor [6], W. Narkeiwicz [11] or Kazuya Kato, Nobushige Kurokawa and Takeshi Saito [8]. The emerging area of computational algebraic number theory is also worth studying. See Henri Cohen [3].

For a splendid treatment of the theory of algebraic numbers, see Paulo Ribenboim [A 5].

13.9. Formally real fields

In chapter 8, Section 8.2, ordered fields were considered. Definition 8.2.4 says that a field K is real, if -1 is not a sum of squares in K.

Definition 13.9.1 : *Let K be a field. The level $s(K)$ of K is the smallest positive integer s such that -1 is a sum of s squares in K.*

If there is no such s (that is, no integer s such that -1 is a sum of s squares in K), we use the convention that $s(K) = \infty$. The notation 's' comes from the German word stufe, meaning 'level'.

Definition 13.9.2 : *Given a field K, $t(K)$ is defined as the smallest positive integer t such that 0_K is non-trivially a sum of t squares in K.*

If there does not exist a number t such that $\sum_{i=1}^{t} a_i^2 = 0 \Rightarrow a_i = 0 (i = 1, 2, \ldots t)$, we write $t(K) = \infty$.

In the case of a finite field $\mathbb{F}_q$ ($q = p^m$; p a prime, $m \geq 1$), it may be verified that $s(\mathbb{F}_q)$ is 1 or 2. Further, $s(\mathbb{F}_q) = 1$ if, and only if, $q \equiv 1 \pmod 4$ or $char(\mathbb{F}_q) = 2$. We note that $s(\mathbb{F}_q) = 1$ whenever $q = 2^m$ ($m \geq 1$).

Definition 13.9.3 : *Given a field K, if $s(K) = \infty$, K is called a formally real field.*

For m square-free, $K = \mathbb{Q}(\sqrt{m})$ is such that $s(K) = \infty$, if $m > 0$. $s(K) = 1$ if $m = -1$. It can be shown that if $m < 0$, and $m \equiv 1 \pmod 8$, $s(K) = 4$. Further, $s(K) = 2$, if m is negative and $m \not\equiv 1 \pmod 8$. See Charles Small [A 7], [A 8].

We note that for m (square-free) > 0, $\mathbb{Q}(\sqrt{m})$ is an example of a formally real field. It is shown by T. Y. Lam [A 4] that formally real fields can be ordered.

Definition 13.9.4 : *Let K be an ordered field. $\alpha \in K$ is called totally positive (that is, $\alpha > 0$), if α is positive in every ordering of K.*

Squares of nonzero elements of an ordered field K are totally positive. So are their finite sums. It is true that if $\alpha > 0$ in an ordered field K, then α is a sum of squares in K.

Let $K = \mathbb{Q}(\theta)$ be an algebraic number field.

Suppose that $[\mathbb{Q}(\theta) : \mathbb{Q}] = n = r + 2s$. Then, r is the number of real conjugates of θ. $\alpha \in K$ is such that $\alpha > 0$ in K if, and only if, $\sigma_i(\alpha) > 0$ for which i, $1 \leq i \leq r$, where $\sigma_i : K \to \mathbb{R}$ is a $\mathbb{Q}$-isomorphism of K into $\mathbb{R}$. In 1902, D. Hilbert conjectured that every element $\beta(> 0)$ in K is a sum of four squares in K. This was proved to be true by C. L. Siegel [A 6] in 1921. In 1941, Maass showed that if $R(5)$ denotes the ring of integer of $\mathbb{Q}(\sqrt{5})$, $\beta(> 0)$ in $R(5)$ is a sum of three squares in $R(5)$. We state without proof.

Proposition 13.9.1 : *(C. L. Siegel (1921)). The only formally real algebraic number fields K in which every totally positive algebraic integer is a sum of squares of algebraic integers in K are the fields $\mathbb{Q}$ and $\mathbb{Q}(\sqrt{5})$.*

For proof, see Juliet Britto [A 1].

13.10. Worked-out examples

a) Let K be an algebraic number field. R_K denotes the ring of integers of K. If $\{\alpha_1, \alpha_2, \ldots, \alpha_n\}$ and $\{\beta_1, \beta_2, \ldots, \beta_n\}$ are two integral bases for R_K, show that

$$\Delta[\alpha_1, \alpha_2, \ldots, \alpha_n] = \Delta[\beta_1, \beta_2, \ldots, \beta_n].$$

Answer: We write β_j $(j = 1, 2, \ldots, n)$ in terms of the elements of the basis $\{\alpha_1, \alpha_2, \ldots, \alpha_n\}$ of R_K. We obtain

(13.10.1) $\qquad \beta_j = a_{j1}\alpha_1 + a_{j2}\alpha_2 + \cdots + a_{jn}\alpha_n, \ (j = 1, 2, \ldots n).$

This yields

$$\begin{bmatrix} \beta_1 \\ \beta_2 \\ \vdots \\ \beta_n \end{bmatrix} = \begin{bmatrix} a_{11} & a_{12} & \cdots & a_{1n} \\ a_{21} & \cdots & \cdots & a_{2n} \\ \cdots & \cdots & \cdots & \cdots \\ a_{n1} & a_{n2} & \cdots & a_{nn} \end{bmatrix} \begin{bmatrix} \alpha_1 \\ \alpha_2 \\ \vdots \\ \alpha_n \end{bmatrix}.$$

Or,

(13.10.2) $\qquad \begin{bmatrix} \beta_1 \\ \beta_2 \\ \vdots \\ \beta_n \end{bmatrix} = A \begin{bmatrix} \alpha_1 \\ \alpha_2 \\ \vdots \\ \alpha_n \end{bmatrix}$

where A is the $n \times n$ matrix $[a_{ij}]$ with $a_{ij} \in \mathbb{Z}$ $(i, j = 1, 2, \ldots, n)$. Let $\sigma_1, \sigma_2, \ldots, \sigma_n$ be the embeddings of K into $\mathbb{C}$.

Applying σ_i to each of the n equations (13.10.2), we obtain the matrix equation

$$[\sigma_i(\beta_j)] = A[\sigma_i(\alpha_j)].$$

Taking determinants and squaring

$$\Delta[\beta_1, \beta_2, \ldots, \beta_n] = (\det A)^2 \Delta[\alpha_1, \alpha_2, \ldots, \alpha_n].$$

As $\det A \in \mathbb{Z}$, $\Delta[\alpha_1, \alpha_2, \ldots, \alpha_n]$ divides $\Delta[\beta_1, \beta_2, \ldots, \beta_n]$ and both have the same sign. Writing α_i ($i = 1, 2, \ldots, n$) in terms of the basis $\{\beta_1, \beta_2, \ldots, \beta_n\}$ and arguing similarly, we note that $\Delta[\beta_1, \beta_2, \ldots, \beta_n]$ divides $\Delta[\alpha_1, \alpha_2, \ldots, \alpha_n]$. So, the discriminants of $\{\alpha_1, \alpha_2, \ldots, \alpha_n\}$ and $\{\beta_1, \beta_2, \ldots, \beta_n\}$ are equal. □

Remark 13.10.1 : The above result justifies the definition of the discriminant of K.

b) As $-17 \equiv 3(\bmod\ 4)$, $\mathbb{Z}[\sqrt{-17}]$ is the ring of integers of the quadratic number field $\mathbb{Q}(\sqrt{-17})$. Find the number of ideals of $\mathbb{Z}[\sqrt{-17}]$ which have norm 18.

Answer: $\mathbb{Z}[\sqrt{-17}]$ is not a UFD, as $18 = 2.3^2 = (1 + \sqrt{-17})(1 - \sqrt{-17})$. However, $\mathbb{Z}[\sqrt{-17}]$ is a Dedekind domain. So, every ideal of $\mathbb{Z}[\sqrt{-17}]$ can be uniquely expressed as a product of prime ideals.

If $P_1 = \langle 2, 1 + \sqrt{-17} \rangle$, $P_2 = \langle 3, 1 + \sqrt{-17} \rangle$ and $P_3 = \langle 3, 1 - \sqrt{-17} \rangle$, it can be shown (after a good deal of calculation) that the ideal generated by 18 has the factorization

(13.10.3) $$\langle 18 \rangle = P_1^2 P_2^2 P_3^2.$$

In fact, we will have

(13.10.4) $$\langle 18 \rangle = (P_1 P_2^2) \cdot (P_1 P_3^2) = \langle 1 + \sqrt{-17} \rangle \cdot \langle 1 - \sqrt{-17} \rangle.$$

Norm of $\langle 18 \rangle$ is 18^2. For I, an ideal of R_K, we know that I divides $N(I)$. So, if I has norm 18, $I | 18$. So, I has the form

$$I = P_1^a \cdot P_2^b \cdot P_3^c$$

which implies that

$$N(I) = 2^a \cdot 3^b \cdot 3^c, \text{ as } N(P_1) = 2, N(P_2) = N(p_3) = 3$$

$N(I) = 18$ gives $a = 1$, $b + c = 2$. One has

$$I = P_1 P_2^2 \text{ or } P_1 P_2 P_3 \text{ or } P_1 P_3^2.$$

Thus, there are 3 ideals (in $\mathbb{Z}[\sqrt{-17}]$) which have norm 18. □

c) (Niven) Let $\alpha = a + 2bi$ be a Gaussian integer, where $b \in \mathbb{Z}$. Show that α can be expressed as a sum of two squares of Gaussian integers if, and only if, not both $a/2$ and b are odd integers.

Answer: $:\Rightarrow$ $\alpha = a + 2bi$ is a Gaussian integer expressible as a sum of two squares of Gaussian integers. We have to prove that not both $a/2$ and b are odd integers. Assume that $a/2$ and b are odd integers.

Write $a = 2a'$ so that a' is an integer.

Let $\alpha = (x+yi)^2 + (s+ti)^2$.

Then, $2a' + 2bi = (x^2 - y^2 + s^2 - t^2) + 2(xy+st)i$

gives

(13.10.5) $x^2 - y^2 + s^2 - t^2 = 2a'$

(13.10.6) $xy + st = b.$

Since b is odd, (13.10.6) implies that exactly two or three of x, y, s, t are odd integers.

If exactly three of x, y, s, t are odd, (13.10.5) will be okay only when all x, y, s and t are odd in which case (13.10.6) will be violated. So, exactly two of x, y, s, t are odd integers. Then, left side of (13.10.5) will be divisible by 4. But then, it will contradict the fact that a' is an odd integer. So, $\alpha = a + 2bi$ is a sum of two squares of Gaussian integers $\Rightarrow$ not both $a/2$ and b are odd.

$\Leftarrow$: Suppose that $\alpha = a + 2bi$ satisfies the condition: not both $a/2$ and b are odd. Four cases arise:

Case (i) a is odd and b even. Then,

$$\alpha + 1 = (a+1) + 2bi$$

$$\frac{\alpha+1}{2} = \left(\frac{a+1}{2}\right) + bi$$

$$\frac{\alpha-1}{2} = \left(\frac{a-1}{2}\right) + bi$$

(13.10.7) $\left(\dfrac{\alpha+1}{2}\right)^2 + i^2\left(\dfrac{\alpha-1}{2}\right)^2 = a + 2bi = \alpha.$

Case (ii) Let $a = 2a'$, where a' is odd and b is even.

Suppose the $\alpha' = a' + bi$.

$$\alpha' + i = a' + (b+1)i,$$

$$\alpha' - i = a' + (b-1)i.$$

a' is odd. $(b+1)$ and $(b-1)$ are odd.

Now, $x + yi$ is such that both x and y are odd integers, then $(1+i)$ divides $x + yi$. For,

$$(1+i)(c+di) = x + yi$$

will give $c - d = x$, $c + d = y$.

$c = \frac{x+y}{2}$, $d = \frac{x-y}{2}$ are integers giving $(c+di)$ as the other factor of $x + yi$. In the same manner $(1+i)$ divides $x - yi$ also when x, y are odd integers. Therefore, $\alpha' \pm i$ is divisible by $(1+i)$. Then,

(13.10.8) $\left(\dfrac{\alpha'+1}{1+i}\right)^2 + i\left(\dfrac{\alpha'-1}{1+i}\right)^2 = \alpha.$

Case (iii) If $a = 4a'$ (where $\frac{a}{2}$ is even) and b is odd, we write $\alpha' = b - 2a'i$, $\alpha' + 1 = (b+1) - 2a'i$. Both $b+1$ and $2a'$ are even. $(1-i)$ divides $\alpha' \pm 1$.

Therefore,

$$(13.10.9) \qquad \left(\frac{\alpha'+1}{1-i}\right)^2 + i^2\left(\frac{\alpha'-1}{1-i}\right)^2 = \alpha.$$

Case (iv) If $a = 4a'$ (so that $2a'$ is even) $\alpha = 4a' + 4b'i$ and b is $2b'$, we write $\alpha' = 2a' + 2b'i$.

$\alpha' \pm 2 = (2a' \pm 2) + 2b'i$ is such that 2 divides both $\alpha' + 2$ and $\alpha' - 2$. Then,

$$(13.10.10) \qquad \left(\frac{\alpha'+2}{2}\right)^2 + i^2\left(\frac{\alpha'-2}{2}\right)^2 = \alpha.$$

We have actually verified that in cases where either $\frac{a}{2}$ or b or both are even, α is a sum of two squares of Gaussian integers. $\qquad\square$

Remark 13.10.2 : The solution has been adapted from W. J. Leahey: A note on a theorem of Niven, Proc. Amer. Math. Soc. **16** (1965) pp 1130–1131.

EXERCISES

1. **Mark the following statements true (T) or false (F) justifying your answer briefly.**
   a) *Let J be an ideal of a number ring R_K. The norm $N(J)$ can be a prime power p^m, (p a prime, $m \geq 1$).*
   b) *Let I, J be nonzero distinct prime ideals of a number ring R_K. Then, $I + J = R_K$ and $I \cap J = \emptyset$.*
   c) *Let $K = \mathbb{Q}(\sqrt{7}, \sqrt{10})$. Let R_K be the ring of integers of K. If $\alpha \in R_K$, it is correct to say that $R_K = \mathbb{Z}[\alpha]$.*
   d) *Let α be a zero of $x^3 - x - 1$. Then,*

   $$\mathbb{Z}[\alpha] = \{a_0 + a_1\alpha + a_2\alpha^2 : a_i \in \mathbb{Z}, i = 0, 1, 2\}$$

   is a Dedekind domain.
   e) *Let $R(m)$ denote the ring of integers of $\mathbb{Q}(\sqrt{m})$ where m is square-free. Let p be a prime dividing m. Then, the ideal $\langle p \rangle$ generated by p in $R(m)$ is given by $\langle p \rangle = \langle p, \sqrt{m} \rangle^2$.*
   f) *The ideal class group of $\mathbb{Z}[\sqrt{-14}]$ is cyclic of order 4.*
2. *In $\mathbb{Z}[\sqrt{-5}]$, show that*

   $$\langle 2 \rangle = \langle 2, 1 + \sqrt{-5} \rangle^2$$

   $$\langle 3 \rangle = \langle 3, 1 - \sqrt{-5} \rangle \langle 3, 1 - \sqrt{-5} \rangle.$$

3. *Show that the number ring $\mathbb{Z}[\sqrt{14}]$ corresponding to the number field $\mathbb{Q}(\sqrt{14})$ is a PID.*
4. *Let $\zeta = \frac{1+\sqrt{-19}}{2}$. Show that the class-number of $\mathbb{Z}[\zeta]$ is 1.*
5. *Show that the class-number of the number ring $\mathbb{Z}[\sqrt{-6}]$ corresponding to the algebraic number field $\mathbb{Q}[\sqrt{-6}]$ is 2.*

6. Let $\xi = \frac{1+\sqrt{-23}}{2}$. Find the class-number of $\mathbb{Z}[\xi]$ corresponding to the algebraic number field $Q[\sqrt{-23}]$.

7. Let $\omega = \exp\frac{2\pi i}{23}$. Show that the class-number of $\mathbb{Z}[\omega]$ is 3.

8. Show that $x^2 - 5y^2 = 4$ has infinitely many solutions.

9. (Esmonde and Murthy) Let U_K denote the group of units in the number ring R corresponding to an algebraic number field K. Show that U_K is finite if, and only if, either $K = Q$ or $K = Q[\sqrt{-m}]$ where m is square-free and greater than or equal to 1.

10. Let $\alpha = \sqrt[3]{2}$ (the real cube-root of 2). Find a fundamental unit in the number ring R corresponding to $\mathbb{Q}(\alpha)$.

11. Show that $15+4\sqrt{14}$ is the fundamental unit of the number ring R corresponding to $K = \mathbb{Q}(\sqrt{14})$. Give the structure of the group U_K of units in R.

12. Illustrate Dirichlet's unit theorem for the number ring $\mathbb{Z}[\zeta_{10}]$ (where $\zeta_{10} = \exp(\frac{2\pi i}{10})$) corresponding to the cyclotomic field $\mathbb{Q}(\zeta_{10})$.

REFERENCES

[1] Ethan D. Bolker: Elementary Number theory–An algebraic approach W. A. Benjamin Inc. (1970) Chapter 6 pp 82–138.

[2] L. Carlitz : A Characterization of algebraic number fields with class number two. Proc. Amer. Math. Soc. 11 (1960) 391–392.

[3] Henri Cohen: A course in computational algebraic number theory Springer GTM No: 138 (first Published 1996) Fourth printing (2000).

[4] Jim Coykendall: Half–factorial domains in quadratic fields, J. Algebra 235 (2001) 417–430.

[5] Jody Esmonde and N. Ram Murthy: Problems in Algebraic Number theory Springer GTM No: 190 (1999). Chapters 3, 4, 6 and 8 pp 25–52, 69–80 and 97–114 (solutions in II for chapters 3, 4, 6 and 8).

[6] A. Fröhlic and M.J. Taylor: Algebraic Number Theory Cambridge studies in Advanced Mathematics No:27 (1993) chapter 1 and 4 pp 8–34 and 152–174.

[7] E. Hecke: Uberdie L-functioneu und den Dirichletsen Prinnizahlsatz für eineu beliebigen Zahlkörper Nachr. Acad.Wiss.Gottingen Math Phys kl II a (1917), pp 299–318.

[8] Kazuya Kato, Nobushige Kurokawa and Takeshi Saito (Translated by Masato Kuwata): Number theory 1: Fermat's Dream and Number Theory 2, Translations of mathematical monographs, vol 186 (1995), Amer. Math. Soc., Providence R.I (Iwanami series in Modern Mathematics). (2000).

[9] H.W. Lenstra Jr: Solving the Pell equation; Notices of Amer. Math. Soc. 49 No. 2 (Feb 2002) pp 182–192.

[10] D.A. Marcus : Number fields, Springer Verlag, Universitext (1977) Chapters 2 and 5 pp 12–54 and 130–157.

[11] W. Narkiewicz: Elementary and Analytic Theory of Algebraic numbers, PWN Polish Scientific publisher, Warzawa 1990. (Original edition) Third revised and expanded edition (English) 2004: Springer Monographs in Mathematics, Chapter 1 pp 1–41.

[12] Rajendran Valieveetil: A study of normed division domains and their analogues with applications to number theory; Ph.D Thesis, University of Calicut (1996).

[13] P. Samuel : Algebraic Theory of Numbers, Hermann Paris Houghton Mifflin Co, Boston (1970) Chapters 2 and 4 pp 27–43 and 53–65.

[14] R. Sridharan: Ancient Indian contributions to quadratic algebra, 'Sulvato Cakravala Science in the West and India: Some historical aspects: Editors B.V. Subbarayyappa and N. Mukunda, Hindustan Publishing House, Bombay, Chapter 11. pp 280–289 (1995).

[15] I. N. Stewart and D. O. Tall: Algebraic Number Theory Chapman and Hall, 2nd Edn (1987) Chapters 2, 3, 6 to 9 and 12, pp 38–69, 135–178 and 211–222.

[16] Rajat Tandon : Roots are not contained in cyclotomic fields, 'Resonance' Vol. 6 No: 4, (2001, April) 78–83.

[17] Andre Weil : Number theory – an approach through history, Birkauser, Boston (1984).

[18] A. Zaks : Half-factorial domains, Bull. Amer. Math. Soc. 82 (1976) 721–723.

[19] A. Zaks : Half-factorial domains, Israel J. Math. 37(1980) 281–302.

ADDITIONAL REFERENCES

[A1] Juliet Britto: On sums of squares, Proc. National seminar on Algebra, Number Theory and Applications to coding and Cryptanalysis, Sep. 16–18, Little Flower college, Guruvayoor, Eds: T. Thrivikraman, Sunny Kuriakose and Lucy V.N, pp 29–35.

[A2] Scott T. Chapman : A single example of non-unique factorization in Integral domains Amer. Math. Monthly 99 (1992) 943–945.

[A3] R. Gilmer : Commutative semigroup rings Chicago Lectures in Mathematics (University of Chicago Press) Chicago IL (1984).

[A4] T. Y. Lam: The algebraic theory of quadratic forms, Benjamin (Reading, Mass.) 1972.

[A5] Paulo Ribenboim : Classical theory of algebraic numbers; Universitext, Springer Verlag, New York Inc., NY (2001).

[A6] C. L. Siegel: Darstellung Total positiver Zahlen durch Quadrate, Math. Zeit, 11 (1921) 246–275.

[A7] Charles Small: Sums of three squares and levels of quadratic number fields, Amer. Math. Monthly, 93 (1986) 276–279.

[A8] Charles Small: Arithmetic of finite fields, Monographs and Textbooks in Pure and App. Mathematics No:148, Marcel Dekker Inc. NY (1991).

Part IV

SOME MORE INTERCONNECTIONS

CHAPTER 14

Rings of arithmetic functions

Historical perspective

It was E. T. Bell (1883–1960) who considered the multiplicative inverse of an arithmetic function. In his paper entitled: An arithmetical theory of certain numerical functions (University of Washington Publications in Mathematical and Physical Sciences Vol No.1 (1915)), Bell shows that an arithmetic function f possesses a Dirichlet inverse if, and only if, $f(1) \neq 0$. In October 1927, R. Vaidyanathaswamy showed the existence of the inverse of a multiplicative function independently. In the Journal of Indian Math. Society (Notes and questions) 17 (1927), 69–73, Vaidynathaswamy established that every multiplicative function of one variable possesses an inverse which is also multiplicative. Only later, when he was in St. Andrews University, (while he was with Professor H. W. Turnbull), he got interested in the papers of E. T. Bell. Vaidynathaswamy refers to the work of Bell in his memoir: 'The theory of multiplicative arithmetic functions' (Trans. Amer. Math. Soc. 33 (1931) 579–662). It is to be remarked that Bell recognized the algebraic foundations of the theory of arithmetical functions and made use of Cauchy composition given by $h(n,r) = \displaystyle\sum_{n \equiv a+b(\bmod r)} f(a)g(b)$ and other techniques. See Bell: Euler Algebra (Trans. Amer. Math. Soc. 25 (1923) 135–154) and Modular interpolation (Bull. Amer. Math. Soc., 37 (1931) 65–68). Since then, the algebraic approach to the theory of arithmetic functions came to be known better resulting in further work by L. Carlitz (1907–1999), Eckford Cohen, P. Kesava Menon and others.

14.1. Introduction

This chapter deals with certain finite dimensional algebras which arise in arithmetic function theory. It was seen that the set $\mathcal{A}$ of arithmetic functions is a Dirichlet algebra over $\mathbb{C}$, the field of complex numbers, multiplication being Dirichlet convolution. (See Remark 4.5.2, chapter 4). It is an example of an infinite dimensional algebra. Let r be a fixed positive integer and F a field of characteristic zero. The set $\mathcal{A}_r F$ of (r, F)-arithmetic functions (See Definition 14.2.2) forms an algebra of dimension r under the rule of Cauchy-composition for multiplication. $\mathcal{A}_r(F)$ is, indeed, a semisimple algebra which is the direct sum of r fields each isomorphic to F. See Proposition 14.2.1. Next, we discuss the algebra of even functions (mod r). The algebra $B_r(\mathbb{C})$ of even functions (mod r), that is, functions f for which $f(n,r) = f(g.c.d(n,r),r)$ is also a semisimple algebra. It is of dimension $d(r)$, the number of divisors of r. The basic technique is by using an orthogonal property of Ramanujan's sum $C(n,r)$, due to Eckford Cohen (see (5.4.4) and (5.4.5), chapter 5). As $B_r(\mathbb{C})$ is a non-nilpotent algebra, a zero divisor in $B_r(\mathbb{C})$ is not nilpotent.

The set $\mathcal{A}'$ of arithmetic functions defined over $\widetilde{\mathbb{Z}}$ has a ring structure when multiplication is via a Lucas product defined in terms of a fixed prime p. We call $\mathcal{A}'$ a Lucas ring. Zero-divisors in the Lucas ring could be nilpotent. An unsettled conjecture of Carlitz says:

Let $(\mathcal{A}', +, *)$ be a Lucas ring of arithmetic functions $f : \widetilde{\mathbb{Z}} \to F$, where F is a field of characteristic zero. $f \in \mathcal{A}'$ is a zero divisor if, and only if, f is nilpotent.

Next, we examine certain norm-preserving transformations of $\mathcal{A}$ considered as a vector space over $\mathbb{C}$. Linear operators help in the derivation of certain number theoretic identities. See Section 14.6.

14.2. Cauchy composition (mod r)

Let r be an arbitrary but fixed positive integer. $\omega = \exp(\frac{2\pi i}{r})$ is an imaginary rth-root of unity. We consider a field F of characteristic zero and assume that F contains the rth-roots of unity. A cyclotomic extension $\mathbb{Q}(\omega)$ (ω an rth root of unity) of $\mathbb{Q}$, the field of rational numbers, is an example of F.

Definition 14.2.1 (Eckford Cohen) : *A function $f : \mathbb{Z} \to F$ is called an (r, F)-arithmetic function, if f is single-valued and for every $a \in \mathbb{Z}$,*

$$f(a') = f(a), \quad whenever\, a' \equiv a\,(\mathrm{mod}\, r).$$

($\widetilde{\mathbb{Z}}$ denotes the set of non-negative integers).

To specify r, we write $f(n)$, $n \in \mathbb{Z}$, as $f(n,r)$.

Example 14.2.1 : ω^n ($n \in \mathbb{Z}$) is an (r, F) arithmetic function, as

$$\omega^m = \omega^n \text{ whenever } m \equiv n\,(\mathrm{mod}\, r).$$

Definition 14.2.2 : *Let f, g be (r, F)-arithmetic functions. The Cauchy product h of f and g is defined by*

$$h(n,r) = (f \cdot g)(n,r) = \sum_{n \equiv a+b(\mathrm{mod}\, r)} f(a,r)g(b,r)$$

where a and b range over elements of a complete residue system (mod r) *such that $n \equiv a + b$ (mod r).*

We introduce the notation

$$(14.2.1) \qquad\qquad e_a(n) = \omega^{an} = \exp(\frac{2\pi i a n}{r})$$

Since

$$\sum_{u(\bmod r)} \exp(\frac{2\pi itu}{r}) = \begin{cases} r, & \text{if } t \equiv 0(\bmod r); \\ 0, & \text{if } t \not\equiv 0(\bmod r) \end{cases}$$

we could write

(14.2.2)
$$\sum_{u(\bmod r)} e_a(u) = \begin{cases} r, & \text{if } a \equiv 0(\bmod r); \\ 0, & \text{otherwise.} \end{cases}$$

Further,

(14.2.3)
$$\begin{cases} e_a(u), & = e_u(a); \\ e_a(s+t), & = e_a(s)e_a(t). \end{cases}$$

Now, e_a is an (r,F) arithmetic function whose value at n is $e_a(n)$. The Cauchy product of e_a and e_b is given by

$$\begin{aligned} (e_a \cdot e_b)(n,r) &= \sum_{n \equiv x+y(\bmod r)} e_a(x)e_b(y) \\ &= \sum_{x(\bmod r)} e_a(x)e_b(n-x) \\ &= e_b(n) \sum_{x(\bmod r)} e_a(x)e_b(-x) \\ &= e_b(n) \sum_{x(\bmod r)} e_a(x)e_{-b}(x), \text{ by } (14.2.3) \\ &= e_b(n) \sum_{x(\bmod r)} e_{a-b}(x). \end{aligned}$$

Using (14.2.2) we deduce that

(14.2.4)
$$(e_a \cdot e_b)(n,r) = \begin{cases} re_b(n), & \text{if } a \equiv b(\bmod r); \\ 0, & \text{otherwise.} \end{cases}$$

Lemma 14.2.1 : *The set $\{e_a : 0 \le a \le r-1\}$ forms a linearly independent set of (r,F)-arithmetic functions over F.*

Proof : Let $\alpha_0, \alpha_1, \ldots \alpha_{r-1}$ be scalars ($\in F$) satisfying

$$\sum_{t=0}^{r-1} \alpha_t e_t(n) = 0.$$

We take the Cauchy product of the above with e_s for a fixed s ($0 \le s \le r-1$). Then, from the left side, we have

$$\sum_{t=0}^{r-1} \alpha_t (e_s \cdot e_t)(n) = \begin{cases} \alpha_s re_s(n), & \text{if } t = s; \\ 0, & \text{otherwise.} \end{cases}$$

This shows that $\alpha_s = 0$ for each s. So, $\{e_a : 0 \le a \le r-1\}$ forms a linearly independent set. $\qquad\square$

We ask whether the set $\{e_a : 0 \le a \le r-1\}$ spans the space of (r,F)-arithmetic functions. The answer is yes. It is easy to check that the set $\mathcal{A}_r(F)$ of (r,F)-arithmetic functions forms a vector space over F.

Theorem 126 (Eckford Cohen (1952)) **:** *The set $\{e_a : 0 \le a \le r-1\}$ is a basis for the vector space $\mathcal{A}_r(F)$. Further, $f \in \mathcal{A}_r(F)$ has the form*

$$(14.2.5) \qquad\qquad f(a,r) = \sum_{s=0}^{r-1} \alpha_s e_s(a),$$

where α_s is given by

$$(14.2.6) \qquad\qquad \alpha_s = \frac{1}{r} \sum_{t(\mathrm{mod}\, r)} f(t,r)e_s(-t).$$

Proof : Assuming (14.2.6), we see that

$$\sum_{s=0}^{r-1} \alpha_s e_s(a) = \frac{1}{r} \sum_{s=0}^{r-1} e_s(a) \sum_{t(\mathrm{mod}\, r)} f(t,r)e_s(-t),$$

$$= \frac{1}{r} \sum_{s=0}^{r-1} \sum_{t=0}^{r-1} f(t,r)e_s(a)e_s(-t),$$

$$= \frac{1}{r} \sum_{t=0}^{r-1} f(t,r) \sum_{s=0}^{r-1} e_a(s)e_t(-s),$$

$$= \frac{1}{r} \sum_{t=0}^{r-1} f(t,r) \sum_{x+y\equiv 0(\mathrm{mod}\, r)} e_a(x)e_t(y),$$

$$= \begin{cases} \frac{1}{r}\sum_{t=0}^{r-1} f(t,r)re_t(0), & \text{if } t \equiv a(\mathrm{mod}\, r), \\ 0, & \text{otherwise.} \end{cases} \quad \text{(by (14.2.4)),}$$

$$= f(a,r), \text{ as } e_a(0) = 1.$$

So, (14.2.6) implies (14.2.5). As $\{e_a : 0 \le a \le r-1\}$ is a linearly independent set, the representation of $f(a,r)$ given in (14.2.5) is unique. $\qquad\qquad \square$

Remark 14.2.1 : Theorem 126 has been adapted from [7].

Corollary 14.2.1 : *If $f(a,r)$ is as given in (14.2.5) and $g(b,r) = \sum_{s=0}^{r-1} \beta_s e_s(b)$, then,*

$$(14.2.7) \qquad\qquad h(n,r) = (f \cdot g)(n,r) = r \sum_{s=0}^{r-1} \alpha_s \beta_s e_s(n).$$

Proof follows from the definition of Cauchy product and from (14.2.5).

In (14.2.1), we gave the notation $e_a(n)$ for $\exp(\frac{2\pi i a n}{r})$. To specify r, we write

$$(14.2.8) \qquad\qquad \varepsilon(an,r) = \exp(\frac{2\pi i a n}{r}).$$

Lemma 14.2.2 (Orthogonality relation) : *For fixed r, let d, δ be two divisors of r. Let $0 \le x < d$, $0 \le y < \delta$. Suppose that*

$$g.c.d\,(x,d) = g.c.d\,(y,\delta) = 1.$$

Then,

(14.2.9) $$\sum_{n \equiv a+b (\mathrm{mod}\ r)} \varepsilon(ax,d)\varepsilon(by,\delta) = \begin{cases} r\varepsilon\,(nx,d) & \textit{if, } x = y,\ \delta = d, \\ 0, & \textit{otherwise.} \end{cases}$$

Proof : We write $dd' = r$, $\delta\delta' = r$. The left side of (14.2.9) is simplified as follows:

$$\varepsilon(ax,d) = \varepsilon(axd',dd') = \varepsilon(axd',r)$$
$$\varepsilon(by,\delta) = \varepsilon(by\delta',\delta\delta') = \varepsilon(by\delta',r)$$

Then,

$$\sum_{n \equiv a+b(\mathrm{mod}\ r)} \varepsilon(axd',r)\varepsilon(by\delta',r) = (e_{xd'}e_{y\delta'})(n,r)$$

Using (14.2.4), we get

$$(e_{xd'} \cdot e_{y\delta'})(n,r) = \begin{cases} re_{xd'}(n) & \text{if } xd' \equiv y\delta'(\mathrm{mod}\ r) \\ 0, & \text{otherwise.} \end{cases}$$

That is, the left side (14.2.9) reduces to zero unless $xd' = y\delta'$. Multiplying both sides by $d\delta$, we get $\delta x = dy$. Since $g.c.d(x,d) = 1$, δ divides d. Also, d divides δ. So $d = \delta$. Then, $x = y$.

As $e_{xd'}(n) = \varepsilon(nxd',r) = \varepsilon(nx,d)$, we have

$$\sum_{n \equiv a+b(\mathrm{mod}\ r)} \varepsilon(ax,d)\varepsilon(by,\delta) = \begin{cases} r \in (nx,d), & \text{if } x = y,\ \delta = d; \\ 0, & \text{otherwise.} \end{cases}$$

If $x = 0$ and $d = 1$, the left side of (14.2.9) is the sum of the δth-roots of unity which is zero unless less $\delta = 1$. But, in that case $y = 0$. So, the sum reduces to r when $x = y = 0$, $d = \delta = 1$.

From lemma 14.2.1 and definition14.2.2, we note that $(\mathcal{A}_r(F),+,\cdot)$ forms a commutative ring with unity

(14.2.10) $$u_0(n,r) = \begin{cases} 1, & \text{if } n \equiv 0(\mathrm{mod}\ r); \\ 0, & \text{otherwise.} \end{cases}$$

(+ denotes ordinary addition and · denotes Cauchy composition.)

From (14.2.2), we also note that

$$u_0(n,r) = \frac{1}{r}\sum_{u(\mathrm{mod}\ r)} e_n(u) = \frac{1}{r}\sum_{u(\mathrm{mod}\ r)} \exp(\frac{2\pi iun}{r}).$$

$\square$

Fact 14.2.1 : As $\mathcal{A}_r(F)$ is an r-dimensional vector space over F, $\mathcal{A}_r(F)$ is a finite dimensional commutative algebra over F.

We digress, for a while, to exhibit the structure of a finite dimensional commutative algebra $\mathcal{A}$ over a field F.

While considering convolution algebras (see Section 11.4, chapter 11), we have defined an F-algebra $\mathcal{A}_F$. We recall that a commutative algebra $\mathcal{A}$ over a field F is such that $\mathcal{A}$ is a vector space over F and $(\mathcal{A}, +, \cdot)$ is a commutative ring where $(\cdot)$ denotes the operation of multiplication. Further, for $\alpha, \beta \in F$ and $a, b \in \mathcal{A}$, one has

$$(14.2.11) \qquad\qquad (\alpha a) \cdot (\beta b) = \alpha\beta(a \cdot b)$$

where αa is obtained from a by doing scalar multiplication by α. Also, $a \cdot b = b \cdot a$ for all $a, b \in \mathcal{A}$. $\mathcal{A}$ is said to have the unity element e_0, if $e_0 \cdot a \cdot e_0 = a$ for all $a \in \mathcal{A}$.

Fact 14.2.2 : If $\mathcal{A}$ is a finite dimensional algebra over F, (a field), $\mathcal{A}$ has no nonzero idempotent elements if, and only if, every element of $\mathcal{A}$ is nilpotent. For proof, see [1].

Since an algebra is also a ring, results which are known for subsets of a ring hold good in connection with subsets of an algebra. In the same manner, as an algebra is a vector space, results which are known for subsets of a vector space are applicable to subsets of an algebra.

A subspace W of an algebra $\mathcal{A}$ is a subalgebra of $\mathcal{A}$ if, and only if, $a \cdot b \in W$ for elements $a, b \in W$. W is, in fact, a subring of the ring $\mathcal{A}$.

Definition 14.2.3 : *A subspace W of an algebra $\mathcal{A}$ is called an ideal of $\mathcal{A}$ if, and only if,*

$$\mathcal{A}W \subset W \text{ and } W\mathcal{A} \subset W.$$

The zero ideal (0) and $\mathcal{A}$ are trivial ideals of $\mathcal{A}$. Any other ideal of $\mathcal{A}$ is said to be non trivial.

Definition 14.2.4 : *A commutative algebra $\mathcal{A}$ (over a field F) is called simple if $\mathcal{A} \neq (0)$ and if $\mathcal{A}$ has no non trivial ideals.*

Definition 14.2.5 : *Let $\mathcal{A}, \mathcal{B}$ be algebras over a field F. A map $\psi : \mathcal{A} \to \mathcal{B}$ is said to be an algebra homomorphism from $\mathcal{A}$ to $\mathcal{B}$, if*

$$\psi(a + b) = \psi(a) + \psi(b),$$
$$\psi(\alpha a) = \alpha\psi(a),$$
$$\psi(a \cdot b) = \psi(a) \cdot \psi(b),$$

for all $a, b \in \mathcal{A}$ and $\alpha \in F$.

ψ is called an isomorphism from $\mathcal{A}$ into $\mathcal{B}$, if ψ is injective. If there exists an isomorphism from $\mathcal{A}$ onto $\mathcal{B}$, $\mathcal{A}$ and $\mathcal{B}$ are said to be isomorphic, written $\mathcal{A} \cong \mathcal{B}$.

Next, we note that an element a of an algebra $\mathcal{A}$ is called idempotent if $a^2 = a$. An element t of an algebra $\mathcal{A}$ is called nilpotent of index m if $t^m = 0$, but $t^{m-1} \neq 0$, for some positive integer m.

Observation 14.2.1 :

(a) A set $\{e_1, e_2, \ldots, e_n\}$ of elements of an algebra $\mathcal{A}$ is said to form pairwise orthogonal idempotents of $\mathcal{A}$ if

$$e_i^2 = e_i \text{ and } e_i e_j = 0 \ (i \neq j)$$

where $i, j = 1, 2, \ldots, n$.

(b) A set $\{e_1, e_2, \ldots, e_n\}$ of non zero pairwise orthogonal idempotent elements of $\mathcal{A}$ is a linearly independent subset of $\mathcal{A}$.

(c) Let $B = [e_1, e_2, \ldots, e_n]$ be the subspace of an algebra $\mathcal{A}$ spanned by n pairwise orthogonal idempotents $e_1, e_2, \ldots, e_n$ of $\mathcal{A}$. Then, B is an n-dimensional subspace of $\mathcal{A}$ and $e = e_1 + e_2 + \cdots + e_n$ is the unity element of B.

(d) Let $\mathcal{A}$ be an algebra and I an ideal of $\mathcal{A}$ with unity. (Recall that I is a subspace of $\mathcal{A}$ such that for $a \in I$, $s \in \mathcal{A}$, $a.s$ and $s.a$ are in $\mathcal{A}$). Then, I is a direct summand of $\mathcal{A}$. That is, $\mathcal{A} = I \oplus J$ for some ideal J of $\mathcal{A}$ and J is unique.

(e) An algebra $\mathcal{A}$ is called reducible if $\mathcal{A}$ is expressible as a direct sum of two of its proper ideals. Otherwise, it is called irreducible.

(f) Let $\mathcal{A}$ be a finite dimensional algebra with a unity element. Then, $\mathcal{A}$ is expressible as a direct sum of irreducible direct summands uniquely except for the order of the direct summands. That is,

(14.2.12) $$\mathcal{A} = \mathcal{A}_1 \oplus \mathcal{A}_2 \oplus \cdots \oplus \mathcal{A}_n$$

In (14.2.12), every direct summand is an ideal of A. The method of derivation is similar to that of rings. See Alexander Abian [1].

Lemma 14.2.3 : *Let $\mathcal{A}$ be a one-dimensional algebra over a field F. Then, $\mathcal{A}$ is either the one-dimensional zero algebra or is a field isomorphic to F.*

Proof : Suppose that $\mathcal{A}$ has a basis $\{e_1\}$ and that the multiplication table for $\mathcal{A}$ is given by $e_1 \cdot e_1 = \alpha e_1$ where α is a fixed element of F. If $\alpha = 0_F$, $\mathcal{A}$ is the one-dimensional zero algebra. If $\alpha \neq 0_F$, we define $\psi : \mathcal{A} \to F$ by $\psi(a) = \alpha\xi$ where $a = \xi e_1, \xi \in F$. If is verified that ψ is an algebra isomorphism onto F. That is, $\mathcal{A} \cong F$.   □

If I is an ideal of an algebra $\mathcal{A}$, we could define the quotient algebra $\mathcal{A}/I$, as we do for a quotient ring or a quotient space. Also, an ideal I of $\mathcal{A}$ is called a nilpotent ideal of index m, if there exists $m \in \mathbb{N}$ such that $I^m = (0)$ and $I^{m-1} \neq (0)$; $m \in \mathbb{N}$.

Fact 14.2.3 : Let $\mathcal{A}$ be a finite dimensional commutative algebra. If I and J are nilpotent ideals of $\mathcal{A}$, so is their sum $I + J$.

Proof is omitted.

Lemma 14.2.4 : *Let A be a finite dimensional commutative algebra with unity element. Then, A has a unique nilpotent ideal of maximal dimension.*

Proof : We consider the collection M of nilpotent ideals of A. As the zero ideal belongs to M, M is non-empty. As A is Noetherian as a ring (or F-module), M has a maximal element. That is, A has a nilpotent ideal of maximal dimension. We call it J. If J' is any other nilpotent ideal of A, by Fact 14.2.3, $J+J'$ is again a nilpotent ideal. But, then, $J \subseteq J+J'$. Since J is of maximal dimension, $J = J+J'$ which implies that $J' \subseteq J$. Also, as $J+J' = J$, $J' \subseteq J+J' = J$. Thus, $J' = J$ or, J is unique.   □

Definition 14.2.6 : *The unique nilpotent ideal of maximal dimension is called the radical of A.*

We remark that an algebra A is nilpotent if, and only if, A is its own radical. Further, when A is nilpotent, every element of A is nilpotent.

When A is a finite dimensional commutative algebra with unity element, A has the structure of an Artinian ring. Looking at A as a ring, the radical of A is the largest nilpotent ideal of A.

Observation 14.2.2 :

(a) Let N be a 2-dimensional algebra whose multiplication table with respect to a basis $\{e_1, e_2\}$ is given by

	e_1	e_2
e_1	e_2	0
e_2	0	0

Then N^2 is a one-dimensional subalgebra and $N^2 \neq (0)$. But $N^3 = 0$. So, N is a nilpotent algebra of index 3.

(b) Let V be a 2-dimensional algebra whose multiplication table with respect to a basis $\{v_1, v_2\}$ is given by

	v_1	v_2
v_1	v_1	v_1
v_2	v_1	v_1

V^2 is a one-dimensional subalgebra and $V^2 = V^3$. V is a nonnilpotent algebra.

(c) An algebra A is simple, if, and only if, A is nonnilpotent and has no nontrivial ideal.

(d) Let J_n be an ideal of $M_n(R)$, the ring of $n \times n$ matrices with entries from a commutative ring R having unity element 1_R. The set I of all entries of all the elements of J_n is an ideal of R. If R is a simple ring with unity 1_R, then $M_n(R)$ is also a simple ring with unity.

(e) An algebra over a field F is called a total $n \times n$ matrix algebra over F, if it is isomorphic to the algebra $M_n(F)$ of all $n \times n$ matrices with entries in F.

(f) *If* $\mathcal{A}$ *is a nonnilpotent algebra, then* $\mathcal{A}$ *has a nonzero idempotent element. In other words,* $\mathcal{A}$ *is nilpotent if, and only if, it has no nonzero idempotent element.*

Definition 14.2.7 : *A nonnilpotent algebra whose radical is* (0) *is called a semisimple algebra . This means that when* $\mathcal{A}$ *is considered as a ring, the intersection of its maximal ideals is (0). In other words, the Jacobson radical (see Section 12.4, chapter 12) of a semisimple algebra* $\mathcal{A}$ *is (0). Obviously, a simple algebra is semisimple.*

Definition 14.2.8 : *Let* $\mathcal{A}$ *be a finite dimensional commutative algebra with unity.* $\mathcal{A}$ *is said to be the (internal) direct sum of its subalgebras* $\mathcal{A}_1, \mathcal{A}_2, \ldots, \mathcal{A}_n$, *if every element a of* $\mathcal{A}$ *is uniquely expressed as a sum of elements* $a_1, a_2, \ldots a_n$ *where* $a_i \in \mathcal{A}_i$ $(i = 1, 2, \ldots n)$ *and for every* $a_i \in \mathcal{A}_i, a_j \in \mathcal{A}_j (i, j = 1, 2, \ldots, n)$, $a_i \cdot a_j = 0$ *for* $i \notin j$, *where 0 is the zero element of the algebra.*

We write $\mathcal{A} = \mathcal{A}_1 \oplus \mathcal{A}_2 \oplus \cdots \oplus \mathcal{A}_n$.

Lemma 14.2.5 : *Let* $\mathcal{A}$ *be a finite dimensional commutative algebra. Suppose that* $\{e_1, e_2, \ldots, e_n\}$ *is a basis of* $\mathcal{A}$ *such that* e_i, e_j *are pairwise orthogonal. That is, whenever* $i \neq j$. $e_i \cdot e_j = 0$. *Then,*

$$(14.2.13) \qquad \mathcal{A} = [e_1] \oplus [e_2] \oplus \cdots \oplus [e_n],$$

when $[e_i]$ *is the one-dimensional subalgebra of* $\mathcal{A}$, *generated by* $e_i (i = 1, 2, \ldots, n)$.

Proof : By definition, $a \in \mathcal{A}$ can be uniquely written as $a = \sum_{i=1}^{n} x_i e_i$, $x_i \in F$ $(i = 1, 2, \ldots, n)$ $x_i e_i \in [e_i]$. As $e_1, e_2, \ldots e_n$ are pairwise orthogonal (14.2.13) holds. $\square$

Corollary 14.2.2 : *By lemma 14.2.3,* $[e_i]$ *is isomorphic to* F, *for each i.*

Fact 14.2.4 : Let $\mathcal{A}$ be a finite dimensional commutative algebra over a field F.

(a) If $\mathcal{A}$ is a semisimple algebra, $\mathcal{A}$ has a unity element. In particular, if $\mathcal{A}$ is simple, $\mathcal{A}$ has a unity element.
(b) If $\mathcal{A}$ is semisimple, $\mathcal{A}$ is irreducible if, and only if, $\mathcal{A}$ is simple.
(c) Suppose that $\mathcal{A}$ has a unity element. If

$$\mathcal{A} = \mathcal{A}_i \oplus \mathcal{A}_2 \oplus \cdot \oplus \mathcal{A}_n,$$

a subalgebra $\mathcal{N}$ of $\mathcal{A}$ is the radical of $\mathcal{A}$ if, and only if, $\mathcal{N} = \mathcal{N}_1 \oplus \mathcal{N}_2 \oplus \cdots \oplus \mathcal{N}_n$ where $\mathcal{N}_i$ is the radical of $\mathcal{A}_i$ and $\mathcal{N}_i = \mathcal{N} \cap \mathcal{A}_i$ $(i = 1, 2, \ldots, n)$. For proofs, see Alexander Abian [1].

Theorem 127 : *Let* $\mathcal{A}$ *be a finite dimensional commutative algebra with unity.* $\mathcal{A}$ *is semisimple if, and only if, it is simple or is expressible uniquely as a direct sum of simple subalgebras, except for the order of the direct summands.*

Proof : If $\mathcal{A}$ is simple, it is nonnilpotent and has radical (0). Thus, $\mathcal{A}$ is semisimple. Conversely, if $\mathcal{A}$ is semisimple either $\mathcal{A}$ is simple or is reducible.

$\Leftarrow$: If $\mathcal{A}$ is reducible and is expressible as a direct sum of irreducible ideals $\mathcal{A}_i$ $(i = 1, 2, \ldots, n)$ uniquely except for the order of the summands, we have

$$\mathcal{A} = \mathcal{A}_1 \oplus \mathcal{A}_2 \oplus \cdots \mathcal{A}_n \text{ (by observation 14.2.1(f))}.$$

Each ideal $\mathcal{A}_i$ is irreducible and so each $\mathcal{A}_i$ is simple (by Fact 14.2.4(b)). By Fact 14.2.4(a), each $\mathcal{A}_i$ has a unity element say u_i $(i = 1, 2, \ldots n)$. Then,

$$u_1 + u_2 + \cdots + u_n = u \text{ is the unity element of } \mathcal{A}.$$

u is an idempotent element. So, $\mathcal{A}$ is non-nilpotent. Since each $\mathcal{A}_i$ is simple, radical N_i of $\mathcal{A}_i$ is (0). By Fact 14.2.4(c), radical of $\mathcal{A}$ is (0). Thus, $\mathcal{A}$ is semisimple
:$\Rightarrow$ Suppose that $\mathcal{A}$ is semisimple. If $\mathcal{A}$ is irreducible, by Fact 14.2.4(b), $\mathcal{A}$ is simple. If $\mathcal{A}$ is reducible, $\mathcal{A}$ is expressible uniquely as

$$\mathcal{A} = \mathcal{A}_1 \oplus \mathcal{A}_2 \oplus \cdots \mathcal{A}_n (\text{except for the order of the summands}).$$

Each $\mathcal{A}_i$ is irreducible and semisimple. So, each $\mathcal{A}_i$ is simple, by Fact 14.2.4(b). This completes the proof of theorem 127. $\square$

Observation 14.2.3 :

(a) Every nonzero ideal of a semisimple algebra is semisimple.
(b) Every nonzero ideal of a semisimple algebra $\mathcal{A}$ has a unity element (different from that of $\mathcal{A}$).
(c) Let $\{e_1, e_2, \ldots, e_n\}$ be a basis of an algebra $\mathcal{A}$ such that

$$e_i.e_j = \begin{cases} e_i, & \text{if } j = i \\ 0, & \text{otherwise.} \end{cases}$$

Then, $\mathcal{A}$ is a commutative semisimple algebra.

Observation 14.2.4 : *A homomorphic image of a semisimple ring need not be semisimple.*

Proof : We recalled that a ring R' is a homomorphic image of a ring R if there exists a subjective homomorphism (epimorphism) from R onto R' (see definition 12.2.5, chapter 12). That is, R' is isomorphic to a quotient ring of R. If I is an ideal of R, we get a ring R/I isomorphic to a homomorphic image of R.

Let $\psi : \mathbb{Z} \to \mathbb{Z}/p^n\mathbb{Z}$ (p a prime, $n > 1$) be defined by $\psi(a) = a + (p^n\mathbb{Z})$, $a \in \mathbb{Z}$. $\mathbb{Z}$ is semisimple. But if $a = kp$, $1 \le k \le p^{n-1}$, $a + (p^n\mathbb{Z})$ is a nilpotent element and so the ideal generated by $p + (p^n\mathbb{Z})$ is a nilideal (see definition 2.3.3, chapter 2) of $\mathbb{Z}/p^n\mathbb{Z}$, denoted by N. N is contained in the Jacobson radical of $\mathbb{Z}/p^n\mathbb{Z}$. So, $\mathbb{Z}/p^n\mathbb{Z}$ is not semisimple for $n > 1$. $\square$

Observation 14.2.5 : *Let R be a Noetherian (Artinian) ring. It is verified that any homomorphic image of R is Noetherian (Artinian). Though chain conditions are not destroyed by homomorphism, the property of 'semisimplicity' is not so, as is seen in observation 14.2.4.*

Next, from lemma 14.2.5 and theorem 127, we deduce that $\mathcal{A}_r(F)$ is the direct sum of the subalgebras $[e_i]$ ($0 \leq i \leq r-1$) and each $[e_i]$ is a field isomorphic to F. That is,

(14.2.14) $$\mathcal{A}_r(F) = [e_0] \oplus [e_1] \cdots \oplus [e_{r-1}].$$

$\mathcal{A}_r(F)$ has a basis containing r elements $e_0, e_1, \ldots e_{r-1}$ and the basis elements are pairwise orthogonal.

If $E_s = \frac{1}{r} e_s$ where $e_s(n) = \exp(\frac{2\pi i n s}{r})$, we note that

(14.2.15) $$E_s \cdot E_t = \frac{1}{r^2}(e_s \cdot e_t) = \begin{cases} \frac{1}{r} e_s(n), & \text{if } s \equiv t \pmod{r}; \\ 0, & \text{otherwise.} \end{cases}$$

That is, we get

(14.2.16) $$E_s \cdot E_t = \begin{cases} E_s, & \text{if } t = s; \\ 0, & \text{otherwise.} \end{cases}$$

So, $\{E_0, E_1, \ldots E_{r-1}\}$ is a basis of $\mathcal{A}_r(F)$ such that (14.2.16) holds. It follows that $\mathcal{A}_r(F)$ is a semisimple algebra. We state this as

Proposition 14.2.1 (Eckford Cohen (1952)) **:** *The ring $\mathcal{A}_r(F)$ of (r, F)-arithmetic functions is a semisimple algebra over F and can be expressed as a direct sum of r fields A_s ($s = 0, 1, 2, \ldots, r-1$) in the form*

(14.2.17) $$\mathcal{A}_r(F) = A_0 \oplus A_1 \oplus \cdots \oplus A_{r-1},$$

where each A_s, ($s = 0, \ldots, (r-1)$) is isomorphic to F. Further, A_s contains $\frac{1}{r} e_s$ as an idempotent generator. Also

$$u_0(n, r) = \begin{cases} 1, & n \equiv 0 \pmod{r}; \\ 0, & \text{otherwise;} \end{cases}$$

serves as the unity element of the algebra.

Remark 14.2.2 : Proposition 14.2.1 has been adapted from Eckford Cohen [7].

Next, we recall the definition of Ramanujan's sum $C(n, r)$ (See (5.1.2) of chapter 5).

$$C(n, r) = \sum_{\substack{h(\text{mod } r) \\ (h, r) = 1}} \exp(\frac{2\pi i n u}{r}) = \sum_{\substack{h=0 \\ (h, r) = 1}}^{r-1} e_h(n),$$

where the summation is over a reduced-residue system (mod r).

Lemma 14.2.6 : *Let d, δ be divisors of r. Then,*

$$\sum_{n \equiv a+b(\text{mod } r)} C(a, d)C(b, \delta) = \begin{cases} rC(n, d) & \text{if } \delta = d \\ 0 & \text{otherwise.} \end{cases}$$

Proof :

$$\sum_{n \equiv a+b(\mathrm{mod}\ r)} C(a,d)C(b,\delta) = \sum_{\substack{(x,d)=1 \\ (y,\delta)=1}} \sum_{n \equiv a+b(\mathrm{mod}\ r)} \varepsilon(ax,d)\varepsilon(by,\delta),$$

where x ranges over non-negative integers $< d$ and prime to d and y ranges over those $< \delta$ and prime to δ. By (14.2.9), the inner sum namely,

$$\sum_{n \equiv a+b(\mathrm{mod}\ r)} \varepsilon(ax,d)\varepsilon(by,\delta) = \begin{cases} r\varepsilon(nx,d) & \text{if } x = y, d = \delta; \\ 0 & \text{otherwise.} \end{cases}$$

So, we have

$$(14.2.18) \qquad \sum_{n \equiv a+b(\mathrm{mod}\ r)} C(a,d)C(b,\delta) = \begin{cases} \sum_{(x,d)=1} r\varepsilon(nx,d), & \text{if } \delta = d; \\ 0, & \text{otherwise.} \end{cases}$$

From (14.2.18) and the definition of $C(n,r)$, the desired result follows. $\qquad \square$

Remark 14.2.3 : Lemma 14.2.6 has been adapted from Eckford Cohen [8]. See [9] also.

We deduce that the set $S = \{\frac{1}{r}C(n,d) : d|r\}$ forms a linearly independent set in which any two distinct elements are mutually orthogonal. For, let $r > 1$. Each element of the set S can be obtained by taking a linear combination of certain basis elements of $B = \{\frac{1}{r}e_s(n) : 0 \leq s \leq (r-1)\}$. We consider a set (having $\phi(d)$ elements) $\frac{1}{r}e_{s_1}(n), \frac{1}{r}e_{s_2}(n) \ldots, \frac{1}{r}e_{s_t}(n)$ where g.c.d $(s_j, r) = \frac{r}{d}$, d being a fixed divisor of r $(j = 1, 2, \ldots t)$. Then,

$$\frac{1}{r}\sum_{j=1}^{t} e_{s_j}(n) = \frac{1}{r} \sum_{\substack{h(\mathrm{mod}\ d) \\ (h,d)=1}} \exp(\frac{2\pi i h(\frac{r}{d})n}{r})$$

$$= \frac{1}{r} \sum_{\substack{h(\mathrm{mod}\ d) \\ (h,d)=1}} \exp(\frac{2\pi h n}{d}),$$

or,

$$\frac{1}{r}\sum_{j=1}^{t} e_{s_j}(n) = \frac{1}{r}C(n,d).$$

Thus, $\{\frac{1}{r}C(n,d) : d|r\}$ forms a basis for a subalgebra $D_r(F)$ (of dimension $d(r)$) of the algebra $\mathcal{A}_r(F)$. To make it down to earth for number theory, we take F to be the field $\mathbb{C}$ of complex numbers. So, then, we have a finite dimensional algebra $D_r(\mathbb{C})$ of dimension $d(r)$ over $\mathbb{C}$. In fact, $D_r(\mathbb{C})$ forms a semisimple subalgebra of the algebra of periodic functions (mod r). Analogous to theorem 126 and proposition 14.2.1, we have

Proposition 14.2.2 : *The set $D_r(\mathbb{C})$ of arithmetic functions of the form*

$$f(n,r) = \sum_{d|r} \alpha(d,r)C(n,d), \quad \alpha(d,r) \in \mathbb{C}, \, r > 1;$$

forms a subalgebra of $\mathcal{A}_r(\mathbb{C})$ with an orthogonal basis $\{\frac{1}{r}C(n,d) : d|r\}$. The unity element of $D_r(\mathbb{C})$ coincides with that of $\mathcal{A}_r(\mathbb{C})$. Further $D_r(\mathbb{C})$ is a direct sum of $d(r)$ fields each isomorphic to $\mathbb{C}$, where $d(r)$ denotes the number of divisors of r.

Proposition 14.2.2 is a consequence of theorem 127 and so, its proof is omitted.

Remark 14.2.4 : (a) $D_r(\mathbb{C})$ is a semisimple algebra.
  (b) The unity element in $D_r(\mathbb{C})$ is given by

$$(14.2.19) \qquad I(n,r) = \sum_{d|r} \frac{1}{r}C(n,d) = \begin{cases} 1, & \text{if } n \equiv 0 \,(\text{mod } r); \\ 0, & \text{otherwise;} \end{cases}$$

which is the same as that of $\mathcal{A}_r(\mathbb{C})$. (See (14.2.10)).

In Section 5.4, chapter 5, we considered the class of even functions (mod r), that is, functions $f : \widetilde{\mathbb{Z}} \times \mathbb{N} \to \mathbb{C}$ which are such that $f(n,r) = f(\text{g.c.d } (n,r),r)$. The arithmetical representation of an even function (mod r) was given in theorem 34.

  If f is even (mod r), it is easy to check that $f(n,r) = f(n',r)$ whenever $n' \equiv n$ (mod r). Therefore, f is even (mod r) $\Rightarrow f$ is an $(r,\mathbb{C})$-arithmetic function.

  By theorem 34, if $B_r(\mathbb{C})$ denotes the set of even functions (mod r), $B_r(\mathbb{C})$ and $D_r(\mathbb{C})$ are subspaces of $\mathcal{A}_r(\mathbb{C})$. Further, $\dim B_r(\mathbb{C}) = \dim D_r(\mathbb{C})$. So, as

$$B_r(\mathbb{C}) \subseteq D_r(\mathbb{C}), B_r(\mathbb{C}) = D_r(\mathbb{C}).$$

See M. Artin [2, Proposition 3.20, chapter 3]. This is also obvious, otherwise.

14.3. The algebra of even functions (mod r)

  By theorem 34 in chapter 5, if f is even (mod r), then,

$$(14.3.1) \qquad\qquad f(n,r) = \sum_{d|r} \alpha(d,r)C(n,d),$$

where

$$(14.3.2) \qquad\qquad \alpha(d,r) = \frac{1}{r}\sum_{t|r} f(\frac{r}{t})C(\frac{r}{d},t)$$

$\alpha(d,r)$ is also expressed by the equivalent formula:

$$(14.3.3) \qquad\qquad \alpha(d,r) = \frac{1}{r\varphi(d)}\sum_{a=1}^{r} f(a,r)C(a,d)$$

For, replacing a (mod r) on the right of (14.3.3) by an equivalent residue system $s = (\frac{r}{t})x$, $t|r$ and g.c.d $(x,t) = 1$, we get

$$\alpha(d,r) = \frac{1}{r\phi(d)} \sum_{t|r} \sum_{\substack{x(\text{mod } t) \\ g.c.d(x,t)=1}} f(\frac{r}{t}x,r)C(\frac{r}{t}x,d)$$

Now, since f and C are even functions (mod r),

$$f(\frac{r}{t}x,r) = f(\frac{r}{t},r) \text{ and } C(\frac{r}{t}x,d) = C(\frac{r}{t},d)$$

By the property of $C(n,r)$,

(14.3.4) $$C(\frac{r}{t},d)\phi(t) = C(\frac{r}{d},t)\phi(d)$$

So, (14.3.3) takes the form

(14.3.5) $$\alpha(d,r) = \frac{1}{r\phi(d)} \sum_{t|r} f(\frac{r}{t},r)\, C(\frac{r}{d},t)\, \phi(d)$$

which reduces to (14.3.2). In obtaining (14.3.5), we have made use of the fact that as x runs through a complete residue system (mod r), we get $\phi(t)$ such systems mod $\frac{r}{t}$.

Next, let $d_1, d_2, \ldots d_q$ be the distinct divisors of r. $q = d(r)$, the number of divisors of r. If the divisors are arranged in ascending order, we could take $d_1 = 1$ and $d_q = r$. We define

(14.3.6) $$\rho_i(n,r) = \begin{cases} 1, & \text{if } (n,r) = d_i, \quad i = 1, 2, \ldots q \\ 0, & \text{otherwise.} \end{cases}$$

$\rho_1(n,r)$ is the familiar Kronecker function ρ.

For $r \geq 1$, g.c.d $(n,r) = d_i$ for some i $(1 \leq i \leq q)$. So, if f is even (mod r),

$$f(n,r) = f(d_i,r), \text{ for some } i(1 \leq i \leq q).$$

So, f is expressible as

(14.3.7) $$f(n,r) = \sum_{i=1}^{q} f(d_i,r)\rho_i(n,r)$$

It is easy to check that the set $E = \{\rho_i(n,r) : 1 \leq i \leq q\}$ is a linearly independent set. By virtue of (14.3.7), $B_r(\mathbb{C})$ is a vector space of dimension $d(r)$ and having E as a basis. See Pentti Haukkanen [19]. (14.3.7) is not that convenient to handle, as it merely says $f(n,r) = f(d_i,r)$ for a divisor d_i of r. $(1 \leq i \leq q)$. The arithmetic representation (14.3.1) of $f(n,r)$ is referred to as a finite Fourier series expansion of f and $\alpha(d,r) : d|r$ are the Fourier coefficients of f. See Remark 5.4.1, in chapter 5.

The Cauchy product of two even functions (mod r) is an even function (mod r). For $f, g \in B_r(\mathbb{C})$, let $\alpha(d,r)$, $\beta(d,r)$ be the Fourier coefficients of f and g respectively, where $d|r$. By theorem 35, chapter 5, the Cauchy product $h = f \cdot g$ has

Fourier coefficients $r\alpha(d,r)\beta(d,r)$. So, $(B_r(\mathbb{C}),+,\cdot)$ is a commutative ring with unity $I(n,r)$ (14.2.19). As

$$\frac{1}{r^2} \sum_{n\equiv a+b(\mathrm{mod}\ r)} C(a,d_i)C(b,d_j) = \begin{cases} \frac{1}{r}C(n,d_i), & \text{if } d_j = d_i; \\ 0, & \text{otherwise;} \end{cases}$$

$\{\frac{1}{r}C(n,d) : d|r\}$ forms a basis of idempotents which are pairwise orthogonal. If $E_i = \frac{1}{r}C(n,d_i)$ $(i = 1,2,\ldots q = d(r))$.

$$\frac{1}{r}\sum_{d|r} C(n,d) = \sum_{i=1}^{t} E_i = \begin{cases} 1, & \text{if } n \equiv 0(\mathrm{mod}\ r); \\ 0, & \text{otherwise.} \end{cases}$$

The basis elements E_i $(i = 1,2,\ldots q)$ of idempotents is such that $E_1+E_2+\cdots+E_t = I(n,r) =$ the unity element in $B_r(\mathbb{C})$. (See (14.2.18) and proof of theorem 127).

Definition 14.3.1 : Let $f \in B_r(\mathbb{C})$. If f is a unit in the ring $(B_r(\mathbb{C}),+,\cdot)$ the Cauchy inverse g of f (if it exists) is defined by the relation:

$$(f \cdot g)(n,r) = (g \cdot f)(n,r) = I(n,r) = \begin{cases} 1, & \text{if } n \equiv 0 \ (\mathrm{mod}\ r); \\ 0, & \text{otherwise .} \end{cases}$$

Theorem 128 : Let $f \in B_r(\mathbb{C})$ have Fourier coefficients $\alpha(d,r) \neq 0$: $d|r$. f possesses a Cauchy inverse if, and only if, $\alpha(d,r) \neq 0$ for all d dividing r. Further, the Fourier coefficients of a Cauchy inverse are $r^{-2}\alpha(d,r)^{-1} : d|r$.

Proof : If g denotes a Cauchy inverse of f, with Fourier coefficients $\beta(d,r) : d|r$,

$$(f \cdot g)(n,r) = r\sum_{d|r} \alpha(d,r)\beta(d,r)C(n,d)$$

As $\frac{1}{r}\sum_{d|r} C(n,d) = I(n,r)$, $(f \cdot g) = I$ if, and only if,

$$r\alpha(d,r)\beta(d,r) = \frac{1}{r} \text{ for each } d \text{ dividing } r.$$

Thus, g is a Cauchy inverse if, and only if, $\beta(d,r) = \frac{1}{r^2}\alpha(d,r)^{-1}$ where $d|r$. $\quad\square$

Remark 14.3.1 : Theorem 128 has been drawn from Haukkanen and Sivaramakrishnan [7].

It is clear from theorem 128 that if any one of the Fourier coefficients of $f \in B_r(\mathbb{C})$ is zero, then f is not a unit in $B_r(\mathbb{C})$. Any such f is a divisor of zero. We point out that the non-units in $\mathbb{B}_r(\mathbb{C})$ are divisors of zero. So, the complement of the group of units in $B_r(\mathbb{C})$ is the set of zero divisors of the ring. Something more can be said about the set of zero divisors of a ring R.

Definition 14.3.2 : Let R be a ring. A non-empty subset S of R is called a multiplicative set if, whenever $a,b \in S$, $ab \in S$.

As an example, we note that the set of all elements in a nonzero ring R with identity which are not divisors of zero is multiplicative. In particular, the set of all nonzero elements of an integral domain is multiplicative. The group of units in a ring with unity is a multiplicative set.

Let R be a commutative ring with unity. If P is a prime ideal of R, P is a multiplicative set and so is the set $S = R \setminus P$.

Theorem 129 : *Let I be an ideal of a commutative ring R with unity 1_R. Suppose that $S \subseteq R$ is a multiplicative set in R, and S and I are mutually disjoint. Then, there exists an ideal P (of R) such that P is maximal in the set of ideals containing I and disjoint from S and P is prime.*

Proof : We consider the family $\mathcal{F}$ of ideals J of R such that $I \subseteq J$ and $J \cap S = \emptyset$. $\mathcal{F}$ is non-empty as I belongs to $\mathcal{F}$. Any chain $\{J_i\}$ of ideals in $\mathcal{F}$ has an upper bound in $\mathcal{F}$. For, $\cup_i J_i$ is an ideal of R contained in $\mathcal{F}$. Also, $\cup_i J_i$ is disjoint from S, since

$$\cup_i J_i \cap S = \cup (J_i \cap S) = U \emptyset = \emptyset$$

Zorn's lemma [20] could be applied to the family and so $\mathcal{F}$ has a maximal element say P. P is maximal in the set of ideals containing I and P does not intersect S.

Claim : P is a prime ideal.

Suppose that for $a, b \in R$, $ab \in P$, but $a \notin P$, $b \notin P$. The ideal (P, a) generated by P and a is strictly larger than P and so contains an element $s \in S$. Similarly, by considering the ideal (P, b), we can find an element $t \in (P, b)$. This shows that

$$st \in (P, a)(P, b) \subseteq (P, ab) \subseteq P, \text{ as } ab \in P.$$

As S is a multiplicative set, $st \in S$. This contradicts the fact that $P \cap S = \emptyset$. Our assumption that $a \notin P$ and $b \notin P$ is wrong. So, either $a \in P$ or $b \in P$. It follows that P is a prime ideal of R. □

Corollary 14.3.1 : *The following statements on the subset S of R are equivalent: [I. Kaplansky [20]]*

(a) S is multiplicative.

(b) The complement of S in R is a set-theoretic union of prime ideals of R.

Proof : $:\Rightarrow$ Given (a), there is a prime ideal P which is maximal with respect to disjointness from S. Let t be an element from $R \setminus S$. We consider the principal ideal (t) generated by t. (t) is disjoint from S. (t) is contained in a prime ideal P which is maximal with respect to disjointness from S. When $t \notin S$, t has been inserted into a prime ideal disjoint from S. So, $R \setminus S$ is a set-theoretical union of prime ideals of R. This is (b).

$\Leftarrow$: Given (b), $t \in R \setminus S$ is contained in a prime ideal P of R. If $a \in S$, $b \in S$, suppose $ab \in R \setminus S$. Then, $ab \in P$, a prime ideal of R. So, either $a \in P$ or $b \in P$. That is, when $ab \notin S$, it can happen that either $a \notin S$ or $b \notin S$. The contrapositive statement yields that $a \in S$ and $b \in S \Rightarrow ab \in S$ or S is a multiplicative set. □

Corollary 14.3.2 : *The set of zero divisors of R is a union of prime ideals.*

Remark 14.3.2 : Corollary 14.3.2 has been adapted from [20].

For, let T be the set of zero divisors of R. Then, $S = R \setminus T$ is a multiplicative set. By Corollary 14.3.1, T is a set-theoretic union of prime ideals of R.

Fact 14.3.1 : A commutative algebra $\mathcal{A}$ over a field F is semisimple if, and only if, it contains no nonzero nilpotent elements. (That is, $a \neq 0$ and $a^m = 0$ for some $m \geq 2$, does not happen).

For proof, see Yu. A. Drozd and V. Y. Kirichenko [13, Chapter 2].
The commutative algebra $B_r(\mathbb{C})$ has no nonzero nilpotent elements and so is semisimple. As it is finite dimensional, it is both Noetherian and Artinian (as a ring). The zero ideal is a product of a finite number of maximal ideals. If we consider the ideals generated by E_i and E_j ($i \neq j$) which are idempotent elements of the basis $\{E_i = \frac{1}{r}C(n, d_i) : d_i|r\}$, we have

$$[E_i][E_j] = (0) \text{ for } i \neq j.$$

So, $B_r(\mathbb{C})$ is Noetherian $\Rightarrow B_r(\mathbb{C})$ is Artinian and conversely. This gives a non-trivial illustration of theorem 94, chapter 12.

Definition 14.3.3 : *Let $f \in B_r(\mathbb{C})$ with Fourier coefficients $\alpha(d, r) : d|r$. The norm of f, written $N(f)$, is defined by*

$$N(f) = r^{d(r)} \prod_{d|r} \alpha(d, r).$$

$N(f) \neq 0$ if, and only if, f is a unit in $B_r(\mathbb{C})$.
It is easy to check that for $f, g \in B_r(\mathbb{C})$.

(14.3.8) $N(f \cdot g) = N(f)N(g).$

Thus, it is possible to make $B_r(\mathbb{C})$ a multiplicatively normed algebra.

It is interesting to note that $B_r(\mathbb{C})$ can be made an inner product space.

Definition 14.3.4 : *For $f, g \in B_r(\mathbb{C})$, the inner product $\langle f, g \rangle$ is defined by*

$$\langle f, g \rangle = \sum_{d|r} f(\tfrac{r}{d}, r)\overline{g(\tfrac{r}{d}, r)}\phi(d)$$

where $\overline{g}$ is the complex conjugate of g.
See Paul J. McCarthy [22]. See also Pentti Haukkanen [19].

That $\langle f, g \rangle$ is an inner product follows from the following observations:

(i) $\langle g, f \rangle = \sum_{d|r} g(\tfrac{r}{d}, r)\overline{f(\tfrac{r}{d}, r)}\phi(d) = \overline{\langle f, g \rangle}$;

(ii) $\langle \alpha f, g \rangle = \alpha \langle f, g \rangle, \quad \alpha \in \mathbb{C}$;

(iii) $\langle f, \alpha g \rangle = \overline{\alpha} \langle f, g \rangle \quad \alpha \in \mathbb{C}$;

(iv) $\langle f + h, g \rangle = \langle f, g \rangle + \langle h, g \rangle; \quad f, g, h \in B_r(\mathbb{C})$.

The norm of f via the inner product is given by

(14.3.9) $< f, f > = \sum_{d|r} f(\tfrac{r}{d}, r)\overline{f(\tfrac{r}{d}, r)}\phi(d) = \sum_{d|r} |f(\tfrac{r}{d}, r)|^2 \phi(d)$

A rigorous definition of inner-product is as given in 15.5.9 of chapter 15. For results relating to even functions (mod r) and applications, see Eckford Cohen [7], [8], [9], [10] and [11]. See also Haukkanen and Sivaramakrishnan [16] and [18].

We remark that in the ring $(B_r(\mathbb{C}), +, \cdot)$, every nonzero non-unit is a divisor of zero. However, as it has no nonzero nilpotent elements, a divisor of zero in $B_r(\mathbb{C})$ is not nilpotent. We examine whether a divisor of zero could be nilpotent in a suitable arithmetical ring. This is considered in Section 14.4 below.

14.4. Carlitz conjecture

In [3], L. Carlitz studies arithmetic functions in an unusual setting. Let F be any field. We consider functions $f : \widetilde{\mathbb{Z}} \to F$ where $\widetilde{\mathbb{Z}}$ is the set of non-negative integers. Let $\mathcal{A}'$ denote the set of functions having domain $\widetilde{\mathbb{Z}}$ and codomain F. $f \in \mathcal{A}'$ is also called an arithmetic function.

For $f, g \in \mathcal{A}'$, the Cauchy product of f and g given by

$$(14.4.1) \qquad (f \odot g)(r) = \sum_{s=0}^{r} f(s)g(r-s)$$

is associative and commutative. Cauchy multiplication distributes addition. It is easily verified that $(\mathcal{A}', +, \odot)$ is an integral domain.

Let p be an arbitrary but fixed prime.
We write

$$r = r_0 + r_1 p + r_2 p^2 + \cdots \quad (0 \le r_i < p)$$
$$s = s_0 + s_1 p + s_2 p^2 + \cdots \quad (0 \le s_i < p)$$

Then, a theorem of Lucas (see theorem 3, chapter 1) says that

$$(14.4.2) \qquad \binom{r}{s} \equiv \binom{r_0}{s_0}\binom{r_1}{s_1}\binom{r_2}{s_2} \cdots \quad (\bmod\ p).$$

$\binom{r}{s}$ is relatively prime to p if, and only if,

$$(14.4.3) \qquad 0 \le s_i \le r_i \quad (i = 0, 1, 2, \ldots)$$

Definition 14.4.1 : *For $f, g \in \mathcal{A}'$, the Lucas product $(f * g)$ of f and g is defined by*

$$(14.4.4) \qquad (f * g)(r) = \sum_{s=0}^{r} {}^{*} f(s)g(r-s)$$

where $\sum^{*}$ is restricted to those values of s for which (14.4.3) *holds* .

(14.4.4) is referred to as Lucas convolution [3].
It is verified that for $f, g, h \in \mathcal{A}'$,

$$(f * g) * h = f * (g * h),$$
$$(f * g) = (g * f),$$
$$f * (g + h) = f * g + f * h.$$

The function

(14.4.5) $z(r) = 0, \quad r \geq 0$

serves as the identity for addition. Also,

(14.4.6) $u(r) = \begin{cases} 1, & r = 0; \\ 0, & r > 0. \end{cases}$

serves as the identity for Lucas multiplication. Further $*$ is commutative. It follows that $(\mathcal{A}', +, *)$ forms a commutative ring with unity u (14.4.6).

Definition 14.4.2 : $f \in \mathcal{A}'$ *is said to be singular, if $f(0) = 0_F$. Otherwise, f is called non-singular.*

Non-singular functions are precisely those for which a Lucas inverse exists. $g \in \mathcal{A}'$ is called a Lucas inverse of f, if $f * g = g * f = u$.

Lemma 14.4.1 : $f \in \mathcal{A}'$ *possesses a Lucas inverse if, and only if, f is non-singular.*

Proof : $:\Rightarrow$ If f possesses a Lucas inverse g, $f * g = u$ yields $f(0)g(0) = 1$ and so, $f(0) \neq 0$.
$\Leftarrow :$ Conversely, suppose that $f(0) \neq 0$. Then, we can construct a function $g \in \mathcal{A}'$ such that $f * g = u$. We take $g(0) = \frac{1}{f(0)}$. We define $g(r)$ recursively from

$$\sum_{s=0}^{r}{}^{*} f(s)g(r-s) = 0, \quad r > 0.$$

That is,

(14.4.7) $g(r) = -\dfrac{1}{f(0)} \displaystyle\sum_{s=1}^{r}{}^{*} f(s)g(r-s) \quad (r \geq 1).$

$g(r)$ is uniquely determined in terms of $g(s)$ ($1 \leq s < r$), where $g(r)$ is to satisfy (14.4.7). Thus, g is determined as a Lucas inverse of f. $\qquad\square$

Remark 14.4.1 : Lemma 14.4.1 is the analogue of the corresponding result for Dirichlet inverse: f has a Dirichlet inverse if, and only if, $f(1) \neq 0$. See the narration after the definition 4.3.1, chapter 4. See also [22].

Definition 14.4.3 : $f \in \mathcal{A}'$ *is a divisor of zero, if $f \neq z$ and there exists $g \neq z$ such that $f * g = z$.*

The ring $(\mathcal{A}', +, *)$ has divisors of zero.
Let $j \in \tilde{\mathbb{Z}}$. Suppose that

(14.4.8)
$$\varphi_j(r) = \begin{cases} 1, & \text{if } r = p^j \\ 0, & \text{otherwise.} \end{cases}$$

Now,

$$(\varphi_r * \varphi_j)(r) = \sum_{s=0}^{r} {}^{*}\varphi_j(s)\varphi_j(r-s)$$

when $r = 2p^j$, $\varphi_j(p^j) = 1$ gives

$$(\varphi_j * \varphi_j)(r) = \begin{cases} 1, & \text{if } r = 2p^j \\ 0, & \text{otherwise.} \end{cases}$$

(14.4.9)
$$(\varphi_j * \varphi_j * \cdots)(r) = \varphi_j^p(r) = \sum_{s=0}^{r} {}^{*}\varphi_j(s)\varphi_j^{p-1}(r-s) = z$$

with p factors on the extreme left of (14.4.9).

(Since $\varphi_j(s) = 1$ if $s = p^j$ and $\varphi_j^{p-1}(r-s) = 1$, only when $r - s = (p-1)p^j$. But, then, $r = p^{j+1}$ and $s = p^j$. So, in $\sum^{*}$ above, $\varphi_j(s)$ and $\varphi_j^{p-1}(r-s)$ do not take the value 1 simultaneously. So, φ_j is a divisor of zero.)

For $f \in \mathcal{A}'$. f^k stands for $f * f * \cdots * f$ (k factors).

Definition 14.4.4 : *Let* $m = a_0 + a_1 p + \cdots + a_t p^t$. $0 \le a_i < p$, $(i = 0, 1, 2 \ldots t)$. *We define the monomial function* ψ_m *as*

(14.4.10) $\psi_m = \varphi_0^{a_0} * \varphi_1^{a_1} * \cdots * \varphi_t^{a_t}$ $(m = 1, 2, 3 \ldots)$

where φ_j *is as given in* (14.4.8).

From (14.4.9), we deduce that

(14.4.11) $\psi_m^p = z$ $(m = 1, 2, \ldots)$

It follows that ψ_m is a nilpotent element in $\mathcal{A}'$. In other words, every monomial function is nilpotent. Further,

if $\binom{m_1 + m_2}{m_1}$ is prime to p, we obtain

(14.4.12) $\psi_{m_1} * \psi_{m_2} = \psi_{m_1 + m_2}$.

If $\binom{m_1 + m_2}{m_1}$ and p are not relatively prime to one another, we get

(14.4.13) $\psi_{m_1} * \psi_{m_2} = z$.

(14.4.12) and (14.4.13) are established as follows:
   Suppose that

$$m_1 = x_0 + x_1 p + \cdots + x_s p^s, \quad 0 \le x_i < p \quad (i = 0, 1, \ldots s)$$
$$m_2 = y_0 + y_1 p + \cdots + y_t p^t, \quad 0 \le y_j < p \quad (j = 0, 1, \ldots t).$$

Then $m_1 + m_2 = b_0 + b_1 p \cdots + b_n p^n$, $0 \le b_i < p$ $(i = 0, 1 \ldots n)$,

$$\binom{m_1 + m_2}{m_1} \equiv \binom{b_0}{x_0}\binom{b_1}{x_1} \cdots \pmod{p}.$$

Next,

$$\psi_{m_1} * \psi_{m_2} = \varphi_0^{x_0} * \varphi_1^{x_1} * \cdots * \varphi_s^{x_s} * \varphi_0^{y_0} * \varphi_1^{y_1} \cdots * \varphi_n^{y_n},$$

or,

$$\psi_{m_1} * \psi_{m_2} = \varphi_0^{x_0 + y_0} * \varphi_1^{x_1 + y_1} * \cdots * \varphi_t^{y_t}, \text{ if } t > s.$$

When $\binom{m_1 + m_2}{m1}$ is relatively prime to p, $\binom{b_j}{x_j}$ and $\binom{b_j}{y_j}$ are relatively prime to p for admissible values of j. So, (14.4.12) holds. When $\binom{m_1+m_2}{m_1}$ and p have g.c.d > 1, p divides $\binom{m_1+m_2}{m_1}$ and contains ψ_j^p as a factor for a suitable value of j and so, $\psi_{m_1} * \psi_{m_2} = z$ which is (14.4.13). (14.4.11) is a special case of (14.4.13), as g.c.d $(mp, p) = p$.

Theorem 130 (L. Carlitz (1966)) : *Let* $\psi_{m_1}, \psi_{m_2} \ldots \psi_{m_k}$ *be monomial functions as in* (14.4.10). *Suppose that* $f_1, f_2, \ldots f_k$ *are arbitrary arithmetic functions such that* $f \in \mathcal{A}'$ *has the form*

(14.4.14) $\qquad f = f_1 * \psi_{m_1} + f_2 * \psi_{m_2} + \cdots + f_k * \psi_{m_k}.$

Then, f *is a nilpotent element in the ring* $(\mathcal{A}', +, *)$. $(m_1, m_2, \ldots, m_k$ *are any set of* k *positive integers)*

Proof : As in (14.4.12) and (14.4.13).

$$\psi_{m_1} * \psi_{m_2} \cdots * \psi_{m_k} = \psi_{m_1 + m_2 + \cdots + m_k} \text{ or } z$$

according as the multinomial coefficient

$$\frac{(m_1 + m_2 + \cdots + m_k)!}{m_1! m_2! \cdots m_k!}$$

is or is not relatively prime to p.

When f is as given in (14.4.14), raising both sides of (14.4.14) to a power λ, we get

$$f^\lambda = \sum_{b_1 + b_2 \ldots + b_k = \lambda} \frac{\lambda!}{b_1! b_2! \cdots b_k!} f_1^{b_1} * \psi_{m_1}^{b_1} * f_2^{b_2} * \psi_{m_2}^{b_2} \cdots * f_k^{b_k} * \psi_{m_k}^{b_k}.$$

For $\lambda > k(p-1)$, at least one $b_j \ge p$ and so, by (14.4.11), $f^\lambda = z$. That is, f is Nilpotent. $\square$

Remark 14.4.2 : The Lucas product of two arithmetic functions could be described in the following manner also:

$f \in \mathcal{A}'$ can be expressed formally in terms of monomial functions ψ_m as

(a) $f = f(0)u + \sum_{m=1}^{\infty} f(m)\psi_m.$

(b) If $g = g(0)u + \sum_{m=1}^{\infty} g(m)\psi_m$, the Lucas product h of f and g is given by

$$h = f * g = f(0)g(0)u + \sum_{m=1}^{\infty} h(m)\psi_m,$$

where $h(m)$ is defined by

(c) $h(m) = f(0)g(m) + f(m)g(0) + \sum_{r=1}^{m-1}{}^{*} f(k)g(m-k).$

In h, one has $\psi_t * \psi_s$ which is ψ_{t+s} or z according as $\binom{t+s}{t}$ is or is not relatively prime to p.

Theorem 131 (Carlitz (1966)) : *Let F be a field of characteristic q, a prime. Then, $f : \tilde{\mathbb{Z}} \to F$ is a zero divisor in $\mathcal{A}'$ if, and only if, f is singular. Further, every zero divisor in $\mathcal{A}'$ is nilpotent.*

Proof : If f is a zero divisor, it is clear that $f(0) = 0$. For, by lemma 14.4.1, if f is non-singular, f is invertible. So, suppose that $f(0) = 0$. Then, f is expressed as

(14.4.15) $f = \sum_{m=1}^{\infty} f(m)\psi_m$

If char $F = q$

$$f * f * \cdots f(q \text{ times }) = \sum_{m=1}^{\infty} (f(m))^q \psi_m^q$$

where $\psi_m^q = \psi_m * \psi_m \cdots \psi_m$ (q factors).

So,

(14.4.16) $f^{q^t} = \sum_{m=1}^{\infty} (f(m))^{q^t} \psi_m^{q^t}$ $(t = 1, 2, 3 \ldots)$

If we choose t such that $q^t \geq p$, as $\psi_m^p = z$, we get $f^{q^t} = z$. If follows that a zero-divisor f is nilpotent, whenever char $F = q$ (a prime).   □

It is shown in [3] that theorem 131 does not stand when F is of characteristic zero. In fact, when char $F = 0$, one can show that a singular function need not be a zero-divisor.

It is likely that in char $F = 0$ case, a function f is a zero-divisor if, and only if, it is of the form given in (14.4.14). If this is true, it will follow that $f \in \mathcal{A}'$ is a zero-divisor if, and only if, f is nilpotent.

14.4.1. CARLITZ CONJECTURE. $(\mathcal{A}', +, *)$ **denotes a Lucas ring of arithmetic functions** $f : \tilde{\mathbb{Z}} \to F$ **where** F **is a field of characteristic zero.** $f \in \mathcal{A}'$ **is a zero divisor of the ring if, and only if,** f **is nilpotent.**

As far as the knowledge of the author goes, this conjecture remains unresolved. See L. Carlitz [3].

14.5. More about zero divisors

Given a commutative ring R with unity 1_R, we examine the set $Z(R)$ of zero divisors of R in the following manner:

Definition 14.5.1 (D. M. Burton) : *Let S be a multiplicative set in R. S is called a saturated multiplicative set if whenever $ab \in S$, both $a \in S$ and $b \in S$.*

See definition 14.3.2 and recall Corollary 14.3.1.

Lemma 14.5.1 : *S is a saturated multiplicative set if, and only if, $R \setminus S$ is a union of prime ideals of R.*

Proof : $:\Rightarrow$ We are given a saturated multiplicative set S in R. Suppose that $a \in R \setminus S$. We consider the ideal J generated by a. That is, $J = (a)$. Then, for $r \in R$, $ra \in J$. If $ra \in S$, it will mean that both r and a are in S. But $a \notin S$. So, $J \cap S = \emptyset$, as $ra \notin S$.

Let $\mathcal{F}$ be the family of ideals I (of R) such that $J \subseteq I$ and $I \cap S = \emptyset$. As J belongs to $\mathcal{F}$, $\mathcal{F}$ is non-empty. For any chain of ideals $\{I_m\}$ in $\mathcal{F}$, $\cup_m I_m$ belongs to $\mathcal{F}$. ($\cup_m I_m$ is an ideal of R). As $J \subseteq I_m$, $J \subseteq \cup_m I_m$. Further,

$$(14.5.1) \qquad (\cup_m I_m) \cap S = \cup(I_m \cap S) = \cup \emptyset = \emptyset.$$

By Zorn's lemma, J has a maximal element P. We claim that P is a prime ideal of R. Suppose that for $a, b \in R$, $ab \in S$. As S is a saturated multiplicative set, both a and b belong to S. Therefore, by multiplicative property of S, $a \in S$, $b \in S \Rightarrow ab \in S$. So, $a \notin P$ and $b \notin P \Rightarrow ab \notin P$. The contrapositive statement gives: $ab \in P \Rightarrow a \in P$ or $b \in P$. Thus, P is a prime ideal of R.

$\Leftarrow$: Conversely, suppose that given a set S contained in R, $R \setminus S$ is a union of prime ideals of R. Given $a \in R \setminus S$, the ideal $J = (a) \subseteq P$, a prime ideal of R. For $a, b \in R$ such that $ab \in P$ implies either $a \in P$ or $b \notin P$. That is, $a \notin P$ and $b \notin P$ imply $ab \notin P$.

Thus, $a \in S$ and $b \in S$ imply $ab \in S$. So, S is a multiplicative set. For $a, b \in S$, if $ab \in S$, $ab \notin P$, a prime ideal of R. This gives the fact that $a \notin P$ and $b \notin P$. For, if $a \in P$ or $b \in P$, we would get $ab \in P$ (as P is an ideal). That is, $ab \in S$ implies $a \in S$ and $b \in S$. That is, S is a saturated multiplicative set. This completes the proof of lemma 14.5.1. $\qquad \square$

Remark 14.5.1 : The set of nonzero-divisors of R is a saturated multiplicative set. For, let S denote the set of nonzero-divisors of R. Then, S consists of units and non-units which are not zero divisors. For $a, b \in R$ and $ab \in S$, $a \in S$ and $b \in S$.

Corollary 14.5.1 : *The set of zero divisors of R along with 0_R is a union of prime ideals of R.*

Proof : By remark 14.5.1, the set S of nonzero-divisors of R is a saturated multiplicative set. So, by lemma 14.5.1, $R \setminus S$ is a union of prime ideals of R. This is the content of Corollary 14.5.1. $\qquad \square$

Next, a nonzero element a in R is nilpotent if there exists a positive integer n such that $a^n = 0_R$. So, a nonzero nilpotent element is a zero divisor, but not conversely, in general. If Carlitz conjecture is true, (see 14.4.1) a Lucas ring of arithmetic functions provides an example of a ring in which every zero divisor is nilpotent.

The set $Z(R)$ of zero divisors consists of
(i) zero divisors which are non-nilpotent and
(ii) zero divisors which are nilpotent.

Definition 14.5.2 : *Let R be a commutative ring with unity 1_R. The prime radical of R denoted by $P(R)$ is the set*

(14.5.2) $P(R) = \cap\{P : P \text{ a prime ideal of } R\}$.

It is verified that $P(R)$ is the ideal containing the nilpotent elements of R. So, if $Z'(R)$ denotes the set of zero divisors of R which are not nilpotent, by Corollary 14.5.1, we obtain

(14.5.3) $$Z'(R) = \bigcup_{\lambda \in \Lambda} P_\lambda \setminus \bigcap_{\lambda \in \Lambda} P_\lambda$$

where $\{P_\lambda\}(\lambda \in \Lambda)$ is the set of prime ideals of R.

It follows that if, in R, every zero divisor is nilpotent, we will arrive at

(14.5.4) $$\bigcup_{\lambda \in \Lambda} P_\lambda = \bigcap_{\lambda \in \Lambda} P_\lambda.$$

Thus, if every zero divisor is nilpotent, there is a unique prime ideal, say N, which contains the nilpotent elements of R.

In order to characterize R for which every zero divisor is nilpotent, we may need more conditions on R such as Noetherian and the like. It will require deeper investigation! In Section 12.10, chapter 12, we have considered primary rings A which contain at least one proper prime ideal. If A is a primary ring, A has a minimal prime ideal which contains all zero divisors. (See Fact 12.10.1). In other words, the set of zero divisors of A along with 0_A coincides with the ideal N of nilpotent elements. Moreover, N is a prime ideal of A. (N is the prime radical of A). It means that A is a primary ring whenever the zero ideal is a primary ideal of A. (See definition 12.3.6). One concludes that a ring A is primary if, and only if, every zero divisor of A is nilpotent. Thus, Carlitz conjecture (14.4.1) boils down to saying that $(\mathcal{A}', +, *)$ is a primary ring. See (b) in Section 14.8.

14.6. Certain norm-preserving transformations

$\mathcal{A}$ denotes the set of arithmetic functions $f : \mathbb{N} \to \mathbb{C}$. $\mathcal{A}$ forms an integral domain under the operations of addition and Dirichlet multiplication. To prove this, we made use of the norm $N(f)$ of $f \in \mathcal{A}$. We recall that $N(f)$ is the least positive integer m such that $f(m) \neq 0$. The norm of a unit in $\mathcal{A}$ is 1, as $f \in A$ possesses a Dirichlet inverse if, and only if, $f(1) \neq 0$. It was shown in corollary 4.5.1, chapter 4 that $\mathcal{A}$ is indeed a UFD. As $\mathcal{A}$ is a vector space over $\mathbb{C}$, $\mathcal{A}$ is an algebra over $\mathbb{C}$ with Dirichlet multiplication for multiplication of elements in $\mathcal{A}$. $\mathcal{A}$ is known as the Dirichlet algebra over $\mathbb{C}$. See Remark 4.5.2 in chapter 4.

The notion of an algebra homomorphism has already been given in definition 14.2.5.

An example of an algebra isomorphism is the isomorphism between the algebra of linear operators $\mathcal{L}(V)$ of an n-dimensional vector space V over a field F and the matrix algebra $M_n(F)$ obtained by assigning to each operator $T : V \to V$, its $n \times n$ matrix with respect to a given basis $\mathcal{B}$ of V.

Next, we consider $\mathcal{A}'$, the Lucas algebra of arithmetic functions defined on $\widetilde{\mathbb{Z}}$. $f \in \mathcal{A}'$ is such that f is a function from $\widetilde{\mathbb{Z}}$ into $\mathbb{C}$. For $f, g \in \mathcal{A}'$,

(14.6.1) $$(f+g)(r) = f(r) + g(r), \quad r \geq 0;$$

(14.6.2) $$(f * g)(r) = \sum_{}^{r} {}^{*} f(s)g(r-s),$$

where the summation is restricted to such s such that $p \nmid \binom{r}{s}$ (p denotes a fixed prime). See definition 14.4.1.

We note that (14.6.2) holds if, and only if,

$$r = r_0 + r_1 p + r_2 p^2 + \cdots \quad 0 \leq r_i < p$$

$$s = s_0 + s_1 p + s_2 p^2 + \cdots \quad 0 \leq s_i < p$$

and

$$0 \leq s_i \leq r_i \quad (i = 0, 1, 2, \ldots)$$

Let $p_0, p_1, p_2 \ldots$ be the sequence of primes in ascending order. If a is not divisible by the pth power of any prime, writing

(14.6.3) $$a = p_0^{a_0} p_1^{a_1} p_2^{a_2} \cdots p_k^{a_k} \cdots \quad (0 \leq a_j < p),$$

where only a finite number of a_j are different from zero; with each a, we may associate the sequence:

(14.6.4) $$\{a_0, a_1, a_2, \ldots \}.$$

Suppose that b has the representation

(14.6.5) $$b = p_0^{b_0} p_1^{b_1} p_2^{b_2} \cdots \quad (0 \leq b_j < p),$$

where only a finite number of b_j are different from zero.

Let $\mathcal{B}_p$ denote the set of arithmetic functions f_p defined on $\mathbb{N}$ such that

(14.6.6) $$f_p(r) = \begin{cases} k_f, & \text{if } r \text{ is not divisible by a } p\text{th-power of a prime,} \\ 0, & \text{if } r \text{ is divisible by a } p\text{th-power of a prime.} \end{cases}$$

(In (14.6.6), k_f is a constant depending on f_p.)

For $f_p, g_p \in B_p$, we define

(14.6.7) $$(f_p \cdot g_p)(r) = \sum_{ab=r}' f_p(a)g_p(b)$$

the summation $\sum'$ is restricted to a and b such that $ab = r$ is not divisible by a pth-power of a prime.

The restriction in $\sum'$ is equivalent to saying that when a, b are as given in (14.6.3) and (14.6.5),

(14.6.8) $a_j + b_j < p \quad (j = 0, 1, 2 \ldots)$

We define $\psi : \mathcal{B}_p \to \mathcal{A}'$ by

(14.6.9) $\psi(f_p) = \overline{f}$

where $f_p(r) = f_p(p_0^{r_0} p_1^{r_1} p_2^{r_2} \cdots)$ $(0 \le r_j < p)$.
(Only a finite number of p_j are nonzero) and $\overline{f}$ is given by

$$\overline{f}(\overline{r}) = \overline{f}(r_0 + r_1 p + r_2 p^2 + \cdots), (\overline{r} = r_0 + r_1 p + r_2 p^2 + \cdots), \quad (0 \le r_j < p).$$

Then, $\overline{f} * \overline{g}$ in $\mathcal{A}'$ corresponds to $f_p \cdot g_p$ in $\mathbb{B}_p$.

The multiplicative identity in $\mathbb{B}_p$ is

$$e_p(r) = \begin{cases} 1, & \text{if } r = 1; \\ 0, & \text{otherwise.} \end{cases}$$

The multiplicative identity in $\mathcal{A}'$ is

$$\overline{u}(\overline{r}) = \begin{cases} 1, & \text{if } \overline{r} = 0; \\ 0, & \text{otherwise.} \end{cases}$$

We check that $\psi(e_p) = \overline{e}$. ψ preserves addition and multiplication. We conclude that there is an algebra isomorphism between $\mathbb{B}_p$ and $\mathcal{A}'$. See L. Carlitz [4].

The notion of transformations of arithmetic functions was first introduced by P. Kesava Menon in [21]. Let

(14.6.10) $r = p_1^{a_1} p_2^{a_2} \cdots p_t^{a_t}. \quad p_i$ primes, $a_i \ge 1, i = 1, 2 \ldots t.$

(14.6.11) $s = p_1^{b_1} p_2^{b_2} \cdots p_t^{b_t} \quad p_i$ primes, $b_i \ge 1, i = 1, 2 \ldots t.$

We define

(14.6.12) $B(s, r) = \begin{cases} 1, & \text{if } s = r; \\ \prod_{i=1}^{t} \binom{a_i - 1}{b_i - 1}, & \text{if } s \text{ divides } r \text{ and } s \text{ contains all} \\ & \text{the prime factors of } r; \\ 0, & \text{otherwise.} \end{cases}$

Definition 14.6.1 : *A transformation* $B : \mathcal{A} \to \mathcal{A}$ *is defined by*

(14.6.13) $B(f)(r) = \sum_{d|r} B(d, r) f(d) : f \in \mathcal{A}.$

It is verified that for $f, g \in \mathcal{A}$

$$B(f + g) = B(f) + B(g)$$

and for $\alpha \in \mathbb{C}$, $B(\alpha f) = \alpha B(f)$. That is, B is a linear operator on the vector space $\mathcal{A}$ over $\mathbb{C}$.

Theorem 132 (Haukkanen and Sivaramakrishnan (1991)) : $B : \mathcal{A} \to \mathcal{A}$ *as defined in* (14.6.13) *is an algebra isomorphism.*

Proof : We have noticed that $B : \mathcal{A} \to \mathcal{A}$ is a linear operator on $\mathcal{A}$.

Claim : B preserves Dirichlet multiplication.
Let $f, g \in \mathcal{A}$.

(14.6.14)
$$[B(f) \cdot B(g)](r) = \sum_{cu\delta=r} f(c)g(u) \sum_{du=\delta} B(c, cd)B(u, uv)$$

Using the combinatorial identity (H. W. Gould [15])
$$\sum_{k=0}^{r} \binom{x+k}{k} \binom{y+r-k}{r-k} = \binom{x+y+r+1}{r},$$
we arrive at
$$\sum_{dv=\delta} B(c, cd)B(u, uv) = B(cu, cus).$$
Or, (14.6.14) is recast as
$$[B(f) \cdot B(g)](r) = \sum_{cus=r} f(c)g(u)B(cu, cus)$$
$$= \sum_{ts=r} B(t, r) \sum_{cu=t} f(c)g(u)$$
$$= \sum_{ts=r} B(t, r)(f \cdot g)(t).$$

Or,
$$B(f \cdot g)(r) = [B(f) \cdot B(g)](r).$$
Next, suppose that for $g \in \mathcal{A}$, we define
$$f(r) = g(r) - \sum_{\substack{d|r \\ d \neq r}} B(d, r)f(d)$$
inductively, then f is such that $B(f) = g$.
This shows that $B : \mathcal{A} \to \mathcal{A}$ is surjective. Therefore, B is a ring homomorphism which is onto.
$$\ker B = \{f \in \mathcal{A} : B(f) = 0\}$$
$$B(f)(r) = 0 \Rightarrow \sum_{d|r} B(d, r)f(d) = 0.$$

For $r = 1$, $f(1) = 0$. Assume that $f(r) = 0$ for $1 \leq r < n$. Then,
$$\sum_{d|n} B(d, n)f(d) = 0 \text{ gives } B(n, n)f(n) = 0. \text{ As } B(n, n) = 1, f(n) = 0.$$

By induction on r, $f(r) = 0$, $r \geq 1$. So, $\ker B = 0$. That is, B is one-one. For $\alpha \in \mathbb{C}$, $B(\alpha f) = \alpha B(f)$. $e_0 = [\frac{1}{r}]$ is the multiplicative identity in $\mathcal{A}$. As B preserves Dirichlet multiplication, $B(f \cdot e_0) = B(f) = B(f) \cdot B(e_0)$ or $B(e_0) = e_0$. Thus, $B : \mathcal{A} \to \mathcal{A}$ is an algebra isomorphism.   □

Theorem 133 (Haukkanen and Sivaramakrishnan (1991)) :

Let $\lambda(r) = (-1)^{\Omega(r)}$ where $\Omega(r)$ denotes the total number of prime factors of r, each being counted according to its multiplicity. Then, the inverse of $B : \mathcal{A} \to \mathcal{A}$ is given by

(14.6.15) $$B^{-1}(f)(r) = \sum_{d|r} \lambda(\frac{r}{d}) B(d, r) f(d).$$

Proof : Suppose that $g \in \mathcal{A}$ is such that $B(g) = f$. Then, $g = B^{-1}(f)$. Then,

(14.6.16) $$\sum_{d|r} \lambda(\frac{r}{d}) B(d, r) f(d) = \sum_{ca=r} g(c) \sum_{e\delta=a} \lambda(\delta) B(ce, r) B(c, ce)$$

By the combinatorial identity (H. W. Gould [15]),

$$\sum_{k=j}^{n} (-1)^k \binom{n}{k} \binom{k}{j} = \begin{cases} (-1)^n, & j = n; \\ 0, & j \neq n. \end{cases}$$

So, we obtain

(14.6.17) $$\sum_{e\delta=a} \lambda(\delta) B(Ce, r) B(c, ce) = \begin{cases} 1, & r = c; \\ 0, & \text{otherwise.} \end{cases}$$

From (14.6.16) and (14.6.17) we obtain $B^{-1}(f)(r) = g(r)$.   □

Remark 14.6.1 : Theorems 132 and 133 have been drawn from [18].

We remark that $B : \mathcal{A} \to \mathcal{A}$ is a norm-preserving linear operator on $\mathcal{A}$. Let $N(f) = m$. For $r < m$,

$$B(f)(r) = \sum_{d|r} B(d, r) f(d) = 0.$$

When $r = m$,

$$B(f)(m) = \sum_{d|m} B(d, m) f(d) = f(m),$$

as $B(m, m) = 1$ and $f(d) = 0$ whenever $d < m$. Therefore, $N(B(f)) = m = N(f)$.

Next, we consider $T : \mathcal{A} \to \mathcal{A}$ given by

(14.6.18) $$T(f)(r) = \sum_{d|r} f((d, \frac{r}{d}))$$

where $(d, \frac{r}{d})$ denotes the g.c.d of d and $\frac{r}{d}$ for each positive divisor d of r. T is a linear operator on $\mathcal{A}$. When $f = e_0$, the unity element in $(\mathcal{A}, +, \cdot)$;

$$T(e_0)(r) = 2^{\omega(r)}$$

where $\omega(r)$ denotes the number of distinct prime divisors of r. The number $2^{\omega(r)}$ gives the number of those divisors d of r for which $g.c.d\,(d,\frac{r}{d}) = 1$. Such divisors are called unitary divisors of r. See Eckford Cohen [12]. As $T(e_0) \neq e_0$, T is not an algebra homomorphism. However, there is an arithmetical representation of $T(f)(r)$ (14.6.18).

Theorem 134 (Rajendran Valiaveetil (1996)) : *For* $f \in A$,

$$(14.6.19) \qquad T(f)(r) = \sum_{k^2|r} f(k) 2^{\omega(r/k^2)}$$

Proof : If t denotes the number of distinct prime divisors of r, where r is expressed as a product $d_1 d_2$ of co-prime factors, as $d_1 d_2 = d_2 d_1$, the number of ways of expressing r as a product of two co-prime factors is $\frac{1}{2} 2^t = 2^{t-1}$ factors. We take $d_1 \cdot d_2 = d_2 \cdot d_1$ whenever $g.c.d\,(d_1, d_2) = 1$. So, the required number is $\frac{1}{2} 2^t = 2^{t-1}$. Suppose that $g.c.d\,(d, \frac{r}{d}) = k$. Then, $d = kd_1$, $\frac{r}{d} = kd_2$ where $g.c.d\,(d_1, d_2) = 1$. Further $r = k^2 d_1 d_2$ and so, k^2 divides r. Thus, if $g.c.d\,(d, \frac{r}{d}) = k$, $k^2|r$. Conversely, if $k^2|r$ and $r = k^2 s$, s can be factored into two co-prime factors s_1 and s_2 in $2^{\omega(s)-1}$ ways. For each of these $2^{\omega(s)-1}$ ways, one has

$$r = k^2 s_1 s_2 = (ks_1)(ks_2) = d(\frac{r}{d}) \text{ with } d = ks_1, \ \frac{r}{d} = ks_2.$$

So, for each k such that $k^2|r$, there exist $2^{\omega(s)-1}$ pairs of divisors $< d, \frac{r}{d} >$ for which $g.c.d\,(d, \frac{r}{d}) = k$. Therefore, the total number of such divisors is $2 \cdot 2^{\omega(s)-1} = 2^{\omega(s)}$ where, $s = \dfrac{r}{k^2}$.

Next, we consider the set $\{d_1 = 1, d_2, \ldots d_n = r\}$ of divisors of r written in ascending order of magnitude. This set is partitioned into mutually disjoint classes:

$$C_1, C_2, \ldots C_m$$

such that the class C_k contains those divisors d of r for which $(d, \frac{r}{d}) = k$, if $k^2|r$. The number of elements in the class C_k is $2^{\omega(r/k^2)}$. We note that C_k is empty if $k^2 \nmid r$. Let $d(r)$ denote the number of divisors of r. Then,

$$(14.6.20) \qquad d(r) = \sum_{k^2|r} 2^{\omega(r/k^2)}$$

Further, $f((d, \frac{r}{d}))$ will appear as $f(k)$ for each d belonging to the class C_k and as pointed out, there are $2^{\omega(r/k^2)}$ elements in the class C_k. Hence,

$$(14.6.21) \qquad T(f)(r) = \sum_{d|r} f((d, \frac{r}{d})) = \sum_{k^2|r} f(k) 2^{\omega(r/k^2)},$$

as claimed in (14.6.19). □

Corollary 14.6.1 (Daniel I. A. Cohen (1965)) : *If* $f(r) = r$, $r \geq 1$, *one gets from* (14.6.21)

$$\sum_{d|r} (d, \frac{r}{d}) = \sum_{k^2|r} k\, 2^{\omega(r/k^2)}.$$

Remark 14.6.2 : The above identity is adapted from [6].

Remark 14.6.3 : $T : \mathcal{A} \to \mathcal{A}$ is not a norm-preserving linear operator.

Analogous to the linear operator T defined in (14.6.18), we define a function $L : \mathcal{A} \to A$ by

$$(14.6.22) \qquad L(f) = \sum_{d|r} f([d, \frac{r}{d}]),$$

where $[d, \frac{r}{d}]$ denotes the l.c.m of d and $\frac{r}{d}$.

Theorem 135 (Rajendran Valiaveetil (1996)) **:** For $f \in A$,

$$(14.6.23) \qquad L(f)(r) = \sum_{k^2|r} f(\frac{r}{k}) 2^{\omega(r/k^2)}.$$

Proof : Follows from that of theorem 134. For, as

$$[d, \frac{r}{d}] = \frac{r}{(d, r/d)}, \text{ where } (d, \frac{r}{d}) = g.c.d \text{ of } d \text{ and } \frac{r}{d},$$

we obtain

$$L(f)(r) = \sum_{d|r} f(\frac{r}{(d, \frac{r}{d})}).$$

Using the identity in (14.6.19), we arrive at (14.6.23). □

Remark 14.6.4 : Theorems 134 and 135 have been drawn from [23].

Corollary 14.6.2 :

$$(14.6.24) \qquad \sum_{d|r}[d, \frac{r}{d}] = \sum_{k^2|r} \frac{r}{k} 2^{\omega(r/k^2)}$$

For, (14.6.24) is a special case of (14.6.23) when $f(r) = r$, $(r \geq 1)$.

We note that $L : \mathcal{A} \to \mathcal{A}$ (14.6.22) is a linear operator and L is norm-preserving. To see this, we proceed as follows:

Let $r = s^2 t$ where t is the greatest square-free divisor of r. Then $k^2|r$ implies $k|s$. Further,

$$\omega(\frac{s^2 t}{k^2}) \leq \omega(\frac{s}{k}) + \omega(t).$$

If $N(f) = m$, then for $1 \leq a < m$, when $a = s^2 b$, b square-free,

$$L(f)(a) = \sum_{k|s} f(\frac{a}{k}) 2^{\omega(sb/k)}.$$

As $f(\frac{a}{k}) = 0$, for k dividing s; $L(f)(a) = 0$. When $a = m$ with $m = s^2 n$ where n is the greatest square-free divisor of m,

$$L(f)(m) = \sum_{k|s} f(\frac{m}{k}) 2^{\omega(\frac{sn}{k})} = f(m) 2^{\omega(m')} \neq 0,$$

where m' is the product of the distinct prime factors of m.
So, $N(L(f)) = N(f)$. That is, L is norm-preserving. However, L is not an algebra homomorphism, as L does not preserve Dirichlet multiplication.

When one confines to the class M of multiplicative functions [22] or [24], one comes across a transformation $R : M \to M$ given by

$$(14.6.25) \qquad R(f) = \sum_{d \mid r^2} f(\frac{r^2}{d}) f(d) \lambda(d)$$

where $\lambda(r) = (-1)^{\Omega(r)}$ and as given in theorem 133. $R(f)$ belongs to M and for $f, g \in M$, $R(f \cdot g) = R(f) \cdot R(g)$. See Haukkanen and Sivaramakrishnan [17].
Let W denote the set of completely multiplicative functions f where f satisfies

$$(14.6.26) \qquad f(r)f(s) = f(rs) \text{ for all pairs of integers } r, s \in \mathbb{N}$$

and for which $f(s) \neq 0$ for every prime p.
W is a subset of M. When $f \in W$ or M, $f(1) = 1$. $\lambda(r) = (-1)^{\Omega(r)}$ belongs to W and $\lambda(p) = -1$ for all primes p.
Fix $h \in W$. Then h is completely multiplicative and $h(s) \neq 0$ for every prime p.

Definition 14.6.2 : $\nu_h : \mathcal{A} \to \mathcal{A}$ is defined by

$$(14.6.27) \qquad \nu_h(f) = hf$$

That is,

$$\nu_h(f)(r) = h(r)f(r)$$

It is known [22, theorem 20] that given $a, b, c \in \mathcal{A}$,

$$c(a \cdot b) = ca \cdot cb$$

if, and only if, c is completely multiplicative. So, $h \in W$ has the property:

$$(14.6.28) \qquad h(f \cdot g) = (hf) \cdot (hg) \text{ for all } f, g \in \mathcal{A}.$$

Theorem 136 : ν_h as defined in (14.6.27) is a norm-preserving algebra isomorphism from $\mathcal{A}$ onto $\mathcal{A}$.

Proof : Let $f \in \mathcal{A}$. $\nu_h(f) = hf$.
As h is completely multiplicative, $N(hf) = N(f)$.
So, ν_h preserves the norm of f. By (14.6.28), ν_h preserves Dirichlet multiplication. For $f \neq g$, $\nu_h(f) \neq \nu_h(g)$. So ν_h is injective. Let $g \in \mathcal{A}$. There exists $f \in A$ such that

$$hf = g.$$

$h(r) \neq 0$ for $r \geq 1$. So, $f = \frac{1}{h}g$ is what is required. For $r \in \mathbb{N}$, we define f by $f(r) = \frac{1}{h(r)}g(r)$. So, ν_h is surjective. Thus, $\nu_h : \mathcal{A} \to \mathcal{A}$ is a norm-preserving algebra isomorphism, as $\nu_h(e_0) = he_0 = e_0$. $\qquad \square$

We remark that there is a lot to learn about rings with zero-divisors. James A. Huckaba makes an interesting study of commutative rings with zero-divisors in [A3]. In [A2], L. Carlitz and M. V. Subbarao have made a study of certain transformations of arithmetic functions f for which $f(1) = 1$. However, results of Section 14.6 are aimed at obtaining certain known arithmetical identities via linear operators.

14.7. Notes with illustrative examples

The study of trigonometric sums in Number theory was first initiated by Gauss. He used these sums for solving problems relating to congruences. Though exponential sums were known earlier, they were handled via the arithmetical representation of a periodic function (mod r) by Eckford Cohen in 1952. He introduced the class of (r, F)-arithmetic functions and studied its properties extensively. A significant theorem of Eckford Cohen says that the set $A_r(F)$ of (r, F)-arithmetic functions forms a commutative semisimple algebra which is a direct sum of r fields each isomorphic to F. See proposition 14.2.1. Analogous results could be given in the context of periodic functions defined over $GF[p^n, x]$ the ring of polynomials in x with coefficients from a finite field $\mathbb{F}_q$, $q = p^n$. Details are in the papers [7], [9], [10]. One could give an analogue of Ramanujan's sum $C(n, r)$ in the polynomial case. It is referred to as Carlitz η-sum. See [7] and [24]. They have applications to congruences.

The three papers [9], [10] and [11] of Eckford Cohen have sparked a lot of interest in research workers for further study in the area of periodic functions (mod r). A major achievement of Eckford Cohen was in the discovery of an orthogonal property for Ramanujan Sums (considered in lemma 14.2.6). It generalizes Carmichael's formula [5]:

$$(14.7.1) \qquad \text{If } r = ef, e \neq f, \quad \sum_{a=1}^{ef} C(a, e)C(a, f) = 0$$

$$(14.7.2) \qquad \text{If } e = f, \quad \sum_{a=1}^{e} c^2(a, e) = e\phi(e)$$

where ϕ is Euler ϕ-function.

The function

$$(14.7.3) \qquad B(n, r) = \sum_{\substack{h(\bmod r) \\ g.c.d(h,r)=\text{a square}}} \exp(\frac{2\pi i h n}{r})$$

where the summation is over a residue system $h(\bmod r)$ such that $g.c.d\ (h, r)$ is a square, is an analogue of $C(n, r)$. It is known [24] that

$$B(n, r) = \sum_{d \mid g.c.d(n,r)} \lambda(\frac{r}{d})d = \lambda(\frac{r}{g})b(g); g = g.c.d\ (n, r)$$

where $\lambda(r) = (-1)^{\Omega(r)}$, $b(r) = B(0, r)$ ($\Omega(r) = $ the total number of prime divisors of r). See [9], [22] also.

Analogous to the orthogonal property of $C(n,r)$, one has the following orthogonal property of $B(n,r)$: If t_1, t_2 are square-free divisors of r,

$$(14.7.4) \qquad \sum_{n \equiv a+b (\text{mod } r)} B(a, \frac{r}{t_1}) B(b, \frac{r}{t_2}) = \begin{cases} rB(n, \frac{r}{t}), & \text{if } t_1 = t_2 = t; \\ 0, & \text{otherwise.} \end{cases}$$

See Rajendran Valiaveetil [23].

Definition 14.7.1 (Cohen [9]) : *A function $f : \mathbb{Z} \times \mathbb{N} \to \mathbb{C}$ is said to be completely even* (mod r) *if there exists a function $F : \widetilde{\mathbb{Z}} \to \mathbb{C}$ such that $f(n,r) = F((n,r))$ where $(n,r) = $ g.c.d of n and r. It can be verified that $\lambda(r)B(n,r)$ is completely even* (mod r).

Proposition 14.7.1 (Rajendran Valiaveetil [21]) : *The set $V_r(\mathbb{C})$ of completely even functions* (mod r) *forms a subspace of $B_r(\mathbb{C})$ having dimension $2^{\omega(r)}$, the number of square-free divisors of r. $V_r(\mathbb{C})$ has an orthonormal basis:*

$$\{\lambda(\tfrac{r}{t})(rb(\tfrac{r}{t}))^{-\frac{1}{2}} B(n, \tfrac{r}{t}) : t \text{ a square-free divisor of } r\}.$$

Proof follows on lines similar to that of proposition 14.2.2. In fact, one has

Proposition 14.7.2 : *The space $B_r(\mathbb{C})$ of even functions* (mod r) *is of dimension $d(r)$, the number of divisors of r and $B_r(\mathbb{C})$ has an orthonormal basis*

$$\{\frac{1}{\sqrt{r\phi(d)}} C(n,d) : d | r\}.$$

See theorem 5 in Haukkanen and Sivaramakrishnan [16].

Analogous to even functions (mod r), Eckford Cohen [12] considers a class of functions called unitary functions mod (r). For $n \in \mathbb{Z}$, $(n,r)_*$ denotes the greatest divisor of n which is a unitary divisor of r, that is, a divisor d of r for which g.c.d $(d, \frac{r}{d}) = 1$. Let $f \in B_r(\mathbb{C})$. f is called an unitary function (mod r), if $f(n,r) = f((n,r)_*, r)$. The unitary analogue $C^*(n,r)$ of Ramanujan's sum could be defined via a semireduced residue system (mod r) [12]:

$$C^*(n,r) = \sum_{(h,r)_* = 1} \exp(\frac{2\pi i h n}{r})$$

where the summation is over $h(\text{mod } r)$ such that $(h,r)_* = 1$.

$C^*(r,r) = \phi^*(r)$, the unitary analogue of Euler's totient. $C^*(n,r)$ has the orthogonal property:

If d_1 and d_2 are unitary divisors of r,

$$(14.7.5) \qquad \sum_{n \equiv a+b (\text{mod } r)} C^*(a,d_1) C^*(b,d_2) = \begin{cases} r\, c^*(n,d), & \text{if } d_1 = d_2 = d; \\ 0, & \text{otherwise.} \end{cases}$$

Proposition 14.7.3 (Eckford Cohen) : *The set $S_r^*(\mathbb{C})$ of unitary functions* (mod r) *forms a subspace of $B_r(\mathbb{C})$ and has dimension $2^{\omega(r)}$, the number of unitary divisors of r. $S_r^*(\mathbb{C})$ has an orthonormal basis*

$$\{\frac{1}{\sqrt{r\phi^*(d)}} C^*(n,d) : d \text{ is a unitary divisor of } r\}.$$

For proof, see [12].

We note that the subspaces $V_r(\mathbb{C})$ and $S_r^*(\mathbb{C})$ of $B_r(\mathbb{C})$ have the same dimension $2^{\omega(r)}$.

There are various ways of constructing transformations of arithmetic functions. For instance, we fix $a \in \mathbb{N}$. Let

$$e_a(r) = \begin{cases} 1, & \text{if } r = a; \\ 0, & \text{otherwise.} \end{cases}$$

Then, for $f \in \mathcal{A}$, the Dirichlet product of f and e_a gives

(14.7.6) $(f \cdot e_a)(r) = \begin{cases} f(\frac{r}{a}), & \text{if } a|r; \\ 0, & \text{otherwise.} \end{cases}$

$T_a : \mathcal{A} \to \mathcal{A}$ given by $T_a(f) = f \cdot e_a$ is a linear operator on $\mathcal{A}$. Generally speaking, transformations are useful in deriving arithmetical identities.

Next, in a finite dimensional algebra A over a field F, every element is either a divisor of zero or a unit (a divisor of multiplicative identity). Also, every element of $\mathcal{A}$ is a root of some polynomial equation $f(x) = 0$ where $f(x) \in F[x]$. For, suppose that a is not a root. As a does belong to $\mathcal{A}$, the subalgebra $F[a]$ will be isomorphic to $F[x]$ which is impossible since $F[x]$ is an infinite dimensional vector space over F.

We wish to add that integers (mod r) form an additive cyclic group of order r. A periodic function (mod r) may be viewed as a function defined on an additive cyclic group $C(r)$ of order r. A subclass of periodic functions consists of those complex-valued functions defined on $C(r)$ which are invariant under automorphisms of $C(r)$. Arithmetically, these functions are the functions $f(n, r) = f(\text{g.c.d } (n, r), r)$ for all n. They are the even functions (mod r).

Remark 14.7.1 : Let D be a division ring. $M_n(D)$ denotes the total n by n matrix ring with entries from D. Then, $M_n(D)$ is a semisimple ring. Further, every semisimple ring is isomorphic to a direct sum of a finite number of total matrix rings over division rings. See J. S. Golan [14].

The structural description of semisimple rings in terms of total matrix rings is one of the major classical results of Wedderburn. See Emil Artin: The influence of J. H. M. Wedderburn on the development of modern algebra, Bull. Amer. Math. Soc., 56 (1950), 65–72.

14.8. Worked-out examples

a) (Sivaramakrishnan (1973)) Kesava Menon defines the norm f^* of a multiplicative function f (see definition 4.3.2, chapter 4) by

$$f^*(r) = \sum_{d|r^2} f(\frac{r^2}{d})\lambda(d)f(d); \quad r \geq 1$$

in which $\lambda(r) = (-1)^{\Omega(r)}$, where $\Omega(r)$ denotes the total number of prime factors of r, each being counted according to its multiplicity and $\Omega(1) = 0$.

Characterize the class of multiplicative functions f for which $f^* = e_0$, the multiplicative identity of the ring $(\mathcal{A}, +, *)$ of arithmetic functions under the operations of addition and Dirichlet convolution.

Answer: We note that $e_0(r) = [\frac{1}{r}]$, $[x]$ being the greatest integer not exceeding x. The definition of f^* gives

$$(14.8.1) \qquad f^*(r) = (f \cdot \lambda f)(r^2). \quad r \geq 1.$$

Let $(M, \cdot)$ denote the group of multiplicative functions. M is a subgroup of the group of units in $\mathcal{A}$. $f \in M$ is known if $f(p^m)$ (p a prime, $m \geq 1$) is known for all primes p. When $f \in M$, so is f^*.

We evaluate $f \cdot \lambda f$ at $r = p^{2m-1}$, p a prime, $m \geq 1$.

$$(f \cdot \lambda f)(p^{2m-1}) = \sum_{i=0}^{2m-1} f(p^{2m-1-i}) \lambda(p^i) f(p^i)$$

$$= \sum_{i=0}^{2m-1} (-1)^i f(p^{2m-1-i}) f(p^i).$$

We take $(f \cdot \lambda f)(p^{2m-1}) = 0$ for all primes p, $m \geq 1$.

Then,

$$(14.8.2) \qquad (f \cdot \lambda f)(p^{2m-1}) = \sum_{i=0}^{2m-1} (-1)^i f(p^{2m-1-i}) f(p^i) = 0.$$

We are interested in characterising the set

$$(14.8.3) \qquad I = \{f \in M : f^* = e_0\}$$

By definition, $f \in I$ if, and only if,

$$(14.8.4) \qquad (f \cdot \lambda f)(p^{2m}) = 0 \text{ for all primes } p, m \geq 1.$$

From (14.8.2) and (14.8.4), we observe that

$$I = \{f \in M : (f \cdot \lambda f) = e_0\}$$

or

$$(14.8.5) \qquad I = \{f \in M : f^{-1} = \lambda f\}.$$

Let $g, h \in I$. Then, $g^{-1} = \lambda g$. $h^{-1} = \lambda h$.

Now,

$$(g \cdot h^{-1})^{-1} = (g \cdot \lambda h)^{-1}$$

$$= (\lambda h)^{-1} \cdot g^{-1}.$$

As λ is completely multiplicative (meaning thereby, $\lambda(rs) = \lambda(r)\lambda(s)$ for all pairs of integers r, s)

$$\lambda(h \cdot h^{-1}) = \lambda h \cdot \lambda h^{-1}.$$

(This is also obvious, as $(\lambda h \cdot \lambda h^{-1})(p^m) = \sum_{i=0}^{m}(-1)^m h(p^{m-i})h^{-1}(p^i) = \lambda(h \cdot h^{-1})(p^m)$).
So, $(\lambda h)^{-1} = \lambda h^{-1}$. Thus,

$$(g \cdot h^{-1})^{-1} = (g \cdot \lambda h)^{-1} = \lambda h^{-1} \cdot g^{-1} = \lambda h^{-1} \cdot \lambda g = \lambda(g \cdot h^{-1}).$$

It follows that $g.h^{-1} \in I$ whenever $g, h \in I$. Therefore, I is a subgroup of M.

From this characterization, it is not clear whether I has any element other than e_0. It has elements other than e_0. For, let f be a multiplicative function given by

$$(14.8.6) \qquad f(p^{2m}) = \frac{1}{2}(-1)^{m+1} f(p^m)^2 + \sum_{i=m+1}^{2m-1}(-1)^{i+1} f(p^i) f(p^{2m-i})$$

(for all prime p, $m \geq 1$), where f has arbitrary values at $f(p), f(p^3), f(p^5) \ldots$.
That is, f takes arbitrary values whenever the argument $= p^{2m-1}$, $(m \geq 1)$.
Then, f given by (14.8.6) belongs to I.   □

Remark 14.8.1 : $\psi : M \to M$ given by $\psi(f) = f^*$ is a homomorphism of M into itself with kernel $= I$. Given $g \in M$, there exists $f \in M$ such that $f^* = g$, that is, $f \cdot \lambda f = g$. Hence ψ is an epimorphism.

Remark 14.8.2 : The solution shown above has been adapted from
Carl Pomerance: Solution to problem 5945, Amer. Math. Monthly 82 (1975) 410–411.

b) Let S be a commutative ring with identity 1_S. S is said to be a primary ring if S has a minimal prime ideal. (see Section 12.10 of chapter 12). Show that a necessary and sufficient condition for S to be a primary ring is that S has a minimal prime ideal which contains all the zero divisors of S.

Answer:
Sufficiency: Let S be a primary ring. The set of zero divisors of S, along with zero, coincides with the ideal N of nilpotent elements. So, N is a prime ideal. But then, N is the prime radical of S, which is the intersection of prime ideals of S. So, N is contained in every prime ideal of S. That is, N is a minimal prime ideal of S.
Necessity: We prove the contrapositive statement of the converse. That is, if $a \in S$ is such that a is non-nilpotent, then, a can never be a zero divisor of S.

Suppose that S has a minimal prime ideal P which contains all zero divisors. $a \in S$, a is non-nilpotent. We consider

$$T = \{ta^n : t \notin P; n \geq 0\}$$

T is a multiplicative set and $1_S \in T$. $0_S \notin T$. Otherwise $0_S \in T \Rightarrow ta^n = 0_S$ with $a^n \neq 0_S$. Then t is a zero divisor and $t \in P$ — a contradiction. Therefore, by theorem 129, the complement of T is S contains a prime ideal P' which is such that $P' \subseteq S \setminus T \subseteq P$, with P a minimal prime ideal. It follows that $S \setminus T = P$.

But, $a \in S$, $a \notin P$. So, a cannot be a zero divisor. That is, a non-nilpotent $\Rightarrow$ a non-zero divisor.

That is, every zero divisor of S is nilpotent. So, if S is such that S has a minimal prime ideal which contains all zero divisors, S has the property that every zero divisor of S is nilpotent. Then, the zero ideal of S is primary. This implies that S is a primary ring. $\qquad\square$

Remark 14.8.3 : A commutative ring S with identity 1_S is a primary ring, if, and only if, every zero divisor of S is nilpotent.

Remark 14.8.4 : The solution of example (b) has been adapted from D. M. Burton [A1, Theorem 8–15, Chapter 8, p 169].

EXERCISES

1. *Mark the following statements true (T) or false (F) justifying your answer briefly.*

   a) *There exist a 2-dimensional algebra A over $\mathbb{R}$, the field of real numbers and having a basis $\{\vec{e}_1, \vec{e}_2\}$ where $(\vec{e}_1)^3 = 0$, $(\vec{e}_1)^2 = \vec{e}_2$.*

   b) *Let A be a 3-dimensional algebra with unity $\vec{e}$ over $\mathbb{R}$ the field of real numbers. Then, A has no singular element (that is, an element $\vec{a}$ which has no multiplicative inverse).*

   c) *Let A be a 2-dimensional algebra over $\mathbb{Z}/3\mathbb{Z}$. It is correct to say that A is commutative, has unity element and every nonzero element of A has a multiplicative inverse.*

   d) *Let $\mathbb{B}_r(\mathbb{C})$ denote the algebra of even functions (mod r). If $f \in \mathbb{B}_r(\mathbb{C})$, there exists a function $g : \mathbb{Z} \times \mathbb{N} \to \mathbb{C}$ such that*

   $$f(n,r) = \sum_{d|g} g(d, \frac{r}{d})$$

   where summation is over divisors d of $g = $ g.c.d (n,r).

   e) *Let r be an arbitrary but fixed positive integer. Consider the matrix $[a_{ij}]$ $(i,j = 1, 2, \ldots, r)$ where $a_{ij} = C(i,j)$, ($C(n,r)$ being Ramanujan's sum). Then,*

   $$det\,[a_{ij}] = r!$$

   f) *Let $f(n,r)$ represent an even function (mod r), possessing a finite Fourier expansion with Fourier coefficients $\alpha(d,r)$. One gets*

   $$\alpha(1,r) = \sum_{a(\text{mod } r)} f(a,r).$$

2. *[Eckford Cohen] Let*

   $$\varepsilon_h(n) = \exp(\frac{2\pi i h n}{r}), \quad r \geq 1.$$

$S_r(F)$ denotes the class of (r, F), arithmetic functions f of the form

$$f(n, r) = \sum_{\substack{h \,(\mathrm{mod}\; r) \\ (h,r) = a\; square}} a(h, r)\varepsilon_h(n), \quad a(h, r) \in F, (1 \leq h \leq r);$$

where the summation is over h (mod r) such that g.c.d $(h, r) = a$ square. Show that $S_r(F)$ is a subalgebra of $\mathcal{A}_r(F)$. Determine the dimension of $S_r(F)$.

3. [Eckford Cohen] With the notation in exercise 1, let $P_r(F)$ denote the class of (r, F)-arithmetic function f of the form

$$f(n, r) = \sum_{\substack{h \,(\mathrm{mod}\; r) \\ h = a\; prime < r}} b(h, r)\varepsilon_h(n), \quad b(h, r) \in F, (1 \leq h \leq r).$$

f is referred to as a Vinogradov sum. Show that $P_r(F)$ is a subalgebra of $\mathcal{A}_r(F)$. Determine the dimension of $P_r(F)$.

4. [Eckford Cohen] Let $N_s(n, r)$ denote the number of solutions of the congruence

$$n \equiv p_1 + p_2 \cdots + p_s \pmod{r},$$

where $p_1, p_2, \ldots, p_s$ are primes, $0 \leq n < r$. Show that

$$N_s(n, r) = \frac{1}{r} \sum_{h=0}^{r-1} \varepsilon_h(n) \Big(\sum_{\substack{k\,(\mathrm{mod}\; r) \\ k = a\; prime < r}} \varepsilon_h(-k) \Big)^s$$

5. [Eckford Cohen] Let J_k denote the Jordan totient given by

$$J_k(r) = \sum_{d \mid r} \mu(\frac{r}{d}) d^k$$

where μ is the Möbius function. $M_s(n, r)$ denotes the number of the solutions of $n \equiv x_1 + x_2 \cdots + x_s$ (mod r) such that g.c.d (g.c.d $(x_1, x_2, \ldots, x_s), r) = 1$.
Show that

$$M_s(n, r) = (\frac{r}{(n, r)})^s J_s((n, r)), \quad where\; (n, r) = g.c.d\;(n, r).$$

(Note that M_s is an even function (mod r))

6. [L. Carlitz] Let F be a field of characteristic zero. Consider the ring L_p of arithmetic functions $f : \widetilde{\mathbb{Z}} \to F$. For $f, g \in L_p$, Lucas multiplication of f and g is given by

$$(f * g)(r) = \sum_{j=0}^{r} {}^* f(j)g(r - j)$$

where summation $\sum^*$ is over j such that $\binom{r}{j}$ is relatively prime to p. $f \in L_p$ is said to be singular if $f(0) = 0_F$. Show by an example that a singular function $g \in L_p$ need not be a divisor of zero.

7. *We use the notation of exercise 6. Let $I(r) = 1$, $r \geq 0$. Show that I is a unit in*
   L_p.
   If $r = r_0 + r_1 p + \cdots + r_s p^s$ $(0 \leq r_j < p; j = 0, \ldots, s)$, define

$$\mu(bp^k) = \begin{cases} 1, & b = 0; \\ -1, & b = 1; \\ 0, & b > 1. \end{cases}$$

   (where $0 \leq b < p$, $k = 0, 1, 2, \ldots s)$ and let

$$\mu(r) = \prod_{j=0}^{\infty} \mu(r_j p^j).$$

   Show that μ is the Lucas inverse of I. Deduce that

$$g(r) = \sum_{j=0}^{r} {}^{*} f(j) \Leftrightarrow F(r) = \sum_{j=0}^{r} {}^{*} f(r-j)\,\mu(j).$$

8. *Let $\mathscr{A}$ be an algebra over a field F. Assume that $\mathscr{A}$ does not possess the multiplicative identity. Suppose that $\bar{\mathscr{A}}$ is defined by*

$$\bar{\mathscr{A}} = \{(a, \alpha) : a \in \mathscr{A}, \alpha \in F\}$$

   Addition and multiplication in $\bar{\mathscr{A}}$ are given by

$$(a, \alpha) + (b, \beta) = (a+b, \alpha+\beta), \quad a, b \in \mathscr{A}, \alpha, \beta \in F;$$
$$r(a, \alpha) = (ra, r\alpha) \quad a \in \mathscr{A}, r, \alpha \in F;$$
$$(a, \alpha) \cdot (b, \beta) = \{ab + \alpha b + \beta a, \alpha\beta\} \quad a, b \in \mathscr{A} : \alpha, \beta \in F.$$

   Show that $\bar{\mathscr{A}}$ is an algebra with the identity element $(0, 1_F)$. Deduce that $\mathscr{A}$ can be embedded in $\bar{\mathscr{A}}$ by showing that elements of the form $(a, 0)$ generate an ideal of $\bar{\mathscr{A}}$ which is isomorphic to $\mathscr{A}$.

9. *Let $\mathscr{A}$ be a two-dimensional algebra over a field F such that every element of $\mathscr{A}$ is an idempotent. Show that the characteristic of $\mathscr{A}$ (as a ring) is two and that $\mathscr{A}$ is a commutative semisimple algebra.*

10. *An element a of a ring R is called Von Neumann regular, if there exists $x \in R$ such that $axa = a$. If every element of R is Von Neumann regular, R is called a regular ring. Show that every regular ring is semisimple. (The ring $\mathbb{Z}$ of integers is semisimple, but not regular).*

11. *Let R be a commutative ring with unity. If R is semisimple and Artinian, show that R is Noetherian.*

REFERENCES

[1] Alexander Abian : Linear Associative algebras, Chapters 3 and 4, pp 65–157, Pergamon Press, Inc., NY (1971).

[2] Michael Artin : Algebra, Chapter 3, pp 78–108, Prentice Hall of India, New Delhi, Reprint (1991).

[3] L. Carlitz : Arithmetic functions in an unusual setting: Amer. Math. Monthly 73 (1966) 582–590.

[4] L. Carlitz : Arithmetic functions in an unusual setting II: Duke Math. J. 34 (1967) 757–759.

[5] R.D. Carmichael : Expansions of arithmetic functions in infinite series, Proc. London Math. Soc. (2) 34 (1932), 1–26.

[6] Daniel I. A. Cohen : Problem No: 5290, Amer. Math. Monthly 72 (1965) p 555.

[7] Eckford Cohen : Rings of Arithmetic functions, Duke Math. J. 19 (1952) 115–129.

[8] Eckford Cohen : A class of Arithmetic functions, Proc. Nat. Acad. Sci. (USA) 41 (1955) 939–944.

[9] Eckford Cohen : Representations of even functions (mod r) I, Arithmetical identities, Duke Math. J. 25 (1958) 401–422.

[10] Eckford Cohen : Representations of even functions (mod r) II, Cauchy products: Duke Math. J. 26 (1959) 165–182.

[11] Eckford Cohen : Representations of even functions (mod r) III, Special topics : Duke Math. J. 26 (1959) 491–500.

[12] Eckford Cohen : Unitary functions (mod r), Duke Math. J. 28 (1961) 475–486.

[13] Yu A. Drozd and V.V. Kirichenko : Finite dimensional algebras, Chapters 1 and 2, pp 1–43, Springer Verlag, Berlin, Heidelberg, (1991).

[14] G.S. Golan : Foundations of Linear Algebra, Kluwer Academic Publishers (1995). (Translation of the book published in Hebrew).

[15] H.W. Gould : Combinatorial Identities, printed by Morgan Printing and Binding Co., (1972).

[16] Pentti Haukkanen and R. Sivaramakrishnan : Cauchy composition and even functions (mod r), Collectanea Mathematica, Universitat de Barcelona, vol XLII Fasc 1, 33–44 (c) (1992).

[17] Pentti Haukkanen and R. Sivaramakrishnan : On certain trigonometric sums in several variables, Collectanea Mathematica, Universitat de Barcelona, 45 (5) (1994) 245–261.

[18] Pentti Haukkanen and R. Sivaramakrishnan : Arithmetic functions in an algebraic setting, Tsukuka Math. J. 15 (1991) 227–234.

[19] Pentti Haukkanen : An elementary linear algebraic approach to even functions (mod r), Nieuw Archief voor Wiskunde Vol 5/2 nr 2 mart 2000 pp 29–31.

[20] I. Kaplansky : Commutative rings : revised edn., Chapters 1 and 2 pp 1–66, The University of Chicago Press, Chicago, USA (1974).

[21] P. Kesava Menon : Transformations of arithmetic functions, J. Indian Math. Soc. 6 (1942) 143–152.

[22] Paul J. McCarthy : Introduction to arithmetical functions, Universitext, Chapters 2 and 3 pp 70–148, Springer Verlag, NY (1986).

[23] Rajendran Valiaveetil : A study of normed division domains and their analogues with applications to number theory, Ph.D Thesis : University of Calicut (India), Oct. 1996.

[24] R. Sivaramakrishnan : Classical theory of arithmetic functions: Chapter 2, pp 25–45, Monographs and Textbooks in Pure and Applied Mathematics No. 126, Marcel Dekker, Inc., NY (1989).

ADDITIONAL REFERENCES

[A1] D. M. Burton: A first course in rings and ideals, Addison-Wesley Pub. Co., Reading, Massachusetts (1970).

[A2] L. Carlitz and M.V. Subbarao : Transformation of arithmetic functions, Duke Math. J. 41 (1975) 949–954.

[A3] James A. Huckaba : Commutative rings with zero-divisors, Monographs and Textbooks in Pure and Applied Mathematics No:117 Marcel Dekker Inc., NY. (1988).

CHAPTER 15

Analogues of the Goldbach problem

Historical perspective

A famous problem concerning prime numbers is known by the name:
Goldbach conjecture. In his letter dated 7th June, 1742 addressed to Leonhard
Euler, Christian Goldbach (1690–1764) mentioned that every even number ≥ 4
is a sum of two primes. As it has remained an unsolved problem for over two
centuries, the conjecture is often referred to as the Goldbach problem. Attempts
to solve the problem were made along with attempts to prove the Prime Number
Theorem which says that if $\pi(x)$ denotes the number of primes not exceeding x,
then,

$$\pi(x) \sim \frac{x}{\log x}, \text{ as } x \to \infty.$$

The Prime Number Theorem was first proved by Jacques Hadamard
(1865–1963) and C. J. de la Vallee Poussin (1866–1962) independently using
Complex Analysis, in 1896. (Probably, it was the first instance of inter-disciplinary
research). An elementary proof of the theorem was given by Atle Selberg (b.1917)
and Paul Erdös (1913–1996) in 1949. An interesting exposition on the Prime
Number Theorem is given by Norman Levinson in [30]. See also H. Diamond
[15] and V. S. Varadarajan [35]. Ralph G. Archibald [2] reports that Rene
Descartes stated without proof that every even number is a sum of 1,2, or 3
primes. E. Waring (1734–1798) also declared that every number is either a prime
or a sum of three primes. See L. E. Dickson (1874–1954) [16]. The statement of
the Goldbach problem may be given as:

(i) BGC: Every even positive integer ≥ 4 can be expressed as a sum of two
primes. (BGC is the short form of Binary Goldbach Conjecture)

(ii) TGC: Every odd number ≥ 7 can be expressed as the sum of three primes.
(TGC is the short form of Ternary Goldbach Conjecture)

Clearly, the truth of BGC implies the truth of TGC. In 1920, Hardy and Littlewood
[25] have shown that under the assumption of a weak version of Generalized
Riemann Hypothesis (GRH), there exists a positive integer M_0 such that TGC
holds for all integers $\geq M_0$. M_0 is referred to as the Hardy-Littlewood Constant.
In 1993, Chen and Wang [10] have shown that M_0 can be chosen as equal to 10^{50}.
In 1997, Zinoviev [40] has shown that assuming GRH, one can make $M_0 = 10^{20}$.
In 1937, I. M. Vinogradov (1891–1983) proved unconditionally that there exists a

positive integer N_o such that TGC holds for all integers $\geq N_o$. More specifically, we state

Vinogradov's Theorem :   *Every sufficiently large odd number is a sum of three primes.*

For proof, see [37].

Experimental verification using algorithms and computations shows that BGC is true for even integers up to $4 \cdot 10^{11}$. See M. K. Sinisalo [34]. New experiments on CRAY C916 super computer and on SGI computer server with 18 R 1000 CPU's carried out by Deshouillers, te Riele and Saouter [14] extend the bound to 10^{14}. The consequences are

(a) TGC holds.

(b) Under the assumption of GRH, every even positive integer can be expressed as the sum of at most four prime numbers. In addition, Deshouillers, te Riele and Saouter [14] have verified the

Goldbach conjecture for all even numbers in the intervals:

(i) $[10^{5i}, 10^{5i+8}]$, $i = 3, 4, 5, \dots, 20$,

(ii) $[10^{10i}, 10^{10i} + 10^9]$, $i = 20, 21, \dots, 30$.

The computation predicts the average number of steps needed to verify Goldbach Conjecture on a given interval. The experimental results are in good agreement with this prediction. This adds to the truth of the Goldbach Conjecture. See A. Granville, J. Van de Lune and H. J. J. te Riele [24].

15.1. Introduction

This chapter gives yet another instance of a number-theoretic situation for which parallel results have been formulated in algebra. We discuss three analogues of the Goldbach problem.

We mention about the Riemann hypothesis which has a bearing on computational aspects in the verification of the Goldbach problem for particular cases.

A finite analogue of the Goldbach problem due to Eckford Cohen is discussed in the context of the quotient ring $R(r) = \mathbb{Z}/r\mathbb{Z}$ where r is any positive composite number. 'Primes' in $R(r)$ are of the form $\pm pm$ where p is a prime and g.c.d $(m, r) = 1$. If r is even, it is shown in theorem 140 that every element of $R(r)$ with the possible exception of primes associated with 2, is expressible as a sum of two primes in $R(r)$.

If $M_n(\mathbb{Z})$ denotes the ring of $n \times n$ matrices ($n \geq 2$) with entries from $\mathbb{Z}$, it is shown in theorems 142 and 143 that for any $A \in M_n(\mathbb{Z})$, there exist matrices $X, Y \in M_n(\mathbb{Z})$ such that

$A = X + Y$ with $\det X = \det Y = q$ (a fixed positive integer) under suitable conditions.

In Section 15.5, we point out an analogue of the Goldbach problem via polynomials over finite fields. It is due to G. W. Effinger and D. R. Hayes [21]. Let

$M(x)$ be a monic polynomial with coefficients from a finite field $\mathbb{F}_q \cdot M(x)$ is called an even polynomial, if $q = 2$ and if x or $(x+1)$ divides $M(x)$. All other polynomials are odd. $M(x) \in \mathbb{F}_q(x)$ is called a 3-primes polynomial, if there exist irreducible monic polynomials $P_1(x), P_2(x)$ and $P_3(x)$ such that $\deg P_1(x) = \deg M(x)$, $\deg P_2(x)$ and $\deg P_3(x)$ are less than $\deg M(x)$ and

(15.1.1) $$M(x) = P_1(x) + P_2(x) + P_3(x).$$

The expressibility of $M(x)$ in the form (15.1.1) has been proved in its generality. We indicate a few particular cases in theorems 146, 151 and 153.

15.2. The Riemann hypothesis

While discussing Dirichlet series (Section 10.6, chapter 10) of an arithmetic function, the ζ-function was introduced as the generating function of $e : \mathbb{N} \to \mathbb{C}$ where $e(r) = 1$, $r \geq 1$. We recall that

$$\zeta(s) = \sum_{r=1}^{\infty} r^{-s} \quad , \text{Re } s > 1.$$

See (10.6.6). Writing $s = \sigma + it$, Riemann conjectured that if $\zeta(s) = 0$ with Res$=\sigma > 0$, then $\sigma = 1/2$. For $s = -2, -4, -6, \cdots, -2m, \cdots, \zeta(s) = 0$. The functional equation [1] for $\zeta(s)$ says:

(15.2.1a) $$\zeta(1-s) = 2(2\pi)^{-s} \cos\left(\frac{\pi s}{2}\right) \zeta(s) \Gamma(1-s),$$

or,

(15.2.1b) $$\pi^{-s/2} \Gamma\left(\frac{s}{2}\right) \zeta(s) = \pi^{-\frac{1}{2}(1-s)} \Gamma\left(\frac{1-s}{2}\right) \zeta(1-s),$$

where $\Gamma(1-s)$ denotes the value of the gamma-function at $1-s$. When $s = 2m+1$ in (15.2.1), $m \geq 1$, $\cos(\frac{\pi s}{2})$ vanishes and so we get $\zeta(-2m) = 0$ for $m \geq 1$. The values $s = -2, -4, -6, \cdots, -2m, \cdots$ are called the trivial zeros of $\zeta(s)$. In 1914, G. H. Hardy proved that an infinite number of values of s can be found for which $\zeta(s) = 0$ and $\sigma = 1/2$. However, it is not known whether all the non-trivial roots of $\zeta(s) = 0$ satisfy Re $s = 1/2$. That is, the Riemann Hypothesis (RH) about the non-trivial zeros of $\zeta(s)$ remains an open problem. In 1974, Norman Levinson [31] showed that at least one-third of the roots of $\zeta(s) = 0$ must lie on Re $s = 1/2$, the so-called critical line. Le Liounair showed in 1983 that Levinson's finding can be sharpened to 40%. It is known that the zeros of $\zeta(s)$ are symmetrically placed about the line $Im\, s = 0$.

Dirichlet's asymptotic formula for the partial sums of the divisor function $d : \mathbb{N} \to \mathbb{C}$ where $d(r) =$ the number of divisors of r is as given below:

Proposition 15.2.1 : *For $x \geq 1$,*

(15.2.2) $$\sum_{r \leq x} d(r) = x \log x + (2\gamma - 1)x + 0(\sqrt{x})$$

where γ denotes Euler's constant.

For proof, see Tom Apostol [1] or Niven, Zuckerman and Montgomery [32]. The error term $0(\sqrt{x})$, (read big 0 of $\sqrt{x}$) has been improved. Assuming the Riemann Hypothesis, the error term is $0(x^\alpha)$ with $\alpha < 1/3$. The determination of the infimum of α such that the error term is $0(x^\alpha)$ is an unsolved problem, known as Dirichlet's divisor problem. H. Ivaniec and C. J. Mozzochi [27] have shown that the error term is $0(x^{\frac{7}{22}+\epsilon})$ (for every $\epsilon > 0$). In the opposite direction, it is known that the error term is infinitely often as large as $x^{1/4}$. That is, inf $\alpha \geq 1/4$. This is mentioned to point out that certain significant results are achieved on the assumption of R.H.

If $\pi(x)$ denotes the number of primes $\leq x(x \geq 1)$ and if $Li(x)$ denotes the logarithmic integral defined by

$$(15.2.3) \qquad Li(x) = \int_2^x \frac{dt}{\log t},$$

R.H is equivalent to the assertion.

$$(15.2.4) \qquad |Li(x) - \pi(x)| \leq c\sqrt{x} \, \log x,$$

where c is a computable constant. See K. Chandrasekharan [9, Chapter III]. See also [8].

R.H was computationally tested and found to be true for the first 2×10^8 zeros by Brent et al. [4], a limit subsequently extended to the first $1.5 \times 10^9 + 1$ zeros by Brent et al. [5]. Brent's calculation covered zeros $\frac{1}{2} + it$ in the region $0 < t < 81,702,130,19$. See also [6] and [7].

Introducing the function $\xi(s)$ as

$$(15.2.5) \qquad \xi(s) = \Gamma(s/2+1)(s-1)\pi^{-s/2} \, \zeta(s),$$

one notes that $\xi(s)$ is an entire function which has as zeros the nontrivial zeros of $\zeta(s)$. $\xi(s)$ is invariant under the transformation $s \to 1-s$. This invariance and the fact that $\xi(s)$ is real for s real imply that $\xi(s)$ is also real on the critical line $s = 1/2 + it, t$ real. In fact,

$$(15.2.6) \qquad \xi(1/2+it) = \left(e^{Re \log(\Gamma(\frac{s}{2}))}\pi^{-1/4}(-t^2 - 1/4)/2\right) \times$$
$$\left(e^{i\,im\,\log(\Gamma(s/2))} \times \pi^{-it/2}\zeta(1/2+it)\right).$$

The first factor on the right of (15.2.6) is always negative. The second factor denoted by $Z(t)$ is real and has the same zeros as $\zeta(s)$ has at $s = 1/2 + it$.

$$(15.2.7) \qquad Z(t) = e^{i\,\vartheta(t)}\zeta(1/2+it)$$

where

$$(15.2.8) \qquad \vartheta(t) = Im \, \log(\Gamma(it/2+1/4)) - t/2 \log \pi.$$

Locating the zeros of $\zeta(s)$ on the critical line amounts to studying $Z(t)$ fot its zeros.

A. Turing (1912–1954) devised an ingenious method of determining the number of zeros on the critical line up to height T. We define $S(T)$ by

$$(15.2.9) \qquad\qquad S(T) = N(T) - \frac{\vartheta(T)}{\pi} - 1,$$

where $N(T)$ is the number of zeros on the critical line between 0 and T. Turing showed that

$$\left| \int_{t_1}^{t_2} S(t)\, dt \right| \leq 2.3 + 0.128 \log(t_2/2\pi),$$

where $168\pi < t_1 < t_2$.

With related numerical computation and more of techniques (leading to the fact that one has to compute $Z(T)$ to know how it changes sign from $+$ to $-$ or $-$ to $+$), much progress has been made in locating zeros of $\zeta(s)$ on the critical line. Van de Lune et al. have shown that the first one and a half billion nontrivial zeros lie on the critical line.

Analogous to $\zeta(s)$, Dirichlet defined an L-function $L(s, \chi)$ as follows:

We consider the ring $\mathbb{Z}/r\mathbb{Z}$ of integers modulo r, U_r denotes the group of units of this ring. $|U_r| = \phi(r)$, the Euler ϕ-function. Let χ be a character of U_r. As usual, $[a]$ denotes the residue class of $a \pmod{r}$. We lift χ (from U_r) to $\mathbb{Z}$ by writing

$$\chi(a) = \begin{cases} \chi([a]), & \text{if g.c.d } (a, r) = 1; \\ 0, & \text{otherwise}. \end{cases}$$

$|\chi(a)| \leq 1$ for all $a \in \mathbb{Z}$.

Definition 15.2.1 : *The function $L(s, \chi)$ is given by*

$$L(s, \chi) = \sum_{n=1}^{\infty} \chi(n) n^{-s}, \; Re\, s > 1.$$

The expression for $L(s, \chi)$ is also the same as

$$(15.2.10) \qquad \log L(s, \chi) = \sum_p \sum_{k=1}^{\infty} \frac{\chi(p^k)}{k p^{ks}}, \qquad Re\, s > 1,$$

where $\sum_p$ denotes summation over all primes p.
Or, equivalently,

$$(15.2.11) \qquad \log L(s, \chi) = \sum_p \frac{\chi(p)}{p^s} + \sum_{k=2}^{\infty} \frac{\chi(p^k)}{k p^{ks}}, \qquad Re\, s > 1.$$

We remark that $L(s, \chi)$ is a tool in proving Dirichlet's theorem on the infinitude of primes in an arithmetic progression. See Apostol [1]. When, $\chi = \chi_0$ the principal character $\pmod{r}$,

$$(15.2.12) \qquad L(s, \chi_0) = \zeta(s) \prod_{p|r} (1 - p^{-s}), \qquad Re\, s > 1,$$

where

$$\chi_0(a) = \begin{cases} 1, & \text{if g.c.d } (a,r) = 1 \\ 0, & \text{otherwise} . \end{cases}$$

Fact 15.2.1 : (a) For Re $s > 1$,

(15.2.13) $\log L(s, \chi_0) = \log \dfrac{1}{s-1} + O\,(\log\log r).$

(b) For $\chi \neq \chi_0$, $L(1, \chi) \neq 0$.

For proofs, see K. Chandrasekharan [9].

15.2.1. GENERALIZED WEAK RIEMANN HYPOTHESIS (R_δ): The real parts of the zeros of all Dirichlet L-functions $L(s, \chi)$ are less than or equal to $1/\delta$.

(R_2) is the Riemann Hypothesis about the zeros of $\zeta(s)$. As remarked earlier, Hardy and Littlewood [25] assumed the generalized weak Riemann Hypothesis (and used their 'Circle method') to prove that there exists a positive integer N such that every odd integer $r \geq N$ is a sum of three primes. That is, TGC holds for all odd numbers $r \geq N$. Vinogradov's theorem establishes the Hardy-Littlewood conclusion without invoking generalized weak R.H. Further, Vinogradov's theorem guarantees a computable value for N.

We mention briefly about Generalized Riemann Hypothesis (GRH) in the following manner :

Riemann Hypothesis (RH) asserts that $\zeta(s) \neq 0$ for any value of s for which $Re\,s > 1/2$. Riemann Hypothesis for $L(s, \chi)$ is that for any character χ modulo r, $L(s, \chi) > 0$ for values of s for which $Re\,s > 1/2$. This is called Generalized Riemann Hypothesis (GRH).

Definition 15.2.2 : *Let U_r denote the group of units in $\mathbb{Z}/r\mathbb{Z}$ $(r \geq 2)$. A Dirichlet character is a homomorphism $\chi : U_r \to \mathbb{C}^*$, the group of nonzero complex numbers under multiplication, with the additional requirement that χ is multiplicative.*

(This agrees with the definition of a character of a finite group given in definition 6.6.3, chapter 6).

It follows that the image of U_r under χ is a subset of the unit circle in $\mathbb{C}$. If $r|t$, χ induces a homomorphism $\varphi \colon U_t \to \mathbb{C}^*$ by composition with the natural map $U_t \to U_r$. So, we would regard χ as being defined modulo t or modulo r. If $[a] \in U_t$, $[a]$ will be the same as $[a'] \in U_r$ $(0 \leq a' < r)$. We take $[a] \mapsto \exp\left(\frac{2\pi i}{r}\right)$. That is, both are essentially the same map.

Definition 15.2.3 : *The minimum number r chosen to define $\chi : U_r \to S_1$ (the unit circle in $\mathbb{C}$) is called the conductor of χ written as f or f_χ.*

Examples 15.2.1 : a) $t = 8$, $r = 4$.
Let $\chi : U_8 \to S_1$ be given by

$$\chi(1) = 1, \ \chi(3) = -1, \ \chi(5) = 1, \ \chi(7) = -1.$$

Since $\chi(a) = \chi(a+4)$, χ is defined on U_4 with $\chi(1) = 1$, $\chi(3) = -1$. So, $f_\chi = 4$.

b) $t = 12$, $r = 6$.

Let $\chi : U_{12} \to S_1$ be given by

$$\chi(1) = 1, \ \chi(5) = -1, \ \chi(7) = 1, \ \chi(11) = -1.$$

We note that $\chi(a) = \chi(a+6)$; $\chi(1) = \chi(7)$ and $\chi(5) = \chi(11)$. 6 is minimal and $f_\chi = 6$.

For defining $L(s, \chi)$ (definition 15.2.1), we lift χ to $\mathbb{Z}$ by taking $\chi : \mathbb{Z} \to \mathbb{C}$ with $\chi(a) = 0$ if g.c.d $(a, f_\chi) \neq 1$. That is, χ is defined modulo its conductor f_χ. Such characters are called primitive characters. χ is periodic modulo f_χ. The notion of a Dirichlet character modulo r is restated fully, in

Definition 15.2.4 (R. A. Mollin) : *A Dirichlet character χ modulo r (an arbitrary, but fixed positive integer) is a function $\chi : \mathbb{Z} \to \mathbb{C}$ satisfying*

a) $\chi(mn) = \chi(m)\chi(n)$; $m, n \in \mathbb{Z}$
b) $\chi(n) = 0$, if g.c.d $(n, r) > 1$
c) $\chi(n) = \chi(m)$, if $n \equiv m(\mathrm{mod}\ r)$
d) $\chi(1) \neq 0$.

χ_0 *given by* (15.2.8) *is called a principal character modulo r.*

Definition 15.2.5 (R. A. Mollin) : *A character χ modulo r is called a primitive character, if, for every divisor r_0 of r, there exists an integer a satisfying $a \equiv 1(\mathrm{mod}\ r_0)$, g.c.d $(a, r) = 1$ and $\chi(a) \neq 1$.*

Fact 15.2.2 : If χ is a Dirichlet character modulo r, then for all $s > 1$

$$L(s, \chi) = \prod_p (1 - \chi(p)p^{-s})^{-1}; \quad Re\ s > 1$$

where $\prod\limits_p$ on the right runs through all primes.

15.2.2. EXAMPLE OF A DIRICHLET CHARACTER. Let $K = \mathbb{Q}(\sqrt{m})$ be a quadratic number field (see Section 3.3, chapter 3), where m is square-free. It is known that given a basis $\mathbb{B} = \{\alpha_1, \alpha_2\}$ for $\mathbb{Q}(\sqrt{m})$ (considered as a vector space of dimension 2 over $\mathbb{Q}$) discriminant of $\mathbb{B}$ is given by

(15.2.14) $$\Delta = \{\det([\sigma_i(\alpha_r)])\}^2 \quad (i, j = 1, 2)$$

where $\sigma_1 : K \to \mathbb{C}$, $\sigma_2 : K \to \mathbb{C}$ are monomorphisms of K into $\mathbb{C}$, keeping every element of $\mathbb{Q}$ fixed. σ_1 is the identity monomorphism and σ_2 is the conjugation map. It is known that Δ is independent of the choice of a basis for K. (See worked-out example (a) and Remark 13.10.1, chapter 13). By theorem 14, chapter 3, the ring $R(m)$ of integers of K is given by

(15.2.15) $$R(m) = \begin{cases} \mathbb{Z}[\sqrt{m}] & \text{if } m \not\equiv 1(\mathrm{mod}\ 4) \\ \mathbb{Z}[\frac{1+\sqrt{m}}{2}], & \text{if } m \equiv 1(\mathrm{mod}\ 4). \end{cases}$$

A basis for $R(m)$, the ring of integers of K is given by

(15.2.16) $$\mathbb{B} = \begin{cases} \{1, \sqrt{m}\}, & m \not\equiv 1 \pmod 4; \\ \{1, \frac{1+\sqrt{m}}{2}\}, & m \equiv 1 \pmod 4. \end{cases}$$

It follows that

(15.2.17) $$\Delta = \begin{cases} 4m, & \text{if } m \not\equiv 1 \pmod 4; \\ m, & \text{if } m \equiv 1 \pmod 4. \end{cases}$$

Thus,

(15.2.18) $$\Delta \equiv 0.1 \pmod 4.$$

Therefore, $K = \mathbb{Q}(\sqrt{m})$ could be expressed as $K = \mathbb{Q}(\sqrt{\Delta})$, where Δ is not a square.

If $n = \prod_{i=1}^{k} p_i^{a_i}$, we write

(15.2.19) $$(\Delta|n) = \prod_{i=1}^{k} (\Delta|p_i)^{a_i}$$

where p_i $(i = 1, 2, \ldots, k)$ are primes and $(\Delta|p_i)$ denotes Legendre symbol. $(\Delta|n)$ (15.2.19) is the Kronecker symbol for Δ (which is similar to Jacobi symbol (Section 6.7, chapter 6) but n could be odd or even).

It is verified that for $m, n \in \mathbb{Z}$.

(15.2.20) $$(\Delta|mn) = (\Delta|m)(\Delta|n)$$

(15.2.21) $$(\Delta|m) = (\Delta|n) \text{ if } m \equiv n \pmod{|\Delta|}.$$

As remarked in [A3], $(\Delta|n)$ is considered as a Dirichlet character modulo Δ.

The Dirichlet L-function in respect of Δ is given by

(15.2.22) $$L(s, (\Delta|n)) = \sum_{n=1}^{\infty} (\Delta|n)n^{-s}; \quad Re\, s > 1.$$

Next, by theorem 44, chapter 6, we know that a finite abelian group G of order n has n distinct characters and that the group $Ch(\mathbb{Z}/n\mathbb{Z}) \cong \mathbb{Z}/n\mathbb{Z}$ (see lemma 6.6.2).

Let G be a finite abelian group. Fix $g \in G$. As χ runs through the elements of $Ch(G)$, $\chi(g)$ runs through the fth roots of unity in $\mathbb{C}$ where $f = o(g)$ (order of g). In fact, $\chi(g)$ runs through all the fth roots of 1 and takes on each value equally many times in the following sense:

We consider the homomorphism $e : Ch(G) \to S_1$ given by the evaluation $e(\chi) = \chi(g)$ at g. Kernel of e consists of those χ which send g to 1. This is the same as the character group of $G/ < g >$, where $< g >$ denotes the subgroup of G generated by g. Therefore, the kernel of the evaluation map has order $|G|/f$. It implies that the image of e consists of all the fth roots of 1. As $e : Ch(G) \to S_1$ is a homomorphism, $\chi(g)$ takes on each fth root of 1 equally many times.

In the context of a cyclotomic field $F = \mathbb{Q}(\omega)$ where $\omega = \exp\left(\frac{2\pi i}{r}\right)$, we identify U_r, (the group of units in $\mathbb{Z}/r\mathbb{Z}$) with the Galois group $G(F/\mathbb{Q})$ of F over $\mathbb{Q}$. Writing $G = Gal(F/\mathbb{Q})$, we note that G is a homomorphic image of U_r. Characters of G can be considered as characters modulo r. Thus, we consider $Ch(G)$ to be a subgroup of $Ch(U_r)$.

Next, let K and L be number fields with $K \subset L$. Suppose that $R = \mathcal{A} \cap K$, $S = \mathcal{A} \cap L$ where $\mathcal{A}$ denotes the ring of algebraic integers. Let P be a prime ideal of R and Q a prime ideal of S. Then, Q divides PS, $Q \supset P$ and $Q \cap R = Q \cap K = P$. We say that Q lies over P. (see definition 13.2.3, chapter 13).

We obtain an embedding $\gamma : R/P \to S/Q$. R/P and S/Q are called residue class fields associated with P and Q. (Note that the ring of integers of a number field is a Dedekind domain and so P, Q are maximal ideals in R, S respectively). Further, R/P and S/Q are finite fields and so S/Q is a finite extension of R/P meaning that degree of extension is finite. Let $f = [S/Q : R/P]$, the degree of extension. Then f is called the inertial degree of Q over P denoted by $f(Q|P)$.

We consider K as a subfield of the cyclotomic field $Q(\omega)$ where $\omega = \exp\left(\frac{2\pi i}{r}\right)$. We can identify U_r, with $Gal(\mathbb{Q}(\omega)/\mathbb{Q})$. Let H be the subgroup of U_r fixing K pointwise.

For a rational prime p not dividing r, let f be the least positive integer such that $[p]^f \in H$, where $[p]$ denotes the congruence class of p (mod r). Let $R' = \mathcal{A} \cap K$. If P is a prime ideal of R', then the prime ideal (p) of $\mathbb{Z}$ is such that P lies over (p). It can be shown [A2] that f equals the inertial degree $f(P|(p))$ of P lying over (p).

Applying definition 13.2.5 for the norm $||I||$ of an ideal I of a number ring $R = \mathcal{A} \cap K$ we make

Definition 15.2.6 : *The Dedekind zeta function ζ_K of a number field K is given by*

$$(15.2.23) \qquad \zeta_K(s) = \sum_I ||I||^{-s}, \quad Re\, s > 1,$$

where the summation is taken over all nonzero ideals I of R, the number ring corresponding to K.

Analogous to the expression for $L(s, \chi)$ given in Fact 15.2.2, by the unique factorization of an ideal I in R, one gets

$$(15.2.24) \qquad \zeta_K(s) = \prod_P (1 - ||P||^{-s})^{-1}, \quad Re\, s > 1$$

where P runs through the prime ideals of R.
  The idea is that when

$$\zeta_K(s) = \prod_P \left(1 + \frac{1}{||P||^s} + \frac{1}{||P||^{2s}} + \cdots\right)$$

formal multiplication gives terms like $||I||^{-s}$ with each I occurring exactly once. Questions of convergence could be tackled using the fact that $\sum | \; ||P||^{-s}| < \infty$ for $Re\, s > 1$.

For each rational prime p, let t_p denote the number of prime ideals P of R lying over (p). The inertial degree $f(P|(p))$ depends only on p. We denote it by f_p. Then,

$$(15.2.25) \qquad \zeta_K(s) = \prod_p (1 - \tfrac{1}{p^{f_p \cdot s}})^{-t_p}, \quad Re\, s > 1.$$

A character χ modulo r is a character of $Gal(\mathbb{Q}(\omega)/\mathbb{Q})$ where $\omega = \exp\left(\frac{2\pi i}{r}\right)$. Let K be the fixed field of the kernel of χ. Then, $K \subseteq \mathbb{Q}(\omega)$. Let r be minimal. That is, $r = f_\chi$. K is called the field belonging to χ.

In general, let G be a finite group of Dirichlet characters. Let r be the l.c.m of the conductors of the characters in G. G is a subgroup of the group of characters of $Gal\left(\mathbb{Q}(\omega)/\mathbb{Q}\right)(\omega = \exp\left(\frac{2\pi i}{r}\right))$. We consider the intersection H of the kernels of these characters.

Let K be the fixed field of H. Then, G is the group of homomorphisms $\chi : Gal(K/\mathbb{Q}) \to \mathbb{C}^\times$, the group of nonzero complex numbers under multiplication. The field K is called the field belonging to G. If G is cyclic, generated by χ, K is the field belonging to χ.

It can be shown that $[K, \mathbb{Q}] = |G|$. In fact,

$$(15.2.26) \qquad\qquad G \cong Gal\,(K/\mathbb{Q}).$$

Example 15.2.2 : Let G be the group of characters χ of U_r for which $\chi(-1) = 1$. The field associated to G is $\mathbb{Q}(\omega + \omega^{-1})$. For, among the monomorphisms of $\mathbb{Q}(\omega)$ into $\mathbb{C}$, the conjugation map $\bar\sigma : \mathbb{Q}(\omega) \to \mathbb{C}$ given by $\bar\sigma(\omega) = \omega^{-1}$, the conjugate of ω is in the kernel of χ. So, the field associated with G is $\mathbb{Q}(\omega + \omega^{-1})$, which is the maximal real subfield of $\mathbb{Q}(\omega)$.

Remark 15.2.1 : (a) It can be shown that if χ is any character, then the field belonging to χ is real if, and only if, $\chi(-1) = 1$.
(b) Example 15.2.2 has been adapted from Lawrence C. Washington [A6].

Fact 15.2.3 : Let G be a group of Dirichlet characters modulo r, with K as the associated field. The Dedekind ζ-function $\zeta_K(s)$ satisfies the identity

$$(15.2.27) \qquad\qquad \zeta_K(s) = \prod_{\chi \in G} L(s, \chi).$$

For proof, see Lawrence C. Washington [A6] or D. A. Marcus [A2].

Fact 15.2.4 : K denotes the field belonging to a characters χ modulo r. $\zeta_K(s)$ has a simple hole at $s = 1$. If χ is of order t

(15.2.28) $$\zeta_K(s) = \zeta(s) \prod_{a=1}^{t-1} L(s, \chi^a).$$

Proof follows on using Fact 15.2.3 and noting that $L(s, \chi_0) = \zeta(s)$. $(Re\ s > 1)$.

Fact 15.2.5 : Kronecker-Weber theorem states that if K is a finite abelian extension of $\mathbb{Q}$ (meaning that K is a normal extension with abelian Galois group), then K is contained in a cyclotomic field. That is, $K \subseteq \mathbb{Q}(\exp(\frac{2\pi i}{r}))$, for some r.

Let K be a quadratic number field given by $K = \mathbb{Q}(\sqrt{\Delta})$ where Δ denotes the discriminant of K. Considering Dirichlet characters χ modulo Δ given by $\chi(n) = (\Delta|n)$, one has

(15.2.29) $$L(s, \chi) = \frac{\zeta_K(s)}{\zeta(s)} \quad (Re\ s > 1).$$

For proof of Kronecker-Weber theorem, see Lawrence C. Washington [A6]. (15.2.29) follows by noting that χ is of order $[K : \mathbb{Q}] = 2$ and by appealing to (15.2.28).

Next, let $h(\Delta)$ denote the class number of $K = \mathbb{Q}(\sqrt{\Delta})$.

Gauss's Conjecture: $h(\Delta) \to \infty$ as $|\Delta| \to \infty$.

This is an open problem.

Remark 15.2.2 : Under the assumption of GRH, one can show that $h(\Delta) \to \infty$ as $\Delta \to -\infty$. That is, if the nontrivial zeros of $L(s, \chi)$ $(\chi(n) = (\Delta|n))$ lie on the critical line $Re\ s = 1/2$, $h(\Delta)$ must grow with $|\Delta|$. L. J. Mordell (1934) proved this, by assuming that RH is false. Hielbronn (1934) showed that if GRH is false, then $h(\Delta) \to \infty$ as $\Delta \to -\infty$. So, under the assumption of either the truth or falsity of GRH, one gets $h(\Delta) \to \infty$, as $\Delta \to -\infty$. It is an example of an unconditional proof.

Gauss's class number one problem is about the fact that the ring $R(m)$ of integers of $\mathbb{Q}(\sqrt{m})$ is a PID for $m = -1, -2, -3, -7, -11, -19, -43, -67$ and -163. This is proved by assuming GRH.

If there could exist a 10th discriminant $\Delta < 0$ with $h(\Delta) = 1$, such existence would be a counterexample to GRH. For more detailed results, see R. A. Mollin [A3].

15.3. A finite analogue of the Goldbach problem

r denotes a composite positive integer > 1. The candidate for obtaining an analogue of BGC is the quotient ring $\mathbb{Z}/r\mathbb{Z}$. We represent the elements of $\mathbb{Z}/r\mathbb{Z}$ by integers belonging to a complete residue system (mod r). We write

(15.3.1) $$r = 2^a p_1^{a_1} p_2^{a_2} \cdots p_k^{a_k} \ (a \geq 0, a_i \geq 1, i = 1, 2, \ldots k);$$

where p_i $(i = 1, 2, \ldots k)$ are odd primes dividing r. Our aim is to obtain an analogue of BGC in the context of the finite ring $\mathbb{Z}/r\mathbb{Z}$. We recall that $u \in \mathbb{Z}/r\mathbb{Z}$ is a unit, if g.c.d $(u, r) = 1$. Two elements m, m' in $\mathbb{Z}/r\mathbb{Z}$ are associates if $m = um'$ for some unit $u \in \mathbb{Z}/r\mathbb{Z}$.

Lemma 15.3.1 : *Let $c_1, c_2 \ldots, c_s$ be arbitrary integers.*

The congruence

(15.3.2) $c_1 x_1 + c_2 x_2 + \cdots + c_s x_s \equiv n (\mathrm{mod}\ r)$

has a solution in $x_i (\mathrm{mod}\ r)(i = 1, 2 \ldots, s)$ where g.c.d $(x_i, r) = 1$ if, and only if, each of the congruences

(15.3.3)
$$\begin{cases} c_1 x_1 + c_2 x_2 + \quad \cdots + c_s x_s \equiv n (\mathrm{mod}\ 2^a), \\ c_1 x_1 + c_2 x_2 + \quad \cdots + c_s x_s \equiv n (\mathrm{mod}\ p_1^{a_1}), \\ \cdots \qquad\qquad \cdots\cdots\cdots \\ \cdots \qquad\qquad \cdots\cdots\cdots \\ c_1 x_1 + c_2 x_2 + \quad \cdots + c_s x_s \equiv n (\mathrm{mod}\ p_k^{a_k}), \end{cases}$$

has a solution in x_i relatively prime to a prime q_i, where q_i equals 2 or p_i $(i = 1, 2, \ldots k)$. Proof is similar to that of the Chinese Remainder theorem and hence omitted.

We recall that by a solution to (15.3.2), we mean an ordered s-tuple of integers $(x_1, x_2, \ldots x_s)$ that satisfies the congruence. Two s-tuples $(x_1, x_2, \ldots, x_s)$ and $(y_1, y_2, \ldots, y_s)$ are counted as the same solution, if $x_j \equiv y_j (\mathrm{mod}\ r)$, $(j = 1, 2, \ldots, s)$. For (15.3.2) to have a solution, it is necessary and sufficient that

g.c.d $(c_1, c_2, \ldots, c_s, r) = d$ divides n.

When solutions exist, let $N(n, r, s)$ denote the number of solutions of (15.3.2). It is easy to check that

(15.3.4) $N(n, r, s) = N(n, 2^a, s) N(n, p_1^{a_1}, s) \cdots N(n, p_k^{a_k}, s)$.

When one guarantees a solution of each of the congruences in (15.3.3), a solution of (15.3.2) is obtained and conversely, when (15.3.2) has a solution, each of (15.3.3) has a solution.

Lemma 15.3.2 : *Every element of $\mathbb{Z}/r\mathbb{Z}$ is expressible in the form*

(15.3.5) $m = 2^b p_1^{b_1} p_2^{b_2} \cdots p_k^{b_k} \cdot n$, *where g.c.d $(n, r) = 1$*

and $0 \leq b \leq a$, $0 \leq b_i \leq a_i$ $(i = 1, 2, \ldots k)$.

Further, the representation (15.3.5) is unique except for the multiple n which is a unit in $\mathbb{Z}/r\mathbb{Z}$.

Proof : We consider the congruences

$$(15.3.6) \qquad \begin{cases} 2^a y & \equiv 2^{a+1}(\bmod\ r), \\ p_1^{a_1} y_1 & \equiv p_1^{a_1+1}(\bmod\ r), \\ \cdots & \cdots \\ \cdots & \cdots \\ p_k^{a_k} y_k & \equiv p_k^{a_k+1}(\bmod\ r). \end{cases}$$

The above congruences are solvable for $y, y_1, y_2, \ldots, y_k$ respectively, where g.c.d $(y, r) = 1$, g.c.d $(y_i, r) = 1$ $(i = 1, 2, \ldots, k)$. Therefore, m given by (15.3.5) is a representative element of $\mathbb{Z}/r\mathbb{Z}$. Further, any representative element of $\mathbb{Z}/r\mathbb{Z}$ has the form (15.3.5). More explicitly, a representative element m in $\mathbb{Z}/r\mathbb{Z}$ has the structure $m = dn$ where $d|r$ and g.c.d $(n, r) = 1$.

Next, suppose that

$$(15.3.7) \qquad m' = 2^t p_1^{t_1} \cdots p_k^{t_k} n', \text{ g.c.d } (n', r) = 1$$

and $0 \le t \le a$, $0 \le t_i \le a_i$ $(i = 1, 2, \ldots k)$. Then, let

$$(15.3.8) \qquad m' = 2^t p_1^{t_1} p_2^{t_2} \cdots p_k^{t_k} \cdot n' \equiv 2^b p_1^{b_1} p_2^{b_2} \cdots p_k^{t_k} \cdot n(\bmod\ r)$$

where g.c.d $(n, r) = $ g.c.d $(n', r) = 1$. Assume that, for some i, $t_i \ne b_i$ or $t \ne b$, $a \ge b > t$ or $a_i \ge b_i > t_i$ (for some fixed i).

From (15.3.8), for $a \ge b > t$, $p_1^{t_1} p_2^{t_2} \cdots p_k^{t_k} \equiv 0(\bmod\ 2)$—a contradiction. Similarly, for $i = 1$, as $a_1 \ge b_1 > t_1$ implies $2^t p_2^{t_2} p_3^{t_3} \cdots p_k^{t_k} \equiv 0(\bmod\ p_1)$—a contradiction.

So, it happens that $t = b$ and $t_i = b_i (i = 1, 2, \ldots k)$, or, the representation of m in (15.3.5) is unique, except for n. □

Next, we examine 'primes' in $\mathbb{Z}/r\mathbb{Z}$. The primes of $\mathbb{Z}/r\mathbb{Z}$ are the elements which are associates of rational primes 2 or p_i $(i = 1, 2, \ldots k)$ dividing r. So, an element $m \in \mathbb{Z}/r\mathbb{Z}$ is a prime in $\mathbb{Z}/r\mathbb{Z}$ if, and only if, it is of the form $m = 2n$ (if 2 divides r) or $m = p_i n'$ where p_i divides r and g.c.d $(n, r) = 1$, g.c.d $(n', r) = 1$. For instance, when $r = 36$, the primes in $\mathbb{Z}/36\mathbb{Z}$ are $2n$ or $3n'$ where g.c.d $(n, 36) = $ g.c.d $(n', 36) = 1$.

In the notation of (14.2.1) of chapter 14, we write $e_a(n) = \exp(\frac{2\pi i n a}{r})$. In (14.1.6), it is shown that

$$(15.3.9) \qquad \sum_{n \equiv x+y(\bmod\ r)} e_a(x)e_b(y) = \begin{cases} re_b(n), & \text{if } a \equiv b \ (\bmod\ r), \\ 0, & \text{otherwise .} \end{cases}$$

We generalize (15.3.9) to the case of a Cauchy product of k functions $e_{a_i}(n)$ $(i = 1, 2, \ldots k)$ $(k \ge 2)$. We state without proof

Lemma 15.3.3 (Eckford Cohen) : *If* $x_1, x_2 \ldots, x_k$ *range independently over numbers of a complete residue system* (mod r) *such that*

$$n \equiv x_1 + x_2 \cdots + x_k(\bmod\ r)$$

then,

(15.3.10) $(e_{a_1}.e_{a_2}\cdots .e_{a_k})(n,r) = \sum_{n \equiv x_1 + x_2 \cdots + x_k (\text{mod } r)} e_{a_1}(x_1)e_{a_2}(x_2)\cdots e_{a_k}(x_k)$

$$= \begin{cases} r^{k-1}e_{a_1}(n), & \text{if } a_1 \equiv \ldots \equiv a_k(\text{mod } r); \\ 0, & \text{otherwise }. \end{cases}$$

For proof, see [11, lemma 3].

Let S denote a least non-negative complete residue system (mod r). $|S| = r$. For $i = 1, 2 \ldots s$, we write P_i for a finite collection of elements chosen from S, with repetitions allowed.

Definition 15.3.1 : *Let $f_r^{(i)}$ be a function from $\widetilde{\mathbb{Z}}$ to $\mathbb{N}$ given by*

$$f_r^{(i)}(a) = t_i,$$

if $a \equiv m(\text{mod } r)$, where $m \in S$ and m appears t_i times in P_i.

It follows that $f_r^{(i)}(b) = f_r^{(i)}(a)$ whenever $b \equiv a(\text{mod } r)$. In the terminology of Section 14.1 of chapter 14, $f_r^{(i)}$ is an $(r, \mathbb{C})$ arithmetic function.

15.3.1. ECKFORD COHEN'S PRINCIPLE. The number $N_s(n,r)$ of ordered sets, including repetitions, of the type $(x_1, x_2, \ldots, x_s)$ such that x_i range over P_i and $n \equiv x_i + x_2 \cdots + x_s (\text{mod } r)$, is given by

$$N_s(n,r) = f_1 \cdot f_2 \cdots f_s(n,r) = \frac{1}{r}\sum_{a=0}^{r-1} e_a(n) \prod_{i=1}^{s} \left(\sum_{u \in P_i} e_a(-u)\right),$$

where u ranges over all numbers in P_i.

This follows from Eckford Cohen's theorem (theorem 126, chapter 14) on the representation of an (r, F)-arithmetic function in terms of the elements of a basis $\{e_a(n) : 0 \le a < r\}$.

To illustrate the principle, we obtain the number $G_s(n,r)$ of solutions in positive primes $< r$ of the congruence

(15.3.11) $n \equiv p_1 + p_2 + \cdots + p_s(\text{mod } r)$

where $o \le n < r$. Let P_i consist of the set of all primes in the interval $(0, r)$; $i = 1, 2, 3, \ldots, s$. Then,

(15.3.12) $G_s(n,r) = \frac{1}{r}\sum_{a=0}^{r-1} e_a(n) \left(\sum_{\substack{p \text{ a prime} \\ 0 < p < r}} e_a(-p)\right)^s.$

Theorem 137 (Eckford Cohen (1954)) **:** *Let p be a prime and $m \ge 1$. $a_1, a_2, \ldots, a_t$, $a_{t+1}, \ldots a_{t+k}$ are integers such that p does not divide a_i $(i = 1, 2, \ldots t)$, whereas p divides a_i $(i = t+1, t+2, \cdots, t+k)$. $t > 0$, $k \ge 0$ and $t + k = s$.*

Then, the number $M(n, p^m, s)$ of solutions of the congruence

(15.3.13) $a_1x_1 + a_2x_2 + \cdots + a_sx_s \equiv n(\text{mod } p^m)$

in x_i prime to p ($i = 1, 2, , \ldots s$) and distinct modulo p^m is given by

(15.3.14) $\qquad M(n, p^m, s) = p^{m(s-1)-s}(p-1)^k\{(p-1)^t + (-1)^t \xi(n, p^m)\},$

where

(15.3.15) $\qquad \xi(n, p^m) = \begin{cases} p-1 & \text{, if } p|n, \\ -1 & \text{if } p \nmid n. \end{cases}$

Proof : With the notation used in Eckford Cohen's principle (15.3.1), we write P_i to denote the collection of integers n_i chosen from a least non-negative complete residue system (mod p^m) and

$$n_i \equiv a_i u(\text{mod } p^m), \quad p \nmid u$$

and each n_i appearing (in P_i) as many times as there are distinct solutions to

$$n_i \equiv a_i u(\text{mod } p^m)$$

in u prime to p. Let $e_a(q) = \exp(\frac{2\pi i a q}{p^m})$. Then, as in (15.3.1),

(15.3.16) $\qquad M(n, p^m, s) = \frac{1}{p^m} \sum_{a=0}^{p^m-1} e_a(n) \prod_{i=1}^{s} \sum_{\substack{1 \le n_i < p^m \\ g.c.d(u,p)=1}} e_a(-a_i u).$

This is simplified using the definition of Ramanujan's sum $C(n, r)$ (5.1.2), chapter 5. Isolating the term corresponding to $a = 0$ in (15.3.16) we obtain

(15.3.17) $\qquad M(n, p^m, s) = \frac{1}{p^m}\left\{ \left(\phi(p^m)\right)^s + \sum_{1 \le a < p^m} e_a(n) \prod_{i=1}^{s} C(aa_i, p^m) \right\}.$

Now, $\phi(p^m) = p^{m-1}(p-1), m \ge 1$;

$$C(a a_i, p^m) = \begin{cases} 0, & \text{whenever } aa_i \text{ contains } p^{m-2} \text{ as a factor ;} \\ (-1)p^{m-1}, & \text{if } p^{m-1} \text{ divides } aa_i. \end{cases}$$

But, $p \nmid a_i (i = 1, 2, \ldots t)$ and $p|a_i$ for $i = t+1, t+2, \ldots, t+k$. We do summation for $a(1 \le a < p^m)$ for which $C(aa_i, p^m) \ne 0$. Writing $a = bp^{m-1}$, $0 < b < p$, and summing over b, we obtain

$$M(n, p^m, s) = \frac{1}{p^m}\{p^{(m-1)s}(p-1)^s + \sum_{0<b<p} \exp(\frac{2\pi i n b}{p}) \prod_{i=1}^{s} C(a_i bp^{m-1}, p^m)\}$$

$$= \frac{1}{p^m}\{p^{(m-1)s}(p-1)^s + \sum_{0<b<p} \exp(\frac{2\pi i n b}{p})(-p^{m-1})^t (p^m - p^{m-1})^k \}.$$

Now,

$$\sum_{0<b<p} \exp(\frac{2\pi i n b}{p}) = \begin{cases} p-1, & \text{if } p|n; \\ -1, & \text{if } p \nmid n. \end{cases}$$

Using the definition of $\xi(n, p^m)$, we get

$$(15.3.18) \quad M(n, p^m, s) = \frac{1}{p^m} \left\{ p^{(m-1)s}(p-1)^s + (-1)^t \xi(n, p^m) p^{(m-1)(t+k)}(p-1)^k \right\}.$$

Simplifying the right side of (15.3.18), we arrive at (15.3.14).   □

Corollary 15.3.1 : *Under the conditions of theorem 137, (15.3.13) is not solvable for x_i prime to p if, and only if, one of the following cases hold:*

(a) p odd, $t = 1$, $p|n$,

(b) $p = 2$, t odd, n even,

(c) $p = 2$, t even, n odd.

For, $(p-1)^t + (-1)^t \xi(n, p^m) = 0$ in each of the cases $(a),(b),(c)$ above. Corollary 15.3.1 is of significance in the results that follow.

We denote the ring $\mathbb{Z}/r\mathbb{Z}$ by $R(r)$. Let $G(r)$ denote the number having the property: Every element of $R(r)$ can be expressed as a sum of $G(r)$ primes in $R(r)$.

Let g denote the minimum value of $G(r)$ when it exists. If $r = p^c$, p a prime, $c \geq 1$, then, for $1 \leq a \leq p^c$, $a = p^d u$ (where $p \nmid u$) is a prime in $R(p^c)$ if, and only if, $d = 1$. Further, no unit in $R(p^c)$ is a sum of two primes in $R(p^c)$. That is, $G(p^c)$ does not exist. So, to find g, we may assume that r has at least two distinct prime factors.

Theorem 138 (Eckford Cohen (1954)) : *Let $r > 1$. There exists a number $G(r)$ such that every element of $R(r)$ is a sum of $G(r)$ primes in $R(r)$ if, and only if, r has at least two distinct prime factors. For such r, the minimum value g of $G(r)$ is given by*

(a) $g = 2$, if r is odd

(b) $g = 3$, if r is even and has at least two distinct odd prime factors or if r is twice on odd prime power and

(c) $g = 4$, if r is of the form $r = 2^b p^k$ where p is an odd prime dividing r, $b > 1$ and $k \geq 1$.

Proof : **Case (i)** r is odd and has at least two distinct prime factors. We write

$$r = p_1^{a_1} p_2^{a_2} \cdots p_h^{a_h} \quad (h \geq 2).$$

By theorem 137 and its corollary 15.3.1,

$$p_1 x_1 + p_2 x_2 \equiv m \pmod{r}$$

is solvable. Further, $m \in R(r)$ is a sum of s primes in $R(r)$ if, and only if,

$$\alpha_1 x_1 + \alpha_2 x_2 \cdots + \alpha_s x_s \equiv m \pmod{r}$$

is solvable in integers x_i prime to r and α_i are rational primes dividing r $(i = 1, 2, \ldots, s)$ or equivalently, if, and only if,

$$\alpha_1 x_1 + \alpha_2 x_2 \cdots + \alpha_s x_s \equiv m \pmod{p_i^{a_i}} \quad (1 \leq i \leq h)$$

is solvable in x_j prime to p_i $(j = 1, 2, \ldots s)$. Every such m is a sum of two primes in $R(r)$.

If m is a non-unit in $R(r)$, then, m has a prime factor say p_1 in common with r. For such m, we consider the congruence:

(15.3.19) $$p_1x_1 + p_2x_2 \equiv m(\bmod\ p_i^{a_i}), \quad i > 1.$$

Now, if p is a prime, the congruence

$$px_1 + px_2 + \cdots + px_s \equiv bt(\bmod\ p^b), \quad b \geq 1, t \in \mathbb{Z}$$

is solvable in x_j prime to p if, and only if, the congruence

$$x_1 + x_2 + \cdots + x_s \equiv t(\bmod\ p^{b-1})$$

has such a solution. So, (15.3.19) is solvable mod $p_i^{a_i}$ in x_j prime to p_i, by theorem 137. So all non-units of $R(r)$ are sums of two primes in $R(r)$.

Case (ii): Suppose that r is even and $r = 2^b p_1^{a_1} p_2^{a_2} \cdots p_h^{a_h}$ ($h \geq 1, b \geq 1$). Not every element of $R(r)$ is expressible as a sum of two primes in $R(r)$. Such an element (not expressible as desired) is $m' = p_1^{a_1} p_2^{a_2} \cdots p_h^{a_h}$. For, let

(15.3.20) $$\pi_1x_1 + \pi_2x_2 \equiv q(\bmod\ r)$$

where π_1, π_2 are two prime divisors of r. By theorem 137, in order that (15.3.20) is solvable modulo 2^b, one of the coefficients π_1 or π_2 must be 2 and the other must be odd. For example, $\pi_1 = p_1$, $\pi_2 = 2$. But, in this case (15.3.20) is not solvable (mod $p_i^{a_i}$). This follows from lemma 15.3.1.

Let $r = 2^b p_i^{a_1}$ ($b \geq 2$). Then, three primes will not suffice. For, if we consider

(15.3.21) $$\pi_1x_1 + \pi_2x_2 + \pi_3x_3 \equiv 4p_1(\bmod\ r),$$

where π_i have values either 2 or p_1; for (15.3.21) to be solvable (mod 2^b), one must have, by corollary 15.3.1, essentially one of the two cases;

(i) $\pi_1 = \pi_2 = \pi_3 = 2$
(ii) $\pi_1 = \pi_2 = p_1$, $\pi_3 = 2$

In case (i), (15.3.21) is not solvable (mod 2^b) by an argument used before. In case (ii) (15.3.21) is not solvable (mod $p_i^{a_i}$) ($i = 1, 2, \ldots h$). So, $4p$ is not a sum of three primes in $R(r)$.

On the other hand, four primes will suffice in this case. First, let r be odd. Then,

(15.3.22) $$p_1x_1 + 2x_2 + 2x_3 + 2x_4 \equiv m(\bmod\ r)$$

is solvable by theorem 137 and Lemma 15.3.1.

Secondly, if r is even, the congruence

(15.3.23) $$p_1x_1 + p_1x_2 + 2x_3 + 2x_4 \equiv m(\bmod\ r)$$

is solvable. So, as before, every element m of $R(r)$ is a sum of four primes of $R(r)$ where $r = 2^b p_1^{a_1}$ ($b \geq 2$).

Case(iii) (remaining cases)

$$r = 2p_1^{a_1} \text{ or } r = 2^b p_1^{a_1} p_2^{a_2} \cdots p_h^{a_h}(b \geq 1, h \geq 2)$$

If m is odd,

(15.3.24) $p_1 x_1 + 2x_2 + 2x_3 \equiv m(\text{mod } r)$

is solvable for either value of r by theorem 137 and lemma 15.3.1. If m is even, we consider

(15.3.25) $2x_1 + 2x_2 + 2x_3 \equiv m(\text{mod } r);$ $(m = 2l, l \text{ odd})$

(15.3.26) $p_1 x_1 + p_2 x_2 + 2x_3 \equiv m(\text{mod } r).$

(15.3.25) is solvable $(\text{mod } 2p_1^{a_1})$ and (15.3.26) is solvable $(\text{mod } 2^b p_1^{a_1} p_2^{a_2} \cdots p_h^{a_h})$. Thus, whether m is odd or even, m is expressible as a sum of three primes in $R(r)$.   □

Theorem 139 (Eckford Cohen (1954)) : *For $r > 1$, $R(r)$ denotes the quotient ring $\mathbb{Z}/r\mathbb{Z}$.*

(a) *Every element of $R(r)$ is expressible as a sum of, at most, three primes in $R(r)$ if, and only if, r has at least two distinct prime factors.*

(b) *Every element of $R(r)$ is a sum of, at most, two primes in $R(r)$ if, and only if, r is odd with at least two distinct prime factors or r is an even number of the form $r = 2^b p$ where $b \geq 1$ and p is an odd prime.*

Proof : Both (a) and (b) of the theorem are established, once we are able to show that

(i) If $r = 2^b p_1^{a_1}$ (p an odd prime, $b \geq 2$, $a_1 \geq 1$) every element of $R(r)$ is a sum of two or three primes in $R(r)$.

(ii) If $r = 2^b p_1^{a_1}$ (p_1 an odd prime, $b \geq 1$, $a_1 \geq 1$), any element of $R(r)$ is a prime or a sum of two primes in $R(r)$.

By theorem 138, it is clear that two primes will not suffice if r is even and not of the form $2^b p_1$ (p_1 an odd prime, $b \geq 1$). For, if π, and π_2 are prime divisors of r,

$$\pi_1 x_1 + \pi_2 x_2 = q \text{ (say)}$$

is neither a prime nor a sum of two primes in $R(r)$.

Step 1 :

(a) Suppose that $r = 2^b p_1^{a_1}$ ($b \geq 2$). Let $m \in R(r)$ be such that m is even and m is not divisible by 4. Then,

$$2x_1 + 2x_2 + 2x_3 \equiv m(\text{mod } r);   (m = 2l, l \text{ odd })$$

is solvable.

(b) If m is even and m is divisible by 4, then the congruence

$$2x_1 + 2x_2 \equiv m(\text{mod } r)$$

is solvable.

(c) If m is odd,

$$p_1 x_1 + 2x_2 + 2x_3 \equiv m(\text{mod } r)$$

is solvable.

(a),(b),(c) are easily checked.

We recall that $m \in R(r)$ is a sum of s primes in $R(r)$ if, and only if, the congruence

$$\alpha_1 x_1 + \alpha_2 x_2 \cdots + \alpha_s x_s \equiv m \,(\mathrm{mod}\ r)$$

is solvable in integers x_i prime to $r\,(i = 1,2,\ldots s)$ and in α_i, rational primes dividing r. $(i = 1,2,\ldots,s)$. It follows that all elements of $R(r)$ are sums of, at most, three primes in $R(r)$.

Step 2: Suppose that $r = 2^b p_1$ (p_1 an odd prime, $b \geq 1$). Then, the congruence

$$p_1 x_1 + 2x_2 \equiv m \,(\mathrm{mod}\ r)$$

is solvable. If r is divisible by 4, that is, $b > 1$, as every element m in $R(r)$ has the form $m = 2^c p_1^d u$ where $0 \leq c \leq b$, $0 \leq d \leq 1$ and g.c.d $(r,u) = 1$,

$$2x_1 + 2x_2 \equiv m \,(\mathrm{mod}\ r)$$

is solvable. Further, if $m \equiv 0(\mathrm{mod}\ 2p_1)$, then,

$$p_1 x_1 + p_1 x_2 \equiv m \,(\mathrm{mod}\ r)$$

is solvable. In all other cases, m is a prime in $R(r)$. It follows that every element of $R(r)$ which is not a prime of $R(r)$ is a sum of two primes in $R(r)$. $\qquad \square$

Next theorem gives a finite analogue of BGC.

Theorem 140 (Eckford Cohen (1954)) : *Let $R(r)$ denote the ring $\mathbb{Z}/r\mathbb{Z}$. If r is even, every element of $R(r)$ with the possible exception of primes associated with 2, is expressible as a sum of two primes in $R(r)$.*

Proof : We have only to recapture some of the steps of proof of theorems 138 and 139.

$$\text{Let } r = 2^b p_1^{a_1} p_2^{a_2} \cdots p_h^{a_h} \quad (b \geq, a_i \geq 1;\ i = 1,2,\ldots h).$$

Let $m \in \mathbb{Z}/r\mathbb{Z}$ and m odd. m is not expressible as a sum of two primes in $R(r)$. So, we are concerned with even elements of $R(r)$. Any associate of 2 is of the form $2u$ where g.c.d $(u,r) = 1$. Such elements are 'even primes'. So, we, next, look at elements of the form $2^c p_1^{b_1} p_2^{b_2} \cdots p_h^{b_h}$, where $0 < c \leq b$, $0 \leq b_i \leq a_i (i = 1,2,\ldots h)$.

If $r = 2^b p_1^{a_1}$ (p_1 an odd prime; $b \geq 1$, $a_1 \geq 1$), any element of $R(r)$ is a prime or a sum of two primes in $R(r)$. So, an even element of $R(r)$ is a sum of two primes in $R(r)$.

Let $r = 2^b p_1^{a_1} p_2^{a_2} \cdots p_h^{a_h}$ (p_i odd primes; $h > 1$). Let $m \in R(r)$. Then, m has the form

(15.3.27) $m = 2^c p_1^{b_1} p_2^{b_2} \cdots p_h^{b_h} n$, g.c.d$(n,r) = 1$;

$(0 \leq c \leq b, 0 \leq b_i \leq a_i\ ;\ i = 1,2,\ldots h)$.

The representation of m in (15.3.27) is unique except for the unit n in the ring $R(r)$ (by lemma 15.3.2). An even element other than an associate of 2 has the form

(15.3.28) $m = 2^c p_1^{b_1} p_2^{b_2} \cdots p_n^{b_n} n$, g.c.d $(n,r) = 1$,

$(1 \leq c \leq b, 0 \leq b_i \leq a_i; i = 1,2,\ldots h)$.

Claim : m as given in (15.3.28) is a sum of two primes in $R(r)$.

If m is a nonzero, non-unit in $R(r)$, m has a prime factor say p_1 in common with r. For such an m, we consider the congruences:

$$(15.3.29) \qquad \begin{cases} p_1x_1 + p_2x_2 \equiv m \pmod{2^b}; \\ p_1x_1 + p_2x_2 \equiv m \pmod{p_i^{a_i}}, i > 1. \end{cases}$$

(15.3.29) is solvable in $x_j(j = 1, 2)$ prime to 2^b as well as prime to $p_i^{a_i}$ $(i = 2, \ldots, h)$. This is a consequence of theorem 137 . So, all the even nonzero non-units of $R(r)$ are sums of two primes of $R(r)$. In the case where $r = 2^b(b > 1)$, for $1 \leq m \leq r$, $m = 2^c u$ where u is odd (that is, $2 \nmid u$) is such that m is a prime in $R(r)$, if, and only if, $c = 1$. Further, no unit of $R(2^b)$ is a sum of two primes of $R(2^b)$. However, the congruence

$$(15.3.30) \qquad 2x_1 + 2x_2 \equiv m \pmod{2^b}$$

is solvable in $x_j(j = 1, 2)$ prime to 2 by theorem 137.

By corollary 15.3.1 (b), given a prime p, integers $a_1, a_2, \ldots a_t, a_{t+1}, \ldots a_{t+k}$ with $p \nmid a_i(i = 1, 2, \ldots t); p \mid a_i$ $(i = t+1, \ldots, s = t+k)$, where t is odd, the congruence:

$$a_1x_1 + a_2x_2 \cdots + a_sx_s \equiv n \pmod{p^q}$$

in x_i prime to $p(i = 1, 2, \ldots, s = t+k)$ has no solution when n is even. That is, (15.3.30) does not fall under any of the cases (a),(b),(c) of corollary 15.3.1. Thus, any even element of $R(r)$ (when r is even) is either a prime associated to 2 or is a sum of two primes of $R(r)$. $\qquad \square$

Remark 15.3.1 : Theorems 137 to 140 have been drawn from Eckford Cohen [12].

15.4. The Goldbach problem in $M_n(\mathbb{Z})$

$M_n(\mathbb{Z})$ denotes the ring of $n \times n$ matrices $(n \geq 2)$ with entries from $\mathbb{Z}$, the ring of integers. In [36], Vaserstein proves that given any integer p and $A \in M_2(\mathbb{Z})$, one can find matrices $X, Y \in M_2(\mathbb{Z})$ such that

$$(15.4.1) \qquad A = X + Y \text{ with } \det X = \det Y = p.$$

(15.4.1) shows the expressibility of A as a sum of two matrices X, Y having the property that $\det X = \det Y = p$ (the given integer). (15.4.1) is an analogue of the Goldbach problem in the context of 2×2 matrices with entries from $\mathbb{Z}$.

For a diagonal matrix $\begin{bmatrix} a & 0 \\ 0 & b \end{bmatrix} = A$, we have

$$(15.4.2) \qquad \begin{bmatrix} a & 0 \\ 0 & b \end{bmatrix} = \begin{bmatrix} a & 1 \\ -p & 0 \end{bmatrix} + \begin{bmatrix} 0 & -1 \\ p & b. \end{bmatrix}$$

A generalization is given by Jun Wang [39] in 1992.

The tool for handling square-matrices, in general, is the process of reduction of a matrix to the 'diagonal form'. Instead of $\mathbb{Z}$, (which is a PID), we consider a principal ideal domain D. We recall that any finite collection of elements of a PID has a g.c.d.

Notation 15.4.1 : Let D be a PID and n, a positive integer. $GL_n(D)$ denotes the group of all invertible elements P in the matrix ring $M_n(D)$ of $n \times n$ matrices with entries from D.

Definition 15.4.1 : *Two $n \times n$ matrices A and B (elements of $M_n(D)$) are said to be equivalent if there exists $P \in GL_n(D)$ such that $B = PAP^{-1}$.*

Let V, V' be two free D-modules each of rank n. Two matrices A, B are equivalent if, and only if, they represent the same D-module homomorphism $\Psi : V \to V'$. The image $\Psi(V)$ of V is a submodule of V'. The rank of the homomorphism Ψ is the rank of the matrix A of Ψ. As in the case of matrices having entries from a field F, the rank of A over D is the maximum number of linearly independent rows (or columns) of A. Further, equivalent matrices have the same rank.

Equivalence of matrices is best understood using 'elementary column or row operations'. The elementary column operations on $A \in M_n(D)$ are

(i) Interchange of any two columns,
(ii) multiplication of the elements of a column by a unit in D,
(iii) addition of d times elements of a column to another column, where $d \in D$.

If one of these is applied to I, the $n \times n$ identity matrix, the resulting matrix is an elementary matrix E. It is an invertible matrix. To apply an elementary column operation on an $n \times n$ matrix A (with entries from D) is to postmultiply A by the corresponding elementary matrix E, that is, A gets transformed into AE by the elementary column operation. Similarly, elementary row operations amount to premultiplication by E. That is, A gets transformed into EA by the elementary row operation.

Observation 15.4.1 : *Any matrix obtained from $A \in M_n(D)$ by elementary row and column operations is equivalent to A; as A gets transformed into the form PAP^{-1} where P is an invertible matrix $\in GL_n(D)$.*

Next, we consider $J \in GL_2(D)$.
Let

$$J = \begin{bmatrix} a & b \\ c & d \end{bmatrix} ; \ a, b, c, d \in D, \quad ad - bc \neq 0_D.$$

We write

(15.4.3)
$$J \oplus I = \begin{bmatrix} a & b & 0_D & 0_D & \cdots & \cdots \\ c & d & 0_D & 0_D & \cdots & \cdots \\ 0_D & 0_D & 1 & 0_D & \cdots & \cdots \\ 0_D & 0_D & \cdots & \cdots & \cdots & 1_D \end{bmatrix}$$

$J \oplus I$ is a $k+2 \times k+2$ matrix, where I is the $k \times k$ unit matrix ($k \geq 1$). Further, $ad - bc$ is a unit in D. $J \oplus I$ is a direct sum square matrix.

Definition 15.4.2 : *[29] $J \oplus I$ as given in* (15.4.3) *is called a secondary matrix. Postmultiplication by $J \oplus I$ is called a secondary column operation.*

Lemma 15.4.1 : *Let A be an $n \times n$ matrix with entries from D. If $A \in M_n(D)$ ($n \geq 2$) is such that the first row of A is $[s \; t \cdots \cdots]$, then A can be transformed into a matrix B with first row $[g \; 0 \; 0 \; \cdots \; 0]$ where $g =$ g.c.d (a,b) by a secondary column operation.*

Proof : The idea of proof is that as D is a PID, a and b have a greatest common divisor g and g is a linear combination of a and b, say,

$$(15.4.4) \qquad\qquad g = as + bt; \quad s, t \in D$$

We could write $s = gs'$, $t = gt'$ where g.c.d $(s', t') = 1_D$. It follows from (15.4.4) that $as' + bt' = 1_D$. Further, $ts' - st' = 0_D$. So, in order to transform the first row of A into $[g \; 0 \cdots 0]$, we do the secondary column operation choosing a direct sum square matrix $J \oplus I$ given by

$$(15.4.5) \qquad\qquad J \oplus I = \begin{bmatrix} a & -t' & 0_D & \cdots & \cdots \\ b & -s' & 0_D & \cdots & \cdots \\ 0_D & 0_D & 1_D & 0_D & \cdots \\ \cdots & \cdots & \cdots & \cdots & \cdots \\ 0_D & 0_D & 0_D & \cdots & 1_D \end{bmatrix}$$

The 2×2 block of $A(J \oplus I)$ appearing in the left-top corner is

$$\begin{bmatrix} s & t \\ - & - \end{bmatrix} \begin{bmatrix} a & -t' \\ b & s' \end{bmatrix} = \begin{bmatrix} sa+tb & -st'+ts' \\ - & - \end{bmatrix} = \begin{bmatrix} g & 0_D \\ - & - \end{bmatrix}$$

Further, $J \in GL_2(D)$, as $as' + bt' = 1_D \neq 0_D$.
This proves the statement of lemma 15.4.1.   □

Lemma 15.4.2 : *Given $A \in M_n(D)$ (A, nonzero) there exists a sequence of elementary row or column operations together with secondary row or column operations transforming A to B where B is of the form*

$$(15.4.6) \qquad\qquad B = \begin{bmatrix} t & 0 \\ 0 & C \end{bmatrix}$$

in which $C \in M_{n-1}(D)$ and t divides every entry in C.

Proof : The use of g.c.d of two elements of D via a secondary column operation (as in lemma 15.4.1) helps. We call the left-upper corner of A with entry $a_{11} \neq 0_D$, the corner of A. Since A is nonzero, we can assume that $a_{11} \neq 0_D$, if necessary, by interchanging rows or columns. Secondary column or row operations (described earlier) will change the first row of A to $[t \; 0 \; 0 \; \cdots 0]$. Analogous secondary row operations reduce entries in the first column below the corner to 0_D. These secondary row operations may spoil some of the entries in the first row. However, using secondary column operations again and then making secondary row operations, we will eventually get the form (15.4.6) for A after a finite number of iterations.

In the matrix B that is obtained, though first row and first column will be okay, t may not divide all the entries in $C \in M_{n-1}(D)$. If t does not divide some entry in the second row, add second row to first row and use secondary row or column operations (as done earlier) so as to produce a new corner t' where $t'|t$. It means that the principal ideal $(t) \subsetneqq (t')$. In case, the new corner fails to divide an entry in some other row, repeat the procedure, one obtains an ascending chain of principal ideals:

$$(t) \subset (t') \subset (t'') \subset \cdots$$

on the successive corner entries. As D is Noetherian, the ascending chain of principal ideals terminates and so we will obtain the matrix B as in (15.4.6) with the stated properties. $\square$

Lemma 15.4.3 : *Let A be a nonzero $n \times n$ matrix with entries from D. A can be reduced to a diagonal matrix.*

$$L = diag\ (t_1, t_2, \ldots, t_n),$$

where $t_i | t_{i+1} (i = 1, 2, \ldots (n-1))$, $t_i \neq 0_D$, $(i = 1, 2, \ldots n)$, by a sequence of elementary and secondary unimodular row and column operations.

Proof : We recall that $U \in M_n(D)$ is unimodular, if $\det U = \pm 1_D$. For $n = 2$, postmultiplication by a secondary matrix takes the form

$$\begin{bmatrix} s & t \\ - & - \end{bmatrix} \begin{bmatrix} a & -t' \\ b & s' \end{bmatrix} = \begin{bmatrix} sa+tb & -st'+ts' \\ - & - \end{bmatrix} = \begin{bmatrix} g & 0_D \\ - & - \end{bmatrix}$$

where $g =$ g.c.d (a,b), $as' + bt' = 1_D \neq 0_D$, $st' - ts' = 0_D$. So, it is the result of a secondary unimodular column operation. Therefore, the result holds for $n = 2$. We apply induction on n. If it is true for $n = k$, it is also true for $n = k+1$, by lemma 15.4.2. This proves lemma 15.4.3. $\square$

Lemma 15.4.4 : *If $P \in GL_n(D)$, P is a product of elementary and secondary matrices.*

Proof : We note that a sequence of elementary and secondary column operations is the same as postmultiplication by elementary and secondary matrices say $E_1, E_2, \ldots E_k$. Similarly, we do premultiplication by elementary and secondary matrices for a sequence of elementary and secondary row operations. Further, P is invertible if, and only if, $\det P$ is a unit in D. So, we obtain

$E'_m E'_{m-1} \cdots E'_1 P E_1 E_2 \cdots E_k = L = diag\ (t_1, t_2, \ldots t_n)$, by lemma 15.4.3. t_i is a unit in D $(i = 1, 2, \ldots n)$. Now, multiplying a column by a unit in D, we can make $L = I$, the $n \times n$ unit matrix. Since the inverse of an elementary matrix is elementary, we get P as a product of elementary and secondary matrices in the form

(15.4.7) $P = E'^{-1}_1\ E'^{-1}_2\ \cdots\ E'^{-1}_m \cdot E^{-1}_k E^{-1}_{k-1}\ \cdots\ E^{-1}_1.$

                                                                        $\square$

Lemma 15.4.5 : *Two matrices A and B (elements of $M_n(D)$) are equivalent if, and only if, there is a sequence of elementary and secondary row and column operations transforming A into B.*

Proof : $\Leftarrow$: Suppose A is transformed into B in the form

$$B = E'^{-1}_1 E'^{-1}_2 \cdots E'^{-1}_m A\, E^{-1}_k E^{-1}_{k-1} \cdots E^{-1}_1,$$

where $E_1, E_2, \ldots E_k, E'_1, \ldots E'_m$ are elementary and secondary matrices. Then, $B = QAP^{-1}$ where P, Q are invertible $n \times n$ square matrices. That is, A and B are equivalent.

:$\Rightarrow$ Given that A and B are equivalent matrices, there exist invertible matrices P, Q such that $B = QAP^{-1}$. We can write P and Q as products of elementary and secondary matrices as in (15.4.7). So, by lemma 15.4.4, Q and P^{-1} are products of elementary and secondary matrices. Therefore, when A and B are equivalent matrices, A can be transformed into B by a sequence of elementary and secondary row and column operations. $\square$

Lemmas 15.4.3 and 15.4.5 together yield the following

Theorem 141 : $A \in M_n(D)$ *is equivalent to a matrix*

$$L = diag\,(t_1, t_2, \ldots t_n)$$

where $t_i (i = 1, 2, \ldots n)$ are nonzero and $t_i \,|\, t_{i+1}\,(i = 1, 2, \ldots (n-1))$.

Remark 15.4.1 : Lemmas 15.4.1 to 15.4.5 and theorem 141 have been adapted from S. MacLane and G. Birkhoff [29, Chapter VIII, section 7].

Definition 15.4.3 : *Given $A \in M_n(D)$, where $A = [a_{ij}]$, we define*

$$d(A) = g.c.d\,(a_{11}, a_{12}, \ldots a_{1n}; a_{21}, \ldots a_{2n}; \ldots; a_{n1}, a_{n2}, a_{nn}).$$

We allow 0_D to divide 0_D and set g.c.d $(0_D, 0_D) = 0_D$. Further, g.c.d $(0_D, a) = a \in D$. By repeated application of lemma 15.4.1 and using theorem 141, $A \in M_n(D)$ is equivalent to a matrix $L = diag\,(t_1, t_2, \ldots, t_n)$, where $t_1 = d(A)$. We state this for the special case $D = \mathbb{Z}$, the PID of rational integers in

Proposition 15.4.1 : *Let $A \in M_n(\mathbb{Z})$ there exist unimodular matrices U, V in $M_n(\mathbb{Z})$ such that*

(15.4.8) $UAV = diag\,(t_1, t_2, \ldots, t_n)$

where $t_1 = d(A)$ and $t_i \,|\, t_{i+1}(i = 1, 2, \ldots, (n-1))$.

See N. Jacobson [28, theorem 3.8, chapter 3, p 176].
By virtue of the above proposition, we note from (15.4.2) that given $A \in M_2(\mathbb{Z})$ and an integer p, A is as shown in (15.4.1).

If $A \in M_n(\mathbb{Z})$, $(n \geq 2)$, we could take $A = diag\,(t_1, t_2, \ldots t_n)$; $t_1 = d(A)$ and $t_i \,|\, t_{i+1}$, $i = 1, 2, \ldots, (n-1)$. When n is even, A is made up of 2×2 blocks. So we deduce

Theorem 142 (Jun Wang (1992)) : *Let n be even. Suppose that q is a given positive integer, there exist $X, Y \in M_n(\mathbb{Z})$ such that*

(15.4.9) $\qquad A = X + Y$ with $\det X = \det Y = q$.

Therefore, an analogue of the Goldbach problem in the form given above is possible for *n* even. The situation, when *n* is odd, is handled in the following

Theorem 143 (Jun Wang (1992)) : *Let $n(> 1)$ be an odd integer and q a fixed positive integer. Then, for any $A \in M_n(\mathbb{Z})$, there exist matrices $X, Y \in M_n(\mathbb{Z})$ such that*

$$A = X + Y \text{ and } \det X = \det Y = q$$

if, and only if, $d(A)$ divides $2q$.

Proof : $\Rightarrow$ Suppose that $A, X, Y \in M_n(\mathbb{Z})$ are such that $A = X + Y$ and $\det X = \det Y = q$. Let $d(A) = d$. Then,

$$\det(A - X) = \det Y = q.$$

But, $\det(A - X) \equiv \det(-X)(\bmod d)$, (by theorem 141). As *n* is odd, $\det(-X) = -\det X$. So, $\det Y \equiv -\det X(\bmod d)$ from which it follows that $2q \equiv 0(\bmod d)$.

$\Leftarrow$: Suppose that $2q \equiv 0(\bmod d)$. We write $2q = kd$, where $k \geq 1$. By theorem 141 and by the validity of (15.4.9) for *n* even, it will suffice if we prove the theorem for $n = 3$. So, we take $A = diag\,(d, a, b)$ where $d \mid a, d \mid b$. Writing

$$(15.4.10) \qquad X = \begin{bmatrix} d & 1 & 0 \\ 0 & a & 1 \\ -q & -k & 0 \end{bmatrix}, Y = \begin{bmatrix} 0 & -1 & 0 \\ 0 & 0 & -1 \\ q & k & b \end{bmatrix}$$

we see that $A = X + Y$ with $\det X = \det Y = q$. Thus, the theorem holds for all odd $n \geq 3$, when $d \mid 2q$. $\qquad \square$

Corollary 15.4.1 : *Let $n(> 1)$ be odd and $A \in M_n(\mathbb{Z})$. Then, for any integer q, there exist matrices $X, Y \in M_n(\mathbb{Z})$ such that*

$$A = X + Y \text{ with } \det X = \det Y = q$$

if $d(A) = 1$ or 2.

For, $d(A) = 1$ or 2 satisfies the condition that $d(A)$ divides $2q$.

Remark 15.4.2 : Theorems 141 to 143 and corollary 14.4.1 have been adapted from Jun Wang [39].

15.5. An analogue of Goldbach theorem via polynomials over finite fields

We examine monic polynomials with coefficients from a finite field $\mathbb{F}_q$ having *q* elements, where $q = p^m$, *p* a prime, $m \geq 1$. We make a restatement of corollary 4.6.1 of chapter 4 under a slightly different notation.

Fact 15.5.1 : Let $\mathbb{F}_q$ be a finite field of characteristic *p* (a prime). Then, for every positive integer *r*, there exists an irreducible polynomial of degree *r* in $\mathbb{F}_q[x]$.

Proof : Suppose that $q = p^m (m \geq 1)$. We denote the algebraic closure of F_q by $\bar{\mathbb{F}}_q$. Then, there exists a field $K \subseteq \bar{\mathbb{F}}_q$ (and containing the prime subfield $\mathbb{Z}/p\mathbb{Z}$) such that K consists of the zeros of $x^{p^{mr}} - x$. Now, every element of $\mathbb{F}_q$ is a zero of $x^{p^m} - x$. That is, if $\alpha \in \mathbb{F}_q$, $\alpha^{p^m} = \alpha$. But, $p^{mr} = p^m \cdot p^{m(r-1)}$. So, $\alpha^{p^{mr}} = (\alpha)^{p^{m(r-2)}} \cdots = \alpha^{p^m} = \alpha$. So, $\alpha \in \mathbb{F}_q \Rightarrow \alpha \in K$. Or, $\mathbb{F}_q \subseteq K$. $[K : \mathbb{F}_q] = r$. So, K is a simple extension of $\mathbb{F}_q$. Therefore, there exists $\beta \in K$ such that $K = \mathbb{F}_q(\beta)$. It follows that the irreducible polynomial of β over $\mathbb{F}_q$ is of degree r. This proves the assertion. □

Definition 15.5.1 : *A monic polynomial $M(x) \in \mathbb{F}_q[x]$ is called even if $q = 2$ and if x or $x + 1$ divides $M(x)$.*

For example, $x^2 + 1 \in \mathbb{F}_2[X]$ is an even polynomial. $x^3 + x^2 + 1 \in \mathbb{F}_2[X]$ is not an even polynomial, though $q = 2$. We note that $x^3 + x^2 + 1$ is irreducible over $\mathbb{F}_2$.

Definition 15.5.2 : *A monic polynomial $M(x) \in \mathbb{F}_q[X]$ which is not an even polynomial, is called an odd polynomial.*

Definition 15.5.3 : *Let $M(x) \in \mathbb{F}_q[X]$ be monic with $\deg M(x) = r$. $M(x)$ is called a 3-primes polynomial, if there exist monic irreducible polynomials $P_1(x)$, $P_2(x)$ and $P_3(x) \in \mathbb{F}_q[x]$ such that $\deg P_1(x) = r$, $\deg P_2(x) < r$, $\deg P_3(x) < r$ and*

(15.5.1) $M(x) = P_1(x) + P_2(x) + P_3(x).$

For instance, $x^4 + x^2 + 1$ is an odd 3-primes polynomial over $\mathbb{F}_2$. For,

$$P_1(x) = x^4 + x^3 + x^2 + x + 1, \quad P_2(x) = x^3 + x^2 + 1, \quad P_3(x) = x^2 + x + 1$$

are irreducible polynomials in $\mathbb{F}_2[x]$ and

$$M(x) = x^4 + x^2 + 1 = P_1(x) + P_2(x) + P_3(x).$$

Lemma 15.5.1 : *(a) When $q = 2^m (m \geq 1)$, $x^2 + a \in \mathbb{F}_q[x]$ is not a 3-primes polynomial.*

(b) When $q = 2$, $g(x) = x(x+1)h(x)$, where $h(x) \in \mathbb{F}_2[x]$ and $g(x) + 1$ is reducible, is not a 3-primes polynomial.

Proof : (a) In $\mathbb{F}_q$, $\Psi : \mathbb{F}_q \longrightarrow \mathbb{F}_q$ given by $\Psi(\alpha) = \alpha^2$, $\alpha \in \mathbb{F}_q$ is an automorphism of $\mathbb{F}_q$, as $\Psi(\alpha + \beta) = (\alpha + \beta)^2 = \alpha^2 + \beta^2 = \Psi(\alpha) + \Psi(\beta)$ and $\Psi(\alpha\beta) = \Psi(\alpha)\Psi(\beta)$ for $\alpha\beta \in \mathbb{F}_q$. Further, $\Psi(0) = 0$ and $\mathbb{F}_q$ is finite. (Ψ is the Frobenius automorphism of $\mathbb{F}_q$).

Let $P_2(x) = x + a_2$, $P_3(x) = x + a_3$ and $P_1(x) = x^2 + (a + a_2 + a_3)$. Then, given $f(x) = x^2 + a$, there exists $\alpha \in \mathbb{F}_q$ such that $\alpha^2 = a + a_2 + a_3$. So, $P_1(x)$ is not irreducible as $P(x) = (x+\alpha)^2$. Evidently, $P_2(x)$ and $P_3(x)$ are first degree polynomials. So, $f(x)$ is not a 3-primes polynomial in $\mathbb{F}_q[x]$.

(b) Let $g(x)$ satisfy the stated conditions. $g(x)$ has an even number of terms if, and only if, it is divisible by $x + 1$, and so, $x + 1$ is the only irreducible polynomial over $\mathbb{F}_2$ with an even number of terms. If $g_1(x)$ and $h_1(x) \in \mathbb{F}_2(x)$ are such that $g_1(x)$ and $h_1(x)$ have an even (odd) number of terms, then $g_1(x) + h_1(x)$ has an even number of terms. If $g_1(x)$ has an odd number of terms and $h_1(x)$ has an even number of terms then $g_1(x) + h_1(x)$ has an odd number of terms.

Let $g(x) = P_1(x) + P_2(x) + P_3(x)$.

As $g(x) = x(x+1)h(x)$, not all of $P_1(x)$, $P_2(x)$ and $P_3(x)$ have an odd number of terms. So, one of them, say, $P_2(x) = x+1$. Not all of them can have 1 as their constant terms. So, as we need only irreducible polynomials, we take $P_3(x) = x$. Then, $P_1(x) = g(x) - (x+1) - x = g(x) - 1 = g(x) + 1$. But, then, $P_1(x)$ is reducible by hypothesis. So, it is impossible to write $g(x) = P_1(x) + P_2(x) + P_3(x)$, where $P_i(x)$ $(i = 1, 2, 3)$ is irreducible. So, $g(x)$ is not a 3-primes polynomial.   □

For $r \geq 1$, $M(x) = x^{4r} + x^{2r}$ is not a 3-primes polynomial, since $M(x) + 1 = x^{4r} + x^{2r} + 1 = (x^{2r} + x^r + 1)^2$ and so $M(x) + 1$ is reducible. Further,

$$M(x) = x^{2r}(x^{2r} + 1) = x(x+1)h(x), \text{ for } h(x) \in \mathbb{F}_2[x].$$

Remark 15.5.1 : D.R. Hayes [26] defines a 3-primes polynomial of degree r as $M(x)$ where $M(x)$ is capable of representation as

(15.5.2) $M(x) = \alpha P_1(x) + \beta P_2(x) + \gamma P_3(x)$

and $\alpha, \beta, \gamma \in \mathbb{F}_q$, $P_i(x) \in \mathbb{F}_q(x)$ is of degree r. $(i = 1, 2, 3)$.

We state, without proof, a theorem of Hayes.

Proposition 15.5.1 (D.R. Hayes (1966)) **:** *For every degree $r \geq 5$, there exists a q_r (depending on r and decreasing as r increases) such that if $q \geq q_r$, then every odd monic polynomial M of degree r in $\mathbb{F}_q[x]$ is a 3-primes polynomial.*

Gove Effinger reports that by Haye's theorem, if $q \geq 6,340,567$, then every monic 5th degree polynomial over $\mathbb{F}_q$ is a 3-primes polynomial. Therefore, $q_5 \leq 6,340,567$. Also, $q_6 \leq 5,297$ and $q_7 \leq 479$. Also, in Hayes' theorem, r is considered to be ≥ 5, and it is an 'asymptotic' result.

For polynomials of low degree, G. W. Effinger [18], [19] gives the 3-primes theorem as follows:

Proposition 15.5.2 :
    (a) If q is odd and if $\deg M(x) = 2, 3, 4, 5$ or 6, then $M(x)$ is a 3-primes polynomial. If $\deg M(x) = 7$ and $q \geq 203$ (and is odd), then $M(x)$ is a 3-primes polynomial.
    (b) Let $M(x)$ be an odd monic polynomial in $\mathbb{F}_q(x)$, where q is even. If degree $M(x)$ is $3, 4, 5$ or 6, then $M(x)$ is a 3-primes polynomial. If $\deg M(x) = 2$ and if $M(x)$ is of the form $x^2 + bx + a$ where $b \neq 0$, then $M(x)$ is a 3-primes polynomial.

15.5.1. THE POLYNOMIAL 3-PRIMES CONJECTURE.
Every odd monic polynomial $M(x) \in \mathbb{F}_q[x]$ is a 3-primes polynomial except for the case q even and $M(x) = x^2 + a \in \mathbb{F}_q[x]$.

The theorems of Hayes and Effinger (propositions 15.5.1 and 15.5.2 above) solve the polynomial 3-primes conjecture except for certain individual values of $\deg M(x)$. The cases not covered are shown in the following table:

For odd monic polynomials of degree r	7	8	9	10	11	12	13	14	15	16
We must still check all fields of order $<$	199*	97	47	29	23	17	13	11	9	8

For odd monic polynomials of degree r	17 – 20	21 – 24	25 – 33	34 – 41
We must still check all fields of order $<$	7	5	4	3

* and the single case $q = 256$.

Thus, the solution of the polynomial 3-primes conjecture is reduced to a finite and apparently tractable calculation as given in the table above. To give a flavour of the method of solution for polynomials of low degree, we discuss a few particular cases.

We need a few results from the theory of finite fields. For $q = p^m$ (p a prime, $m \geq 1$) $\mathbb{F}_q$, denoting a finite field of q elements, contains $\mathbb{Z}/p\mathbb{Z}$ as its prime subfield. $\mathbb{F}_q$ is a finite extension of $\mathbb{Z}/p\mathbb{Z}$ of degree m. For $r \geq 2$, $\mathbb{F}_{q^r}$ is a finite extension of $\mathbb{F}_q$ of degree r over $\mathbb{F}_q$. We could write $\mathbb{F}_{q^r} = \mathbb{F}_q(\alpha)$ for some $\alpha \in \mathbb{F}_{q^r}$. The irreducible polynomial of α over $\mathbb{F}_q$ is denoted by irr $(\alpha, \mathbb{F}_q)$. It is also called the minimal polynomial of α, as it is the monic polynomial of minimal degree having α as a zero. It is known [23] that every field F has an algebraic closure, that is, an algebraic extension $\bar{F}$ which is algebraically closed. If E is a finite extension of F, the number of isomorphisms of E into $\bar{F}$ leaving F fixed is denoted by $\{E : F\}$, called the index of E over F. $\{\mathbb{F}_q(\alpha) : \mathbb{F}_q\}$ gives the number of distinct zeros of irr $(\alpha, \mathbb{F}_q)$. If $\{\mathbb{F}_q(\alpha) : \mathbb{F}_q\} = [\mathbb{F}_q(\alpha) : \mathbb{F}_q](= r)$, $\mathbb{F}_q(\alpha)$ is called a separable extension of $\mathbb{F}_q$. An element α in $\bar{\mathbb{F}}_q$ is separable over $\mathbb{F}_q$, if $\mathbb{F}_q(\alpha)$ is a separable extention of $\mathbb{F}_q$. A field F is perfect if every finite extension of F is a separable extension. Every finite field is perfect.

Now, $\mathbb{F}_q$ has characteristic p. $E = \mathbb{F}_{q^r}(r \geq 2)$ is a finite extension of $\mathbb{F}_q$. Analogous to the primitive element theorem [23] (which says that a finite separable extension E of an infinite field F contains an element α such that $E = F(\alpha)$), we note that $\mathbb{F}_{q^r}$ contains primitive elements α such that $\mathbb{F}_{q^r} = \mathbb{F}_q(\alpha)$.

We denote by E, a field which is an algebraic extension of a field F. Suppose that $\alpha \in E$. β is said to be a conjugate of α, if the irreducible polynomial of α has β as a zero also. If $F \subseteq E \subseteq \bar{F}$ (the algebraic closure of F), an automorphism σ of $\bar{F}$ which leaves F fixed, maps $\alpha(\epsilon E)$ onto some conjugate β of α over F. The collection of all automorphisms of E leaving F fixed forms a group $G(E/F)$. Let $F \subset E \subseteq \bar{F}$. E is called a splitting field over F, if it is the splitting field of a specified set of polynomials in $F[x]$. E is a splitting field over F, if all isomorphisms of E into $\bar{F}$ leaving F fixed are automorphisms of E. That is, if $E \subseteq \bar{F}$ is a splitting field over F, then every irreducible polynomial in $F[x]$ having a zero in E splits in E (factors into a product of linear factors in $E[x]$). Further, if E is a finite extension of F and is a separable splitting field over F, then, $|G(E/F)| = [E : F]$. In such a situation, E is referred to as a finite normal extension of F.

Definition 15.5.4 : *Let E be a finite normal extension of a field F. For $\alpha \in E$, the norm of α over F, written $N_{E/F}(\alpha)$, is given by*

$$N_{E/F}(\alpha) = \prod_{\sigma \in G(E/F)} \sigma(\alpha).$$

The trace of α over F is given by

$$T_{E/F}(\alpha) = \sum_{\sigma \in G(E/F)} \sigma(\alpha).$$

It is verified that $N_{E/F}(\alpha)$ and $T_{E/F}(\alpha)$ are elements of F. Further, when $E = F(\alpha)$, if

$$irr(\alpha, F) = x^n + a_{n-1}x^{n-1} + \cdots + a_1 x + a_0,$$

then,

(15.5.3) $\qquad\qquad\qquad N_{E/F}(\alpha) = (-1)^n a_0$

(15.5.4) $\qquad\qquad\qquad \mathrm{Trace}_{E/F}(\alpha) = T_{E/F}(\alpha) = -a_{n-1}.$

In what follows, we consider only monic polynomials with coefficients from a finite field $\mathbb{F}_q$.

Definition 15.5.5 : *Given $P(x) = x^r + a_1 x^{r-1} + a_2 x^{r-2} + \cdots + a_r$ where $a_i \in \mathbb{F}_q$ ($i = 1, 2, \ldots r$), a_1 is called the first or trace coefficient of $P(x)$, a_2 the second coefficient, $\ldots, a_r$ the rth coefficient.*

In order to solve the 3-primes problem for polynomials of low degree, the general procedure is as follows:

Let $M(x)$ be a given polynomial of degree r. We seek an irreducible polynomial $P_1(x)$ of degree r such that $M(x) - P_1(x)$ is monic and is of as low a degree as possible. Then, we find an irreducible polynomial $P_2(x)$ such that $\big(M(x) - P_1(x)\big) - P_2(x)$ is monic and irreducible. We call this $P_3(x)$. It follows that

(15.5.5) $\qquad\qquad\qquad M(x) = P_1(x) + P_2(x) + P_3(x)$

making $M(x)$ a 3-primes polynomial.

For instance, if $\deg M(x) = 3$, it will suffice if $P_1(x)$ can be obtained so that $M(x) - P_1(x)$ is a (monic) quadratic polynomial and if $P_2(x)$ is obtained so that $\big(M(x) - P_1(x)\big) - P_2(x)$ is (monic) linear. If $\deg M(x) = 5$, we find $P_1(x)$ with $\deg P_1(x) = 5$ and $P_2(x) = M(x) - P_1(x)$ is (monic) cubic. Then, $\big(M(x) - P_1(x)\big) - P_2(x)$ is to be a (monic) linear polynomial.

Theorem 144 : *Let $a_1 \in \mathbb{F}_q$ and $r \geq 2$. Then, there exists an irreducible polynomial $P(x)$ of degree r whose first coefficient is a_1 except for the case q even, $r = 2$, $a_1 = 0$.*

Proof : Case 1. Let $p = $ char $\mathbb{F}_q$. Suppose that $p \nmid r$. Pick any t_0, a primitive element of $\mathbb{F}_{q^r}$ (an extension of $\mathbb{F}_q$). Let

$$t = t_0 - \frac{a_1 + \text{trace } (t_0)}{r}$$

be also a primitive element of $\mathbb{F}_{q^r}$. Then,

$$\text{trace } (t) = \text{trace } (t_0) - \text{trace } (\frac{a_1 + \text{trace } (t_0)}{r}) = \text{trace } (t_0) - \frac{ra_1}{r} - \text{trace } (t_0) = -a_1.$$

Let $f_t(x)$ be the irreducible polynomial of t over $\mathbb{F}_q$. Then $f_t(x)$ has the first coefficient a_1.

Case 2. Suppose that char $\mathbb{F}_q = p$ and p divides r. Since $\mathbb{F}_{q^r}$ is a separable extension of $\mathbb{F}_{q^r}$, the trace function is not identically zero. Select $t_0 \in \mathbb{F}_{q^r}$ with trace $(t_0) \neq 0$. Then, for any $t \in \mathbb{F}_q$, $t = (\frac{-a_1}{\text{trace } (t_0)})t_0$ has trace $(t) = -a_1$. So, $\Psi : \mathbb{F}_{q^r} \to \mathbb{F}_q$ where $\Psi(t) = $ trace (t), $t \in \mathbb{F}_{q^r}$ is onto. But, Ψ is an additive homomorphism. So, for every $t \in \mathbb{F}_q$,

$$\#\{t \in \mathbb{F}_{q^r} : \text{trace } (t) = -a_1\} = \frac{q^r}{q} = q^{r-1}.$$

If $r > 2$, $r - 1 > \frac{r}{2}$. But, there are at the most $q^{r/2}$ elements (in $\mathbb{F}_{q^r}$) which are not primitive elements. So, there exists a primitive element t with trace $(t) = -a_1$. Then, $f_t(x)$ is the irreducible polynomial, as desired.

If $r = 2$ and $q = 2^m (m \geq 1)$, ker $\Psi = \{t \in \mathbb{F}_{q^2} : \text{trace } (t) = 0\}$ is given by ker $\Psi = \mathbb{F}_q$. So, there are no primitive elements t in $\mathbb{F}_{q^2}$ with trace$(t) = 0$. $\square$

Let $P_i(x)$ be an irreducible polynomial of degree r with coefficients from $\mathbb{F}_q$. $(i = 1, 2, \ldots n)$. Suppose that g.c.d $(q, r) = 1$. If n distinct elements of $\mathbb{F}_q$ appear as the second coefficient in $P_i(x)(i = 1, 2, \ldots n)$ where trace coefficient of $P_i(x) = 0$ for each $i = 1, 2, \ldots n$, then, the n distinct elements will appear as the trace coefficient for every irreducible polynomial $P(x) \in \mathbb{F}_q[x]$, where $\deg P(x) = r$. We say that all trace coefficients are obtained by 'translation'. This is justified in

Theorem 145 (Translation lemma) : *Suppose that for $a \in \mathbb{F}_q$, there are $n_0(a)$ irreducible polynomials of degree r with trace coefficient $= 0$ and second coefficient a, where g.c.d $(q, r) = 1$. Let $a_1 \in \mathbb{F}_q$ be such that there are $n_1(a)$ irreducible polynomials of degree r with trace coefficient a_1 and second coefficient a. Then,*

$$(15.5.6) \qquad n_1(a) = n_0\left(a - \binom{r}{2}\frac{a_1^2}{r^2}\right).$$

That is, for any $a_1 \in \mathbb{F}_q$, the set $\{n_1(a) : a \in \mathbb{F}_q\}$ is just a permutation of the set $\{n_0(a) : a \in \mathbb{F}_q\}$.

Proof :

Suppose that $t \in \mathbb{F}_{qr}$ satisfies the irreducible polynomial $x^r + ax^{r-2} + \cdots + a_r$. Assume that t is a primitive element of $\mathbb{F}_{qr}$. Then, $t - \frac{a_1}{r}$ is also primitive in $\mathbb{F}_{qr}$ and satisfies

$$(x + \frac{a_1}{r})^r + a(x + \frac{a_1}{r})^{r-2} + \cdots = (x^r + a_1 x^{r-1} + \binom{r}{2} \frac{a_1^2}{r^2} x^{r-2} + \cdots)$$

$$+ a(x^{r-2} + \frac{r-2}{r} a_1 x^{r-3} + \cdots).$$

Or,

$$(x + \frac{a_1}{r})^r + a(x + \frac{a_1}{r})^{r-2} + \cdots = x^r + a_1 x^{r-1} + \{\binom{r}{2} \frac{a_1^2}{r^2} + a\} x^{r-2} + \cdots.$$

From the above, (15.5.6) follows. □

Remark 15.5.2 : When $\deg P(x) = r$ and char $\mathbb{F}_q = p$ is such that $p \nmid r$, theorem 145 says that we can obtain all trace coefficients by 'translation'. So, it suffices to study polynomials with trace coefficient $= 0$. This is a technique used by Effinger [18], [19], [21] in his investigations. See also [22].

Next, using theorems 144 and 145, we obtain

Theorem 146 (G. W. Effinger (1988)) :

  a) If q is odd, then every quadratic and cubic polynomial over $\mathbb{F}_q$ are 3-primes polynomials.

  b) If q is even, every monic quadratic or a cubic polynomial over $\mathbb{F}_q$ is a 3-primes polynomial, except for $x^2 + a$, $a \in \mathbb{F}_q$.

Proof : Case 1, Quadratic case :

  Let $M(x) = x^2 + a_1 x + a$. $a_1, a \in \mathbb{F}_q$, $a \neq 0$.
Then, by theorem 144, there exists an irreducible polynomial $P_1(x)$ of degree 2 whose first coefficient is a_1, except for q even, $r = 2$ and $a_1 = 0$. So we take $P_1(x) = x^2 + a_1 x + a'$, $P_2(x) = x + a$, $P_3(x) = x + a'$, so that $M(x) = P_1(x) + P_2(x) + P_3(x)$, when q is even.

  Next, suppose that q is odd. Then $\deg M(x)$ is relatively prime to $p = $ char $\mathbb{F}_q$. Theorem 144 guarantees that there exists an irreducible polynomial $P_1(x) = x^2 + (a_1 - 2)x + a'$ for some $a' \in \mathbb{F}_q$. Let $P_2(x) = x + a$ and $P_3(x) = x - a'$. Then, $M(x) = P_1(x) + P_2(x) + P_3(x)$ as required.

Case 2, Cubic Case:

  (i) Char $\mathbb{F}_q = 2$. Let $M(x) = x^3 + a_1 x^2 + ax + b$. If $q > 2$, there must exist at least two distinct coefficients a_1' and a_2' such that

$$P_{1,1}(x) = x^3 + (a_1 + 1)x^2 + a_1' x + b_1',$$

$$P_{1,2}(x) = x^3 + (a_1 + 1)x^2 + a_2' x + b_2'.$$

Both $P_{1,1}(x)$ and $P_{1,2}(x)$ are irreducible for some b'_1, b'_2. If this is not the case, then, $(a_1 + 1) = a'_0$ (say). If b' ranges over all the nonzero elements of $\mathbb{F}_q$, there would exist at the most $(q-1)$ irreducible polynomials with first coefficient $(a_1 + 1)$. But, by theorem 144, the $\frac{q^3-q}{3}$ irreducible polynomials of degree 3 are equally divided among the q possible trace values. So, we notice that there exist $\frac{q^2-1}{3}$ irreducible polynomials of degree 3, with first coefficient $(a_1 + 1)$. In order that

$$\frac{q^2-1}{3} \leq (q-1),$$

we must have $q + 1 \leq 3$. That is $q \leq 2$. So, there exist at least two polynomials $P_{1,1}(x)$ and $P_{1,2}(x)$ with the stated properties.

So, when $q > 2$, we select $P_1(x) = x^3 + (a_1 + 1)x^2 + a'x + b'$ with $a_1 + a' + 1 \neq 0$. By theorem 144, we may select

$$P_2(x) = x^2 + (a_1 + a' + 1)x + b''$$

and $P_3(x) = x + (b + b' + b'')$. Then, $M(x) = P_1(x) + P_2(x) + P_3(x)$ as required.

Next, when $q = 2$, there are 8 monic cubic polynomials over $\mathbb{F}_2$. We display the '3-primes representations' for each using the fact that $x^3 + x^2 + 1$, $x^3 + x + 1$ and $x^2 + x + 1$ are irreducible polynomials.

   (i) $x^3 = (x^3 + x^2 + 1) + (x^2 + x + 1) + x$
   (ii) $x^3 + 1 = (x^3 + x^2 + 1) + (x^2 + x + 1) + (x + 1)$
   (iii) $x^3 + x = (x^3 + x + 1) + (x + 1) + x$
   (iv) $x^3 + x + 1 = (x^3 + x + 1) + x + x$
   (v) $x^3 + x^2 = (x^3 + x^2 + 1) + (x + 1) + x$
   (vi) $x^3 + x^2 + 1 = (x^3 + x^2 + 1) + x + x$
   (vii) $x^3 + x^2 + x = (x^3 + x + 1) + (x^2 + x + 1) + x$
   (viii) $x^3 + x^2 + x + 1 = (x^3 + x + 1) + (x^2 + x + 1) + (x + 1)$.

We have already noted that $x^2 + a$, $(a \in \mathbb{F}_q)$ is not a 3-primes polynomial when $q = 2^m$. See Lemma 15.5.1(a).

(ii) Char $\mathbb{F}_q = p$, odd. Let $M(x) = x^3 + a_1 x^2 + ax + b$. The irreducible polynomials $P_1(x)$, $P_2(x)$ and $P_3(x)$ are given by

$$P_1(x) = x^3 + (a_1 - 1)x^2 + a'x + b',$$

$$P_2(x) = x^2 + (a - a' - 1)x + b'',$$

$$P_3(x) = x + b - b' - b''.$$

So, every monic polynomial of degree 3 over $\mathbb{F}_q$ (q odd) is a 3-primes polynomial.

This completes the proof of theorem 146.

$\square$

The first or trace coefficient was needed to obtain a 3-primes representation of a quadratic or a cubic polynomial in $\mathbb{F}_q[x]$. For polynomials of degree 4 or 5, we examine the second coefficient of the polynomial.

Now, given a field F, a map $\eta : F \to F$ given by $\eta(x) = x^2$, $x \in F$; defines a 'homogeneous' quadratic form, in the sense that

(15.5.7) $\qquad \eta(x+y) - \eta(x) - \eta(y) = (x+y)^2 - x^2 - y^2 = 2xy.$

Let V, V' be finite dimensional vector spaces over a field F. An F-bilinear function $\xi : V \times V' \to W$, a finite dimensional vector space over F, is given by

(15.5.8) $\qquad \begin{cases} \xi(\alpha_1 \vec{v_1} + \alpha_2 \vec{v_2}, \vec{v}') = \alpha_1 \xi(\vec{v_1}, \vec{v}') + \alpha_2 \xi(\vec{v_2}, \vec{v}'); \\ \xi(\vec{v}, \alpha_1 \vec{v_1}' + \alpha_2 \vec{v_2}') = \alpha_1 \xi(\vec{v}, \vec{v_1}') + \alpha_2 \xi(\vec{v}, \vec{v_2}'), \; (\alpha_1, \alpha_2 \in F), \end{cases}$

where $\vec{v}, \vec{v_1}, \vec{v_2}$ are elements of V and $\vec{v}', \vec{v_1}', \vec{v_2}'$ are those of V'. A bilinear form on the finite dimensional vector spaces V, V' (over the same field F) is a bilinear function $\xi : V \times V' \to F$ satisfying relations of the type (15.5.8). Suppose that $\dim V = m$, $\dim V' = n$. Let $\mathcal{B}, \mathcal{B}'$ be bases for V, V' respectively. We define the matrix of the form $\xi : V \times V' \to F$ to be the $m \times n$ matrix $[a_{ij}]$ with entries from F (relative to bases $\mathcal{B}, \mathcal{B}'$) given by

(15.5.9) $\qquad a_{ij} = \xi(\vec{b}_i, \vec{b}_j'), \quad i = 1, 2, \ldots m; j = 1, 2, \ldots n;$

where $\mathcal{B} = \{\vec{b}_1, \vec{b}_2, \ldots, \vec{b}_m\}$, $\mathcal{B}' = \{\vec{b}_1', \vec{b}_2' \ldots, \vec{b}_m'\}$. In fact, for $v = \sum x_i \vec{b}_i$, $v' = \sum y_j \vec{b}_j'$;

$$\xi(\vec{v}, \vec{v}') = \sum_{i=1}^{m} \sum_{j=1}^{n} x_i \xi(\vec{b}_i, \vec{b}_j') y_j.$$

To each bilinear form ξ, there corresponds an $m \times n$ matrix $A = [a_{ij}]$ given in (15.5.9). It is an isomorphism of the vector space of bilinear forms $\xi : V \times V' \to F$ to the vector space $M_{m,n}(F)$ of $m \times n$ matrices with entries from F. If $A \in M_{m,n}(F)$ corresponds to ξ, the rank of A is referred to as the rank of the bilinear form ξ. Two $m \times n$ matrices represent the same bilinear form relative to two different bases in V and V' if, and only if, they are equivalent. We know that a matrix A of rank r is equivalent to one in canonical form with r entries 1_F on the main diagonal and all other entries zero.

Definition 15.5.6 : *A bilinear form $\xi : V \times V \to F$ (where V is a finite dimensional vector space over F) is said to be symmetric if $\xi(\vec{v}, \vec{v}') = \xi(\vec{v}', \vec{v})$ for all $\vec{v}, \vec{v}' \in V$.*

We note that a bilinear form $\xi : V \times V \to F$ is symmetric if, and only if, its matrix A (relative to any one basis $\mathcal{B}$ of V) is symmetric.

Observation 15.5.1 : *If $\xi : V \times V' \to F$ is a bilinear form of rank r, there exist vectors $\vec{v}, \vec{v}'$ in V and V' respectively with $\vec{v} = (x_1, x_2, \ldots x_m)$, $\vec{v}' = (y_1, y_2, \ldots, y_n)$ and*

$$\xi(\vec{v}, \vec{v}') = x_1 y_1 + x_2 y_2 + \cdots + x_r y_r.$$

Definition 15.5.7 : *Let V be a finite dimensional vector space over a field F (of characteristic $\neq 2$). A quadratic form $\theta : V \to F$ is such that $\theta(-\vec{v}) = \theta(\vec{v})$ for all $\vec{v} \in V$ and*

$$2\xi(\vec{u}, \vec{v}) = \theta(\vec{u} + \vec{v}) - \theta(\vec{u}) - \theta(\vec{v})$$

gives a bilinear form $\xi : V \times V \to F$. The rank of ξ is accepted as the rank of θ.

It can be verified that for $\vec{u}, \vec{v}, \vec{w} \in V$,

$$(15.5.10) \quad \theta(\vec{u}+\vec{v}+\vec{w}) - \theta(\vec{u}+\vec{v}) - \theta(\vec{v}+\vec{w}) - \theta(\vec{w}+\vec{u}) + \theta(\vec{u}) + \theta(\vec{v}) + \theta(\vec{w}) = 0.$$

Taking $\vec{u} = \vec{v} = \vec{w} = 0$ in (15.5.10), we obtain $\theta(0) = 0$.

Let $\vec{v} = \vec{u}$ and $\vec{w} = -\vec{u}$ in (15.5.10). We see that

$$(15.5.11) \quad \theta(\vec{u}) - \theta(2\vec{u}) - \theta(0) - \theta(0) + \theta(\vec{u}) + \theta(\vec{u}) + \theta(-\vec{u}) = 0.$$

As $\theta(-\vec{u}) = \theta(\vec{u})$, $\vec{u} \in V$; we get

$$(15.5.12) \quad \theta(2\vec{u}) = 4\theta(\vec{u}).$$

The result that is of relevance to polynomials over $\mathbb{F}_q$ is the following:

Theorem 147 : *Let V be a finite dimensional vector space over a field F of characteristic $\neq 2$. Then, each symmetric bilinear form $\xi : V \times V \to F$ defines a quadratic form $\theta : V \to F$ given by the equation.*

$$\theta(\vec{v}) = \xi(\vec{v}, \vec{v}), \quad \vec{v} \in V.$$

Further, θ is the only quadratic form satisfying

$$(15.5.13) \quad 2\xi(\vec{u}, \vec{v}) = \theta(\vec{u}+\vec{v}) - \theta(\vec{u}) - \theta(\vec{v}).$$

Proof : $:\Rightarrow$ As $\theta(\vec{v}) = \xi(\vec{v}, \vec{v})$ for all $\vec{v} \in V$,

$$\theta(-\vec{v}) = \xi(-\vec{v}, -\vec{v}) = (-1)^2 \xi(\vec{v}, \vec{v}) = \theta(\vec{v}).$$

Further, ξ is bilinear and symmetric. So,

$$\begin{aligned}
\theta(\vec{u}+\vec{v}) - \theta(\vec{u}) - \theta(\vec{v}) &= \xi(\vec{u}+\vec{v}, \vec{u}+\vec{v}) - \xi(\vec{u}, \vec{u}) - \xi(\vec{v}, \vec{v}) \\
&= \xi(\vec{u}, \vec{u}) + \xi(\vec{u}, \vec{v}) + \xi(\vec{v}, \vec{u}) + \xi(\vec{v}, \vec{v}) - \xi(\vec{u}, \vec{u}) - \xi(\vec{v}, \vec{v}) \\
&= \xi(\vec{u}, \vec{v}) + \xi(\vec{v}, \vec{u}) \\
&= 2\xi(\vec{u}, \vec{v}), \text{ by symmetry of } \xi.
\end{aligned}$$

$\Leftarrow$: Suppose that ξ and θ are such that (15.5.13) holds.

From (15.5.13), we note that $2\xi(\vec{u}, \vec{u}) = \theta(2\vec{u}) - 2\theta(\vec{u})$. But, by (15.5.12), $\theta(2\vec{u}) = 4\theta(\vec{u})$ and $\theta(0) = 0$, as θ is a quadratic form. Therefore, $2\xi(\vec{u}, \vec{u}) = 2\theta(\vec{u})$. Or, $\xi(\vec{v}, \vec{v}) = \theta(\vec{v})$, for all $\vec{v} \in V$. Thus, $\theta : V \to F$ is such that $\xi(\vec{v}, \vec{v}) = \theta(\vec{v})$. Or, the symmetric bilinear form ξ defines θ, as stated in theorem 147. $\qquad\square$

Remark 15.5.3 : Theorem 147 has been adapted from MacLane and Birkhoff [29, chapter IX, §3, theorem 2, pp 382–384].

Definition 15.5.8 : *Let q be odd. If t is a primitive element of $\mathbb{F}_{q^r}$ with $P_t(x)$, the irreducible polynomial of t (over $\mathbb{F}_q$), the second coefficient of $P_t(x)$ is denoted by $A(t)$.*

We note that $\deg P_t(x) = r$ and

$$(15.5.14) \quad P_t(x) = (x-t)(x-t^q) \cdots (x-t^{q^{r-1}}).$$

The second coefficient in $P_t(x)$ is given by

(15.5.15) $$A(t) = t^{1+q} + t^{1+q^2} + \cdots + t^{q^{r-2}+q^{r-1}} = \sum_{\substack{0 \le i,j < r \\ i < j}} t^{q^i + q^j}.$$

Theorem 148 : *Suppose that* $t, s \in \mathbb{F}_{q^r}$ *and trace* $(t) = 0$. *Then,*

(15.5.16) $$A(t+s) = A(t) + A(s) - \text{trace } (ts).$$

Proof :

$$\text{trace } (t) = T_{E/F}(t) = \sum_{\sigma \in G(E/F)} \sigma(t),$$

where $E = \mathbb{F}_{q^r}$, $F = \mathbb{F}_q$ and $G(E/F)$ is the group of automorphisms of E leaving F fixed.

$$A(t+s) = \sum_{\substack{0 \le i,j < r \\ i < j}} (t+s)^{q^i + q^j}$$

$$= (t+s)(t+s)^q + (t+s)(t+s)^{q^2} + \cdots + (t+s)^{q^{r-2}}(t+s)^{q^{r-1}};$$

$$= (t+s)(t^q + s^q) + (t+s)(t^{q^2} + s^{q^2}) + \cdots + (t^{q^{r-2}} + s^{q^{r-2}})(t^{q^{r-1}} + s^{q^{r-1}})$$

$$= (t^{1+q} + ts^q + t^q s + s^{1+q}) + (t^{1+q^2} + ts^{q^2} + t^{q^2}s + s^{1+q^2}) + \cdots$$

$$\qquad + (t^{q^{r-2}}s^{q^{r-1}} + t^{q^{r-1}}s^{q^{r-2}} + t^{q^{r-1}}s^{q^{r-2}} + s^{q^{r-2}+q^{r-1}});$$

$$= A(t) + A(s) + s(t^q + t^{q^2} + \cdots + t^{q^{r-1}}) + s^q(t + t^{q^2} + \cdots + t^{q^{r-1}}) + \cdots$$

$$\qquad + s^{q^{r-1}}(1 + t^q + \cdots + t^{q^{r-2}});$$

$$= A(t) + A(s) - ts - t^q s^q - \cdots - t^{q^{r-1}}s^{q^{r-1}}, \text{ since trace } (t) = 0.$$

Or,

$$A(t+s) = A(t) + A(s) - \text{trace } (ts).$$

□

Theorem 149 : *If* $\alpha \in \mathbb{F}_q$, *then,* $A(\alpha t) = \alpha^2 A(t)$.

Proof : $A(\alpha t) = (\alpha t)^{1+q} + (\alpha t)^{1+q^2} + \cdots + (\alpha t)^{q^{r-2}+q^{r-1}}$. We know that

$$\alpha^{q-1} = 1, \text{ if } \alpha \ne 0.$$

So, $\alpha^{1+q} = \alpha^2 \cdot \alpha^{q-1} = \alpha^2$. Also, $\alpha^{q^i+q^j} = \alpha^{q^j(1+q^{i-j})}$, if $j \le i$. Further, $\alpha^{1+q^{i-j}} = \alpha^2 \cdot \alpha^{q^{i-j}-1} = \alpha^2$, if $i - j \ge 1$. So,

$$A(\alpha t) = \alpha^2(t^{1+q} + t^{1+q^2} + \cdots + t^{q^{r-2}+q^{r-1}}) = \alpha^2 A(t).$$

□

Corollary 15.5.1 : *If* $t \in \mathbb{F}_{q^r}$ *has trace*$(t) = 0$, *then,*

(15.5.17) $A(t) = -\dfrac{1}{2}$ *trace* (t^2).

For, $4A(t) = A(2t) = A(t+t) = A(t)+A(t) -$ *trace* (t^2), *by* (15.5.16). (15.5.17) *is immediate.*

Next, we observe that (15.5.17) allows us to tackle the second coefficient of an irreducible polynomial via trace(t^2). We have noticed that $\Psi : \mathbb{F}_{q^r} \to \mathbb{F}_q$ given by $\Psi(t) =$ trace (t), is onto $\mathbb{F}_q$. So, if V denotes kernel of Ψ, by rank-nullity theorem.

$$\dim(Im\Psi) + \dim(V) = \dim \mathbb{F}_{q^r} = r$$

So, V is an $(r-1)$-dimensional vector space over $\mathbb{F}_q$.

Definition 15.5.9 : *Let* V *be a vector space over a field* F. *By an inner product* $\vec{u} * \vec{v}$ *of vectors* $\vec{u}$, $\vec{v} \in V$, *we mean an* F-*bilinear function* $\eta : V \times V \to F$. *The inner product is said to be symmetric if* $\vec{u} * \vec{v} = \vec{v} * \vec{u}$.

In the case of $\mathbb{R}^n$, considered as a vector space over $\mathbb{R}$, the inner product $\vec{u} \cdot \vec{v}$ of two vectors $\vec{u} = (u_1, u_2, \dots u_n)$ and $\vec{v} = (v_1, v_2, \dots v_n)$ given by

(15.5.18) $\vec{u} * \vec{v} = u_1 v_1 + u_2 v_2 \cdots + u_n v_n$

is $\mathbb{R}$-linear in $\vec{u}$ for each $\vec{v} \in \mathbb{R}^n$ and is $\mathbb{R}$-linear in $\vec{v}$ for each $\vec{u} \in \mathbb{R}^n$. $\vec{u} * \vec{u}$ is an $\mathbb{R}$-bilinear function from $\mathbb{R}^n \times \mathbb{R}^n$ to $\mathbb{R}$. Further, as $\vec{u} * \vec{v} = \vec{v} * \vec{u}$, the inner product is symmetric.

Definition 15.5.10 : *Let* V *be a vector space (over a field* F) *endowed with a symmetric inner product.* V *is said to be nonsingular if there does not exist a vector* $\vec{u_0} \in V$ *such that* $\vec{u_0} * \vec{v} = 0$ *for all* $\vec{v} \in V$.

We have noted that $\Psi : \mathbb{F}_{q^r} \to \mathbb{F}_q$ given by $\Psi(t) =$ trace t $(t \in \mathbb{F}_{q^r})$ is such that $V = \ker \Psi$ is an $(r-1)$-dimensional vector space. Now, for $t, s, \in V$, the product $t * s =$ trace (ts) serves as an inner-product on V. When $t * s =$ trace (ts), the relation (15.5.16) shows that $A(t) = -\frac{1}{2}$ trace t^2 is the quadratic form associated with the inner product. (See (15.5.16) and (15.5.17)). Further, if $0 \neq t \in V$ is such that trace$(ts) = 0$ for all $s \in V$, then $tV \subseteq V$, by the definition of V. Since $\#\{tV : t$ fixed $\} = \#\{V : V = \ker \Psi\}$ and the number of elements in each set is finite, we have $tV = V$. So we must have $1 \in V$, which is not true as $\dim V < \dim \mathbb{F}_{q^r}$. So, V is nonsingular with respect to the inner product $t * s =$ trace (ts). Therefore, if $t \in V$ and t is a primitive element of $\mathbb{F}_{q^r}$, then, the minimal polynomial $P_t(x)$ of t is of degree r, the trace coefficient of $P_i(x)$ is zero and the second coefficient $A(t) = -\frac{1}{2}$ trace $(t^2) = -\frac{1}{2}(t * t)$.

Let $r \geq 2$. $\mathbb{F}_{q^r}$ is obtained as a finite extension of degree r over $\mathbb{F}_q$. For $a \in \mathbb{F}_{q^r}$, there exists a monic irreducible polynomial $f_a(x)$ of degree r with co-efficients from $\mathbb{F}_q$. Further, $\mathbb{F}_q$ is realized as a simple extension of $\mathbb{F}_q$ obtained by adjourning a to $\mathbb{F}_q$. The degree of $irr(a, \mathbb{F}_q)$ is r and $irr(a, \mathbb{F}_q) = f_a(x)$. The conjugates of a are the zeros of $f_a(x)$. $\mathbb{F}_{q^r}$ can be obtained by adjoining any of the r conjugates of a to $\mathbb{F}_q$. So corresponding to a monic irreducible polynomial of degree r (with coefficients from $\mathbb{F}_q$), there exist r primitive elements in $\mathbb{F}_{q^r}$. For instance, to a cubic irreducible polynomial $g_a(x) \in \mathbb{F}_q(x)$, there corresponds 3 primitive elements in $\mathbb{F}_{q^3}$. Elements of $\mathbb{F}_{q^r}$ which do not generate $\mathbb{F}_{q^r}$ (as a vector space of dimension r over $\mathbb{F}_q$) are non-primitive elements in $\mathbb{F}_{q^r}$. For example, if $f_a(x) = x^3 + a_1 x + a_2$ $(a_1, a_2 \in \mathbb{F}_q)$ and if $f_a(x)$ is irreducible and having a as a zero, a is not a primitive element of $\mathbb{F}_{q^4}$. In fact, there are q^2 such irreducible polynomials. So, there corresponds q^2 non-primitive elements of $\mathbb{F}_{q^4}$ all having trace = 0.

Observation 15.5.2 : *Since there are q^2 non-primitive elements of $\mathbb{F}_{q^4}$ all having trace = 0, there exist $\frac{q^3-q^2}{4}$ irreducible polynomials of degree 4 over $\mathbb{F}_q$ with trace = 0 and $\frac{q^3}{4}$ irreducible polynomials with trace = a_1 for each $a_1 \neq 0$.*

Next, let r denote the degree of a monic irreducible polynomial $P_t(x) \in \mathbb{F}_q(x)$. By theorem 145, if it is a primitive element of $\mathbb{F}_{q^r}$, and if r is odd, it suffices to study $P_t(x)$ with trace coefficient equal to zero. In what follows, we take $q = 2^m (m \geq 1)$. That is, char $\mathbb{F}_q = 2$. Quadratic and cubic polynomials have been considered in theorem 146. In order to tackle the cases $r = 4$ and $r = 5$, we need to know more about cubic irreducible polynomials.

Theorem 150 (G. W. Effinger) : *Among the irreducible cubic polynomials with trace coefficient = 0,*

(a) there are $\frac{2(q-1)}{3}$ cubic polynomials with second coefficient zero and $\frac{q-1}{3}$ with each nonzero second coefficient, provided $q = 2^m$ with m even.

(b) there are none with second coefficient zero and $\frac{q+1}{3}$ with each nonzero second coefficient, provided $q = 2^m$ with m odd.

Proof : The linear functional $\Psi : \mathbb{F}_{q^3} \to \mathbb{F}_q$ given by $\Psi(t) =$ trace (t), $t \in \mathbb{F}_{q^3}$ is such that $V_3 = \ker \Psi$ is a vector space of dimension 2 over $\mathbb{F}_q$. V_3 is non-singular with respect to the inner product $t * s =$ trace (ts) where $t, s \in \mathbb{F}_{q^3}$. Let $\{t_1, t_2\}$ be a basis for V_3. Then, $t \in V_3$ has the representation.

(15.5.19) $\qquad t = \alpha_1 t_1 + \alpha_2 t_2 ; \quad \alpha_i \in \mathbb{F}_q \ (i = 1, 2)$

It is known [17] that if $P_t(x)$ denotes the minimal polynomial of t over $\mathbb{F}_q$, the second coefficient $A(t)$ in $P_t(x)$ is given by

(15.5.20) $\qquad A(t) = \beta \alpha_1^2 + \alpha_1 \alpha_2 + \beta \alpha_2^2$

where either $\beta = 0$ or $\beta x^2 + x + \beta \in \mathbb{F}_q[x]$ is irreducible.

In (15.5.20), α_1, α_2 vary over the elements of $\mathbb{F}_q$. We shall count the number of solutions $\{\alpha_1, \alpha_2\}$ of (15.5.20) for fixed $A(t)$. All nonzero elements t in V_3 are primitive elements in $\mathbb{F}_{q^3}$, since g.c.d $(3, q) = 1$, by theorem 145. Two cases arise:

Case(i) $\beta = 0$. Then, $A(t) = \alpha_1 \alpha_2$. If $A(t) = 0$, either $\alpha_1 = 0$ or $\alpha_2 = 0$. If $\alpha_2 = 0$, there are $(q-1)$ choices for α_1. As $\alpha_1 = \alpha_2 = 0$ is excluded, there are $2(q-1)$ nonzero solutions $\{\alpha_1, \alpha_2\}$ for (15.5.20). If $A(t) = \alpha_0 \neq 0$, there are $(q-1)$ solutions $\{\alpha_1, \alpha_2\}$.

Case(ii) $\beta \neq 0$. Then, $\beta x^2 + x + \beta$ is irreducible over $\mathbb{F}_q$. So, $A(t) = 0$ has no nonzero solutions. If $A(t) = \alpha_0 \neq 0$, we examine $\beta \alpha_1^2 + \alpha_1 \alpha_2 + \beta \alpha_2^2 = \alpha_0$. If $\alpha_2 = 0$, there is one solution obtained from $\beta \alpha_1^2 = \alpha_0$. If $\alpha_2 \neq 0$, we get

$$\beta(\frac{\alpha_1}{\alpha_2})^2 + (\frac{\alpha_1}{\alpha_2}) + \beta = \frac{\alpha_0}{\alpha_2^2},$$

or,

(15.5.21) $(\frac{\alpha_1}{\alpha_2})^2 + \frac{1}{\beta}(\frac{\alpha_1}{\alpha_2}) + (1 + \frac{\alpha_0}{\beta \alpha_2^2}) = 0$, as char $\mathbb{F}_q = 2$.

We count the number of irreducible quadratic polynomials with trace coefficient $\neq 0$. There are $(q-1)$ nonzero elements of $\mathbb{F}_q$ which can serve as a trace coefficient. In (15.5.21), we get different equations for different elements $\alpha_2 \neq 0$. The number of irreducible quadratic polynomials with trace coefficient $\neq 0$ is $\frac{q^3 - q^2}{2}$, as each nonzero element of V_3 is a primitive element of $\mathbb{F}_{q^3}$. For the total of $(q-1)$ nonzero traces, there are $(\frac{q}{2} - 1)$ irreducible quadratic polynomials and $\frac{q}{2}$ reducible quadratic polynomials. (There are $(q-1)$ choices for α_2 in $1 + \frac{\alpha_0}{\beta \alpha_2^2}$). Since a reducible quadratic polynomial has two solutions, we obtain a total of $2(\frac{q}{2}) = q$ solutions. Thus, there are a total of $(q+1)$ solutions for a given $A(t)$.

Now, each irreducible cubic polynomial corresponds to 3 primitive elements of $\mathbb{F}_{q^3}$. If $q = 2^m$ (m odd), $q + 1 = 2^m + 1$ is divisible by 3. If m is even, $2^m - 1$ is divisible by 3.

Therefore, there are $2(\frac{q-1}{3})$ cubic polynomials with second coefficient $A(t)$ when $q = 2^m$, m even. When $q = 2^m$, m odd or even, $A(t) = 0$ shows that there are no cubic polynomials (case(ii) above with $\beta \neq 0$). If $q = 2^m$ with m odd, there are $\frac{q+1}{3}$ cubic polynomials with second coefficient $A(t) \neq 0$.   □

Corollary 15.5.2 : *Among the irreducible cubic polynomials in $\mathbb{F}_q[x]$ (where $q = 2^m$) with trace coefficient a_1, there exist q distinct second coefficients, if m is even and $(q-1)$ distinct second coefficients, if m is odd.*

For, by theorem 145 (translation lemma), we need only to look into irreducible cubic polynomials with trace coefficient zero. As indicated in the proof of theorem 150, all elements of $\mathbb{F}_q$ could be a second coefficient, if m is even. However, only the $(q-1)$ nonzero elements of $\mathbb{F}_q$ could occur, if m is odd.

Now, we move on to the case $r = 4$.

Theorem 151 (G. W. Effinger (1988)) : *Every odd fourth degree polynomial over $\mathbb{F}_q$ (q even) is a 3-primes polynomial.*

Proof : By observation 15.5.2, we note that there are q^2 non-primitive elements in $\mathbb{F}_{q^4}$. If $q = 2^m$, $\mathbb{F}_{q^4}$ is a finite extension of $\mathbb{F}_q$ such that $[\mathbb{F}_{q^4} : \mathbb{F}_q] = 4$. For $\alpha \in \mathbb{F}_{q^4}$, $\alpha^{2^{4m}} = \alpha$. A non-primitive element β in $\mathbb{F}_{q^4}$ is of the form $\beta = \alpha^{2^{2m}}$ for some $\alpha \in \mathbb{F}_{q^4}$. The Galois group $G(\mathbb{F}_{q^4}, \mathbb{F}_q)$ of automorphisms of $\mathbb{F}_{q^4}$, leaving every element of $\mathbb{F}_q$ fixed, is cyclic of order 4. Therefore, the non-primitive elements of $\mathbb{F}_{q^4}$ are of the form $\alpha^{q^2}, \alpha^{2q^2}, \ldots, \alpha^{q \cdot q^2}, \alpha^{q^3 + q^2}, \alpha^{q^3 + 2q^2}, \ldots, \alpha^{q^4 - q^2}$. They are $(q^2 - 1)$

in number. Including 1 in the set of non-primitive elements, we see that as $G(\mathbb{F}_{q2}, \mathbb{F}_q)$ is cyclic of order 2, the trace of β (where β is non-primitive) is equal to $\sigma_1(\beta) + \sigma_2(\beta)$ where $\sigma_1 : \mathbb{F}_{q2} \to \mathbb{F}_{q2}$ is the identity map and $\sigma_2 : \mathbb{F}_{q2} \to \mathbb{F}_{q2}$ is given by $\sigma_2(\beta) = \beta^{2^{2m}}$. We get $\sigma_1(\beta) + \sigma_2(\beta) = \alpha + \alpha = 0$, as $\beta = \alpha^{2^{2m}}$ for some $\alpha \in \mathbb{F}_{q4}$. Thus, trace of a non-primitive element is zero.

Therefore, there must be $\frac{q^3 - q^2}{4}$ irreducible polynomials of degree 4 over $\mathbb{F}_q$ with trace = 0 and $\frac{q^3}{4}$ with trace = a_1 for each a_1 ($\neq 0$) $\in \mathbb{F}_q$. Two cases arise:

Case 1: Suppose that $q = 2^m$, where m is even. Let

$$M(x) = x^4 + a_1 x^3 + a x^2 + b x + c.$$

We select $P_1(x) = x^4 + (a_1 + 1)x^3 + dx^2 + ex + f$. Using corollary 15.5.2, we select $P_2(x) = x^3 + (a + d)x^2 + (b + e + 1)x + g$. Finally, let $P_3(x) = x + (e + f + g)$. We get $M(x) = P_1(x) + P_2(x) + P_3(x)$.

Case 2: Suppose that $q = 2^m$ where m is odd. Suppose that among the fourth degree irreducible polynomials with trace $a_1 + 1$, every second coefficient which appears is matched with a single third coefficient. Then, there can be at the most $q(q - 1)$ irreducible polynomials with trace coefficient $a_1 + 1$. But, we know that there are at least $\frac{q^3 - q^2}{4}$ such irreducible polynomials. So, $\frac{q^3 - q^2}{4} \leq q(q - 1)$. That is, $q \leq 4$. So, we must have $q = 2$ (as m is odd).

If we suppose that $q > 2$, we know that there exists some second coefficient a such that

$$P_{1,1}(x) = x^4 + (a + 1)x^3 + a'x^2 + b_1'x + c'$$

and

$$P_{1,2}(x) = x^4 + (a + 1)x^3 + a'x^2 + b_2'x + e'$$

are both irreducible polynomials and $b_1' \neq b_2'$. By corollary 15.5.2, we now select

$$P_2(x) = x^3 + a'x^2 + (b_i' + 1)x + f' \text{ where } b_i' = b_1' \text{ or } b_i' = b_2^1.$$

Further, if $P_3(x) = x + (c' + e' + f')$, we obtain

$$M(x) = P_{1,i}(x) + P_2(x) + P_3(x), \text{ as desired, } (i = 1 \text{ or } 2).$$

Case 3: For the remaining case $q = 2$, we observe that there are even polynomials which are not 3-primes polynomials. For instance, $x^4 + x^2$ is one such. Suppose that $M(x) \in \mathbb{F}_2(x)$ is an odd monic polynomial. Then the constant term in $M(x)$ is 1. We may choose $P_1(x)$ as an irreducible 4th degree polynomial so that $M(x) - P_1(x) = x^3 + tx^2 + sx$. We establish that four polynomials of this type can be written as a sum of two irreducible polynomials. But,

$$x^3 = (x^3 + x + 1) + (x + 1),$$

$$x^3 + x = (x^3 + x^2 + 1) + (x^2 + x + 1),$$

$$x^3 + x^2 = (x^3 + x + 1) + (x^2 + x + 1),$$

$$\text{and } x^3 + x^2 + x = (x^3 + x^2 + 1) + (x + 1).$$

As we work with the field $\mathbb{F}_2$, in $x^3 + tx^2 + sx$, one has either $t = 0$, $s = 0$ or $t = 0$, $s = 1$ or $t = 1$, $s = 0$ or $t = 1$, $s = 1$. So, an odd 4th degree polynomial $M(x)$ over $\mathbb{F}_2$ is a sum of three irreducible polynomials.

This proves theorem 151.   □

Next, we move on to the case $r = 5$. The argument is similar to that of case 3. We obtain an analogue of theorem 150.

Theorem 152 (G. W. Effinger (1988)) : *Among the irreducible fifth degree polynomials with trace coefficient equal to zero, (over $\mathbb{F}_q$, $q = 2^m$)*;

(a) *there are $(q-1)(q+1)^2/5$ with second coefficient = 0 and $(q-1)q(q+1)/5$ with each nonzero second coefficient, provided m is even,*

(b) *there are $(q-1)(q^2+1)/5$ with second coefficient = 0 and $q(q^2+1)/5$ with each nonzero second coefficient, provided m is odd.*

Proof : We work with $\mathbb{F}_{q^5}$ and the linear functional $\Psi : \mathbb{F}_{q^5} \to \mathbb{F}_q$ given by $\Psi(t) =$ trace (t). Let V_5 denote ker Ψ. Then, dim $V_5 = 4$ over $\mathbb{F}_q$.

It is known [17, p 34] that there is a basis for V_5 say $\{t_1, t_2, t_3, t_4\}$ such that $t \in V_5$ has the representation $t = \alpha_1 t_1 + \alpha_2 t_2 + \alpha_3 t_3 + \alpha_4 t_4$, where

$$(15.5.22) \qquad A(t) = \alpha_1\alpha_2 + \beta\alpha_3^2 + \alpha_3\alpha_4 + \beta\alpha_4^2$$

with either $\beta = 0$ or $\beta x^2 + x + \beta \in \mathbb{F}_q[x]$ is irreducible. ($A(t)$ being the second coefficient in the irreducible polynomial of t over $\mathbb{F}_q$). We have to count the number of solutions in α_i ($i = 1, 2, 3, 4$) of (15.5.22) for fixed values of $A(t)$. Let $t \neq 0$. $t \in V_5$ is primitive in $\mathbb{F}_{q^5}$ (5 being odd). The following cases arise:

Case 1: $\beta = 0$.

We have $A(t) = \alpha_1\alpha_2 + \alpha_3\alpha_4$. First, suppose that $\alpha_1\alpha_2 + \alpha_3\alpha_4 = 0$. We compute the number of nonzero solutions by breaking the total number into summands by the number of zeros among the $\alpha_i's$.

(a) 3 out of four $\alpha_i's$ equal to zero : $\binom{4}{3}(q-1) = 4(q-1)$,

(b) 2 out of four $\alpha_i's$ equal to zero : $\binom{2}{1}\binom{2}{1}(q-1)^2 = 4(q-1)^2$,

(c) 1 out of four $\alpha_i's$ equal to zero : no solutions,

(d) all $\alpha_i's$ nonzero : $(q-1)(q-1)(q-1) = (q-1)^3$.

So, the total number of solutions equals $4(q-1) + 4(q-1)^2 + (q-1)^3 = (q-1)(q+1)^2$.

Next, suppose that $\alpha_1\alpha_2 + \alpha_3\alpha_4 = \alpha_0$ ($\neq 0$).

The possibilities are

(a) 3 out of four $\alpha_i's$ zero : no solution,

(b) 2 out of four $\alpha_i's$ zero : $2(q-1)$ solutions,

(c) 1 out of four $\alpha_i's$ zero : $4(q-1)^2$ solutions,

(d) all $\alpha_i's$ nonzero : $(q-1)^2(q-2)$ solutions.

So, the total number of solutions equals $2(q-1) + 4(q-1)^2 + (q-1)^2(q-2)$ which is $(q-1)q(q+1)$.

To verify the above results, we have only to note that the number of nonzero elements with trace zero, occurring in $\mathbb{F}_{q^5}$ is the number of ways of choosing b_i

$(i = 1, 2, 3, 4) \in \mathbb{F}_q$ in the quintic: $x^5 + b_1 x^3 + b_2 x^2 + b_3 x + b_4$ and that is equal to $q^4 - 1$.

For, we check that # zero solutions $+(q-1)$ (# nonzero solutions)

$$= (q-1)(q+1)^2 + (q-1)(q-1)q(q+1)$$
$$= (q-1)(q+1)\{(q+1) + (q-1)q\} = (q^2-1)(q^2+1)$$
$$= (q^4 - 1), \text{ as required .}$$

Case 2: $\beta \neq 0$, $\beta x^2 + x + \beta$ is irreducible over $\mathbb{F}_q$.

We have $A(t)$ as given in (15.5.22). If $A(t) = 0$, one gets

$$\alpha_1 \alpha_2 + \beta \alpha_3^2 + \alpha_3 \alpha_4 + \beta \alpha_4^2 = 0$$

(a) $\alpha_3 = 0 = \alpha_4 \Rightarrow \alpha_1 \alpha_2 = 0$. There are $2(q-1)$ solutions.

(b) At least one of $\alpha_3, \alpha_4 \neq 0$. Assume that $\alpha_3 \neq 0$, $\alpha_4 = 0$. Then, $\beta \alpha_3^2 + \alpha_1 \alpha_2 = 0$. There are $(q-1)$ choices for α_1 and the same number for α_2. So, one part of solution gives $(q-1)^2$. Similarly, $\beta \alpha_4^2 + \alpha_1 \alpha_2 = 0$ has $(q-1)^2$ solutions. When $\alpha_3 \neq 0$, $\alpha_4 \neq 0$, from $\beta \alpha_3^2 + \alpha_1 \alpha_2 + \alpha_3 \alpha_4 + \beta \alpha_4^2 = 0$, one gets $(q-1)^3$ solutions for $\alpha_1, \alpha_2, \alpha_3$. (Once $\alpha_1, \alpha_2, \alpha_3$ are known, α_4 is fixed by the equation $\beta \alpha_3^2 + \alpha_1 \alpha_2 + \alpha_3 \alpha_4 + \beta \alpha_4^2 = 0$). So, the total number of solutions is

$$2(q-1)^2 + (q-1)^3 = (q-1)^2\{2 + q - 1\} = (q-1)^2(q+1) = (q^2-1)(q-1).$$

So, altogether, total number of solutions is

$$2(q-1) + (q^2-1)(q-1) = (q-1)(q^2+1).$$

Next, suppose that $\alpha_1 \alpha_2 + \beta \alpha_3^2 + \alpha_3 \alpha_4 + \beta \alpha_4^2 = \alpha_0 \neq 0$. As in the proof of theorem 150, we break up the total:

$$\beta \alpha_3^2 + \alpha_3 \alpha_4 + \beta \alpha_4^2 = \alpha_0 : (q+1)\big(2(q-1)+1\big) = (q+1)(2q-1)$$
$$\beta \alpha_3^2 + \alpha_3 \alpha_4 + \beta \alpha_4^2 \neq \alpha_0, \neq 0 : (q+1)(q-2)(q-1)$$
$$\beta \alpha_3^2 + \alpha_3 \alpha_4 + \beta \alpha_4^2 = 0 : 1(q-1).$$

The total number $= (q+1)(2q-1) + (q+1)(q-2)(q-1) + (q-1)$

$$= (q+1)\{q^2 - q + 1\} + (q-1)$$
$$= q(q^2 + 1).$$

This is verified by checking that (# zero solutions) $+(q-1)$ (# nonzero solutions)

$$= (q-1)(q^2+1) + (q-1)q(q^2+1)$$
$$= (q-1)(q^2+1)(1+q)$$
$$= (q^2-1)(q^2+1)$$
$$= (q^4 - 1), \text{ as required.}$$

Now, there are five primitive elements of $\mathbb{F}_{q^5}$, corresponding to each irreducible fifth degree polynomial over $\mathbb{F}_q$. If $q = 2^m$, where m is even, 5 divides $(q^2 - 1)$. If

$q = 2^m$, where m is odd, 5 divides $q^2 + 1$.
So, the assertion of the theorem 152 is true. $\qquad\qquad\qquad\qquad\qquad$ $\square$

Corollary 15.5.3 : *Let $q = 2^m$ where m is odd. Then given $a_1, a \in \mathbb{F}_q$ there exist at least $(q-1)(q^2+1)/5$ irreducible fifth degree polynomials with trace coefficient a_1 and second coefficient a.*

For, by theorem 145 (translation lemma) we have only to look for irreducible fifth degree polynomials with trace coefficient 0 and zero or nonzero second coefficient. As $\frac{(q-1)(q^2+1)}{5} < \frac{q(q^2+1)}{5}$, the number of such irreducible polynomials is either $\frac{(q-1)(q^2+1)}{5}$ or more.

Theorem 153 (G. W. Effinger (1988)) **:** *Every odd monic fifth degree polynomial over any finite field of characteristic 2 is a 3-primes polynomial.*

Proof : Let $q = 2^m$. Take $M(x) = x^5 + a_1 x^4 + a x^3 + b x^2 + c x + d$.

Case 1: Let m be even. By theorems 145 and 152, there exists an irreducible polynomial $P_1(x) = x^5 + a_1 x^4 + (a+1)x^3 + b'x^2 + c'x + d'$. By corollary 15.5.2, there exists an irreducible polynomial

$$P_2(x) = x^3 + (b+b')x^2 + (c+c'+1)x + d''.$$

Taking $P_3(x) = x + (d + d' + d'')$, we obtain

$$M(x) = P_1(x) + P_2(x) + P_3(x), \text{ as desired.}$$

Case 2: Let m be odd. According to corollary 15.5.2, one second coefficient is 'missing' among cubic irreducible polynomials with any fixed trace. We claim that if $q > 2$, among fifth degree irreducible polynomials of fixed trace and second coefficient, there exists a third coefficient for which there must exist at least two distinct fourth (that is, the term containing x) coefficients. Were this not the case, there could exist at most $q(q-1)$ irreducible polynomials with fixed first two coefficients. But, by corollary 15.5.3, there are at least $(q-1)(q^2+1)/5$ such polynomials. Therefore,

$$\frac{(q-1)(q^2+1)}{5} \le q(q-1).$$

That is, $q^2 - 5q + 1 \le 0$, from which we infer that $q = 2$, since m is odd. So, we first assume that $q > 2$. Then, by the above argument, there exists $b' \in \mathbb{F}_q$ such that

$$P_{1,1}(x) = x^5 + a_1 x^4 + (a+1)x^3 + b'x^2 + c'x + d_1'$$

and

$$P_{1,2}(x) = x^5 + a_1 x^4 + (a+1)x^3 + b'x^2 + c_2'x + d_2'$$

are both irreducible polynomials with $c_1' \ne c_2'$. By the corollary 15.5.2, there exists an irreducible polynomial

$$P_2(x) = x^3 + (b+b')x^2 + (c+c_k'+1)x + d'',$$

where $c_k' = c_1'$ or $c_k' = c_2'$. We write

$$P_3(x) = x + (d + d_k' + d'') \quad ; d_k' = d_1' \text{ or } d_2'.$$

Then, $M(x) = P_{1,k}(x) - P_2(x) + P_3(x)$, where $k = 1$ or 2.

The remaining case is $q = 2$. Here, there are even polynomials which are not 3-primes polynomials. For example,

$$x^5 + x^4 + 1 = (x^3 + x + 1)(x^2 + x + 1)$$

and so $x^5 + x^4 + 1$ is reducible. By lemma 15.5.1(b),

$$M(x) = x^5 + x^4 = x^4(x + 1)$$

is not a 3-prime polynomial. Suppose, then, that $M(x)$ is odd and so the constant term in $M(x)$ is 1. By theorem 145 and theorem 152, we observe that there exists an irreducible polynomial $P_1(x)$ of degree 5 so that $M - P_1(x) = x^3 + a'x^2 + b'x$.

As a', b' take values 0 or 1 only, as in the last step (case 3) of proof of theorem 151, we see that the four cubic polynomials $x^3, x^3 + x, x^3 + x^2$ and $x^3 + x^2 + x$ are sums of two irreducible polynomials.

Thus, a fifth degree polynomial over $\mathbb{F}_q$ (q even) is a 3-primes polynomial. $\square$

For solving the 3-primes problem for polynomials of degree 6 (over $\mathbb{F}_q$) one needs to study the 3rd coefficient as well. This is systematically developed by G. W. Effinger in [19] for q even.

Fact 15.5.2 : Among irreducible fourth degree polynomials over $\mathbb{F}_q$ (q even) which have fixed nonzero trace coefficient a_1 and fixed arbitrary second coefficient a_0, there are least $(q - 1)$ distinct third coefficients.

For proof see [18].

Fact 15.5.3 : If $q > 8$ in $\mathbb{F}_q$ (q even), then for every fixed first and second coefficient, there exists a fourth coefficient such that there are irreducible sixth degree polynomials with that fourth coefficient and with at least two distinct third coefficients and for each of those third coefficients, there exist at least two distinct fifth coefficients.

For proof, see [19].

Fact 15.5.4 : Every odd monic sixth degree polynomial over $\mathbb{F}_q$ (a finite field of characteristic 2) is a 3-primes polynomial.

For proof, see [19].

Combining the above with theorems 146(b), 151 and 153, we obtain the assertion in Proposition 15.5.2(b) given at the beginning of this section.

Incidentally, we remark that analytical number theory makes use of a good dose of complex analysis. For a study of number theory involving complex analysis, see Anatolij A. Karatsuba [A1].

15.6. Notes with illustrative examples

The experimental method of verifying the Goldbach conjecture is to find two sets P, Q of primes such that

$$P + Q = \{p + q : p, q \text{ primes belonging to } P \text{ and } Q \text{ respectively}\}$$

covers all even numbers in a given interval $[a, b]$. It is computationally verified that GBC is true for all even numbers up to $4 \cdot 10^{11}$ [34]. All the work of an algorithmic nature does give evidence of the truth of BGC. So far, no even number which is not a sum of two odd primes has been found out. If the given even number is not big enough, the easiest way to check BGC is with a table of primes.

Eckford Cohen's analogue of BGC is via the (finite) residue class ring $\mathbb{Z}/r\mathbb{Z}$ where r is even > 2. Theorem 140 says that when r is even, every element of $\mathbb{Z}/r\mathbb{Z}$ considered as belonging to a least non-negative complete residue system (mod r) is either a prime or a sum of two primes. For instance, in the case $r = 24$,

(15.6.1) $\mathbb{Z}/24\mathbb{Z} = \{0, 1, 2, 3, \dots, 23\}$

The primes in $\mathbb{Z}/24\mathbb{Z}$ are in the set $\{2, 10, 14, 18, 22; 3, 5, 21\}$ as a prime is either of the form $2k$ or $3k'$ where g.c.d $(k, 24) =$ g.c.d $(k', 24) = 1$. It is easy to check that

$0 = 3 + 21$ or $14 + 10$	$11 = 14 + 21$
$1 = 3 + 22$ or $15 + 10$	$12 = 2 + 10$
$4 = 2 + 2$	$13 = 3 + 10$
$5 = 2 + 3$	$16 = 2 + 14$
$6 = 3 + 3$	$17 = 2 + 15$
$7 = 10 + 21$	$19 = 21 + 22$
$8 = 10 + 22$	$20 = 10 + 10$
$9 = 15 + 18$	$23 = 2 + 21$

We observe that an extension of the Goldbach problem to the ring of integers of an algebraic number field is possible. See Eckford Cohen [13]. Let K be an algebraic number field with D_K, the ring of integers of K. Let A denote a proper ideal of D_K. It is known (see Remark 12.6.6(ii), of chapter 12) that the quotient ring D_K/A is finite. Let $R(A)$ denote the quotient ring of A in D_K. The 'primes' of $R(A)$ are suitably defined.

Proposition 15.6.1 : *There exists an $s \geq 1$ such that all elements of $R(A)$ are expressible as a sum of s primes in $R(A)$ under suitably chosen conditions.*

For details, see Eckford Cohen [13].

In the case of the polynomial analogue of BGC, the starting point is Hayes' asymptotic theorem (analogous to Vinogradov's theorem) which says:

For every degree $r \geq 5$, there exists a q_r, depending on r and decreasing, as r increases so that if $q \geq q_r$, then every odd monic polynomial of degree r over $\mathbb{F}_q (q = p^m, p$ a prime $, m \geq 1)$ is a 3-primes polynomial. Moreover, $q_r = 2$, if r is sufficiently large.

G. W. Effinger [20] proves results in the form of theorems (for q odd or even; q even considered in theorems 146(b), 151 and 153) which reduce the cases not covered by the asymptotic theorem to a finite, tractable number. Applications of these theorems reduces the polynomial 3-primes problem to a situation where one has to consider 85 separate combinations of q and r. For instance, $q = 256$, $r = 5$; $q = 199$, $r = 4, 5, \ldots$; $q = 2$, $r = 25$ and so on.

One needed to check that every monic polynomial (except for odd polynomials when $q = 2$) with first coefficient 0 and the second coefficient 0, 1 and (for odd q) a fixed quadratic non-residue is a sum of two monic irreducible polynomials. This involved laborious computation with a powerful computer. G. W. Effinger programmed the *IBM*3090 Super Computer at the Cornell National Super Computing Facility to check those remaining cases. Algorithms were designed to

(a) generate lists of irreducible polynomials and
(b) check off the sums of appropriate pairs of irreducibles.

For (a), both the Berlekamp Factorization Algorithm for $\mathbb{F}_q[x]$ and an *extension field* algorithm were employed. For (b), extensive indexing was used. The details of the algorithm design are given in Effinger [20]. See also [21].

On December 19, 1989, the *IBM*3090 completed the list of 85 cases which needed to be checked. A total of 64.8 hours of central processing was needed. This culminated in the complete solution to the 3-primes problem and the credit for this achievement goes to D. R. Hayes and G. W. Effinger. See [22].

Before conclusion, we make a mention about Bertrand's postulate. In a paper published in 'j.de l'Ecole royale polytechnique XVIII Pt 30(1845) pp 123–140, J. Bertrand (1822–1903) asked for the occurrence of a prime p lying in the interval.

$$n/2 < p \leq n-2, \text{ where } n > 6.$$

He had verified it for numbers n less than 6 million. P. L. Chebyshev (1821–1894) was the first to prove 'Bertrand's Postulate' in 1850. See Mem'oires pre'sente's a l' Academie Imperiale des Sciences de St. Petersburg VII (1854) pp 15–33 and Journal de mathematiques pures et appliques (Paris) XVII (1852) pp 366–390. Chebyshev proved that given $\epsilon > 1/5$, there exists a real number ξ such that for $n \geq \xi$, at least one prime satisfies

$$n < p \leq (1+\epsilon)n$$

J. J. Sylvester (1814–1897) and others showed that ϵ can be taken as a positive quantity, however small. From this, it would follow that if p_m denotes the mth prime,

$$\lim_{m \to \infty} \frac{p_{m+1}}{p_m} = 1$$

In simple terms, Bertrand's postulate says that if n is a positive integer, there exists a prime p such that $n < p \leq 2n$. S. Sivasankaranarayana Pillai (1901–1950) gave a proof of Bertrand's postulate without using Stirling's formula for $\Gamma(n)$ and his proof reduced the number of verifications to a minimum.

See K. Chandrasekharan [9]. Defining Chebyshev's function ϑ by

$$\vartheta(x) = \sum_{p \le x} \log p, \ x > 0, \ p \text{ a prime},$$

Pillai showed that $\vartheta(2n) - \vartheta(n) > 0$ for all $n \ge 2^6$ and verified the inequality directly for $1 \le n < 2^6$.

For an update of the Goldbach problem, see Wang Yuan [38].

15.7. A variant of Goldbach conjecture:

Following D. Shanks [A4], we give a variant of Goldbach conjecture as shown below:

(15.7.1)   Every integer of the form $4k + 2 \, (k \ge 2)$ is a sum of two primes

of the form $4t + 1 \, (t \ge 1)$.

It follows that by Fermat's Two squares theorem, every integer of the form $4k + 2(k \ge 2)$ is a sum of four squares. When $4k + 2 = 14$, one writes

$$14 = 1 + 13 = 0^2 + 1^2 + 2^2 + 3^2,$$

where 1 is considered as a prime for small values of $4k + 2$.

If $P(2r)$ denotes the number of solutions of

(15.7.2) $2r = p_1 + p_2$

where p_1 and p_2 are primes and $p_1 \equiv p_2 \equiv 1 \pmod 4$, one can obtain an asymptotic result for $P(2r)$ (due to Hardy and Wright) in the form

(15.7.3) $P(2r) \sim f(2r) \prod_{q \mid r} \left(\frac{q-1}{q-2} \right),$

where

(15.7.4) $f(r) = 1.3203236 \int_0^r \frac{dx}{(\log x)^2}$

and the product $\prod\limits_{q \mid r}$ on the right of (15.7.3) is taken over all odd primes q if any,

that divide r. (15.7.3) has been verified for values of r up to $r = 10^5$ [A5]. The constant on the right side of (15.7.4) is not an empirical constant, but, the infinite product

(15.7.5) $2 \prod_{p, \text{ an odd prime}} \left(1 - \frac{1}{(p-1)^2} \right)$

equals $1.3203236 \ldots$ ($\prod\limits_{p}$ running through all odd primes). (See D. Shanks [A4]).

If (15.7.1) is true, we obtain a simple proof of Lagrange's four squares theorem. For, given a prime p, if $2p$ is a sum of four squares, p is also a sum of four squares. So, any positive odd integer is the sum of four squares. If $r \equiv 0 \pmod 4$, either

$r = m^2s$ or $r = n^2 \cdot 2s'$, where m, n are even powers of 2 and s, s' are odd integers. In all of these cases, r is a sum of four squares, yielding Lagrange's theorem.

EXERCISES

1. **Mark the following statements true (T) or false (F) justifying your answer briefly.**

    a) *Every positive integer ≥ 12 is a sum of two composite integers.*

    b) *Given integers a, b, $1 < b < a$, let $p_1, p_2, \ldots p_b, p_{b+1}, \ldots p_n$ be a set of n primes arranged in ascending order of magnitude. We choose a set of b primes from $S = \{p_1, p_2, \ldots, p_n\}$ and take M to be their product. If N denotes the product of the remaining primes in S, it is correct to say that $M + N$ is divisible by all the primes of the set S.*

    c) *It is possible that a prime p could be of the form $a^4 - b^4$ where $a, b \in \mathbb{Z}$.*

    d) *There exists a composite integer $r > 1$ such that the nonzero non-units of $\mathbb{Z}/r\mathbb{Z}$ are primes in the ring.*

    e) *Let*
    $$A = \begin{bmatrix} 1 & 2 & 3 \\ 2 & 3 & 1 \\ 3 & 1 & 2 \end{bmatrix}.$$
    There exist matrices $X, Y \in M_3(\mathbb{Z})$ such that $A = X + Y$ and $\det X = \det Y = 9$.

    f) *$M(X) = X^5 + 1 \in \mathbb{F}_4[X]$ is a 3-primes polynomial.*

2. *Let p be a prime greater that a positive integer n. If p divides $N = \frac{(2n)!}{(n!)^2}$, show that p^2 does not divide N.*

3. *(Euler) If $4k + 3$ and $8k + 7$ are both primes, show that*
    $$2^{4k+3} - 1$$
    is divisible by $8k + 7$.

4. *(Euler) Let m be a positive integer. For positive integers x, y; show that every divisor of*
    $$x^{2^m} + y^{2^m}$$
    is of the form $2^{m+1}k + 1$, where $k \geq 1$.

5. *(Jacobi) Let p be a prime of the form $4k + 3$. Show that*
    $$(2k+1)! + (-1)^\eta \equiv 0 \pmod{p}$$
    where η is the number of quadratic non-residues of p which are less that $p/2$.

6. *Following S. Ramanujan (1887–1920) [33], a positive integer m is called highly composite, if $m(> 1)$ has more number of divisors than any preceding positive integer. For instance, 36 is highly composite. Show that every prime is a divisor of some highly composite number.*

    Analogously, highly composite ideals could be defined in the context of algebraic number fields. See Ralph G. Archibald [3].

7. *Find all positive integers m for which $m(m+30)$ is a perfect square. ($m = 2$ is the least among such numbers).*

8. *If p_n denotes the nth prime, show that*
   (a) $p_n < 2^{2^n}$;
   (b) $p_{n+1} < p_n^n + 1$, $(n > 1)$.

9. *Let $p = 4^k + 1$(an integer) for $k \geq 1$. Show that p is a prime if, and only if,*

$$3^{\frac{p-1}{2}} \equiv -1 (\mod p).$$

10. *[Eckford Cohen] Let $r > 1$ be an even integer. Show that 2 and its associates are not sums of two primes in $\mathbb{Z}/r\mathbb{Z}$ if, and only if, $r \equiv 0 (\mod 4)$ and r has at most one distinct odd prime divisor.*

11. *[Eckford Cohen] Let $r = 2^b p_1^{a_1} p_2^{a_2} \cdots p_h^{a_h}$ ($h \geq 1$) where $p_1, p_2, \ldots, p_h$ are odd primes. Show that an odd integer m belonging to $\mathbb{Z}/r\mathbb{Z}$ is not expressible as a sum of two primes in $\mathbb{Z}/r\mathbb{Z}$ if, and only if, m is divisible by every odd prime dividing r.*

12. *[Eckford Cohen] Let $r = 2^b p_1^{a_1}$ ($b \geq 2$, p_1 an odd prime, $a_1 \geq 1$). Show that a number m belonging to $\mathbb{Z}/r\mathbb{Z}$ cannot be represented as a sum of three primes of $\mathbb{Z}/r\mathbb{Z}$ if, and only if, m is of the form $2^c p_1^t q$, where $c \geq 2$, $t \geq 1$ and g.c.d $(q, r) = 1$.*

13. *Let $M(x) = x^5 + x^4 + x^3 + x^2 + x + 1 \in \mathbb{Z}_2(x)$. Show that $M(x)$ is a 3-primes polynomial.*

14. *Let $M(x) = x^4 - x^3 + x^2 + 1 \in \mathbb{Z}_3(x)$. Show that $M(x)$ is a 3-primes polynomial.*

15. *Let $M(x) = x^6 + 1 \in \mathbb{Z}_3[x]$. Express $M(x)$ as $P_1(x) + P_2(x) + P_3(x)$ where $P_1(x)$ is an irreducible 6th degree polynomial and $P_2(x), P_3(x)$ are irreducible 5th degree polynomials.*

REFERENCES

[1] Tom M. Apostol, Introduction to Analytic Number theory, Narosa Publishing House Pvt. Ltd, New Delhi, (1996).

[2] Ralph G. Archibald, An Introduction to the theory of Numbers, Charles E. Merril Publishing Co., Columbus, Ohio, (1970).

[3] Ralph G. Archibald, Highly composite ideals, Transactions of Royal Society of Canada, Third Series, XXX, Sec III, (1936), 41–47.

[4] Brent, Van de Lune, H.J.J. te Riele and Winter, On the zeros of the Riemann ζ-function in the critical strip, I, Math. Comp., 33, (1979), 1361–1372.

[5] Brent, Van de Lune, H.J.J. te Riele and Winter, On the zeros of the Riemann ζ-function in the critical strip, II, Math. Comp., 39, (1982), 681–682.

[6] Brent, Van de Lune, H.J.J. te Riele and Winter, On the zeros of the Riemann ζ-function in the critical strip, III, Math. Comp., 41, (1983), 759–767.

[7] Brent, Van de Lune, H.J.J. te Riele and Winter, On the zeros of the Riemann ζ-function in the critical strip, IV, Math. Comp., 46, (1986), 667–681.

[8] K. Chandrasekharan, Introduction to Analytic Number Theory, Chapter VII, pp 71–76, Springer-Verlag, N.Y. Inc, (1968).

[9] K. Chandrasekharan, Arithmetical functions, Chapter III pp 54–87, Springer-Verlag NY, Berlin, Heidelberg, (1970).

[10] J.R. Chen and T.Z. Wang, On the odd Goldbach problem, Acta Math. Sinica 32, (1989), 702–718, (in Chinese).

[11] Eckford Cohen, Rings of arithmetic functions, Duke Math. J., 19, (1952), 115–129.

[12] Eckford Cohen, A finite analogue of the Goldbach problem, Proc. Amer. Math. Soc. 5, (1954), 478–483.

[13] Eckford Cohen, The finite Goldbach problem in algebraic number fields, Proc. Amer. Math. Soc. 7, (1956) 500–506.

[14] J.M. Deshoillers, H.J.J. te Riele and Y. Saouter, New experimental results concerning the Goldbach Conjecture; Algorithmic Number Theory, J.P. Buhler (Ed) ANTS:III, Springer Verlag: Berlin, Heidelberg, New York, (1988) 204–215.

[15] H. Diamond, Elementary methods in the study of the distribution of prime numbers, Bull. Amer. Math. Soc. 7, (1982), 553–589.

[16] L.E. Dickson, History of the theory of numbers Vol I, Chelsea Publishing Co., Reprint (1971).

[17] J. Diedonné, La geometric des groups classiques, Springer Verlag, Berlin (1955).

[18] G.W. Effinger, A Goldbach theorem for polynomials of low degree over odd finite fields, Acta Arithmetica, 42, (1983), 329–365.

[19] G.W. Effinger, A Goldbach 3-primes theorem for polynomials of low degree over finite fields of characteristic 2, J. Number Theory, 29, (1988), 345–363.

[20] G.W. Effinger, The polynomial 3-primes conjecture, Computer Assisted Analysis and Modelling on the IBM 3090, M.I.T Press, Cambridge, MA, (1992).

[21] G.W. Effinger and D.R. Hayes, Additive Number Theory of polynomials over a finite field, Oxford University Press, London (1992).

[22] G.W. Effinger and D.R. Hayes, A complete solution to the Polynomial 3-primes problem, Bulletin (New Series) of the Amer. Math. Soc., 24, (1991) 363–369.

[23] J.B. Fraleigh, A first course in Abstract Algebra Chapters 38–46 pp 312–379, 5th Edition (International Student Edn), Addison-Wesley Longman, Inc, (1999), Reading, Mass.

[24] A. Granville, J. Van de Lune and H.J.J. te Riele, Checking the Goldbach Conjecture on a vector computer, Number Theory and Applications, R.A. Mollin (Ed), Kluwer, Dordrecht, (1983) pp 423–433.

[25] G.H. Hardy and J.E. Littlewood, Some problems in 'Partitio Numerorium' III, On the expansion of a number as a sum of three primes, Acta Math, 44, (1922/23) pp 1–70.

[26] D.R. Hayes, The Expression of a polynomial as a sum of three irreducibles, Acta Arith XI, (1966) 461–488.

[27] H. Ivaniec and C. J. Mozzochi, On the divisor and circle problems, J. Number Theory, 29, (1988), 60–93.

[28] Nathan Jacobson, Basic Algebra Vol I, Hindustan Publishing Corporation, New Delhi, Reprint (1985).

[29] S. Mac lane and G. Birkhoff, Algebra, Chapter XI, pp 382–386, The Macmillan Co. Collier-Macmillan Ltd., London, (1967).

[30] Norman Levinson, A motivated account of an elementary proof of the Prime Number Theorem, Amer. Math. Monthly, 76 (1969), 225–244.

[31] Norman Levinson, More than one-third of the zeros of Riemann Zeta function are on $\sigma = 1/2$, Advance Math, 13, (1974), 383–436.

[32] I. Niven, H.S. Zuckerman and H.L. Montgomery, An Introduction to the theory of Numbers, Chapter 8, Section 8.3, pp 389–401. John Wiley and Sons (Asia) Pvt. Ltd, 5th Edition (2000).

[33] S. Ramanujan, Highly composite numbers, Proc. London Math. Soc. Ser 2xiv, (1915) 347–409.

[34] M.K. Sinisalo, Checking Goldbach Conjecture up to 4.10^{11}, Math. Comp., 61, (1993) 931–934.

[35] V.S. Varadarajan, Some remarks on the analytic proof of the Prime Number Theorem, Nieuw Archief voor Wiskunde Vierde Serie Dee 16, No. 3, (Nov 1998), 153–160.

[36] L.N. Vaserstein, Non-commutative number theory, Contemporary Math, 83, (1989) 445–449.

[37] I.M. Vinogradov, Representation of an odd number as a sum of three primes, Comptus Rendues (Doklady), de 1' Academic des Sciences de 1', URSS, 15, (1937) 291–294.

[38] Wang Yuan (Ed), Goldbach Conjecture, Series in Pure Mathematics Vol. 4, World Scientific Publishing Co Pte Ltd, (2002) Singapore, 2nd Edn.

[39] Jun Wang, Goldbach problem in the ring $M_n(\mathbb{Z})$, Amer. Math. Monthly 99, (1992), 856–857.

[40] Zinoviev, On Vinogradov's Constant in Goldbach's ternary problem, J. Number Theory, 65, (1997), 334–358.

ADDITIONAL REFERENCES

[A1] Anatolij A. Karatsuba: Complex analysis in Number Theory. CRC Press, Boca Raton, Florida 33431, USA (1995).

[A2] Daniel A. Marcus: Number fields, Universitext, Springer Verlag, New York (1977).

[A3] Richard A. Mollin: Quadratics, CRC Press, Boca Raton, Florida, 33431, USA (1996).

[A4] D. Shanks: Solved and unsolved problems in Number Theory, Chelsea Publishing Co, New York (1977).

[A5] D. Shanks: Review of Stein and Stein UMT 36, Math. Comp. (1975) 427–434.

[A6] Lawrence C. Washington: Introduction to Cyclotomic Fields, GTM No. 83, Springer-Verlag, New York (1982).

CHAPTER 16

An epilogue: More interconnections

Introduction

Thus far, a genuine attempt has been made to highlight some interconnections between Number Theory and Algebra. This concluding chapter gives a sum-up of the salient features of commutative rings. It is a significant fact that the integral domain $\mathbb{Z}$ of rational integers is characterised by its double-remainder property. Something more is said about integral domains.

Krull-Zorn theorem (see Proposition 2.4.1, chapter 2) is about the existence of a maximal ideal in a commutative ring with unity. But, does there exist a commutative ring without a maximal ideal? The answer is yes. See theorem 156. We mention that many more interesting results could be given in the context of commutative rings. In Section 16.3, an analogue of Euclid's theorem on the infinitude of primes is given for a PID. See theorem 158. Corollary 16.4.2 gives a connection of the number of units in a ring with Mersenne primes. In theorem 162 (the last one), a quadratic reciprocity law (due to William Duke and Kimberly Hopkins) for a finite group is stated and proved.

16.1. On commutative rings

Commutative rings are in plenty. They have two subdivisions (i) Noetherian rings (ii) Non-Noetherian rings. (See figure 16): The class of Noetherian rings

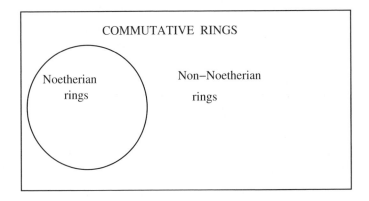

COMMUTATIVE RINGS

Noetherian rings

Non–Noetherian rings

Figure 16

could be subdivided further. However, if we confine to the class of integral domains, we have the following subclasses (see figure 17):

   i. Half-factorial domains
  ii. Unique factorization domains
 iii. Principal ideal domains
 iv. Euclidean domains
  v. The Euclidean domain having double remainder property ($d.r.p$) which is $\mathbb{Z}$, the ring of integers. (See Theorem 21, chapter 3). Figure 17 below illustrates the unique position of $\mathbb{Z}$, among integral domains.

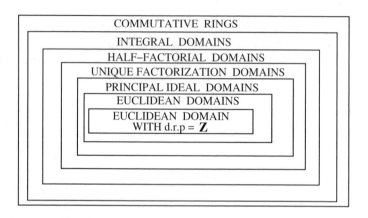

Figure 17

In Section 8.4 of chapter 8, fields with valuation were considered. We recall definition 8.4.10.

Definition 16.1.1 : *Let F be a field. A valuation on F is a map $\nu : F \to G \cup \{\infty\}$ where $(G,+)$ is a totally ordered abelian group and ∞ is a symbol which is greater than any element of G such that for all $a, b \in F$,*

(i) $\nu(a \cdot b) = \nu(a) + \nu(b)$
(ii) $\nu(a) = \infty$, *if, and only if, $a = 0$ and*
(iii) $\nu(a+b) \geq \min\{\nu(a), \nu(b)\}$.

    G is called the value-group of the valuation ν. Let

(16.1.1) $R = \{x : x \in F, \nu(x) \geq 0\}$.

It is verified that R is a subring of F. If $t \in F$, $\nu(t) = 0$, if $t = 1_F$. Also, $\nu(t^{-1}) = -\nu(t)$. Therefore, when $t \in F$, either $\nu(t) \geq 0$ or $\nu(t^{-1}) \geq 0$. This is an interesting property of R.

Definition 16.1.2 : *Let R be an integral domain with field of quotients F. R is called a valuation domain, if it satisfies either of the following equivalent conditions:*

(i) *For any two elements a, b in R, either $a|b$ or $b|a$.*

(ii) *For any element t in K, either $t \in R$ or $t^{-1} \in R$.*

See definition 8.3.3 and lemma 8.3.2, chapter 8. R given by (16.1.1) is, thus, a valuation domain. Further,

(16.1.2) $$M = \{x : x \in F \text{ and } \nu(x) > 0\}$$

is the unique maximal ideal of R. For this reason, R is a quasi-local ring. It is possible that we may get the same valuation domain by using two or more distinct valuation maps. Recalling definition 12.6.6, chapter 12, we have

Theorem 154 : *If R is a valuation domain, R is integrally closed.*

Proof : Let K be the field of fractions of R. Suppose that $a = \dfrac{s}{t} \in F$ where $s, t \in R$ with $t \neq 0$. As R is a valuation domain, for $s, t \in R$, either $s|t$ or $t|s$. If $t|s$, $\dfrac{s}{t} \in R$. So, if $\dfrac{s}{t}$ is integral over R, $\dfrac{s}{t} \in R$ whenever $t|s$. Suppose that $s|t$ and $a = \dfrac{s}{t}$ is integral over R. Then, $\dfrac{s}{t}$ is a root of a monic polynomial equation:

(16.1.3) $$x^n + a_1 x^{n-1} + \cdots + a_{n-1}x + a_n = 0, \; q_i \in R, i = 1, 2, \ldots, n.$$

If $s|t$, we write $t = sb$ where $0 \neq b \in R$. $\dfrac{s}{t} = \dfrac{1}{b}$ satisfies (16.1.3). That is,

$$1 + a_1 b + \cdots + a_{n-1}b^{n-1} + a_n b^n = 0$$

or,

$$b(a_1 + a_2 b + \cdots + a_n b^{n-1}) = -1.$$

This shows that b is a unit in R. That is, s and t are associates in R. Thus, when $s|t$, $\dfrac{s}{t}$ is a unit belonging to R. Hence, any element $a = \dfrac{s}{t}$ of the field of quotients of R which is integral over R belongs to R. $\qquad \square$

As remarked in [14], valuation domains exist in abundance. Let R be an integral domain contained in a field K. Assume that R has a maximal ideal M (say). Then, we can construct an integral domain D contained in K and containing R such that K is the field of quotients of D and for any $x \in K$, either x or x^{-1} belongs to D. Then, D is a valuation domain having unique maximal ideal M' such that $M' \cap R = M$.

Fact 16.1.1 : If R is a Noetherian valuation domain, then, the value group of R is isomorphic to $(\mathbb{Z}, +)$, the additive group of rational integers.

For proof, see I. Kaplansky [15].

Definition 16.1.3 : *An integral domain D is called a Prüfer domain, if every finitely generated nonzero ideal of D is invertible.*

We note that Prüfer domains are integrally closed [11]. A PID is Prüfer, as principal ideals are invertible. A Noetherian Prüfer domain is a Dedekind domain. We have to show that every nonzero prime ideal of the domain is maximal.

Definition 16.1.4 : *Let R be a commutative ring with unity* 1_R. *A prime ideal P of R is said to be of height n, written* $ht(P) = n$, *if there exists an ascending chain*

$$P_0 \subset P_1 \subset \ldots \subset P_n = P$$

of prime ideals of R.

In particular, a prime ideal is of height 0, if it contains no other prime ideals.

Definition 16.1.5 : *An commutative ring R with unity* 1_R *is said to be of Krull dimension r, written,* $\dim(R) = r$ *(possibly infinite) if r = the supremum of* $ht(P)$ *for all prime ideals P of R.*

For example, Z has Krull dimension 1. In fact, a PID has Krull dimension 1. Let D be a Noetherian domain. Suppose that $x \in D$ is a non-unit. We call P a prime ideal minimal among all prime ideals containing (x), if there does not exist a prime ideal P' such that $(x) \subset P' \subset P$. Then, $ht(P) \leq 1$. Further, if x is nilpotent (that is, there exists $m \in \mathbb{N}$ such that $x^m = 0_D$), then $ht(P) = 0$, as D has no zero divisors. If x is not a zero divisor, $ht(P) = 1$. This is the content of Krull's Principal ideal theorem. For details and a more general result, see I. Kaplansky [15].

We remark that a Noetherian domain D may have $\dim(D) = \infty$, whereas non-Noetherian domains D can be found such that $\dim(D) < \infty$.

Let D be an integral domain with unity element 1_D. A nonzero element $x \in D$ is called a principal prime, if the ideal (x) is a prime ideal. Any principal prime element is irreducible. However, there exist irreducible elements which are not principal primes. In any domain with ACCP (ascending chain condition on principal ideals), there exist irreducible elements (see Section 4.8, chapter 4). But, in a Noetherian domain, it is possible that principal prime ideals may not exist.

In a unique factorization domain (UFD), every prime ideal of height 1 is principal. Also, if an integral domain D is Noetherian and height 1 prime ideals are principal, then, D is a UFD. This is what makes $\mathbb{Z}$ a UFD. In other words, a Noetherian domain is a UFD, if, and only if, every prime ideal of height 1 is principal. We remark that a Noetherian valuation domain is one-dimensional. But, not all one-dimensional valuation domains are Noetherian. See R. Gilmer [11] or N. Bourbaki [3].

Next, we give a characterization of Dedekind domains (see Section 12.6, chapter 12).

An integral domain D is a Dedekind domain if, and only if,

(i) D is Noetherian, integrally closed and one-dimensional or
(ii) every nonzero ideal of D is invertible, or
(iii) every proper ideal of D is a product of prime ideals (unique, except for order).

Definition 16.1.6 : *Let D be an integral domain with unity* 1_D. *D is called a Bézout domain (Etienne Bézout (1730–1783)), if every finitely generated ideal of D is principal.*

We note that a valuation domain is a Bézout domain. Now, as principal ideals of an integral domain are invertible, a Bézout domain is also a Prüfer domain. Further, there exist Bézout domains which are not valuation domains as also there are Prüfer domains which are not Bézout domains.

We have

a) a Dedekind domain $\subset$ a Noetherian domain $\subset$ a 1-dimensional domain which is integrally closed.

b) a Dedekind domain $\subset$ a Prüfer domain $\subset$ a domain which is integrally closed.

c) a Principal Ideal Domain $\subset$ a Bézout domain $\subset$ a Prüfer domain.

For a more detailed account of commutative rings, see Harry C. Hutchinson [14] and M. F. Atiyah and I. G. Macdonald [2].

16.2. Commutative rings without maximal ideals

In $\mathbb{Z}$, the principal ideal (6) is contained in the maximal ideal (2) or (3). If $n \in \mathbb{Z}$ has k distinct prime factors, the ideal (n) is contained in k maximal ideals generated by the k distinct prime factors of n. However, there exist commutative rings without maximal ideals. The idea is to look for the so-called 'divisible groups'.

Definition 16.2.1 : *Let $(G,+)$ be an abelian group. G is said to be a divisible group, if given $a \in G$ and $n(\neq 0) \in \mathbb{Z}$, there exists $b \in G$ such that $nb = a$.*

Example 16.2.1 : If $\mathbb{Q}$ denotes the set of rational numbers, $(\mathbb{Q},+)$ is a divisible group. For, given $\frac{a}{b} \in \mathbb{Q}, (b \neq 0)$, there exists $\frac{c}{d} \in \mathbb{Q}$ such that for $n \in \mathbb{Z}$,

$$\frac{nc}{d} = \frac{a}{b} \ (a,b,c,d \in \mathbb{Z}; \quad b,d \neq 0).$$

It is enough if we make $(\frac{1}{n})(\frac{a}{b}) = \frac{c}{d}$.

However, $\mathbb{Z}$ is not a divisible group. For, given $a \in \mathbb{Z}$, $n(\neq 0) \in \mathbb{Z}$, there does not exist $b \in \mathbb{Z}$ for which $nb = a$, when $n > 1$.

Fact 16.2.1 :

a) An abelian group G is divisible, if $mG = \{mg : g \in G\} = G$ for every positive integer m.

b) A direct sum of divisible groups is divisible; if, and only if, each summand is divisible.

c) A homomorphic image of a divisible group is divisible. (In contract, a homomorphic image of a semisimple ring need not be semisimple. see observation 14.2.4, chapter 14).

Proofs are omitted.

Theorem 155 (Peter Malcolmson and Frank Okoh (2000)) : *A divisible abelian group has no maximal subgroups* .

Proof : Assume the contrary. Suppose that M is a maximal subgroup of a divisible abelian group G. Then, G/M is abelian and simple. Therefore, G/M is cyclic. As $(\mathbb{Z},+)$ is not a divisible group, G/M is not infinite cyclic. So, $G/M \cong \mathbb{Z}/p\mathbb{Z}$, for some prime p. p is such that $pa = 0$ for every element a in G/M. That is, $p(G/M) = 0$ from which it follows that $pG \subseteq M$. By Fact 16.2.1 (a), as G is divisible, $pG = G$. It follows that $G \subseteq M$, a contradiction to the fact that M is maximal (proper) subgroup of G. Hence, G has no maximal subgroups. □

Lemma 16.2.1 : *A torsion-free abelian group* $(G,+)$ *is divisible if, and only if, G is a vector space over* $\mathbb{Q}$, *the field of rationals.*

Proof : $\Leftarrow$: Suppose that $(G,+)$ forms a vector space over $\mathbb{Q}$. Given $a \in G$ and $n\, (\neq 0) \in \mathbb{Z}$, $(\frac{1}{n})\, a \in G$. So there exists $b \in G$ such that $nb = a$. For, we have only to choose b as $(\frac{1}{n})a \in G$. No nonzero element of G is of finite order. Therefore, G is a divisible group.

$:\Rightarrow$ Assume that $(G,+)$ is abelian, torsion-free and divisible. If $n(\neq 0) \in \mathbb{Z}$ and $a \in G$ are given, there exists $b \in G$, for which $nb = a$. We denote b by $(\frac{1}{n})a$.

For $m, n \in \mathbb{Z}$, $n \neq 0$, we define $(\frac{m}{n})\, a = m(\frac{1}{n})\, a$. Scalar multiplication is associative and distributes addition. No element of G is of finite order. So, scalar multiplication is nontrivial. Moreover $1(a) = a$. So, G is a vector space over $\mathbb{Q}$. □

Next, let R be a commutative ring. By definition (see definition 2.3.2, chapter 2)

$$(16.2.1) \qquad R^2 = \{ \sum_{\text{finite}} a_{i1}a_{i2} \, : \, a_{ij} \in R \cdot j = 1, 2 \}.$$

R^2 is an ideal generated by products of elements of R. Two cases arise
(i) $R \neq R^2 \neq (0_R)$,
(ii) $R = R^2 \neq (0_R)$.
(Where 0_R is the zero element in R). We note that a particular case of $R = R^2$ happens when R has a unity element.

It was mentioned in theorem 25 (chapter 4) that given F a field of characteristic 0, the ring $F[[x]]$ of formal power series in x is a PID and on the basis of definition of a quasilocal ring (see definition 4.5.3, chapter 4). $F[[x]]$ is quasilocal with the unique maximal ideal $(x) = xF[[x]]$. Writing

$$(16.2.2) \qquad\qquad R = xF[[x]],$$

we note that $R^2 = x^2 F[[x]]$ and $R \neq R^2 \neq (0_R)$ (case (i) above).

Theorem 156 (Peter Malcolmson and Frank Okoh (2000)) : *Let F be a field of characteristic zero. If R is as given in* (16.2.2), *R has no maximal ideals.*

Proof : Assume the contrary. Suppose that M is a maximal ideal of R. By Remark 4.5.1, chapter 4, if $f \in F[[x]]$, f has an inverse, if the constant term in f is nonzero.

Claim : $R^2 \subseteq M$.

If $M \subseteq R^2$, $M = R^2$, as M is maximal. Otherwise, M contains an element xf in R such that $xf \notin R^2$. When $xf \notin R^2$, f has a nonzero constant term and so f has an inverse. Then, if $x^2g \in R^2$, one could write $x^2g = (xf)(xf^{-1}g)$, where $g \in F[[x]]$. As $xf \in M, x^2g \in M$. That is, $R^2 \subseteq M$. If $a + R^2$, $b + R^2 \in R/R^2$,

$$ab + R^2 \in R^2 \text{ and so, } R/R^2 \text{ has 'trivial' multiplication.}$$

We consider $(R/R^2, +)$. It is an abelian group. Given $a \in G$. $n \in \mathbb{N}$, na is well defined. If, by any chance, $na = 0$, na is a product of two elements of R. So, there exists $b = cd$ (say) such that $na = b$. If $na \neq 0$, na is an element of R/R^2 and so $na = c$ (say). In either case, we could define an element of R/R^2 as $(\frac{1}{n})a$. For negative integer n, writing $n = -n'$, one could define $-(\frac{1}{n'})a \in R/R^2$. Then, $(\frac{m}{n})a$ is given by $m(\frac{1}{n})a$. Thus, scalar multiplication by $\frac{m}{n} \in \mathbb{Q}$ works out alright. Scalar multiplication distributes addition. As in the proof of lemma 16.2.1, we note that $(R/R^2, +)$ is a divisible group. Next, as M is designated as a maximal ideal of R, M/R^2 is a maximal ideal of R/R^2. By theorem 155, the divisible group R/R^2 has no maximal subgroups. The situation giving M/R^2 as a maximal ideal of R/R^2 contradicts the fact that $(R/R^2, +)$ has no maximal subgroups. So, R has no maximal ideals. □

Remark 16.2.1 : R is the unique maximal ideal of $F[[x]]$. $R \neq R^2$. Then, R/R^2 is shown to have no maximal ideals in theorem 156. It is mentioned in [17] that if R is a commutative ring with unity and R has a unique maximal ideal M with $M = M^2$, then, the ring M has no maximal ideals.

We proceed to consider case (ii): $R = R^2 \neq (0_R)$. We look for a subring of A_ω = the polynomial ring in countably many indeterminates $x_1, x_2, \ldots$ over a commutative ring A subject to the condition:

$$(16.2.3) \qquad x_i^2 = x_{i-1}, \ i = 2, 3, 4, \ldots.$$

Theorem 157 (Peter Malcolmson and Frank Okoh (2000)) : *Let R be a subring of A_ω (the ring of polynomials in $x_1, x_2, x_3, \ldots$ satisfying (16.2.3) such that R consists of those polynomials in $x_1, x_2, \ldots$ having constant term equal to 0_A). Then, R has no maximal ideals.*

Proof : Because of the relation (16.2.3), $R = R^2 \neq (0_A); 0_A$ being the zero element in A. We consider a proper ideal I of R. Suppose that $x_n (n \in \mathbb{N})$ belongs to I. For $a \in A$, we get

$$ax_n = (ax_{n+1})x_{n+1} \in I.$$

Therefore, for $m \geq n$, $x_m \in I$. Since I is proper, we choose the least value $t \in \mathbb{N}$, such that $x_t \notin I$. Denoting the ideal generated by x_t and I by $< I, x_t >$, we see that $J = < I, x_t >$ is a proper ideal of R. If $J = R$, we will have

$$(16.2.4) \qquad x_{t+1} = a + sx_t + gx_t$$

where $a \in I, g \in R$ and s is an integer. Multiplying both sides of (16.2.4) by x_t, we obtain $x_t x_{t+1} \in I$, since $x_t^2 = x_{t+1} \in I$. Next, multiplying both sides of (16.2.4) by x_{t+1}, we see that $x_t \in I$, since $x_{t+1}^2 = x_t$. This contradicts the fact that $x_t \notin I$. So, $J \neq R$. That is, J is a proper ideal of R. This implies that given I a proper ideal of R, we can always find another proper ideal J of R such that J contains I properly. Hence, R has no maximal ideals. $\qquad \square$

Remark 16.2.2 :

(a) Theorems 155,156 and 157 have been adapted from Peter Malcolmson and Frank Okoh [17].

(b) The rings R considered in theorems 156 and 157 are such that R does not possess the unity element (multiplicative identity).

For more examples of rings having no maximal ideals, see [17].

16.3. Infinitude of primes in a PID

An analogue of Euclid's theorem on the infinitude of primes was shown in theorem 92, chapter 12. It says that a PID is semisimple if, and only if, either it is a field or it has an infinite number of maximal ideals. In theorem 9, chapter 2, we noted that if D is a PID, an irreducible element is also a prime. Further, a principal ideal generated by a prime is a prime ideal. That is, p is a prime in D if, and only if, the principal ideal (p) is a prime ideal. When D has an infinite number of maximal ideals, D has an infinite number of irreducible elements that are primes in D. Is there a way to characterise a PID which has an infinite number of irreducibles? It is possible to get this by noting that given an integral domain D, the polynomial ring $D[x]$ is semisimple (see definition 12.4.2, chapter 12).

We recall that the Jacobson radical $J(R)$ of a commutative ring R with unity 1_R is defined as the intersection of maximal (prime) ideals of R. In the notation of definition 12.3.3, chapter 12, $\sqrt{0_R}$ is the nilradical of the zero ideal. It was observed that the nilradical of R consists of the nilpotent elements of R.

Lemma 16.3.1 : *Given a commutative ring R with unity 1_R, the nilradical $\sqrt{0_R}$ of R is the intersection of prime ideals of R.*

Proof : Let

(16.3.1) $J = \cap \{P : P \text{ is a prime ideal of } R\}$.

Suppose that $a \in R$ and $a \notin \sqrt{0_R}$. Then, a is non-nilpotent. So, $S = \{a^n : n \in \mathbb{N}\}$ does not intersect $\sqrt{0_R}$. Since S is a multiplicative set (see definition 14.3.2, chapter 14), by Corollary 14.3.1, the complement of S in R is a set-theoretic union of prime ideals of R. So, there exists a prime ideal P not containing a. That is, a does not belong to the intersection of prime ideals of R. Or, $a \notin J$. That is, $a \notin \sqrt{0_R} \Rightarrow a \notin J$. Therefore,

(16.3.2) $$J \subseteq \sqrt{0_R}.$$

If $b \in R$ and $b \notin J$, there exists a prime ideal not containing b. So, no power of b belongs to P. It means that b is not nilpotent. That is, $b \notin \sqrt{0_R}$. Thus,

(16.3.3) $\sqrt{0_R} \subseteq J.$

From (16.3.2) and (16.3.3), we arrive at $\sqrt{0_R} = J$. $\square$

Remark 16.3.1 : The nilradical of R is also the prime radical of R. See definition 14.5.2, chapter 14.

Lemma 16.3.2 : *If R is an integral domain, $R[x]$ is semisimple.*

Proof : Assume the contrary. Let $0 \neq f(x) \in J(R[x])$. By Corollary 12.4.1, chapter 12, $xf(x) + 1_R$ is a unit. This is evident, even otherwise. For, if $xf(x) + 1_R$ is not a unit, $xf(x) + 1_R$ will belong to a maximal ideal M of $R[x]$. Since $f(x) \in M$, $xf(x) \in M$. Then, $1_R \in M$, a contradiction. Now, R is an integral domain, so is $R[x]$. When $xf(x) + 1_R$ is a unit, $xf(x) + 1_R$ is a constant polynomial, forcing $f(x)$ to be the zero polynomial. So, $J(R[x]) = (0_R)$. $\square$

Theorem 158 (Fabrizio Zanello (2004)) **:** *R denotes a PID. R has an infinite number of pairwise nonassociated irreducible elements if, and only if, every maximal ideal of $R[x]$ has height 2.*

Proof : $\Rightarrow$

Let M be a maximal ideal in $R[x]$. We write $P = M \cap R$. Then, P is a prime ideal of R. Suppose that $P = (0)$. We select an element $g(x)$ in M of lowest degree which is possible, since $R[x]$ is a UFD. Now, M is a prime ideal of $R[x]$. So, we could assume that $\deg g(x) > 0$ and $g(x)$ is an irreducible in $R[x]$.

Let K denote the quotient field of R. If $f(x) \in M$, as division algorithm holds in $K[x]$, we obtain

(16.3.4) $f(x) = g(x)q_1(x) + r_1(x)$

where either $r_1(x) = 0$ or $\deg r_1(x) < \deg g(x)$. From (16.3.4), we have

(16.3.5) $f(x) = g(x)\dfrac{q(x)}{a} + \dfrac{r(x)}{a}$

where a denotes the l.c.m of denominators of the coefficients (which are from K) of $q_1(x)$ and $a \neq 0_R$. Or,

(16.3.6) $af(x) = g(x)q(x) + r(x)$

where $q(x)$ and $r(x) \in R[x]$. This implies that $r(x) \in M$. This leads to the fact that $r(x)$ is the zero polynomial, as $\deg r(x) \geq \deg g(x)$ by the choice of $g(x)$ as an irreducible polynomial. So, $af(x) \in (g(x)) \subset M$. Therefore, $f(x)$ belongs to the principal ideal $(g(x))$. But $(g(x))$ is a prime ideal. So, either $a \in (g(x))$ or $f(x) \in (g(x))$. But, prime ideals in $R[x]$ cannot contain nonzero constants. Thus, $M = (g(x))$ which implies that $ht(M) = 1$ in case $P = (0_R)$.

M is an arbitrary maximal ideal of $R[x]$. If $P = M \cap R = (0_R)$, $M = (g(x))$ where $g(x)$ is an irreducible polynomial in $R[x]$. For every irreducible p in R, g.c.d $(g(x), p) = a$ unit in R. Therefore, if $J = $ the ideal generated by $g(x)$ and p, written $(g(x), p)$, we have

(16.3.7) $R[x]/J = (0_R)$.

Now, $R/(p)$ is a field say F. There exist polynomials $u(x)$, $v(x)$ in $R[x]$ such that $u(x)g(x) + v(x)p = 1_R$. So, in $F[x]$, as $u(x)g(x) \equiv 1_R \pmod{I}$, where $I = (p)$, the ideal generated by p in $R[x]$, $g(x)$ is a constant in $F[x]$, for each such p. So, every coefficient of $g(x)$ other than the constant term is divisible by every irreducible p in R. This is a contradiction to the hypothesis that there are infinitely many pairwise non-associated irreducible elements in R and the fact that R, being a PID, is a UFD. Therefore, $P = M \cap R \neq (0_R)$. So, $P = (p)$ where p is an irreducible element of R.

Now, M is not a principal ideal. For, if it were, M would coincide with (p). But, then, $R[x]/(p) \cong F[x]$, where $F = R/(p)$ and $F[x]$ is not a field. Therefore, M strictly contains the prime ideal (p). So, height of M is at least 2. But, as R is Noetherian,

(16.3.8) $\dim R[x] = \dim R + 1$.

For proof, see G. Karpilovsky [A2]. Here, $\dim R = 1$. Therefore, $\dim R[x] = 2$. Thus, height of M is exactly 2. That is, if R has infinitely many pairwise nonassociated irreducibles, height of a maximal ideal M of $R[x]$ is 2.

$\Leftarrow$: Conversely, assume that every maximal ideal M of $R[x]$ has height 2. Every such M must contain some nonzero constant and so must contain at least one irreducible element of R. For, if $P = M \cap R$ and $P = (0)$, it was shown already that M has height 1.

Suppose that there are only a finite number of pairwise non-associated irreducible elements in R. Let the irreducibles be $p_1, p_2, \ldots, p_m$ (say). Then, $p_1 \cdot p_2 \cdot \ldots \cdot p_m$ belongs to every maximal ideal M in $R[x]$. Therefore, the product $p_1 \cdot p_2 \cdots p_m$ belongs to the Jacobson radical $J(R[x])$ of $R[x]$. Since R is an integral domain, by lemma 16.3.2, $J(R[x]) = (0_R)$, a contradiction to the fact that the product $p_1 \cdot p_2 \cdots p_m$ of irreducibles belongs to $J(R[x])$. This proves that R has an infinite number of non-associated irreducible elements. $\qquad \Box$

Remark 16.3.2 : The proof given above has been adapted from [20].

Remark 16.3.3 : Theorem 158 catches the infinitude of primes in a PID R for which every maximal ideal of $R[x]$ has height 2. Of course, we use the fact that $R[x]$ is semisimple.

Remark 16.3.4 : Let $P = (p)$, (p a prime) be a prime ideal of $\mathbb{Z}$. Then, $P[x]$ is a prime ideal of $\mathbb{Z}[x]$. $\mathbb{Z}[x]$ is semisimple and by Theorem 92 (Burton's theorem), $\mathbb{Z}[x]$ has an infinite number of maximal ideals. Not all prime ideals of $\mathbb{Z}[x]$ are of the form $P[x]$. For, (x), the ideal generated by x is a prime ideal of $\mathbb{Z}[x]$ and it is not a maximal ideal, though. As mentioned in chapter 12, $\mathbb{Z}[x]$ is not a Dedekind domain.

Next, we make the following

Observation 16.3.1 : *Let P, Q be prime ideals of a Noetherian domain R such that $P \subset Q$. Suppose that ht $P = n$, ht $Q = n + 1$ ($n \geq 0$). Then, there is no prime ideal in $R[x]$ which lies strictly in between $P[x]$ and $Q[x]$. See [A2].*

Observation 16.3.2 : *Given a prime ideal P of a Noetherian domain R such that ht $P = n$ (≥ 1), if J is a prime ideal of $R[x]$ which contracts to P (meaning that $P = J \cap R$) and contains $P[x]$ properly, then, in $R[x]$,*

$$(16.3.9) \qquad ht\ (P[x]) = n, \quad ht\ J = n + 1.$$

For proof, see [A2].

We deduce that if P is a prime ideal (generated by a prime p) in $\mathbb{Z}$, as $(0) \subset P$ forms an ascending chain of prime ideals of $\mathbb{Z}$, ht $P = ht\ (P[x]) = 1$. $P[x]$ is contained in a maximal ideal of $\mathbb{Z}[x]$. If M denotes the ideal of $\mathbb{Z}[x]$ which contracts to P and contains $P[x]$, ht $M = 1 + 1 = 2$. It implies that every maximal ideal of $\mathbb{Z}[x]$ has height 2. Theorem 158 confirms that $\mathbb{Z}$ has an infinite number of non-associated irreducible elements that are primes.

16.4. On the group of units of a commutative ring

Dirichlet's unit theorem (see theorem 124, chapter 13) determines the structure of the group of units in the ring of integers of an algebraic number field. As a partial converse, given a cyclic group G, one could determine a finite commutative ring R for which the given group G is the group of units, see Robert Gilmer [10]. Going a step further, K. E. Elridge and I. Fisher [9] determine an artinian ring (see definition 12.5.4, chapter 12) whose group of units is a given cyclic group. In a way, a ring structure is determined by a given group G which will serve as the group of units [8]. In [5], the following questions are posed.

Question 1: Which groups can be reckoned as the group of units of a commutation ring with unity?
Question 2: Which numbers u can serve as the number of units of a commutative ring with unity?

S. Z. Ditor [5], answers questions 1 and 2 for finite groups of odd order. The needed background picture is presented below:

Fact 16.4.1 : In a finite dimensional algebra A with unity 1_A, over a field F
(i) Every left divisor of zero is a right divisor of zero and vice versa.
(ii) Every left unit (a left divisor of 1_A) is also a right unit (a right divisor of 1_A).
(iii) A divisor of zero in A is never a unit in A.
These are obvious.

Definition 16.4.1 : *A finite dimensional algebra A without divisors of zero is called a division algebra.*

Definition 16.4.2 : *A representation of an algebra A over a field F is a homomorphism $T : A \to \mathcal{L}(V)$, where $\mathcal{L}(V)$ denotes the F-algebra of linear operators on some vector space V (over F). In other words, to define a representation T of A is to assign to every element a in A, a linear operator $T_a : V \to V$ such that for $a, b \in A$, $\alpha \in F$,*

$$T_{a+b}(v) = (T_a + T_b)(v),$$
$$T_{\alpha a}(v) = \alpha T_a(v),$$
$$T_{a \cdot b}(v) = T_a(T_b(v)),$$
$$and \quad T_{1_A}(v) = I(v),$$

where $v \in V$ and I denotes the identity operator on V.

If V is finite dimensional, its dimension is called the dimension of the representation T. The image of the representation T (the set of linear operators of the form $T_a : a \in A$, forms a subalgebra of $\mathcal{L}(V)$. Image of T is denoted by Im T.

Definition 16.4.3 : *Let T be a representation of an algebra A. If $T : A \to \mathcal{L}(V)$ is a monomorphism, Im T which is isomorphic to A is called a faithful representation of A.*

Lemma 16.4.1 (Cayley) : *Every algebra admits a faithful representation. That is, every algebra is isomorphic to a subalgebra of $\mathcal{L}(V)$, for some vector space V over F.*

Proof : Given A, if $a \in A$, we could define $T_a : A \to \mathcal{L}(V)$ by $T_a(x) = ax$, $x \in A$. T_a is a linear operator on the space A over F.

$$T_{a+b}(x) = (T_a + T_b)(x),$$
$$T_{\alpha a}(x) = \alpha T_a(x)$$
$$T_{a \cdot b}(x) = T_a(T_b(x))$$
$$and \quad T_{1_A}(x) = I_A(x) = x = I(x);$$

where $x \in A$ and I denotes the identity operator on V. T is a representation of A and for $a, b \in A$ whenever $a \neq b$ $1_A(a) \neq 1_A(b)$ or T is a faithful representation of A. $\square$

Definition 16.4.4 : *The representation given in lemma 16.4.1 is called the regular representation of A. When A is finite dimensional, the dimension of the regular representation is equal to the dimension of A.*

When T is finite dimensional, we may choose a basis of V and assign to each operator $T_a \in \mathcal{L}(V)$ its matrix $[T_a]$. $T : A \to \mathcal{L}(V)$ gives a homomorphism of A into the algebra $M_n(F)$, where $n = \dim T$. Such a homomorphism is a matrix representation of the algebra A. So, a representation T of A also gives rise to a matrix representation of A.

Definition 16.4.5 : *Let A be an algebra over a field F. A left A-module M is a vector space over F together with a function* $\eta : A \times M \to M$ *such that if* $\eta(a,m)$ *is denoted by* $am (a \in A, m \in M)$, *and*

$$a(m_1 + m_2) = am_1 + am_2,$$

$$(a + b)m = am + bm,$$

$$a(\alpha m) = \alpha a(m) = \alpha(am), \alpha \in F,$$

$$a.b(m) = a(bm),$$

$$1_A m = m,$$

where $a, b \in A, m, m_1, m_2$ *are in M.*

If $T : A \to \mathcal{L}(V)$ is a representation of A, we define $T_a(V)$ as av for $a \in A, v \in V$. Then, V becomes a left A-module. This module corresponds to T. Conversely, given M as a left A-module, for $a \in A$, we can consider $T_a : M \to M$ by defining $T_a(m) = am$. Then, T_a becomes a linear operator on M. That is, we get a representation of A corresponding to the module M. In particular, there corresponds a regular module corresponding to a regular representation of A. In fact, when $M = A$, ma is nothing but the product of elements m and a in A.

Next, we consider A-module homomorphisms. Let M, N be left A-modules. A linear map $\psi : M \to N$ satisfying $\psi(am) = a\psi(m)$. $a \in A, m \in M$ is called a homomorphism. If ψ is both a homomorphism and epimorphism, it is called an isomorphism. Further, M and N are said to be isomorphic. We state without proof:

Fact 16.4.2 : The representations of an algebra A corresponding to isomorphic modules are similar and modules corresponding to similar representations are isomorphic.

Homomorphisms of A-modules can be multiplied. The idea is that given $\psi : M \to N, \eta : N \to L$, the product $\eta \psi : M \to L$ is given by $\eta \psi(m) = \eta(\psi(m))$ for all $m \in M$. $\eta \psi$ is also a homomorphism. Further, if ψ_1, ψ_2 are homomorphisms from M to N,

$$(\psi_1 + \psi_2)(m) = \psi_1(m) + \psi_2(m)$$

and

$$(\alpha \psi_i)(m) = \alpha(\psi_i(m)), \alpha \in F, i = 1, 2; m \in M;$$

multiplication is associative and distributes addition. The set $Hom_A(M, N)$ denotes the set of all homomorphisms and it is considered as a vector space over the field F. In particular, $Hom_A(M, M)$ is such that homomorphisms from M to M can be multiplied. It forms an F-algebra called the algebra of endomorphisms of the module M and is denoted by $E_A(M)$. Its elements are endomorphisms of M. We note that an A-module M is called simple if its only submodules are (O_M) and M.

Lemma 16.4.2 (Schur lemma) : *If M and N are simple A-modules, every nonzero homomorphism* $\psi : M \to N$ *is an isomorphism.*

Proof : $ker\,\psi$ and $Im\,\psi$ are submodules of M and N respectively. ψ is a nonzero homomorphism implies that $ker\,\psi \neq M$ and $Im\,\psi \neq (O_N)$. As M is simple, $ker\,\psi = (O_M)$. So, ψ is a monomorphism. As N is simple, $Im\,\psi = N$. That is, ψ is an epimorphism. Thus, ψ is an isomorphism.   □

Corollary 16.4.1 : *A regular A-module is simple if, and only if, A is a division algebra.*

Proof : We observe that $A \cong \mathcal{L}(A)$. So, when A has no nontrivial left ideals (sub-spaces N' which absorb products from left), A has no zero divisors. That is, A is a division algebra. As a left ideal N' of A is realized as the kernel of a linear operator, when A is a division algebra, the regular A-module is simple.   □

Next, we remark that commutative semisimple algebras have been considered in chapter 14. By theorem 127 or by Schur lemma, one obtains

Fact 16.4.3 : (Weierstrass-Dedekind theorem) A commutative semisimple algebra is isomorphic to a direct sum of fields. Conversely, a direct sum of fields is a semisimple algebra (see theorem 127, chapter 14).

Proof is omitted.

Let D denote a finite dimensional division algebra over a field F. Suppose that V is a finite dimensional left D-module. We call V a finite dimensional vector space over the division algebra D.

Fact 16.4.4 :

a) V is a semisimple D-module.

b) Every vector space over D is isomorphic to a direct sum of n copies of D. $(n = \dim V)$

c) If $A = M_n(D)$, V, as an A-module, is simple.

d) $A = M_n(D)$ is simple.

e) An algebra A is semisimple if, and only if, there exists a faithful semisimple A-module.

For proofs, see Yu. A Drozd and V. V. Kirichenko [6].

Theorem 159 (Wedderburn-Artin theorem (1908)) : *Every semisimple algebra is isomorphic to a direct sum of matrix algebras over division algebras. Conversely, a direct sum of matrix algebras over division algebras is a semisimple algebra.*

Proof : $\Rightarrow$ Let A be a semisimple algebra. Suppose M is a regular A-module. By Fact 16.4.4 (e), M is regular $\Rightarrow$ M is faithful and so M is semisimple.

$$M \cong n_1 M_1 \oplus n_2 M_2 \oplus \cdots \oplus n_k M_k$$

is a decomposition of M into a direct sum of simple modules M_i $(i = 1, 2, \ldots, k)$. $M_i \neq M_j$, for $i \neq j$. So, we get

$$A \cong S_1 \oplus S_2 \ldots \oplus S_k \text{ where } S_i = n_i M_i.$$

By Schur lemma, $Hom_A(M_i, M_j) = (0)$ for $i \neq j$.

Then, $A = \oplus \sum_{i=1}^{k} A_i$, where $A_i = E_A(S_i)$, as $E_A(M) \cong \oplus \sum_{i=1}^{n} E_A(S_i)$. As $S_i = n_i M_i$
$(i = 1, 2, \ldots k)$, $A_i \cong M_n(D_i)$ with a division algebra $D_i = E_A(M_i)$.

$\Leftarrow$: If $A = \oplus \sum_{i=1}^{k} A_i$ where $A_i = M_n(D_i)$, the regular A-module M can be decomposed

into a direct sum $\oplus \sum_{i=1}^{k} M_i$ with M_i as regular A_i-modules. As every module over
$A = M_n(D)$ is semisimple, M_i are semisimple modules. That is, A, being a direct
sum of semisimple modules, is semisimple. $\qquad \square$

Let G be a group. A representation of G over a field F is a homomorphism
$T : G \rightarrow A(V)$, where $A(V)$ denotes the group of invertible linear operators on
V, a vector space (over F). The terminology is similar to that of representations
of an algebra. G acts on V as a group of F-linear automorphisms of V by the
action $g \cdot v = T_g(v)$, for $g \in G, v \in V$. V becomes a left G-module. This module
corresponds to T. Conversely, given M as a left G-module, for $g \in G$ we make
$T_g : M \rightarrow M$ by defining $T_g(m) = g \cdot m$. Then, T_g becomes a linear operator on M.
That is, we get a representation of G corresponding to the module M. As for the
case of algebras, we can talk about the representation module of group G.

Remark 16.4.1 : Theorem 159 has been included to give a flavour of 'semisimplicity' via representation theory.

Next, given a finite group G, we consider the group algebra FG (see definition
11.4.2, chapter 11). We need

Fact 16.4.5 : (Maschke's theorem) Let G be a finite group of order n. Given a field
F of arbitrary characteristic (0 or a prime p), the group algebra FG is semisimple
if, and only if, the characteristic of F does not divide n, the order of G.

For proof, see Yu. A. Drozd and V. V. Kirichenko [6].

Theorem 160 (S. Z. Ditor (1971)) :
$\quad$ *a) A finite group of G of odd order is the group of units of some ring if, and
only if, G is abelian and is the finite product of cyclic groups G_i where order of
each G_i is of the form $2^{k_i} - 1$.*
$\quad$ *b) If G is the group of units of a ring R and if G is finite and of odd order,
then the subring $[G]$ (of R) generated by G is a finite direct sum of Galois fields
of characteristic 2, namely,*

$$[G] = \oplus \sum_{i=1}^{r} GF(2^{k_i}).$$

Proof : We prove (b) first.
$\quad$ Since G is of odd order, $-1_R = 1_R$. Otherwise $\{-1_R, 1_R\}$ will be a subgroup of
G of order 2. Therefore, the subring $[G]$ (of R) generated by G is a finite dimensional algebra over a field $GF(2)$ of characteristic 2. Now, $[G]$ is a representation

module of G over $GF(2)$. Also the characteristic 2 of $GF(2)$ does not divide $o(G)$ (being odd). By Fact 16.4.5 (Maschke's theorem), $[G]$ is semisimple. By theorem 159, $[G]$ is the finite direct sum of rings A_i where each A_i is the full ring of $n_i \times n_i$ matrices (for some n_i) over a division ring D_i; $i = 1, 2, \ldots, r$. By Wedderburn's theorem (which says that a finite division ring is a field), the division rings D_i are fields. Since $-1_R = 1_R$ in $[G]$, each D_i is a Galois field of characteristic 2.

Now, if $M_n(F)$ is the ring of $n \times n$ matrices over a finite field F having s elements. $M_n(F)$ has precisely

(16.4.1) $(s^n - 1)(s^n - s) \cdots (s^n - s^{n-1})$

units (matrices whose rows are linearly independent). When s is of the form $2^k (k \geq 1)$, the number of units obtained from (16.4.1) is odd if, and only if, $n = 1$. Hence

$$[G] = \oplus \sum_{i=1}^{r} GF(2^{k_i}).$$

Proof of (a) (using (b)). Since the multiplicative group of $GF(2^{k_i})$ is cyclic of order $2^{k_i} - 1$, the group of units of R is a direct product of cyclic groups of order $2^{k_i} - 1 (i = 1, 2, \ldots, r)$ and so G is cyclic and hence abelian.

Conversely, if a finite group G is abelian and is a finite direct product of cyclic groups G_i, where $o(G_i) = 2^{k_i} - 1, (i = 1, 2, \ldots r)$, considering $\{0\} \cup G = S$ and obtaining a ring generated by S, say $[S]$, we see that G is the group of units of $[S]$ and G is of odd order.   □

Corollary 16.4.2 : *Let* $t = p^m$ *where* p *is a prime and* $m \geq 1$. *Then,* t *is equal to the number of units of a ring* R *if, and only if,* $p = 2$ *or* p *is Mersenne prime* $M_q = 2^q - 1$ *where* q *is a prime.*

Proof : If $p = 2$, the direct sum of m fields $GF(2)$ has 2^m units.

If $p = M_q = 2^q - 1$ (a Mersenne prime) the direct sum of m fields $GF(2^q)$ has p^m units. Conversely, let G be a group of odd p^m and suppose that G is the group of units of a ring R. By theorem 160, G is a direct product of cyclic groups of order $2^{k_i} - 1$ $(i = 1, 2, \ldots, r)$. So, $p^m = \prod_{i=1}^{r} (2^{k_i} - 1)$.

So, there exist integers n and k such that $p^n = 2^k - 1$. When n is even,

$$p^n - 1 = (p - 1)(p^{n-1} + p^{n-2} + \cdots + p + 1)$$

and $p^n - 1$ is divisible by 4. If $k \neq 1$, $2^k - 2$ is not divisible by 4. So $p^n = 2^k - 1$ is not feasible, when n is even. Therefore, n is odd and $p^n + 1 = 2^k$. Then, as $(p + 1)$ divides $p^n + 1$, $p + 1$ has to be a power of 2. Thus, $p = 2^q - 1$, making p a Mersenne prime.   □

16.5. Quadratic reciprocity in a finite group

In this section, we revisit the law of quadratic reciprocity (see theorem 42, chapter 6) in the context of a finite group. In [7], William Duke and Kimberly Hopkins give a nice generalization via group characters.

Let p be an odd prime. Then $\mathbb{F}_p^* = \mathbb{Z}/p\mathbb{Z} \setminus [0]$ is a cyclic group of order $(p-1)$. The associated Dirichlet character is the Legendre symbol $(\cdot \mid p)$. In 1872, Zolotarev [7] referred to the sign of a permutation of the elements of $G = \mathbb{Z}/p\mathbb{Z}$ induced by multiplication by a where $p \nmid a$.

Definition 16.5.1 : *Let α be a permutation of a finite set S. Sign α, written $sgn\,\alpha$, is given by*

(16.5.1)
$$sgn\,\alpha = \begin{cases} 1, & \text{if } \alpha \text{ is an even permutation} \\ -1, & \text{otherwise.} \end{cases}$$

Since α is expressed as a product of s transpositions (say), $sgn\,(\alpha) = (-1)^s$. If β is another permutation expressed as a product of t transpositions,

(16.5.2)
$$sgn\,(\alpha \circ \beta) = (-1)^{s+t} = (sgn\,\alpha)\,(sgn\,\beta).$$

Thus, sgn is a mapping of S_r (the symmetric group on r symbols) into the multiplicative group $\{1,-1\}$ which preserves products. In fact, $sgn : S_r \to \{1,-1\}$ is a homomorphism of S_r into $\{1,-1\}$ with kernel A_r, the alternating group.

In the case of $\mathbb{F}_p^*$ which is a cyclic group of order $(p-1)$, the map $\pi : \mathbb{F}_p^* \to \mathbb{F}_p^*$ given by $\pi(a) = a^2$ is such that image of $\mathbb{F}_p^*$ under π is the unique subgroup of order $\frac{p-1}{2}$ and $ker\,\pi$ is the unique subgroup of order 2 and is equal to $\{1,-1\}$. This unique subgroup of order 2 identifies the Legendre symbol $(\cdot \mid p)$. We also observe that a generator of $\mathbb{F}_p^*$ induces a $(p-1)$-cycle which is an odd permutation.

In what follows, we take G to be a finite group of order r. We recall that if F is a field, the general linear group $GL_n(F)$ is given by

(16.5.3) $\quad GL_n(F) = \{A : A \text{ is an invertible } n \times n \text{ matrix with entries from } F\}$.

Definition 16.5.2 : *An n-dimensional matrix representation of G is a homomorphism $\psi : G \to GL_n(F)$.*

We denote the image of $g \in G$ made ψ by $\psi(g)$ and it is an invertible matrix, as an element of $GL_n(F)$. For $g, h \in G$,

$$\psi(g.h) = \psi(g)\psi(h).$$

Notation 16.5.1 : Let V be a finite dimensional vector space over F. $GL(V)$ denotes the group of invertible linear operators on V. The choice of a basis for V determines an isomorphism of $GL(V)$ with $GL_n(F)$.

We recall

Definition 16.5.3 : *A representation of G on V is a homomorphism $\rho : G \to GL(V)$. The dimension of the representation ρ is defined as the dimension of the vector space V (over F).*

As $V \cong F^n$, matrix representation of G can be considered as representations of G on the space F^n of column vectors. As remarked in M. Artin [1], all representations of G on finite dimensional vector spaces can be reduced to matrix representation, once we decide to choose a basis.

Definition 16.5.4 : *Two representations* $\rho : G \to GL(V)$ *and* $\rho' : G \to GL(V')$ *of a group G are called isomorphic (or equivalent) if there is an isomorphism of vector spaces V and V' given by* $T : V \to V'$ *which is compatible with the group operation in the sense that for* $v \in V$, $g \in G$,

(16.5.4) $gT(v) = T(gv)$

or

(16.5.5) $\rho'(g)(T(v)) = T(\rho(g)v)$.

If $\mathbb{B}$ is a basis of V and if $\mathbb{B}' = T(\mathbb{B})$ is the corresponding basis of V', the associated matrix representations $\psi(g)$ and $\psi(g')$ are equal, where $g' \in G$.

Next, let V be an n-dimensional vector space order $\mathbb{C}$ (the held of complex numbers).

Definition 16.5.5 : *The character* χ *of a representation* ρ *for a choice of a basis of V is the map* $\chi : G \to \mathbb{C}$ *defined by*

(16.5.6) $\chi(g) = trace\ (\rho(g))$

If ψ *is the matrix representation obtained from* ρ *for a choice of a basis of V,*

(16.5.7) $\chi(g) = trace\ (\psi(g)) = \lambda_1 + \lambda_2 \cdots + \lambda_n$

where $\lambda_i (i = 1, 2 \cdots n)$ *are the eigenvalues of* $\psi(g)$.

Let ρ be a representation of G on a finite dimensional vector space V. Given ρ, it is known [1] that for each $g \in G$, there is a basis of V so that the matrix of $\rho(g)$ is diagonal. We will look for a basis which will diagonalize $\rho(g)$ for all $g \in G$. If G is a finite abelian group, every matrix representation ψ of G is diagonalisable. That is, there exists $\rho \in GL_n(\mathbb{C})$ such that $\rho\ \psi(g)\rho^{-1}$ is a diagonal matrix for all $g \in G$.

Definition 16.5.6 : *Let* ρ *be a representation of G on a vector space V. A subspace W of V is called G-invariant if* $\rho(g)w \in W$ *for all* $w \in W$ *and* $g \in G$. *That is,* $\rho(g)W \subseteq W$ *for all* $g \in G$.

Definition 16.5.7 : *If a representation* ρ *of a (finite) group G on a vector space V has no proper G-invariant subspaces, it is called an irreducible representation.*

We remark that the character of an irreducible representation is called an irreducible character.

Fact 16.5.1 : Given a (finite) group G, let $\rho_1, \rho_2, \cdots$ be the distinct isomorphism classes of irreducible representations of G. Let χ_i be the character of ρ_i.

a) *orthogonality relations*: The characters χ_i are orthonormal. That is,

(16.5.8) $\langle \chi_i, \chi_j \rangle = \dfrac{1}{|G|} \displaystyle\sum_{g \in G} \overline{\chi_i(g)}\ \chi_j(g) = \begin{cases} 1, & \text{if } j = i \\ 0, & \text{otherwise.} \end{cases}$

b) There are finitely many isomorphism classes of irreducible representations, the same number as the number of conjugacy classes in the group.

For proof, see Michael Artin [1, chapter 9 sections 1,2,4,5 and 9].

Next, if $f : G \to \mathbb{C}$ is such that f is a constant on each conjugacy class of G, f is called a class-function on the set of conjugacy classes. The set of class-functions defined on G forms a vector space $\mathscr{C}$ (say) over $\mathbb{C}$. It is verified that the set of irreducible characters of G forms an orthonormal basis of $\mathscr{C}$. Thus, $\mathscr{C}$ is a finite dimensional complex vector space.

If χ is a character of G, $\chi(g)$, $g \in G$, is called a character value. As $\mathscr{C}$ has a basis consisting of a finite number of distinct irreducible characters, the character values of all representations of G could be found out. For let $\chi_1, \chi_2, \cdots, \chi_m$ be the distinct irreducible characters of G. Then, $\chi \in \mathscr{C}$ is expressible as

(16.5.9) $$\chi = k_1 \chi_1 + k_2 \chi_2 + \cdots + k_m \chi_m$$

where $k_i = \langle \chi, \chi_i \rangle$ $(i = 1, 2, \cdots m)$, (see (16.5.8)).

Now, if $\{g_1, g_2 \cdots, g_m\}$ is a set of representatives of conjugacy classes in G, we form a table, which is referred to as the character table of G.

	g_1	g_2	g_j	g_m
χ_1	$\chi_1(g_1)$	$\chi_1(g_2)$	$\chi_1(g_j)$	$\chi_1(g_m)$
χ_2	$\chi_2(g_1)$	$\chi_2(g_2)$	$\chi_2(g_j)$	$\chi_2(g_m)$
$\vdots$	$\cdots$	$\cdots$	$\cdots$	$\cdots$
χ_i	$\chi_i(g_1)$	$\chi_i(g_2)$	$\chi_i(g_j)$	$\chi_i(g_m)$
	$\cdots$	$\cdots$	$\cdots$	$\cdots$
χ_m	$\chi_m(g_1)$	$\chi_m(g_2)$	$\chi_m(g_j)$	$\chi_m(g_m)$

Table 1

Next, we denote the order of G by r. Suppose that $a \in \mathbb{N}$ and g.c.d $(a, r) = 1$. We denote the set of m conjugacy classes in G by

(16.5.10) $$\Gamma = \{C_1, C_2, \cdots, C_m\}.$$

$m = r$ if, and only if, G is abelian. a induces a permutation of Γ, if we define $\psi : \Gamma \to \mathbb{F}$ by $\psi(g) = g^a$, for all $g \in G$. It means that

$$C_j^a = \{g^a : g \in C_j\}.$$

(16.5.11) ψ takes C_j to C_j^a $(j = 1, 2, \cdots m)$.

Definition 16.5.8 : *The quadratic symbol for G at any nonzero integer a is defined by*

$$(a|G) = \begin{cases} 0, & \text{if g.c.d } (a,r) \neq 1; \\ 1, & \text{if } \psi \text{ is even;} \\ -1, & \text{if } \psi \text{ is odd} \end{cases}$$

(where ψ is as given in (16.5.11)).

If we take the l.c.m k of orders of elements of G, $(a|G)$ is a Dirichlet character modulo k (see definition 15.2.3). In the special case when $G = \mathbb{Z}/p\mathbb{Z}$, p an odd prime

(16.5.12) $(a|G) = (a|p)$, the Legendre symbol.

Definition 16.5.9 : $C_j \in \Gamma$ (16.5.10) *is said to be a real conjugacy class, if* $C_j^{-1} = C_j$. *If* $C_j^{-1} \neq C_j$, C_j *is said to be a complex conjugacy class.*

We note that C_j^{-1} is obtained from Γ by taking the conjugacy class of g^{-1} where g determines C_j. Complex conjugacy classes occur in pairs $\{C_j, C_j^{-1}\}$ with $|C_j| = |C_j^{-1}|$. Then, as is done in the case of number fields, we could write $|\Gamma| = m = s + 2t$ where t is half the number of complex conjugacy classes.

Notation 16.5.2 : Given a finite group G, if $m = |\Gamma|$ and $m = s + 2t$ (s being the number of real conjugacy classes), we write

(16.5.13) $$d = d(G) = (-1)^t |G|^s \prod_{j=1}^{s} |C_j|^{-1}.$$

We recall that for $a \in G$, the normalizer of a in G is given by

$$N(a) = \{g \in G : g \cdot a = a \cdot g\}.$$

$N(a)$ is a subgroup of G and if $a \in C_j$, $|C_j| = \frac{|G|}{|N(a)|}$. (See I. N. Herstein [13, chapter 2, section 11 pp 69–77]). So, $|C_j|^{-1} = \frac{N(a)}{|G|}$. Therefore d given in (16.5.13) is an integer.

Remark 16.5.1 : It will be shown in theorem 161 that

$$d = 0 \text{ or } 1 (\text{mod } 4).$$

In a sense, d resembles Δ, the discriminant of a quadratic number field. (See Section 15.2, chapter 15). Therefore, it is appropriate to call d (given in (16.5.13)), the discriminant of G.

Let b be an odd positive integer. Suppose that $b = \prod_{i=1}^{k} p_i$ (p_i primes not necessarily distinct). In (6.7.1), chapter 6, we have seen the Jacobi symbol $(a|b)$ expressed as

(16.5.14) $$(a|b) = \prod_{i=1}^{k} (a|p_i)$$

where $(a|p_i)$ is Legendre symbol. $(a|b) = 1$, if $b = 1$.

Fact 16.5.2 : The following statements are valid for the Jacobi symbol.
  (i) If $b > 0$, b odd and $a \equiv a' (\text{mod } b)$ with g.c.d $(a,b) = 1$

(16.5.15) $$(a|b) = (a'|b).$$

(ii) If $b > 0$, b odd; $b' > 0$, b' odd and g.c.d (a, b) = g.c.d (a, b') = 1

(16.5.16) $$(a|b)(a|b') = (a|bb').$$

(iii) If $b > 0$, b odd

(16.5.17) $$(-1|b) = (-1)^{\frac{b-1}{2}}.$$

(iv) If $b > 0$, b odd

(16.5.18) $$(2|b) = (-1)^{\frac{b^2-1}{8}}$$

(v) Let a, b be odd and g.c.d (a, b) = 1. Then,

(16.5.19) $$(a|b)(b|a) = \begin{cases} -(-1)^{\left(\frac{a-1}{2}\right)\left(\frac{b-1}{2}\right)}, & \text{if} \quad a < 0 \text{ and } b < 0 \\ \\ (-1)^{\left(\frac{a-1}{2}\right)\left(\frac{b-1}{2}\right)}, & \text{if} \quad a > 0 \text{ and } b > 0, \text{ or} \\ & \qquad a < 0 \text{ and } b > 0 \text{ or} \\ & \qquad a > 0 \text{ and } b < 0. \end{cases}$$

For proofs, see E. Landau [16].

Kronecker symbol (15.2.19) was already used in Section 15.2, chapter 15. We need to look into it more closely: Following Landau [16], Kronecker symbol is introduced as given below:

Let $d \equiv 0$ or $1 \pmod 4$. d is not a perfect square.

(For instance, $d = 5, 8, 12, 13, 17, 20, 21, \ldots$ or $-3, -4, -7, -8, \cdots$). Let $a > 0$.

Definition 16.5.10 : *The symbol $(d|m)$ is always given a meaning by means of the following:*

$$(d|p) = 0, \ \textit{if } p|d \ (p \textit{ a prime}),$$

$$(d|2) = \begin{cases} 1, & \textit{if } d \equiv 1 \pmod 8, \\ -1, & \textit{if } d \equiv 5 \pmod 8. \end{cases}$$

It means that $(d|2)$ = the Jacobi symbol $(2|d)$ for 2 not dividing d.

$$(d|p) = \textit{Legendre symbol, if } p > 2 \textit{ and } p \nmid d$$

$$(d|a) = \prod_{i=1}^{k} (d|p_i), \textit{ where } a = \prod_{i=1}^{k} p_i$$

and $(d|1) = 1$.

For those integers d and a for which Kronecker and Jacobi symbols are defined namely, odd $a > 0$ and g.c.d $(d, a) = 1$, both definitions agree. Further, if g.c.d $(d,a) > 1$, $(d\,|a) = 0$. If g.c.d $(d,a) = 1$, $(d\,|a) = \pm 1$ (for the above-mentioned d).

Fact 16.5.3 : If $a > 0$, $a' > 0$, then

$$(d\,|aa') = (d\,|a)\,(d\,|a').$$

Next, let $|d| = D$.

Fact 16.5.4 : Let $a > 0$ and g.c.d $(d,a) = 1$. If d is odd, $(d\,|a) = (a\,|D)$ (the Jacobi symbol). If d is even and $d = 2^q m$ (m odd), given $|m| = M$, then,

(16.5.20) $$(d\,|a) = (2\,|a)^q(-1)^{(\frac{m-1}{2})(\frac{a-1}{2})}(a\,|M).$$

(symbols on the right are Jacobi symbols.)

For proofs, see E. Landau [16].

As mentioned in [16], the Kronecker symbol $(d\,|a)$ when expressed as a function (of a) gives a number-theoretic function $f_d : \mathbb{N} \to \mathbb{C}$ taking values in $\{0, \pm 1\}$.

Fact 16.5.5 : The function $(d\,|a)$ has the following properties:

(i) $(d\,|a) = 0$ if g.c.d $(d,a) > 1$.
(ii) $(d\,|1) = 1$.
(iii) $(d\,|a_1 a_2) = (d\,|a_1)(d\,|a_2)$, $a_1 > 0$, $a_2 > 0$.
(iv) $(d\,|a_1) = (d\,|a_2)$ whenever $a_1 \equiv a_2 (\bmod D)$; $D = |d|$.
(v) If d is odd and g.c.d $(D,a) = 1$ writing $d = p^r m$, where p is odd, r odd, p does not divide m we can choose a quadratic nonresidue $s(\bmod p)$ so that $a \equiv s \,(\bmod p)$, $a \equiv 1(\bmod m)$ and as g.c.d $(D,a) = 1$

$$(d\,|a) = (a\,|D) = (a\,|p)^r(a\,|m)$$
$$= (s\,|p)^r(1\,|m)$$
$$= (-1)^r$$

or

$$(d\,|a) = -1, \text{ when } r \text{ is odd}.$$

(vi) If d is even, so that $d = 2^q m$, where m is odd, choose q odd and $a \equiv 5 \,(\bmod 8)$, $a \equiv 1(\bmod |m|)$. Then, g.c.d $(D,a) = 1$ and by (16.5.20)

$$(d\,|a) = (2\,|a)^q(-1)^{(\frac{m-1}{2})(\frac{a-1}{2})}(a|\,|m|).$$

Or,

$$(d\,|a) = (2\,|a) \cdot 1 \cdot (1|\,|m|) = -1.$$

(vii) If d is even and q is even, m is not a perfect square. If g.c.d $(D,a) = 1$ and $a > 0$

$$(d\,|a) = (-1)^{(\frac{m-1}{2})(\frac{a-1}{2})}, \ (a|\,|m|)$$

If $m = 3(\bmod\ 4)$, choose $a \equiv -1(\bmod\ 4)$, $a \equiv 1(\bmod\ |m|)$. As g.c.d $(D,a) = 1$, $a > 0$,

$$(d\,|a) = (-1)^{\frac{m-1}{2}}(1\,|\,|m|) = -1.$$

If $m \equiv 1(\bmod\ 4)$, when g.c.d $(D,a) = 1$, $a > 0$,

$$(d\,|a) = (a|\,|m|) = -1.$$

(viii) From (vi) and (vii), one gets $(d\,|a) = -1$ for a suitable value of a. Details of proof of (i) to (iv) are in [16].

Lemma 16.5.1 (Landau (1927)) : *Given $d \equiv 0$ or $1(\bmod\ 4)$ and $D = |d|$*

$$(16.5.21) \qquad (d|D-1) = \begin{cases} 1, & \text{if } d > 0 \\ -1, & \text{if } d < 0. \end{cases}$$

Proof :
Case (i): d an odd integer.
By Fact 16.5.4, $(d\,|D-1) =$ the Jacobi symbol $(D-1|D)$.
But,

$$(D-1|D) = (-1|D) = (-1)^{\frac{D-1}{2}} = \begin{cases} 1 & \text{if } d > 0, \\ -1 & \text{if } d < 0. \end{cases}$$

Case (ii): d even and $d = 2^q m$, m odd. Then, by Fact 16.5.4

$$(d|D-1) = (2|D-1)^q(-1)^{\frac{m-1}{2}}(D-1|M)$$

where $M = |m|$. Now, $(2|D-1)^q = 1$, since it is true for $q = 2$, and for $q \geq 3$, as $D-1 \equiv 7(\bmod\ 8)$, $(2|D-1)^q = 1$.
    Next,

$$(-1)^{\frac{m-1}{2}}(D-1\,|M) = (-1)^{\frac{m-1}{2}}(-1\,|M)$$

$$= (-1)^{\frac{m-1}{2}}(-1)^{\frac{M-1}{2}}.$$

Or,

$$(-1)^{\frac{m-1}{2}}(D-1\,|M) = \begin{cases} 1, & d > 0 \\ -1 & d < 0. \end{cases}$$

$\square$

Theorem 161 (Landau (1927)) : *For $a > 0$, $b > 0$ and $a \equiv -b(\bmod\ |D|)$*

$$(16.5.22) \qquad (d\,|a) = \begin{cases} (d\,|b), & \text{if } a > 0 \\ -(d\,|b) & \text{if } d < 0. \end{cases}$$

Proof :

$$(d\,|a) = (d\,|Db-b) = (d\,|b(D-1))$$

$$= (d\,|b)(d\,|D-1)$$

which yields (16.5.22) using (16.5.21). $\square$

If d is any integer and a a positive odd integer, we apply Jacobi's generalized reciprocity law (16.5.19) to obtain

$$(16.5.23) \qquad (d \,|\, a)((-1)^{\frac{a-1}{2}} a \,|\, d) = 1.$$

It implies that

$$(16.5.24) \qquad (d \,|\, a) = ((-1)^{\frac{a-1}{2}} a \,|\, d)$$

which is an equivalent form of the quadratic reciprocity law.

As a is odd, $(-1)^{\frac{a-1}{2}} a \equiv 1 \pmod 4$ and so $a^* = (-1)^{\frac{a-1}{2}} a$ is a discriminant. That is.

$$(d \,|\, a) = (a^* \,|\, d)$$

is the type of relation, we think of proving, given a finite group G as the premise.

Theorem 162 (William Duke and Kimberly Hopkins (2005)) **:** *Let G be a finite group having discriminant d* (16.5.13)*. Then,*

(i) $d \equiv 0$ or $1 \pmod 4$

(ii) $(a \,|\, G) = (d \,|\, a)$.

Proof : The set Γ (16.5.10) of conjugacy classes is such that if χ is a character of G, $\chi(C_j) = \chi(g)$, where $g \in C_j$. Let $\chi_1 = 1, \chi_2, \cdots, \chi_m$ be the irreducible characters of G. The character table of G is the $m \times m$ matrix.

$$(16.5.25) \qquad M = \begin{bmatrix} \chi_1(C_1) & \chi_1(C_2) & \cdots & \chi_1(C_m) \\ \chi_2(C_1) & \chi_2(C_2) & \cdots & \chi_2(C_m) \\ \cdots & \cdots & \cdots & \cdots \\ \chi_m(C_1) & \chi_m(C_2) & \cdots & \chi_m(C_m) \end{bmatrix}.$$

We denote the conjugate transpose of M by M^*. Then, the (i,j)-entry in M^*M is given by

$$\sum_{k=1}^{m} \bar{\chi}_i(C_k)\chi_k(C_j) = \begin{cases} |C_i|^{-1}|G|, & \text{if } j = i; \\ 0, & \text{if } j \neq i. \end{cases}$$

(as $\chi(C_j) = \chi(g)$ for all $g \in C_j$).

Thus,

$$(16.5.26) \qquad M^*M = \begin{bmatrix} |G||C_1|^{-1} & 0 & \cdots & 0 \\ 0 & |G||C_2|^{-1} & \cdots & 0 \\ & & & \\ 0 & & \cdots & |G||C_m|^{-1} \end{bmatrix}$$

M^*M is a diagonal matrix. Since $\chi(C_j^{-1}) = \overline{\chi(C_j)}$ for any character χ and any conjugacy class C_j, we obtain

$$(16.5.27) \qquad \det \bar{M} = (-1)^t \det M,$$

where t is as given in (16.5.13). From (16.5.26) and (16.5.27), we deduce that

$$(16.5.28) \qquad (\det M)^2 = q^2 d \text{ for some integer } q.$$

Each entry $\chi_i(C_j)$ of M is an algebraic integer in the cyclotomic field $Q(\zeta_r)$ where $\zeta_r = \exp(\frac{2\pi i}{r})$ and $r = |G|$. $Q(\zeta_r)$ is a finite extension of Q whose Galois group Gal $(Q(\zeta_r)/Q)$ is isomorphic to $((\mathbb{Z}/r\mathbb{Z})^*, \cdot)$ by the map

$$\sigma_a : \text{Gal}(Q(\zeta_r)/Q) \to (\mathbb{Z}/r\mathbb{Z})^*$$

given by

(16.5.29) $$\sigma_a(\zeta_r) = \zeta_r^a.$$

Then,

(16.5.30) $\qquad \sigma_a(\chi_j(g)) = \chi_j(g^a)$ for any χ_j and $g \in G$. $\quad (j = 1, 2, \cdots m)$.

By the definition of a determinant [12, chapter 6 section 9, pp 279–294],

$$\det M = \sum_{\sigma \in S_m} sgn\,(\sigma)\chi_1(C_{\sigma(1)}) \cdots \chi_m(C_{\sigma(m)}),$$

where the summation is over all $\sigma \in S_m$ and $sgn\,(\sigma) = \pm 1$, according as σ is even or odd.

We write

$$\det M = A - B,$$

where A is the sum of even permutations in S_m and B, the sum of odd permutation in S_m. By (16.5.30), $A + B$ and AB are invariant under the Galois group of automorphisms of $\mathbb{Q}(\zeta_r)$ leaving $\mathbb{Q}$ fixed.

Then,

$$q^2 d = (A - B)^2 = (A + B)^2 - 4AB \equiv (A + B)^2 \,(\text{mod } 4).$$

Thus,

$$q^2 d \equiv 0 \text{ or } 1(\text{mod } 4)$$

from which it follows that $d \equiv 0$ or $1(\text{mod } 4)$.

This proves (i).

Next, from (16.5.25) and (16.5.29), we note that

$$\sigma_a(\det M) = (a\,|G)\det M.$$

From (16.5.28)

(16.5.31) $$\sigma_a(\sqrt{d}) = (a\,|G)\sqrt{d}.$$

To show that $(a\,|G) = (d\,|a)$ we prove it for $a = p$, a prime such that p does not divide $|G|$ and for $a = -1$. If $p \nmid |G|$, we use the Frobenius automorphism σ_p. It is known [18] that p splits in $\mathbb{Z}[\zeta_r]$ if, and only if, σ_p fixes $\mathbb{Z}[\zeta_r]$ pointwise. So, p splits in the ring of integers of $\mathbb{Q}(\sqrt{d})$ if, and only if, $\sigma_p(\sqrt{d}) = \sqrt{d}$. Further, the Kronecker symbol $(d\,|p)$ has the fundamental property that p splits in the ring of integers of $\mathbb{Q}(\sqrt{d})$ if, and only if, $(d\,|p) = 1$. See P. Samuel [18]. Thus, from (16.5.31), we see that

$$(p\,|G) = (d\,|p).$$

For $a = -1$, since $\det(\bar{M}) = (-1)^t \det M$, by (16.5.13),

$$(-1\,|G) = (-1)^t = (d\,|-1).$$

This completes the proof of (ii). $\qquad\qquad\qquad\qquad\qquad\qquad\qquad\qquad$ $\square$

Example 16.5.1 : If $|G|$ is odd, C_1 is the only real conjugacy class. (See W. Burnside [4]). For, if $g \in G$ and g is in a real conjugacy class, we have $h^{-1}gh = g^{-1}$ for some h. Then, $h^{-2}g\,h^2 = g$. So, $h^2 \in N(g)$, the normalizer of g. Since $|G|$ is odd, order of h is odd, say $2k+1$. It follows that $h = (h^2)^{k+1}$ implying that $h \in N(g)$. Thus, $h^{-1}gh = (h^{-1}h)g = g = g^{-1}$. Since the order of g is odd, g is the identity element in G. So, then, in $m = s+2t, s = 1$. Therefore, from the equation (16.5.13) for d, we have

$$d = (-1)^{\frac{m-1}{2}} r.$$

As $d \equiv 0$ or $1 \pmod 4$, and r is odd, we have

$$(16.5.32) \qquad\qquad d = (-1)^{\frac{r-1}{2}} r = r^*$$

holds for any group of odd order.

Remark 16.5.2 : When $|G|$ is odd, W. Burnside [4] shows that $|G|$ and m are related by the congruence

$$|G| \equiv m \pmod{16}.$$

Example 16.5.2 : Let $\mathbb{F}_q$ denote a finite field of order $q = 2^k$ $(k \geq 1)$.

$$SL(2, \mathbb{F}_q) = \{A : A \text{ is a } 2 \times 2 \text{ matrix with entries from } \mathbb{F}_q \text{ and det } A = 1\}.$$

I. Schur [19] has shown that $|SL(2, \mathbb{F}_q)| = q(q^2 - 1)$ and $m = s = q+1$. Further,

$$(16.5.33) \qquad\qquad d = q^2(q+1)(q^2-1)^{q/2}.$$

It is known that d is a perfect square if, and only if, $q = 2^3 = 8$. When $q = 4$, $SL(2, \mathbb{F}_4) \cong A_5$ (a nonabelian group of order 60). Then, $(a|A_5) = (5|a)$. For $d = 16 \times 5 \times 15^2, d \equiv 0 \pmod 4$. So, $(a|A_5) = (d|a) = (5|a)$.

16.6. Worked-out examples

(a) Let R be a ring with the property that for every $a \in R$, there exists an integer $n = n(a) \geq 4$ such that

$$a + a^2 + a^3 = a^n + a^{n+1} + a^{n+2}.$$

Show that
(i) $a^{3n(a)-2} = a$, for every $a \in R$.
(ii) R is a commutative ring.
(iii) every element of R has finite additive order.

Answer: (i) More generally, suppose that for each $a \in R$, there exist integers $k = k(a)$ and $n = n(a)$ with $n > k \geq 1$ such that

$$(16.6.1) \qquad\qquad a + a^2 \cdots + a^k = a^n + a^{n+1} + \cdots + a^{n+k-1}.$$

To prove (i), we make a
Claim:

(16.6.2) $$a^{k(a)n(a)-k(a)+1} = a, \text{ for all } a \in R.$$

Step I: We first prove that R has no nonzero nilpotent elements. On the contrary, suppose that $x \in R$ is such that $x^r = 0_R$ for some $r > 1$ and r is minimal.
Using (16.6.1), we note that

(16.6.3) $$x+x^2\cdots+x^k = (x+x^2\cdots+x^k)x^{n-1}.$$

Iterating (16.6.3) m times, where $(n-1)m \geq r$, we obtain

$$x+x^2\cdots+x^k = 0_R.$$

Multiplying both sides by x^{r-2}, we get

$$(x+x^2\cdots+x^k)x^{r-2} = x^{r-1}+x^r+\cdots+x^{k+r-2} = 0_R.$$

It follows that $x^{r-1} = 0_R$ which contradicts the minimality of r. So, R has no nonzero nilpotent elements.
Step II: We rewrite (16.6.1) as

(16.6.4) $$(a+a^2+\cdots+a^k)(a^{n-1}-1) = 0_R.$$

Here, '1' is a formal symbol. It means that we get (16.6.1) when we multiply out the factor on the left side of (16.6.4). It is made clear that we are not assuming the existence of a multiplicative identity 1_R in R.
It follows that multiplying both sides of (16.6.4) by $(a-1)$, we obtain

$$(a-1)(a+a^2\cdots+a^k)(a^{n-1}-1) = (a^{k+1}-a)(a^{n-1}-1) = 0_R.$$

Or,

(16.6.5) $$a(a^k-1)(a^{n-1}-1) = 0_R.$$

We consider two polynomials $f(a)$ and $g(a)$ as defined below:

(16.6.6) $$f(a) = \sum_{i=0}^{n-2} a^{ik}, \quad g(a) = \sum_{j=0}^{k-1} a^{j(n-1)}$$

Then, we get

$$a(a^k-1)f(a) = a(a^{n-1}-1)g(a) = a^{kn-k+1}-a.$$

Therefore,

$$(a^{kn-k+1}-a)^2 = \{a(a^k-1)f(a)a(a^{n-1}-1)g(a)\}.$$

Or,

$$(a^{kn-k+1}-a)^2 = a(a^k-1)(a^{n-1}-1)af(a)g(a) = 0_R.$$

As R has no nonzero nilpotent elements, we conclude that

$$a^{kn-k+1}-a = 0_R.$$

Thus, (16.6.2) holds for all $a \in R$.

(ii) A theorem of Jacobson [13] states that if R is a ring such that for each $a \in R$, there exists an integer $m(a) > 1$ for which $a^{m(a)} = a$, then R is commutative. By (i), $m(a) = k(a)n(a) - k(a) + 1 > 1$ makes $a^{m(a)} = a$ for all $a \in R$. That is, R is commutative.

(iii) We need to prove that the additive order of any $a(\neq 0) \in R$ is finite. Let $a \in R$. $(a \neq 0)$. Then, $a + a = 2a \in R$. Using (i), we see that there exists integers s and t such that

(16.6.7) $$x^s = x, \quad (2x)^t = 2x.$$

Considering a^{s-1}, we see that

$$(a^{s-1})^2 = a^{2s-2} = a^s \cdot a^{s-2} = a \cdot a^{s-2} = a^{s-1}.$$

Thus, a^{s-1} serves as an idempotent element. In the same manner, we show that $(2a)^{t-1}$ is also an idempotent.
Therefore,

(16.6.8) $$a^{(s-1)(t-1)+1} = (a^{s-1})^{t-1} \cdot a = a^{s-1} \cdot a = a.$$

Likewise, $(2a)^{(s-1)(t-1)+1} = 2a$. Writing $(s-1)(t-1) + 1 = q$ (say), we see that

$$2a = (2a)^q = 2^q.a^q = 2^q a, \text{ by } (16.6.8).$$

Hence $(2^q - 2)a = 0_R$. Thus, a is of finite additive order.
The solution is complete. □

Remark 16.6.1 : The above worked-out example is due to Erwin Just. The solution given has been adapted from Charles Lanski: Solution to problem 10841, Amer. Math. Monthly 109 (2002) p 858. A nontrivial example of a ring possessing the property stated is yet to be found. For any set X if $\mathbb{P}(X)$ denotes the power set of X, $(\mathbb{P}(X), \Delta, \cap)$ is a ring. (Δ being symmetric difference and $\cap$, set-intersection). It is true that for any $A \in \mathbb{P}(X)$, $A \Delta A^2 \Delta A^3 = A = A^n \Delta A^{n+1} \Delta A^{n+2}, n \geq 4$. We know, already, that every element in $\mathbb{P}(X)$ is of additive order 2.

(b) We recall that a Mersenne number is one of the form $M_n = 2^n - 1$. It is known that if $n > 1$ and $a^n - 1$ is a prime, then $a = 2$ and n is a prime. Let p be a prime. Lucas (1876) has given a method of checking whether M_p is a prime or not. In fact, he showed that M_{127} is a prime.

Let p be an odd prime. Suppose that q is a positive integer such that $q < p, r = qp + 1$ or $qp^2 + 1$ and $2^q \equiv 1 \pmod{r}$. If $2^{r-1} - 1 \equiv 0 \pmod{r}$, show that r is a prime.

Answer: We write $r = qp^i + 1$, where $i = 1$ or 2. Suppose that the element [2] in $U(r)$ (the group of units in $\mathbb{Z}/r\mathbb{Z}$) has order d. That is, d is the least positive integer such that $[2]^d = [1]$. In other words, d is the least positive integer such that $2^d \equiv 1 \pmod{r}$. Then, $d \nmid q$. But $d \mid (r-1)$. That is, $d \mid qp^i$. There exists

d' such that $dd' = qp^i$. As $d \nmid q$, $p \mid d$. Now, $d \mid \phi r)$. So $p \mid \phi(r)$. Further, if
$r = \prod_{i=1}^{k} p_i^{a_i}$ (p_i, primes. $a_i \geq 1$; $i = 1, 2 \ldots k$)

(16.6.9) $\phi(r) = p_1^{a_1-1} p_2^{a_2-1} \cdots p_k^{a_k-1} (p_1-1)(p_2-1) \cdots (p_k-1)$.

Some $p \nmid r$, p divides at least one of $(p_i - 1)$ $(i = 1, 2 \ldots k)$. So, r has a prime factor s satisfying $s \equiv 1 (\text{mod } p)$. Let $r = s r'$. Since $r \equiv 1 \equiv s(\text{mod } p)$, $r' \equiv 1 (\text{mod } p)$. If $r' > 1$,

(16.6.10) $r = (mp+1)(m'p+1), \quad 1 \leq m \leq m'$.

Then,

$$qp^i + 1 \equiv (mm'p^2 + (m+m')p + 1,$$

or

(16.6.11) $qp^{i-1} = mm'p + (m+m')$.

If $i = 1$, $q = mm'p + (m+m')$ and so,

$$p \leq mm'p < q < p - \text{a contradiction}.$$

If $i = 2$, $qp = mm'p + (m+m')$. So

$$p \mid (m+m'), p < (m+m').$$

Therefore, $mm' < q$ and $q < p$ (given). Moreover, $2m' \geq (m+m') > p$. So, $mm' < q < p$. Or, $mm' \leq (p-2)$ and $m' > p/2$. That is,

$$m \leq \frac{(p-2)}{m'} < \frac{2(p-2)}{p} < 2.$$

Therefore, $m = 1$. But then, $m' > p-1, mm' \geq (p-1)$ — a contradiction. So, (16.6.10) is impossible, unless $r' = 1$. This forces r to be equal to s, a prime $\equiv 1(\text{mod } p)$.   □

We examine the factorability of $M_p = 2^p - 1$.

(c) (Euler) Let p be a prime of the form $4k+3$ and > 7. Show that $2p+1$ is a prime if, and only if,

(16.6.12) $2^p \equiv 1 (\text{mod } (2p+1))$,

(when $2p+1$ is a prime, $M_p = 2^p - 1$ is composite).

Answer: $:\Rightarrow$ Let $2p+1$ be a prime.
Then, $2p+1 = t$ is a prime $\equiv 7 (\text{mod } 8)$. 2 is a quadratic residue of s. So,

(16.6.13) $2^p = 2^{\frac{t-1}{2}} \equiv 1 (\text{mod } t)$.

So (16.6.12) is a necessary condition.
As $p > 7$, $k > 1$, $M_p = 2^p - 1 > s = 2p+1$, a prime. So, $M_p = 2^p - 1$ is a composite number.
$\Leftarrow$: Let (16.6.12) be given, when $p = 4k+3$. In worked-out example (b), taking $q = 2$, $r = 2p+1$, we note that $q < p$ and $2^q = 4 \not\equiv 1 (\text{mod } r)$. Now,

$$2^{r-1} = 2^{2p} \equiv 1 (\text{mod } r).$$

By worked-out example (b), r is a prime. Therefore, the condition (16.6.12) is sufficient. □

Remark 16.6.2 : Worked-out examples (b) and (c) have been adapted from Hardy & Wright [A1, chapter VI pp 78–80]. Worked-out example (c) gives a criterion for testing the primality of Mersenne numbers M_p where p is a prime $\equiv 3 \pmod 4$ and $p > 7$. For instance, for $p = 11$, 23 is a factor of M_{11}. For $p = 23$, 47 is a factor of M_{23}.

Remark 16.6.3 : The Fibonacci type of a primality test is the following:
Let $F_0 = F_1 = 1$ and $F_{n+1} = F_n + F_{n-1} (n \geq 1)$. If F_n is divisible by N for $n = N + 1$, but not for $n = \frac{N+1}{p}$, where p ranges over the prime factors p of $N + 1$, then N is a prime.

REFERENCES

[1] M. Artin: Algebra, Prentice Hall of India Private Ltd. New Delhi 110 001 (1994) Eastern Economy Edn.

[2] M. F. Atiyah and I. G. Macdonald: Introduction to Commutative Algebra, Addison-Wesley Publishing Co., Inc Reading, Mass. (1969).

[3] N. Bourbaki: Commutative Algebra, Chapter VI, Houghton Mifflin Co, (1972) Boston.

[4] W. Burnside: Theory of groups of finite order, 2nd edn., Cambridge Univ. Press, Cambridge (1911) Dover reprint (1937).

[5] S. Z. Ditor: On the group of units of a ring, Amer. Math. Monthly 78 (1971) 522–523.

[6] Yu. A. Drozd and V. V. Kirichenko: Finite dimensional algebras (translated from the Russian by Vlastimil Dlab) Springer Verlag, Berlin, Heidelberg, New York (1994).

[7] William Duke and Kimberly Hopkins: Quadratic reciprocity in a finite group, Amer. Math. Monthly 112 (2005) 251–256.

[8] K. E. Elridge: On ring structures determined by groups, Proc. Amer. Math. Soc. 23 (1969) 472–477.

[9] K. E. Elridge and I. Fisher: D. C. C. wing with a cyclic group of units, Duke Math. J. 34 (1967) 243–248.

[10] Robert Gilmer: Finite wings having a cyclic multiplicative group of units, Amer. J. Math 85 (1963) 447–452.

[11] Robert Gilmer: Multiplicative Theory of ideals Chap III. Marcel Dekker Inc, (1972) New York.

[12] I. N. Herstein: Topics in algebra, Blaidsell Pub Co. NY (1965) Third printing.

[13] I. N. Herstein: Noncommutative rings, Carus Math. monographs, No. 15 MAA, (1968) p 73.

[14] Harry C. Hutchinson: Examples of commutative rings, chapter 2 pp 12–22, Polynomial Publishing House, (1981) 80 Passaic Ave, Passaic NJ 07055.

[15] I. Kaplansky: Commutative rings, Allyn & Bacon Inc, (1970) Boston.

[16] E. Landau: Elementary number thoery, Chelsea Publishing Co. N. Y., 2nd Edn. (1966) reprint, Part I chapter VI pp 53–75.

[17] Peter Malcolmson and Frank Okoh: Rings without maximal ideals, American Math. Monthly 107 (2000) 61–66.

[18] P. Samuel: Algebraic theory of numbers : (trans. A. J. Silberger) Houghton Mifflin, Boston (1970).

[19] I. Schur : Untersuchungen über die Darstellung der endliche Gruppen durch gebrochene lineare Substitutionen, J. Reine. Angewant. Math. 132 (1907) 85–137; and also in Gesammelte Abhandlungen, vol. I, Springer-Verlag, Berlin, 1973, pp. 198–250.

[20] Fabrizio Zanello: When are there infinitely many irreducible elements in a principal ideal domain? Amer. Math. Monthly, 111 (2004) 150–152.

ADDITIONAL REFERENCES

[A1] G. H. Hardy and E. M. Wright: An introduction to the theory of numbers, Oxford University Press, UK (1937), Reprint, 5th Edn. (1979).

[A2] G. Karpilovsky: Commutative group algebras, Monographs and Textbooks in Pure and Applied Mathematics No: 78, Marcel Dekker Inc., New York (1983) chapter 3, section 5 pp 75–79.

True/False statements : Answer key

	(a)	(b)	(c)	(d)	(e)	(f)
chapter 1	T	T	T	T	T	F
chapter 2	T	T	F	F	F	T
chapter 3	T	F	T	T	T	T
chapter 4	T	F	T	T	T	T
chapter 5	F	T	T	T	T	T
chapter 6	T	T	F	T	F	T
chapter 7	T	T	T	F	T	F
chapter 8	T	T	T	T	F	T
chapter 9	F	T	F	F	T	T
chapter 10	T	T	T	F	F	T
chapter 11	T	F	F	T	T	F
chapter 12	T	F	T	F	T	T
chapter 13	T	F	F	T	T	T
chapter 14	T	F	T	T	T	T
chapter 15	T	F	T	F	F	T

Index of some selected structure theorems/results

20 The ring $\mathcal{A}_r(F)$ of (r,F)-arithmetic functions is a semisimple algebra over F and can be expressed as a direct sum of r fields each isomorphic to F.

Proposition 14.2.1   *p 493*

21 CARLITZ CONJECTURE : $(\mathcal{A}', +, *)$ denotes the Lucas ring of arithmetic functions $f : \widetilde{\mathbb{Z}} \to F$, where F is a field of characteristic zero. $f \in \mathcal{A}'$ is a zero divisor if, and only if, f is nilpotent.

Section 14.4.1   *p 504*

22 The ring $B_r(\mathbb{C})$ of even functions (mod r) is a semisimple algebra of dimension $d(r)$, the number of divisors of r.

Proposition 14.7.2   *p 515*

23 THE POLYNOMIAL 3-PRIMES CONJECTURE: Every odd monic polynomial $M(x) \in \mathbb{F}_q[x]$ is a 3-primes polynomial except for the case q even and $M(x) = x^2 + a \in \mathbb{F}_q[x]$.

Subsection 15.5.1   *p 551*

24 Every odd monic fifth degree polynomial over any finite field of characteristic 2 is a 3-primes polynomial

Theorem 153   *p 566*

25 If $R = xF[[x]]$, R has no maximal ideals.

Theorem 156   *p 582*

26 Suppose that R is a Noetherian ring in which all maximal ideals are principal, then R is a principal ideal ring (PIR).

Worked-out example (b)   *p 430*

27 A commutative ring S with identity 1_S is a primary ring, if, and only if, every zero divisor of S is nilpotent.

Remark 14.8.3   *p 519*

28 (Fabrizio Zanello) R denotes a PID. R has an infinite number of pairwise nonassociated irreducible elements if, and only if, every maximal ideal of $R[x]$ has height 2.

Theorem 158   *p 585*

29 (S. Z. Ditor) A finite group G of odd order is the group of units of some ring if, and only if, G is abelian and is the finite product of cyclic groups G_i where order of each G_i is of the form $2^{k_i} - 1$.

Theorem 160(a)   *p 591*

30 (S. Z. Ditor) Let $t = p^m$ where p is a prime and $m \geq 1$. Then, t is equal to the number of units of a ring R if, and only if, $p = 2$ or p is Mersenne prime $M_q = 2^q - 1$ where q is a prime.

Corollary 16.4.2   *p 592*

Index of symbols and notations

$\mathbb{Z}[i]$	the ring of Gaussian integers	p. 8
$\phi(r)$	the value of Euler ϕ-function at r	p. 5
$[a]$	the congruence class of a or the greatest integer not exceeding a (to be understood from the context)	p. 5, 281
$\binom{a}{b}$	a choose $b = \frac{a!}{b!(a-b)!}$, if $b < a$	p. 10
$R[x]$	polynomial ring in x with coefficients from R	p. 75
$R[t]$	an R-module generated by t	p. 395
$F[x_1, x_2, \ldots, x_n]$	polynomial ring in $x_1, x_2, \ldots, x_n$ with coefficients from a field F	p. 83
g.c.d $(a_1, a_2, \ldots, a_s)$	greatest common divisor of $a_1, a_2, \ldots, a_s$.	p. 18
$\mathbb{R}^m$	$\{(x_1, x_2, \ldots, x_m) : x_i \in \mathbb{R}, i = 1, 2, \ldots, m\}$	p. 9
(a)	the principal ideal generated by a	p. 33
$\mathbb{F}_q[x]$	ring of polynomials in x with coefficients from the finite field $\mathbb{F}_q$ having q elements	p. 419
$R[[x]]$	seq $R = \{a_0, a_1, a_2, \ldots\}$ where $a_i \in R$, $i = 0, 1, 2, \ldots$	p. 85
$F[[x]]$	ring of formal power series in x with coefficients from F	p. 86
$\mathbb{C}_l$ ($\mathbb{C}[[x_1, x_2, \ldots, x_l]]$)	ring of formal power series in $x_1, x_2, \ldots, x_l$ over $\mathbb{C}$	p. 90
$\mathbb{C}_\omega$	ring of formal power series in countably infinite number of indeterminates over $\mathbb{C}$	p. 87
$I + J$	$\{a + b : a \in I, b \in J\}$	p. 32
IJ	$\{\sum_{\text{finite}} a_i b_i : a_i \in I, b_i \in J\}$	p. 32
$I : J$	$\{a \in R : aJ \subseteq I\}$ where I, J are ideals of R, a commutative ring with unity 1_R.	p. 34
RH	Riemann Hypothesis	p. 527
GRH	Generalized Riemann Hypothesis	p. 525
$ht(P)$	height of prime ideal P	p. 580
$\mathbb{Z}[\sqrt{-5}]$	$\{a + b\sqrt{-5} : a, b \in \mathbb{Z}\}$	p. 35
$\mathbb{Q}(\alpha) = \mathbb{Q}[\alpha]$	$\{a_0, a_1\alpha + a_2\alpha^2 \ldots + a_{n-1}\alpha^{n-1} : a_i \in \mathbb{Q}, i = 0, 1, 2, \ldots, (n-1)\}$.	p. 49, 50
$\mathbb{Q}(\sqrt{m})$	$\{a + b\sqrt{m} : a, b \in \mathbb{Q}\}$	p. 50
$R(m)$	the number ring corresponding to the algebraic number field $\mathbb{Q}(\sqrt{m})$	p. 51
$N(\alpha)$	norm of α	p. 52
$N(I) = \|I\|$	norm of the ideal I in a number ring	p. 437
$L(I)$	length of the ideal I of an integral domain	p. 422
$U(r)$	group of units in $\mathbb{Z}/r\mathbb{Z}$	p. 5
PID	principal ideal domain	p. 8
UFD	unique factorization domain	p. 37
HFD	half factorial domain	p. 446
$F.T.A$	fundamental theorem of arithmetic	p. 3

F-T.A	functional-theoretic algebra	p. 366
FLT	Fermat's Last Theorem	p. 29
a.c.c	ascending chain condition	p. 376
d.c.c.	descending chain condition	p. 390
NDD	normed division domain	p. 230
ACCP	ascending chain condition on principal ideals	p. 39
BGC	binary Goldbach conjecture	p. 525
TGC	ternary Goldbach conjecture	p. 525
$\mathcal{A}$	the Dirichlet algebra of arithmetic functions	p. 94
$\mathcal{A}^*$	ring of arithmetic functions under addition and Cauchy multiplication	p. 323
$\mathcal{A}'$	the Lucas ring of arithmetic functions	p. 501
$\mathcal{D}(\mathcal{A})$	the ring of Dirichlet series of arithmetic functions $f \in \mathcal{A}$.	p. 323
$\mathcal{A}(R)$	the algebra of functions $f : \tilde{\mathbb{Z}} \to R$	p. 366
$\mathcal{B}(R)$	the subalgebra of functions $f : \tilde{\mathbb{Z}} \to R$ which vanish except for a finite subset of $\tilde{\mathbb{Z}}$	p. 367
$\mathcal{A}_1 \oplus \mathcal{A}_2 \oplus \cdots \oplus \mathcal{A}_n$	direct sum of $\mathcal{A}_1, \mathcal{A}_2, \ldots \mathcal{A}_n$	p. 491
d.r.p	double-remainder property	p. 60
$\mathbb{F}_q$	a finite field having q elements	p. 97
$GF(2^k)$	Galois field of characteristic 2, having 2^k elements	p. 591
D^*	$D \setminus \{0_D\}$ where D is an integral domain	p. 48
aRm	a is a quadratic residue of m	p. 139
aNm	a is a quadratic non-residue of m	p. 139
$(a|p)$	Legendre symbol	p. 139
$(a|b)$	Jacobi symbol for b an odd positive integer	p. 166
$(\Delta|n)$	Kronecker symbol for Δ	p. 532
$(\alpha, \pi)_3$	the cubic character of α modulo π	p. 154
ω	an imaginary root of unity	p. 435
$\mathbb{Z}[\omega]$	$\{a_0 + a_1\omega + \ldots + a_{n-2}\omega^{n-2} : \omega = exp(\frac{2\pi i}{n});$ $a_i \in \mathbb{Z}, i = 0, 1, 2, \ldots, (n-2)\}$	p. 442
$\chi(\alpha, \pi)$	a notation for $(\alpha, \pi)_3$	p. 155
$spl.f(x)$	$\{p : p$ is a prime and $f(x)$ splits into a product of linear factors in $\mathbb{F}_p[x]\}$	p. 169
$p(n)$	value of partition function at n	p. 178
$Z(G)$	centre of G, where G is a group	p. 181
D_n	dihedral group of order $2n$	p. 187
$[E : F]$	degree of the field extension E over the field F	p. 553
$\{E; F\}$	index of the field extension E over a field F	p. 552
$G(E/F)$	Galois group of E over F	p. 552
(D, g)	Euclidean domain R with the associated function g	p. 48
$\mathbb{Z}_p$ (p is a prime)	the integral domain consisting of rational numbers of the form $\frac{m}{n}$ where $\frac{m}{n}$ is in its lowest terms and $p \nmid n$.	p. 214

(X, Δ)	weak partially ordered set	p. 230
poset	partially ordered set	p. 233
$(X, \leq)$	the poset X with partial order $\leq$	p. 233
$(L, \vee, \wedge)$	a lattice L	p. 237
$M_n(D)$	the total matrix ring of $n \times n$ matrices with entries from D	p. 545
$GL_n(D)$	$\{[a_{ij}] : [a_{ij}]$ is an invertible $n \times n$ matrix with entries from $D\}$	p. 545
$GL(V)$	the group of invertible linear operators on V	p. 593
$\|a\|_\infty$	Q-absolute value of $a \in F$, a field	p. 253
$d(r)$	the number of divisors of r	p. 80
$\sigma_k(n)$	sum of the kth powers of the divisors of r	p. 304
$S(n, k)$	Stirling numbers of first kind	p. 309
$G(-n, k)$	Stirling numbers of second kind	p. 309
B_n	Bernoulli numbers	p. 311
$V_n(q)$	vector space of dimension n over $\mathbb{F}_q$, a finite field having q elements	p. 262
$L(V_n(q))$	Lattice of k-dimensional subspaces of $V_n(q)$	p. 262
$\binom{n}{k}_q$	number of k-dimensional subspaces of $V_n(q)$.	p. 279
A_P	the incidence algebra of functions defined on $P \times P$ where P is a locally finite poset.	p. 263
$\chi : G \to \mathbb{C}$	a semicharacter (character) of a semigroup (group) G	p. 344, (p. 157)
$ch(G)$	the group of characters of G	p. 159
FG	group algebra of a group G over a field F	p. 355
$\mathcal{A}(k)$	the set of arithmetic functions f for which $\sum_{n=1}^{\infty} f(n) n^{-s}$ converges absolutely where $Re\, s > k - \delta$	p. 364
$\sqrt{I}$	nilradical of I	p. 382
$J(R)$	Jacobson radical of R	p. 384
C_K	ideal class group of an algebraic number field K	p. 439
$\mathcal{A}_r(F)$	the algebra of (r, F) arithmetic functions	p. 486
$B_r(\mathbb{C})$	the algebra of even functions (mod r)	p. 483
$C(n, r)$	Ramanujan's sum	p. 106
$C^*(n, r)$	the unitary analogue of Ramanujan's sum	p. 515
$\zeta(s)$	Riemann ζ-function at s	p. 291
$\tau(n)$	Ramanujan's τ-function at n	p. 292
$\pi(x)$	the number of primes less than or equal to x	p. 291
$Li(x)$	the logarithmic integral	p. 528
$L(s, \chi)$	Dirichlet's L-function at s, χ	p. 529
$\vartheta(x)$	Chebyshev's ϑ-function at x	p. 570
$\xi(s)$	function related to $\zeta(s)$	p. 528
$Z(t)$	function related to $\zeta(1/2 + it)$	p. 528

BIBLIOGRAPHY

The list of books/journal articles given below is meant to supplement the references which were shown at the end of each chapter. Though each chapter from 1 to 16 is self-contained, the following references provide the material for co-lateral reading.

[1] Adhikari S. D., Katre S. A. and Dinesh Thakur: Cyclotomic fields and related topics, Proc. Summer School, June 7–30, 1999, Organized by the Department of Math., Univ. Pune, Bhaskaracharya Prathishtana, Pune, Printed at Mudra, 383, Narayan Path, Pune 411030, India.

[2] Anderson, Daniel D. (Ed.) : Factorization in Integral Domains, Lecture Notes in Pure and Applied Mathematics, Marcel Dekker Inc. N. Y. (2001).

[3] A. Baker: Imaginary quadratic fields with class number two; Ann of Math 94(1971) 139–152.

[4] Bambah R. P., Dumir V. C. and Hans-Gill R. J.: Number Theory, Hindustan Book Agency and Indian National Science Academy (2000), P 19, Green Park Extension, New Delhi 110016, India.

[5] M. Barnabai, A. Brini and G. C. Rota: The theory of Möbius functions, Russian Math. Surveys 41:3 (1986) 135–188.

[6] Barnes F. W. : On the stufe of an algebraic number field, J. Number Theory, 4, (1972), 474–476.

[7] John A. Beechy: Introductory Lectures on rings and modules, London Math. Society Student Texts 47, Cambridge University Press (1999).

[8] Beiter, M : The mid-term coefficient of the cyclotomic polynomial, Amer. Math. Monthly, 71, (1964), 769–770.

[9] Nicholas Bourbaki: Elements of the History of Mathematics (translated from the French by John Meldrum), Springer-Verlag, Berlin, Heidelberg (1994).

[10] A. N. Chatters and C. R. Hajarnavis: An introductory Course in Commutative Algebra, Oxford University Press (1998), London.

[11] Paul-Jean Cahen and Jean-Luc Chabert: Integer-valued polynomials, Math Surveys and Monographs No.48, Chapters I to VI, pp 1–121, Amer. Math. Soc., Providence, RI, USA, (1996).

[12] L. Carlitz : The number of terms in the cyclotomic polynomial $F_{pq}(x)$, Amer. Math. Monthly, 73, (1966), 979–981.

[13] L. Carlitz : Kloosterman sums and finite field extensions, Acta Arithmetica, XVI, (1969), 179–193.

[14] Child Lindsay N: A Concrete Introduction to Higher Algebra, UTM, Springer Verlag NY Inc., (1995), Second Edition.

[15] Harvey Cohn : Class-number formula for the quadratic case, Chapter 10, Advanced Number Theory, Dover Publication, (1962) Reprint.

[16] P. M. Cohn : Algebra III, Second Edition, John Wiley and Sons, NY, (1991).

[17] Leo Corry : (The Cohn Institute for history and philosophy of science and ideas, Tel Aviv University, Ramat Aviv 69978, Israel): Modern algebra and the rise of mathematical structures, Science Networks/Historical Studies, Birkhäuser, 2nd revised Edn (2004).

[18] David A. Cox : Primes of the form $x^2 + ny^2$, Fermat, Class-field Theory and complex multiplication, John Wiley and Sons NY Inc., (1989).

[19] Henri Darmon, Fred Diamond and Richard Taylor: Fermats' Last Theorem, pp 1–154 in Current Developments in Mathematics, 1995, International Press incorporated, Boston (1994), P O Box 2872, Cambridge MA 02238-2892. USA.

[20] P. G. L. Dirichlet: Lectures in Number Theory (Supplement by Dedekind), History of Mathematics Sources vol 16, (translated by John Stillwell), Amer. Math. Soc. (and London Math. Soc.), PO Box 6248, Providence RI 02940-0268 USA.

[21] John L. Drost : A shorter proof of the Ramanujan Congruence modulo 5, Amer. Math. Monthly, 104, (1997), 963–964

[22] P. Eakin and W. Heinzer : More non-Euclidean PID's and Dedekind domains with prescribed class-group, Proc. Amer. Math. Soc. 40 (1973) 66–68.

[23] F. Faltin, N. Metropolis, B. Ross and G. C. Rota: The real numbers as a wreath product, Advances in Mathematics 16 (1975), 278–304, in Surveys in Applied Mathematics (Essays dedicated to S M Ulam) pp 271–297. Academic Press, New York (1976).

[24] R. Gilmer: On polynomial and power series rings over a commutative ring, Rocky Mountain J. Math. 5 (1975) 157–175.

[25] Pierre Antoine Grillet: Algebra, Wiley Interscience, John Wiley and Sons Inc., NY (1999).

[26] Richard K. Guy: Unsolved problems in Number Theory, section C1 pp 159–164, Springer-Verlag, New York, LLC (2004).

[27] Helmut Hasse: Number Theory, Springer Verlag, Berlin, Heidelberg (1980).

[28] Pentti Haukkanen: On the l_p norm of GCD and related matrices, Journal of Inequalities in Pure and Applied Mathematics, Vol 5, Issue 3, Article 61, 2004, pp 1–7.

[29] Pentti Haukkanen, Jun Wang and Juha Sillanpää: On Smith's Determinant, Linear Algebra and its Applications 258: 251–269 (1997).

[30] Pentti Haukkanen: An abstract Möbius inversion formula with number-theoretic applications, Discrete Mathematics 142 (1995) 87–96.

[31] Pentti Haukkanen: Some binomial inversions in terms of ordinary generating functions, Publications Mathematicae (Debrecen), Tomus 47 (1995) Fasc 1-2, pp 181–191.

[32] E. Hecke: Lectures on the theory of Algebraic Numbers, GTM No 77, Springer Verlag, NY Inc. (1980).

[33] L. K. Hua: Introduction to Number Theory, Springer-Verlag, Berlin, Heidelberg, New York (1982).

[34] A. Ivic : On the number of abelian groups of a given order and on certain multiplicative functions, J. Number Theory, 16, (1983), 119–137.

[35] Bernard L. Johnston and Fred Richman: Numbers and symmetry, An Introduction to Algebra, CRC Press Inc Boca Raton NY, London and Tokyo, (1997).

[36] G. A. Jones and J. Mary Jones: Elementary Number Theory, Springer Undergraduate Mathematics Series (SUMS), Springer Verlag NY Inc. (1998).

[37] S. A. Katre and Sangitha A. Khule: A discriminant criterion for matrices over orders in algebraic number fields to be sums of squares, Proc. of Symposium on Algebra and Number Theory, Cochin University of Science and Technology, Kochi, July (1990), pp 31–38.

[38] S. A. Katre and Sangitha A. Khule: Matrices over orders in algebraic number fields as sums of kth-powers, Proc. Amer. Math. Soc. 128, (1999), 671–675.

[39] Helmut Koch: Algebraic Number Theory, Springer Verlag, Berlin, Heidelberg (1997).

[40] John Knopfmacher and Wen Bin Zhang: Number Theory arising from finite fields (Analytic and Probabilistic Theory) Monographs and Textbooks in Pure & Applied Mathematics No. 241, Marcel Dekker, NY (2001).

[41] A. O. Kostrikin and I. R. Shaverevich : Encyclopaedia of Mathematical Sciences, No 11, Algebra I, Basic notions of algebra, Chapters 1–14, pp 4–124, Springer Verlag, Berlin (1990).

[42] A. O. Kostrikin and I. R. Shaverevich : Encyclopaedia of Mathematical Sciences, No 18, Algebra II, Non-commutative rings-Identities, Chapter 1–17, pp 4–106, Springer Verlag, Berlin, (1991).

[43] M. Laczkovich : On Lambert's proof of the irrationality of π, Amer. Math. Monthly, 104 (1997), 439–443.

[44] I. S. Luthar and I. B. S. Passi: Algebra Vol 2, Rings, Narosa Publishing House Pvt Ltd, (1999), 22, Daryaganj, Delhi Medical Association Road, New Delhi 110002, India.

[45] Olberding, Bruce: Factorization into prime and invertible ideals, J. London Math. Soc. (2) 62 (2000) 336–344.

[46] McCarthy, Paul J. : Arithmetical rings and multiplicative lattices, Annali di Matematica pura ad applicata (iv), Vol LXXXII, pp 267–274 (1969).

[47] T. T. Moh: Algebra, Series in University Mathematics Vol 5, World Scientific, Singapore (1995), Reprint.

[48] Richard A. Mollin: Fundamental Number Theory with applications, CRC Press, Boca Raton, Florida (1998).

[49] K. Nageswara Rao and R. Sivaramakrishnan: Ramanujan's sum and its applications to some combinatorial problems, Proc. Tenth Manitoba Conf. Numer. Math and Comput. Vol II, Congr. Numer 31 (1981), 205–239.

[50] W. Narkeiwicz: The development of Prime Number Theorem from Euclid to Hardy and Littlewood, Springer Monographs in Mathematics December (2000).

[51] M. B. Nathanson: Elementary methods in Number Theory, GTM No 195, Springer Verlag, Berlin, Heidelberg (2000).

[52] Donald J. Newman: Analytic Number Theory GTM No. 177. Springer, NY Corrected Second printing (2000).

[53] Takashi Ono: An Introduction to Algebraic Number Theory, Plenum Press, NY and London (1990), (a subdivision of Plenum Publishing Corporation, 233, Spring Street, NY 10013, USA).

[54] D. S. Passman: A course in ring theory, Wadsworth and Brooks/Cole, Pacific Grove, California (1991).

[55] M. Pohst and H. Zassenhaus: Algorithmic Algebraic Number Theory, Encyclopaedia of Mathematics and it applications, G C Rota (Ed), Cambridge University Press (1989).

[56] Richard S. Price: Associative Algebras, GTM No 88, Springer-Verlag, NY (1981).

[57] M. Ram Murthy: Prime numbers and irreducible polynomials, Amer. Math. Monthly, 109 (2002) 452–458.

[58] John Scherk: Algebra, a computational introduction, Adv. Studies in Mathematics Series, Chapman and Hall/CRC, Boca Raton, Florida, USA (July 2000).

[59] W. M. Schmidt: Equations over finite fields: an elementary approach, Lecture Notes in Mathematics Vol 536, Springer Verlag, NY Inc (1976).

[60] J. P. Serre: A Course in Arithmetic, Springer Verlag NY Inc. (1973), Reprint, Narosa Publishing House Pte Ltd, 22 Daryaganj, Delhi Medical Association Road, New Delhi 110002, India.

[61] Harold N. Shapiro: On the convolution ring of arithmetic functions, Communications on Pure and Applied Mathematics, XXV (1972) 287–336, (John Wiley and Sons Inc., 1972).

[62] Kalheinz Spindler: Abstract Algbra with applications, Vol II, Rings and Fields, Marcel Dekker Inc., NY (1994).

[63] V. Srinivas: Diophantine Equations, Expository lecture delivered at Pune University, 23–26 (February 1994), Lecture Notes, TIFR Publication.

[64] Richard P. Stanley: Combinatorics & Communicative Algebra, Progress in Mathematics No 41, Birkhauser, Boston, Basel, Stutgart (1983).

[65] John Stillwell: Elements of algebra, geometry, numbers, equations, UTM, Springer-Verlag NY Inc. (1994).

[66] B. Sury: The congruence subgroup problem, An elementary approach aimed at applications, Texts and Readings in Mathematics No 24, Hindustan Book Agency, India, P 19, Green Park Extension, New Delhi 110016 (2003) India.

[67] Sudesh K. Khanduja and Jayanthi Saha: The prime and maximal ideals in $R[x]$: R a one-dimensional Prüfer domain, Indian J. Pure & Appl. Math. 29 (12): 1275–1279 (Dec. 1998).

[68] R. J. Valenza: Elasticity of factorization in Number Fields, J. Number Theory, 36 (1990) 212–218.

[69] V. S. Varadarajan: Algebra in ancient and modern times, Texts and Readings in Mathematics No.14, (1997) Hindustan Book Agency, India, P 19, Green Park Extension, New Delhi 110016 (India).

[70] A. Weil: Number of solutions of equations in a finite field, Bull. Amer. Math. Soc. 55 (1949) 497–508.

[71] L. Weisner, Some properties of prime-power groups, Trans. Amer. Math. Soc. 38 (1935) 485–492.

[72] W. C. Winnie Li: Number theory with applications, World Scientific, Singapore, NJ, London, Hongkong (1996).

[73] D. Zagier: Newman's short proof of the Prime Number Theorem, Amer. Math. Monthly, 104 (1997) 705–708.

Subject index

Index of names